The Wiley Blackwell Handbook of the Psychology of the Internet at Work

Wiley Blackwell Handbooks in Organizational Psychology

Series Editor: Jonathan Passmore

The aim of the Wiley Blackwell Handbooks in Organizational Psychology is to create a set of uniquely in-depth reviews of contemporary research, theory and practice across critical sub-domains of organizational psychology. Series titles will individually deliver the state-of-the-art in their discipline by putting the most important contemporary work at the fingertips of academics, researchers, students and practitioners. Over time, the series will grow in into a complete reference for those seeking to develop a comprehensive understanding of the field.

Published

The Wiley-Blackwell Handbook of the Psychology of Coaching and Mentoring
Edited by Jonathan Passmore, David B. Peterson and Teresa Freire

The Wiley-Blackwell Handbook of the Psychology of Leadership, Change and Organizational Development
Edited by H. Skipton Leonard, Rachel Lewis, Arthur M. Freedman and Jonathan Passmore

The Wiley Blackwell Handbook of Psychology of Training, Personal Development and E-Learning
Edited by Kurt Kraiger, Jonathan Passmore, Nuno Rebelo dos Santos and Sigmar Malvezzi

The Wiley-Blackwell Handbook of the Psychology of Occupational Safety and Workplace Health
Edited by Sharon Clarke, Tahira M. Probst, Frank W. Guldenmund and Jonathan Passmore

The Wiley Blackwell Handbook of the Psychology of the Internet at Work

Edited by Guido Hertel,
Dianna L. Stone, Richard D. Johnson,
and Jonathan Passmore

WILEY Blackwell

This edition first published 2017

Registered Office(s)
John Wiley & Sons, Inc., 111 River Street, Hoboken, NJ 07030, USA
John Wiley & Sons Ltd, The Atrium, Southern Gate, Chichester, West Sussex, PO19 8SQ, UK

Editorial Office
The Atrium, Southern Gate, Chichester, West Sussex, PO19 8SQ, UK

For details of our global editorial offices, customer services, and more information about Wiley products visit us at www.wiley.com.

Wiley also publishes its books in a variety of electronic formats and by print-on-demand. Some content that appears in standard print versions of this book may not be available in other formats.

Library of Congress Cataloging-in-Publication Data
Names: Hertel, Guido, editor.
Title: The Wiley Blackwell handbook of the psychology of the Internet at work / [edited by] Guido Hertel, University of Muenster, Munster, [and three others].
Description: First edition. | Hoboken, NJ : John Wiley & Sons, Inc., [2017] | Series: Wiley Blackwell handbooks in organizational psychology ; 7696 | Includes index. |
Identifiers: LCCN 2017013705 (print) | LCCN 2017028341 (ebook) | ISBN 9781119256175 (pdf) | ISBN 9781119256168 (epub) | ISBN 9781119256144 (cloth)
Subjects: LCSH: Information technology—Management. | Telematics. | Online social networks in business. | Virtual reality in management. | Organizational effectiveness. | Psychology, Industrial.
Classification: LCC HD30.2 (ebook) | LCC HD30.2 .W5278 2017 (print) | DDC 658/.054678—dc23
LC record available at https://lccn.loc.gov/2017013705

Cover image: © busypix/Gettyimages
Cover design by Wiley

Set in 9.5/11, Galliard by SPi Global, Pondicherry, India

10 9 8 7 6 5 4 3 2 1

Contents

About the Editors vii
About the Contributors ix
Foreword xv

1 The Psychology of the Internet @Work 1
Guido Hertel, Dianna L. Stone, Richard D. Johnson, and Jonathan Passmore

Part I Individual Perspectives 19

2 Digitized Communication at Work 21
Nicole C. Krämer and Stephan Winter

3 Ergonomics of Information Technologies at Work 39
Benjamin V. Hanrahan and John M. Carroll

4 Competencies for Web-Based Work and Virtual Collaboration 61
Stefan Krumm and Julian Schulze

5 User Experience, Gamification, and Performance 79
Meinald T. Thielsch and Jörg Niesenhaus

6 Trust in Virtual Online Environments 103
Sirkka L. Jarvenpaa, Celeste Cantu, and Shi Ying Lim

7 Workplace Cyberdeviance 131
Steven D. Charlier, Gary W. Giumetti, Cody J. Reeves, and Lindsey M. Greco

8 Blended Working 157
Nico W. Van Yperen and Burkhard Wörtler

9 Flexwork, Work–Family Boundaries, and Information and Communication Technologies 175
Ronald E. Rice

10 Mobile Computing and Hand-Held Devices at Work 195
Humayun Zafar

Part II Organizational Perspectives **211**

11 E-Recruiting: Using Technology to Attract Job Applicants 213
Derek S. Chapman and Anna F. Gödöllei

12 Social Networking Sites, Search Engines, and the Employment Process 231
Kimberly M. Lukaszewski and Andrew F. Johnson

13 The Evolution of E-Selection 257
David N. Dickter, Victor Jockin, and Tanya Delany

14 E-Leadership 285
Surinder Kahai, Bruce J. Avolio, and John J. Sosik

15 Virtual Teams 315
M. Travis Maynard, Lucy L. Gilson, Nicole C. Jones Young, and Matti Vartiainen

16 Online Employee Surveys and Online Feedback 347
Bernad Batinic and Carrie Kovacs

17 E-Learning 369
Richard D. Johnson and Kenneth G. Brown

Part III Societal and Cross-Sectorial Perspectives **401**

18 Robots in the Digitalized Workplace 403
Jochen J. Steil and Günter W. Maier

19 Social Issues Associated with the Internet at Work 423
Dianna L. Stone, Dianna Krueger, and Stephen Takach

20 Employee Age Differences in Using Internet-Based Tools at Work 449
Gabriela Burlacu, Donald M. Truxillo, and Talya N. Bauer

21 The Future of Work 481
Stela Lupushor and Alex Fradera

Index 509

About the Editors

Guido Hertel Guido is a Professor of Organizational and Business Psychology at the University of Münster, Germany. His research addresses emerging trends and challenges in work organizations, such as electronic human resource management, demographic changes, and synergy effects in cooperation and negotiations. He has published more than 100 chapters and journal papers, for instance, in the *Journal of Applied Psychology*, *Journal of Organizational Behavior*, *Journal of Management*, *Journal of Occupational and Organizational Psychology*, *Journal of Personality and Social Psychology*, *Psychological Bulletin*, and *Research Policy*. Guido is Associate Editor of *Group & Organization Management*, and member of the editorial boards of *Organizational Psychology Review*, *Small Group Research*, and the *International Journal of Internet Science*. He is also member of the founding editorial board of *Work, Aging, & Retirement*, and has served as guest editor of special issues for various journals such as the *European Journal of Work and Organizational Psychology* and the *Journal of Managerial Psychology*. In 2013, Guido was host and program committee chair of the sixteenth Congress of the European Association of Work and Organizational Psychology (EAWOP), and he received the Innovation Award of the German Association of Psychologists, Section Work, Organization and Business Psychology in 2015. In addition to his academic activities, Guido works as speaker, trainer, and consultant for work and business organizations.

Dianna L. Stone, PhD Dianna received her PhD from Purdue University, and is currently a Visiting Research Professor at the University at Albany and an Affiliate Professor at Virginia Tech. Her research focuses on a variety of issues including diversity in organizations, cross-cultural issues, electronic human resource management (eHRM), e-recruiting, e-selection, and privacy. Results of her research have been published in the *Journal of Applied Psychology, Personnel Psychology*, the *Academy of Management Review, Human Resource Management Review*, *Organizational Behavior and Human Decision Processes*, and the *Journal of Management*. She is an Associate Editor of *Human Resource Management Review*, and has served as guest editor for several special issues of the journal. She is also the Editor of *Research in Human Resource Management*, and was the Editor for the *Journal of Managerial Psychology* from 2007 to 2014. In 2012 she was awarded the Sage Scholarly Achievement Award for Research on Gender and Diversity, and received the

Janet Chusmir Sage Service Award in 2013 from the same division. She also received the Lead Editor Award from Emerald Publishing in 2014 and was honored with the Trailblazer Award from the PhD Project for her research on diversity and work with minority doctoral students. Dianna is a Fellow of the Society for Industrial and Organizational Psychology, the American Psychological Association, and the Association for Psychological Science.

Richard D. Johnson, PhD Richard received his PhD from the University of Maryland, USA, and is an Associate Professor of Management, Department Chair, and Director of the Human Resource Information Systems (HRIS) program at the University at Albany, State University of New York. He has published more than 50 journal articles and book chapters on topics such as human resource technology, the psychological impacts of computing, training and e-learning, and issues surrounding the digital divide. His research has been published in outlets such as *Information Systems Research*, *Journal of the Association for Information Systems*, *Human Resource Management Review*, and the *International Journal of Human Computer Studies*. Richard is a Past Chair of Association for Information Systems Special Interest Group on Human–Computer Interaction (AIS SIGHCI) and is a Senior Editor at Data Base and an Associate Editor at *AIS Transactions on Human-Computer Interaction*. He is also an editor of the book, *Human Resource Information Systems: Basics, Applications and Future Directions*.

Jonathan Passmore, D.Occ.Psych Jonathan is a Professor of Psychology at the University of Evora, Portugal, and Director for the Centre of Coaching & Behavioural Change, Henley Business School, UK. Prior to this, he has worked for PriceWaterhouseCoopers, IBM and OPM. He is a chartered psychologist, holds five degrees and has an international reputation for his work in coaching and leadership. He has published over 20 books on the themes of leadership, personal development and change, including editing the *Association for Coaching* series of coaching titles. He speaks widely at conferences across the world from the United States to Europe and Asia and has published more than 100 journal papers and book chapters. He was awarded the Association for Coaching Global Coaching Award for his contribution to practice and research in 2010, the British Psychology Society Research Award for his research into safety coaching in 2012, and the Association of Business Psychologist Chairman's Award for Excellence in 2015. He sits on the editorial board of several journals including *Coaching: An International Journal* and the *International Coaching Psychology Review*.

About the Contributors

Bruce J. Avolio, PhD Bruce is the Mark Pigott Chair in Business Strategic Leadership at Michael G. Foster School of Business, University of Washington, USA. Bruce has published 11 books and more than 150 articles. His latest book, *The First Principle of Organizational Transformation*, covers organizations that have undergone successful and sustainable transformation.

Bernad Batinic Bernad is Head of Department of Work, Organizational and Media Psychology at the Johannes Kepler University, Austria. He was a founding member of the German Society for Online Research and involved in the development of several large online panels. His current research focuses on job characteristics and wellbeing, as well as the use and impact of media and new technologies.

Talya N. Bauer, PhD Talya is a Cameron Professor of Management in the Business School at Portland State University, USA. Her well-published research focuses on recruitment, applicant reactions to selection, onboarding, and leadership. In addition, she has co-authored three textbooks and co-edited the *Oxford Handbook of Leader-Member Exchange*.

Kenneth G. Brown, PhD Ken is an Associate Dean and Ralph L. Sheets Professor of Management and Organizations at the Tippie College of Business, University of Iowa, USA. He has a PhD in Psychology, and has served as a Fulbright Scholar at Seoul National University, South Korea. His research interests include workplace learning and development, and science-practice connections in human resource management.

Gabriela Burlacu, PhD Gabriela is a researcher at SAP SuccessFactors, where she focuses on identifying, understanding, and communicating ways companies can effectively support their human resource management (HRM) processes using Cloud-based solutions. She brings a primarily academic background to this role, having done extensive research on workforce age differences and HRM.

Celeste Cantu Celeste is a doctoral student in Industrial Engineering at the Tecnológico de Monterrey, Mexico. She has a degree in Electronical Engineering from Tecnológico de Monterrey and a master's in Media Management from KTH Royal Institute of Technology, Sweden. She is experienced in implementing radiofrequency identification (RFID), telehealth, and physical security systems. She also has an interest in trust issues and adoption of emerging technologies.

John M. Carroll John is a Distinguished Professor of Information Sciences and Technology at Pennsylvania State University, USA. Trained as an experimental psychologist, his research is in internet tools for collaborative learning and problem-solving, and sustaining community. John has published 24 books and 600 papers, and received 10 fellowship/lifetime awards and an honorary doctorate in engineering.

Derek S. Chapman, PhD Derek is an Associate Professor of Industrial and Organizational Psychology at the University of Calgary, Canada. His research interests include recruiting, person–organization fit, personnel selection, and technology use in human resources.

Steven D. Charlier, PhD Steven is an Associate Professor in the Department of Management at Georgia Southern University, USA. He received his PhD in Management and Organizations from the University of Iowa. His research focuses on the impact of technology on various aspects of the modern work environment, including leadership, teams, training, and job attitudes.

Tanya Delany, PhD Tanya leads selection and onboarding for IBM, as well as talent acquisition metrics and analytics. She received her doctorate in Industrial Organizational Psychology. Tanya has 15 years' experience in organizational development and change management, surveys, job analysis, competency modeling, test development, training system design, and program evaluation.

David N. Dickter, PhD David is a Director of Interprofessional Education Research and Strategic Assessment at Western University of Health Sciences, USA. He earned his PhD in Industrial Organizational Psychology from Ohio State University. David has spent more than 20 years developing and validating psychometric assessments in internal and external consulting roles.

Alex Fradera, PhD Alex is a science writer for the British Psychological Society (BPS) Research Digest (http://digest.bps.org.uk), a consultant in workplace psychology since 2007, and the head of research for wiserbydesign.com, an organization dedicated to making sense of the world. He holds a doctorate in cognitive psychology from University College London.

Lucy L. Gilson, PhD Lucy is a Professor and Head of the Management Department at the University of Connecticut, USA. Her research focuses on individual and team creativity, why managers should want employees to be creative, team effectiveness, and virtual teams. Lucy has been published in the *Academy on Management Journal*, *Journal of Applied Psychology*, *Journal of Management*, *Leadership Quarterly*, and *Group & Organization Management*.

Gary W. Giumetti, PhD Gary is an Associate Professor in the Department of Psychology at Quinnipiac University, USA, where he teaches courses in and directs the concentration in Industrial Organizational Psychology. He received his PhD in Industrial Organizational

Psychology from Clemson University. Gary's research focuses on organizational justice and occupational stress and health.

Anna F. Gödöllei Anna is a master's student in Industrial Organizational Psychology at the University of Calgary, Canada. Her interests include the validity and fairness of selection assessments and applicant reactions to assessment methods. Specifically, she studies gamification and how game-based assessments may be used in the recruitment and selection of employees.

Lindsey M. Greco, PhD Lindsey is an Assistant Professor in the Department of Management at Oklahoma State University, USA. She received her PhD in Management and Organizations from the University of Iowa. Her research interests center around individual differences in counterproductive work behavior, social identity at work, and research methods.

Benjamin V. Hanrahan, PhD Benjamin is a Research Associate at the Information Sciences and Technology department at Pennsylvania State University, USA. Previously, he worked as a research scientist in the Work Practice Technology group at Xerox Research Centre Europe. He received his PhD in Computer Science from Virginia Polytechnic State University.

Sirkka L. Jarvenpaa Sirkka is the Bayless/Rauscher Pierce Refsnes Chair in Business Administration and Professor of Information Systems at the University of Texas at Austin, USA, where she serves as the Director of the Center for Business, Technology and Law and the Director of the Information Management program in the Department of Information, Risk and Operations Management.

Victor Jockin, PhD Victor earned his PhD in Psychology from the University of Minnesota, USA. As Senior Manager of Assessment Solutions at PSI Services, Victor has developed a broad range of pre-employment assessment products. He was also instrumental in the development of PSI's online testing platform and much of its proprietary scoring code.

Andrew F. Johnson, PhD Andrew is an Assistant Professor of Management at Texas A&M University-Corpus Christi, USA. He earned his PhD in Business Administration from the University of Texas at San Antonio. In addition to work in social networking and online learning, he studies corporate political activity.

Nicole C. Jones Young, PhD Nicole is an Assistant Professor of Management at the Franklin & Marshall College, USA. She earned her PhD in Organizational Behavior from the University of Connecticut. Her current research focuses on marginalized populations and social class as related to the selection process and organizational inclusion.

Surinder Kahai, PhD Surinder is an Associate Professor at Binghamton University, USA. Surinder's research attempts to understand how information and communication technologies mediate leadership, collaborative work, and learning. He has employed his research, which has been published in prestigious journals and presented at selective conferences, to coach many business leaders.

Carrie Kovacs, PhD Carrie is a Research Assistant at the Department of Work, Organizational and Media Psychology, Johannes Kepler University, Austria. Her current research extends her longstanding interest in research methods to the work context, with a focus on job satisfaction, worker wellbeing, and the use of wearable technologies in organizational research.

Nicole C. Krämer, PhD Nicole is a Professor for Social Psychology: Media and Communication at the University of Duisburg-Essen, Germany. Her research addresses human–computer interaction as well as computer-mediated communication, with a particular focus on social psychological processes and social effects.

Dianna Krueger, PhD Dianna received her PhD from the University of Texas at San Antonio, and is an Assistant Professor of Management at Tarleton State University, USA. Her research focuses on weight-based bias, diversity, and Hispanic work issues. She has published the results of her research in the *Journal of Managerial Psychology* and the *Business Journal of Hispanic Research.*

Stefan Krumm Stefan is the Head of Department of Psychological Assessment, Differential and Personality Psychology at the Freie Universität Berlin, Germany. Among other research interests, his current focus is on knowledge, skills, and abilities as predictors of success in digital collaboration and virtual teamwork.

Shi Ying Lim Shi Ying is a PhD candidate in Information Systems at the University of Texas at Austin, USA. Her research interests include health information technology design, digital innovation, information systems strategy, and entrepreneurship.

Kimberly M. Lukaszewski, PhD Kimberly is an Associate Professor of Management at Wright State University, USA. She holds an MBA in Human Resources Information Systems (HRIS), and a PhD in Organizational Studies from the University at Albany, State University of New York. Her research focuses on HR technology, e-recruitment, and social media.

Stela Lupushor Stela is the founder of Reframe.Work Inc. and a thought leader on the future of work and workforce analytics topics. She is consulting Fortune 100 companies on how to future-proof their business by reframing their workforce, workplace, and work strategies and practices. She holds a diploma in Mathematics and Computer Science, has a patent pending for a social sentiment analysis tool, and is a sought-after speaker on the topics of future of work and workforce analytics.

Günter W. Maier Günter is a Professor for Work and Organizational Psychology at Bielefeld University, Germany. He focuses on questions in the field of personnel selection, personality at work, leadership, organizational justice, innovation and creativity, personal work goals, and digitalized work. He has published more than 70 articles and chapters.

M. Travis Maynard, PhD Travis is an Associate Professor in the Department of Management at Colorado State University, USA. He received his PhD from the University of Connecticut and has conducted extensive research in the area of organizational team effectiveness. Specifically, his research interests include the role that team context has on team interactions and outcomes.

Jörg Niesenhaus, PhD Jörg is a Branch Manager at Centigrade, where he supports clients in creating software products of high usability, visual attractiveness, and technical elegance. He uses playful design elements to enhance user satisfaction and motivation. He holds a PhD in Computer Science and has worked in the games and gamification industry since 1997.

Cody J. Reeves, PhD Cody is an Assistant Professor in the Department of Organizational Leadership and Strategy at the Marriott School of Management, Brigham Young University, USA. He received his PhD in Management and Organizations from the University of Iowa. Cody's research investigates entry into teams and organizations and the forces that influence the entry process.

Ronald E. Rice, PhD Ronald is the Arthur N. Rupe Chair in Social Effects of Mass Communication at University of California, Santa Barbara, USA. He has published 13 books and more than 125 articles and 70 chapters on public communication campaigns, computer-mediated communication, organizational communication, information science, and social networks.

Julian Schulze Julian is a Research Assistant and PhD candidate at the Department of Psychological Assessment, Differential and Personality Psychology at Freie Universität Berlin, Germany. His research interests include competencies for virtual teamwork and communication skills in face-to-face and computer-mediated interactions.

John J. Sosik, PhD John is a Professor of Management and Organization and Professor-in-Charge of the Master of Leadership Development program at the Great Valley School of Graduate Professional Studies, Pennsylvania State University, USA. His current research interests include character and leadership development, multi-level leadership, and e-leadership.

Jochen J. Steil Jochen is a Professor for Robotics and Process Control at Technische Universität Braunschweig, Germany. He focuses on robot learning, human–robot interaction, programming by demonstration, and applications in production and control systems. Jochen Steil has published more than 150 papers in cognitive robotics, neural networks, and learning systems.

Stephen Takach, PhD Stephen earned his PhD in Business Administration from the University of Texas at San Antonio, and is a Lecturer at the University of New Mexico-Valencia, USA. His research focuses on a variety of issues including justification of organizational performance, cognitive biases in strategic decision making, culturally bound firm specific advantage creation and dissemination, and social issues in work organizations.

Meinald T. Thielsch, PhD Meinald is an Akademischer Rat (Assistant Professor) at the Department of Psychology, University of Münster, Germany. He earned his PhD and habilitation in Psychology. His main work and research interests are human–computer teraction, user experience, applied research, and science–practice transfer.

Donald M. Truxillo, PhD Donald is a Professor of Psychology at Portland State University, USA, where he has done extensive research examining issues associated with older workers, including job design, age stereotypes, and work ability. He has published more than 70 peer-reviewed journal articles and book chapters, and serves on numerous journal editorial boards.

Nico W. Van Yperen, PhD Nico is a Professor of Organizational Psychology at the University of Groningen, the Netherlands. His research interests include achievement motivation, competence, blended working, and talent development.

Matti Vartiainen, PhD Matti is a Professor of Work and Organizational Psychology at the Department of Industrial Engineering and Management, Aalto University, Finland. With his research teams, he is studying organizational innovations, new ways of working such as digital, mobile, and multi-locational work and distributed teams and organizations, reward systems, knowledge and competence building, and e-learning systems.

Stephan Winter, PhD Stephan is an Assistant Professor of Persuasive Communication at the University of Amsterdam, The Netherlands. His research interests include selective exposure and opinion formation in online contexts, science and crisis communication, as well as self-presentation and self-disclosure in social media.

Burkhard Wörtler Burkhard is a PhD student at the Department of Organizational Psychology, University of Groningen, The Netherlands. His research interests include psychological needs in the workplace, blended working, and leadership in organizations.

Humayun Zafar Humayun is an Associate Professor of Information Security and Assurance in the Department of Information Systems at Kennesaw State University, USA. He is also a Research Fellow at the Distance Learning Center. He received his doctorate from the University of Texas in San Antonio.

Foreword

The digital revolution, sparked by the development of the Internet, affects all of us 24/7/365. Given that two key objectives of industrial organizational psychology are to improve productivity as well as the quality of work–life for men and women, this edited volume on the psychology of the internet at work is especially timely. To understand the wide-ranging implications of the digital revolution in the context of work requires multi-level perspectives, and this book delivers. Indeed, the chapters are organized around three levels of analysis. At the micro level, primary concern is with the experience and behavior of the individual worker. This section of the book examines the effects of internet-based technologies on classical industrial organizational topics such as work motivation, performance, the analysis of work to identify required competencies, and workplace health and wellbeing.

The second section of the book focuses on the effects of Internet-based technologies at the meso level – work organizations. Here the focus is on topics such as recruiting, that is, finding talent, selecting it, developing it, leading it, and promoting effective teamwork, whether co-located or virtual. The critical issues of trust and distrust in e-commerce and in virtual teamwork are addressed explicitly (see Chapter 6), as are multilevel perspectives on trust that include other persons, as well as teams, organizations, brands, and even the Internet itself.

The final section of the book considers macro-level phenomena – societies as a whole. To their credit, the editors and chapter authors explicitly recognize that the effects of internet-based technologies have both positive and negative implications. On the one hand, these technologies have spawned many new business ideas to generate wealth, to grow the economy, and to improve the quality of life (e.g., Amazon, Google, Facebook, eBay, and millions of "apps"). On the other hand, the Internet has eroded personal privacy, it has enabled scammers, hackers, and thieves, and technologies such as machine learning and robots have led to the displacement and unemployment of many workers. Challenges wrought by the Internet at the micro, meso, and macro levels will engage industrial organizational scientist-practitioners for years to come. Whether you are just beginning your career or are well established in it, you will find a fascinating array of intriguing ideas in every chapter of this volume. I invite you to explore each one in detail.

Wayne F. Cascio
Robert H. Reynolds Chair in Global Leadership
University of Colorado, Denver, USA

1

The Psychology of the Internet @ Work

Guido Hertel, Dianna L. Stone, Richard D. Johnson, and Jonathan Passmore

Introduction

The Internet has radically changed the way we work, and the way work is organized. Similar to other core technologies in the past, such as steam engines, electricity, or computer technologies, the Internet influences not only singular work activities or discrete branches, but affects nearly all aspects of work in a striking speed of time. For industrial and organizational psychologists, it is therefore essential to understand the implications of the Internet at work from different levels of analysis. At the micro level, Internet-based technologies have significant implications not only for the experience and behavior of the individual worker, for her or his work motivation, trust experience, and individual performance, but also for health and wellbeing and for required competencies at work. At the meso level, the Internet offers new strategies for work organizations, for recruiting and selection of employees, for leadership and teamwork, and for training and development. At the macro level, Internet-based work has implications for societies as a whole, creating new opportunities for economically underdeveloped regions and for the integration of disadvantaged workers, but also for new conflicts and legal problems in increasingly global workplaces. This edited handbook covers these three perspectives in an integrative way, providing state-of-the-art reviews of existing research, guidance for future work, and suggestions for practitioners.

In this first chapter, we start with a brief history of the Internet at work to understand the specific characteristics of Internet-based technologies that underlie different qualitative shifts in working conditions. In doing so, we identify five core characteristics of Internet-based work that might clarify and structure the still divergent usage of concepts in this emerging field. Moreover, such a task-oriented approach might help to analyze more precisely the various implications of Internet-based technologies for work-related experiences and behaviors as the main focus of work and organizational psychologists. We provide

The Wiley Blackwell Handbook of the Psychology of the Internet at Work,
First Edition. Edited by Guido Hertel, Dianna L. Stone, Richard D. Johnson, and Jonathan Passmore.

initial suggestions and examples for this idea, and illustrate general opportunities and risks of the core characteristics at the three main levels introduced above, that is, the individual worker, work organizations, and societies. After these more general observations, we provide a preview of the other chapters in this handbook. In addition to newest findings from scientific research in the different fields, the book provides best practices for the usage and optimization of these rapidly evolving technologies for different sectors and industries.

A Brief History of the Internet at Work

The high prevalence of Internet-based technologies at work is reflected in a multitude of labels, such as "online" (e.g., online assessment), "web" (e.g., web-based working), "virtual" (e.g., virtual teams), "e-" (e.g., e-leadership), "cyber" (e.g., cyber loafing), "tele-" (e.g., telework), or "digital" (e.g., digital divide). Originating in different research traditions and disciplines, these different labels might be confusing when considering what the core attributes of Internet-based work might be. For instance, Internet-based work implies more than just electronically mediated information, and can also be relevant when people collaborate at short distances in the same building.

The Internet is a global communication system that connects private, public, academic, business, and government networks using a broad and constantly developing array of electronic, wireless, and optical technologies (Internet, n.d., retrieved November 15, 2016, from https://www.britannica.com/technology/Internet; see also Internet, n.d., retrieved November 15, 2016 from https://en.wikipedia.org/wiki/Internet). Thus, in addition to electronic computer technologies, global accessibility and interactivity are core characteristics of the Internet. Moreover, the Internet functions without a central governing body, with different autonomous networks being voluntarily interconnected.

The Internet emerged from early computer networks developed in the 1960s, such as the Advanced Research Projects Agency Network (ARPANET; e.g., Leiner et al., 1997). These communication networks were expanded to the 1980s in order to provide efficient communication for science and military needs. For instance, based on the ARPANET, the first email was sent 1971, and the File Transfer Protocol (FTP) specification was defined in 1973. However, the birth of the Internet as we know it today can be dated in 1990 when the ARPANET was opened for commercial usage. This was followed quickly by the first web page in 1991, and the first live stream (of a lab coffee pot) in 1993 (Trojan Room coffee pot, n.d., retrieved November 15, 2016, from https://en.wikipedia.org/wiki/Trojan_Room_coffee_pot). In addition, early forms of crowdsourcing started at that time, such as the Linux operating system software development project, one of the most successful Open Source software development projects. Open Source projects such as Linux, which had no strong financial incentives or organizational structures, demonstrated the potential of collaborative work via the Internet (e.g., Hertel, Niedner, & Hermann, 2003; Saxena, Deodhar, & Ruohonen, 2017).

In 1995, the Internet was globally commercialized, and Internet-based (e-commerce) enterprises such as eBay and Amazon were founded. In the remaining years of the twentieth century, there was tremendous excitement about Internet-related businesses, with pundits predicting that brick and mortar stores would be replaced by click-and-order markets. This development led to an exploding stock market, and ultimately to the so-called "dotcom crisis" in 2000, when the hype surrounding Internet businesses finally burst, with many early e-commerce innovators becoming bankrupt. Following this hype-and-bust cycle, Internet-based e-commerce stabilized and set the stage for further, more solid developments (e.g., Fenn & Raskino, 2008). In addition, new developments in multimedia and collaboration tools emerged in the mid-2000s that allowed for individuals to not only consume content but to also interactively and collaboratively create new content ("Web 2.0"). Wikipedia and other knowledge communities expanded on the idea of crowdwork, from rather specialized communities such as software developers in Open Source software projects, to the whole population (e.g., Schroer & Hertel, 2009). Moreover, social media tools such as Facebook, Twitter, YouTube, LinkedIn, Viadeo, and Skype supported both organizational and non-organizational communities to develop around common interests. The introduction of Internet-connected mobile "smart" phones (e.g., the first Apple iPhone in 2007) was a major step towards permanent accessibility at work, allowing users to not only make phone calls but to also access emails, webpages, and Internet-based data files. Somewhat ironically, the first email-free workday was also introduced in 2007 in some companies as a result of workers' reactions towards the high number of emails to be attended every day.

Only a few years later, the idea of Internet-based collaboration was picked up by commercial or semi-commercial initiatives in so-called "sharing" communities, providing new services for transportation (Uber), housing (Airbnb), or labor leasing (Amazon Mechanical Turk). Notably, workers and service providers in these communities often lack conventional labor rights protection and insurances, raising various legal and ethical questions with these new conceptualizations of work.

More recently, the increasing use of sensor technologies has enabled again a new level of interactivity of the Internet, including direct connections between artifacts and machines ("Internet-of-Things"), which has further increased the speed of work and commerce. Although automatic driving, smart clothes, and smart offices provide many facilitations at work, these innovations can also cause risks and strain, and have been accused of destroying workplaces in various industries. For instance, use of the global positioning system (GPS) in monitoring in transportation businesses, while facilitating logistics and security issues, also puts pressure and strain on drivers, and automatic driving might even replace human drivers. Moreover, artificial intelligence might substitute important leadership tasks such as decision making in complex scenarios (Parry, Cohen, & Bhattacharya, 2016). Thus, Internet-based innovations at work come both with opportunities and challenges for the individual worker, for work organizations, and for societies (Cascio & Montealegre, 2016; Stone & Dulebohn, 2016).

We do not know exactly what will come next. One of the exciting aspects of doing research in this field is the innovative potential of the Internet for work and work organization over the coming decades. New technologies, tools, and business ideas are continually invented, some being picked up immediately, some only after a delay of months or even years, and some perhaps never. However, in order to understand the implications of the Internet at work, it might be useful to abstract from discrete tools and phenomena, and reconsider basic characteristics of the Internet from a task-oriented perspective.

Core Characteristics of Internet-Based Work

Internet-based work has been associated with many different attributes, for instance with high team diversity (e.g., Hoch & Kozlowski, 2014) or more autonomy at work (e.g., Gajendran & Harrison, 2007). However, some of these attributes are less central to or not unique for Internet-based work in general. Based on the definition of the Internet as a global system of interconnected computer systems, we consider the following characteristics of Internet-based work as most central and distinct from traditional work forms: accessibility, interactivity, reprocessability, automatization, and boundary crossing.

Accessibility of information. Connecting billions of autonomous computers throughout the world (with numbers still growing), the Internet provides easy and often un-delayed access to myriads of data both within organizational networks and beyond. In addition to increasing speed and amount of information available for work processes and decisions, high accessibility of information also supports the mobility of work because information (and people as information carriers) is neither restricted to certain locations nor to certain hours or days. Working "anywhere and anytime" has been made possible by this high level of information accessibility. However, this development has also increased the expectation that workers are accessible regardless of time or location. In fact, ubiquity has emerged as a major stressor at work, blurring traditional forms of separation between occupational and private life (e.g., Derks, Bakker, Peters, & van Wingerden, 2016). Moreover, high information accessibility comes with many ethical and legal issues about data ownership and data privacy, for instance, when organizations recruit and select new employees.

Interactivity of communication. A second major characteristic of interconnected computer systems is the opportunity to directly interact with other users regardless of where they are. Individuals can not only passively retrieve information or contact other persons, but can also send or post information and reply to others, enabling multidirectional exchange and collaboration. This high degree of potential interactivity is the backbone of many collaborative processes at work today. Users of Internet-based technologies can share and exchange information with other workers (e.g., in virtual teams) or with organization representatives (e.g., during recruitment processes) regardless whether or not the others are currently co-present. Whereas interactivity was mainly text-based and asynchronous in the early years of the Internet, the rapid development of technologies now includes synchronous audio and visual communication (e.g., web conferencing), and even digital exchange of physical products (e.g., three-dimensional [3D] printing).

Reprocessability of information. The mediated nature of working via Internet enables automatic storage of many working steps. As a consequence, related information can be retrieved and reprocessed on demand. Such documentation and reprocessability of work (Dennis, Fuller, & Valacich, 2008) has many advantages, such as reducing the risk of misunderstandings in complex (e.g., international) collaborations, or reducing the need for trust in virtual teams (e.g., Breuer, Hüffmeier, & Hertel, 2016). However, the fact that automatic storage and reprocessability makes "forgetting" considerably more difficult can also cause problems, such as information overload, stigmatization of individuals due to past behavior or rumors about past behavior ("cyber-mobbing"), and data protection issues. In light of the constantly increasing data volumes accessible for individuals and organizations, developing intelligent algorithms for handling "big data" (e.g., George, Haas, & Pentland, 2014; George, Osinga, Lavie, & Scott, 2016) and for "intentional forgetting" (e.g., Niederee, Kanhabua, Gallo & Logie, 2015) are important challenges for future knowledge management at work.

Automatization. A fourth core characteristic of interconnected computer systems is the potential to automatize processes in these networks, for instance, by using computer routines to send information at certain pre-programmed times, to scan (billions of) websites for information, or to monitor and control workers or machines from a distance (GPS monitoring). Automatization has the potential to disburden the individual at work in many respects, freeing resources for other tasks. Moreover, automatization in computer networks highly increases the opportunities for information seeking and control.

However, Internet-based automatization can also create undesired costs (e.g., unsolicited and undesired "spam" emails). Although automatization is also used in other work forms, its potential is greatly increased in interconnected computer networks.

Boundary crossing. High boundary-crossing capabilities are suggested as a fifth core characteristic of the Internet at work. In interconnected computer systems, information and products are communicated in a standardized "language" that not only connects different work processes at different geographic or organizational locations (e.g., cross-company collaboration in automotive industries) but can also integrate non-work domains (e.g., usage of social media networks at work; e.g., van Iddekinge, Lanivich, Roth, & Junco, 2016). Moreover, individuals not only collaborate remotely with others but also with machines ("manufacturing 2.0," "industry 4.0"), and machines can directly communicate with other machines ("Internet-of-Things"). The more pervasive computer technology becomes in our (work) lives, the higher the potential interconnections of these computers will grow. This boundary-crossing potential can dissolve work structures such as hierarchies or demarcations of knowledge exchange within organizations. Moreover, required competencies for knowledge management, communication, and leadership are changing due to the boundary-crossing capabilities of the Internet.

These five core characteristics might help to structure research on psychological implications of the Internet at work, enabling more specific and perhaps precise predictions based on psychological theories. We illustrate this idea in the next section with initial examples.

Psychological Implications of the Internet at Work

When reviewing the existing literature on psychological implications of the Internet at work, we found that studies often focused on quite different aspects and outcomes, making comparisons of research results and literature summaries difficult. For instance, some research on Internet-based "virtual" teams has focused on the (lack of) synchronicity or interactivity of collaboration, while other studies have focused on difficulties in accessing and reprocessing data, and yet others on cultural diversity (see, for instance Gilson, Maynard, Jones Young, Vartiainen, & Hakonen, 2015, for a review). The suggested core characteristics of Internet-based work might help to structure existing and future research, and to connect technological phenomena with psychological theories. In addition, considering core characteristics of Internet-based work might provide helpful guidance for empirical research, for instance, suggesting appropriate baselines when examining specific effects of Internet-based work.

One of the basic questions in this respect is, which processes and requirements truly change when using the Internet at work, and which processes and requirements remain the same as in traditional work settings. For instance, at the person level, psychological theories related to high information accessibility include dual process models of information processing (e.g., Chaiken, Liberman, & Eagly, 1989; Petty & Cacioppo, 1986), predicting more systematic information processing of individual workers when more information is accessible. At the same time, theories from cognitive psychology also cover consequences of information overload on workers' decision accuracy and strain experience (e.g., Speier, Valacich, & Vessey, 1999). At the organizational level, effects of information availability on individual decision processes are complemented by theories on organizational knowledge management, including the motivation and coordination of knowledge generation, memory, and dissemination as well as decision-making routines. Theoretical approaches relevant for these themes include networking and knowledge transfer theories (e.g., Foss, Husted, & Michailova, 2010) as well as conceptualizations of potential information overload at the organizational level (e.g., Niederee et al., 2015). Finally, at the societal level, high information accessibility due to Internet-based technologies increases opportunities to create new workplaces even in areas with low infrastructure and for workers with reduced mobility. Moreover, high information accessibility about work organizations affect how these organizations are perceived by job applicants or customers, which is described, for example, in theories of trust in organizations (e.g., Fulmer & Gelfand, 2012), employer branding (e.g., Rupp, Ganapathi, Aguilera, & Williams, 2006; Walker, Feild, Giles, Bernerth, & Short, 2011), or customer expectations (e.g., Barley, 2015).

Psychological theories relevant for increased interactivity due to Internet-based work include communication and feedback theories at the level of individual workers (e.g., Kluger & DeNisi, 1996), and theories on motivation for Enterprise Social Networks (e.g., Chin, Evans, & Choo, 2015; Leftheriotis & Giannakos, 2014) as well as social impact and minority influences (e.g., Latané & L'Herrou, 1996) at the level of organizations and societies. In addition, much of the field of electronic human resources depends on the ability of technology to expand communication and interaction opportunities among customers, employers, employees, retirees, and prospective employees (Johnson, Lukaszewski, & Stone, 2016). Reprocessability effects due to Internet-based work can be addressed, for instance, based on communication theories (e.g., Dennis et al., 2008), models of trust (e.g., Mayer, Davis, & Schoorman, 1995) and fairness perceptions (e.g., Colquitt & Zipay, 2015), as well as theories on error management (e.g., Frese & Keith, 2015).

Psychological effects of Internet-based automatization are explained, for instance, by theories on workers' need for control (e.g., Karasek & Theorell, 1990), highlighting potential resource saving but also aversive and even threatening effects of Internet-based surveillance and monitoring. In addition, the change to automated human resource (HR) systems may evoke psychological reactance among applicants and employees, affecting individuals' job acceptance rates or attraction to organizations. Internet-based recruiting has the potential to invade personal privacy, and may have an adverse impact on protected group members. As a result, psychological models of reactance (Brehm, 1966), person-organization fit (Chatman, 1989), privacy (Stone & Stone, 1990), and self-efficacy (Bandura, 1986) are relevant to explain the extent to which individuals accept these new systems and feel comfortable to use them. Together with research from sociology, human computer interaction, and information systems, these psychological theories are relevant for how individuals might respond to the replacement of human workers by Internet-based technologies (Hess, Fuller, & Campbell, 2009; Johnson, et al., 2006).

Finally, boundary-spanning aspects of Internet-based work are related to psychological theories on work-life balance and organizational segmentation norms (e.g., Derks et al., 2016) as well as basic approaches of job design (e.g., Hackman & Oldham, 1980) and job crafting (e.g., Wrzesniewski & Dutton, 2001). Based on these psychological theories, potential benefits and risks of Internet-based work can be derived more precisely. Initial examples are listed in Table 1.1 and Table 1.2.

Table 1.1 Potential benefits of Internet-based work.

		Level of analysis		
		Individual worker	*Work organization*	*Society*
Core characteristics	Accessibility	Breadth of available information Decision quality Flexible work times	Processing speed Decision quality Global presence and activities	New workplaces in underdeveloped regions Employability of disadvantaged persons Global knowledge repositories
	Interactivity	Feelings of agency and voice Frequent exchange with colleagues	Decisions speed and quality Speed of product development	Reduction of cultural distance Global connection of those with similar interests
	Reprocessability	Perceived transparency and fairness Reduced risks of misunderstanding	Reduced risks of process errors Increased knowledge sharing	Transparency of business processes Analyses of accidents and error management
	Automatization	Work safety Strain relief	Cost reduction Breadth of information available	Health management Reduction of business barriers
	Boundary crossing	Job crafting Work–life balance	Breadth of available information Cross-company collaboration	Opportunities for electronic job freelancing

Table 1.2 Potential risks of Internet-based work.

		Level of analysis		
		Individual worker	*Work organization*	*Society*
Core characteristics	Accessibility	Information overload Ubiquitous working	Data security	Privacy issues
	Interactivity	Challenges to work–life balance Escalating aggression ("flaming")	Cyber-loafing	Post-truth thinking and conspiracy beliefs
	Reprocessability	Suability Stigmatization	Challenge to rebuild negative reputation	Ability to interfere with political processes
	Automatization	Monotony Lack of control	Layoffs and reduction in job opportunities	Loss of workplaces
	Boundary crossing	Blurred boundaries between work and non-work	Cyber-loafing	Legal issues

A Research Agenda for the Psychology of the Internet at Work

Research on the psychological implications of Internet-based technologies at work is still in its infancy. In addition to different foci and conceptualizations of Internet-based implications, published studies span multiple journals, books, conference proceedings, and research communities (see also Cascio & Montealegre, 2016). Often, valuable research is done by academics outside of work and organizational psychology, such as by computer or communication scientists. At the same time, theoretical work on the psychological implications of Internet-based work is still underdeveloped. Consequently, it is difficult to maintain a decent overview about research findings in specific fields of interest, for example, e-leadership, and it is even more difficult to integrate findings from different fields into a more complete picture of the Internet at work. For instance, research on ergonomics, cyber-loafing, Internet-based trust, and virtual teamwork is all relevant when developing models of e-leadership. Recognizing the implications of Internet-based work at different levels and for different processes is important in order to truly understand its psychological effects on experience and behavior at work.

This edited title is the first international handbook that integrates the various sources and fields of evidence-based psychological research on the Internet at work in one volume. As such, the handbook provides useful information for researchers and practitioners alike. The handbook is structured in three parts, reflecting the micro, meso, and macro perspectives referred to in the introduction to this chapter. Although the different chapters are sorted into these three parts according to their predominant perspective, most chapters are not restricted to only one perspective but cover various levels of analysis. Together, the chapters combine what are currently the most important findings, conceptual frameworks, and practices related to the psychology of the Internet at work.

Part I: Micro-level implications

The first part of the handbook addresses implications of Internet-based work at the level of the individual person (micro perspective). The aim of this part is to present theoretical concepts and empirical literature on individual reactions to Internet-based work, including both positive and negative consequences. In addition to individuals' motivation and performance, this part covers implications of Internet-based work for health and wellbeing as well as deriving implications for individual competencies.

The first two chapters of this part set the stage by presenting more general frameworks in this field. In Chapter 2, Nicole C. Krämer and Stephan Winter provide a comprehensive overview of theoretical frameworks on communication with digital technologies from a social science perspective. Given that Internet-based work usually includes some kind of mediated communication, the described constructs and mechanisms are at the heart of many of the following chapters, for instance, when considering required competencies or leadership strategies. Moreover, the chapter also includes excellent examples of psychological theories relevant for core characteristics of Internet-based work. The authors review both classic theories of computer-mediated communication, for instance, of the selection of specific communication tools, as well as more recent developments, such as communication in enterprise social media networks. In doing so, the chapter provides both illustrative description of existing concepts as well as promising new directions for future work.

Complementing the social science perspective of Chapter 2, Chapter 3 introduces a more technical perspective and presents the general principles of the ergonomics of human–computer interaction. Summarizing the wealth of findings in this very active and creative community could easily fill a full volume instead of only a single chapter. Therefore, Benjamin V. Hanrahan and John M. Carroll focus on more general processes of ergonomic analyses of requirements and implications of Internet-based technologies at work. Rather than reviewing micro issues in ergonomics (e.g., usability, human–machine interface), this chapter takes a macro approach in viewing how Internet-based technologies are embedded within broader work practices, how they impact how work is conducted, how power is structured in organizations, and how relationships are maintained. Moreover, the authors provide instructional case examples at different levels, including motivational implications of ergonomics in web-based customer service centers, online social networks in organizations, and Internet-based crowdworking. Thus, well in line with the general approach of this handbook, the authors demonstrate that in addition to individual-level effects, principles of human–computer interaction are also relevant for the meso and macro levels of Internet-based work.

Following and building on the introduction of basic concepts, Chapter 4 derives specific affordances for individuals during Internet-based work. Stefan Krumm and Julian Schulze review new research on competencies for Internet-based work, finding only low convergence among existing studies to date, despite many individual competencies having been proposed in the literature. Moreover, existing taxonomies provide few explicit links to specific challenges of Internet-based work, illustrating the already mentioned lack of theoretical connections between core characteristics of Internet-based technologies and existing psychological theories. In addition to a closer analysis of discrete task specifics, the authors argue for more rigorous empirical tests of whether assumed competencies are truly unique for Internet-based work. In their chapter, the authors provide initial examples of this quest. Moreover, the authors inspect existing theories on digital communication and virtual collaboration in order to derive potentially overlooked competencies. Together, this chapter sets the stage for more fine-grained theories of work competencies that integrate different aspects of task, organizational context, and mediating technology.

Switching from competencies to individual experience, in Chapter 5 Meinald T. Thielsch and Jörg Niesenhaus review the emerging literature on the impact of user experience in Internet-based work and its effects on motivation, learning, and performance. In doing so, the authors particularly address gamification as an increasingly popular aspect of the user experience. Building on general concepts of human–computer interaction, the authors summarize empirically established design principles and provide informative research examples. Moreover, they also discuss best practices for practical applications, and provide instructive cases for motivating design. In general, the individual experience of Internet-based technologies has been shown to considerably affect motivation and performance at work. At the same time, more integrative theories are needed that particularly consider mediating psychological mechanisms in this respect, such as attention, cognition, or emotion.

Chapter 6 also addresses motivational processes, focusing on trust and distrust as a consequence of Internet-based work. Indeed, difficulties to build and maintain trust are among the most prominent topics in the existing literature on Internet-based collaboration, and can be related to different core characteristics such as accessibility or reprocessability of information. Sirkka L. Jarvenpaa, Celeste Cantu, and Shi Ying Lim start with a comprehensive review of different conceptualizations of trust in the literature, before applying different trust models to the contexts of electronic commerce and virtual teamwork. Interestingly, trust in Internet-based context not only refers to other persons but also to larger entities such as teams, organizations, or brands. Moreover, trust can refer to specific technologies and even to the Internet per se. Thus, even though psychologists usually consider trust as a state experienced by individuals, multilevel modelling of its determinants is required. Finally, the authors address important findings on the development, maintenance, and repair of trust in Internet-based contexts, considering current technologies and potential future developments.

Further exploring motivational implications of Internet-based work, Chapter 7 addresses potential counterproductive behavior summarized under the term of "cyber-deviance." In their review, Steven D. Charlier, Gary W. Giumetti, Cody J. Reeves, and Lindsey Greco structure the existing research on cyber-deviance into behaviors that are intrapersonally focused (e.g., visiting Internet-based entertainment or shopping sites during work hours), interpersonally focused (e.g., using digital resources to harass others), and organizationally focused (e.g., posting defamatory comments about a company in publicly available Internet sites). The authors summarize both antecedents and consequences of cyber-deviant behaviors as well as potential countermeasures. Interestingly, not all cyber-deviant behaviors are intentional, some might be rather based on carelessness and inattention, and some deviant behaviors can even have positive consequences in the long term.

The following two chapters address consequences of Internet-based work for health and wellbeing, implications that are particularly related to high accessibility and boundary-crossing aspects of the technologies. Research on this topic has strongly grown in the past years as a consequence of increased complains of strain due to ubiquity expectations at work. Although this certainly is a concern for organizations as a whole, the psychological processes of strain reactions and wellbeing are located at the level of individual persons. In Chapter 8, Nico W. Van Yperen and Burkhard Wörtler present the research on "blended working," a relatively new concept describing the opportunities for more flexibility at work due to Internet-based technologies. When discussing both potential benefits and risks of blended working, the authors stress that these consequences are not automatically given but depend on specific moderating factors, such as workers' personality, type of task, and organizational policies.

In Chapter 9, Ronald E. Rice focuses on the implications of Internet-based flexible work on work–life balance and related outcomes. This chapter provides an excellent example of the need to integrate findings from disparate research streams into a cohesive framework. The author draws on research from communications, health, information systems, management, and psychology as he reviews and integrates the work on communication technologies, flexwork, and work–life balance. Specifically, he develops a theoretically driven model to shape research on flexwork and work–life balance to inform those researching in this area.

Today, a large portion of Internet traffic is driven by mobile devices, increasing accessibility as core characteristic of Internet-based work even further. This can have profound implications for how work is conducted and how employees connect and interact with each other and with organizational resources. Despite the centrality of mobile computing to work practices, only limited research has begun to investigate its employee and organizational implications. In Chapter 10, Humayun Zafar first discusses the evolution of the use of mobile computing in organizations. He then reviews the research on both the positive and negative implications of mobile computing in organizations, and provides guidance for future researchers. Finally, he discusses how mobile computing and the "Internet-of-Things" will potentially transform the workplace of the future. Similar to other topics in this handbook, research on mobile computing is in an early stage and there are opportunities for scholars from various domains to use their theories to contribute to this phenomenon.

Part II: Meso-level implications

Part II addresses implications of the Internet at work at the organizational level, covering typical topics of "electronic human resource management" (e.g., Stone, Deadrick, Lukaszewski, & Johnson, 2015). The different chapters review the current literature on how organizations (can) use the Internet for the management of their employees, including recruiting and assessment, leadership, teamwork, and personnel development.

In Chapter 11, Derek S. Chapman and Anna F. Gödöllei focus on Internet-based recruiting, and describe how technology has dramatically altered the recruitment process in the past decades, modifying employer practices and applicant expectations. With respect to the introduced core characteristics of Internet-based technologies, e-recruiting particularly benefits from high accessibility of information and interactivity of communication technologies. Moreover, automatization of data collection and analyses increasingly plays a role. Discussing different strategies of e-recruiting, the authors not only consider finding and attracting job applicants but also how to keep them interested during the selection process. In addition to reflecting both potential benefits and risks of using e-recruiting in organizations, the authors review the existing theories and empirical research on the topic and introduce a new model of e-recruiting with interesting new directions for future research.

In Chapter 12, Kimberly M. Lukaszewski and Andrew F. Johnson consider the advantages and disadvantages of using information from social network sites and search engines in the employment decision-making process (e.g., selection, termination). They argue that organizations are using these sites to determine whether individuals have the knowledge, skills, and abilities to perform the job, and ensure that they are trustworthy and conscientious. In spite of the increased use of social network sites and search engines, the authors maintain that many individuals believe that the use of these sites in the employment process is unfair. One reason for this is that social network sites were designed for communication with friends and family, and were not meant to be viewed by employers. The authors begin their chapter with a review of literature on individuals' reactions to using social network sites and search engines. Next, they apply a model of organizational justice and Leventhal's fairness principles to understand the factors affecting individuals' perceptions of unfairness. Finally, the authors offer interesting directions for future research on these issues.

David N. Dickter, Victor Jockin, and Tanya Delany present a review of the literature on e-selection in Chapter 13, describing the current state of research and its applicability to organizations. Similar to e-recruiting, main core aspects of Internet technologies relevant for e-selection are the high accessibility of information, the opportunity of interactive communication despite geographic distance, as well as automatization in data acquisition and analyses. The chapter begins by highlighting the problems and solutions associated with e-selection including deployment, data security, test equivalence, and the use of unproctored (unsupervised) tests. Next the authors discuss the landscape of e-selection tests and assessments and offer some considerations involved in their development and use. They describe issues associated with the implementation and emphasize that multiple stakeholders are involved in the e-selection process (e.g., applicants, employees, managerial decision makers). They also consider contextual factors affecting the e-selection process including international laws, policies and data-privacy standards that apply to all candidates as well as protected groups, delivering tests under unproctored conditions, and communicating and training. Finally, the authors offer suggestions for areas for future research in this rapidly-evolving field.

Switching from recruitment and selection to the management of workers, Surinder Kahai, Bruce J. Avolio, and John J. Sosik argue in Chapter 14 that the spread of information technology has influenced how leadership is transmitted and received in organizational contexts. In this context, all five core characteristics of Internet-based work are involved. E-leadership is defined as the "social influence process embedded in ... multiple contexts and mediated by advanced information technology (AIT), that produces changes in attitudes, feelings ... behavior, and performance (Avolio, Sosik, Kahai, & Baker, 2014, p. 107). In their chapter, the authors consider the various themes related to e-leadership in the literature, and discuss whether information technology changes leadership in three different contexts: media context, virtual teams, and social media. In doing so, the authors describe the different dynamics between information technologies and leadership, especially how information technologies change leadership, how leadership might change information technologies, and how leaders must be aware of these changes in the new sociotechnical context. The authors end by offering instructive directions for future research.

Also focusing on the management of Internet-based work, Chapter 15 provides a review on the vibrant literature on virtual teams. After summarizing the existing theoretical work and methodological paradigms, Travis Maynard, Lucy L. Gilson, Nicole C. Jones Young, and Matti Vartiainen move forward and outline promising future research perspectives, including longitudinal designs and network analyses in existing teams, as well as mobile devices for data collection. While discussing the different conceptualizations of team virtuality in the literature, the authors argue for a more fine-grained examination of the different dimensions of team virtuality (e.g., geographic dispersion, technology usage), a call that fits well to the overall perspective taken also in the current chapter. Finally, the authors present a comprehensive overview of empirical research findings on the different determinants of virtual team effectiveness.

The last two chapters of Part II are devoted to personnel development. In Chapter 16, Bernad Batinic and Carrie Kovacs maintain that the use of Internet-based technology has been widely used to conduct and report the results of employee surveys in organizations. These processes are particularly related to faster and more economic access to information and to documentation and reprocessability of communication. Internet-based technologies are used not only to plan surveys, collect and analyze data, and communicate the results, but also to implement and evaluate new programs. In the beginning of the chapter, the authors argue that this new technology offers many advantages (e.g., ease, speed of use) while presenting several challenges. Next, the authors explore the technology applied to using the Internet throughout the survey process, for example, project management, use of online employee surveys, and electronic reporting of results. In doing so, they also discuss specific issues associated with using electronic surveys, including data quality, response rates, survey design, and benchmarking. Finally, the authors discuss the risks and opportunities of using online employee surveys, and offer directions for future research.

Chapter 17 provides a comprehensive review on the wide and vibrant field of e-learning at work. High accessibility, increasing interactivity and boundary-crossing aspects of technologies are central in this field. Richard D. Johnson and Kenneth G. Brown argue that technology has transformed how training and education are delivered in organizations, and summarize the considerable research on effectiveness and acceptance of e-learning processes at work. They contend that research has consistently revealed that a well-designed e-learning program can be as effective as face-to-face programs. After reviewing and discussing several e-learning frameworks, the authors present a new integrative framework to guide research on the topic. Based on a review of the literature from multiple disciplines (e.g., management, education, psychology) they also offer suggestions for how to design effective e-learning initiatives, and consider directions for research to move the field forward.

Part III: Societal-level implications

The third section further broadens the scope of analysis to the societal level. Indeed, the introduction of Internet-based technologies has allowed changes at the macro-level of work with both positive and negative implications. On the one hand, the Internet has ignited many new powerful business ideas, creating new job perspectives and wealth. Many of the newly created products and services facilitate our lives (e.g., e-commerce), protect the environment (e.g., decreased commuting due to flexwork and blended working) and our health (e.g., tele-medicine), and increase our knowledge (e.g., free online encyclopedias) and wellbeing (e.g., Internet-based entertainment). On the other hand, the Internet has changed the way work is distributed among social groups, and new forms of work organizations have negatively affected workers' rights and social security. In addition, digitalization and automatization of work, along with a general transition from manufacturing to knowledge and service economies in many industrialized countries, have resulted in the displacement and unemployment of many workers. These developments may also have political implications because alienation and a lack of occupational perspectives contribute to feelings of social exclusion, political radicalization, and xenophobia. Part III addresses this complex field in three exemplary chapters that focus on psychological implications of automatization at work, on potential conflicts between social groups due to different access to the Internet, and on potential age-related opportunities and conflicts at work. The fourth and final chapter concludes this part, summarizing more general opportunities and visions based on the current development of the Internet, creating new ways of working and organizing work.

In Chapter 18, Jochen J. Steil and Günter W. Maier focus particularly on automatization and boundary-crossing aspects of Internet-based technologies, and review the current research on robots in digitalized industries, including the implications for human workers having to interact with robots. Particularly in light of more recent technological trends known as "manufacturing 2.0," "cyber-physical systems," and "Internet-of-Things," potential qualitative shifts in the interaction and decision-making power between human workers and machines have become an important issue, for example, who is taking the lead in the work process. The authors first describe standard as well as novel and advanced robot technology that are particularly relevant to the interaction with human workers. This is followed by an in-depth discussion of implications both from the user and the developer/system designer perspective. Although these considerations are mostly applied in the field of industrialized production where the prevalence of robots is highest to date, the authors also discuss potential usages of digital robot technologies in other work contexts.

In Chapter 19, Dianna L. Stone, Dianna Krueger, and Stephen Takach focus on social conflicts that might occur as unintended consequences of the growing use of the Internet at work. In doing so, they particularly address potential implications of different access to information as well as boundary-crossing aspects of Internet-based technologies. In the first part of the chapter, the authors outline potential disparities between social groups (gender, ethnic groups, etc.) in the access to the Internet and/or differences in the usage of Internet-based technologies that might lead to discrimination in job opportunities, career, and socio-economic status. Understanding the psychological dynamics that uphold such negative processes is important for developing effective countermeasures. Next, the authors discuss potential privacy invasion (and perceptions thereof) as another important source of conflicts during Internet-based work. They introduce a psychological model of privacy perception, and review the existing empirical research in light of this framework.

Moreover, the authors compare the obtained central factors with international privacy laws and suggest instructional directions for future research as well as guidelines for organizations and legislators to develop sound privacy protection policies.

Also addressing potential inequalities between social groups when adopting Internet-based technologies, Gabriela Burlacu, Donald M. Truxillo, and Talya N. Bauer discuss in Chapter 20 potential implications of chronological age for Internet-based work. The chapter starts with a review of empirical research on age differences at work, followed by a more general review of age-related changes in cognition, personality, motivation, and societal context, as well as age differences in Internet use overall. Based on this foundation, the authors develop specific predictions for age differences in the various domains of electronic HR management, such as e-recruiting and e-selection, e-leadership, virtual teamwork, and e-learning. In light of the current lack of empirical research here, these sound theoretical predictions help to structure this emerging field, and might already provide initial guidance for age-differentiated management of Internet-based work facing an increasingly age-diverse workforce in many countries today.

Finally, in the concluding Chapter 21, Stela Lupushor and Alex Fradera outline rich and fascinating visions for the future of Internet-based work. In doing so, the authors consider all five core characteristics of Internet-based work, and categorize their anticipated developments into three major trends: "datafication," describing the conversion of work-related activities, interactions, and relationships into a common language that facilitates new forms of analytics of workflow and work dynamics; "digitization" describing the transformation of working activities and even the workplace itself through the use of digital media and devices, providing new opportunities to integrate physical and digital aspects of work; and "disintermediation," describing the decomposition of work value chains and the removal of intermediaries, which reconfigures work more closely around the consumer's experience. The authors illustrate these processes in various work contexts, including data analytics, automatization and robotics, block-chain technology, and virtual reality applications and show how drastically the Internet and related technologies can change the world of work as well as impacting the wider society. In their outlook of potential future developments, the authors stress both bright and dark aspects, underlining again the need for continuous research to support the positive and prevent the negative consequences of the Internet at work.

Together, the different parts and chapters of this handbook cover the most prominent research topics of psychological implications of the Internet at work. Of course, this selection is not conclusive. Some potential topics currently lack the critical mass of empirical studies required for a literature review. Other topics are at least partly addressed, for instance, e-commerce and electronic bargaining is discussed in Chapter 6. The large amount of research on online communities and online knowledge management is partly addressed in Chapter 3. The emerging field of virtual coaching and mentoring is covered in another volume of this handbook series (see Ghods & Boyce, 2012).

We hope that this handbook will be a useful resource for postgraduate researchers, students, and academics looking for a comprehensive review of the literature as a starting point for their own research. In addition to stimulating research in the different fields, the collection of topics and the cross-references between the chapters might also ignite new perspectives, connecting different fields and levels of analysis for a more complete understanding of the Internet at work. Last but not least, scientist-practitioners might gain a fuller understanding of the depth and scope of the literature in their area of interest, finding innovative ideas and guidelines for their practical work.

Conclusion

Editing a handbook is probably the best way to obtain a decent overview about a research field. These are the conclusions we have taken from this fascinating and insightful process: without doubt, more research by work and organizational psychologists is needed in this rapidly evolving field. In addition to established topics (e.g., e-recruiting, e-leadership, e-learning) and tools (e.g., email communication), more consideration of current technological trends is required. As noted by Gilson and colleagues, "research appears not to be keeping up with practice" (Gilson et al., 2015, p. 14). Notably, the ongoing technological innovations also provide fascinating new tools for doing research, i.e., collecting continuous and rich data from work processes that still have to be better integrated (e.g., George et al., 2016; Kozlowski, 2015). In addition, theoretical frameworks are central to connecting technological features with specific psychological mechanisms. This not only provides a better understanding of different effects of technologies, but might also enable psychologists to take the lead in future technological developments to design both effective and healthy work conditions. Indeed, adapting technological developments to human needs is a still growing occupational field where psychologists not only benefit from sound theories and methods, but also from social and communicative skills when taking the role of a mediator, translator, or knowledge broker in interdisciplinary research groups. In light of these prospects, and given the wide and ever-increasing prevalence of Internet-based technologies in nearly all aspects of work today, psychological implications of the Internet at work should receive more acknowledgment in the regular teaching curricula of work and organizational psychology, in addition to the classic issues of human factors and human–computer interaction. We hope that the current handbook provides useful insights and material for all of these processes.

References

Avolio, B. J., Sosik, J. J., Kahai, S. S., & Baker, B. (2014). E-leadership: Re-examining transformations in leadership source and transmission. *The Leadership Quarterly, 25*, 105–131.

Bandura, A. (1986). *Social foundations of thought and action: A social cognitive theory.* Englewood Cliffs, NJ: Prentice-Hall.

Barley, S. R. (2015). Why the Internet makes buying a car less loathsome: How technologies change role relations. *Academy of Management Discoveries, 1*, 31–60.

Brehm, J. W. (1966). *A theory of psychological reactance.* New York: Academic Press.

Breuer, C., Hüffmeier, J., & Hertel, G. (2016). Does trust matter more in virtual teams? A meta-analysis of trust and team effectiveness considering virtuality and documentation as moderators. *Journal of Applied Psychology, 101*, 1151–1177.

Cascio, W. F., & Montealegre, R. (2016). How technology is changing work and organizations. *Annual Review of Organizational Psychology and Organizational Behavior, 3*, 349–375.

Chaiken, S., Liberman, A., & Eagly, A. H. (1989). Heuristic and systematic information processing within and beyond the persuasion context. In J. S. Uleman & J. A. Bargh (Eds.), *Unintended thought* (pp. 212–252). New York: Guilford Press.

Chatman, J. A. (1989). Improving interactional organizational research: A model of person-organization fit. *Academy of Management Review, 14*, 333–349.

Chin, C. P.-Y., Evans, N., & Choo, K.-K. R. (2015). Exploring factors influencing the use of Enterprise Social Networks in multinational professional service firms. *Journal of Organizational Computing and Electronic Commerce, 25*, 289–315.

Colquitt, J. A., & Zipay, K. P. (2015). Justice, fairness, and employee reactions. *Annual Review of Organizational Psychology and Organizational Behavior, 2*, 75–99.

Dennis, A. R., Fuller, R. M., & Valacich, J. S. (2008). Media, tasks, and communication processes: A theory of media synchronicity. *MIS Quarterly, 32*, 575–600.

Derks, D., Bakker, A. B., Peters, P., & van Wingerden, P. (2016). Work-related smartphone use, work–family conflict and family role performance: The role of segmentation preference. *Human Performance, 69*, 1045–1068.

Fenn, J., & Raskino, M. (2008). *Mastering the hype cycle: How to choose the right innovation at the right time.* Stamford, CT: Gartner.

Foss, N. J., Husted, K., & Michailova, S. (2010). Governing knowledge sharing in organizations: Levels of analysis, governance mechanisms, and research directions. *Journal of Management Studies, 47*, 455–482.

Frese, M., & Keith, N. (2015). Action errors, error management, and learning in organizations. *Annual Review of Psychology, 66*, 661–687.

Fulmer, C. A., & Gelfand, M. J. (2012). At what level (and in whom) we trust across multiple organizational levels. *Journal of Management, 38*, 1167–1230.

Gajendran, R. S., & Harrison, D. A. (2007). The good, the bad, and the unknown about telecommuting: Meta-analysis of psychological mediators and individual consequences. *Journal of Applied Psychology, 92*, 1524–1541.

George, G., Haas, M. R., & Pentland, A. (2014). Big data and management. *Academy of Management Journal, 57*, 321–325.

George, G., Osinga, E. C., Lavie, D., & Scott, B. A. (2016). Big data and data science methods for management research. *Academy of Management Journal, 59*, 1493–1507.

Ghods, N., & Boyce, C. (2012). Virtual coaching & mentoring. In J. Passmore, D. Peterson, & T. Freire (Eds.), *The Wiley-Blackwell handbook of the psychology of coaching and mentoring* (pp. 501–523). Chichester: John Wiley & Sons, Ltd.

Gilson, L. L., Maynard, M. T., Jones Young, N., Vartiainen, M., & Hakonen, M. (2015). Virtual teams research: Ten years, ten themes, and ten opportunities. *Journal of Management, 41*, 1313–1337.

Hackman, J. R., & Oldham, G. R. (1980). *Work redesign.* Reading, MA: Addison-Wesley.

Hertel, G., Niedner, S., & Herrmann, S. (2003). Motivation of software developers in Open Source projects: An internet-based survey of contributors to the Linux kernel. *Research Policy, 32*, 1159–1177.

Hess, T., Fuller, M., & Campbell, D. (2009). Designing interfaces with social presence: Using vividness and extraversion to create social recommendation agents. *Journal of the Association for Information Systems, 10*, 889–919.

Hoch, J. E., & Kozlowski, S. W. J. (2014). Leading virtual teams: Hierarchical leadership, structural supports, and shared team leadership. *Journal of Applied Psychology, 99*, 390–403.

Johnson, R. D., Lukaszewski, K. M., & Stone, D. L. (2016). The evolution of the field of human resource information systems: Co-evolution of technology and HR processes. *Communications of the Association for Information Systems, 38*, 533–553.

Johnson, R. D., Marakas, G. M., & Palmer, J. W. (2006). Differential social attributions toward computing technology: An empirical investigation. *International Journal of Human-Computer Studies, 64*, 446–460.

Karasek, R. A., & Theorell, T. (1990). *Health work.* New York, NY: Basic Books.

Kluger, A. N., & DeNisi, A. (1996). The effects of feedback interventions on performance: Historical review, a meta-analysis and a preliminary feedback intervention theory. *Psychological Bulletin, 119*, 254–284.

Kozlowski, S. W. (2015). Advancing research on team process dynamics: Theoretical, methodological, and measurement considerations. *Organizational Psychology Review, 5*, 270–299.

Latané, B., & L'Herrou, T. (1996). Spatial clustering in the conformity game: Dynamic social impact in electronic groups. *Journal of Personality and Social Psychology, 70*, 1218–1230.

Leftheriotis, I., & Giannakos, M. N. (2014). Using social media for work: losing your time or improving your work? *Computers in Human Behavior, 31*, 134–142.

Leiner, B. M., Cerf, V. G., Clark, D. D., Kahn, R. E., Kleinrock, L., Lynch, D. C., Postel, J., Roberts, L. G., & Wolff, S. S. (1997). The past and future history of the Internet. *Communications of the ACM, 40*, 102–108.

Mayer, R. C., Davis, J. H., & Schoorman, F. D. (1995). An integrative model of organizational trust. *Academy of Management Review*, *20*, 709–734.

Niederee, C., Kanhabua, N., Gallo, F., & Logie, R. H. (2015). Forgetful digital memory: Towards brain-inspired long-term data and information management. *ACM SIGMOD Record*, *44*, 41–46.

Parry, K., Cohen, M., & Bhattacharya, S. (2016). Rise of the machines: A critical consideration of automated leadership decision making in organizations. *Group & Organization Management*, *41*, 571–594.

Petty, R. E., & Cacioppo, J. T. (1986). The elaboration likelihood model of persuasion. In L. Berkowitz (Ed.), *Advances in experimental social psychology*, *Vol. 19* (pp. 124–205). New York, NY: Academic Press.

Rupp, D. E., Ganapathi, J., Aguilera, R. V., & Williams, C. A. (2006). Employee reactions to corporate social responsibility: An organizational justice framework. *Journal of Organizational Behavior*, *27*, 537–543.

Saxena, K. B. C., Deodhar, S. J., & Ruohonen, M. (2017). History of the penguin: Emergence of Open Source Software paradigm. In K. B. C. Saxena, S. J. Deodhar, & M. Ruohonen (Eds.), *Business model innovation in software product industry* (pp. 79–93). New Delhi: Springer India.

Schroer, J., & Hertel, G. (2009). Engagement in an open web-based encyclopedia: Wikipedians and why they do it. *Media Psychology*, *12*, 96–120.

Speier, C., Valacich, J. S., & Vessey, I. (1999). The influence of task interruption on individual decision making: An information overload perspective. *Decision Sciences*, *30*, 337–360.

Stone, D. L., Deadrick, D. L., Lukaszewski, K. M., & Johnson, R. (2015). The influence of technology on the future of human resource management. *Human Resource Management Review*, *25*, 216–231.

Stone, D. L., & Dulebohn, J. H. (2016). Challenges that may prompt changes in hrm theory and research. In D. L. Stone & J. H. Dulebohn (Eds.), *Research in human resource management: HRM theory and research for the new employment relationship*. Charlotte, NM: Information Age.

Stone, E. F., & Stone, D. L. (1990). Privacy in organizations: Theoretical issues, research findings and protection mechanisms. In G. Ferris & K. Rowland (Eds.), *Research in personnel and human resource management* (*Vol. 8*) (pp. 349–411). Greenwich, CT: JAI Press.

van Iddekinge, C. H., Lanivich, S. E., Roth, P. L., & Junco, E. (2016). Social media for selection? Validity and adverse impact potential of a Facebook-based assessment. *Journal of Management*, *42*, 1811–1835.

Walker, H. J., Feild, H. S., Giles, W. F., Bernerth, J. B., & Short, J. C. (2011). So what do you think of the organization? A contextual priming explanation for recruitment web site characteristics as antecedents of job seekers' organizational image perceptions. *Organizational Behavior and Human Decision Processes*, *114*, 165–178.

Wrzesniewski, A., & Dutton, J. E. (2001). Crafting a job: Revisioning employees as active crafters of their work. *Academy of Management Review*, *26*, 179–201.

Part I
Individual Perspectives

2

Digitized Communication at Work

Nicole C. Krämer and Stephan Winter

Introduction

The emergence of the Internet has fundamentally altered the opportunities of communication and connections around the world, for instance by enabling asynchronous or synchronous exchange over a variety of channels and platforms from e-mail to social networking sites. These changes have not only sustainably affected people's private interactions but also changed the way people exchange information in their corporate lives. Within organizations, employees initiate videoconferences with their distant colleagues, collaborate with co-workers in digital rooms such as Google Docs, or exchange news within Enterprise Social Media platforms. In order to provide a basis for understanding how these new ways of interaction change traditional processes within organizations, we discuss theories and models that explain the selection of specific channels as well as the characteristics and psychological effects of computer-mediated communication (CMC).

Retracing the developments of technological changes in CMC, we will distinguish between two phases of research: first, an early phase in which communication in the world wide web and its derivates was primarily one-sided and interaction was largely conducted via e-mail or anonymous forums (and in which "classic" theories of CMC emerged) and, second, a more recent phase in which the Web 2.0 (O'Reilly, 2005) and the so-called "social media" platforms were established. This second phase was characterized by an opening and flexibilization of the Internet into a more participatory medium in the sense that users started to actively generate content of their own – rendered possible by clearing technical barriers. Internet users were no longer mere consumers but additionally became producers of information – which inspired the term prosumer (a combination of the words producer and consumer; Ritzer & Jurgenson, 2010). In parallel, users started to self-disclose large amounts of personal information and connect to other people behaving likewise. Related to these changes, expectations were raised that people's new willingness to engage in information exchange and knowledge production would also transfer to corporate life and organizational processes if social media would be incorporated in organizations. We will therefore, in the second part of the chapter, also review approaches that model relevant factors for digitized communication in the age of social media and information diversity.

The Wiley Blackwell Handbook of the Psychology of the Internet at Work,
First Edition. Edited by Guido Hertel, Dianna L. Stone, Richard D. Johnson, and Jonathan Passmore.

Classic Theories – How Beneficial is Computer-Mediated Communication?

Before the Internet gained wider popularity among the general public, universities and pioneers among companies and organizations were already employing digitized communication in the 1970s and 1980s. Soon, questions about the specifics and effects of this new medium emerged: how can people relate to one another, exchange opinions, or work with each other when they communicate via e-mail, bulletin boards, or chats? The lack of nonverbal cues was regarded as the most striking difference from face-to-face (FtF) communication and even to telephone communication (which also lacked cues about the other's appearance and her or his proxemic and kinesic behavior, although it at least transmitted vocalic cues). Furthermore, most CMC applications (apart from chats) were asynchronous. In the following, we will review theoretical approaches that describe the motivations why people selected these new channels, and the effects of CMC on people's interpersonal and intergroup impressions as well as their task fulfillment.

Selection of computer-mediated communication

Daft and Lengel (1986) argued that media can be categorized according to their richness, which is affected by the diversity of channels that are available, the opportunities for feedback, and the degree to which they are able to convey personal feelings and emotions. Typically, FtF is evaluated as richer than telephone and e-mail. Furthermore, Daft and Lengel (1986) distinguished between different types of task when working within an organization along the characteristics of uncertainty and equivocality (how ambiguous a situation is and how difficult it is to find a common ground). Media richness theory posits that a matching between tasks and the richness of a medium leads to better outcomes in terms of efficiency: rich media are appropriate for equivocal tasks, whereas simple tasks should lead to the selection of a lean medium. Such a matching would avoid overcomplication (if a rich medium is selected for a simple task) as well as oversimplification (if a lean medium is selected for an ambiguous task). In a survey study, Daft, Lengel, and Trevino (1987) showed that managers choose richer channels for more equivocal task scenarios, and those who more often select matching media to the specific task were more likely "high performing managers." However, this finding only represents an indirect test of the hypothesized relation between appropriate media choices and performance outcomes. An experimental study did not find that a match between task equivocality and the richness of a medium (e.g., video vs. text-based CMC) leads to better decision outcomes (Dennis & Kinney, 1998). Additionally, results of a later investigation regarding decisions to select rich versus lean media (here: voice mail vs. e-mail) were not in line with the predictions of media richness theory (El-Shinnawy & Markus, 1997).

Dennis and Valacich (1999) extended the approach of media richness theory and highlighted the importance of enabling the process of simultaneously working together. The authors proposed the immediacy of feedback, the number of potential simultaneous conversations (parallelism) as well as the rehearsability and reprocessability of messages and the variety of cues as determinants of media synchronicity. With regard to tasks, simple scenarios of transmitting information (conveyance) are distinguished from more complex situations in which co-workers have to find a compromise or choose between contradictory options (convergence). According to media synchronicity theory (Dennis & Valacich, 1999), synchronous channels are selected for convergence and less synchronous channels for conveyance tasks. In an experimental study (Dennis, Valacich, Speier,

& Morris, 1998), participants were asked to generate ideas (conveyance) as well as select the best idea (convergence), either in a written exchange (low levels of feedback, low synchronicity) or in an FtF (high levels of feedback, high synchronicity). Participants generated more ideas when communicating in written form, but the task of reaching a consensus was solved more quickly in FtF sessions. This finding is in line with the basic assumption of media synchronicity theory but focuses only on the aspect of feedback and not on media-related aspects (such as CMC scenarios). However, empirical research on the other factors of the theory remained scarce.

While the models mentioned here assume that people make rational decisions about the most appropriate medium for a given situation, Fulk, Steinfield, Schmitz, and Power (1987) argued that the perception of media is not only influenced by objective characteristics but also – and to a stronger extent – by subjective impressions (e.g., one's own competence with regard to the medium) and social factors. Their social influence model of technology use holds that media use is affected by perceived task requirements and attitudes toward communication media, both of which are shaped by social influence. In this line, a survey study with employees of a research center (Schmitz & Fulk, 1991) showed that the use of e-mail was related to the opinions toward and usage of this communication channel among co-workers and supervisors. The social influence perspective can therefore be regarded as an important amendment to the more rational models concerning media richness and synchronicity. From a practical point of view, it may help to identify situations in which social norms or preferences of superiors lead organizations and employees to use communication systems that are not appropriate for specific tasks.

Effects of computer-mediated communication

Deficit theories Regarding the question of how beneficial or satisfying the new medium might be, early theories highlighted the deficits of CMC in comparison with FtF and argued that the lack of cues would lead to lower socioemotional quality of the formed impressions and relations. Culnan and Markus (1987) critically named these assumptions as the cues-filtered-out approach. Short, Williams, and Christie (1976) were among the first researchers who coined this view: their influential social presence theory posits that electronic media have a lower capacity to transmit nonverbal information and that fewer channels would lead to less attention that people pay to the presence of others. Their original work referred to videoconferencing (restricting proxemic and kinesic cues) and audioconferencing (filtering any visual cues) – later, they were applied to computer-mediated conferencing and messaging systems (e.g., Rice & Case, 1983; Rice & Love, 1987).

Building on this research, the lack of social cues hypothesis (Kiesler, Siegel, & McGuire, 1984; Sproull & Kiesler, 1986) specified that particularly information about social background of the interaction partner and the context of the exchange is filtered out in CMC. For instance, viewing the other helps to detect the social group to which he or she belongs (gender, age, ethnicity, social status), and viewing the facial expression as well as the physical environment contextualizes the situation. However, this was thought to be impossible in CMC. It was therefore assumed that people have more difficulties in inferring communicative norms. This may have positive consequences in that CMC can become more egalitarian when differences of hierarchy, gender, or ethnicity are less important. Sproull and Kiesler (1986) argued that electronic messages in organizations likely foster new ideas, and it was hoped that digitized communication would generally lead to a less restricted exchange of opinions (see Papacharissi, 2002). In the same vein, Joinson (2001) found higher levels of spontaneous self-disclosures among strangers using CMC than FtF, which can have a liberating effect. However, the reduced amount of social control may lead to violation of norms and negative outcomes of disinhibition, such as flaming or other forms of incivility (Gervais, 2015; Lapidot-Lefler & Barak, 2012). In their pioneer survey study of e-mail communication in organizations, Sproull and Kiesler (1986) found that people focused more on themselves, overestimated their own contributions, and behaved more irresponsibly than in FtF.

The cues-filtered-out approach was a dominant position in the early days of the Internet. Similarly, first studies showed negative consequences of Internet usage such as increased loneliness (Kraut et al., 1998) while later studies by the same research group found that these tendencies had largely disappeared (Kraut et al., 2002)). Although the assumption that specific cues are missing is adequate, the deficit approaches can be criticized due to their restricted view of CMC users and an idealization of FtF communication (see Döring, 2003). Indeed, these considerations have paid less attention to the opportunities to compensate for channel limitations as well as to the fact that FtF is also restricted in several ways (for instance, with regard to time and space). Meanwhile, anecdotal reports emerged about people who develop online friendships or even fall in love with their chat partners, or about virtual communities with a lot of active participation that called the early deficit theories of CMC into question.

Adaptation theories In the light of the developments in the early 1990s, researchers began to analyze the potentials of CMC and the mechanisms by which people are still able to form rich impressions and relationships (which may sometimes even exceed those of offline encounters). These approaches, which Walther (2006) called adaptation theories, also refer to the metaphor of channels. While accepting the limitation of visual cues they describe situations in which users creatively overcome these restrictions.

In his social information processing theory, Walther (1992) posits that people develop impressions of and relations with people, no matter what medium they use. He questions the notion that CMC lacks socioemotional qualities and instead proposes a functional approach of communication. In this view, there are no fixed roles for verbal and nonverbal cues (for instance, in the sense that verbal information would be necessary to convey content and nonverbal cues would always provide affective information). This leads to the conclusion that nonverbal cues do not have a monopoly on fostering interpersonal impressions and relations and that this may also happen through written text. As one major difference from FtF, it is assumed that more time is required to develop impressions in CMC. However, given sufficient time and messages, there should be similar impressions and relations.

These assumptions received support in empirical studies that compared people who worked together FtF or via CMC over several weeks. Walther (1993) found that, at first, impressions developed more strongly in the FtF groups (consistent with the assumptions of the cues-filtered-out approach). But over time, impressions gradually increased in CMC, and at the end, they were on a similar level as in FtF (which is no longer consistent with the cues-filtered-out approach). The interchangeability of verbal and nonverbal cues in their capacity to arouse interpersonal impressions was demonstrated by Walther, Loh, and Granka (2005): in their experiment, participants were prompted to create a positive versus negative relationship with their interaction partner, either in a virtual chat or in a FtF meeting. The instructions did not specify how to achieve this goal and thereby left open the cues, expressions, and behaviors used by participants to make a positive versus negative impression on their interaction partner. Results showed that participants' behavior led to impressions of affinity that were consistent with the experimental manipulation of positive versus negative goals, regardless of the channel. This suggests that it is possible to induce liking/disliking in different modes and therefore that both channels can lead to equivalent affective outcomes. An analysis of participants' behavior showed that the differences in liking in FtF were mainly due to vocalic and kinesic cues; there were only few differences in the verbal content that was expressed; in contrast, differences in the CMC condition were strongly based on verbal information (e.g., statements of disagreement). These findings suggest that CMC users, who are restricted in their opportunity to convey liking or disliking nonverbally, find other ways to express their affinity.

The development of social information processing theory marked a fundamental departure from the deficit-oriented models as it described how users overcome the lack of nonverbal cues. In 1996, Walther proposed a model that extended these assumptions and focused on the advantages "that accompany the flexibility of communicating via language and only language" (Walther, Van der Heide, Ramirez, Burgoon, & Pena, 2015, p. 4). According to the hyperpersonal model (Walther, 1996), CMC can lead to more positive and more intense impressions of others than FtF – thereby, CMC might sometimes be even more personal than the FtF mode, which has typically been regarded as the gold standard of relational communication. The model names four characteristics of CMC that foster such hyperpersonal impressions: senders, channel, feedback, and receivers. Senders have more control and more time for self-presentation than in FtF and are thereby able to carefully select positive aspects of themselves. This is stimulated by characteristics of the channel such as the opportunity to edit messages before sending them. In this regard, the act of publishing a message on a public CMC channel has been shown to strengthen the attitude or the self-presentational view that was conveyed in this message. In a study by Gonzales and Hancock (2008), participants were asked to write a public versus private CMC message in which they portrayed themselves as extrovert versus introvert. Those who posted messages on a public channel internalized their self-presentation, suggesting that composing and sending public messages can lead to identity shifts. These changes in one's self-views are likely to be reinforced if the sender's messages receive positive comments (Walther et al., 2011). Receivers of CMC messages are likely to overestimate the limited number of given cues and/or to interpret them based on their prior impression of the sender – in cases of favorable impressions, this may lead to an idealization of the interaction partner. Although the predictions and empirical tests about the receiver aspect of the model are less specific than for the processes of selective self-presentation, there is some support for the assumption of stronger interpersonal attributions in CMC. In an experiment by Jiang, Bazarova, and Hancock (2011), participants engaged in an FtF versus CMC discussion with a confederate who showed either high or low levels of self-disclosure. Results showed that high self-disclosure led to greater intimacy, and this association was stronger in CMC, which was explained by the assumption that participants attributed the confederates' self-disclosure more strongly to the interpersonal relationship than they did in an FtF interaction. Taken together, these processes may lead to hyperpersonal dynamics such that "when a receiver comes upon a selectively self-presented message and idealizes its source, that individual may respond in such a way that reciprocates and reinforces the partially modified personae, reproducing, enhancing, and potentially exaggerating them" (Walther et al., 2015, p. 16).

In a somewhat similar fashion, compensatory adaptation theory (Kock, 1998, 2005) highlights the potential of CMC to produce "better" results than FtF but provides different explanations derived from theories of human evolution. It is assumed that the biological communication system has been optimized for FtF and therefore modes of communication with other characteristics increase cognitive efforts. At the same time, humans are motivated to overcome obstacles, and this (over)compensation may even lead to better outcomes than in a more convenient situation without obstacles. Empirically, Kock (1998) analyzed groups that discussed how to optimize work processes in their university, either in an FtF setting or via e-mail. Results showed that e-mail was perceived to be more effective in terms of group success and quality, indicating that people can overcome the initial limitations of CMC and exploit the actual advantages (such as the opportunity to communicate asynchronously and from different locations).

While the hyperpersonal model and social information processing theory mainly focused on relational and interpersonal communication, another influential theory captured the communication patterns of groups in computer-mediated settings: the social identity model of deindividuation effects (SIDE; Postmes, Spears, & Lea, 1998; Reicher, Spears, & Postmes, 1995). SIDE proposes that visual anonymity and the salience of a group identity leads to depersonalization and deindividuation. These notions are based on social identity and self-categorization theories (e.g., Tajfel, Billig, Bundy, & Flament, 1971), holding that individuals who see themselves as members of a certain group (e.g., students of university X, employees of company Y, fans of football club Z) are more susceptible to social influence if the specific identity is activated. According to the SIDE model, CMC's characteristic of visual anonymity reinforces this process. When other members of the group are not visible, individual differences are masked, and therefore, the interaction partners likely appear more similar than in FtF interactions (depersonalization), where one would see unique characteristics of individual persons, some of which arouse more positive and some more negative impressions. Additionally, people may feel unidentifiable and therefore shift their attention to the group and away from self-presentational concerns (deindividuation). As a result, deindividuated CMC users are seen as even more susceptible to peer influence than in offline settings, which may have positive or negative consequences, depending on the norms that are provided in the salient group.

Classic experimental investigations of the SIDE model (e.g., Spears, Lea, & Lee, 1990) typically manipulate the degree of visual anonymity and the focus of the instructions. Participants taking part in a CMC discussion are either in the same or in different rooms (visible versus anonymous, respectively), and instructions either address participants as individuals (personal identity) or as group members (social identity). Results show higher degrees of identification and social attraction when people are visually anonymous and the group identity is salient. This leads to a stronger adherence to group norms and stronger persuasive influence of group members (e.g., Lee, 2004). Similar patterns, that is, consistent with the predictions of SIDE, have also been found outside the laboratory: Sassenberg (2002) analyzed chat protocols of common identity groups (sharing a common interest or hobby) and common bond groups (based on personal connections) and found that chat groups with a common identity showed a stronger similarity in their language use.

Although all of these adaptation theories highlight that CMC can be as social as FtF, or even more so, there are notable differences. Whereas the social information processing theory predicts that people get to know each other as individuals (even when working together in a task-oriented manner), the SIDE view focuses on CMC users who see others as group members. A tentative resolution may lie in the boundary conditions of the theories: according to Walther and Carr (2010), SIDE more adequately explains settings of restricted interaction without anticipation of future cooperation, and the social information processing propositions become relevant when people expect to cooperate and communicate further with an interaction partner or members of a group.

In the realm of industrial and organizational psychology, the discussion on effects of CMC has been taken one step further to the question of whether CMC applications alter the existing structures of organizations. Here, constructivist and deterministic assumptions can be distinguished (Brotheridge, Neufeld & Dyck, 2015). While constructivist assumptions would claim that CMC is used in a way that it reinforces existing structures

(Arrow, 1997; Ducheneaut, 2002), the deterministic perspective suggests that technology may facilitate change in organizations especially by altering hierarchical communication (Lemke, 1999). Since the latter assumptions heavily rely on the lab experiments depicted above (Spears & Lea, 1992), it can be questioned whether they are generalizable to organizational life. More recently, the construct of affordances has been introduced to this area in order to resolve the constructivism–determinism debate (Boudreau & Robey, 2005; Brotheridge et al., 2015; Hutchby, 2001).

Originally, Gibson (1979) introduced the concept of affordances in a more general way to explain how species perceive and react to objects in their environment. The reactions are seen as determined by the perception of the possibilities for actions that the objects themselves offer (e.g., a chair's affordance character is to sit down on it). This notion has been transferred to technology (Kaptelinin & Nardi, 2012) in the sense that different technologies likewise include different affordances of how to use them. Therefore, an affordance lens "directs attention to what users perceive themselves as being able to do with technology" (Larsen, 2015, p. 65). In line with this, Brotheridge et al. (2015) propose that hierarchical communication in an organization is informed by what a communication medium allows. They suggest six attributes which might be helpful to identify the affordances of different communication media: communications frequency, feedback immediacy, message length, information seeking, idea generation and socioemotional expression. An affordance perspective might also help in spanning the gap between the old days of CMC and new platforms and applications within the area of social media, which are characterized by partly new and different affordances (see Treem & Leonardi, 2012). In conclusion, the affordance assumptions might be beneficial as a theoretical framework that helps to focus on overarching questions and results in the sense of general mechanisms instead of merely addressing specific features and phenomena of specific platforms.

Digitized Communication in the Age of Social Media and Information Diversity

While most of the theories and assumptions discussed in the previous section were developed and tested in a media landscape in which CMC was common, Web 2.0 platforms that were first developed around the year 2005 enabled communication that is – at least with regard to some aspects – radically different from classic technologies for CMC. In the second decade of the twenty-first century, social media are ubiquitous and enable people to broadcast easily via the Internet (or Intranet, that is, Internet-like infrastructures within organizations) information about oneself, one's interests and occupations to large groups of people. Users can communicate via various online applications: blogs and micro-blogging services, social networking sites (SNS) such as Facebook, news boards, rating platforms such as Tripadvisor, wikis, and video platforms offer a variety of opportunities for communication, self-presentation, and information exchange. Still similar to the possibilities of the 1990s is the fact that social media can facilitate social interaction and make interpersonal communication more efficient by overcoming limitations in space and time and that people still tend to communicate predominantly via plain text (Walther et al., 2015). Some of the differences, however, are that people communicate to larger groups of people and networks and that at least the most popular social media platforms, such as Facebook, require an account with the user's real name so that people's identity is more salient compared with the chat-rooms or multi-user dungeons/domains (MUDs) that were commonly used 25 years ago. Many of the classic theories, moreover, start out from the assumption that the persons who communicate online do not know each other personally, while more recently people seem to use social media applications first and foremost to interact with people they already know and are friends with. Also, information about the network itself (for example who is connected to whom) and that shared by others is displayed to everyone.

In the past years, these possibilities have also entered the field of organizations – especially fueled by the observation that people willingly engage in producing content in social media platforms. Therefore, numerous companies have provided possibilities to use social media: what has been termed "Enterprise Social Media" (Leonardi, Huysman & Steinfield, 2013). Another term used in the literature is "Organizational Social Media," defined as follows:

> *Organizational social media are technology artefacts, both material and virtual, that support various intra- and extra-organizational actors – including management, employees and external stakeholders – in a multiplicity of organizational communication activities for producing user-generated content, developing and maintaining social relationships, or enabling other computer-mediated interactions and collaborations in the context of a specific organization and its environment.*
>
> Van Osch and Coursaris (2013, p. 703)

This definition focuses on the functions and possibilities the technology provides and also includes the notion that social media are used by organizations for communication with external parties (which has been studied more commonly, see Leonardi et al., 2013). However, other definitions zoom in on the opportunities of employing social media for internal communication and center around the specific activities of employees which Enterprise Social Media enable – they are:

> *web-based platforms that allow workers to (1) communicate messages with specific coworkers or broadcast messages to everyone in the organization; (2) explicitly indicate or implicitly reveal particular coworkers as communication partners; (3) post, edit, and sort text and files linked to themselves or others; and (4) view the messages, connections, text, and files communicated, posted, edited and sorted by anyone else in the organization at any time of their choosing.*
>
> Leonardi et al. (2013, p. 2)

This latter definition especially highlights several aspects that make social media unique and distinct from earlier technologies. As decisive difference, the authors stress the fact that everyone can see both the connections and the communication occurring between those who are not (yet) their communication partners. In order to enable this, most Enterprise Social Media mimic popular SNS and additionally provide blogs, wikis, social tagging, and document sharing. Organizations thereby have the choice of three different forms of Enterprise Social Media (Leonardi et al., 2013). They can either utilize publicly available sites such as Facebook, provide a private implementation of Open Source or proprietary software (e.g., Microsoft's Sharepoint, IBM's Connections, Yammer, Jive) by which strictly internal wikis, blogs, or SNS are enabled or an in-house proprietary solution (prototypes that might later be sold to other companies, e.g., Beehive at IBM). In order to differentiate functions of Enterprise Social Media, Leonardi et al. (2013) illustrated three metaphors which can be used to describe the effects of Enterprise Social Media. The notion of Enterprise Social Media as a leaky pipe incorporates first and foremost the aspect of visibility of what others are communicating, which might be advantageous for learning and knowledge sharing and to build social capital. Enterprise Social Media, however, can also be understood as echo chambers, providing room for reciprocal verification of like-minded people with like-minded thoughts, which might increase the danger of groupthink. Finally, Enterprise Social Media are described as social lubricants, giving the opportunity to connect easily to others, either when getting to know each other or when working together. Enterprise social media are expected to ease the building of social capital by giving people information that they can use as a starting point in their conversations. First studies indeed show that several of the assumed effects can be demonstrated empirically. For example, Jackson, Yates, and Orlikowski (2007) found that blogging in an organization can lead to both social and informational benefits. With regard to the former, employees reported feeling more as a part of the community, and concerning the latter, feedback on ideas was given. Similarly, Steinfield, Dimicco, Ellison, and Lampe (2009) described how usage of an Enterprise Social Media leads to increased social capital, for instance by providing access to new people and expertise. Leidner, Koch, and Gonzalez (2010) conclude that Enterprise Social Media can be especially beneficial for newcomers by cultivating a sense of belonging, of being part of a family.

In the following, we will focus on three different aspects, which are important parts of organizational activities and which have been influenced heavily by the emergence of social media platforms. We address two fields that concern not only the individual but also the organization as a whole, namely new opportunities for building knowledge and meta-knowledge as well as building relationships. The third field focusses the realm of cooperation in small groups.

Building knowledge and meta-knowledge/visibility

According to Treem and Leonardi (2012), visibility constitutes a central affordance of Enterprise Social Media. The authors argue that the display of the network structure (Who is friends with whom?) and the public form of communication via wall posts and so on make "communication network connections that were once invisible (or at least very hard to see) visible to others in the organization" (Treem and Leonardi, 2012, p. 150). Without Enterprise Social Media, the visibility of work processes in contemporary jobs is decreased as there are less physical manifestations of work than in the age of industrialization. Knowing what others are doing is possible when observing others (although this is probably less informative in computer-based tasks), overhearing colleagues' conversations, or asking them – but this is all relatively unlikely or is restricted to close colleagues due to department structures or dispersed geographic locations. In line with the leaky pipe metaphor (Leonardi et al., 2013), Enterprise Social Media are able to re-establish visibility in organizations, encourage knowledge sharing, and enable co-workers to gather information and acquire knowledge by observing. Majchrzak, Faraj, Kane, and Azad (2013) emphasize the importance of features by which users can follow threads that are interesting for them, for instance, with notifications (triggered attending) that help to stay up to date and at the same time reduce information overload.

Against this background, Leonardi (2014) has proposed a theory of communication visibility, positing that the visibility of communication between co-workers enhances meta-knowledge, that is, knowledge "about who knows what and who knows whom within an organization" (p. 797). According to this theory, message transparency (public access to messages between co-workers or routine communication with customers) is crucial for the knowledge of who knows what. Furthermore, the display of connections is assumed to elicit network translucence. This information tends to be weaker and not fully transparent but still useful to increase the knowledge of who knows whom ("clear enough to show that a relationship exists, but not clear enough to show the scope and/or nature of that relationship" [Leonardi, 2014, p. 805]). The theory predicts that these aspects of visibility lead to a reduction of double work (in the sense that people more likely notice if other employees are working on the same question), with increases in productivity, and also foster innovation (as the combination of given knowledge becomes easier). A qualitative field study in a financial services organization yielded results that were largely consistent with these optimistic propositions (Leonardi, 2014). In a more recent facet of their work, Leonardi and Meyer (2015) posited that the fact that Enterprise Social Media make others' communications visible to everyone leads to so-called ambient awareness, which then helps knowledge seekers to lubricate own interactions with knowledge sources. This ties in with the findings of Breuer, Hüffmeier, and Hertel (in press), who demonstrated that the effectiveness of teams is either fueled by trust or can be compensated by documentation of the interaction.

Visibility may also result in greater likelihood of knowledge sharing. In related studies on contributions to wikis, Moskaliuk, Kimmerle, and Cress (2009) showed that high levels of perceived importance of the topic and moderate incongruences between the current state of the wiki and one's own view help to support knowledge contribution. On the one hand, visibility of Enterprise Social Media may mitigate the concern that one's own contribution is not properly recognized when a valuable idea is posted (see Cress & Kimmerle, 2008). On the other hand, the large audience may foster self-presentational behavior (Leonardi, 2014) that is less concerned with the true merits of a solution but rather aims to convey the impression of an active employee. Furthermore, employees might be afraid of the opportunities of surveillance that are connected to high levels of visibility (Leonardi et al., 2013), which could restrict creativity.

Building and maintaining relationships

As the results by Leonardi and Meyer (2015) already show, it is not only important to know who communicates what and with whom but this meta-knowledge can also be used to build new relationships which – in turn – enable more direct access to knowledgeable and important people. This notion is in line with the original idea of the term social capital, which has been defined as:

> *the sum of the resources, actual or virtual, that accrue to an individual or a group by virtue of possessing a durable network of more or less institutionalized relationships of mutual acquaintance and recognition.*
>
> Bourdieu and Wacquant (1992, p. 119)

By establishing and maintaining relationships, people can gain a number of emotional, structural, and economic advantages. Numerous theories postulate different forms of social capital. For example, Granovetter (1973) distinguished between strong and weak ties. Based on this, Putnam (2000) described the attributes of weak ties and strong ties as bridging and bonding social capital. Weak ties (for instance, colleagues and acquaintances) primarily provide bridging social capital and information that is not yet known and might broaden horizons, whereas strong ties are associated with bonding social capital in that they provide emotional and social support. Relationships with weak ties are characterized by breadth rather than depth since they consist of people from different backgrounds connecting (i.e., bridging) over social networks to provide access to novel information. Recent research on online social capital takes up the early notion of Granovetter (1973) about the "strength of weak ties" and suggests that weak ties and their power to provide new information (e.g., about potentially interesting job offers) are what makes it worthwhile to use SNS such as Facebook.

Ellison, Steinfield, and Lampe (2007) were the first to apply empirically the concept of social capital to SNS. In a survey study among students, they demonstrated that the intensity of Facebook use is positively associated with individuals' perceived bridging social capital as well as their perceived bonding social capital. The extent to which social capital is beneficial seems to be influenced by the way the features of SNS are used: Burke, Marlow, and Lento (2010) demonstrated that active (vs. passive) engagement with Facebook leads to more beneficial outcomes of SNS use in terms of building social capital. Ellison, Wohn, and Greenhow (2014) focused on the relationship between specific Facebook communication behaviors and perceived bridging social capital. Their results suggest that social grooming and attention-signaling activities are important to maintain access to resources in one's network as measured by bridging social capital. This is in line with prior research by Ellison, Vitak, Steinfield, Gray, and Lampe (2011), who found that self-disclosure is necessary to benefit from the network.

As regards organizations and the use of Enterprise Social Media, it seems obvious to tune in to the notion put forward by Granovetter (1973) that weak ties might be more important than strong ties. This is in line with Leonardi's visibility approach, which also stresses that by increasing the visibility of what happens in the network of ties, people might receive knowledge they would otherwise not receive. Numerous empirical studies have supported Granovetter's thesis in that the diversity of a weak-tie network leads to greater gains. Similarly, workers in organizations benefit from using weak-tie connections via CMC (Pickering & King, 1995). Also, a large-scale field experiment using social network technologies showed that weak ties are superior in information dissemination (Bakshy, Rosenn, Marlow, & Adamic, 2012). While stronger ties were individually more influential, weak ties were responsible for the propagation of novel information within the social media environment. More recent research, however, has shown that at least with regard to subjectively perceived benefits, weak ties are perceived to contribute not only less emotional support but also less informational support compared with strong ties (Krämer et al., 2014). As this has been demonstrated in the area of private interactions, it does not necessarily hold true for organizational interactions and Enterprise Social Media. Depending on the way such a site is designed (in the sense of how closely it resembles the typical structure of Facebook and also includes self-presentation and impression management, see Krämer & Winter, 2008), the specific effects of weak and strong ties haveto be considered carefully. Related to this, the blurring of boundaries between private and public ties can be disadvantageous (Leonardi et al., 2013) as such a "context collapse" (Marwick & boyd, 2010) can lead to problematic encounters.

Organizing cooperation

With regard to the general question whether CMC enhances or hinders cooperation in groups, there is a large body of research that has been summarized in previous sections of this chapter. The interesting question to be discussed here is whether the new communication opportunities provided by Enterprise Social Media will lead to specific advantages or disadvantages with regard to cooperation and communication within small working groups. Contractor and Monge (2002) suggest that networked organizations have an influence on the way small-group communication networks function. Drawing on Bavelas' (1948; Leavitt, 1951) theory on small group communication networks, one can conclude that due to the visibility and meta-knowledge afforded by CMC nowadays, there are no longer (more or less) centralized communication forms such as chain, circle, and wheel. Rather, all communication networks are necessarily "comcon" (completely connected) and therefore maximally decentralized, as everyone interacts with everyone. Based on the results of empirical studies (Shaw, 1964) that demonstrate centralized organizations are more efficient for routine tasks whereas decentralized networks are more advantageous for tasks that involve collaborative problem solving, it can be assumed that groups cooperating via Enterprise Social Media will also experience advantages with regard to more creative tasks. Moreover, more satisfaction with the work process can be expected. As other research shows, these beneficial outcomes can be maximized if the virtual groups adhere to the six simple rules (Walther & Bunz, 2005) derived from the social information processing theory (discussed earlier in this chapter) and its assumptions of how to overcome the lack of visual cues: get started right away, communicate frequently, organize multitasking and do substantive work simultaneously, acknowledge that you have read one another's messages, be explicit about what you are thinking and doing, set deadlines and stick to them. These rules were tested in a study in which students were assigned to work in virtual teams and to write papers collaboratively. Although the results did not consistently show an influence of (self-reported) adherence on the performance of the groups, following the rules led to greater trust, social attraction, and liking between team members. As the first studies more specifically testing the effects of using SNS and Enterprise Social Media for small-group collaboration emerge, there is large potential for future research (see next section; see also Maynard, Gilson, Jones Young, & Vartiainen, 2017, Chapter 15 in this volume). For example, Carter, Asencio, Wax, DeChurch, and Contractor (2015) argue that new big data approaches will lead to rapid progress with regard to theory of teams. Not only will this enable researchers to test theories (e.g., the input-process-outcome model suggesting that team processes are the mediating link between inputs and outcomes [McGrath, 1984; Steiner, 1972]) in a better way than meta-analyses, but it will also render the inclusion of new constructs (e.g., virtual proximity) and the application to new phenomena (voluntary virtual communities such as Wikipedia) (Keegan, Gergle, & Contractor, 2012) possible. Especially with a view to these new forms of organizations (but probably also for common organizations employing Enterprise Social Media), additionally analyses on the emergence of leadership have to be conducted. Early research demonstrates that besides the influence of traditional factors such as formal role and centrality, people who post many positive and concise messages with simple language that is familiar to other participants are perceived as leaders (Johnson, Safadi, & Faraj, 2015).

Future Research

Interestingly, the current literature on new applications such as Enterprise Social Media focuses more on opportunities and chances than on potential disadvantages and drawbacks. A necessity for future research therefore is to provide a more complete and realistic account of the effects of using Enterprise Social Media in organizations. Specifically, the potential drawbacks on an individual as well as on an organizational level have to be considered. With regard to the latter, and in line with the echo chamber metaphor (Leonardi et al., 2013), it has to be analyzed to what degree Enterprise Social Media will merely lead to recursive processes in which like-minded people are connected and conflicting ideas are suppressed. These recursive processes might also lead to more fragmentation of different communities and the increased likelihood of groupthink in subgroups. Also, Leonardi et al. (2013) warned against relying too much on analyses of the Enterprise Social Media structures in order to diagnose the state and processes of the organization. The structures visible in the Enterprise Social Media might not match the processes in the organization as the most active Enterprise Social Media users might not be the most active or knowledgeable members offline.

For the individual user especially the potential social overload can be detrimental. Here, the social components of Enterprise Social Media might introduce interruptions in corporate life that can hinder productivity (Agarwal & Karahanna, 2000) and be perceived as annoying. Additionally, the surveillance aspect (the fact that people know that their actions and communication are visible) may lead members of the organization to avoid this form of communication altogether and search for other possibilities that better preserve their anonymity and autonomy (Leonardi et al., 2013).

Although first studies and insights have been presented, van Osch and Coursaris (2013) argue that research on Enterprise Social Media needs to be intensified. They especially argue that research focuses too much on the individual user as the level of analysis instead of relating users' behavior to higher-level units of analysis (group and organizational levels). Here, new methodological approaches of network analysis within organizations (see Contractor & Monge, 2002) might be helpful to jointly analyze individuals, groups, the organization, and external partners. Current developments and changes in organizational landscapes as particularly fueled by globalization need to be investigated by appropriate methods. In addition, it will be more important to analyze emergent instead of formal networks as they are reflected in the organizational chart (Contractor & Monge, 2002). This emergence will not only be more important in formal organizations as they employ new communication platforms such as Enterprise Social Media but also in work communities that do not (yet) have a formal structure (e.g., Open Source software communities).

This chapter's focus on theory and general assumptions reflects the sparsity of practice-based research and theory testing. However, both are difficult to accomplish. On the one hand, practice-based research is difficult to organize, because, given the nature of the phenomenon of Enterprise Social Media, whole organizations would have to be involved. Theory-based research, on the other hand, also needs additional prerequisites. The assumptions presented so far need to be concretized further to enable testing of fundamental mechanisms in systematically controlled conditions. Therefore, future endeavors need to consider further theory development as well as empirical investigations. In sum, while the outcomes of employing Enterprise Social Media in organizations have to be uncovered yet, it can already be concluded that the data gathered in Enterprise Social Media will lead to innovative opportunities for research. The new methods of big data analysis will be helpful in combining theories, social science methods and digital data streams in a way that will lead to "unprecedented and unparalleled" opportunities for organizational research (Carter et al., 2015, p. 553).

Conclusion

The present chapter has provided an overview on classic and current approaches to CMC. What should have become apparent is that the "classic" theories and assumptions partly no longer fit the new forms of CMC as these do not conceal people's identity, make communication structures visible, and often do not distinguish between personal and job-related information. The new approaches, however, are not sufficiently mature and cannot yet be denoted as "theory." This demands much research and theoretical modeling within the next years, but there is reason to be optimistic that such attempts will be fruitful. This is especially because new phenomena not only enable new kinds of communication but, at the same time, they provide unique opportunities for employing innovative methods that could help delineate relevant factors and build new theories.

References

Agarwal, R., & Karahanna, E. (2000). Time flies when you're having fun: Cognitive absorption and beliefs about information technology usage. *MIS Quarterly*, 665–694. DOI:10.2307/3250951

Arrow, H. (1997). Stability, bistability, and instability in small group influence patterns. *Journal of Personality and Social Psychology*, *72*, 75–85.

Bakshy, E., Rosenn, I., Marlow, C., & Adamic, L. (2012, April). The role of social networks in information diffusion. Paper presented at the Association of Computing Machinery World Wide Web Conference, Lyon, France. Retrieved April 3, 2017, from http://arxiv.org/pdf/1201.4145v2.pdf

Bavelas, A. (1948). A mathematical model for group structures. *Human Organization*, *7*, 16–30.

Boudreau, M.-C., & Robey, D. (2005). Enacting integrated information technology: A human agency perspective. *Organization Science*, *16*, 3–18.

Bourdieu, P., & Wacquant, L. J. D. (1992). *An invitation to reflexive sociology*. Chicago: University of Chicago Press.

Breuer, C., Hüffmeier, J., & Hertel, G. (2016). Does trust matter more in virtual teams? A meta-analysis of trust and team effectiveness considering virtuality and documentation as moderators. *Journal of Applied Psychology*, *101*, 1151–1177.

Brotheridge, C. M., Neufeld, D., & Dyck, B. (2015). Communicating virtually in a global organization. *Journal of Managerial Psychology*, *30*, 909–924.

Burke, M., Marlow, C., & Lento, T. (2010). Social network activity and social well-being. In *Proceedings of the SIGCHI Conference on Human Factors in Computing Systems (CHI '10)* (pp. 1909–1912). New York: ACM.

Carter, D. R., Asencio, R., Wax, A., DeChurch, L. A., & Contractor, N. S. (2015). Little teams, big data: Big data provides new opportunities for teams theory. *Industrial and Organizational Psychology*, *8*, 550–555.

Contractor, N. S., & Monge, P. R. (2002). *Theories of communication networks*. USA: Oxford University Press.

Cress, U., & Kimmerle, J. (2008). A systemic and cognitive view on collaborative knowledge building with wikis. *International Journal of Computer-Supported Collaborative Learning*, *3*, 105–122.

Culnan, M. J., & Markus, M. L. (1987). Information technologies. In F. M. Jablin, L. Putnam, K. H. Roberts, & L. W. Porter (Eds.), *Handbook of organizational communication: An interdisciplinary perspective* (pp. 420–443). Newbury Park, CA: Sage.

Daft, R. L., & Lengel, R. H. (1986). A proposed integration among organizational information requirements, media richness, and structural design. *Management Science*, *32*, 191–233.

Daft, R. L., Lengel, R. H., & Trevino, L. K. (1987). Message equivocality, media selection, and manager performance: Implications for information systems. *MIS Quarterly*, *11*, 355–366.

Dennis, A. R., & Kinney, S. T. (1998). Testing media richness theory in the new media: The effects of cues, feedback, and task equivocality. *Information Systems Research*, *9*, 256–274.

Dennis, A. R., & Valacich, J. S. (1999). Rethinking media richness: Toward a theory of media synchronicity. In *Proceedings of the 32nd Hawaii International Conference on System Sciences* (pp. 1017–1026). New York, NY: IEEE.

Dennis, A. R., Valacich, J. S., Speier, C., & Morris, M. G. (1998). Beyond media richness: An empirical test of media synchronicity theory. In *Proceedings of the 31st Annual Hawaii International Conference on System Sciences* (pp. 48–57). New York, NY: IEEE.

Döring, N. (2003). *Sozialpsychologie des Internets* [The social psychology of the Internet]. Göttingen, Germany: Hogrefe.

Ducheneaut, N. B. (2002). The social impacts of electronic mail in organizations: A case study of electronic power games using communication genres. *Information, Communication & Society*, *5*, 153–188.

Ellison, N. B., Steinfield, C., & Lampe, C. (2007). The benefits of Facebook "friends": Social capital and college students' use of online social network sites. *Journal of Computer-Mediated Communication*, *12*, 1143–1168.

Ellison, N. B., Vitak, J., Steinfield, C., Gray, R., & Lampe, C. (2011). Negotiating privacy concerns and social capital needs in a social media environment. In S. Trepte & L. Reinecke (Eds.), *Privacy online: Perspectives on privacy and self- disclosure in the social web* (pp. 19–32). Berlin: Springer.

Ellison, N. B., Wohn, D. Y., & Greenhow, C. M. (2014). Adolescents' visions of their future careers, educational plans, and life pathways: The role of bridging and bonding social capital experiences. *Journal of Social and Personal Relationships, 31*, 516–534.

El-Shinnawy, M., & Markus, M. L. (1997). The poverty of media richness theory: Explaining people's choice of electronic mail vs. voice mail. *International Journal of Human-Computer Studies, 46*, 443–467.

Fulk, J., Steinfield, C. W., Schmitz, J., & Power, J. G. (1987). A social information processing model of media use in organizations. *Communication Research, 14*, 529–552.

Gervais, B. T. (2015). Incivility online: Affective and behavioral reactions to uncivil political posts in a web-based experiment. *Journal of Information Technology & Politics, 12*, 167–185.

Gibson, J. J. (1979). *The ecological approach to visual perception*. Boston, MA: Houghton Mifflin.

Gonzales, A. L., & Hancock, J. T. (2008). Identity shift in computer-mediated environments. *Media Psychology, 11*, 167–185.

Granovetter, M. S. (1973). The strength of weak ties. *American Journal of Sociology, 78*, 1360–1380.

Hutchby, I. (2001). Technologies, texts and affordances. *Sociology, 35*, 441–456.

Jackson, A., Yates, J., & Orlikowski, W. (2007). Corporate blogging: Building community through persistent digital talk. In *Proceedings of the 40th annual Hawaii International Conference on System Sciences (HICSS '07)* (p. 80). New York, NY: IEEE

Jiang, L. C., Bazarova, N. N., & Hancock, J. T. (2011). The disclosure-intimacy link in computer-mediated communication: An attributional extension of the hyperpersonal model. *Human Communication Research, 37*, 58–77.

Johnson, S. L., Safadi, H., & Faraj, S. (2015). The emergence of online community leadership. *Information Systems Research, 26*, 165–187.

Joinson, A. N. (2001). Self-disclosure in computer-mediated communication: The role of self-awareness and visual anonymity. *European Journal of Social Psychology, 31*, 177–192.

Kaptelinin, V., & Nardi, B. A. (2012). Affordances in HCI: Toward a mediated action perspective. In *Proceedings of the SIGCHI Conference on Human Factors in Computing Systems (CHI '12)* (pp. 967–976). New York, NY: ACM.

Keegan, B., Gergle, D., & Contractor, N. (2012). Do editors or articles drive collaboration? Multilevel statistical network analysis of Wikipedia co-authorship. *Proceedings of the ACM 2012 Conference on Computer Supported Cooperative Work* (pp. 427–436). New York, NY: ACM.

Kiesler, S., Siegel, J., & McGuire, T. W. (1984). Social psychological aspects of computer-mediated communication. *American Psychologist, 39*, 1123–1134.

Kock, N. (1998). Can communication medium limitations foster better group outcomes? An action research study. *Information & Management, 34*, 295–305.

Kock, N. (2005). Compensatory adaptation to media obstacles: An experimental study of process redesign dyads. *Information Resources Management Journal, 18*, 41–67.

Krämer, N. C., Rösner, L., Eimler, S. C., Winter, S., & Neubaum, G. (2014). Let the weakest link go! Empirical explorations on the relative importance of weak and strong ties on social networking sites. *Societies, 4*, 785–809.

Krämer, N. C., & Winter, S. (2008). Impression management 2.0. Self-presentation on social networking sites and its relationship to personality. *Journal of Media Psychology, 20*, 106–116.

Kraut, R., Kiesler, S., Boneva, B., Cummings, J., Helgeson, V., & Crawford, A. (2002). Internet paradox revisited. *Journal of Social Issues, 58*, 49–74.

Kraut, R., Patterson, M., Lundmark, V., Kiesler, S., Mukhopadhyay, T., & Scherlis, W. (1998). Internet paradox: A social technology that reduces social involvement and psychological well-being? *American Psychologist, 53*, 1017–1031.

Lapidot-Lefler, N., & Barak, A. (2012). Effects of anonymity, invisibility, and lack of eye-contact on toxic online disinhibition. *Computers in Human Behavior, 28*, 434–443.

Larsen, S. (2015). Videoconferencing in business meetings: An affordance perspective. *International Journal of e-Collaboration, 11*, 64–79.

Leavitt, H. J. (1951). Some effects of certain communication patterns on group performance. *The Journal of Abnormal and Social Psychology, 46*, 38–50.

Lee, E.-J. (2004). Effects of visual representation on social influence in computer-mediated communication: Experimental tests of the social identity model of deindividuation effects. *Human Communication Research, 30*, 234–259.

Leidner, D., Koch, H., & Gonzalez, E. (2010). Assimilating generation Y IT new hires into USAA's workforce: The role of an Enterprise 2.0 system. *MIS Quarterly Executive, 9*, 229–242.

Lemke, J. L. (1999). Discourse and organizational dynamics: Website communication and institutional change. *Discourse and Society, 10*, 21–47.

Leonardi, P. M. (2014). Social media, knowledge sharing, and innovation: Toward a theory of communication visibility. *Information Systems Research, 25*, 796–816.

Leonardi, P. M., Huysman, M., & Steinfield, C. (2013). Enterprise social media: Definition, history, and prospects for the study of social technologies in organizations. *Journal of Computer-Mediated Communication, 19*, 1–19.

Leonardi, P. M., & Meyer, S. R. (2015). Social media as social lubricant: How ambient awareness eases knowledge transfer. *American Behavioral Scientist, 59*, 10–34.

Majchrzak, A., Faraj, S., Kane, G. C., & Azad, B. (2013). Contradictory influence of social media affordances on online communal knowledge sharing. *Journal of Computer-Mediated Communication, 19*, 38–55.

Marwick, A. E., & boyd, d. (2010). I tweet honestly, I tweet passionately: Twitter users, context collapse, and the imagined audience. *New Media & Society, 13*, 114–133.

Maynard, M. T., Gilson, L. L., Jones Young, N. & Vartiainen, M. (2017). Virtual teams. In G. Hertel, D. Stone, R. Johnson, & J. Passmore (Eds.), *The Wiley Blackwell handbook of the psychology of the internet at work* (pp. 315–346). Chichester: John Wiley & Sons, Ltd.

McGrath, J. E. (1984). *Groups: Interaction and performance*. Englewood Cliffs, NJ: Prentice-Hall.

Moskaliuk, J., Kimmerle, J., & Cress, U. (2009). Wiki-supported learning and knowledge building: Effects of incongruity between knowledge and information. *Journal of Computer Assisted Learning, 25*, 549–561.

O'Reilly, T. (2005). *What is web 2.0 – design patterns and business models for the next generation of software*. Retrieved April 3, 2017, from http://oreilly.com/pub/a/web2/archive/what-is-web-20.html?page=1

Papacharissi, Z. (2002). The virtual sphere: The internet as a public sphere. *New Media & Society, 4*, 9–27.

Pickering, J. M. & King, J. L. (1995). Hardwiring weak ties: Interorganizational computer-mediated communication, occupational communities, and organizational change. *Organizational Science, 6* (4), 479–486.

Postmes, T., Spears, R., & Lea, M. (1998). Breaching or building social boundaries?: SIDE-effects of computer-mediated communication. *Communication Research, 25*, 689–715.

Putnam, R. (2000). *Bowling alone: The collapse and revival of American community*. New York, NY: Simon and Schuster.

Reicher, S. D., Spears, R., & Postmes, T. (1995). A social identity model of deindividuation phenomena. *European Review of Social Psychology, 6*, 161–198.

Rice, R. E., & Case, D. (1983). Electronic message systems in the university: A description of use and utility. *Journal of Communication, 33*, 131–152.

Rice, R. E., & Love, G. (1987). Electronic emotion: Socioemotional content in a computer-mediated communication network. *Communication Research, 14*, 85–108.

Ritzer, B., & Jurgenson, N. (2010). Production, consumption, prosumption. The nature of capitalism in the age of the digital "prosumer." *Journal of Consumer Culture, 10*, 13–36.

Sassenberg, K. (2002). Common bond and common identity groups on the Internet: Attachment and normative behavior in on-topic and off-topic chats. *Group Dynamics: Theory, Research, and Practice, 6*, 27–37.

Schmitz, J., & Fulk, J. (1991). Organizational colleagues, media richness, and electronic mail. A test of the social influence model of technology use. *Communication Research, 18,* 487–523.

Shaw, M. E. (1964). Communication networks. *Advances in Experimental Social Psychology, 1,* 111–149.

Short, J., Williams, E., & Christie, B. (1976). *The social psychology of telecommunications.* London, UK: John Wiley & Sons, Ltd.

Spears, R., & Lea, M. (1992). Social influence and the influence of the social in computer-mediate communication. In M. Lea (Ed.), *Contexts of computer-mediated communication* (pp. 30–65). New York: Harvester Wheatsheaf.

Spears, R. Lea, M. & Lee, S. (1990). De-individuation and group polarization in computer-mediated communication. *British Journal of Social Psychology, 29,* 121–134.

Sproull, L., & Kiesler, S. (1986). Reducing social context cues: Electronic mail in organizational communication. *Management Science, 32,* 1492–1512.

Steiner, I. D. (1972). *Group process and productivity.* New York, NY: Academic Press.

Steinfield, C., Dimicco, J. M., Ellison, N. B., & Lampe, C. (2009). Bowling online: Social networking and social capital within the organization. *Proceedings of the Fourth International Conference on Communities and Technologies (C&T '09)* (pp. 245–254). New York, NY: ACM.

Tajfel, H., Billig, M. G., Bundy, R. P., & Flament, C. (1971). Social categorization and intergroup behavior. *European Journal of Social Psychology, 1,* 149–178.

Treem, J. W., & Leonardi, P. M. (2012). Social media use in organizations: Exploring the affordances of visibility, editability, persistence, and association. *Communication Yearbook, 36,* 143–189.

Van Osch, W., & Coursaris, C. K. (2013). Organizational social media: A comprehensive framework and research agenda. *Proceedings of the 2013 46th Hawaii International Conference on System Sciences* (pp. 700–707). New York, NY: IEEE.

Walther, J. B. (1992). Interpersonal effects in computer-mediated interaction: A relational perspective. *Communication Research, 19,* 52–90.

Walther, J. B. (1993). Impression development in computer-mediated interaction. *Western Journal of Communication, 57,* 381–398.

Walther, J. B. (1996). Computer-mediated communication: Impersonal, interpersonal, and hyperpersonal interaction. *Communication Research, 23,* 3–43.

Walther, J. B. (2006). Nonverbal dynamics in computer-mediated communication, or: (and the net: ('s with you, :) and you:) alone. In V. Manusov & M. L. Patterson (Eds.), *Handbook of nonverbal communication* (pp. 461–479). Thousand Oaks, CA: Sage.

Walther, J. B., & Bunz, U. (2005). The rules of virtual groups: Trust, liking, and performance in computer-mediated communication. *Journal of Communication, 55,* 828–846.

Walther, J. B., & Carr, C. T. (2010). Internet interaction and intergroup dynamics: Problems and solutions in computer-mediated communication. In H. Giles, S. Reid, & J. Harwood (Eds.), *The dynamics of intergroup communication* (pp. 209–220). New York, NY: Peter Lang.

Walther, J. B., Liang, Y., DeAndrea, D. C., Tong, S. T., Carr, C. T., Spottswood, E. L. & Amichai-Hamburger, Y. (2011). The effect of feedback on identity shift in computer-mediated communication. *Media Psychology, 14,* 1–26.

Walther, J. B., Loh, T., & Granka, L. (2005). Let me count the ways: The interchange of verbal and nonverbal cues in computer-mediated and FtF affinity. *Journal of Language and Social Psychology, 24,* 36–65.

Walther, J. B., Van Der Heide, B., Ramirez, A. Jr., Burgoon, J. K., & Pena, J. (2015). Interpersonal and hyperpersonal dimensions of computer-mediated communication. In S. S. Sundar (Ed.), *The handbook of psychology and communication technology* (pp. 3–22). New York, NY: Wiley-Blackwell.

3

Ergonomics of Information Technologies at Work

Benjamin V. Hanrahan and John M. Carroll

Introduction

In this chapter, we discuss the ergonomics of information technology (IT) at work, particularly Internet-based technology. As Wilson (2014) notes, ergonomics is a scientific discipline concerned with the understanding of interactions among humans and a system in such a way that human wellbeing and overall system performance are optimized. Work and work systems are a major component of modern human life and activity. The Internet has clearly, broadly, and significantly affected work in many ways. To highlight the ergonomics of IT at work we focus on research that has sought to impact the practice of work, how work is accomplished, and how each of these themes affect these practices.

Despite the broad, systems focus, of ergonomics, many researchers have limited their focus to the ergonomics of technology design. These researchers have focused on the fit between humans and machines. For example, researchers have focused on how to best design computer workstations and other technical devices (Kroemer, Kroemer, & Kroemer-Elbert, 2001; see also Thielsch & Niesenhaus, 2017, Chapter 5 in this volume), optimize human–computer performance (Wickens & Hollands, 2000; see also Zafar, 2017, Chapter 10 in this volume), decompose tasks (Hackos & Redish, 1998), support offloading of cognition to tools (Hutchins & Klausen, 1996), and produce safe operating environments (Vicente, 1999). However, these micro views on ergonomics miss the broader context and system in which work is conducted, the social interactions necessary for collaboration at work, and the overall meaning of work. This system-oriented view of ergonomics is necessary to fully understand the ergonomics of IT at work, especially Internet-based work. Therefore, our chapter examines ergonomics in a broad sense; one which considers the social connections critical to work. We seek to examine how well Internet-based technology "fits" with the individual, and how well it "fits" the work practices and collaborative nature of today's work environments.

The topics that we cover in this chapter emerged from a review of the literature. For each of the topics we give a broad overview of the current research and choose a specific

The Wiley Blackwell Handbook of the Psychology of the Internet at Work,
First Edition. Edited by Guido Hertel, Dianna L. Stone, Richard D. Johnson, and Jonathan Passmore.

study that illustrates the ergonomic issues for each respective topic. We first outline work practices in more detail, and then associate them with several closely related practices and topics. Specifically, we focus on three main parts of work systems: community, relationships, and communication. We discuss the particular complexities of supporting community development within the workplace, because some form of community is necessary for collaboration. After reviewing community, we move into the closely related topic of how IT can support relationships in the workplace because these relationships form the bedrock of collaborative activities. Finally, in order for relationships to form and flourish there needs to be some form of communication. We conclude with several challenges for the future workplace and propose directions for future research. Our conclusions will focus on how organizational IT systems both embody and enforce their policies and philosophies.

Our Perspective

As noted previously, our focus on ergonomic is macro, or system, oriented. Our perspective of what constitutes work is drawn from several places. The first, and primary source that we draw on is the work of Button on IT work systems (Button, 1993, 2003; Button & Dourish, 1996; Button & Sharrock, 2009). Button and colleagues argued strongly that bad systems are rooted in a poor understanding of how work is done. One of their critiques is that poorly designed methods result in poorly implemented systems, and, more often than not, in over-formalized and inflexible work practices (Button & Sharrock, 2009, p. 362). Button and colleagues' study of work drew on the work of Suchman (1987) and particularly that of Garfinkel (1967). Suchman (1987) emphasized the context, particularly the social context, of action and the use of computer systems. One of her critical arguments was that plans are resources that guide action, rather than prescriptions. She described actions as situated within the overall context of work, whereas plans are too schematic and inflexible to properly describe said actions. Garfinkel (1967) and his disciples insisted that in order to understand work we must focus on the social order that constitutes work.

Another source from which we draw our broader view of work is the French tradition of *ergonomie*, which outlines several dimensions or scopes of work (Daniellou, 2005), specifically that of the collective. In particular, the collective is meant to widen the scope of investigations into work, and builds on the individual activity focused dimension. The last of the major line of research that framed our perspective is based on the tradition of participatory design (Holtzblatt & Jones, 1993; Muller & Kuhn, 1993; Schuler & Namioka, 1993). This tradition focuses on the granting of agency and the involvement of different stakeholders from the work environment. From this research, we draw on issues of power in the work setting.

Finally, we draw on work practices research. Understanding work practices from various perspectives can help leverage our understanding of the context into which technology will be, or has been, deployed. After discussing how ergonomics might be understood through the lens of work practices, we discuss how researchers have focused on supporting some specific practices of work. We frame the discussion on how work gets done in terms of the practices around collaboration, because work rarely happens without collaboration. The Internet highlights the value of understanding the work context, because it is always accessible and has implications for the broad geographic reach of teams.

Analytic Outlook

In this chapter, we use ergonomics as a criterion of practice. That is, how well a given technology fits into a workplace is dictated in part by how ergonomic it is, not just from a human–machine perspective, but also how it is embedded within broader work practices. Therefore, we review research on different technologies and how they impact work practices. Measuring the impact of a technology can be generally viewed in terms of the extent to which it impacts work practices. To a degree, the extent to which technology impacts the workplace is a result of how well it designed to fit human use (e.g., is ergonomic). One tool that has certainly had a large effect on work practices is email, one of the most successful applications deployed within the workplace, and often used to the point of being problematic (Dabbish & Kraut, 2006; Whittaker, Bellotti, & Gwizdka, 2007; Whittaker & Sidner, 1996). Email has both wide reach and deep contribution rates. In other words, nearly everyone in the organizational context has an email account, and almost everyone sends and receives emails regularly.

Email can be compared to the more advanced Web 2.0 tools that have received greater press but whose organizational impact and use is much lower. With great fanfare, Web 2.0 tools such as wikis, blogs, and social networking tools have been widely deployed in corporate intranets, with some suggesting that they are key to corporate knowledge management (Levy, 2009). But these tools have reported contribution rates of 3% in major companies such as IBM and Microsoft (Dugan, Geyer, & Millen, 2010; Efimova & Grudin, 2007). Even simple discussion forums are used at a rate of 10–28 times that of Web 2.0 tools (Brzozowski, Sandholm, & Hogg, 2009). In other words, the embeddedness and use of technology within work practices can be viewed as its ergonomics. That said, our review is less concerned about the tool's impact and more concerned about how the tool impacts work practices.

Studying Work and Work Practices

Several lines of research have highlighted the importance of work practices, each of which have different strengths. These lines of research have all served to avoid formalization of work processes and restore agency (to varying degrees) to the workers who use these systems. For example, task analysis (Hackos & Redish, 1998) focuses on potential users' work goals, and how they are accomplished, or could be accomplished with IT. To achieve any work goal there has to be a certain amount of formalization of the work processes through technology and an understanding of how to improve these processes. Likewise, cognitive task analysis focuses on improving the performance of a task (Chipman, Schraagen, & Shalin, 2000) or formalizing how a user accomplishes goals such as sensemaking (Pirolli & Card, 2005). Although task performance is certainly an important aspect of work (particularly to more time-critical work), performance or goal achievement does not provide a complete view of work practices. In fact, efficiency of a user interface does not necessarily even correlate with work effectiveness (Frøkjær, Hertzum, & Hornbæk, 2000).

Suchman (1987) highlighted this and other key problems in designing IT systems that successfully support work. Her research has important implications for framing the study of work practices because it focuses on the action of work instead of the objects of work (e.g. the product). Also, Suchman's major contribution centers on the validity and utility of forming plans to perform actions. That is, IT systems that support work or work flows easily fall into the trap of formalizing a plan of action to accomplish a specific task or build a specific product. Suchman argued that these plans should be resources for action and a basis for improvisation of said action as opposed to dictating them. This is important for the practice of work, because work cannot be understood simply by examining the produced product. Instead work is best understood by examining the action of work. Studying work practice rather than outcomes helps to avoid the pitfalls of over-formalization. That is, when examining a work product, it is easy to assume that the practices involved in producing the product are as formalized as the work product itself.

An illustrative example of this concept is provided by Button and Harper (1996), who described the failings of an IT system meant to facilitate crime reporting by police officers. In their example, they outline how the designers of the system treated existing reports as "literal descriptions of the work they documented," instead of taking into consideration how the reports were produced. One particular aspect that Button and Harper's (1996) work highlights is the rhythm of reporting that the IT system sought to impose on the officers. That is, the IT system expected an officer to produce a complete report of the crime immediately after they had finished dealing with the crime. However, actual police work consists of a variety of activities that have different priorities. For instance, some work must be performed right away (e.g., responding to a call from the public), but other work can be done by the officer at a later point in time (e.g., completing a report). Thus, officers are constantly juggling different priorities in an ad-hoc manner, and their schedule cannot be specified ahead of time. As a result, it is not always clear that they will be able to fill out a report immediately after a crime. This is a central failing of assuming that the formal report that an officer produces represents the practices that are part of producing the report.

In fact, the difference between work products and their production is a central theme in Button and colleagues' research (Button, 1993, 2003; Button & Dourish, 1996; Button & Sharrock, 2009). They argue that to understand work, one should study the practice of work, not the objects or work products. In order to judge how well an IT system fits with the workplace, a researcher must look at how well it supports workplace practices, and the consequences of its deployment. Button's proposed approach to study this topic draws on ethnomethodology (Garfinkel, 1967), which focuses on the way that the people organize and make sense of their actions.

The final approach to studying work that we utilize is participatory design. Participatory design takes a different approach to a similar end. Although there are some differences between the specific applications of participatory design (Bjerknes, Ehn, Kyng, & Nygaard, 1987; Björgvinsson, Ehn, & Hillgren, 2010; Muller & Kuhn, 1993), all approaches incorporate the various organizational stakeholders in some way. In our review we draw specifically on the Scandinavian tradition, which derives from the collective resource framing by Ehn and Kyng (1987). In this tradition, the system and the processes around them are redesigned with a specific focus on the involvement of users in the design of work systems. Drawing from these traditions, the ergonomics of an Internet technology system at work is not only about how the system supports the work practices, but also how these systems broker power throughout the system and how they affect organizational stakeholders.

The combination of these perspectives lets us highlight the aspects of work that we see as constituting its 'ergonomics.' That is, we take a broader look than just performance and look at the practices of how work gets done and how work systems impact these practices.

Embodiment of work practices

As noted by Suchman, Button, and the participatory design tradition, the embodiment of work practices is not simply a description of work processes or products, but also reflects the meaning of work, the structure of power in work relationships, and the social norms which drive situated work practices. We briefly describe a case study of the design of an IT system within the organizational context (Colombino, Hanrahan, & Castellani, 2014). This study used a combination of work practices and participatory design approaches and highlights the importance of looking more broadly at the impact of deploying technology, even if the context and practices are well understood. This is especially important when the technology is meant to reconfigure the work practices and balance of power within the workplace.

The embodiment in question was built at Xerox Research Centre Europe in the Work Practice Technologies group. The focus of the research was on the customer service agents at outsourced call centers. The job of these agents was to field calls from customers (the majority of the call centers were for major US telecommunications companies) and to assist customers with problem resolution, or, at the very least, to assuage their frustrations. The agents' pay was in large part based on a series of key performance indicators (KPIs). The primary tension that arose from these KPIs was that the agents needed to optimize the amount of time they spent with an individual customer (typically in the order of 5–15 minutes), and that customer's satisfaction. The various IT systems were deployed over the Internet, and the centers were networked with each other.

Colombino et al. (2014) began their research with a series of ethnographic studies, in which they observed various employees in a call center performing their jobs (e.g., call center agents fielding calls and doing their administrative work, low-level managers attending performance meetings with agents, support staff monitoring call performance, and mid- to high-level management conducting strategic meetings with their peers and customers). These studies all had an ethnomethodological outlook, and the researchers soon discovered that the agents did not have access to the data that served as the basis for their pay (e.g., time spent talking on the phone, customer satisfaction, time spent documenting the call). In fact, in some call centers, the agents only received these numbers in the form of weekly updates from their managers. Without these data, it was difficult for the workers to determine their pay or track their current status in order to adjust their performance strategies. Instead, they relied on workarounds such as keeping a manual tally of different metrics to keep an inexact view of their current performance. Through a series of interviews with the workers and their managers, the researchers discussed and confirmed that the lack of data was indeed a problem for the agents. The research team then engaged in a participatory design process to support the agents and improve their ability to self-regulate and monitor their performance, thereby enfranchising the workers. The result of this process was a dashboard system that gave real-time feedback on the set of KPIs that served as the basis for pay rates. Initial studies of the prototypes revealed that the research team's system successfully bridged the gap and it seemed that it would be an unmitigated success.

These successes led the company to deploy the new system to several thousand users, and through this effort they hoped to improve the ergonomics of the Internet-based systems at these call centers. However, when Colombino et al. (2014) studied the consequences of this deployment, they learned of several additional, unseen factors associated with work practices. In particular, they learned that the IT infrastructure that supported the call center was inadequate. For instance, it was essentially impossible to provide agents with their data in real time. The reason for this was that when the system was being built and designed, the designers did not consider the need to report the agents' KPI data to the agents. In addition, data were not reported in real time. Instead, the primary purpose of the IT infrastructure was to report gathered data to the customer organization every few weeks. They also observed that the agents' feature requests were not prioritized when features were designed, as opposed to manager and customer organization requests. This was because each decision was considered in isolation, and the highest priority was always for features that supported reporting data to customers. In addition, Colombino et al. (2014) also found other forces at work that affected deployment in this work context. For example, upper-level managers were reticent to let workers know when they were nearing the next "pay level" (i.e., when workers reached certain benchmarks their hourly rate for the entire month would increase).

This case provides a strong example of how complex work practices are, and how continued engagement is needed to study the impact of the deployed technology on work practices. If Colombino et al. (2014) had stopped at measuring the task efficiency of the agents checking their performance data, they would have missed the rich tapestry of work practices and priorities of the many stakeholders that this relatively small technology impacted. Some of the pitfalls of this system are especially relevant to Internet-based work platforms, as designers are rarely (if ever) embedded within the context of the work that they serve. That is, it is easy for designers to lose sight of the actual practice of the work, as happened in the case of the call-center agents.

Community

Communities are a key component affecting the accomplishment of work. There has been much research over centuries on how communities form, function correctly, and maintain productivity. A particular thread of thinking that is especially relevant to work contexts is the tension between community and self-interests. In the following section we review the various perspectives on this tension and apply the resulting lens to specific examples of communities within IT.

In de Tocqueville's (1835/2002) discussions of democracy in America, he described that in order for democracies to function, individuals needed to be engaged with their communities in a direct way. However, de Tocqueville also outlined that the trait of individualism may prove to be a threat to the functioning of these communities. For example, when individualism is combined with direct involvement with community, the result may prevent collective action, because individuals would be working to improve their own situation rather than that of the community. This threat, or tension, was somewhat evident in Tönnies' (2012) anchoring of the opposing concepts of *Gemeinschaft* and *Gesellschaftö*. *Gemeinschaft*, directly translated as community, "emphasized bonds of kindship, friendship and place" (Carroll, 2014, p. 3). *Gemeinschaft* is somewhat similar to de Tocqueville's *habits of the heart*, in that, having bonds with the members of one's community is, in large part, what motivates one to be engaged with the community. Conversely, *Gesellschaft* emphasized the formalities of living in a community or an individual "viewing other people instrumentally" (Carroll, 2014, p. 3). These aspects of *Gesellschaft* move to isolate members from the other members of the community, a threat to a well-functioning democracy. These two concepts give additional depth to the discussion of de Tocqueville, where *Gemeinschaft* is the types of relationship that a community relies on and *Gesellschaft* represents the types of individualism that threatens them. More recently, scholars have found de Tocqueville's work to be prophetic. Bellah, Madsen, Sullivan, Swidler, and Tipton (2007) further specified the tension described by de Tocqueville and Tönnies in terms of the individual, namely as expressive individualism and utilitarian individualism. Expressive individualism represents individuals' intuition and their ability to identify (or merge) with others to form communities (i.e. *Gemeinschaft*). In contrast, utilitarian individualism is more instrumental in nature, in that it represents the protection of self-interests. Bellah et al. (2007) saw individualism as having become a negative, cancerous force in American society. Putnam's (1995) found in the later part of the twentieth century, that people were becoming increasingly isolated and disengaged with their community, certainly points to the effects of this trend.

These same tensions between individual interests and that of the community are directly relatable to the work context. Communities are important in the work context in much the same way that they are in broader society. Within broader societies, communities are necessary for the functioning of democracies and healthy, meaningful lives, whereas, in the work context they are necessary for producing work of any magnitude. Just as democracies are predicated on individuals' direct participation and interests, they also serve to draw individuals out and provide perspective on their place within a community:

> *When the members of a community are forced to attend to public affairs, they are necessarily drawn from the circle of their own interests, and snatched at times from self-observation. As soon as a man begins to treat of public affairs in public, he begins to perceive that he is not so independent of his fellowmen as he had at first imagined, and that, in order to obtain their support, he must often lend them his co-operation.*
>
> De Tocqueville (1835/2002)

This explanation has a direct parallel in the work context, as work is predicated on the accomplishments of individuals, but each individual must also understand their place within the broader team context and the specific value of their contribution. A work community is most effective when its members are drawn out of their individual work, and they can see how the value of their work fits with the value of others' work to create a meaningful *gestalt*. In psychological literature, these two seemingly opposing concepts are termed as rational self-interest and other orientation (Meglino & Korsgaard, 2004). De Dreu (2006) however, calls into question that rational self-interest and other orientation have a negative relationship such that they represent end points on a continuum. Instead, he proposes that they are independent, multilevel constructs. In fact, coupling this with De Dreu's (2006) reframing of rational self-interest as self-concern, brings further insight into how individuals balance their own interests with that of their communities or workplaces. De Dreu's analysis shows that situations can occur where self-concern and other orientation can be synergistic instead of diametrically opposed, that is, a worker may both satisfy their self-concern and other orientation in the same activity. For instance, one can imagine an individual working on a project that is motivated by both expectancy value variables (e.g., a bonus for project success), as well as to fulfill group membership responsibilities. In this way, the two cannot be diametrically opposed, because they are entirely separate constructs.

An oft-mentioned example of community is communities of practice (Wenger, 1998, 1999; Wenger & Snyder, 2000). Communities of practice are centered on a shared goal of learning and development, and are more of an example of expressive individualism than utilitarianism. That is, these communities are more than just instrumental, they are ways that members help to construct their identity through work (Carter, 2015) or signal their value through making their contributions visible to themselves and others (DiMicco & Millen, 2007). Ideally, they also serve to draw individuals out and force them to realize how they are situated within the said community.

Researchers have recognized the importance of communities at work, particularly in the context of the Internet, and have undertaken several activities to support their construction and maintenance. We discuss many of these in the following section.

Embodiment of community

Beehive, the social network developed and deployed at IBM (Steinfield, DiMicco, Ellison, & Lampe, 2009) serves as a particularly illuminating example of a community network in the workplace. In reporting their findings about the different motivations of their users, DiMicco et al. (2008) provided support for the tension noted in the work of de Tocqueville (1835/2002), Tönnies (2012), and Bellah et al. (2007). DiMicco et al. created three categories with which to analyze the different motivations that users reported:

- *Caring*: motivations in this category were primarily focused on connecting with others, seeing a different personal side to colleagues, or strengthening weak ties and sharing information about oneself.
- *Climbing*: motivations in this category were around employees using the site to advance their careers.
- *Campaigning*: where users were trying to gain support for their different projects.

Using the lens of Bellah et al. (2007), the user motivation of caring might be seen as expressive individualism, or the exhibition of characteristics closer to *Gemeinschaft* than *Gesellschaft* (Tönnies, 2012). That is, they are not treating other members of the community instrumentally, they are connecting on a personal level with members of the community. However, the reported motivations of climbing and campaigning treat the other members of the community as more instrumental, which is closer to utilitarian individualism or *Gesellschaft*. Put more plainly, the users of the system are looking at the community in terms of how it can benefit their personal interests (e.g., promoting a personal project or increasing their visibility). This serves as a particularly good example of the tension of being part of a community at work, expressive individualism versus utilitarian individualism. The tension is perhaps even more salient within the work context as aspects of utilitarian individualism are expressed within the community more publicly than perhaps in the broader societal context. DiMicco et al. (2009) further explained the more instrumental uses of the social network site. Specifically, users were more information-centric, and used the site to help determine information about colleagues, including their position in the enterprise, connections, interests, or job functions. DiMicco et al. (2009) termed this people sense-making, and found that users were spending slightly more time looking at people they did not know relative to those they did know.

Interestingly, it seems that at IBM, online social networks did not necessarily reflect actual social networks in the ways that one might expect (Jacovi, Guy, Kremer-Davidson, Porat, & Aizenbud-Reshef, 2014; Mark, Guy, Kremer-Davidson, & Jacovi, 2014). For example, the number of followers (or "friends") within social media was not a reflection of reputation within the enterprise (Jacovi et al., 2014). In fact, the number of "friends" on a social network seemed to be inversely proportional to the "likability" of a person and their content within the network (Mark et al., 2014). Although Mark et al. (2014) did not find a conclusive reason for this relationship, they hypothesized that this behavior was not recognized as important in the enterprise, or that perhaps there were many other ways to connect with co-workers. Perhaps these social networks were only a tool and could not be considered a reflection of the actual social networks and structure within an organization. Alternatively, collecting online "friends" may have been a way to compensate for the lack of a personal network or an attempt to expand it.

When we take the work practices perspective and examine the impact of deploying Beehive within IBM, its relationships to measures such as productivity or effectiveness appear inconclusive (DiMicco et al., 2008; Steinfield et al., 2009). It seems that these systems, at the very least, could help people to perform some of their softer job responsibilities (DiMicco et al., 2008) and serve as a resource for co-workers (DiMicco et al., 2009). It also seems the purpose of the social network site was less about reinforcing the existing communities at work and more about fostering and forming weak contacts within the enterprise. There is of course a benefit from having weak contacts in terms of access to a more diverse set of expertise (DiMicco et al., 2009). In fact, Skeels and Grudin (2009) came to a similar conclusion in their work examining the use of Facebook and LinkedIn at work. Their findings suggested that the biggest benefit of these tools was to maintain weak ties. They hypothesize that the use of social network sites will increase much in the same way as email. One issue with using social media at work, which has been echoed by both email and instant messaging, is that boundary crossing can occur over the enterprise firewall and within hierarchies. The authors posit that over time new social norms will evolve, much like how email communication broke down boundaries across the workplace hierarchy. Dabbish, Stuart, Tsay, and Herbsleb (2012) provide some insight into this in their investigation into how the social interactions on Github (https://github.com/) have evolved to increase the transparency of what others are doing on the platform.

Github is the most popular source control platform for software development, particularly for open source projects. On top of standard source control capabilities, Github supports numerous social functionalities to promote interaction between its users. Dabbish et al. (2012) found that software developers (e.g., Github's users) were making rich inferences about others' activities, which helped them to coordinate work and collaborate more effectively, for example, the number of users who indicated that they were "watching" a specific software project. Members of the community inferred that the number of users who were watching a project was an indicator of the importance of that project. Thus by simply viewing the most watched projects, Github users quickly understood the set of projects important to that community and the likely direction of development. Due to the success of the widely used social features on Github, an open question was why the adoption of these features was so widespread. Perhaps a bit of insight can be gained from Ducheneaut and Bellotti's (2001) framing of email as a habitat; that is, they argued that email was not just a communication tool, but a place in which work was done. Similarly, Github was not just a social network, but a place where much of the work in managing software projects occurred. Perhaps, community tools would experience more use if they are embedded within the work practices for that community.

Looking at the possibilities of social media within organizations more broadly, the potential impact is large. Treem and Leonardi (2012) suggested that there are four affordances of social media, visibility, persistence, editability, and association. *Visibility* is "the ability [for users] to make their behaviors, knowledge, preferences, and communication network connections" easily available to others (Treem and Leonardi, 2012, p. 150), an example being the various types of status updates provided by social networking sites. *Persistence* reflects a permanency of contributions and is illustrated as when "a poster to a blog or SNS [social networking site] logs out, that information remains available to users and does not expire or disappear" (Treem and Leonardi, 2012, p. 155). *Editability* is the ability for the author of the content to revise their content before or after it is made available (Treem and Leonardi, 2012, p. 159). *Association* reflects "established connections between individuals, between individuals and content, or between an actor and a presentation" (Treem and Leonardi, 2012, p. 162). Treem and Leonardi suggest that these various affordances may be able to influence several processes around socialization, information sharing, and power distribution. They also provided a research direction for communication within the enterprise, and their work mostly serves as a "thought exercise." That said, there is evidence that some of the new behaviors that the authors envision are manifesting themselves. Zhao and Rosson (2009) outlined several ways that employees were leveraging microblogs to share information. They found that Twitter was being used as a replacement for water-cooler talk. There were also some newer behaviors, such as expanding one's social network, encountering new information, and keeping in more personal contact with others. A direction of research that seems quite necessary and promising is acknowledging that people are combining their social tools (Matthews, Whittaker, Badenes, & Smith, 2014), and that we need to study the entire constellation of tools instead of focusing on a particular one when we want to study how communities are formed.

Relationships

We contend that forming communities at work is important to accomplishing work. In other words, communities are a precondition for getting work of any magnitude done. We see relationships as the vessel through which these communities do work. Indeed, when we examine how relationships, as opposed to communities, have been framed and examined in the literature on human–computer interaction we see that research into relationships is more concretely associated with actual work and its products. This is in particular contrast to the work in communities, where it is assumed that healthier communities will eventually lead to work being done. When one considers that communities are the nesting ground for relationships, the reason for assuming that healthier, more connected communities will lead to productivity becomes clearer. Moreover, communities comprise a broad set of relationships.

In this section, we discuss two primary threads of research: first, how systems have supported the forming of relationships, and second, the issues and implications of software systems within these (sometimes instantiated) relationships. We then discuss crowdwork, an interesting embodiment of computer-supported relationships, where both the systems to find work and their implications highlight several pertinent issues.

Expertise finding

There are several organizational efforts in expert finding, starting in earnest in the computer-supported cooperative work field. For example, McDonald and Ackerman (2000) developed Expertise Recommender based on their studies (McDonald & Ackerman, 1998) at a medical software company. Expertise Recommender was meant to make it easier to locate the person who had the required or desired expertise. The system was based on several behaviors that the authors saw as important to future expertise-finding systems, namely, expertise identification and selection, as well as escalation mechanisms for each. An expertise system needs to know who has the needed expertise, whether they are available, and whether they would fit the project requirements. Expertise Recommender met some of these goals by leveraging work products to construct profiles that suited expertise finding. However, constructing profiles has since been analyzed to determine their appropriateness for locating expertise, and their potential negative outcomes (e.g., privacy concerns). In fact, Ackerman, Dachtera, Pipek, and Wulf (2013) point out that how to best construct these profiles is quite "contentious" (p. 555).

The effort of constructing profiles and its ensuing contentiousness is clearly evident in McDonald and Ackerman's (2000) research on Expert Recommender. Whereas Expert Recommender emphasized using public work products for profile construction (e.g., change logs in source control systems to identify the expertise of developers), Ehrlich, Lin, and Griffiths-Fisher (2007) leveraged chat logs and email messages to construct expertise profiles. They stressed that in their system, SmallBlue, users were not required to explicitly construct their profiles and that they protected user (e.g., employee) privacy. Guy et al. (2008) continued this trend with SONAR, a system that integrated several different social media feeds to understand and visualize how employees were interacting with one another. Later, Guy et al. (2013) leveraged similar techniques to combine eight social feeds to mine expertise. Interestingly, the feeds that were more associated with work products (e.g., wiki entries) were less useful than feeds that had less content but contextually situated users within the organization (e.g., belonging to a community or tags). We would argue that this contextually situated information is more informative on the practice of the work (e.g., who collaborated with whom or who knows what) than work products, which are formalizations of the practice of work.

Despite all of the work and progress that has gone into these expertise finding systems, when employees seek expertise they still ask their own co-workers for advice (Balog, Fang, de Rijke, Serdyukov, & Si, 2012). Like us, Balog et al. (2012) question the efficacy of quantifying expertise through work products, because there are many different documents and it is unclear what textual work actually says about a person's expertise. In addition, several facets of work may not lend particularly well to the verbalization of tacit knowledge. Perhaps, taking from some of the findings of Guy et al. (2013), we can find proxies for the work practices instead of mining the work products.

Trust

Trust is obviously an important part of work relationships. IT can directly affect levels of trust between members in a workplace and is an important consideration for properly supporting its ergonomics. Trust has many positive benefits, it can help to increase performance (Robert, 2016) and it can be a "catalyst" for performance (Knowles et al., 2014). Trust is also not just binary, instead trust reflects the degree, and quality of trust directly shapes and impacts the work practices of teams and workplaces (Knowles et al., 2015, see also Jarvenpaa, Cantu, & Lim, 2017, Chapter 6 in this volume).

One particular aspect of trust that Internet-based systems that affect their operation is whether and how they implement the monitoring of their users. Robert (2016) examined several strategies for monitoring of virtual teams. For example, he found a nuanced and complex relationship between monitoring, trust, and team performance. One aspect of trust that Robert highlighted was the difference between cognitive trust, the trust in someone's abilities, and affective trust, which is derived from "strong interpersonal emotional connections" (2016, p. 245). In fact, this is a longstanding distinction made in trust research (McAllister, 1995). For example, Robert (2016) found that when teams were not monitored and had a high level of cognitive trust, they performed significantly better than teams that were monitored and had a similar level of trust. Affective trust had a negative relationship with performance for teams that did not perform their own team monitoring, but a strong positive relationship when they did engage in their own monitoring. Interestingly, this relationship was reversed when external monitoring (e.g., by those not on the team) was considered. Clearly, more research is needed to fully understand the complex interactions among trust, monitoring, and performance in virtual teams.

In addition, Knowles et al. (2014) borrowed from privacy research and focused on strategies for increasing trust in deliberate ways, suggesting a trustworthy by design approach. In support of this they developed five design patterns to improve the degree and quality of trust between the users of the system (Knowles et al., 2015). These patterns centered around co-workers being able to communicate both successes (e.g., when people exceeded work goals) and failure, as opposed to only communicating failure. They also focused on empowering individuals to influence organizational change, and incorporating social translucence to help avoid "filling in the gaps" and harboring distrust between parties.

Power

Within the work context, power or agency has been a longstanding focus of research and is an important part of relationships at work. The participatory design tradition has focused rather prominently on design methodologies that give agency within the design process and beyond to workers, particularly those who are disenfranchised (Bjerknes et al., 1987; Björgvinsson et al., 2010; Muller & Kuhn, 1993; Schuler & Namioka, 1993). The Florence project (Bjerknes & Bratteteig, 1988) provides key examples of these strategies. The researchers built an IT system to support nurses. Their position was that nurses should be included in the design of the tools that were purportedly being built to support them. That is, the designers recognized the mutual learning that must take place between the two parties (nurses and designers). This mutual learning was achieved by including the nurses in the design process and granting them agency over the design directions. By giving a voice to their needs and expertise, this agency in the design process had a long-term positive effect on their work and work relationships.

More recently, Hinds, Retelny, and Cramton (2015) studied how power impacts globally distributed teams and their performance. They outlined how lack of a voice or representation within power negatively impacted distributed team performance. In contrast, feeling that one's voice is being heard and having access to executives and decision makers has a positive effect on performance.

Embodiment of relationships

In this section we discuss several aspects of relationships at work that are affected or mediated by Internet-based systems. We examine two different Internet-based embodiments of relationships: first, we look at Open Source software development communities and second, we look at crowdsourcing.

Open Source relationships Bach, DeLine, and Carroll's (2009) investigation into the experience of designers who contributed to Open Source software projects found that software developer contributions were limited by several barriers. The decentralized construction of the teams caused problems such as "feature bloat," because anybody on the project could add to the project. In addition, the fast, iterative development cycles did not leave time for user-centered design at the start of the project. The decentralized structure of the collaboration meant that a majority of the team needed to "buy-in" to the philosophy of user-centered design. This was difficult because of developer perceptions (e.g., users being equally technical or that usability problems were trivial and unchallenging). The structure and makeup of the collaboration made it virtually impossible for the designer's contributions to be incorporated and valued. This is contrasted by another role in the Open Source community, the curator. Wu, Kropczynski, Prates, and Carroll (2015) detailed how different curation projects were highly valued in the Open Source community. These curation projects are "where developers systematically put efforts to select, evaluate, and organize existing artifacts" (, p. 46). These projects received a high degree of contributions and attention from the communities, even from those who were not part of the core work activity, and who served more of a supporting role in terms of documentation or gathering resources. The structure of the collaboration in this case was completely self-selected, that is, the only people that worked on a curation project were the ones who valued it. Those who thought that it was not a valuable contribution simply ignored it. This contrast highlights some of the benefits and pitfalls of Internet-based, volunteer projects.

Crowd labor relationships A newer trend that provides an interesting illustration of Internet-based work relationships is crowd labor (Felstiner, 2011). There are several forms of crowd or crowd-like labor: comparatively low-skill labor (e.g., ridesharing services such as Uber (Raval & Dourish, 2016), micro task services such as Amazon Mechanical Turk (AMT) (Irani, 2015a), or skilled labor such as software developers (e.g. TopCoder [www.topcoder.com; Archak, 2010]). In our embodiment section we discuss how relationships operate and are affected by AMT.

AMT was launched around 2006 when the chief executive officer of Amazon, Jeff Bezos, announced the system as a stopgap solution for the shortcomings of artificial intelligence (Irani, 2015a, pp. 3–5). As the research community began to investigate and utilize this new form of labor, researchers began to notice that AMT and its community of workers should be taken seriously, and researchers must be responsible for how they utilize and represent the crowd (Fort, Adda, & Cohen, 2011; Martin, Hanrahan, O'Neill, & Gupta, 2014; Salehi, Irani, & Bernstein, 2015). Crowd labor, particularly as realized on AMT, provides a salient case for how issues around power in the workplace can cause significant problems for all stakeholders when not properly considered. Issues around power in these new ITs, such as AMT, are more important as these systems do not operate within an existing social structure (e.g., email, knowledge management systems, or even workflow systems). Instead they facilitate the instantiation of a new social structure that did not exist previously. In this way, the system and its features/functionalities have a direct impact on the actions available to the stakeholders in the system (Hanrahan, Willamowski, Swaminathan, & Martin, 2015). Silberman, Ross, Irani, and Tomlinson (2010) outlined problems around unfair rejection of work, uncertain payment, low wages, and little to no communication.

Also of interest are the legal ramifications of the power dynamics, particularly in terms of worker protections, forms of labor, and some of the difficulties entailed for this new system (Bederson & Quinn, 2011; Felstiner, 2011; Wolfson & Lease, 2011). Researchers analyzed the crowd in terms of different types of labor (Irani, 2015b). Others have used the setting of crowd labor to examine the impact of the invisible algorithms that manage the workers (Lee, Kusbit, Metsky, & Dabbish, 2015; Raval & Dourish, 2016). This research led to a number of questions about the ethics and efficacy of current crowd labor practices (Bederson & Quinn, 2011; Felstiner, 2011; Kittur et al., 2013; O'Neill & Martin, 2013; Silberman et al., 2010).

Some of the consequences above can be linked directly to the implementation of the AMT platform. AMT does not display or prioritize the work tasks in a way that reflects the needs or desires of the workers. This results in a large amount of unpaid work in finding work tasks, difficulty in maintaining awareness in the face of market volatility, and difficulty with the onboarding process (Martin et al., 2014). Lack of functionality and limited information makes many decisions more difficult, such as which employers are trustworthy, which jobs pay a reasonable wage, or which skills would be more profitable to acquire.

In large part, many of these problems could be ameliorated by providing reciprocal actions to the various stakeholders. In the case of AMT, for example, if the system designers believe that the work providers need to block worker access to their tasks, perhaps workers should be able to block work providers from being displayed in search results (Hanrahan et al., 2015). In addition, if work providers need access to the reputation of different workers, the workers may need information about the reputation of the work providers (Silberman et al., 2010). When the actions as defined by the platform shape the interactions available between stakeholders, there must be good reason, and great care must be taken when granting an ability to one stakeholder and not to the other.

Communication

Communication is a vital part of forming relationships and collaborating at work. In this section we deviate slightly from the structure of the previous sections and instead, contrast two different types of communication technologies, email and wikis. In this way, we seek to illustrate some of the ergonomics around communication technologies in the workplace.

Email

Email has been called one of the most historically successful computer applications (Dabbish & Kraut, 2006; Whittaker & Sidner, 1996; Whittaker et al., 2007). It is a common, if not critical, part of work for many (Dabbish et al., 2005; Ducheneaut & Bellotti, 2001; Fisher, Brush, Gleave, & Smith, 2006; Mackay, 1988; Wainer, Dabbish, & Kraut, 2011; Whittaker et al., 2011). As such it has a long history of research in its 40 years of existence and, even with its success, many modifications have been proposed by the research community (Cselle, Albrecht, & Wattenhofer, 2007; Faulring et al., 2010; Freed et al., 2008; Nardi et al., 2002; Venolia & Neustaedter, 2003; Whittaker et al., 2004). Considerable research has centered on email triage, however, there have also been recent calls for better triaging tools (Vacek, 2014).

Email is clearly a place to do work. Along these lines, researchers have both studied and proposed several re-engineering of the tasks that are contained within the inbox (Bellotti, Ducheneaut, Howard, & Smith, 2003; Gwizdka, 2000, 2002). Another line of research has dealt with email overload (Whittaker & Sidner, 1996), finding that increased engagement with email increases feelings of overload (Sumeckia, Chipulua, & Ojiakoa, 2011). Other studies have identified that email volume contributes to email stress, as well as worrying about email (Jerejian, Reid, & Rees, 2013).

Researchers have also proposed changes to email practices rather than email tools. Chase and Clegg (2011) proposed that users should develop ways to manage sender expectations. Others have called for organizational training courses (McMurtry, 2014), or putting corporate or individual limits on email (Hemp, 2009). Bradley, Brumby, Cox, and Bird (2013) found evidence for the efficacy of these types of approaches in reducing email stress, finding that infrequent checking is better than frequent checking. This is somewhat in contrast to Dabbish and Kraut (2006), who found that frequently checking email reduced email-related stress, but they also hypothesized that letting email messages pile up creates a high degree of stress. Thus, checking new messages as they arrive relieves the tension of not knowing what is waiting for you in the inbox. However, we would argue that this is not necessarily a contradiction, and both of these results can be true. Specifically, if sender expectations are not managed or there are no mandated or endorsed limitations, individually disengaging from email may not decrease email-related stress. However, a group of collaborators disengaging from email altogether may decrease email-related stress.

These numerous research activities tell us much about the nature of communication through email, the research activities that have succeeded, the features that have been added to email and adopted by its users, and the studies into email use. We will discuss these insights when we contrast email with wikis.

Enterprise wikis

We now discuss the research around how knowledge is shared among coworkers (Davenport & Prusak, 1998; Grudin & Poole, 2010). In our knowledge economy the ability to efficiently share and reuse knowledge among workers is a key advantage. Enterprise wikis offer collaborative authoring mechanisms as a way to develop and organize such knowledge. Using wikis, workers can share content such as project updates, organized lists of tools or links, calls for participation, reports, publications, and frequently asked questions. However, much of the current sharing among knowledge workers continues to occur via email, which remains the central content management tool (Ducheneaut & Bellotti, 2001). This leads to two problems. First, email provides an inefficient channel to share, as it does not enable reuse across the organization. In fact, useful knowledge is trapped in personal inboxes, making it unavailable to co-workers and making its reuse difficult. This problem is exacerbated when a worker leaves an organization, rendering the information lost. Second, as more media are piped into email (e.g., updates from social networks) workers increasingly experience email overload, in the absence of suitable tools for offloading content from email (Whittaker & Sidner, 1996). Researchers have investigated this adoption problem through examination of specific deployments of enterprise wikis and Web 2.0 tools (O'Reilly, 2007). These studies point to both the intrinsic limitations within an enterprise and the deficiencies in the design and deployment of these new tools. One such limitation is a very small population willing to edit and maintain enterprise wiki pages (see power law distributions in Brzozowski et al., 2009), which results in less shared content in comparison to the same tools deployed outside enterprises (e.g., Wikipedia, see Wilkinson, 2008).

In our previous work (Hanrahan et al., 2011), we argued that the design of an Enterprise 2.0 tool, such as a corporate wiki, must be grounded in understanding the sharing ecology in organizations. The sharing ecology is fundamentally different between an enterprise, where a popular topic might have 10–100 possible contributors, and the iconic example of Wikipedia, where a popular topic will have thousands more possible contributors. In both examples, the percentage of potential contributors who actually contribute follows the power law. However, within the enterprise this is insufficient and new content is not authored and these technologies suffer disuse. Our approach to this problem was to construct a convincing adoption path. In line with this vision, drawing on our observations and prior research, we built the Mail2Wiki system to reduce interaction costs and integrate wikis with email (Hanrahan et al., 2011). Mail2Wiki accomplished this by embedding relevant wiki pages directly into the email client. If a user wanted to provide a contribution from an email message, they could select the text and drag it into the article's structure. We designed the system to enable incremental adoption, where using the simplest functionalities requires the simplest interactions at the lowest cost. This built-in adoption example accentuates the amelioration of interaction costs and tool integration. Note that while these pro-adoption solutions are embodied here with enterprise wikis, they might be later re-applied to other common Enterprise 2.0 tools.

Future Research

In our analysis we reviewed a number of ways to study work, oriented more towards the practices and contexts of work. Through this lens we identified the different aspects of work that researchers have tried to influence with their work. These threads of research were the basis of our frame for the preconditions of work. We would urge future researchers to continue to look, not only at the activity or tasks of work, but the conditions and context that surround it.

Much of the research into the ergonomics of IT and the psychology of work has been in more traditional work contexts. Looking toward the future, Internet-based systems will not merely be tools used at work, instead they will help to shape and define work. One way to research this theme is to look at other structures of work, as this may have a broad impact on how to support the future of work, and may reshape many practices. One such emerging work structure occurs when labor is mediated by a third-party platform. Crowd labor is a key example and several important identifying features that seem to have inspired new platforms. Some of these features are transformational to their industry and sometimes quite contentious. These key features are: a labor market that is maintained by a central party; the involvement of several stakeholders (e.g., the worker–employer relationship is somewhat complex and involves multiple entities); loose ties between the worker and employer (be it the work provider or the platform maintainer); and finally, the classification of workers as independent contractors. These aspects are being used in other contexts already, for instance, Uber, the ride-sharing company. The basic structure of Uber is that they provide and maintain a platform where drivers and riders can locate one another to coordinate rides. There are several stakeholders (e.g., Uber the company, the driver and the rider) involved in this exchange of labor, but few of them are under the direct control of the Uber organization. There are loose ties between the different stakeholders, for instance, the driver and rider have extremely loose ties, and the relationship between the driver and Uber is quite loose as well. Finally, driving a car is a relatively low-skill job, and the drivers are categorized as independent contractors. Of course the same can be said for competitors of Uber. There are many other examples of such third-party brokers of labor: from low-skilled labor such as TaskRabbit, where individuals coordinate primarily home-related tasks (e.g. packing boxes or running errands) to high-skilled labor such as TopCoder (e.g., where programmers or designers complete highly technical work). Given the realized value of this new type of workplace and its numerous benefits for both workers and employers, the model of independent contracting of labor or hyper-freelancing will only grow.

Conclusion

In this chapter we discussed the ergonomics of Internet technology at work by discussing what we identified as major research themes around the practice of work. We broadly took an analytic outlook of how IT impacts work practices within the contexts in which it is deployed. The themes that we identified represent some of the requirements or preconditions for getting work done. There must be some practice for how the work gets done, there must be some kind of community to provide support and meaning to the work, these communities must have relationships for them to function, and communication is necessary for these relationships to form and flourish. The community, particularly the human–computer interaction and computer-supported cooperative work community has recognized these aspects as fundamental to work, and much of the research literature can be viewed in terms of the impact of one or more of these items.

In addition, the Internet highlights the importance of a broader view of ergonomics in that it considers the systems that constitute work and how work is accomplished. As a result of ubiquitous Internet access, a number of new phenomena have appeared such as continually connected workers, offshoring/outsourcing, crowd labor, and highly distributed teams. This means that aspects of how work gets done, which used to be more implicit to the workplace, now need to be given more explicit consideration and support in the tools that we develop to support workers and work environments.

References

Ackerman, M. S., Dachtera, J., Pipek, V., & Wulf, V. (2013). Sharing knowledge and expertise: The CSCW view of knowledge management. *Computer Supported Cooperative Work, 22*(4):531–573.

Archak, N. (2010). Money, glory and cheap talk: Analyzing strategic behavior of contestants in simultaneous crowdsourcing contests on topcoder.com. In *Proceedings of the 19th International Conference on World Wide Web (WWW '10)* (pp. 21–30). New York, NY: ACM.

Bach, P. M., DeLine, R., & Carroll, J. M. (2009). Designers wanted: Participation and the user experience in open source software development. In *Proceedings of the SIGCHI Conference on Human Factors in Computing Systems* (pp. 985–994). Boston, MA: ACM.

Balog, K., Fang, Y., de Rijke, M., Serdyukov, P., & Si, L. (2012). Expertise retrieval. *Foundations and Trends in Information Retrieval, 6*, 127–256.

Bederson, B. B., & Quinn, A. J. (2011). Web workers unite! Addressing challenges of online laborers. In *CHI 2011 Extended abstracts on human factors in computing systems* (pp. 97–106). Boston, MA: ACM.

Bellah, R. N., Madsen, R., Sullivan, W. M., Swidler, A., & Tipton, S. M. (2007). *Habits of the heart: Individualism and commitment in American life.* Berkeley, CA: University of California Press.

Bellotti, V., Ducheneaut, N., Howard, M., & Smith, I. (2003). Taking email to task: the design and evaluation of a task management centered email tool. In *Proceedings of the SIGCHI Conference on Human Factors in Computing Systems* (pp. 345–352). New York, NY: ACM.

Bjerknes, G., & Bratteteig, T. (1988). The memoirs of two survivors – or evaluation of a computer system for cooperative work. In *Proceedings of the Second Conference on Computer Supported Cooperative Work* (pp. 167-177). Portland, OR: ACM.

Bjerknes, G., Ehn, P., Kyng, M., & Nygaard, K. (1987). *Computers and democracy: A Scandinavian challenge.* Aldershot, England: Avebury.

Björgvinsson, E., Ehn, P., & Hillgren, P.-A. (2010). Participatory design and "democratizing innovation." In *Proceedings of the 11th Biennial Participatory Design Conference (PDC '10)* (pp. 41–50). New York, NY: ACM.

Bradley, A., Brumby, D. P., Cox, A. L., & Bird, J. (2013). How to manage your inbox: Is a once a day strategy best? In *Proceedings of the 27th International BCS Human Computer Interaction Conference (BCS-HCI '13)* (pp. 20:1–20:6). Swinton, UK: British Computer Society.

Brzozowski, M. J., Sandholm, T., & Hogg, T. (2009). Effects of feedback and peer pressure on contributions to enterprise social media. In *Proceedings of the ACM 2009 International Conference on Supporting Group Work* (pp. 61–70). Sanibel Island, FL: ACM.

Button, G. (1993). *Technology in working order: Studies of work, interaction, and technology.* New York, NY: Routledge.

Button, G. (2003). *Studies of work in human-computer interaction. HCI models, theories and frameworks: Toward a multidisciplinary science* (pp. 357–380). San Francisco, CA: Morgan Kaufmann.

Button, G., & Dourish, P. (1996). Technomethodology: Paradoxes and possibilities. In *Proceedings of the SIGCHI Conference on Human Factors in Computing Systems, (CHI '96)* (pp. 19–26). New York, NY: ACM.

Button, G., & Harper, R. (1996). The relevance of 'work-practice' for design. *Computer Supported Cooperative Work, 4*, 263–280.

Button, G., & Sharrock, W. (2009). Studies of work and the workplace in HCI: Concepts and techniques. *Synthesis Lectures on Human-Centered Informatics, 2*(1), 1–96.

Carroll, J. M. (2014). *The neighborhood in the Internet: Design research projects in community informatics.* New York, NY: Routledge.

Carter, M. (2015). Me, myself, and I(T): Conceptualizing information technology identity and its implications. *MIS Quarterly, 39*(4), 931–957.

Chase, N., & Clegg, B. (2011). Effects of email utilization on higher education professionals. *International Journal of Technology and Human Interaction, 7*(4), 31–45.

Chipman, S. F., Schraagen, J. M., & Shalin, V. L. (2000). Introduction to cognitive task analysis. In *Cognitive task analysis* (pp. 3–23). Québec, Canada: St Joseph Ottawa/Hull.

Colombino, T., Hanrahan, B., & Castellani, S. (2014). Lessons learnt working with performance data in call centres. In *Proceedings of the 11th International Conference on the Design of Cooperative Systems (COOP)* (pp. 277–292). Cham, Switzerland: Springer International Publishing.

Cselle, G., Albrecht, K., & Wattenhofer, R. (2007). Buzztrack: Topic detection and tracking in email. In *Proceedings of the 12th International Conference on Intelligent User Interfaces (IUI)* (pp. 190–197). New York, NY, USA. ACM.

Dabbish, L., Stuart, C., Tsay, J., & Herbsleb, J. (2012). Social coding in Github: Transparency and collaboration in an open software repository. In *Proceedings of the ACM 2012 Conference on Computer Supported Cooperative Work (CSCW)* (pp. 1277–1286). New York, NY: ACM.

Dabbish, L. A., & Kraut, R. E. (2006). Email overload at work: an analysis of factors associated with email strain. In *Proceedings of the 2006 20th Anniversary Conference on Computer Supported Cooperative Work (CSCW)* (pp. 431–440). New York, NY: ACM.

Dabbish, L. A., Kraut, R. E., Fussell, S., & Kiesler, S. (2005). Understanding email use: Predicting action on a message. In *Proceedings of the SIGCHI Conference on Human Factors in Computing Systems (CHI)* (pp. 691– 700). New York, NY: ACM.

Daniellou, F. (2005). The French-speaking ergonomists' approach to work activity: cross-influences of field intervention and conceptual models. *Theoretical Issues in Ergonomics Science, 6*(5), 409–427.

Davenport, T. H., & Prusak, L. (1998). *Working knowledge: How organizations manage what they know.* Boston, MA: Harvard Business Press.

De Dreu, C. K. W. (2006). Rational self-interest and other orientation in organizational behavior: A critical appraisal and extension of Meglino and Korsgaard (2004). *Journal of Applied Psychology, 9*(6), 1245–1252.

de Tocqueville, A. (2002). *Democracy in America* (Reeve, H., Trans.). Washington, DC: Regnery. (Original translated work published 1889, original work published 1835).

DiMicco, J., Millen, D. R., Geyer, W., Dugan, C., Brownholtz, B., & Muller, M. (2008). Motivations for social networking at work. In *Proceedings of the ACM 2008 Conference on Computer Supported Cooperative Work (CSCW)* (pp. 711–720), New York, NY: ACM.

DiMicco, J. M., Geyer, W., Millen, D. R., Dugan, C., & Brownholtz, B. (2009). People sensemaking and relationship building on an enterprise social network site. In *Hawaii International Conference on System Sciences (HICSS)* (pp. 1–10). Washington DC: IEEE.

DiMicco, J. M., & Millen, D. R. (2007). Identity management: multiple presentations of self in Facebook. In *Proceedings of the 2007 International ACM Conference on Supporting Group Work (GROUP)* (pp. 383–386). New York, NY: ACM.

Ducheneaut, N., & Bellotti, V. (2001). E-mail as habitat: an exploration of embedded personal information management. *Interactions, 8*(5), 30–38.

Dugan, C., Geyer, W., & Millen, D. R. (2010). Lessons learned from blog muse: Audience-based inspiration for bloggers. In *Proceedings of the SIGCHI Conference on Human Factors in Computing Systems (CHI)* (pp. 1965–1974). New York, NY: ACM.

Efimova, L., & Grudin, J. (2007). Crossing boundaries: A case study of employee blogging. In *Hawaii International Conference on System Sciences (HICCS)* (pp. 86–86). Washington DC: IEEE.

Ehn, P., & Kyng, M. (1987). The collective resource approach to systems design. In *Computers and democracy – A Scandinavian challenge* (pp.17–58). Aldershot, England: Avebury.

Ehrlich, K., Lin, C.-Y., & Griffiths-Fisher, V. (2007). Searching for experts in the enterprise: Combining text and social network analysis. In *Proceedings of the 2007 International ACM Conference on Supporting Group Work (GROUP)* (pp. 117–126). New York, NY: ACM.

Faulring, A., Myers, B., Mohnkern, K., Schmerl, B., Steinfeld, A., Zimmerman, J., ... Siewiorek, D. (2010). Agent-assisted task management that reduces email overload. In *Proceedings of the 15th International Conference on Intelligent User Interfaces (IUI)* (pp. 61–70). New York, NY: ACM.

Felstiner, A. (2011). Working the crowd: Employment and labor law in the crowdsourcing industry. *Berkeley Journal of Employment and Labor Law, 32*(1), 143–203.

Fisher, D., Brush, A. J., Gleave, E., & Smith, M. A. (2006). Revisiting Whittaker & Sidner's "Email Overload" ten years later. In *Proceedings of the 2006 Conference on Computer Supported Cooperative Work (CSCW)* (pp. 309–312). New York, NY: ACM.

Fort, K., Adda, G., & Cohen, K. B. (2011). Amazon mechanical turk: Gold mine or coal mine? *Computational Linguistics, 37*(2), 413–420. MIT.

Freed, M., Carbonell, J. G., Gordon, G. J., Hayes, J., Myers, B. A., Siewiorek, D. P., ... Tomasic, A. (2008). Radar: A personal assistant that learns to reduce email overload. In *Proceedings of the 23rd National Conference on Artificial Intelligence (AAAI)* (pp. 1287–1293). Chicago, Illinois: AAAI.

Frøkjær, E., Hertzum, M., & Hornbæk, K. (2000). Measuring usability: Are effectiveness, efficiency, and satisfaction really correlated? In *Proceedings of the SIGCHI Conference on Human Factors in Computing Systems (CHI)* (pp. 345–352). New York, NY: ACM.

Garfinkel, H. (1967). *Studies in ethnomethodology.* Upper Saddle River, NJ: Prentice-Hall.

Grudin, J., & Poole, E. S. (2010). Wikis at work: success factors and challenges for sustainability of enterprise wikis. In *Proceedings of the 6th International Symposium on Wikis and Open Collaboration* (pp. 1–8). New York, NY: ACM.

Guy, I., Avraham, U., Carmel, D., Ur, S., Jacovi, M., & Ronen, I. (2013). Mining expertise and interests from social media. In *Proceedings of the 22nd International Conference on World Wide Web* (pp. 515–526). New York, NY: ACM.

Guy, I., Jacovi, M., Shahar, E., Meshulam, N., Soroka, V., & Farrell, S. (2008). Harvesting with sonar: The value of aggregating social network information. In *Proceeding of the 26th Annual SIGCHI Conference on Human Factors in Computing Systems (CHI)* (pp. 1017–1026). New York, NY: ACM.

Gwizdka, J. (2000). Timely reminders: A case study of temporal guidance in pim and email tools usage. In *Extended Abstracts on Human Factors in Computing Systems* (pp. 163–164). New York, NY: ACM.

Gwizdka, J. (2002). Reinventing the inbox: Supporting the management of pending tasks in email. In *Extended Abstracts on Human Factors in Computing Systems* (pp. 550–551). New York, NY: ACM.

Hackos, J. T. & Redish, J. (1998). *User and task analysis for interface design.* New York, NY: John Wiley & Sons, Inc.

Hanrahan, B., Bouchard, G., Convertino, G., Weksteen, T., Kong, N., Archambeau, C., & Chi, E. H. (2011). Mail2wiki: low-cost sharing and early curation from email to wikis. In *Proceedings of the 5th International Conference on Communities and Technologies* (pp. 98–107). New York, NY: ACM.

Hanrahan, B. V., Willamowski, J. K., Swaminathan, S., & Martin, D. B. (2015). Turkbench: Rendering the market for Turkers. In *Proceedings of the 33rd Annual ACM Conference on Human Factors in Computing Systems (CHI)* (pp. 1613–1616). New York, NY: ACM.

Hemp, P. (2009). Death by information overload. *Harvard Business Review, 87*(9), 82–89.

Hinds, P., Retelny, D., & Cramton, C. (2015). In the flow, being heard, and having opportunities: Sources of power and power dynamics in global teams. In *Proceedings of the 18th ACM Conference on Computer Supported Cooperative Work (CSCW)* (pp. 864–875). New York, NY: ACM.

Holtzblatt, K., & Jones, S. (1993). Contextual inquiry: A participatory technique for system design. In Schuler, D. & Nomioka, A. (Eds.), *Participatory design: Principles and practices* (pp. 177–210). Hillsdale, NJ: Lawrence Erlbaum Associates.

Hutchins, E., & Klausen T. (1996). Distributed cognition in an airline cockpit. In Engeström, Y. & Middleton, D. (Eds.), *Dognition and communication at work.* New York, NY: Cambridge University Press.

Irani, L. (2015a). The cultural work of microwork. *New Media & Society, 17*(5), 720–739. Sage.

Irani, L. (2015b). Difference and dependence among digital workers: The case of Amazon Mechanical Turk. *South Atlantic Quarterly, 114*(1), 225–234. Durham, NC: Duke University Press.

Jacovi, M., Guy, I., Kremer-Davidson, S., Porat, S., & Aizenbud-Reshef, N. (2014). The perception of others: Inferring reputation from social media in the enterprise. In *Proceedings of the 17th ACM Conference on Computer Supported Cooperative Work* (pp. 756–766). New York, NY: ACM.

Jarvenpaa, S., Cantu, C. & Lim, S. Y. (2017). Trust in virtual online environments: Toward a contextualized understanding of trust. In G. Hertel, D. L. Stone, R. D. Johnson, & J. Passmore (Eds.), *Wiley Blackwell handbook of the psychology of the internet at work* (pp. 103–130). Chichester: John Wiley & Sons, Ltd.

Jerejian, A., Reid, C., & Rees, C. S. (2013). The contribution of email volume, email management strategies and propensity to worry in predicting email stress among academics. *Computers in Human Behavior, 29*(3), 991–996.

Kittur, A., Nickerson, J. V., Bernstein, M., Gerber, E., Shaw, A., Zimmerman, J., ... Horton, J. (2013). The future of crowd work. In *Proceedings of the 16th ACM Conference on Computer Supported Cooperative Work* (pp. 1301–1318). New York, NY: ACM.

Knowles, B., Harding, M., Blair, L., Davies, N., Hannon, J., Rouncefield, M., & Walden, J. (2014). Trustworthy by design. In *Proceedings of the 17th ACM Conference on Computer Supported Cooperative Work and Social Computing* (pp. 1060–1071). New York, NY: ACM.

Knowles, B., Rouncefield, M., Harding, M., Davies, N., Blair, L., Hannon, J., ... Wang, D. (2015). Models and patterns of trust. In *Proceedings of the 18th ACM Conference on Computer Supported Cooperative Work and Social Computing* (pp. 328–338). New York, NY: ACM.

Kroemer, K. H., Kroemer, H. B., & Kroemer-Elbert, K. E. (2001). *Ergonomics: how to design for ease and efficiency.* Lebanon, IN: Pearson College Division.

Lee, M. K., Kusbit, D., Metsky, E., & Dabbish, L. (2015). Working with machines: The impact of algorithmic and data-driven management on human workers. In *Proceedings of the 33rd Annual ACM Conference on Human Factors in Computing Systems* (pp. 1603–1612). New York, NY: ACM.

Levy, M. (2009). WEB 2.0 implications on knowledge management. *Journal of Knowledge Management, 13*(1), 120–134.

Mackay, W. E. (1988). Diversity in the use of electronic mail: A preliminary inquiry. *ACM Transactions on Information Systems, 6*(4), 380–397.

Mark, G., Guy, I., Kremer-Davidson, S., & Jacovi, M. (2014). Most liked, fewest friends: Patterns of enterprise social media use. In *Proceedings of the 17th ACM Conference on Computer Supported Cooperative Work & Social Computing* (pp. 393–404) New York, NY: ACM.

Martin, D., Hanrahan, B. V., O'Neill, J., & Gupta, N. (2014). Being a turker. In *Proceedings of the 17th ACM Conference on Computer Supported Cooperative Work & Social Computing* (pp. 224–235). New York, NY: ACM.

Matthews, T., Whittaker, S., Badenes, H., & Smith, B. (2014). Beyond end user content to collaborative knowledge mapping: Interrelations among community social tools. In *Proceedings of the 17th ACM Conference on Computer Supported Cooperative Work & Social Computing* (pp. 900–910). New York, NY: ACM.

McAllister, D. J. (1995). Affect-and cognition-based trust as foundations for interpersonal cooperation in organizations. *Academy of Management Journal, 38*(1), 24–59.

McDonald, D. W., & Ackerman, M. S. (1998). Just talk to me: A field study of expertise location. In *Proceedings of the ACM Conference on Computer Supported Cooperative Work* (pp. 315–324). New York, NY: ACM.

McDonald, D. W., & Ackerman, M. S. (2000). Expertise recommender: a flexible recommendation system and architecture. In *Proceedings of the 2000 ACM Conference on Computer Supported Cooperative Work* (pp. 231–240). New York, NY: ACM.

McMurtry, K. (2014). Managing email overload in the workplace. *Performance Improvement, 53*(7), 13–37.

Meglino, B. M., & Korsgaard, A. (2004). Considering rational self-interest as a disposition: organizational implications of other orientation. *Journal of Applied Psychology, 89*(6), 946.

Muller, M. J., & Kuhn, S. (1993). Participatory design. *Communications of the ACM, 36*(6), 24–28.
Nardi, B. A., Whittaker, S., Isaacs, E., Creech, M., Johnson, J., & Hainsworth, J. (2002). Integrating communication and information through contactmap. *Communications of the ACM, 45*(4), 89–95.
O'Neill, J., & Martin, D. (2013). Relationship-based business process crowdsourcing. In *Human-Computer Interaction 14th IFIP TC 13 International Conference (INTERACT)* (pp. 429–446). Berlin: Springer.
O'Reilly, T. (2007). What is web 2.0: Design patterns and business models for the next generation of software. *Communications & Strategies, 65*(1), 17–37.
Pirolli, P., & Card, S. (2005). The sensemaking process and leverage points for analyst technology as identified through cognitive task analysis. In *Proceedings of International Conference on Intelligence Analysis* (pp. 2–4). McLean, VA: MITRE.
Putnam, R. D. (1995). Bowling alone: America's declining social capital. *Journal of Democracy, 6*(1), 65–78.
Raval, N., & Dourish, P. (2016). Standing out from the crowd: Emotional labor, body labor, and temporal labor in ridesharing. In *Proceedings of the 19th ACM Conference on Computer-Supported Cooperative Work & Social Computing* (pp. 97–107). New York, NY: ACM.
Robert, Jr., L. P. (2016). Monitoring and trust in virtual teams. In *Proceedings of the 19th ACM Conference on Computer-Supported Cooperative Work & Social Computing* (pp. 245–259). New York, NY: ACM.
Salehi, N., Irani, L. C., & Bernstein, M. S. (2015). We are dynamo: Overcoming stalling and friction in collective action for crowd workers. In *Proceedings of the 33rd Annual ACM Conference on Human Factors in Computing Systems* (pp. 1621–1630). New York, NY: ACM.
Schuler, D., & Namioka, A. (1993). *Participatory design: Principles and practices.* Hillsdale, NJ: Lawrence Erlbaum Associates.
Silberman, M., Ross, J., Irani, L., & Tomlinson, B. (2010). Sellers' problems in human computation markets. In *SIGKDD Workshop on Human Computation* (pp. 18–21). New York, NY: ACM.
Skeels, M. M., & Grudin, J. (2009). When social networks cross boundaries: A case study of workplace use of Facebook and LinkedIn. In *Proceedings of the ACM 2009 International Conference on Supporting Group Work* (pp. 95–104). New York, NY: ACM.
Steinfield, C., DiMicco, J. M., Ellison, N. B., & Lampe, C. (2009). Bowling online: Social networking and social capital within the organization. In *Proceedings of the Fourth International Conference on Communities and Technologies* (pp. 245–254). New York, NY: ACM.
Suchman, L. A. (1987). *Plans and situated actions: the problem of human machine communication.* New York, NY: Cambridge University Press.
Sumeckia, D., Chipulua, M., & Ojiakoa, U. (2011). Email overload: Exploring the moderating role of the perception of email as a 'business critical' tool. *International Journal of Information Management, 31*, 407–414.
Thielsch, M. T., & Niesenhaus, J. (2017). User experience, gamification, and performance. In G. Hertel, D. L. Stone, R. D. Johnson, & J. Passmore (Eds.), *Wiley Blackwell handbook of the psychology of the internet at work* (pp. 79–102). Chichester: John Wiley & Sons, Ltd.
Tönnies, F. (2012). Community and society (*Gemeinschaft und Gesellschaft*) (Loomis, C. P. Trans.). In J. Lin & C. Mele (Eds.), *The urban sociology reader.* Berlin: Springer. (Original translated work published 1963, original work published 1877).
Treem, J. W., & Leonardi, P. M. (2012). Social media use in organizations: Exploring the affordances of visibility, editability, persistence, and association. *Communication Yearbook, 36*:143–189.
Vacek, M. (2014). How to survive email. In *9th IEEE International Symposium on Applied Computational Intelligence and Informatics* (pp. 49–54). Washington DC: IEEE.
Vicente, K. (1999). *Cognitive work analysis: Toward safe, productive, and healthy computer-based work.* Mahwah, NJ: Lawrence Erlbaum Associates.
Venolia, G. D., & Neustaedter, C. (2003). Understanding sequence and reply relationships within email conversations: a mixed-model visualization. In *Proceedings of the SIGCHI Conference on Human Factors in Computing Systems* (pp. 361–368). New York, NY: ACM.

Wainer, J., Dabbish, L., & Kraut, R. (2011). Should I open this email? Inbox level cues, curiosity and attention to email. In *Proceedings of the 2011 Annual Conference on Human Factors in Computing Systems* (pp. 3439–3448). New York, NY: ACM.

Wenger, E. (1998). Communities of practice: Learning as a social system. *Systems Thinker*, *9*(5), 2–3.

Wenger, E. (1999). *Communities of practice: Learning, meaning, and identity*. New York, NY: Cambridge University Press.

Wenger, E. C., & Snyder, W. M. (2000). Communities of practice: The organizational frontier. *Harvard Business Review*, *78*(1), 139–146.

Whittaker, S., Bellotti, V., & Gwizdka, J. (2007). Chapter everything through email. In W. Jones & J. Teevan (Eds.), *Personal information management*. Seattle, WA: University of Washington Press.

Whittaker, S., Jones, Q., Nardi, B., Creech, M., Terveen, L., Isaacs, E., & Hainsworth, J. (2004). Contactmap: Organizing communication in a social desktop. *ACM Transactions on Computer Human Interaction*, *11*(4), 445–471.

Whittaker, S., Matthews, T., Cerruti, J., Badenes, H., & Tang, J. (2011). Am I wasting my time organizing email? A study of email refinding. In *Proceedings of the 2011 Annual Conference on Human Factors in Computing Systems* (pp. 3449–3458). New York, NY: ACM.

Whittaker, S., & Sidner, C. (1996). Email overload: exploring personal information management of email. In *Proceedings of the SIGCHI Conference on Human Factors in Computing Systems* (pp. 276–283). New York, NY: ACM.

Wickens, C., & Hollands, J. (2000). *Engineering psychology and human performance* (3rd ed.). Upper Saddle River, NJ: Prentice Hall.

Wilkinson, D. M. (2008). Strong regularities in online peer production. In *Proceedings of the 9th ACM Conference on Electronic Commerce* (pp. 302–309). New York, NY: ACM.

Wilson, J. R. (2014). Fundamentals of systems ergonomics/human factors. *Applied Ergonomics*, *45*(1), 5-13.

Wolfson, S. M., & Lease, M. (2011). Look before you leap: Legal pitfalls of crowdsourcing. In *Proceedings of the American Society for Information Science and Technology*, *48*(1), 1–10.

Wu, Y., Kropcznyski, J., Prates, R., & Carroll, J. M. (2015). The rise of curation on GitHub. In *Third AAAI Conference on Human Computation and Crowdsourcing* (pp. 46-47). Chicago, IL: AAAI.

Zafar, H. (2017). Mobile computing and hand held devices. In G. Hertel, D. L. Stone, R. D. Johnson, & J. Passmore (Eds.), *Wiley Blackwell handbook of the psychology of the internet at work* (pp. 195–210). Chichester: John Wiley & Sons, Ltd.

Zhao, D., & Rosson, M. B. (2009). How and why people twitter: the role that micro-blogging plays in informal communication at work. In *Proceedings of the ACM 2009 International Conference on Supporting Group Work* (pp. 243–252). New York, NY: ACM.

4

Competencies for Web-Based Work and Virtual Collaboration

Stefan Krumm and Julian Schulze

Introduction

During the last decade, the world has experienced a steady rise in web-based technologies. Mobile device usage, as an example, has quintupled in the period 2010–2015 in the developed world (Statista, 2016). Although web-based technologies are already pervading almost all aspects of today's life (e.g., work, education, entertainment, health and social services), the steep increase in its usage at the end of the twentieth and at the beginning of the twenty-first century suggests that this trend is far from over and it is safe to say that new developments will continue to shape the way we work and communicate in the future. In light of the rate of change, the competencies required for web-based work and virtual collaboration – the former being defined as work that is predominantly based on web-based applications and the latter as collaboration predominantly based on digital communication media (*cf.* Hertel, Geister, & Konradt, 2005) – are key competencies of the twenty-first century (OECD, 2013).

The aim of this chapter is twofold. First, we review the literature on competencies in web-based work and digital collaboration. Second, we seek to widen the picture and review theories on digital communication and collaboration with respect to their aptness to derive further competencies. Specifically, we review theories pertaining to media choice, computer-mediated communication (CMC), and digital collaboration. Each section ends with a list of competencies that the reviewed theories imply are relevant to these areas. The chapter ends by suggesting future research directions.

A Brief Review of Competency Modeling Approaches

While several authors have sought to identify individual qualities required in web-based work and virtual collaboration (e.g., Hertel, Konradt, & Voss, 2006; Krumm & Hertel, 2013; Schulze & Krumm, 2017, Wang & Haggerty, 2011), research has not yet

---

The Wiley Blackwell Handbook of the Psychology of the Internet at Work,
First Edition. Edited by Guido Hertel, Dianna L. Stone, Richard D. Johnson, and Jonathan Passmore.

reached a consensus concerning which competencies are particularly relevant for certain forms of digital work. In fact, a review by Gilson, Maynard, Young, Vartiainen, and Hakonen (2015) conceded that "virtual competence . . . has been explored only minimally" (p. 13). Although the rich body of research on face-to-face collaboration has suggested ample potentially relevant competencies, one has to acknowledge that "this list is neither totally transferable nor complete in the virtual environment" (Furst, Blackburn, & Rosen, 1999, p. 257).

When aiming to identify competencies for web-based work and virtual collaboration, it is a key prerequisite to define the term *competency*. As Shippmann et al. (2000) noted, the term competency has a variety of meanings, both in research as well as in practice. For the purpose of this chapter, we adopt the notion of competencies as "any individual characteristic . . . that can be shown to differentiate superior from average performers" (Spencer, McClelland, & Spencer, as cited in Shippmann et al., 2000, p. 706). According to this definition, the scope of predictors for effective web-based work and virtual collaboration may include all kinds of individual difference constructs, such as motives, attitudes, personality traits, knowledge in relevant domains, as well as cognitive and noncognitive skills.

A competency modeling approach comes with not only a large spectrum of potentially relevant individual difference constructs but also a variety of methods. These methods include – in their simplest forms – having subject matter experts to infer competencies from global job descriptions, having job incumbents rate or sort competencies according to their relevance, task and behavior analytic approaches (such as the critical incident technique; Flanagan, 1954), comparisons of high- and low-performing groups, as well as regression analytic approaches relating self- and peer-rated competencies to performance criteria (where the more scientifically rigorous among these methods are usually referred to as job analysis rather than as competency modeling, *cf.* Shippmann et al., 2000). Research on competencies in several important work-related domains (e.g., leadership) has benefited from using the aforementioned methods and, as a result, has produced several comprehensive competency models (e.g., Hollenbeck, McCall, & Silzer, 2006).

One of the most generic competency models is the great eight competency framework (Bartram, 2005), derived from a synthesis of organizational competency models. This framework includes eight broad competency domains, further subdivided into 20 smaller competency dimensions. Among those, only one dimension explicitly targets technology. It is dubbed "applying expertise and technology" and is part of the broader domain of "analyzing and interpreting." As technology has pervaded many aspects of work, it seems surprising that only one out of 20 dimensions explicitly addresses technology, and we thus agree with Furst et al. (1999) that more competencies need to be considered to capture the requirements of web-based work and digital collaboration.

The same conclusion might be drawn when inspecting less generic competency models, such as those explicitly created for face-to-face collaboration/teamwork (e.g., Cannon-Bowers, Tannenbaum, Salas, & Volpe, 1995; Hoegl & Gemuenden, 2001; Loughry, Ohland, & Moore, 2007; Stevens & Campion, 1994). Krumm and Hertel (2013) summarize the overlap between the available competency models for face-to-face collaboration as follows:

> *a good team-player is willing to do a fair share of the team's work, communicates relevant information effectively, has a good understanding of the overall team goals and monitors the team's progress towards these goals, has high performance standards, has an overview of other team members' tasks, can coordinate task completion appropriately, and, finally, is well accepted by the other team members.*
>
> (p. 82)

Again, however, no obvious links to technology-related challenges and corresponding requirements are provided in these models.

We now turn to competency modeling approaches developed specifically for web-based work and digital collaboration. Some of these approaches are purely theoretical. Shin (2004), for instance, employed a theoretical person–environment fit approach, focusing on organizational, cultural, temporal, and spatial dispersion as virtual environment dimensions, positing that virtual organizations vary on these environment dimensions. The main assumption of this model is that these dimensions moderate the extent to which certain individual qualities lead to perceived person–environment fit. For example, individuals who value autonomy highly will perceive a greater person–environment fit when the organization is characterized by a high degree of temporal and spatial dispersion. This is because spatially and temporally dispersed work is characterized by flexible timeframes and less monitoring by supervisors and managers, which is likely to increase an individual's responsibilities and autonomy (Shin, 2004). Furthermore, Shin also theorized that computer literacy, willingness to trust, trustworthiness, lateral skills (the skills to cooperate with and learn from individuals with a variety of backgrounds), and virtual communication skills are crucial antecedents of person–environment fit perceptions within all four virtual environment dimensions.

As one form of an empirical competency modeling approach, Blackburn, Furst, and Rosen (2003) interviewed 25 experts in digital collaboration. They found the following set of skills on the individual level: to behave proactively and manage themselves, to select the most appropriate transmission medium and to seek and interpret feedback (as media differ in their appropriateness to transfer information and feedback cannot be gathered instantly), to be culturally sensitive (as one might collaborate frequently with individuals from other cultures), to build and maintain trust (as one cannot observe others' efforts), and – of course – to be confident and competent in using the relevant technology.

Lee-Kelley and Sankey (2008), as an example of another competency modeling approach, compared efficient and inefficient virtual collaborators. They conducted semi-structured interviews with 11 managers of virtual project teams. The interviews were analysed thematically to identify key issues of virtual project teamwork, which in turn allowed for the derivation of the necessary competencies to handle such challenges. The authors conclude that personal project management competencies, appropriate use of technology and networking, willingness for self-management, and cultural and interpersonal awareness are crucial for virtual team project success (Lee-Kelley & Sankey, 2008). Krumm, Kanthak, Hartmann, and Hertel (2016) also sought to identify competencies for virtual collaboration by comparing two groups. They asked members of virtual and traditional teams to rate the importance of competencies organized along the great eight competency framework (Bartram, 2005). The results revealed that competencies related to "leading and deciding" (e.g., including skills to work autonomously, to act on one's own initiative) and "analyzing and interpreting" (e.g., including skills to communicate in writing, to use communication media effectively) were considered more important in digital forms of collaboration, that is, in virtual teams.

Finally, several studies have used a regression analytic approach to derive competencies for web-based work and digital collaboration. For instance, Hertel et al. (2006) used a set of theoretically derived competencies to predict performance in virtual teams. From the initial set of predictors, conscientiousness, integrity, loyalty, and cooperativeness (on the individual level), as well as creativity, learning orientation, and independence (on the team level) significantly predicted performance in virtual teamwork. Wang and Haggerty (2011), another example among regression analytic approaches, created the umbrella term "individual virtual competence" and posited that it comprised virtual self-efficacy, virtual media skills, and virtual social skills. They defined virtual self-efficacy as an individual's self-concept about their ability to meet the challenges imposed by virtual work. Virtual media skills denote the skills to use "technologies to communicate in virtual settings to their full potential" and virtual social skills refer to the "ability to build social relationships with others in virtual settings" (Wang & Haggerty, 2011, pp. 304–305). Empirical results from their online survey among alumni of a North American business school (n = 199) working in industries such as consulting, trading, and financial services, confirmed that the broader construct of individual virtual competence significantly predicted work performance and job satisfaction. However, the individual virtual competence latent variable was mainly built by virtual media skills and by remote work self-efficacy but not by virtual social skills or computer self-efficacy. Whereas the construct of remote work self-efficacy addresses one's confidence to perform well in the virtual work context, computer self-efficacy is conceptualized as a belief in one's own ability to complete a job with new software under various conditions (i.e., with and without support; with a manual as the only reference). According to the authors, the insignificant contribution of computer self-efficacy and virtual social skills to latent virtual competence might have been because the participants showed rather high levels and, as a consequence, reduced variation in these skills (Wang & Haggerty, 2011).

A concept closely related to self-efficacy is also frequently considered in research on technology acceptance. The prominent technology acceptance model (Davis, 1989) rests on assumptions outlined in the theory of reasoned action (Fishbein & Ajzen, 1975) and posits that individual intention to use technology is not only determined by the perceived usefulness of a technology but also by the perceived ease of use. Perceived ease of use is defined as the beliefs held by an individual regarding "the degree to which using a system would be free of effort" (Davis, 1989, p. 320), and thus it is conceptually closely related to technology-related self-efficacy (the latter even being a predictor of perceived ease of use in extensions of the technology acceptance model [e.g., Venkatesh & Bala, 2008]). In fact, perceived ease of use plays an important role in the technology acceptance model, as it not only predicts intentions to use technology but also individual perceptions of the usefulness of a technology. The technology acceptance model has been extensively tested in various contexts, such as social networks (e.g., Braun, 2013), health technologies (e.g., Yu, Li, & Gagnon, 2009), and e-learning (e.g., Cho, Cheng, & Lai, 2009), and has also been confirmed by several meta-analyses (King & He, 2006; Ma & Liu, 2004; Schepers & Wetzels, 2007; Yousafzai, Foxall, & Pallister, 2007).

Spitzberg (2006), our final example of the regression analytic approach, developed a competency model specifically for CMC. In this model, CMC competence is defined through three broad components: motivation, knowledge, and skills. Specifically, Spitzberg posited that one must have an *affinity* to use digital media, be *knowledgeable* about how and when to use which media, and possess the *skills* to communicate effectively through digital media. In his CMC competence model, skills are further specified as including attentiveness, expressiveness, composure, coordination, and selectivity. Attentiveness refers to the skills to show interest and empathy in conversations, expressiveness denotes the skills to convey emotions and use adequate substitutes for face-to-face nonverbal communication cues such as emoticons, composure relates to displaying certainty and assertiveness, and coordination refers to the skills to stay on topic and prioritize media traffic. The CMC competence model additionally specifies a set of communication outcomes: co-orientation, appropriateness, effectiveness, satisfaction, efficiency, and relational development. Co-orientation denotes the convergence of a sender's intention and the interpretation of the receiver; appropriateness refers to how appropriate the communication is perceived in a certain context; effectiveness describes the extent of goal achievement through communication; satisfaction refers to positive affect when communicating; efficiency addresses the economy with which objectives are achieved; and relational development refers to the quality of maintaining or building relationships (Spitzberg, 2006, p. 648). When using the skills specified in the CMC competence model to predict (most of) the digital communication outcomes, Schulze, Schultze, West, and Krumm (2017) showed that the components of the model (assessed with self- and peer-reports) significantly predicted the digital communication outcomes of co-orientation (these authors refer to it as clarity), appropriateness, and effectiveness of the communication.

When reviewing these competency modeling approaches, it becomes obvious that their results overlap only marginally. Although there seems to be a consensus that any kind of work conducted with technologies requires at least a minimum level of literacy in using the technology as well as a user's conviction that he or she can (learn to) handle the technology, other findings on competencies for web-based work and digital collaboration vary substantially across studies. This variation can be explained by differences in aims of studies (e.g., identifying competencies for digital communication generally vs. for virtual teamwork specifically; Hertel, Konradt, & Voss, 2006; Spitzberg, 2006), by different work contexts (e.g., computer-mediated collaboration generally vs. teamwork in cross-cultural teams specifically; Krumm, Terwiel, & Hertel, 2013; Wang & Haggerty, 2011), and by different sets of initially assumed individual characteristics as predictors (e.g., personality vs. competencies; Blackburn et al., 2003; Hertel et al., 2006).

Moreover, identifying competencies for web-based work and virtual collaboration involves the particular challenge of identifying competencies that are relevant only because of the digital aspect of work. In other words, when competency models for virtual collaboration include competencies also required for face-to-face collaboration (e.g., cooperativeness, conscientiousness), it is difficult to gauge how specific the competencies are for the virtual or digital aspect of the work. Some authors argue that such competencies may still be specific for virtual aspects of work, that is, the more virtual the work, the more important are these competencies (e.g., Hoegl, Ernst, & Proserpio, 2007). Yet others call for more rigorous tests of competencies with unique relevance in virtual work contexts (e.g., Krumm & Hertel, 2013; Schulze & Krumm, 2017).

Such rigorous tests are difficult to conduct outside the laboratory and exceed the methods usually employed for competency modeling. It is therefore not surprising that few studies have pursued such rigorous tests. As a noteworthy exception, Schulze et al. (2017) investigated the interrelationship of face-to-face communication competence and CMC competence. They used exactly the same items but framed them to address either the domain of face-to-face communication or the domain of CMC and used correlated trait-correlated methods models (Eid, 2000) to inspect the convergence between these contextualized competencies. In correlated trait-correlated methods models, traits and method factors are specified independently. In other words, variance is decomposed into trait variance, method variance, and also error variance (Eid, 2000). In Schulze et al.'s (2017) study, the face-to-face communication context served as the reference method that was contrasted with the CMC context, thus allowing for the inspection of convergence between the two contextualized communication skills. The results revealed that competencies from both domains (face-to-face and computer-mediated) should be treated as distinct constructs, therefore pointing to the necessity of virtuality-specific and contextualized construct definitions and assessments (Schulze et al., 2017; Spitzberg, 2006).

We suggest more research should be conducted along these lines and that a viable theoretical approach to identifying competencies for web-based work and digital collaboration is to scrutinize the theories developed specifically in this realm. We next outline theories pertaining to media choice, CMC, and digital collaboration, and discuss their relevance for deriving competencies (see also Krämer & Winter, 2017, Chapter 2 in this volume).

Selected Theories of Technology Choice and Related Competencies

Technologies differ in the extent to which their capabilities fit certain tasks or communication purposes (Kirkman & Mathieu, 2005). For instance, a digitally conducted negotiation between a vendor and a customer will benefit from using media that allow transferring emotional cues (e.g., to observe the customer's reaction to a price offer) and enable a parallel instead of time-displaced exchange (e.g., enable immediate inquiries and clarification). We now discuss two theories that emphasize appropriate media choice as being pivotal to effective digital work: media synchronicity theory (Dennis, Fuller, & Valacich, 2008) and information and communication technology succession theory (Stephens, 2007).

Media synchronicity theory assumes that media can be described by five capabilities that make them more or less synchronous: immediacy of feedback (the capability to provide rapid feedback), symbol variety (the capability to convey information in different formats, such as verbal and nonverbal), parallelism (the capability to convey conversations simultaneously), rehearsability (the capability to fine tune information before sending it), and reprocessability (the capability to reprocess information) (DeLuca & Valacich, 2006, p. 326). Synchronous media characteristically provide immediate feedback as well as high symbol variety and low parallelism. Asynchronous media, on the contrary, have the benefit of being rehearsable and reprocessable.

In addition to specifying capabilities of synchronous and asynchronous media, media synchronicity theory distinguishes two communication processes that are relevant for all tasks, specifically "convergence" and "conveyance" processes (Dennis et al., 2008). Convergence processes describe the sharing of individual interpretations of a situation in order to create at a common understanding. Conveyance processes, in contrast, are characterized by the transmission of information. Conveyance is about analyzing incoming information, sense-making, and finally the development of an individual's mental model (Dennis et al., 2008). Media synchronicity theory posits that all tasks require both conveyance and convergence processes but to different extents. For example, a problem-solving activity may first need information distribution, that is, conveyance; however, convergence on a shared meaning may be the focus of subsequent conversations. Generally, low synchronicity media are considered more appropriate when conveyance is a priority, whereas a high synchronicity medium is a better choice when convergence is the primary focus.

The assumptions of media synchronicity theory have been empirically tested and confirmed in several studies (e.g., Lam, 2016; Niinimäki, Piri, Lassenius, & Paasivaara, 2012). For example, Lam (2016) provided students with media synchronicity training, which resulted in a change in media-fit behavior and subsequently improved communication quality. Niinimäki et al. (2012) found support for media synchronicity theory in global software development projects. Their research revealed that when media choices were aligned with the theoretical framework, communication was more effective and efficient compared with nonaligned media choices.

As a noteworthy addition to media synchronicity theory and other theories addressing single media choice, information and communication technology succession theory also considers sequential use of media (Stephens, 2007). The main proposition of this theory is that communication can be improved by using complementary successive technologies (e.g., a face-to-face meeting or a videoconference followed by an email exchange) as opposed to successive singular or single message transmission (e.g., videoconference followed by videoconference) (Stephens & Rains, 2011; Stephens, Sørnes, Rice, Browning, & Sætre, 2008). Drawing from between-channel redundancy (Hsia, 1977) and dual coding (Paivio, 1990), information and communication technology succession theory posits that sequential communication over multiple channels leads to modality augmentation (e.g., visual, textual), which should reduce errors and information overload in a communication interaction (Stephens, 2007). The assumptions of information and communication technology succession theory have been tested by, for example, Stephens et al. (2008). These authors found that experts in digital communication indeed reported the delivery of repeated messages with a complementary modality. Two of the most frequent follow-up communication modes were email and meeting face-to-face. Further research confirmed that complementary information technology use significantly decreased perceptions of information overload and therefore improved information processing compared with the repeated use of the same information technology (Stephens & Rains, 2011).

In sum, the competence to know when and why to choose a specific technology seems to be key in web-based work and digital collaboration (Schulze & Krumm, 2017). Thus, competent users will be those who have an awareness of media capabilities, how they augment each other, and how they are best exploited for the communication purposes at hand; and, as a result, competent technology users will flexibly choose the most appropriate available technology (Shachaf & Hara, 2007).

Selected Theories on Computer-Mediated Communication and Related Competencies

The so-called cues-filtered-out theories have a longstanding tradition in CMC research. These theories build on the premise that mediated communication lacks the nonverbal cues available in face-to-face interaction (*cf.* Walther, 2011). Several theories, such as social presence theory (Short, Williams, & Christie, 1976) or the lack of social context cues theory (Siegel, Dubrovsky, Kiesler, & Mcguire, 1986) basically argue that the absence of social cues in mediated interaction leads to less warm and interpersonally involved and more deindividuated, normless, and self-focused interactions (for a comprehensive and controversial discussion of these and related theories, see Walther, 2011).

Importantly, another line of theorizing, which preceded cues-filtered-out theories and focused on how communicators can adapt to specifics of digital media, is particularly insightful for the purpose of the current chapter. Three examples are presented here. Social information processing theory (Walther, 1992) acknowledges – like cues-filtered-out theories – that CMC does not offer the same amount of nonverbal communication cues as face-to-face interaction. But in contrast to classical cues-filtered-out theories, the framework proposes that individuals are intrinsically motivated to form interpersonal relationships and therefore adapt their communication style to the medium at hand (e.g., by language content, by providing social information on a text basis; Walther, 1992). Notably, this adaptation process can lead to the same quality of interpersonal relationships as in classical face-to-face interaction, although it should take more time to develop (Walther, 2011). Several studies have confirmed this assumption, that individuals can adapt their communication style to form interpersonal relations over CMC but that this process takes more time (e.g., Peter, Valkenburg, & Schouten, 2005; Walther, Loh, & Granka, 2005). Much research on virtual teams also corroborates the presumptions of social information processing theory (e.g., Jarvenpaa & Leidner, 1998).

It follows that competent users can form appropriate interpersonal relations by explicitly sending relational development messages and by sharing socioemotional cues at an early phase of the communication (Beranek & Martz, 2005). Specifically, Hancock, Landrigan, and Silver (2007) identified that, compared with expressers of negative emotions, computer-mediated communicators use three distinct strategies to convey positive emotions: they agree more with their communication partner, they use more punctuation, and respond more quickly. Negations and exclamation marks are particularly useful cues in detecting the emotional state of others.

Electronic propinquity theory (Korzenny, 1978) focuses on differences in perceived psychological closeness between face-to-face and computer-mediated interactions. Electronic propinquity is defined as a psychological feeling of closeness, involvement, and nearness to other persons that individuals experience when using different communication devices (Walther & Bazarova, 2008). Concepts that share some similarity with the propinquity construct have been labeled "subjective distance" (e.g., Siebdrat, Hoegl, & Ernst, 2013) or "perceived proximity" (e.g., Prikladnicki, 2012). The traditional term "social presence" (e.g., Bente, Rüggenberg, Krämer, & Eschenburg, 2008; Short, Williams, & Christie, 1976; Walther & Bazarova, 2008) also points to the degree of involvement and nearness people can perceive while communicating digitally. Korzenny (1978) primarily addressed media characteristics that may influence feelings of propinquity (e.g., bandwidth, mutual directionality, communication rules). For the purpose of this chapter, however, we focus on individual characteristics that might influence the same. Specifically, Walther and Bazarova (2008) showed that individuals with high communication skills perceived media with different bandwidths (e.g., audio, text-based chat) as equal in propinquity and equally satisfactory in a group setting. These authors conceptualized communication skills as a six-dimensional construct consisting of emotional expressivity, emotional sensitivity, emotional control, social expressivity, social sensitivity, and social control. They concluded that these "communication skills appear to help communicators address the challenges of lower bandwidth and of relative bandwidth deprivation" (Walther & Bazarova, 2008, p. 640).

Channel expansion theory (Carlson & Zmud, 1999) also assumes that individual characteristics can shape the perception of digital media: individuals experienced in communicating via digital media develop more detailed and fine-grained knowledge about how to encode and decode information in CMC. As a result, they perceive CMC as less restrictive or richer than inexperienced users. Carlson and Zmud (1999) provided empirical evidence to support the main assumption of the channel expansion theory. Experience with the medium was positively and significantly associated with richness perceptions of that medium (i.e., email). D'Urso and Rains (2008) replicated these findings for email and also extended them to instant messaging. Thus, the channel expansion theory supports the notion that knowledge about encoding and decoding information in a particular medium constitutes an important personal characteristic for digital collaboration.

In sum, the following competencies seem important for successful digital communication: the ability to adapt the communication style to overcome restrictions of the medium at hand (i.e., to form interpersonal relationships, to create propinquity) and possessing the relevant knowledge to encode and decode (task-related and social) information according to the specifics of a certain medium.

Selected Theories on Digital Collaboration and Related Competencies

In this section, we present theories that seem particularly apt to derive competencies for digital collaboration (for an overview on virtual teams as a frequent form of digital collaboration, see Maynard, Gilson, Jones Young, & Vartiainen, 2017, Chapter 15 in this volume). Specifically, we focus on social impact theory (Latané, 1981) and a preliminary model of virtual collaboration presented by Peters and Manz (2007).

Although social impact theory (Latané, 1981) was not specifically developed for digital collaboration, several authors underline its importance for this domain (Blaskovich, 2008; Chidambaram & Tung, 2005; Driskell, Radtke, & Salas, 2003; Furst et al., 1999). Social impact is defined as "social force fields that impinge upon us, pushing us to think or behave in a particular way" (Jackson, 1987, p. 111). Social impact theory posits that the larger the team and the more distant the team members, the less the social impact, which in turn increases the likelihood of social loafing (Blaskovich, 2008; i.e., "the tendency to reduce one's effort when working collectively compared with individually on the same task," [Williams & Karau, 1991, p. 570]). Chidambaram and Tung (2005) found evidence for increased social loafing in larger virtual teams and Blaskovich (2008) provides evidence that support the predictions of social impact theory with respect to team member distribution.

The traditional team literature offers several suggestions as to how to address social loafing tendencies. A meta-analytic review concluded:

> *providing individuals with feedback about their own performance or the performance of their work group, monitoring individual performance or making such performance identifiable . . . might all serve, under some conditions, to reduce or eliminate social loafing.*
>
> Karau and Williams (1993, p. 700)

One might assume that such activities reduce social loafing in virtual settings as well, in so far as they can increase the perceived social impact between team members. In line with this presumption, Hertel et al. (2006) point out that self-management skills are of particular importance in virtual teams, especially when supervisory or team member control is less pronounced. Specifically, independence, persistence, as well as intrinsically motivated learning (Hertel et al., 2006) should help diminish the effect of lower social impact. We thus posit that monitoring, giving feedback, and self-management are relevant competencies in digital collaboration as they can counteract the effects assumed by social impact theory.

Peters and Manz (2007) proposed a preliminary model of virtual collaboration which, in particular, suggests three important and interrelated antecedents of "the degree of virtual collaboration that can be achieved" (p. 119). These three antecedents are: trust, shared understanding, and depth of relationships among team members. Furthermore, depth of relationships among team members is seen as a precursor of trust and shared understanding, while trust and shared understanding mutually influence each other.

Trust among team members is frequently considered as key in digital collaboration (e.g., Duarte & Snyder, 2001; see also Jarvenpaa, Cantu, & Lim, 2017, Chapter 6 in this volume). Digital collaboration environments not only make casual meetings and chatting less likely (Arnison & Miller, 2002) and thus the creation of trust and cohesion more difficult (Jarvenpaa & Leidner, 1998), but also, perhaps more importantly, reduce team members' ability to observe each other directly during task completion (e.g., Harvey, Novicevic, & Garrison, 2004). According to this argument, team members thus need to trust that everyone is making the same amount of task-related effort. Other authors, however, argue that a main advantage of digital collaboration is electronic communication can be saved and retrieved at any time, thus enabling team members to track task completion. Consequently, trust is less pivotal in virtual work environments (*cf.* Breuer, Hüffmeier, & Hertel, 2016). Indeed, meta-analytic evidence confirms this presumption: although trust was more important in virtual as compared with traditional teams, this effect was moderated by documentation possibilities (Breuer et al., 2016). Specifically, when digital communication was stored and could be retrieved by others, trust became less important.

Peters and Manz define shared understanding as "a clear sense of strategic direction for all team members" (2007, p. 122) as well as the coordination of knowledge among team members and knowing who knows what (also referred to as transactive memory systems; Wegner, 1987). They argue that digital collaboration often takes place among individuals with different areas of expertise and cultural backgrounds. Thus, for effective digital collaboration, it is vital that these individuals share a mutual understanding of their goals, the ways to achieve their goals, and the distribution of responsibilities. Indeed, developing such knowledge was significantly associated with subsequent successful performance of teams collaborating on a web-based business simulation (Kanawattanachai & Yoo, 2007). Maynard, Mathieu, Rapp, and Gilson (2012), who analyzed global virtual supply chain teams, also found support for the crucial role of transactive memory systems in virtual team effectiveness. In line with this, Siebdrat et al. (2009) found that high-performing virtual teams not only showed more effort and mutual support, but they also showed better coordination between team members.

The relevance of shared understanding is particularly highlighted in research on cross-cultural virtual teams. As digital forms of collaboration have the advantage of bringing experts together regardless of their physical location, many virtual teams comprise individuals with different cultural backgrounds (e.g., Ellingson & Wiethoff, 2002). This can have complex effects on team effectiveness (*cf.* Webber & Donahue, 2001): for instance, creating shared understanding among culturally diverse collaborators who communicate predominantly through CMC is certainly challenging. Indeed, several studies identify cross-cultural communication skills as pivotal in digital collaboration environments (e.g., Anawati & Craig, 2006; Grosse, 2002; Olaniran, 2007; Scott & Wildman, 2015; Zaugg & Davies, 2013). Specifically, collaborators should use language clearly and precisely, avoid jargon or slang, and display reliable and rule-adherent patterns of behavior (e.g., Krumm et al., 2013). In line with aforementioned theories positing that CMC may lead to less warm interactions and less interpersonal closeness compared with face-to-face interaction, Peters and Manz (2007) argue in their preliminary model of virtual collaboration that deep interpersonal relationships can alleviate the negative effects of working remotely and communicating over digital media. This reasoning is supported by studies showing that developing cohesion over a distance is more difficult (e.g., Jarvenpaa & Leidner, 1998) and that virtual teams benefit from face-to-face meetings and getting to know each other at early phases of the team life cycle (e.g., Maznevski & Chudoba, 2000; Montoya-Weiss, Massey, & Song, 2001).

Thus, in addition to the competencies regarding monitoring, giving feedback, and managing oneself, the preliminary model of virtual collaboration and related evidence suggest that digital collaborators should be willing to trust and be trustworthy and have the skills to develop a shared understanding, also among culturally diverse collaborators, as well as close interpersonal relations.

Synthesis

The current chapter has addressed the question "What are the relevant competencies for web-based work and virtual collaboration?" from two main angles: first, a review of existing (theoretical and empirical) studies on competencies for web-based work and virtual collaboration; and second, an inspection of theories of media choice, digital communication, and digital collaboration with regard to their usefulness to derive competencies. The resulting insights are summarized below.

Low convergence of studies on competencies for web-based work and virtual collaboration. The low convergence among existing studies – that technology-related work requires self-efficacy and/or literacy in handling the technology – is in stark contrast to the plethora of individual qualities proposed. For instance, Krumm and Hertel (2013) counted around 50 different constructs considered relevant in virtual teamwork. While we have outlined potential reasons for such inconsistencies (e.g., different work contexts, different sets of initially assumed predictors), one take-home message for readers might be to consider competencies as relevant for their setting only if the original studies were conducted in similar settings. The comprehensive theory building on a person–environment fit approach by Shin (2004) might provide further guidance when delineating competencies, as it specifies which individual qualities might be most relevant under certain virtual environment dimensions (see also Schulze & Krumm, 2017 for a similar approach). Another general conclusion might be to consider the spectrum of competencies identified so far as a point of departure for conducting additional requirement analyses. Finally, we reiterate that when digital communication is of primary interest, the comprehensive and well-validated model by Spitzberg (2006) is available.

Using theories to derive competencies enables a more fine-grained understanding of when and why competencies are relevant. The reviewed theories on media choice, digital communication, and virtual collaboration emphasize different aspects of digital work. When a variety of media are available and communicators are free to choose among them, they will likely benefit from having knowledge about media capabilities and how they are best used separately or in combination with each other to achieve the best possible communication outcome. These skills may also be understood as part of the broader construct of media literacy. The same is true for competencies derived from theories pertaining to digital communication. The ability to adapt one's communication style to the medium at hand and to possess the relevant knowledge to encode and decode information conveyed by a certain medium might also be considered as media literacy. When it is particularly important to collaborate digitally, both theories as well as empirical evidence stress the relevance of interpersonal skills such as giving feedback, trusting each other, and developing a shared understanding. Hence, the theories reviewed here and the conclusions drawn not only converge with empirical evidence but also clarify why some competencies are identified as relevant in some studies but not in others.

A greater fine-grained understanding of when and why competencies are relevant is needed. Although there is ample empirical evidence regarding the relevance of certain competencies, none of the existing theoretical frameworks can explain the "why" and "when" of their relevance. Thus, our present review also calls for developing a more fine-grained network of theories that address the various aspects of web-based work and digital collaboration simultaneously. More specifically, we propose developing a matrix combining all relevant aspects such as, for example, dimensions of virtuality, task type, team processes (e.g., communication, collaboration, trust), as well as environmental aspects (e.g., industry, culture). This could also be understood as a facet theory approach (Borg & Shye, 1995) including a virtuality facet, a task type facet, a team process facet, and an environmental facet. Ideally, in such an approach, every configuration (or cell of a matrix) could subsequently be filled with theory and empirical evidence, to ultimately guide selection and development in the realm of web-based work and digital collaboration.

Future Research

Our review indicates future research on competencies for web-based work and digital collaboration should focus on providing more clear-cut insights as to when and why specific competencies are relevant. This would overcome the problem of heterogeneity among prevailing findings. While we do not consider the recommendations below as exhaustive, these approaches offer viable next steps in this line of research.

Include a broad range of potentially relevant competencies. Many researchers are interested in a specific aspect of web-based work or digital collaboration and therefore, rightly, gear their research design to enable in-depth analyses with regard to this specific aspect (e.g., Maynard et al., 2012). One can, of course, collate all the single findings to create a broader picture concerning competencies required. However, the empirical relevance of competencies as included in single studies is contingent on many methodological aspects, which may vary across studies. For instance, an initial study (1) examines the relevance of competency A with highly reliable assessments of both competency A and performance outcomes. Let us assume the sample exhibited high variation in competency A and the assessed performance outcomes. A follow-up study (2) investigated the predictive power of competency B with less reliable assessments, but using a sample providing little variation in competency B and the assessed performance outcomes. Obviously, it would be premature to conclude that competency A is relevant for web-based work and digital collaboration and competency B is not. Although there are ways to account for the methodological differences between studies (1) and (2), there remains an additional problem: many competencies are not well defined (Shippmann et al., 2000). Therefore, many reviews will also be confronted with the question whether competencies A and B as assessed in studies (1) and (2) are empirically distinguishable or highly correlated. We thus call for more studies assessing a broad range of competencies (e.g., Schulze et al., 2017). In such studies, methodological aspects can be kept constant and the theoretically assumed nomologic network of competencies can be empirically tested and subsequently applied to predict performance outcomes.

Establish the factorial structure of typical instruments. In light of the many competencies considered relevant for web-based work and digital collaboration, future research would benefit from establishing higher-order factors on which several measurement instruments of competencies load. This would help uncover the underlying structure of digital work competencies and enable more parsimonious approaches in subsequent studies (e.g., by only including competencies with high loadings on a relevant factor). For instance, Krumm et al. (2013) examined cross-cultural virtual and traditional teams, including altogether 29 different concepts. They found that these concepts formed only four factors, which they labeled working conscientiously, coping with stress and ambiguity, openness and perspective taking, and knowledge about other cultures. Future studies along these lines will not only help advance our understanding of the underlying structure of competencies, but also help to (i) better understand previous results (e.g., similar result patterns for competencies loading on one factor), (ii) identify redundant competencies, and, thereby, (iii) conduct more parsimonious studies in the future.

Consider (technology-related) moderators. Some findings regarding the relevance of personal qualities in the realm of digital collaboration overlap with results obtained for nondigital, face-to-face collaboration (e.g., Hertel et al., 2006; Siebdrat, Hoegl, & Ernst, 2009). Hence, these findings remain somewhat open whether the relevance of an identified competency is due to the digital aspect of the work or is due to some other aspect of work (e.g., the collaborative aspect). We need more studies including samples that are heterogeneous with regard to the form of digital work (e.g., the number of technologies used, their information value, the asynchronicity of their interaction; *cf.* Kirkman & Mathieu, 2005). If some of the assessed personal qualities are relevant for performance outcomes only when one aspect of the digital work is given, the research can make a better point that these personal qualities can account for the challenges raised by (a specific aspect of) the digital work (e.g., asynchronicity). Such studies might even consider interactions of two or more moderators and assess their competency requirements (e.g., the competencies required in asynchronous, cross-cultural virtual collaboration).

Moreover, research would benefit from more studies examining nontechnology-related boundary conditions. That is, some competencies for web-based work and digital collaboration might be particularly relevant to junior personnel or in the early phases of the collaboration process (see Dennis et al., 2008; Jarvenpaa & Leidner, 1998) but become irrelevant later on. In the same vein, leadership and management strategies may render some otherwise relevant competencies unnecessary. Certainly, more research is needed to account for the complex conditions of web-based work and digital collaboration in the field.

Assess competencies in a facetted form. Schulze et al.'s (2017) study is a rare example of research assessing and comparing the same competencies (i.e., communication competencies) in two different contexts (face-to-face and computer-mediated). Such a study provides a test of the context specificity of the competencies, that is, whether face-to-face communication competencies and CMC competencies can be distinguished empirically. We suggest that future research should test the context dependency of other competencies considered relevant in the realm of digital collaboration (e.g., self-management, intercultural competence). This might provide greater insight as to whether the traditional assessment (framed in a face-to-face context) can be aptly applied to the computer-mediated context. In fact, some competencies might not have emerged as relevant for web-based work and digital collaboration in previous studies simply because they were assessed in a version originally designed for the face-to-face context.

Provide causal evidence by manipulating competencies. Finally, we also suggest planning experimental research designs that will provide causal evidence for the relevance of competencies. That is, we suggest manipulating competencies by training individuals in applying them. Causal evidence about their relevance can be tested by 2×2 (trained vs. untrained individuals and computer-mediated vs. face-to-face) designs, in which one would expect interaction between both factors. Examples of early studies comparing trained and untrained individuals include Beranek and Martz (2005) and Lam (2016).

Conclusion

To date, a variety of different forms of web-based work and digital collaboration exist. The required competencies are likely contingent on the specific form of digital work (Krumm & Hertel, 2013; Schulze & Krumm, 2017). That is, highly asynchronous communication imposes distinct challenges compared with communication over media with low information value. Thus, the so-called "one-size-fits-all" or generic competency models are unable to capture the complexity of digital work environments. Although taxonomies exist that describe and categorize digital forms of collaboration along the lines of virtuality dimensions (e.g., Kirkman & Mathieu, 2005), competency models that can differentiate between various virtuality dimensions have not yet been developed. In light of this paucity, we conclude that, first, personnel selection and training for web-based work and digital collaboration should be based on job analyses conducted specifically for the job at hand.

Second, we posit that job analyses use the competencies required for the "non–web-based" or "nondigital" version of the job at hand as a starting point. This conclusion rests on the overlap between competency modeling in digital and in face-to-face environments. For instance, Krumm et al. (2016) compared virtual and traditional teams along the lines of the great eight competency framework (Bartram, 2005). They found that virtual team members considered only two of the eight competency domains more important in virtual as compared with traditional teamwork. Similarly, Hertel et al. (2006) found that several requirements of traditional collaboration were also relevant in digital forms of collaboration (such as cooperation and conscientiousness).

Third, there seems to be a consensus that individuals should be confident and competent in choosing and using the relevant technology (e.g., Blackburn et al., 2003; Lee-Kelley & Sankey, 2008; Wang & Haggerty, 2011) and communicating adequately with digital media (e.g., Spitzberg, 2006). Moreover, several theories maintain that the drawbacks of some media can be overcome as users gain experience and competence in using these media (see social information processing theory [Walther, 1992]; electronic propinquity theory [Korzenny, 1978]).

Fourth, relevance of other competencies may be more contingent on the specifics of web-based work and digital collaboration. Self-management skills, for example, may frequently emerge as relevant because digital work environments are often associated with less monitoring and feedback (e.g., Harvey, Novicevic, & Garrison, 2004). However, if web-based work or digital collaboration follows strict rules and is closely monitored, self-management skills may become less relevant. The same may be true for willingness to trust and trustworthiness. If individuals have formed deep interpersonal relationships, mutual trust may be established regardless of each individual's trust-related personal qualities (Peters & Manz, 2007). Similarly, mutual trust is less relevant for performance of virtual teams in which interaction and cooperation is documented (Breuer et al., 2016).

Finally, we reiterate that research on competencies for web-based work and digital collaboration has created a rather heterogeneous picture. On the one hand, this might reflect the diversity of digital work in the field. On the other hand, we might need more differentiated research designs to address the many challenges imposed by various forms of web-based work and digital collaboration. Such research would contribute to improving personnel selection and training in this work domain and would also help human resource management to design web-based work and digital collaboration in ways that impose fewer challenges to individuals.

References

Anawati, D., & Craig, A. (2006). Behavioral adaptation within cross-cultural virtual teams. *IEEE Transactions on Professional Communication*, *49*(1), 44–56.

Arnison, L., & Miller, P. (2002). Virtual teams: A virtue for the conventional team. *Journal of Workplace Learning*, *14*(4), 166–173.

Bartram, D. (2005). The great eight competencies: A criterion-centric approach to validation. Journal of Applied Psychology, 90(6), 1185–1203.

Bente, G., Rüggenberg, S., Krämer, N. C., & Eschenburg, F. (2008). Avatar-mediated networking: Increasing social presence and interpersonal trust in net-based collaborations. *Human Communication Research*, *34*(2), 287–318.

Beranek, P. M., & Martz, B. (2005). Making virtual teams more effective: Improving relational links. *Team Performance Management*, *11*(5/6), 200–213.

Blackburn, R., Furst, S., & Rosen, B. (2003). Building a winning virtual team. In C. B. Gibson & S. G. Cohen (Eds.), *Virtual teams that work: Creating conditions for virtual team effectiveness* (pp. 95–120). San Francisco: John Wiley & Sons, Inc.

Blaskovich, J. L. (2008). Exploring the effect of distance: An experimental investigation of virtual collaboration, social loafing, and group decisions. *Journal of Information Systems*, *22*(1), 27–46.

Borg, I., & Shye, S. (1995). *Facet theory: Form and content*. London: Sage.

Braun, M. T. (2013). Obstacles to social networking website use among older adults. *Computers in Human Behavior, 29*(3), 673–680.

Breuer, C., Hüffmeier, J., & Hertel, G. (2016). Does trust matter more in virtual teams? A meta-analysis of trust and team effectiveness considering virtuality and documentation as moderators. *Journal of Applied Psychology, 101*(8), 1151–1177.

Cannon-Bowers, J. A., Tannenbaum, S. I., Salas, E., & Volpe, C. E. (1995). Defining competencies and establishing team training requirements. In R. A. Guzzo & E. Salas (Eds.), *Team effectiveness and decision making in organizations* (pp. 333–381). San Francisco: Jossey-Bass.

Carlson, J. R., & Zmud, R. W. (1999). Channel expansion theory and the experiential nature of media richness perceptions. *Academy of Management Journal, 42*(2), 153–170.

Chidambaram, L., & Tung, L. L. (2005). Is out of sight, out of mind? An empirical study of social loafing in technology-supported groups. *Information Systems Research, 16*(2), 149–168.

Cho, V., Cheng, T. E., & Lai, W. J. (2009). The role of perceived user-interface design in continued usage intention of self-paced e-learning tools. *Computers & Education, 53*(2), 216–227.

Davis, F. D. (1989). Perceived usefulness, perceived ease of use, and user acceptance of information technology. *MIS quarterly, 13*(3), 319–340.

DeLuca, D., & Valacich, J. S. (2006). Virtual teams in and out of synchronicity. *Information Technology & People, 19*(4), 323–344.

Dennis, A. R., Fuller, R. M., & Valacich, J. S. (2008). Media, tasks, and communication processes: A theory of media synchronicity. *MIS quarterly, 32*(3), 575–600.

Driskell, J. E., Radtke, P. H., & Salas, E. (2003). Virtual teams: Effects of technological mediation on team performance. *Group Dynamics: Theory, Research, and Practice, 7*(4), 297–323.

Duarte, D. L., & Snyder, N. T. (2001). *Mastering virtual teams: Strategies, tools, and techniques that succeed.* San Francisco: Jossey-Bass.

D'Urso, S. C., & Rains, S. A. (2008). Examining the scope of channel expansion: A test of channel expansion theory with new and traditional communication media. *Management Communication Quarterly, 21*(4), 486–507.

Eid, M. (2000). A multitrait-multimethod model with minimal assumptions. *Psychometrika, 65*(2), 241–261.

Ellingson, J. E., & Wiethoff, C. (2002). From traditional to virtual: Staffing the organization of the future today. In R. L. Henemann & D. B. Greenberger (Eds.), *Human resource management in virtual organizations* (pp. 141–177). Greenwich, CT: Information Age Publishing.

Fishbein, M., & Ajzen, I. (1975). *Belief, attitude, intention and behavior: An introduction to theory and research.* Reading, MA: Addison-Wesley.

Flanagan, J. C. (1954). The critical incident technique. *Psychological Bulletin, 51*(4), 327–357.

Furst, S., Blackburn, R., & Rosen, B. (1999). Virtual team effectiveness: A proposed research agenda. *Info Systems Journal, 9*(4), 249–269.

Gilson, L. L., Maynard, M. T., Young, N. C. J., Vartiainen, M., & Hakonen, M. (2015). Virtual teams research: 10 years, 10 themes, and 10 opportunities. *Journal of Management, 41*, 1313–1337.

Grosse, C. U. (2002). Managing communication within virtual intercultural teams. *Business Communication Quarterly, 65*(4), 22–38.

Hancock, J. T., Landrigan, C., & Silver, C. (2007). Expressing emotion in text. In *Proceedings of the SIGCHI Conference on Human Factors in Computing Systems (CHI '07)* (pp. 929–932). New York, NY: Association for Computing Machinery.

Harvey, M., Novicevic, M. M., & Garrison, G. (2004). Challenges to staffing global virtual teams. *Human Resource Management Review, 14*(3), 275–294.

Hertel, G., Geister, S., & Konradt, U. (2005). Managing virtual teams: A review of current empirical research. *Human Resource Management Review, 15*, 69–95.

Hertel, G., Konradt, U., & Voss, K. (2006). Competencies for virtual teamwork: Development and validation of a web-based selection tool for members of distributed teams. *European Journal of Work and Organizational Psychology, 15*(4), 477–505.

Hoegl, M., Ernst, H., & Proserpio, L. (2007). How teamwork matters more as team member dispersion increases. *Journal of Product Innovation Management, 24*(2), 156–165.

Hoegl, M., & Gemuenden, H. G. (2001). Teamwork quality and the success of innovative projects: A theoretical concept and empirical evidence. Organization Science, *12*(4), 435–449.

Hollenbeck, G. P., McCall, M. W., & Silzer, R. F. (2006). Leadership competency models. *The Leadership Quarterly, 17*(4), 398–413.

Hsia, H. (1977). Redundancy: Is it the lost key to better communication? *AV Communication Review, 25*(1), 63–85.

Jackson, J. M. (1987). Social impact theory: A social forces model of influence. In B. Mullen & G. R. Goethals (Eds.), *Theories of group behavior* (pp. 111–124). New York, NY: Springer.

Jarvenpaa, S. L., Cantu, C. & Lim, S. Y. (2017). Trust in virtual online environments. In G. Hertel, D. L. Stone, R. D. Johnson, & J. Passmore (Eds.), *The Wiley Blackwell handbook of the psychology of the internet at work* (pp. 103–130). Chichester: John Wiley & Sons, Ltd.

Jarvenpaa, S. L., & Leidner, D. E. (1998). Communication and trust in global virtual teams. *Journal of Computer-Mediated Communication, 3*(4), 1–36.Kanawattanachai, P., & Yoo, Y. (2007). The impact of knowledge coordination on virtual team performance over time. *MIS Quarterly, 31*(4), 783–808.

Karau, S. J., & Williams, K. D. (1993). Social loafing: A meta-analytic review and theoretical integration. *Journal of Personality and Social Psychology, 65*(4), 681–706.

King, W. R., & He, J. (2006). A meta-analysis of the technology acceptance model. *Information & Management, 43*(6), 740–755.

Kirkman, B. L., & Mathieu, J. E. (2005). The dimensions and antecedents of team virtuality. *Journal of Management, 31*(5), 700–718.

Korzenny, F. (1978). A theory of electronic propinquity: Mediated communication in organizations. *Communication Research, 5*(1), 3–24.

Krämer, N. C., & Winter, S. (2017). Digitized communication at work. In G. Hertel, D. L. Stone, R. D. Johnson, & J. Passmore (Eds.), *The Wiley Blackwell handbook of the psychology of the internet at work* (pp. 21–38). Chichester: John Wiley & Sons, Ltd.

Krumm, S., & Hertel, G. (2013). Knowledge, skills, abilities and other characteristics (KSAOs) for virtual teamwork. In A. Bakker & D. Derks (Eds.), *The psychology of digital media and work* (pp. 80–99). East Sussex, UK: Psychology Press.

Krumm, S., Kanthak, J., Hartmann, K., & Hertel, G. (2016). What does it take to be a virtual team player? The knowledge, skills, abilities, and other characteristics required in virtual teams. *Human Performance, 29*(2), 123–142.

Krumm, S., Terwiel, K., & Hertel, G. (2013). Challenges in norm formation and adherence: The knowledge, skills, and ability requirements of virtual and traditional cross-cultural teams. *Journal of Personnel Psychology, 12*(1), 33.

Lam, C. (2016). Improving technical communication group projects. An experimental study of media synchronicity theory training on communication outcomes. *Journal of Business and Technical Communication, 30*(1), 85–112.

Latané, B. (1981). The psychology of social impact. *American Psychologist, 36*(4), 343–356.

Lee-Kelley, L., & Sankey, T. (2008). Global virtual teams for value creation and project success: A case study. *International Journal of Project Management, 26*, 51–62.

Loughry, M. L., Ohland, M. W., & Moore, D. D. (2007). Development of a theory-based assessment of team member effectiveness. *Educational and Psychological Measurement, 67*(3), 505–524.

Ma, Q., & Liu, L. (2004). The technology acceptance model: A meta-analysis of empirical findings. *Journal of Organizational and End User Computing, 16*(1), 59–72.

Maynard, M. T., Gilson, L. L., Jones Young, N., & Vartiainen, M. (2017). Virtual teams. In G. Hertel, D. L. Stone, R. D. Johnson, & J. Passmore (Eds.), *The Wiley Blackwell handbook of the psychology of the internet at work* (pp. 315–346). Chichester: John Wiley & Sons, Ltd.

Maynard, M. T., Mathieu, J. E., Rapp, T. L., & Gilson, L. L. (2012). Something(s) old and something(s) new: Modeling drivers of global virtual team effectiveness. *Journal of Organizational Behavior, 33*(3), 342–365.

Maznevski, M. L., & Chudoba, K. M. (2000). Bridging space over time: Global virtual team dynamics and effectiveness. *Organization Science, 11*(5), 473–492.

Montoya-Weiss, M., Massey, A. P., & Song, M. (2001). Getting it together: Temporal coordination and conflict management in global virtual teams. *Academy of Management Journal, 44*(6), 1251–1262.

Niinimäki, T., Piri, A., Lassenius, C., & Paasivaara, M. (2012). Reflecting the choice and usage of communication tools in global software development projects with media synchronicity theory. *Journal of Software: Evolution and Process*, *24*(6), 677–692.
OECD (2013). *OECD skills outlook 2013: First results from the survey of adult skills*, OECD Publishing. Retrieved April 5, 2017, from https://dx.doi.org/10.1787/9789264204256-en
Olaniran, B. A. (2007). Culture and communication challenges in virtual workspaces. In K. St-Amant (Ed.), *Linguistic and cultural online communication issues in the global age* (pp. 79–92). Hershey, PA: Information Science Reference (IGI Global).
Paivio, A. (1990). *Mental representations: A dual coding approach*. New York, NY: Oxford University Press.
Peter, J., Valkenburg, P. M., & Schouten, A. P. (2005). Developing a model of adolescent friendship formation on the Internet. *Cyber Psychology & Behavior*, *8*(5), 423–430.
Peters, L. M., & Manz, C. C. (2007). Identifying antecedents of virtual team collaboration. *Team Performance Management: An International Journal*, *13*(3/4), 117–129.
Prikladnicki, R. (2012). Propinquity in global software engineering: Examining perceived distance in globally distributed project teams. *Journal of Software: Evolution and Process*, *24*(2), 119–137.
Schepers, J., & Wetzels, M. (2007). A meta-analysis of the technology acceptance model: Investigating subjective norm and moderation effects. *Information & Management*, *44*(1), 90–103.
Schulze, J., & Krumm, S. (2017). The "Virtual Teamplayer": A review and initial model of knowledge, skills, abilities, and other characteristics for virtual collaboration. *Organizational Psychology Review*, *7*(1), 66–95.
Schulze, J., Schultze, M., West, S.G., & Krumm, S. (2017). The knowledge, skills, abilities, and other characteristics required for face-to-face versus computer-mediated communication: Similar or distinct constructs? *Journal of Business and Psychology*, *32*(3), 283–300.
Scott, C. P., & Wildman, J. L. (2015). Culture, communication, and conflict: A review of the global virtual team literature. In J. L. Wildman & R. L. Griffith (Eds.), *Leading global teams* (pp. 13–32). New York, NY: Springer.
Shachaf, P., & Hara, N. (2007). Behavioral complexity theory of media selection: A proposed theory for global virtual teams. *Journal of Information Science*, *33*(1), 63–75.
Shin, Y. (2004). A person-environment fit model for virtual organizations. *Journal of Management*, *30*(5), 725–743.
Shippmann, J. S., Ash, R. A., Batjtsta, M., Carr, L., Eyde, L. D., Hesketh, B., . . . Sanchez, J. I. (2000). The practice of competency modeling. *Personnel Psychology*, *53*(3), 703–740.
Short, J., Williams, E., & Christie, B. (1976). *The social psychology of telecommunications*. London: John Wiley & Sons, Ltd.
Siebdrat, F., Hoegl, M., & Ernst, H. (2009). How to manage virtual teams. *MIT Sloan Management Review*, *50*(4), 63–68.
Siebdrat, F., Hoegl, M., & Ernst, H. (2013). Subjective distance and team collaboration in distributed teams. *Journal of Product Innovation Management*, *31*(4), 765–779.
Siegel, J., Dubrovsky, V., Kiesler, S., & McGuire, T. W. (1986). Group processes in computer-mediated communication. *Organizational Behavior and Human Decision Processes*, *37*(2), 157–187.
Spitzberg, B. H. (2006). Preliminary development of a model and measure of computer-mediated communication (CMC) competence. *Journal of Computer-Mediated Communication*, *11*(2), 629–666.
Statista (2016). Number of worldwide internet users from 2000 to 2015 (in millions). Retrieved April 5, 2017, from www.statista.com/statistics/273018/number-of-internet-users-worldwide/
Stephens, K. K. (2007). The successive use of information and communication technologies at work. *Communication Theory*, *17*(4), 486–507.
Stephens, K. K., & Rains, S. A. (2011). Information and communication technology sequences and message repetition in interpersonal interaction. *Communication Research*, *38*, 101–122.
Stephens, K. K., Sørnes, J. O., Rice, R. E., Browning, L. D., & Sætre, A. S. (2008). Discrete, sequential, and follow-up use of information and communication technology by experienced ICT users. *Management Communication Quarterly*, *22*(2), 197–231.
Stevens, M. J., & Campion, M. A. (1994). The knowledge, skill, and ability requirements for teamwork: Implications for human resource management. *Journal of Management*, *20*(2), 503–530.

Venkatesh, V., & Bala, H. (2008). Technology acceptance model 3 and a research agenda on interventions. *Decision Sciences, 39*(2), 273–315.
Walther, J. B. (1992). Interpersonal effects in computer-mediated interaction: A relational perspective. *Communication Research, 19*(1), 52–90.
Walther, J. B. (2011). Theories of computer-mediated communication and interpersonal relations. In M. L. Knapp & J. A. Daly (Eds.), *The SAGE handbook of interpersonal communication* (pp. 443–479). Thousand Oaks, CA: Sage Publications.
Walther, J. B., & Bazarova, N. N. (2008). Validation and application of electronic propinquity theory to computer-mediated communication in groups. *Communication Research, 35*(5), 622–645.
Walther, J. B., Loh, T., & Granka, L. (2005). Let me count the ways the interchange of verbal and nonverbal cues in computer-mediated and face-to-face affinity. *Journal of Language and Social Psychology, 24*(1), 36–65.
Wang, Y., & Haggerty, N. (2011). Individual virtual competence and its influence on work outcomes. *Journal of Management Information Systems, 27*(4), 299–334.
Webber, S. S., & Donahue, L. M. (2001). Impact of highly and less job-related diversity on work group cohesion and performance: A meta-analysis. *Journal of Management, 27*(2), 141–162.
Wegner, D. M. (1987). Transactive memory: A contemporary analysis of the group mind. In B. Mullen & G. R. Goethals (Eds.), *Theories of group behavior* (pp. 185–208). New York, NY: Springer.
Williams, K. D., & Karau, S. J. (1991). Social loafing and social compensation: The effects of expectations of co-worker performance. *Journal of Personality and Social Psychology, 61*(4), 570–581.
Yousafzai, S. Y., Foxall, G. R., & Pallister, J. G. (2007). Technology acceptance: A meta-analysis of the TAM: Part 2. *Journal of Modelling in Management, 2*(3), 281–304.
Yu, P., Li, H., & Gagnon, M. P. (2009). Health IT acceptance factors in long-term care facilities: A cross-sectional survey. *International Journal of Medical Informatics, 78*(4), 219–229.
Zaugg, H., & Davies, R. S. (2013). Communication skills to develop trusting relationships on global virtual engineering capstone teams. *European Journal of Engineering Education, 38*(2), 228–233.

5

User Experience, Gamification, and Performance

Meinald T. Thielsch and Jörg Niesenhaus

Introduction

Computers and the Internet are used to master a broad variety of tasks. One major challenge is to make user interfaces usable, enabling effective and efficient work processes (see Hanrahan & Carroll, 2017; Chapter 3 in this volume). Another significant challenge for researchers is understanding and improving human–computer interaction based on an experiential perspective: users' subjective impressions and feelings, and resulting consequences for users' work performance (see Bargas-Avila & Hornbæk, 2011; Hassenzahl & Tractinsky, 2006). This chapter focuses on the experiential perspective to answer emerging questions such as:

- How do subjective experiences affect work performance?
- Are appealing and aesthetic website designs beneficial or harmful?

We distinguish between the indirect effect of an appealing design on learning and working performance and a more direct effect of motivating people with game-based concepts applied to working interfaces. Game-based approaches are an aspect of user experience capable of directly enhancing performance. Researchers have only recently begun to study the best ways to enhance work performance via appealing or motivating interfaces, and we provide case examples of best practices in the field, discuss practical implications, and point to directions for the greatly needed research in this area.

User Experience

Many businesses highly value digitalization and Internet user experiences. In part 210 of the ISO standard 9241, titled "Ergonomics of Human–System Interaction," the International Organization for Standardization broadly defines user experience as users' perceptions and

The Wiley Blackwell Handbook of the Psychology of the Internet at Work,
First Edition. Edited by Guido Hertel, Dianna L. Stone, Richard D. Johnson, and Jonathan Passmore.

responses toward actual or anticipated use of interactive products, systems, or services (ISO, 2009). The term covers the use or expected use of a wide range of digital products and interactive systems, such as software and Internet tools. Moreover, user experience is a multifaceted construct influenced by several factors, such as design factors, interaction characteristics, and subjective appraisals (e.g., Lee & Koubek, 2012; van der Heijden, 2003). The components of user experience model (CUE model) (Thüring & Mahlke, 2007) helpfully categorizes these factors (Figure 5.1) by describing how interaction with a system (which is influenced by the system's properties) and user and task/context characteristics might influence key user experience outcomes. Instrumental qualities such as usability, and non-instrumental qualities such as aesthetics, both evoke emotional reactions and influence appraisals. The CUE model is derived from experimental data (see Thüring & Mahlke, 2007) and supports the idea that user experience depends on the interplay between emotions, perceptions, and evaluations during interactions with technical systems or interactive products. From an organizational perspective, the consequences for work behavior, performance, and outcomes are an important interest.

The CUE model categorizes the main dimensions of the human–computer interaction within any technical system. With respect to the World Wide Web, three core constructs are essential to the current research on user experiences: content, usability, and aesthetics (e.g., Cober, Brown, Levy, Cober, & Keeping, 2003; Schenkman & Jönsson, 2000; Tarasewich, Daniel & Griffin, 2001; Thielsch, Blotenberg & Jaron, 2014).

Content is of prime importance when using the web, especially in work settings. ISO standard 9241–151 defines web content as "a set of content objects" on a web user interface, and describes a content object as an "interactive or non-interactive object containing information represented by text, image, video, sound or other types of media" (ISO, 2006, p. 3). Besides such objective characteristics (see Thielsch & Hirschfeld, under review), subjective experiences of web content are also essential, especially in business settings (Huizingh, 2000; Palmer, 2002). These subjective perceptions of web content depend on characteristics of the reader, purpose of website use, and a website's domain and specific content.

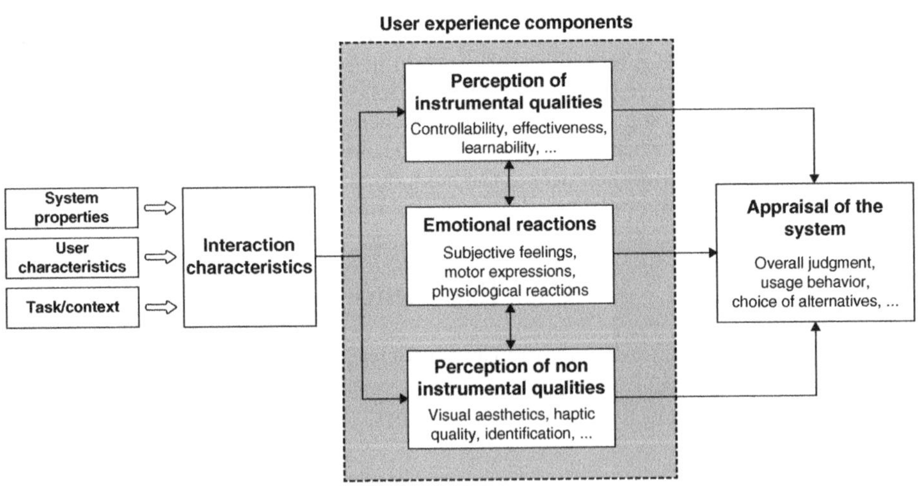

Figure 5.1 Components of the user experience model (CUE-Model). *Source:* Adapted from Thüring & Mahlke (2007).

Usability is a well-known construct of web user experiences (see Shneiderman & Plaisant, 2009; with respect to measurement issues, see Hornbæk, 2006). It is defined, based on ISO 9241–11, as the "extent to which a product can be used by specified users to achieve specified goals with effectiveness, efficiency and satisfaction in a specified context of use" (ISO, 1998, p. 2). Usability and ergonomics of Internet-based tools are discussed in Chapter 3 of this handbook (see Hanrahan & Carroll, 2017). Here we focus on an experiential perspective on the Internet and digital products at work (see Bargas-Avila & Hornbæk, 2011; Hassenzahl & Tractinsky, 2006).

Aesthetics, beauty, and pleasure are among the non-instrumental experiential factors of website experience (Lavie & Tractinsky, 2004; Moshagen & Thielsch, 2010). The generic term aesthetics is often described as an immediate pleasurable subjective experience (Leder, Belke, Oeberst, & Augustin, 2004; Moshagen & Thielsch, 2010; Reber, Schwarz, & Winkielman, 2004). Users can quickly perceive, process, and evaluate the aesthetics of a web site – often within a split second (e.g., Lindgaard, Fernandes, Dudek, & Browñ, 2006; Thielsch & Hirschfeld, 2012; Tractinsky, Cokhavi, Kirschenbaum, & Sharfi, 2006; Tuch, Presslaber, Stöcklin, Opwis, & Bargas-Avila, 2012). Aesthetic evaluations influence several outcomes such as preferences, trust, and even user urges to buy impulsively or intentions to revisit websites (see Moshagen & Thielsch, 2010). High aesthetics lead to user satisfaction (e.g., Lavie & Tractinsky, 2004; Lindgaard & Dudek, 2003; Tractinsky, Katz, & Ikar, 2000) and influence subjective perceptions of usability (e.g., Thielsch, Engel, & Hirschfeld, 2015; Tractinsky et al., 2000; Tuch, Roth, Hornbæk, Opwis, & Bargas-Avila, 2012).

The content of an Internet tool or a given website is often determined by the target group and is task related, as well as being determined by a specific context. Thus, most approaches on Internet content involve specific situations or specific aspects of content (e.g., its actuality, availability or reliability, see Thielsch & Hirschfeld, under review). Consequently, developers must analyze and optimize content in relation to individual circumstances. The same applies to usability, which is at best optimized in a way that the user experiences no problems. The fact that enhanced functionality of an interactive tool increases performance is rather trivial. However, how do non instrumental qualities (such as aesthetics) affect work performance? Can game-based design enhance user performance? In the following sections, we discuss current understandings of these issues. We use the term *aesthetics* to refer to the appeal, beauty, and attractiveness of, and the pleasure experienced when visiting a website or using an interactive system.

Effects of user experiences on performance

How do Internet users' subjective experiences affect work performance? Do emotions and moods interfere with organizational performance goals? Can aesthetics harm performance? Early user experience research stressed that workplace computer systems are meant to enhance efficiency, not to give pleasure. Some authors even argued that aesthetic designs interfere with work goals (e.g., Andre & Wickens, 1995; Hollnagel, 2003) and objected to favoring aesthetics over usability (Norman, 1988). Further research has altered that perspective. For example, Norman (2002) argued that aesthetics positively influence problem-solving cognition and performance by changing emotional states.

We conducted a systematic literature search, applying several search terms (such as "user experience," "aesthetics," "design," and "emotional design," combined with the term "performance") in several scientific databases (e.g., the Association of Computing Machinery [ACM] digital library, Google Scholar, Web of Science). Additional publications were found by searching the cited references in the studies retrieved through the systematic search. Publication lists of prominent researchers working in the field were also individually checked for papers on the subject. Studies were only included in the systematic review if distinct performance measures had been applied; other publications were dropped (e.g., when participants only subjectively rated their performance and no objective measure was used). Thus, our review finally included 46 studies on non-instrumental user experience qualities and performance (Table 5.1, Table 5.2, and Table 5.3). Most of the studies have included student samples; field studies, systematic reviews, and meta-analyses are as yet lacking. However, extant research has tested a broad variety of variables and tasks, and here we summarize the current state of knowledge in this area.

Research provides mixed findings about performance effects of design variables. At present, we lack wide-ranging advice on enhancing Internet workers' performance, but many studies offer promising suggestions for specific tasks and situations (Table 5.1, Table 5.2, and Table 5.3). Only a few studies suggest design variables deteriorate performance (i.e., Gnambs, Appel, & Batinic, 2010; Sauer & Sonderegger, 2011; Sonderegger, Uebelbacher, Pugliese, & Sauer, 2014; van Schaik & Ling, 2009). Thus, there seems to be no general conflict around designing websites and Internet tools that are both usable and enjoyable.

Many studies have focused on design variables of interactive systems and websites that indirectly affect user behavior (however, this chapter will also describe a more direct approach in the context of game-based approaches). To date, there is little evidence regarding which theory best explains the effects. Researchers have provided a range of explanations, such as mediation by positive affects (Norman, 2002), increased motivation due to prolonged joyful experiences (e.g., Sonderegger & Sauer, 2010), or effects caused by reduced cognitive effort in processing (e.g., Szabo & Kanuka, 1998). Additionally, different effects might interact in determining user experiences.

Furthermore, researchers distinguish learning performance from task performance. Users pursuing learning goals focus on increasing their competence and knowledge; users pursuing performance goals focus on outcomes, favorable judgments, or high scores (see Elliott & Dweck, 1988; Locke & Latham, 2006; Seijts & Latham, 2005). Users with learning goals are expected to gain deeper understandings, take their time, and make mistakes as a natural and important part of learning. In contrast, work situations usually require employees to find the best and fastest task solutions, without making mistakes. Although not all work situations are stressful or are meant to be a test experience, regular work days often leave little time for learning or making mistakes. Nevertheless, learning is important for job performance, especially for working smart through acquired knowledge rather than working hard (e.g., Seijts & Latham, 2005). The effects of design variables in learning scenarios are quite positive, as we will see in the next section.

Table 5.1 Overview of studies addressing user experience and learning performance.

Effects	*Authors*	*Typical design variables*	*Typical performance measures and effects*
No effect	Hall & Hanna (2004)	Color	Retention
Partial positive effects	Heidig et al. (2015), Plass et al. (2014), Szabo & Kanuka (1998)	Balance, color, shape, unity, general appeal/aesthetics	Comprehension (+), recalled information, response time, retention, transfer
Positive effects	Miller (2011), Pomales-García, Liu, & Mendez (2005), Strebe (2016), Um et al. (2012)	Background, shape, icons, color, contrast	Comprehension, pages retrieved, time spent learning (+), recalled information, transfer

(+): Positive effects on a specific performance measure in more than one study.

Table 5.2 Overview of studies addressing user experience and task performance.

Effects	*Authors*	*Typical design variables*	*Typical performance measures and effects*
Negative effects	Sauer & Sonderegger (2011), Sonderegger et al. (2014)	Color, symmetry, general appeal/aesthetics	Task completion time, number of commands needed to fulfill task
No effect	Ben-Bassat, Meyer, & Tractinsky (2006), Hartmann et al. (2007), Ilmberger et al. (2008), Katz (2010), Lee & Koubek (2010, 2011), Nakarada-Kordic & Lobb (2005), Nordeborn (2013), Schmidt et al. (2009), Sonderegger et al. (2012), Thielsch et al. (2009), Thüring & Mahlke (2007), Tractinsky et al. (2000), Tuch et al. (2012)	Background, color, font, layout, shape, symmetry, texture, general appeal/aesthetics	Task completion time, number of clicks/inputs, number of errors
Partial positive effects	Bonnardel, Piolat & Le Bigot (2011), Quinn & Tran (2010), Reinecke & Bernstein (2011), Sauer & Sonderegger (2009), Sonderegger, Sauer, & Eichenberger (2014)	Color, general appeal/esthetics	Task completion time, number of clicks/inputs, number of errors, amount of used information
Positive effects	Douneva et al. (2016), Sonderegger & Sauer (2010)	Color, pictures, text-picture-ratio, general appeal/aesthetics	Number of correct answers, task completion time, number of commands needed to fulfill task, number of errors

Table 5.3 Overview of studies addressing user experience and performance in testing tasks and under stress.

Effects	*Authors*	*Typical design variables*	*Typical performance measures and effects*
Negative effects	Gnambs et al. (2010), van Schaik & Ling (2009)	Color	Task completion, test score (−), number of correct answers
No effect	Chawda et al. (2005), Douneva et al. (2015), Ling & van Schaik (2006), Salimun (2013)	Color, general appeal/aesthetics	Task duration, number of errors, number of tasks solved
Partial positive effects	Cawthon & Vande Moere (2007), McDougall et al. (2016), Moshagen et al. (2009), Reppa et al. (2008), Reppa & McDougall (2015), Salimun (2013), Salimun et al. (2010)	Color, shape, visual complexity, general appeal/aesthetics	Response time (+), number of errors
Positive effects	Tuch et al. (2009), van Schaik & Ling (2008)	Visual complexity, general appeal/aesthetics	Reaction time, number of correct answers

(+): Positive effects on a specific performance measure in more than one study.
(−): Negative effects in more than one study.

Learning performance User learning requires time for acquiring new information, understanding how to use a system, or finding solutions to a specific problem. Eight publications on user experience and learning in our review investigated reading and response times, comprehension, recall, retention, and knowledge transfer. Most found at least partial positive effects of interface design variables (Table 5.1). Besides general appeal, the use of specific colors leads to positive outcomes. Pleasant designs encourage users to spend more time reading information material, which increases their retention and comprehension. Several possible reasons have been discussed, for example, effects of emotions and motivation (e.g., Heidig, Müller, & Reichelt, 2015; Plass, Heidig, Hayward, Homer, & Um, 2014; Um, Plass, Hayward, & Homer, 2012) or the idea that good designs enable automatic processing (Szabo & Kanuka, 1998), reducing learners' mental effort and perceived stress (Miller, 2011). Thus, interactive and web-based learning environments should be created in a way that facilitates positive user experiences.

Task performance Our search strategy yielded 23 studies on typical working tasks with computer applications, systems, or websites. The tasks were somewhat comparable with everyday work tasks: researchers asked participants to acquire information or to use an application or interactive system for data entry, communication, or system queries. Two studies clearly had negative results and two had positive results (Table 5.2). Most other studies found interface design variables to have partially positive or no effects on task performance. Color or general appeal were the most commonly studied variables, but no clear effect pattern emerged. Several studies discussed a possible weak manipulation of investigated design variables; others conjectured there may be moderating factors that are currently unknown. Additionally, some of these studies did not focus on performance or examined only one aspect of performance, sometimes including merely a manipulation check. In sum, aesthetic and appealing designs usually do not harm task performance. Instead, well-designed digital work environments can enhance task performance even in everyday tasks such as search, use, or information input.

Performance in testing and under stress Many digital workplaces have stressful work demands requiring rapid and competitive work. We found 15 publications investigating effects of pleasant interface designs in such tasks (Table 5.3). Study participants were often required to work as fast as possible or to answer test questions. Again, only a few studies found negative effects of design variables while several publications reported at least partially positive results. For instance, the color red has been found to detrimentally affect the performance of men in tests of general knowledge (Gnambs et al., 2010). The authors discuss the possibility of stereotyped threats in achievement contexts. In contrast, several studies on icon and interface design found that low visual complexity or applying general aesthetic design principles positively affected response times. Possible explanations included performance gains caused by positive emotions according to mood mediation theory (Norman, 2002) or reduced cognitive effort caused by more aesthetic designs. In conclusion, color effects are essential when designing for stressful or test situations, and well-made designs can support job performance. Regarding icon design, McDougall, Reppa, and colleagues provide helpful insights (see, e.g., McDougall, Reppa, Kulik, & Taylor, 2016; Reppa & McDougall, 2015). These authors present several icon examples and empirical findings on appeal, familiarity and visual complexity, which can be directly applied for improving existing icons or the creation of new ones.

Exemplary study: Positive effects of optimized user experience

Miller (2011) used an online tool designed to help instructors assess student learning in American Sign Language (ASL) to investigate whether an aesthetically optimized design would influence learning and performance in e-assessment environments. Students were asked to perform tasks such as re-telling a short story or describing a photo or a picture. They had to perform the tasks in sign language and present the solution on their webcam. The online software recorded the answers. Students voluntarily self-assessed their recorded performance using the original media and several self-evaluation items. This learning environment was designed to address the educational and technical needs of assessments within the ASL curriculum (see Miller, Hooper, & Rose, 2005). Miller (2011) aesthetically optimized the original online tool without changing functions, task objectives, or media. For example, saturated colors were used for buttons, texts, and time bars; the background was changed from white to neutral gray; drop shadows were placed behind screen content. Miller based the design changes on Norman's (2004) visceral, behavioral, and reflective levels of emotional design. Basic color and background changes aligned with the visceral level. Animated time bars and text tickers rather than numerical timers addressed the behavioral and reflective design levels.

Miller (2011) randomly assigned 66 students with some ASL experience to the original or the redesigned version of the online tool. Both groups performed the same tasks and answered the same questions, leading to several interesting findings. First, the students evaluated the redesigned tool as more aesthetic but not more usable. That is, the aesthetic manipulation was effective but without change or decrease in experienced usability. Second, the more aesthetic version decreased perceived mental effort, stress, and task demands. Third, students who worked with the more aesthetic version reported greater satisfaction and willingness to continue using the tool. All findings were associated with at least medium but mostly large effect sizes. The self-assessment time was nearly doubled among students using the aesthetic version, and these students also showed significantly higher ASL task performance scores (given by external raters). Qualitative student interviews confirmed that aesthetics had positive effects, especially with respect to satisfaction. Furthermore, students who used the original version often complained about the difficulty and inappropriateness of tasks for their level of ASL experience. Students using the optimized aesthetic version experienced tasks and level of difficulty as appropriate.

In sum, Miller (2011) showed how an optimized aesthetic design not only influences subjective perceptions and satisfaction of users but also impacts on voluntary learning time and objective performance scores. Thus, this research exemplifies best practice in designing online learning tasks in educational settings, in personnel development, and in professional training. However, all the studies and results discussed so far have relied mostly on design variables to indirectly affect performance through positive user experiences. In contrast, it might be preferable to motivate users directly through, for instance, "gamification." That is, gamification can be understood as one specific aspect of user experience dedicated to motivate users through game elements.

Gamification

Gamification, which is growing in popularity, applies game elements to non-game contexts to make computerized and nondigital systems, services, and activities more enjoyable and motivating (Deterding, Dixon, Khaled, & Nacke, 2011). In the service industry, gaming experiences create overall value for users (Huotari & Hamari, 2012). Gamification, also called game-thinking (Marczweski, 2015), emphasizes game-oriented designs rather than game mechanics (Niesenhaus, 2014a). Game-based learning (Prensky, 2001), serious games (Michael & Chen, 2005), games with a purpose (von Ahn, 2006), and gamification all use game elements and technology to generate benefits beyond pure entertainment. However, the design processes are challenging. Many projects fail because serious applications are often incompatible with games, and achieving compatibility is sometimes commercially nonviable.

Rather than building whole games, gamification puts playful mechanics into non-game contexts in a more flexible and cost-efficient way. Although gamification means that game elements are integrated into non-game applications, unfortunately, often the only game elements used are virtual awards through point lists, high-score tables, or badges attached to established products or processes. In many applications, simply adding game elements has not yielded the expected benefits. The market research company Gartner estimates that 80% of gamified applications fail to meet their business objectives primarily because of poor design (Pettey & van der Meulen, 2012). Further challenges include diverse user needs and motivation, as shown by research into play personas (an approach based on fictional characters of typical users, see, e.g., Bartle, 1996; Canossa & Drachen, 2009) and research into user preferences, needs, and frustrations (Marczweski, 2015). Furthermore, gamification frameworks have focused on motivations and needs in relation to typical game elements. The Octalysis framework (Chou, 2015) offers eight core drivers for human motivation such as meaning, accomplishment, and unpredictability. For example, a game awards points, badges, progress bars, and leaderboards to fulfill users' desire to progress, develop skills, and overcome challenges. Werbach and Hunter (2012) use a pyramidal gamification framework: low-level components are at the base, followed by mechanics, while higher-level dynamics are at the apex. Components include tangible mechanics such as points, leaderboards, boss fights, and virtual goods. Mechanics move the system forward and generate player engagement (e.g., challenge, competition, feedback, and rewards), whereas dynamics are the highest level of abstraction and comprise fundamentals such as emotions, narratives, and progressions. Such frameworks have provided initial explanations of user motivation and game elements, but the research community is still in need of a reliable and validated model to describe gamification effects. The early approaches also lack a closer connection to motivational psychology theory, which makes it difficult to validate them. Closing this gap will be one of the most important and challenging tasks of future gamification research.

However, gamification potentially optimizes participation, engagement, and commitment (Herger, 2014; Schering, Niesenhaus, & Schmidt, 2014). In business and working environments, gamification can increase the attractiveness of companies and their products, enhance process efficiency, and optimize customer relations (e.g., Werbach & Hunter, 2012; Zichermann & Linder, 2010). Consequently, the number of published gamified business applications has considerably increased over the past years, and includes the team-based challenges to find errors in Microsoft Office (McDonald, Musson, & Smith, 2008) and a community network at SAP (Kumar & Herger, 2013).

Effects of gamification on work performance

Researchers have studied effects of gamification on work performance in areas such as education (Denny, 2013; McDaniel, Lindgren, & Friskics, 2012), online communities and social networks (Cramer, Rost, & Holmquist, 2011; Thom, Millen, & DiMicco, 2012), health and wellness (Cafazzo Casselman, Hamming, Katzman, & Palmert, 2012; Fuchslocher et al., 2011; Stinson et al., 2013), sustainability (Gnauk et al., 2012; Y. Liu, Alexandrova, & Nakajima, 2011), orientation (Depura & Garg, 2012; Fitz-Walter, Tjondronegoro, & Wyeth, 2012), and marketing (Downes-Le Guin, Baker, Mechling, & Ruyle, 2012). Seaborn and Fels (2015) discussed theoretical foundations used in gamification frameworks, including self-determination theory (Ryan & Deci, 2000a), intrinsic and extrinsic motivation (Ryan & Deci, 2000b), situational relevance (Wilson, 1973), situated motivational affordance (Deterding et al., 2011), universal design for learning (Rose & Meyer, 2002), transtheoretical model of behavior change (Prochaska & Marcus, 1994), and Norman's user-centered design (2002). Motivation is assumed to have the most positive effect (see Hamari, Koivisto, & Sarsa, 2014).

As gamification is an emerging field, currently few studies have examined gamification effects on work performance. Although businesses are increasingly applying gamification projects, they often keep their efforts confidential, hence again few company-related studies are publicly available (Seaborn & Fels, 2015). Work-related gamification examples thus come from software development (Farzan et al., 2008; Kumar & Herger, 2013; McDonald et al., 2008), software evaluation (Eickhoff, Harris, de Vries, & Srinivasan, 2012), and crowdsourcing (Y. Liu et al., 2011; Mason, Michalakidis, & Krause, 2012) applications. Most of those work-related gamification systems are based on standard elements such as points, badges, and leaderboards, which are easily applied but represent only a small selection of potential gamification elements (Deterding et al., 2011). Table 5.4 summarizes findings on work-related gamification effects, and lists the applied gamification elements and performance measures.

Microsoft's popular Windows Language Quality Game illustrates how gamification positively affects work performance. The game encouraged native language speakers within the company to perform the expensive work of traditional software localizers; more than 900 voluntary players found 170 bugs across all 36 native language editions of Microsoft Windows, thereby providing a cost-effective way to improve product quality (McDonald et al., 2008). Unfortunately, the benefits were not compared with benefits of non-gamified testing methods.

SAP's community network is an often-cited case of fostering engagement in the customer and developer communities by the introduction of basic gamification mechanisms such as points and levels (see Kumar & Herger, 2013). SAP used active community members to provide feedback for fine-tuning its gamification system. In 2013, two months after SAP introduced the changes, it reported a 1113% increase in comments on content creation and a 250% rise in community feedback (Kumar & Herger, 2013). Even before SAP applied gamification elements to its community network, IBM gamified "Beehive," its internal social networking website, with a point and status interaction. Employees were highly motivated to increase contributions, but the effects were not sustained (Farzan et al., 2008).

Bagley (2012) used gamification elements such as points, ranks, and badges to encourage users to participate in a crowdsourcing task. The gamification elements did not attract all users in the same way: older users who had less general interest in gaming reported fewer positive experiences. Eickhoff et al. (2012) asked users to participate in a relevance assessment by relating keywords to concepts. Limited game rounds, a points system, progress visualization, high scores, and leaderboards gamified the application and yielded significantly higher annotation efficiency. Mason et al. (2012) used points and badges to engage users in a human computation task and reported mixed results: the gamified application motivated participation and thereby improved recognition of algorithms, but the gamification elements interfered with the quality of results. Other studies have also

Table 5.4 Overview of gamification studies by effect, applied gamification elements, and performance measures.

Effects	*Authors*	*Gamification elements*	*Performance measures and effects*
No effect	Y. Liu et al. (2011)	Points and scoring system	No significant changes in user behavior
	Witt et al. (2011)	Points and activity counter	Similar participation as in prior studies, lower impact than expected because of unclear presentation
Partial positive effects	Bagley (2012)	Points, ranks and badges	Encouraged participation (+); performance varies depending on age and game interest (−)
	Mason et al. (2012)	Points and badges	Engaged users and improved algorithms (+); interfered with quality of results (−)
	McDonald et al. (2008)	Point and scoring system, team contest	Positive results in participation and error reporting (+) but not in efficiency or effectivity compared with a similar non-gamified process (−)
	Grant & Betts (2013)	Badges	Positive effects on user participation in community network (+)
	Farzan et al. (2008)	Points and status system	Higher motivation to contribute and increased contribution (+), reduced effects over time (−)
Positive effects	Eickhoff et al. (2012)	Limited game rounds, points, progress visualization, leaderboards	Higher annotation efficiency and quality (+)
	Kumar & Herger (2013)	Ranks, points, progress visualization	Significant increase of community network activity (+)
	Niesenhaus (2014b)	Team scoring, progress visualization, playful time management	Improved process efficiency, higher rate of problem reporting and problem-solving (+)
	Flatla et al. (2011)	Playful visualization	Similar to standard calibration procedures but more enjoyable and strongly preferred (+)

(+): Positive effects on a specific performance measure.
(−): Negative effects.

found ambiguous effects of gamified elements. Y. Liu et al. (2011) implemented a point and scoring system in two software applications but found no significant changes in user behavior. Witt, Scheiner, and Robra-Bissantz (2011) found a tendency towards positive changes in behavior but lower-than-expected impact of gamification elements because the presentation was unclear.

Production and management scenarios offer further promising applications that may have high economic impact. Next, we present a case study highlighting the use of gamification elements in an industrial production setting (Niesenhaus, 2014b).

Case study: Gamifying industrial production

This case study was conducted in a lamp-producing facility. Industrial assembly requires that machines are always functional. Operators and maintenance personnel must collaborate to achieve optimal throughput, high-quality products, and minimal downtimes. Production requires both flexibility and quality control. For this reason, most gamification mechanisms presented in this case study focus on operators and their activities, although gamification interventions are also appropriate in other departments such as management, sales, and plant engineering. A confidentiality agreement with the client dictated use of stylized plastic models to describe the scenario rather than images of the real production facility (Figure 5.2 and Figure 5.3). Our ability to show the screen contents of the operator stations was also restricted. Each operator had at least one screen at their station to show the next production step, and offered interactive elements to flag production errors, and incorporated gamification elements.

The operator stations were aligned with the production machines, so the process flow was streamlined, and offered several constraints. First, the low agency among the employees was a limitation in the implementation of motivational and playful elements. The design team considered these constraints and looked for games that share qualities with the industrial application. They were inspired by games that have simple and repetitive interaction, such as "FarmVille," the physical interaction with little freedom in "Dance Dance Revolution," and the simple touch gestures with maximal impact of "Infinity Blade." The team devised several motivational and playful elements to maximize employee autonomy and agency and to avoid potential harm from direct competition and control (Niesenhaus, 2014b). The gamification design follows the Lean UX interaction design approach (Gothelf & Seiden, 2013), an agile software engineering approach that focuses on personas and scenarios, and generated assumptions according to their criticality for product viability. To minimize the risk of failure, small experiments or questionnaires are recommended to validate the most relevant assumptions (Gothelf & Seiden, 2013). This design approach allows fast iterations of gamified software prototypes and adapts well to the agile development process used by most software companies.

Figure 5.2 Stylized model of a shift team watching their daily score (Niesenhaus, 2014b). *Source:* With kind permission of Niesenhaus.

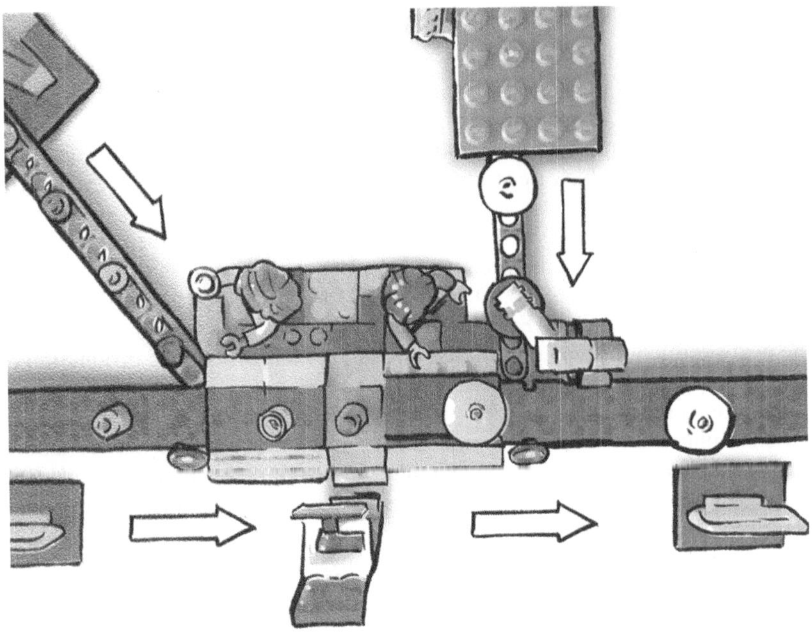

Figure 5.3 Stylized model of optimized collaboration in a lamp production unit, achieved by synchronizing the movement of the various operators (Niesenhaus, 2014b). *Source:* With kind permission of Niesenhaus.

Team score The design team observed and interviewed production team members and learned that they had strong social bonds and were willing to help each other. The team chose to use a team-oriented score system rather than introducing potentially harmful competitive elements (Figure 5.2). Each production line operator generates team points by performing actions such as boxing lamp parts, suggesting ways to optimize the workflow, and checking quality. The team score encourages efficiency and supports other team members. When team members log out of their shifts, they are asked, for example, to choose the most valuable colleague of the shift, or identify which station is the most difficult to handle. Answering voluntary questions gives the team extra points.

However, the teams do not compete with each other. Instead, each team views daily team scores compared with past scores and all-time high scores. We observed that operators reacted positively; they discussed their reasons for and ways to improve their daily scores. Once or twice a year, the company connects teams from other company locations in different countries for a week-long competition including daily videoconferences between locations so that operators around the world socialize with each other and exchange knowledge. The competition is between company branches, not within a local branch, and is designed to encourage socialization and efficiency.

Idle time Gaming elements also managed work breaks. As 10–12 operators work in parallel, if four or more take work breaks at the same time, production is significantly lowered. Operators can use their touch-screen terminals to request break times. They have the option to take immediate breaks but can also adhere to the system's suggestions for optimal beginning and ending time to minimize impacts on the assembly line. When operators agree to suggested break times, they receive 3–15-minute bonuses, which are directly added to their break allowance. The bonus is based on an algorithm that calculates potential reduction in production time when an operator takes an immediate break. The higher the potential reduction, the greater is the bonus time that can be earned.

Team rhythm Some assembly processes required two or more operators to collaborate. The design team observed that two operators work together to assemble three lamp parts. The operator who is second in line must adjust to their partner's working rhythm. At one step (Figure 5.3), the second operator makes a 90° turn to check the quality of the lampshade before assembly, momentarily losing track of their partner's progress. They then waited for their partner, which frequently generated inefficient idle times.

The team evaluated various types of visual and acoustic feedback tools to provide the second worker with immediate feedback about the first operator's production progress. They finally chose to use an LED-powered progress bar that showed the assembly progress in three steps: start, in assembly, nearly ready for take-over. The display was based on optical sensors that recognized hand movements. The progress feedback minimized waiting times. Another positive side effect was that the tool revealed which operators shared similar rhythms or better fitted the roles of first or second operator. The team leader used the information to choose better placements, making the process more efficient and enhancing workers' role satisfaction.

Results Early findings of an analysis of long-term effects of the gamification interventions were promising (Niesenhaus, 2014b). Over one month, data were collected on efficiency (throughput, average duration of downtimes), quality rate (number of rejected products during quality control) and user feedback (based on questionnaires) of one production line with a gamification system and one without the intervention. Although operators of the gamified production line could choose to use the system or deactivate it at each operator terminal (with no impact on team score), 85% used it. The gamified production line had 6% higher average throughput and 12% lower downtimes in comparison with the non-gamified production line. Quality rate tended toward positive, although the results were statistically insignificant. User feedback based on short questionnaires during terminal logout revealed significantly higher motivation, knowledge exchange, and social interaction among the operators.

Psychological Processes Affecting User Performance

So far no distinct evidence is available to show which model or theory best explains the observed findings. Early user experience research suggested that good design enhances performance by reducing cognitive processing efforts (Szabo & Kanuka, 1998), allowing faster detection of visual objects, caused by less complexity and more coherence in good designs. Good designs promote automatic processing, whereas bad designs induce less-efficient manual processing. In addition, perceptions of content quality are driven by halo effects of good designs, leading to greater attention by or higher motivation in users (Szabo & Kanuka, 1998). Building on these ideas, researchers have discussed attentional effects of good design (e.g., Reppa, Playfoot, & McDougall, 2008) and other cognitive effects in website perception, such as mental models, bottom-up perception processes, visual complexity and prototypicality (e.g., Douneva, Jaron, & Thielsch, 2016; Tuch, Bargas-Avila, Opwis, & Wilhelm, 2009). Tractinsky and colleagues (2000) introduced the "what is beautiful is usable" approach to human–computer interaction research and proposed attitude effects such as halo effects, stereotypes and affective responses to aesthetic designs as driving processes. Several studies were in agreement (for an overview, see Tuch, Roth et al., 2012), but others have argued that aesthetic pleasure is the result, not the cause, of processing dynamics. Processing fluency theory (Reber et al., 2004) assumed that the more fluently a user can process an object, the more positive is his or her aesthetic response. Thus, "what is beautiful is usable" may have a reversed connection under certain conditions (see Tuch, Roth, et al., 2012). Human–computer interaction research must still solve this kind of chicken-or-egg causality dilemma.

Norman (2002, 2004) proposed a mood mediation model: good design and aesthetics influence cognition by evoking positive emotions. Thus, aesthetics can improve performance and compensate for usability problems, particularly in creative and problem-solving tasks. Users who feel good about the system will overlook design flaws. Several researchers have followed Norman's theory (e.g., Moshagen, Musch, & Göritz, 2009; Quinn & Tran, 2010; Reppa & McDougall, 2015). In addition, emotional design has positive effects in learning contexts (e.g., Plass et al., 2014; Um et al., 2012). But evoked emotions might be relatively weak or short-lived (e.g., Douneva, Haines, & Thielsch, 2015; Katz, 2010) and Norman's theory still lacks validation.

Motivational effects are often discussed, in particular, in learning or gaming scenarios. Game elements are especially known to generate enjoyable and interesting settings that engage users for long periods (Plass, Homer, & Kinzer, 2015). Thus, several user experience researchers adhere to the prolonged joy/increased motivation hypothesis (e.g., Heidig et al., 2015; Sauer & Sonderegger, 2011; Sonderegger & Sauer, 2010). That is, users prolong their enjoyment with aesthetically appealing products (partly, instead of directly solving given tasks) and perform better by entering flow situations (see Csikszentmihalyi, 1977) in which they perceive their skills to be congruent with challenges in engrossing activities. In gaming, they experience flow when their attention is focused such that they feel a compelling sense of being present in a mediated virtual environment (Liu & Chang, 2012). Furthermore, motivation and flow are essential preconditions for engagement. The engagement model captures the close relation between user motivation and cognitive, affective, behavioral, and sociocultural engagement, largely depending on the context, the user, and the game (see Plass et al., 2015). As research has little focused on engagement, the model is based on the INTERACT model of learner activity (Domagk, Schwartz & Plass, 2010; [INTERACT, Integrated Model of Multimedia Interactivity]). Future application of such models in user experience research appears promising.

In sum, research is just beginning to unravel the non-instrumental aspects of interface design and effects on performance. Causes and effects such as attention, cognition, and emotion may act in combination. Applying playful elements might increase task motivation due to higher engagement and flow experiences. Nevertheless, existing research already highlights several practical implications.

Practical Implications

To optimize user experiences with Internet tools at work and to allow decision-makers to choose among the available products, first we need to know how users experience a product. Research has provided several reliable and valid measures for such an inquiry, especially regarding general user experiences: The AttrakDiff questionnaire (Hassenzahl, 2004; Hassenzahl, Burmester, & Koller, 2003) employs 28 pairs of adjectives to measure pragmatic quality, identification, stimulation, and appeal of an interactive product. Evidence of reliability and validity of the tool is available (for further information, see www.attrakdiff.de). Another measure using a semantic differential is the User Experience Questionnaire ([UEQ], Laugwitz, Held, & Schrepp, 2008; Laugwitz, Schrepp, & Held, 2006), which consists of 26 items representing attractiveness, perspicuity, efficiency, dependability, stimulation, and novelty. (For information about reliability and validity see www.ueq-online.org.) A more recent measure is the meCUE questionnaire (Minge & Riedel, 2013; Minge, Riedel & Thüring, 2013) based on the CUE model (Thüring & Mahlke, 2007). The instrument has 34 items in four modules: instrumental and non-instrumental product perceptions, emotions, consequences, and overall evaluations. Evidence regarding reliability and validity has been provided (see www.mecue.de).

Few scientifically validated measures are available for measuring aesthetics. Lavie and Tractinsky (2004) developed the first standardized multiscale measure for assessing subjective aesthetics of websites. The instrument includes two five-item scales reflecting classical and expressive aesthetics. The authors found evidence for reliability and for convergent, divergent, and concurrent validity. More recently published, the Visual Aesthetics of Websites Inventory ([VisAWI], Moshagen & Thielsch, 2010) is an 18-item questionnaire that measures a general aesthetics factor consisting of four facets: simplicity, diversity, color, and craftsmanship. The authors found evidence for high reliability as well as convergent, divergent, discriminative, concurrent and experimental validity. A short four-item version, the VisAWI-S (Moshagen & Thielsch, 2013) is also available as well as information about optimal cut points (Hirschfeld & Thielsch, 2015), and a manual (for further information, see www.visawi.de). A sound measure would assist in decisions regarding whether to invest in new tools, systems, or websites, or to build new ones or relaunch existing ones. Table 5.1, Table 5.2, and Table 5.3 might help determine specific design factors that can account for performance improvements. But research has yet to elucidate causal relationships between design variables and performance outcomes. Thus, ultimately, a well-skilled trusted designer is essential (Chevalier & Ivory, 2003; Park, Choi, & Kim, 2004).

Before developing a gamified system, in particular, developers should set a clear mission statement, which can be recalled if needed. Will the game elements be used to strengthen customer loyalty, to increase user motivation, or to make processes more efficient? Almost as important is choosing how to measure achievement. When and how will a given goal be achieved, and can it be quantified? Questionnaires and interviews with customers and employees, pre/post studies comparing efficiency data, or investigating data quality are reasonable methods for examining achievement (Herger, 2014; Niesenhaus, 2014b).

Generally, game elements can be used to create high situational interest (Rotgans & Schmidt, 2011), but not to enhance unattractive mechanics (Plass et al., 2015). That is, game elements cannot enrich every process. Sometimes the underlying process has to be redesigned before game elements can be integrated. All stakeholders should participate in a redesign to represent different perspectives and implications. Prototypes should be available early to investigate functionality of game elements (Fullerton, Swain, & Hoffman, 2004; Salen & Zimmerman, 2004; Schell, 2014). Experiences from game development highlight the importance of testing game mechanics as soon as possible. Most companies do not have professional game designers and hire external expertise for process, analysis, and idea generation. Individuals who are experienced in game design and who possess additional skills in user experience design and/or interaction design can aid balancing game mechanics and the general user experience.

Future Research

Researchers are just beginning to study the effects of experiential variables on user performance. Further studies should identify main variables and investigate causal relationships. Additionally, existing results should be summarized in systematic reviews and meta-analyses. Objective design factors should be systematically linked to subjective perceptions (e.g., Miniukovich & de Angeli, 2015; Seckler, Opwis & Tuch, 2015) and underlying perception processes (e.g., Leder & Nadal, 2014; Thielsch & Hirschfeld, 2012). Research on performance effects should be embedded in a theoretical framework, ordering performance goals, tasks, application characteristics, and design variables. Published reports of field approaches are also urgently needed to verify research results under real conditions of practice. The same is true for gamification studies, especially in the work environment. The number of publications is increasing, but most applications and studies are in areas with no direct impact on daily work routines (Hamari et al., 2014; Seaborn & Fels, 2015). Some insights have been gained about gamified systems in education (Denny, 2013; McDaniel et al., 2012), health and wellness (Cafazzo et al., 2012; Stinson et al., 2013), and orientation (Depura & Garg, 2012; Fitz-Walter et al., 2012). We can transfer those findings to work scenarios, but our knowledge of gamification effects on work performance requires more application to and studies from real work environments. Furthermore, the research community needs more mature theoretical models and frameworks on gamification effects. Although we have frameworks for game mechanics (Werbach & Hunter, 2012; Yu-Kai Chou, 2015) and early approaches to user types (Marczweski, 2015), developing comprehensive models and frameworks is essential to broaden future research.

In addition, it will be interesting to investigate further and extend results of existing user studies into the areas of performance and user experience. For example, some interfaces can automatically adapt to users' cultural preferences (see Reinecke & Bernstein, 2013), which should have consequences for the workplace. Good interface design may contribute not only to satisfaction or performance but also directly to wellbeing (see Thieme, Wallace, Meyer, & Olivier, 2015), which is promising. Finally, technical innovations have triggered significant developments in the Internet in the past (e.g., Engholm, 2002) and will remain important for future developments and research, especially considering the rise of smart and connected products (e.g., Porter & Heppelmann, 2014).

Conclusion

User experience and gamification can enhance performance in Internet-based working contexts. When the right designs and appropriate gamification approaches are chosen, users will benefit from pleasant and joyful interactions with technical systems. Especially in learning contexts, positive user experiences relate to learning outcomes. Gamification can enhance productivity in classic online community environments (e.g., Kumar & Herger, 2013) and in production settings (e.g., Niesenhaus, 2014b). However, more basic and applied research is urgently needed to investigate underlying processes and causes. The Internet and related technologies are moving and developing targets, and the same applies to user experience and gamification research. The research reviewed in this chapter suggests a promising outlook for future possibilities in this field.

References

Andre, A. D., & Wickens, C. D. (1995). When users want what's not best for them. *Ergonomics in Design: The Quarterly of Human Factors Applications, 3*(4), 10–14.

Bagley, K. S. (2012). Conceptual mile markers to improve time-to-value for exploratory search sessions. (Unpublished doctoral thesis.) University of Massachusetts Lowell, Ann Arbor, MI.

Bargas-Avila, J. A., & Hornbæk, K. (2011). Old wine in new bottles or novel challenges: A critical analysis of empirical studies of user experience. In *Proceedings of the 2011 Annual Conference on Human Factors in Computing Systems (CHI '11)*, (pp. 2689–2698). New York, NY: ACM.

Bartle, R. (1996). Hearts, clubs, diamonds, spades: Players who suit MUDs. *Journal of MUD Research, 1*(1), 19.

Ben-Bassat, T., Meyer, J., & Tractinsky, N. (2006). Economic and subjective measures of the perceived value of aesthetics and usability. *ACM Transactions on Computer-Human Interaction, 13*(2), 210–234.

Bonnardel, N., Piolat, A., & Le Bigot, L. (2011). The impact of colour on Website appeal and users' cognitive processes. *Displays, 32*(2), 69–80.

Cafazzo, J. A., Casselman, M., Hamming, N., Katzman, D. K., & Palmert, M. R. (2012). Design of an mHealth app for the self-management of adolescent type 1 diabetes: A pilot study. *Journal of Medical Internet Research, 14*(3), e70.

Canossa, A., & Drachen, A. (2009). Patterns of play: Play-personas in user-centered game development. In *Breaking new ground: Innovation in games, play, practice and theory: Proceedings of the 2009 Digital Games Research Association Conference* (p. 11). London: Brunel University.

Hanrahan, B., & Carroll, J. M. (2017). Ergonomics of information technologies at work. In G. Hertel, D. L. Stone, R. D. Johnson, & J. Passmore (Eds.), *The Wiley Blackwell handbook of the psychology of the internet at work* (pp. 39–60). Chichester: John Wiley & Sons, Ltd.

Cawthon, N., & Vande Moere, A. (2007). The effect of aesthetic on the usability of data visualization. In *Proceedings of the International Conference on Information Visualisation* (pp. 637–645). Washington DC: IEEE.

Chawda, B., Craft, B., Cairns, P., Heesch, D., & Rüger, S. (2005). Do "attractive things work better"? An exploration of search tool visualisations. *(Proceedings) Human Computer Interaction*, 46–51.

Chevalier, A., & Ivory, M. Y. (2003). Web site designs: Influences of designer's expertise and design constraints. *International Journal of Human–Computer Studies, 58*, 57–87.

Chou, Y.-K. (2015). *Actionable gamification – beyond points, badges, and leaderboards.* Freemont, CA: Octalysis Media.

Cober, R. T., Brown, D. A., Levy, P. E., Cober, A. B., & Keeping, L. M. (2003). Organizational web sites: Web site content and style as determinants of organizational attraction. *International Journal of Selection and Assessment, 11*(2/3), 158–169.

Cramer, H., Rost, M., & Holmquist, L. E. (2011). Performing a check-in: Emerging practices, norms and "conflicts" in location-sharing using foursquare. In *Proceedings of the 13th International Conference on Human Computer Interaction with Mobile Devices and Services* (pp. 57–66). New York, NY: ACM.

Csikszentmihalyi, M. (1977). *Beyond boredom and anxiety.* San Francisco, CA: Jossey-Bass.

Denny, P. (2013). The effect of virtual achievements on student engagement. In *Proceedings of the SIGCHI Conference on Human Factors in Computing Systems* (pp. 763–772). New York, NY: ACM.

Depura, K., & Garg, M. (2012). Application of online gamification to new hire onboarding. In *Proceedings of the 2012 Third International Conference on Services in Emerging Markets (ICSEM)* (pp. 153–156). Washington DC: IEEE.

Deterding, S., Dixon, D., Khaled, R., & Nacke, L. (2011). From game design elements to gamefulness: Defining gamification. In *Proceedings of the 15th International Academic MindTrek Conference: envisioning future media environments* (pp. 9–15). New York, NY: ACM.

Domagk, S., Schwartz, R. N., & Plass, J. L. (2010). Interactivity in multimedia learning: An integrated model. *Computers in Human Behavior, 26*(5), 1024–1033.

Douneva, M., Haines, R., & Thielsch, M. T. (2015). Effects of interface aesthetics on team performance in a virtual task. In *Proceedings of the 23rd European Conference on Information Systems (ECIS 2015)*. Atlanta, GA: AIS.

Douneva, M., Jaron, R., & Thielsch, M. T. (2016). Effects of different website designs on first impressions, aesthetic judgments, and memory performance after short presentation. *Interacting with Computers, 28* (4), 552–567.

Downes-Le Guin, T., Baker, R., Mechling, J., & Ruyle, E. (2012). Myths and realities of respondent engagement in online surveys. *International Journal of Market Research, 54*(5), 1–21.

Eickhoff, C., Harris, C. G., de Vries, A. P., & Srinivasan, P. (2012). Quality through flow and immersion: Gamifying crowdsourced relevance assessments. In *Proceedings of the 35th International ACM SIGIR Conference on Research and Development in Information Retrieval* (pp. 871–880). New York, NY: ACM.

Elliott, E. S., & Dweck, C. S. (1988). Goals: an approach to motivation and achievement. *Journal of Personality and Social Psychology, 54*(1), 5–12.

Engholm, I. (2002). Digital style history: The development of graphic design on the Internet. *Digital Creativity, 13*(4), 193–211.

Farzan, R., DiMicco, J. M., Millen, D. R., Dugan, C., Geyer, W., & Brownholtz, E. A. (2008). Results from deploying a participation incentive mechanism within the enterprise. In *Proceedings of the SIGCHI Conference on Human Factors in Computing Systems* (pp. 563–572). New York, NY: ACM.

Fitz-Walter, Z., Tjondronegoro, D., & Wyeth, P. (2012). A gamified mobile application for engaging new students at university orientation. In *Proceedings of the 24th Australian Computer-Human Interaction Conference* (pp. 138–141). New York, NY: ACM.

Flatla, D. R., Gutwin, C., Nacke, L. E., Bateman, S., & Mandryk, R. L. (2011). Calibration games: Making calibration tasks enjoyable by adding motivating game elements. In *Proceedings of the 24th annual ACM symposium on User interface Software and Technology* (pp. 403–412). New York, NY: ACM.

Fuchslocher, A., Niesenhaus, J., & Krämer, N. (2011). Serious games for health: An empirical study of the game "Balance" for teenagers with diabetes mellitus. *Entertainment Computing, 2*(2), 97–101.

Fullerton, T., Swain, C., & Hoffman, S. (2004). *Game design workshop: Designing, prototyping, & playtesting games*. Boca Raton, FL: CRC Press.

Gnambs, T., Appel, M., & Batinic, B. (2010). Color red in web-based knowledge testing. *Computers in Human Behavior, 26*(6), 1625–1631.

Gnauk, B., Dannecker, L., & Hahmann, M. (2012). Leveraging gamification in demand dispatch systems. In *Proceedings of the 2012 Joint EDBT/ICDT workshops* (pp. 103–110). New York, NY: ACM.

Gothelf, J., & Seiden, J. (2013). *Lean UX: Applying lean principles to improve user experience*. Sebastopol, CA: O'Reilly Media.

Grant, S., & Betts, B. (2013). Encouraging user behaviour with achievements: An empirical study. In *Proceedings of the 10th Working Conference on Mining Software Repositories* (pp. 65–68). IEEE.

Hall, R. H., & Hanna, P. (2004). The impact of web page text-background colour combinations on readability, retention, aesthetics and behavioural intention. *Behaviour & Information Technology, 23*(3), 183–195.

Hamari, J., Koivisto, J., & Sarsa, H. (2014). Does gamification work? – A literature review of empirical studies on gamification. In *Proceedings of the Annual Hawaii International Conference on System Sciences* (pp. 3025–3034). Piscataway, NJ: IEEE.

Hartmann, J., Sutcliffe, A., & De Angeli, A. (2007). Investigating attractiveness in web user interfaces. In *Proceedings of the SIGCHI Conference on Human Factors in Computing Systems* (pp. 387–396). New York, NY: ACM.

Hassenzahl, M. (2004). The interplay of beauty, goodness, and usability in interactive products. *Human Computer Interaction, 19*, 319–349.

Hassenzahl, M., Burmester, M., & Koller, F. (2003). AttrakDiff: Ein Fragebogen zur Messung wahrgenommener hedonischer und pragmatischer Qualität [AttrakDiff: A questionnaire for the measurement of perceived pragmatic and hedonic quality]. In J. Ziegler, & G. Szwillus (Eds.), *Mensch & Computer* (pp. 187–196). Teubner, Stuttgart.

Hassenzahl, M., & Tractinsky, N. (2006). User experience – a research agenda. *Behaviour & Information Technology, 25,* 91–97.

Heidig, S., Müller, J., & Reichelt, M. (2015). Emotional design in multimedia learning: Differentiation on relevant design features and their effects on emotions and learning. *Computers in Human Behavior, 44,* 81–95.

Herger, M. (2014). *Enterprise gamification: Engaging people by letting them have fun.* North Charleston, SC: CreateSpace Independent Publishing Platform.

Hirschfeld, G., & Thielsch, M. T. (2015). Establishing meaningful cut points for online user ratings. *Ergonomics, 58*(2), 310–320.

Hollnagel, E. (2003). Is affective computing an oxymoron? *International Journal of Human Computer Studies, 59*(1–2), 65–70.

Hornbæk, K. (2006). Current practice in measuring usability: Challenges to usability studies and research. *International Journal of Human-Computer Studies, 64*(2), 79–102.

Huizingh, E. (2000). The content and design of web sites: an empirical study. *Information & Management, 37*(3), 123–134.

Huotari, K., & Hamari, J. (2012). Defining gamification: A service marketing perspective. In *Proceeding of the 16th International Academic MindTrek Conference* (pp. 17–22). New York, NY: ACM.

Ilmberger, W., Schrepp, M., & Held, T. (2008). Cognitive processes causing the relationship between aesthetics and usability. In *HCI and Usability for Education and Work* (pp. 43–54). Berlin: Springer.

ISO. (1998). *ISO 9241: Ergonomic requirements for office work with visual display terminals (VDTs) – Part 11: Guidance on usability.* Geneva: International Organization for Standardization.

ISO. (2006). *ISO 9241: Ergonomics of human-system interaction – part 151: Guidance on world wide web interfaces.* Geneva: International Organization for Standardization.

ISO. (2009). *9241-210: Ergonomics of human system interaction-Part 210: Human-centred design for interactive systems.* Geneva: International Organization for Standardization.

Katz, A. (2010). Aesthetics, usefulness and performance in user-search engine interaction. *Journal of Applied Quantitative Methods, 5*(3), 424–446.

Kumar, J. M., & Herger, M. (2013). *Gamification at work: Designing engaging business software.* Interaction Design Foundation.

Laugwitz, B., Held, T., & Schrepp, M. (2008). Construction and evaluation of a user experience questionnaire. In A. Holzinger (Ed.), *HCI and Usability for Education and Work. USAB 2008. Lecture Notes in Computer Science, vol 5298* (pp. 63–76). Berlin: Springer.

Laugwitz, B., Schrepp, M., & Held, T. (2006). Konstruktion eines Fragebogens zur Messung der User Experience von Softwareprodukten [Construction of a questionnaire to measure the user experience of software products]. In A. M. Heinecke & H. Paul (Eds.), *Mensch & Computer 2006 – Mensch und Computer im Strukturwandel* (pp. 125–134). Munich: Oldenbourg Verlag.

Lavie, T., & Tractinsky, N. (2004). Assessing dimensions of perceived visual aesthetics of web sites. *International Journal of Human–Computer Studies, 60,* 269–298.

Leder, H., Belke, B., Oeberst, A., & Augustin, D. (2004). A model of aesthetic appreciation and aesthetic judgements. *British Journal of Psychology, 95,* 489–508.

Leder, H., & Nadal, M. (2014). Ten years of a model of aesthetic appreciation and aesthetic judgments: The aesthetic episode – developments and challenges in empirical aesthetics. *British Journal of Psychology, 105*(4), 443–464.

Lee, S., & Koubek, R. J. (2010). Understanding user preferences based on usability and aesthetics before and after actual use. *Interacting with Computers, 22*(6), 530–543.

Lee, S., & Koubek, R. J. (2011). The impact of cognitive style on user preference based on usability and aesthetics for computer-based systems. *International Journal of Human-Computer Interaction, 27*(11), 1083–1114.

Lee, S., & Koubek, R. J. (2012). Users' perceptions of usability and aesthetics as criteria of pre- and post-use preferences. *European Journal of Industrial Engineering, 6*(1), 87–117.

Lindgaard, G., & Dudek, C. (2003). What is this evasive beast we call user Satisfaction? *Interacting with Computers, 15,* 429–452.

Lindgaard, G., Fernandes, G., Dudek, C., & Browñ, J. (2006). Attention web designers: You have 50 milliseconds to make a good first impression! *Behaviour & Information Technology, 25*(2), 115–126.

Ling, J., & Van Schaik, P. (2006). The influence of font type and line length on visual search and information retrieval in web pages. *International Journal of Human Computer Studies, 64*(5), 395–404.

Liu, C. C., & Chang, I. C. (2012). Measuring the flow experience of players playing online games. In *Proceedings of the Pacific Asia Conference on Information Systems (PACIS 2012)* (n.p.). AIS.

Liu, Y., Alexandrova, T., & Nakajima, T. (2011). Gamifying intelligent environments. In *Proceedings of the 2011 International ACM Workshop on Ubiquitous Meta User Interfaces* (pp. 7–12). New York, NY: ACM.

Locke, E. A., & Latham, G. P. (2006). New directions in goal-setting theory. *Current Directions in Psychological Science, 15*(5), 265–268.

Marczweski, A. (2015). *Even ninja monkeys like to play: Gamification, game thinking & motivational design*. San Francisco, CA: Blurb Inc.

Mason, A. D., Michalakidis, G., & Krause, P. J. (2012). Tiger nation: Empowering citizen scientists. In *Digital Ecosystems Technologies (DEST), 2012 6th IEEE International Conference on* (pp. 1–5). Washington DC: IEEE.

McDaniel, R., Lindgren, R., & Friskics, J. (2012). Using badges for shaping interactions in online learning environments. In *Professional Communication Conference (IPCC), 2012 IEEE International* (pp. 1–4). Washington DC: IEEE.

McDonald, M., Musson, R., & Smith, R. (2008). Using productivity games to prevent defects. In M. McDonald, R. Musson, and R. Smith (Eds.), *The practical guide to defect prevention* (pp. 79–95). Redmond: Microsoft Press.

McDougall, S., Reppa, I., Kulik, J., & Taylor, A. (2016). What makes icons appealing? The role of processing fluency in predicting icon appeal in different task contexts. *Applied Ergonomics, 55*, 156–172.

Michael, D. R., & Chen, S. L. (2005). *Serious games: Games that educate, train, and inform*. New York, NY: Muska & Lipman/Premier-Trade.

Miller, C. (2011). Aesthetics and e-assessment: The interplay of emotional design and learner performance. *Distance Education, 32*(3), 307–337.

Miller, C., Hooper, S., & Rose, S. (2005). Avenue ASL: Developing an environment for assessing American Sign Language learner performance. *Advanced Technology for Learning, 2*(3), 140–147.

Minge, M., & Riedel, L. (2013). meCUE-Ein modularer Fragebogen zur Erfassung des Nutzungserlebens [meCUE – a modular user experience questionnaire]. In S. Boll, S. Maaß & R. Malaka (Eds.), *Mensch und Computer 2013: Interaktive Vielfalt* (pp. 89–98). München: Oldenbourg Verlag.

Minge, M., Riedel, L., & Thüring, M. (2013). Und ob du wirklich richtig stehst ... Zur diskriminativen Validität des User Experience Fragebogens "meCUE" [Are you really right ... Discriminant validity of the user experience questionnaire meCUE]. In S. Boll, S. Maaß & R. Malaka (Eds.), *Mensch und Computer 2013 – Workshopband* (pp. 137–144). München: Oldenbourg Verlag.

Miniukovich, A., & De Angeli, A. (2015). Computation of interface aesthetics. In *Proceedings of the 33rd Annual ACM Conference on Human Factors in Computing Systems (CHI '15)* (pp. 1163–1172).

Moshagen, M., Musch, J., & Göritz, A. S. (2009). A blessing, not a curse: Experimental evidence for beneficial effects of visual aesthetics on performance. *Ergonomics, 52*, 1311–1320.

Moshagen, M., & Thielsch, M. T. (2010). Facets of visual aesthetics. *International Journal of Human-Computer Studies, 68*, 689–709.

Moshagen, M., & Thielsch, M. (2013). A short version of the visual aesthetics of websites inventory. *Behaviour & Information Technology, 32*(12), 1305–1311.

Nakarada-Kordic, I., & Lobb, B. (2005). Effect of perceived attractiveness of web interface design on visual search of web sites. In *Proceedings of the 6th ACM SIGCHI New Zealand Chapter's International Conference on Computer-Human Interaction Making CHI Natural (CHINZ '05)* (pp. 25–27). New York, NY: ACM.

Niesenhaus, J. (2014a). *Best of both worlds: Game Design meets User Interface Design. UP14-Talks*. Stuttgart: German UPA. Retrieved April 7, 2017, from http://dl.mensch-und-computer.de/handle/123456789/3941

Niesenhaus, J. (2014b). *Labour's love won? Gamifying industrial production.* Gamification Research Network. Retrieved February 12, 2016, from http://gamification-research.org/2014/12/gamification-of-industrial-production/

Nordeborn, G. (2013). The effect of color in website design: Searching for medical information online. (Unpublished master's thesis). Lund University, Lund, Sweden.

Norman, D. A. (1988). *The psychology of everyday things.* New York, NY: Basic Books.

Norman, D. A. (2002). Emotion and design: Attractive things work better. *Interactions Magazine, ix*(4), 36–42.

Norman, D. A. (2004). Emotional design: Why we love (or hate) everyday things. New York, NY: Basic Books.

Palmer, J. W. (2002). Web site usability, design, and performance metrics. *Information Systems Research, 13*(2), 151–167.

Park, S., Choi, D., & Kim, J. (2004). Critical factors for the aesthetic fidelity of web pages: Empirical studies with professional web designers and users. *Interacting with Computers, 16*, 351–376.

Pettey, C., & van der Meulen, R. (2012). Gartner says by 2014, 80 percent of current gamified applications will fail to meet business objectives primarily due to poor design. Gartner Newsroom. Retrieved February 12, 2016, from www.gartner.com/newsroom/id/2251015.

Plass, J. L., Heidig, S., Hayward, E. O., Homer, B. D., & Um, E. (2014). Emotional design in multimedia learning: Effects of shape and color on affect and learning. *Learning and Instruction, 29*, 128–140.

Plass, J. L., Homer, B. D., & Kinzer, C. K. (2015). Foundations of game-based learning. *Educational Psychologist, 50*(4), 258–283.

Pomales-García, C., Liu, Y., & Mendez, D. (2005). Web-based distance learning technology: Does appearance matter? *Proceedings of the Human Factors and Ergonomics Society Annual Meeting, 49*(7), 755–758.

Porter, M. E., & Heppelmann, J. E. (2014). How smart, connected products are transforming competition. *Harvard Business Review, 92*(11), 64–88.

Prensky, M. (2001). Fun, play and games: What makes games engaging. *Digital Game-Based Learning, 5*, 1–05.

Prochaska, J. O., & Marcus, B. H. (1994). The transtheoretical model: Applications to exercise. *Medicine & Science in Sports & Exercise, 26*(11), 1400–1404.

Quinn, J. M., & Tran, T. Q. (2010). Attractive phones don't have to work better: Independent effects of attractiveness, effectiveness, and efficiency on perceived usability. In *Proceedings of the SIGCHI Conference on Human Factors in Computing Systems (CHI '10)* (pp. 353–362). New York, NY: ACM.

Reber, R., Schwarz, N., & Winkielman, P. (2004). Processing fluency and aesthetic pleasure: Is beauty in the perceiver's processing experience? *Personality and Social Psychology Review, 8*(4), 364–382.

Reinecke, K., & Bernstein, A. (2011). Improving performance, perceived usability, and aesthetics with culturally adaptive user interfaces. *ACM Transactions on Computer-Human Interaction, 18*(2), 1–29.

Reinecke, K., & Bernstein, A. (2013). Knowing what a user likes: A design science approach to interfaces that adapt to culture. *MIS Quarterly, 37*(2), 427–453.

Reppa, I., & McDougall, S. (2015). When the going gets tough the beautiful get going: Aesthetic appeal facilitates task performance. *Psychonomic Bulletin & Review, 22*(5), 1243–1254

Reppa, I., Playfoot, D., & McDougall, S. J. P. (2008). Visual aesthetic appeal speeds processing of complex but not simple icons. *Proceedings of the Human Factors and Ergonomics Society Annual Meeting, 52*(18), 1155–1159.

Rose, D. H., & Meyer, A. (2002). *Teaching every student in the digital age: Universal design for learning.* Alexandria, VA: Association for Supervision and Curriculum Development.

Rotgans, J. I., & Schmidt, H. G. (2011). Situational interest and academic achievement in the active-learning classroom. *Learning and Instruction, 21*(1), 58–67.

Ryan, R. M., & Deci, E. L. (2000a). Self-determination theory and the facilitation of intrinsic motivation, social development, and well-being. *American Psychologist, 55*(1), 68.

Ryan, R. M., & Deci, E. L. (2000b). Intrinsic and extrinsic motivations: Classic definitions and new directions. *Contemporary Educational Psychology*, *25*(1), 54–67.

Salen, K., & Zimmerman, E. (2004). *Rules of play: Game design fundamentals*. Cambridge, MA: MIT Press.

Salimun, C. (2013). The relationship between visual interface aesthetics, task performance, and preference (PhD). University of Glasgow, Glasgoow, U.K..

Salimun, C., Purchase, H. C., Simmons, D. R., & Brewster, S. A. (2010). The effect of aesthetically pleasing composition on visual search performance. In *Proceedings of the 6th Nordic Conference on Human-Computer Interaction Extending Boundaries (NordiCHI '10)* (p. 422. New York, NY: ACM.

Sauer, J., & Sonderegger, A. (2009). The influence of prototype fidelity and aesthetics of design in usability tests: Effects on user behaviour, subjective evaluation and emotion. *Applied Ergonomics*, *40*(4), 670–677.

Sauer, J., & Sonderegger, A. (2011). The influence of product aesthetics and user state in usability testing. *Behaviour & Information Technology*, *30*(6), 787–796.

Schell, J. (2014). *The art of game design: A book of lenses*. Boca Raton, FL: CRC Press.

Schenkman, B. N., & Jönsson, F. U. (2000). Aesthetics and preferences of web pages. *Behaviour & Information Technology*, *19*(5), 367–377.

Schering, S., Niesenhaus, J., & Schmidt, R. (2014). Press Play? Praxisorientierte und wissenschaftliche Sicht auf Gamification [Press Play? Practical and scientific viewpoints on gamification]. In M. Koch, A. Butz, & J. Schlichter (Eds.), *Mensch & Computer 2014–Workshopband*. Berlin: De Gruyter Oldenbourg, pp. 355–359.

Schmidt, K. E., Liu, Y., & Sridharan, S. (2009). Webpage aesthetics, performance, and usability: Design variables and their effects. *Ergonomics 52* (6), 631–643.

Seaborn, K., & Fels, D. I. (2015). Gamification in theory and action: A survey. *International Journal of Human-Computer Studies*, *74*, 14–31.

Seckler, M., Opwis, K., & Tuch, A. N. (2015). Linking objective design factors with subjective aesthetics: An experimental study on how structure and color of websites affect the facets of users' visual aesthetic perception. *Computers in Human Behavior*, *49*, 375–389.

Seijts, G. H., & Latham, G. P. (2005). Learning versus performance goals: When should each be used? *Academy of Management Perspectives*, *19*(1), 124–131.

Shneiderman, B., & Plaisant, C. (2009). *Designing the user interface: Strategies for effective human–computer-interaction* (5th ed.). Boston, MA: Addison-Wesley.

Sonderegger, A., & Sauer, J. (2010). The influence of design aesthetics in usability testing: Effects on user performance and perceived usability. *Applied Ergonomics*, *41*, 403–410.

Sonderegger, A., Sauer, J., & Eichenberger, J. (2014). Expressive and classical aesthetics: Two distinct concepts with highly similar effect patterns in user–artefact interaction. *Behaviour & Information Technology*, *33*(11), 1180–1191.

Sonderegger, A., Uebelbacher, A., Pugliese, M., & Sauer, J. (2014). The influence of aesthetics in usability testing. In *Proceedings of the 32nd Annual ACM Conference on Human Factors in Computing Systems (CHI '14)* (pp. 21–30). New York, NY: ACM.

Sonderegger, A., Zbinden, G., Uebelbacher, A., & Sauer, J. (2012). The influence of product aesthetics and usability over the course of time: a longitudinal field experiment. *Ergonomics*, *55*(7), 713–730.

Stinson, J. N., Jibb, L. A., Nguyen, C., Nathan, P. C., Maloney, A. M., Dupuis, L. L., ... Portwine, C. (2013). Development and testing of a multidimensional iPhone pain assessment application for adolescents with cancer. *Journal of Medical Internet Research*, *15*(3), e51.

Strebe, R. (2016). Aesthetics on the web: Effects on approach and avoidance behaviour. *Behaviour & Information Technology*, *35*(1), 4–20.

Szabo, M., & Kanuka, H. (1998). Effects of violating screen design principles of balance, unity, and focus on recall learning, study time, and completion rates. *Journal of Educational Multimedia and Hypermedia*, *8*, 23–42.

Tarasewich, P., Daniel, H. Z., & Griffin, H. E. (2001). Aesthetics and web site design. *Quarterly Journal of Electronic Commerce*, *2*(1), 67–81.

Thielsch, M. T., Blotenberg, I., & Jaron, R. (2014). User evaluation of websites: From first impression to recommendation. *Interacting with Computers, 26*(1), 89–102.

Thielsch, M. T., Engel, R., & Hirschfeld, G. (2015). Expected usability is not a valid indicator of experienced usability. *PeerJ Computer Science, 1*, e19.

Thielsch, M. T., & Hirschfeld, G. (2012). Spatial frequencies in aesthetic website evaluations – explaining how ultra-rapid evaluations are formed. *Ergonomics, 55*(7), 731–742.

Thielsch, M. T., & Hirschfeld, G. (under review). Facets of website content.

Thielsch, M. T., Nikolaeva, D., & Kebeck, G. (2009). Beeinflusst Ästhetik die Performanz bei Suchaufgaben im Web? [Does aesthetics influence performance in web search tasks?] In H. Wandke, S. Kain & D. Stroove (Hrsg.), *Mensch & Computer 2009* (p. 487–490). München: Oldenbourg Verlage.

Thieme, A., Wallace, J., Meyer, T. D., & Olivier, P. (2015). Designing for mental wellbeing: Towards a more holistic approach in the treatment and prevention of mental illness. In *Proceedings of the 2015 British HCI Conference* (pp. 1–10). New York, NY: ACM.

Thom, J., Millen, D., & DiMicco, J. (2012). Removing gamification from an enterprise SNS. In *Proceedings of the ACM 2012 Conference on Computer Supported Cooperative Work* (pp. 1067–1070). New York, NY: ACM.

Thüring, M., & Mahlke, S. (2007). Usability, aesthetics and emotions in human–technology interaction. *International Journal of Psychology, 42*(4), 253–264.

Tractinsky, N., Cokhavi, A., Kirschenbaum, M., & Sharfi, T. (2006). Evaluating the consistency of immediate aesthetic perceptions of web pages. *International Journal of Human–Computer Studies, 64*(11), 1071–1083.

Tractinsky, N., Katz, A. S., & Ikar, D. (2000). What is beautiful is usable. *Interacting with Computers, 13*(2), 127–145.

Tuch, A. N., Bargas-Avila, J. A., Opwis, K., & Wilhelm, F. H. (2009). Visual complexity of websites: Effects on users' experience, physiology, performance, and memory. *International Journal of Human-Computer Studies, 67*(9), 703–715.

Tuch, A. N., Presslaber, E. E., Stöcklin, M., Opwis, K., & Bargas-Avila, J. A. (2012). The role of visual complexity and prototypicality regarding first impression of websites: Working towards understanding aesthetic judgments. *International Journal of Human-Computer Studies, 70*(11), 794–811.

Tuch, A. N., Roth, S. P., Hornbæk, K., Opwis, K., & Bargas-Avila, J. A. (2012). Is beautiful really usable? Toward understanding the relation between usability, aesthetics, and affect in HCI. *Computers in Human Behavior, 28*(5), 1596–1607.

Um, E., Plass, J. L., Hayward, E. O., & Homer, B. D. (2012). Emotional design in multimedia learning. *Journal of Educational Psychology, 104*(2), 485–498.

Van der Heijden, H. (2003). Factors influencing the usage of websites: The case of a generic portal in the Netherlands. *Information & Management, 40*(6), 541–549.

Van Schaik, P., & Ling, J. (2008). Modelling user experience with web sites: Usability, hedonic value, beauty and goodness. *Interacting with Computers, 20*(3), 419–432.

Van Schaik, P., & Ling, J. (2009). The role of context in perceptions of the aesthetics of web pages over time. *International Journal of Human-Computer Studies, 67*, 79–89.

Von Ahn, L. (2006). Games with a purpose. *Computer, 39*(6), 92–94.

Werbach, K., & Hunter, D. (2012). *For the win: How game thinking can revolutionize your business.* Philadelphia, PA: Wharton Digital Press.

Wilson, P. (1973). Situational relevance. *Information Storage and Retrieval, 9*(8), 457–471.

Witt, M., Scheiner, C., & Robra-Bissantz, S. (2011). Gamification of online idea competitions: Insights from an explorative case. *Informatik schafft Communities, 192.*

Zichermann, G., & Linder, J. (2010). *Game-based marketing: inspire customer loyalty through rewards, challenges, and contests.* Hoboken, NJ: John Wiley & Sons, Inc.

6

Trust in Virtual Online Environments

Sirkka L. Jarvenpaa, Celeste Cantu, and Shi Ying Lim

Introduction

Work has become increasingly mediated by information and communication technologies. In light of the changes, questions have been raised as to how much we can trust other people and organizations in virtual computer-mediated environments, and how much we can trust the technology itself as it replaces or augments social actors (Bonabeau, 2003; McAfee, 2013; Sullivan, 2015). As simulation software became more precise, Ducati's designers started to trust the simulation software so much that when motorcycle test results disagreed with a simulation, the first reaction was to mistrust the physical test results (McAfee, 2013). Answers to questions about whether trust matters, what forms trust takes, and how it functions in virtual environments have serious consequences for virtual work environments, as well as for our daily lives. However, the specifics about the technologies and the circumstances surrounding their use have been sparsely addressed in conversations about trust.

We begin this chapter by defining trust. We then discuss trust research in two virtual contexts that have received significant attention: electronic commerce (e-commerce) and virtual teams. We examine commonly used trust models in these two contexts. We then turn to two perspectives that are expected to have increasing importance in conversations about trust: trust in technology and contextualized models of trust. We conclude with proposed directions for future research. In the remainder of the chapter, we use the terms "online," "virtual," and "technology" interchangeably.

The Wiley Blackwell Handbook of the Psychology of the Internet at Work,
First Edition. Edited by Guido Hertel, Dianna L. Stone, Richard D. Johnson, and Jonathan Passmore.

Defining Trust and Why Trust Matters

In work environments, trust commonly is defined as the:

> *willingness to be vulnerable to the actions of another party, based on the expectation that the other will perform a particular action important to the trustor, irrespective of the ability to monitor or control that other party.*
>
> Mayer, Davis, and Schoorman (1995, p. 712)

This interpersonal view of trust is echoed in another widely used organizational definition from Rousseau, Sitkin, Burt, and Camerer (1998): "a psychological state comprising the intention to accept vulnerability based upon positive expectations of the intentions or behavior of another" (p. 395). Both definitions view trust as the psychological state in which a human or a collective (e.g., a team, organization, or a community comprising humans) participates. Traditional conceptualizations of trust assume volition and choice, even if trust is considered in the aggregate at the collective level (Zaheer, McEvily, & Perrone, 1998). Hence, not just an individual, but an organization, team, institution, or community can be associated with a state of trust (Rousseau et al., 1998; Zaheer et al., 1998). Technology also can be a trustee and a trustor, as we later argue.

In discussions about trust in online and virtual environments, psychological perspectives prevail. Trust can be viewed as a disposition or trait directed toward others (Rotter, 1967). However, most commonly it is considered a state of being (Mayer et al., 1995). Sociological views of trust also exist, as the following definition suggests: "an objective social reality not reducible to individualistic psychological factors" (Lewis & Weigert, 2012, p. 25). According to Luhmann (2000), trust is inherent in most human relationships and experiences, and is often confused with familiarity, confidence, and power. Each of these four relational states can reduce the complexity of events and result in positive expectations. In terms of psychological theories, trust research builds on social exchange, social influence, psychological contracts, fairness theories, and the theory of reasoned action, among others (Culnan & Armstrong, 1999; Komiak & Benbasat, 2008; Pavlou & Gefen, 2005). In addition, to examine trust transference in the context of hyperlinks on the Internet, Stewart (2003) and Stewart and Zhang (2003) leveraged balance theory (Heider, 1958) and priming effects (Higgins, Bargh, & Lombardi, 1985). Trust research also combines psychological perspectives with economic theories. For example, using agency theory, Bansal, Zahedi, and Gefen (2010) examined personal dispositions and antecedents to trust in a health website. Liang, Laosethakul, Lloyd, and Xue (2005) used transaction cost economics to understand trust effects on online prescription filling.

Trust matters in areas where risk and interdependencies cannot be reduced and where vulnerabilities prevail. McEvily (2011) argued that trust and risk, which is a probabilistic cost–benefit calculation, inevitably co-occur and overlap. Trust reflects heuristics or intuitive judgments that are based on frames or schemes embedded in the social systems (McEvily, 2011). Meanwhile, risk perception refers to the "trustor's belief about the likelihood of gains and losses outside of the considerations that involve the relationships with the particular trustee" (Mayer et al., 1995, p. 726).

Concerns about risk are central in online environments because such environments are characterized by incomplete or asymmetric information about the parties, weak legal protections, and delayed remedies. The inherent risks generate feelings of "uncertainty," "discomfort," and "anxiety" (Dowling & Staelin, 1994), and increase the sense of a lack of control (Pavlou, 2003), and hence negatively affect the overall online experience (Forsythe &

Shi, 2003). For trust to emerge and to result in particular outcomes, some form of interdependence or dependence has to be present between parties. That is, one party cannot satisfy its own interests or expectations without the participation of the other party. In the context of e-commerce, D. J. Kim (2014) described the dependence and risk as follows:

> *(1) goods and money usually are not exchanged simultaneously, (2) consumers have to provide information that is recorded by vendors, and (3) the risk exists that the consumer might not get goods that fit the on-screen description*
>
> (Kim, 2014, p. 11)

Next, we briefly review how trust has been examined commonly in relation to e-commerce and to virtual teams and then consider trust models that were influential in these contexts.

Trust Research in E-Commerce and Virtual Teams

Two frequently studied contexts of trust in virtual environments include online transactional environments and new organizing work forms. Prior to the 1990s, research on trust was relatively scarce and generally concerned about whether trust can exist or be cultivated online, whether it matters, and what is distinctive about online trust.

The Birth of the Commercial Internet

The early Internet technologies – particularly the worldwide web technologies – gave birth to online merchants and online transactions. One of the first online merchants was a science fiction bookstore in Palo Alto in the United States (Jarvenpaa & Ives, 1994). The commercial use of the Internet and the hyperlinked worldwide web made trust relevant because of the high levels of uncertainty in engagement. Until then, computer-based technologies served highly routine and structured transactional functions in organizations and in work environments. Users had few choices or discretion in using technologies, relying primarily on word processing and accounting systems. Systems were driven by individual operators in batched modes, and they did not interact with each other in real time. By the mid-1990s, web technologies and new information technology (IT) architectures had entered organizations that led to unstructured, networked, and real-time processing and communication. The new technologies gave birth not only to web storefronts but also to new organizing forms, such as virtual teams and virtual organizations (Jarvenpaa & Ives, 1993).

From early on, researchers were skeptical about whether trust was present and could be cultivated online. Nohria and Eccles (1992) wrote: "you cannot build network organizations on electronic networks alone ... . If so, we will probably need an entirely new sociology of organizations" (pp. 304–305). Handy (1995) reflected on the hype in the business press about virtual teams and cautioned that "one cannot function effectively without frequent face-to-face interaction; that is, 'trust needs touch'" (p. 46). *The Economist* (1997) echoed the sentiment: "How do you know whom to trust? In a virtual world, the issue of trust gets magnified" (n.p.). In 1993, *The New Yorker* published a cartoon that showed two dogs – one sitting in front of a computer. The caption read, "On the Internet, nobody knows you're a dog" (Steiner, 1993). As of 2011, this cartoon was the most frequently reproduced of *The New Yorker*'s cartoon archive. The cartoon spoke to the anonymity of users and to the possibility that users were not who they claimed to be. Trust in the online environment was seen as weak or fragile. Meanwhile, offline, the organizational theories of trust argued that trust becomes more critical in environments that are decentralized, less formal, less hierarchical, more inclusive, and open (Rousseau et al., 1998; Whitener, Brodt, Korsgaard, & Werner, 1998). Paradoxically, then, the trust research for offline environments highlighted the lack of trust in online environments – the very trust that was essential for any meaningful online work activities.

The concerns of weak trust originated at least partially from the impersonal nature of online relationships. Gutek (1995) discussed how social actors (trustors and trustees) would become interchangeable in service encounters. Organizations would not care about who did the work, as long as the supply of willing workers was sufficient and workers could be accessed on demand. Similarly, the online shops would not care about the customers' identity or whether they trusted the "store," as long as enough of them shopped in the online store. Such transactions were perceived to be devoid of trust – trust was irrelevant. Earlier researchers had underscored that virtual relationships needed affective bonds and personal engagement to render trust as feasible (Kiesler, Siegel, & McGuire, 1984). As a solution, Gutek (1995) discussed how these impersonal transactions could be turned into pseudorelationships by tracking customer actions and preferences and relying on this information for future encounters to provide consistent and predictable service. This consistency would convey that the providers had their customers' interests in mind; customers would develop trust and keep returning.

Privacy concerns propel research on online trust

In response to the impersonal nature of online interactions, an active line of research on personalization developed. But personalization in turn intensified tensions around privacy, and these tensions were seen to undermine trust (Alsop, 1998). Already in 1995, scholars, such as Gutek (1995), were concerned with the implications for online trust from disclosure of personal information. Personalization required the use of personal data, which raised privacy concerns (Culnan & Armstrong, 1999). Personal data allowed a merchant or a provider to know its customers' identities, interests, and needs and so to tailor the relationship development process and the offerings to increase customer satisfaction and convenience – as well as their own profits. The availability and sale of personal information became the engine of growth in Internet business. Customers were concerned about vendors using the data or the supplied information in an exploitative manner. In the late 1990s, some sources predicted that within the next 6–8 months, most Internet users would experience identity theft first-hand (Jarvenpaa & Tiller, 2001).

The first wave of online trust research hence centered on privacy issues. Trust was considered a relevant factor given the risks and vulnerabilities inherent in online activities. The relationship with privacy concerns was theorized both as an effect and as an outcome (Culnan & Armstrong, 1999; Smith, 1994). The sharing of personal information to complete an online transaction was a "second exchange" that required a social contract between individuals and impersonal online web sites (Culnan & Armstrong, 1999). For people to enter the social contract, trust needed to exist. Studies followed that investigated whether merchants could develop privacy practices to build consumer trust or whether government regulation of privacy was a better way to build consumers' confidence in the Internet (e.g., Milberg, Smith, & Burke, 2000). To date, privacy continues to be an active area of research (see also Stone, Krueger, & Takach, 2017, Chapter 19 in this volume), and trust has many roles in that research; it is depicted as antecedent, mediator, moderator, and outcome of privacy concerns (Smith, Dinev, & Xu, 2011).

Trust and electronic commerce Since the early days of commercial Internet activity, e-commerce comprised many types of commercial transactions and relationships, including Internet shopping. In addition to privacy risks, many other sources of risk prevailed in Internet shopping, including economic, social, performance, and personal risk (Jarvenpaa & Todd, 1996). How socially acceptable was shopping online? Early on, a flurry of studies addressed varied risks in relation to trust (e.g., Ba & Pavlou, 2002; Lee & Turban, 2001; Nambisan, 2002; Pavlou, 2003; Wallage, Christiaanse, & Noteberg, 2003). How trust operated in early online stores was not strikingly different from what one would have expected of brick-and-mortar stores. Trust in a virtual salesperson (recommendation agent) had many parallels with the trust placed in a real sales person (Komiak, Wang, & Benbasat, 2005). The merchant's size and reputation were found to affect consumers' assessment of trustworthiness (Jarvenpaa, Tractinsky, & Vitale, 2000). Over time, researchers focused on why trust and risk emerge and on how websites can be built to decrease risk and increase trust. Li, Pieńkowski, Van Moorsel, and Smith (2012) summarized many of the interventions: third-party certificates, reputation systems, security policies, website design (e.g., graphic design, navigation, presentation of sellers, products), and communication strategies, among others. Reichheld and Schefter (2000) gave the example of the company that intentionally incorporated security features that made it difficult for customers to access its website but protected customers' account information. This safeguard, although inconvenient, enhanced customer trust.

Just as in physical stores, institutional mechanisms played an important part in the study of online marketplaces (e.g., Pavlou & Dimoka, 2006). Institutional mechanisms involved "third-party safeguarding mechanisms, such as online credit card guarantees, escrow services, and privacy protection" which existed "to protect [consumers] against potential risks in the e-commerce environment" (Fang et al., 2014, p. 409). However, the importance of such institutional mechanisms might be a function of the organizations involved. Recently, Botsman (2015) contended that platform-based business models enabled by technology and the Internet have transformed how trust is built, lost, and repaired. Botsman (2015) gave the example of how she would be more conscious of the condition in which she left her Airbnb rental as opposed to a hotel room because it could have repercussions for her online rating and her availability of choices as a consumer thereafter. Consequently, trust development and maintenance are no longer limited to a dyadic relationship between the vendor and the customer but are entangled in a network or web of stakeholders.

Many of the early studies focused on adoption and purchase intentions as outcomes of trust (M. K. O. Lee & Turban, 2001; McKnight, Kacmar, & Choudhury, 2004; Ratnasingam & Pavlou, 2003). However, as e-commerce became increasingly widespread and routinized, later research has taken a longer term perspective and focused on continuing use of the technology, including repurchase intentions and satisfaction (Balasubramanian et al., 2003; Cyr, 2008; Fang et al., 2014; Gefen, 2002; D. J. Kim, 2014; Li et al., 2012). Research also has taken a multilevel view of online trust (Kim, 2014). Figure 6.1 summarizes the key patterns in research on trust in e-commerce and how the outcomes of trust shifted as e-commerce developed and diffused.

New forms of organizing work: virtual teams

Another online context that has received much attention is the virtual team. Also called distributed work, online groups or telework, virtual teams offered new and flexible ways of structuring work that leveraged lateral as opposed to command-and-control structures (Jarvenpaa & Ives, 1993). They might be either standing or ad hoc teams

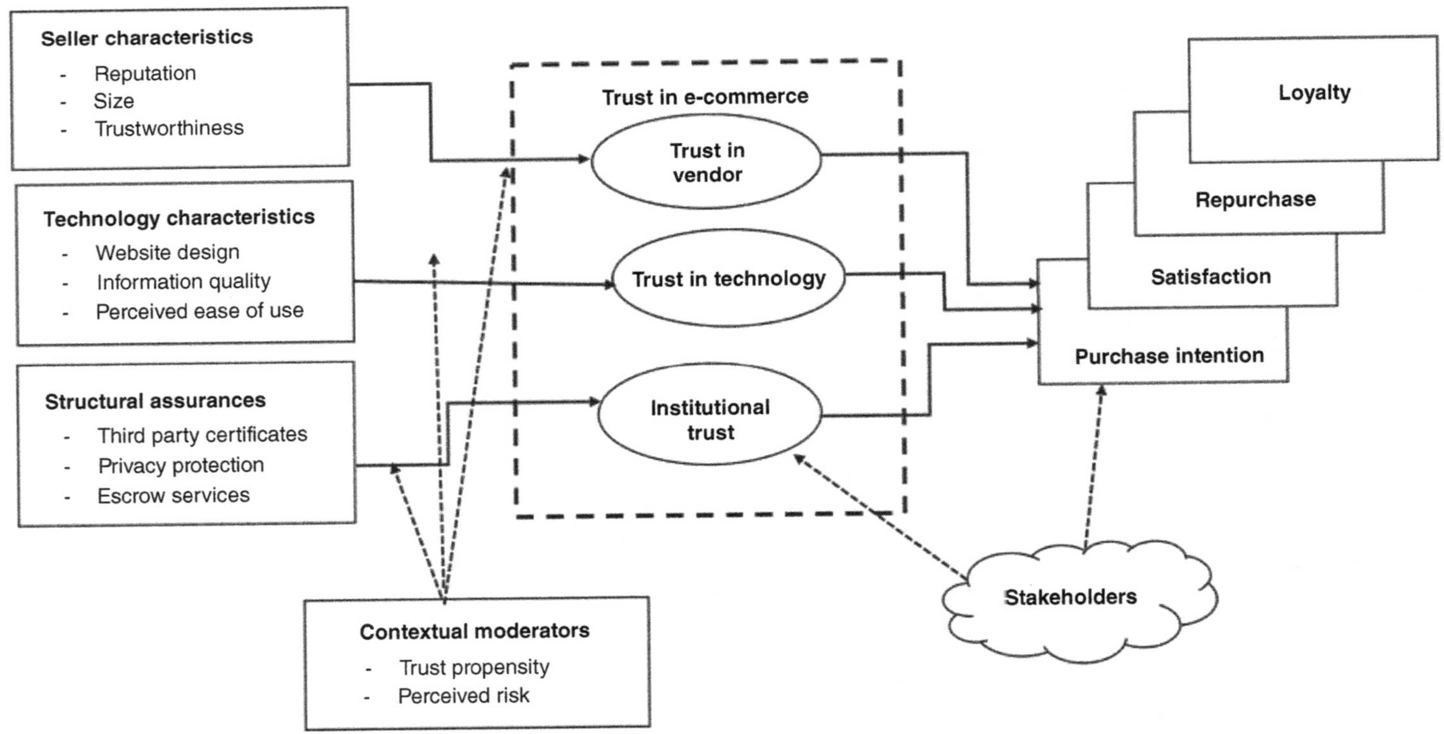

Figure 6.1 Overview of research on trust in e-commerce.

(Bakker, 2010). In an ad hoc virtual team, team members in different parts of the world communicate and collaborate with each other for a short period of time to solve a problem, but without necessarily spending face-to-face time with each other (Jarvenpaa, Ives, & Pearlson, 1995). With regard to standing teams, trust was "[the] glue of the global workplace" (O'Hara-Devereaux & Johansen, 1994: 243). Online environments lacked the customary institutions, norms, and rules that gave the familiar permanency, reliability, and consistency to the offline environment (Mowshowitz, 1997). Knoll and Jarvenpaa (1995) provided an early example of how trust can be built online – but not easily. Others cautioned about the importance of using virtual settings only to a limited extent or only after initial face-to-face encounters (De Meyer, 1991; Maznevski & Chudoba, 2000). Indeed, research continues to emphasize the need for extended site visits by team members in long-standing virtual teams (Cramton & Hinds, 2014).

Researchers have also sought to understand how the degree of virtuality affects trust building and maintenance, as well as the relationship between trust and performance, although without much consensus (De Jong, Dirks, & Gillespie, in press). Some studies have found less trust and more conflict in virtual teams than in face-to-face teams (e.g., Hinds & Mortensen, 2005), while others have suggested that as long as trust is high, increased virtuality does not have detrimental effects (Peñarroja, Orengo, Zornoza, & Hernández, 2013). Distributed isolates (i.e., members who have no teammates at their site) can be more trusting toward their other colleagues who are collocated than collocated teammates toward each other or toward their distributed isolated colleagues (O'Leary & Mortensen, 2010). Broader national and organizational cultures and governance structures are also likely to affect how virtuality influences trust (Lowry, Zhang, Zhou, & Fu, 2010).

A set of studies has examined the effect of a team's configuration on trust in virtual teams. Polzer, Crisp, Jarvenpaa, and Kim (2006) found that virtual teams that have large, collocated subgroups have the highest level of conflict and the lowest level of trust. Webster and Wong (2008) found that trust was lowest in hybrid teams (i.e., teams having both local and dispersed team members), compared with teams that were fully dispersed or fully local. Polzer et al. (2006) and Webster and Wong (2008) revealed that the subgroups within teams created real or perceived divisions so-called fault-lines. Chiu and Staples (2013) found that self-disclosure and information elaboration reduced perceived faultlines, even when real divisions were strongly evident in a team.

The effects from both virtuality and configuration are likely influenced by communication. Richer communication media increase trust as well as performance (Burgoon et al., 2002). Spontaneous communication also can increase trust, or at least can decrease distrust (Hinds & Mortensen, 2005). Frequent feedback (Henttonen & Blomqvist, 2005; Jarvenpaa & Leidner, 1998) and predictable communication (Jarvenpaa & Leidner, 1998; Warkentin & Beranek, 1999) have positive effects on trust. Sarker, Ahuja, Sarker, and Kirkeby (2011) found that communication is a mediator between trust and performance: communication by trustworthy team members increases performance, while more communication by those who are not trusted can backfire. Kanawattanachai and Yoo (2002) found that enhanced performance improved trust, and this improved trust further improved performance. These authors also noted that task-oriented communication early on among team members was important for trust, but at the end of the project more implicit coordination based on expertise location (i.e., awareness of who knows what) was important for trust and performance. In their meta-analysis of trust and team effectiveness, Breuer, Huffmeier, and Hertel (2016) found that when technology was used to document interactions and commitments, trust had less effect on performance.

In research on virtual teams, outcome variables have included task performance, team process, and team performance, similar to Hackman's (1974) definition of team effectiveness. Both subjective and objective measures have been used. Measures of task performance include decision quality (Kanawattanachai & Yoo, 2002) and stock price (Kanawattanachai & Yoo, 2007). Other team-dependent variables used to measure team effectiveness include cohesiveness (e.g. Jarvenpaa, Shaw, & Staples, 2004), status and authority relations, and counter-normative behaviors (Driskell, Radtke, & Salas, 2003). Research on team processes has looked at process improvement and customer satisfaction (Kirkman, Rosen, Tesluk & Gibson, 2004). Trust has been observed to be an antecedent to subjective outcomes, such as team performance as rated by team members, but it has been shown to act as a moderator to the relationship between functional diversity and team performance, as rated by managers (e.g., Peters & Karren, 2009). Figure 6.2 summarizes some general patterns in research on trust in virtual teams. Noteworthy is that e-commerce research is generally positive about trust and its consequences. For example, the online context in e-commerce is theorized to speed up trust formation (Ou, Pavlou, & Davison, 2014). Yet, negative views are not absent in e-commerce research. The online context is theorized to delay trust development in virtual teams when members are geographically distributed (Wilson, Straus, & McEvily, 2006).

Trust models in virtual online environments

Trust research in both e-commerce and virtual teams has leveraged and built on trust models from organizational studies, including the following: (1) Mayer et al.'s (1995) integrative model of organizational trust, (2) Lewicki and Bunker's (1995) developmental

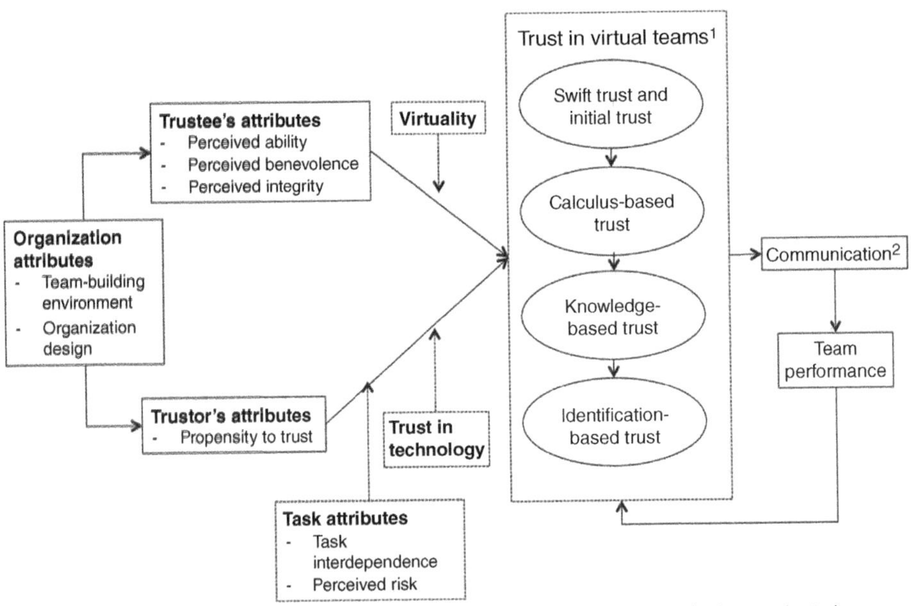

[1] Other than its role as an antecedent, mediator, and dependent variable, trust has also been examined as a moderator in virtual team contexts.

[2] Similarly, communication has also been examined both as an antecedent to trust (e.g. as a part of team building) and mediator between trust and team performance.

Figure 6.2 Overview of research on trust in virtual environments.

model, (3) Meyerson, Weick, and Kramer's (1996) swift trust, and (4) McKnight, Cummings, and Chervany's (1998) initial trust. We review these models and briefly highlight related research in virtual environments.

Integrative view of trust (Mayer et al., 1995)

Mayer et al.'s (1995) model integrated the fragmented trust literature that had developed across several disciplines and presented trust not as a trait or disposition, but as the psychological state of individuals in a relationship. In the Mayer et al. (1995) model, a trustor trusts a trustee based on the perceived trustworthiness in a specific relationship. Perceived trustworthiness comprises ability, integrity, and benevolence. Whether trust leads to risk-taking in a relationship depends on whether a perceived risk is present. As a relational model, the model is based on social exchange theory and reciprocity (Schoorman, Mayer, & Davis, 2007). Recent works have advanced the relational model beyond reciprocity to include mutual and asymmetric trust (Korsgaard, Brower, & Lester, 2014). The model has been found to be empirically robust at the individual, team, organizational, and interorganizational levels (Fulmer & Gelfand, 2012). The Mayer et al. (1995) model has served as the basis for numerous studies on e-commerce (e.g., Gefen, 2002; McKnight, Choudhury, & Kacmar, 2002; Ridings, Gefen, & Arinze, 2002) and virtual teams (for a review, see Martins, Gilson, & Maynard, 2004; Breuer et al., 2016).

Development view of trust (Lewicki and Bunker, 1995)

The developmental view sees trust development as intertwined with relationship development processes (Lewicki & Bunker, 1995; Sheppard & Tuchinsky, 1996). Trust is a result of history-bounded interactions that determine the perceived costs and benefits of interdependence (Kramer, 1999). Direct personal communication and interactions, shared experiences, and completed promises become the basis for calibrating and updating trust. Lewicki, Stevenson, and Bunker (1997) proposed three forms of trust. First, calculus-based trust relies on punishment or deterrence mechanisms in the relationship. Second, knowledge-based trust assumes socialization or courtship types of interaction, in which people try to learn about the other and establish an interpersonal relationship. Third, identification-based trust is based on shared goals, empathy, and common values, to the point that one person is able to act as an agent for the other. The three forms of trust were depicted as a unitary stage model, in which calculus-based trust formed first and then enabled development of knowledge-based trust followed by identification-based trust (Lewicki, Tomlinson, & Gillespie, 2006).

Further research has acknowledged the existence of these three (or related) forms of trust in virtual contexts (e.g., Ba & Pavlou, 2002; Kuo & Yu, 2009; Liang et al., 2005; Polzer et al., 2006). However, the notion of linear progression of trust development in virtual environments has not received empirical support. Kuo and Yu (2009) found that both calculus-based and knowledge-based trust existed early on in a team relationship. Members imported knowledge from prior relationships to substitute for a lack of knowledge in the current relationship. Identification-based trust also developed initially on the basis of members' past experiences, but this trust form remained weak over time and was not associated with the knowledge-based trust in the virtual teams (Kuo & Yu, 2009). In sum, empirical research suggests that the three types of trust are not necessarily linked in the sequential manner theorized by Lewicki and Bunker (1995).

Swift trust (Meyerson et al., 1996)

Although Mayer et al.'s (1995) model and Lewicki and Bunker's (1995) model assume the existence of the shadows of past and future, Meyerson et al. (1996) developed the concept of swift trust for temporary systems such as film crews. Temporary systems are akin to ad hoc groups or teams that, by definition, are fluid and without fixed organizational structures (Bakker, 2010). Temporary systems are found in existing organization structures – even in groups and teams inside large organizations (Valentine, Nembhard, & Edmondson, 2015). According to Meyerson et al. (1996), trust was quickly and swiftly built and maintained by action that demonstrated competent execution of role schemes and scripts. Although initially trust was influenced by dispositions, role expectations, and institutional structures, action gave swift trust its unique form. Swift trust is a unique, contextualized trust that occurs in "short-lived, transient, and fleeting" social structures (Meyerson et al., 1996, p. 179).

In addition to its application to virtual teams (Iacono & Weisband, 1997; Jarvenpaa & Leidner, 1998), swift trust has been applied in virtual communities of practice (Gammelgaard, 2010), interorganizational emergency teams (Majchrzak, Jarvenpaa, & Hollingshead, 2007), and military groups (Ben-Shalom, Lehrer, & Ben-Ari, 2005), as well as to online learning environments (Erez, Lisak, Harush, Glikson, Nouri, & Shokef, 2013) and mobile commerce platforms (Ruf & Back, 2014). In the context of global virtual teams, Crisp and Jarvenpaa (2013) suggested that swift trust can be seen as consisting of both cognitive components, which set the initial expectations, and normative components, and reinforce trust through action. Swift trust is "trust but verify" (Meyerson et al., 1996, p. 187). Other researchers have similarly underscored the importance of follow-up or monitoring to maintain trust (Zolin, Hinds, Fruchter, & Levitt, 2004). Swift trust also has close cousins in related concepts such as "fast trust" (Blomqvist, 2005; Blomqvist, Hurmelinna-Laukkanen, Nummela, & Saarenketo 2008; Henttonen & Blomqvist, 2005), and active trust (Möllering, 2005).

Initial trust (McKnight et al., 1998)

Like swift trust, initial trust complements the developmental and relationship models. The model was created to explain why some people (e.g., MBA students) who do not know each other have high levels of trust toward each other. Trust can be high at the outset, even before the parties have any chance to interact.

Rather than being based on action, as swift trust is, initial trust depends heavily on social embeddedness. Initial trust is a result of categorical and institutional factors: unit grouping, faith in humanity, structural assurance, and situation normality. It is highly dependent on initial stable structures outside the particular relationship, such as guarantees, regulators, and legal resources. External normative structures prevail, such as privacy rules. For example, initial trust was used to explain why consumers can have trust in an unknown web merchant with whom they have never interacted before (McKnight, Choudhury, & Kacmar, 2002).

In e-commerce, researchers often borrow from both swift trust and initial trust models to explain why shoppers trust an unfamiliar web merchant. Initial trust has been important in providing grounding for trust in structural and institutional mechanisms in e-commerce, as well as in mobile commerce (Wang, Shen, & Sun, 2013). In mobile environments, if trust is not instantly present, merchants will miss opportunities for transactions and relationships. Table 6.1 summarizes the four key models of trust that have been used in research in virtual environments.

Table 6.1 Key models of trust.

	Integrative model of trust	*Development view of trust*	*Swift trust*	*Initial trust*
	(Mayer et al., 1995)	(Lewicki & Bunker, 1995)	(Meyerson et al., 1996)	(McKnight et al., 1998)
Foundation	Relationship	History-bounded interactions in a relationship	Ad hoc and temporary systems	Social embeddedness
Sources of trust	Trustee's ability, benevolence, and integrity	Direct personal communication and interactions, shared experiences, and completed promises	Action, role scripts	Unit grouping, faith in humanity, structural assurance, and situation normality; all can be external to the relationship between trustor and trustee
Key mechanisms	Reciprocity, risk	Time, deterrence, familiarity, identity	Heuristics	Categorization processes
Virtual environment research	Supported in virtual environments	Three types of trust found in virtual environments, but not sequential interdependence	Supported in virtual environments	Supported in virtual environments

Other key trust concepts in virtual environments

Several other trust concepts have had a major influence on trust research in virtual environments, including transference, distrust, and trust repair. We briefly review the concepts and relate them to selective research in virtual environments.

Transference of trust Transference of trust examines how trust travels from one target to another. Hyperlinks on websites make facilitation of trust transfer possible. An unknown website can include links to a well-known offline or online merchant. In an early paper on trust transference in an online merchant context, Stewart (2003) examined whether trust is "transferred across hypertext links based on the perceived interaction and similarity of the linked organizations," and whether "institution-based trust is transferred from the traditional shopping channel to a Web-based organization based on evidence that the Web-based organization has a physical store" (p. 5). The study found support for the proposition that trust is transferred from a better known site to an unknown site via a hyperlink. The reverse relationship, from the unknown site to the better known site, was not studied. Many other studies replicated the results, showing positive transference of trust from an offline context to an online context (Lee, Kim, & Ahn, 2011; Ye, 2010). Studies have also found positive transference from an online context to a mobile context (Wang et al., 2013) and have examined online trust transference in different countries (Badrinarayanan, Becerra, Kim, & Madhavaram, 2012; Pentina, Zhang, & Basmanova, 2013). The US context facilitated a lesser amount of trust transfer compared with other countries such as South Korea and Ukraine. Some researchers attributed the finding to the emphasis on relationships and the high context characters of the non-US counterparts (Pentina et al., 2013).

Distrust in online environments Distrust is commonly evoked in research on virtual environments but is rarely studied systematically. One reason for the lack of studies on distrust might be the fundamental divide over whether distrust is a separate concept or is synonymous with low trust or lack of trust. This divide transcends the trust literature in general. Some believe that trust and distrust are opposite ends of the same continuum (Singh & Sirdeshmukh, 2000; Worchel, 1979). Others argue that they are separate constructs with unique antecedents (Lewicki, McAllister & Bies, 1998; Sitkin & Roth, 1993).

Rousseau et al. (1998) defined distrust as negative expectations of others' intentions and motives. These negative expectations are most often seen to imply negative consequences. Trust, representing positive expectations of others' intentions and motives, is most often associated with positive consequences. However, in both offline and online contexts, exceptions have been identified. Grazioli and Jarvenpaa (2000) demonstrated that high levels of online trust caused even technology-savvy online consumers to become victims of fraud.

Colesca (2009) describes how low levels of trust in Internet technology inhibit citizens' participation in e-government, despite high trust in the government; furthermore, she states that bad personal experiences and reported failures with the technology "reinforce distrust or reduce a high level of trust in the Internet and in the agencies that use them" (p. 34). Similarly, Adjei (2013) uses the concepts of low trust and distrust interchangeably, suggesting that low privacy concerns are signs of perceived trustworthiness, whereas high privacy concerns reflect low trust, or distrust, in the identity management system. When distrust is treated as the opposite end of the same continuum as trust, the factors that increase trust should reduce distrust.

In the context of virtual recommendation agents, Komiak and Benbasat (2008) established that trust and distrust are two separate processes with their own antecedents. Differences in the antecedents suggest significant differences in how distrust and trust operate. Komiak and Benbasat (2008) noted that competence helped most to build trust and that the revelation of something previously unknown to them helped most to reduce distrust. They concluded that focusing solely on trust provides an incomplete picture and might lead to the oversight of high levels of distrust. In the case of mobile services, Ho and Chau (2013) reported coexistence of both trust (in integrity) and distrust.

For the most part, researchers who have attended to both trust and distrust still associate trust with positive outcomes and distrust with negative outcomes. That is, technology designers are told to focus on factors that increase trust and decrease distrust. However, Jarvenpaa and Majchrzak (2010) argue that in highly vulnerable situations, high trust and low distrust can create mindlessness. Consequently, both high trust and high distrust are needed for vigilant behavior in highly vulnerable situations.

Trust repair in online environments In offline environments, trust repair is often studied as an aftermath of breach (Tomlinson, Dineen, & Lewicki, 2004). Forms or means for repairing trust include apologies, reparations and penance, and forgiveness. In offline contexts, Kramer and Lewicki (2010) argue that the timing of a breach can be important. Trust violations that occur early in a relationship when trust is building are far more harmful to the future of the relationship than when trust is violated later (Lount, Zhong, Sivanathan, & Murnighan, 2008). Both the nature of the underlying relationship and whether the breach is seen as an isolated event or as habitual and repetitive matter for trust repair. An isolated event is likely to be associated with competence failure, whereas a habitual breach is possibly associated with integrity loss.

Breaches in trust have been studied in online environments in terms of disrespectful behaviors, unmet expectations, unwillingness to acknowledge performance issues, incongruence with core values, structural issues (e.g., misalignment of job duties and authority), and transactions that do not go as planned in the e-commerce setting (Choi & Nazareth, 2014). However, little research has examined what can be done to counteract low trust or distrust. Some researchers (e.g., Choi & Nazareth, 2014) speculate that in the online context, or in virtual teams, actions to repair trust might not achieve their intended effect because face-to-face interaction is not possible and because communication could be misinterpreted when trust is already low or distrust is already high. Choi and Nazareth (2014) also suggest that breaches decrease knowledge-based trust or, if interpreted as inconsistent behavior, can decrease calculus-based trust. In addition, in the online context, the other parties might not know that a breach has occurred until too much time has passed to make timely and fair action possible. However, we were not able to find any studies, including studies in an online security context, that had considered the extent to which the vendor's acknowledgment of the incident, apology, and appropriate restorative action shapes the consumers' willingness to reconcile. Chen, Wu, and Chang (2013) found that positive mood is an important mediator in trust repair. They also shared Choi and Nazareth's (2014) view that repairing trust is especially difficult in online settings because of the impoverished communication channel and the narrow time window associated with each interaction. Bansal and Zahedi (2015) elucidated the harmful effects that privacy violation in the forms of hacking and unauthorized sharing can have on the trust violation and repair process. Apology emerged as a universally effective response, although its reparative power was far lower in situations that involved unauthorized sharing than it was in hacking situations.

Trust in virtual environments generally has built on models and concepts from research in organizational studies. In their review of online transactions, Li et al. (2012) concluded that "online trust is, first and foremost, about trust" (p. 87) but that the online context brings important new distinctions. In the effort to develop a better understanding of the effects of the virtual context, we discuss (1) the role of technology in trust, and (2) contextualized models of trust. In considering these two lines of research, we have aimed to increase our understanding of the effects of the material elements of the technology on the actions, interactions, and goals of social entities: individuals, groups, and organizations.

Technology Trust

The arrival of intelligent systems introduced human-like technologies. Not only do intelligent systems have to be trusted by humans, but humans also must be trusted by the systems (e.g., robots in a nuclear power station) – and, perhaps more importantly, humans can become distrusted by systems in special high-stress situations (Markiewicz & Adamus, 2013). The implication is that technology can be a trustor, trustee, or a context of trust. First, algorithmic capabilities give technologies, as trustors, both social and moral agency, as technology responds to our actions, talks back to us, and is associated with provisional, inscrutable, and consequential actions (Orlikowski, 2015). Second, a technological artifact as a trustee can involve mundane online transactions in an e-commerce site or complex technology manifestations, such as a mobile app with complex algorithmic capabilities that matches organ donors and their recipients (Orlikowski, 2015). Third, technology is and can create a context for interpersonal or interorganizational trust in online or mobile merchants, online or mobile marketplaces, virtual teams, virtual organizations, enterprise social media, virtual worlds, telework, and so on.

In the trust models in organizational studies, the trustee is an individual, team, or organization. Trust in technology assumes that the trustee is technology, an inanimate object (McKnight, 2005). We trust cars, airplanes, and smart phones. Inanimate objects do not have the volition that usually is assumed with trustees, but vulnerability and uncertainty still are present (McKnight, Carter, Thatcher, & Clay, 2011). The literature on trust in technology has had four somewhat distinct but clearly overlapping themes: (1) whether trust in technology is distinguishable from trust in social actors; (2) which processes and features drive trust in technology; (3) broadening the conceptualization of trust in technology beyond technology acceptance and adoption; and (4) exploring trust and risk in emergent technologies. We briefly discuss each theme.

Distinguishing trust in technological actors from social actors

Authors introducing the technology trust concept made the case against those who argued that no trust relationship exists between a person and a technology because of a lack of volition and of correspondence between the parties (Friedman, Khan, & Howe, 2000; Shneiderman, 2000). Initial papers that tackled trust in online commerce concentrated on both the merchant and the technology medium as a trustee (Jarvenpaa et al., 2000; McKnight et al., 2004; Yousafzai, Pallister, & Foxall, 2003). In these papers efforts were focused on making a distinction between interpersonal trust and technology trust. For example, in the arena of Internet shopping in particular, differences were highlighted between trust in the merchant and trust in the medium – namely, the Internet (Corbitt, Thanasankit, & Yi, 2003; M. K. O. Lee & Turban, 2001; Ratnasingam & Pavlou, 2003). The results often proved that, indeed, technology trust beyond trust in the merchants was linked to user behaviors in online technologies.

Research has examined technology from a broad category perspective, such as "the Internet" or "online shopping," as well as considering trust in specific websites. The studies underscore risk concerns. A reliable vendor needs to make sure that the technology is functioning with the appropriate infrastructure and that it engages in safe procedures (Ratnasingam & Pavlou, 2003). However, studies have failed to establish a link between trust in technology and risk, stating instead that users can choose to trust the Internet in general, even if they identify high levels of risk (Corbitt et al., 2003). However, most studies did not consider the possible effects of technology maturity, familiarity, or scope on the perception of risk and its corresponding effect on trust in technology.

In studies that have focused on a more specific target (e.g., a website, a web vendor, or an online store), the institutional factors were salient in dealing with a never-before-seen vendor or website. However, institutional factors also tended to include infrastructure and security structures that were encompassed by a broader definition of the technology referent (Ratnasingam & Pavlou, 2003; Corbitt et al., 2003). Even given a more specific segmentation of institutional factors and technology-intrinsic characteristics, the differentiation between websites and vendors becomes ambiguous (Lim et al., 2006; McKnight et al., 2004). Also, experiential evidence in the form of web experience (McKnight et al., 2002) and website attractiveness (Wakefield, Stocks & Wilder, 2004) were significant in evaluating trust in technology for these studies. The direct relationships between website trust and purchase behavior were not as significant as expected, suggesting that other mechanisms had to be accounted for, particularly if the trustee (technology) was no longer novel (Lim et al., 2006; Wakefield et al., 2004).

Processes and features that drive trust development

In trust-in-technology research, traditional antecedents of trust (e.g., benevolence, competence, and integrity) have been contested and compared with more artifact-specific conceptualizations, such as helpfulness, capability, and reliability (McKnight, 2005; Lippert & Swiercz, 2005; Ratnasingam, 2005). Different antecedents are needed because of the lack of volition. Still, some studies argue that technology takes on human-like attributes, focusing on virtual salespersons (Komiak et al., 2005), an online auction marketplace (Pavlou & Dimoka, 2006), and an online bookstore (Lim et al., 2006).

Broadening the conceptualization of the information technology artifact

Studies incorporate descriptions of the technology in terms of interfaces, procedures, and physical characteristics that influence trust in the technology (Charki & Josserand, 2008; Cyr, 2008; Komiak & Benbasat, 2008; Vance, Elie-Dit-Cosaque, & Straub, 2008). Over time, the target of trust has broadened to encompass agents, technologies, and services behind the technology, as in the case of e-government (Teo, Srivastava, & Jiang, 2008), Facebook (Lankton & McKnight, 2008), mobile banking (G. Kim, Shin, & Lee, 2009), and online auction marketplaces (Dimoka, 2010). Such broadening has created confusion in terms of where the technology artifacts end and social actors begin (Stenmark, 2013).

The boundaries of the artifact's scope and the context of use have also generated concern and study (Charki & Josserand, 2008; Teo et al., 2008). Montague, Winchester, and Kleiner, (2010) suggested the need for more active engagement by patients to increase trust in the obstetrics work system. The recommendations included increasing trust in many stakeholders, including nurses, doctors, and technicians. Others have also found that organizations that collect data about their users through their technology-enabled services, such as Pandora or Amazon, should also seek to cultivate trust among their users (especially with respect to their data use and giving users control over the data use) so that users volunteer data (Morey, Forbath, & Schoop, 2015). These trust-building efforts become essential because voluntary sharing of data is what enables organizations to provide more personalized services. Trust-building efforts thus extend beyond the technology itself to include the organization and the education of its users.

More recent studies have responded to the call to further conceptualize the information technology artifact and to incorporate more sophisticated discussions on the personal versus impersonal features of the technology. General versus specific trust (Thatcher, Carter, Li, & Rong, 2013) and system-like versus human-like trust (Lankton, McKnight, & Tripp, 2015) are frameworks that have been used to segment different attributes of a technology and to analyze the specific trust effects of each. Studies draw on affordance, social appraisal, and cognitive appraisal theory in an effort to reconcile both the social and the material conceptualizations of technology and to account for emotionality in the trust relationship (Deng, 2012; Lankton et al., 2015; Thatcher et al., 2013).

Evaluation of trust in technology at the macro level also requires adopting a longer time horizon. The focus has shifted to technology's continued use, rather than remaining on its adoption (Lankton et al., 2015; Lankton, McKnight, & Thatcher, 2012). In addition, with technologies such as knowledge management systems, the use of which is mandated, the emphasis also has shifted away from adoption. Instead, the focus is on the user's intention to explore the technology, which later could translate into continued use

(Thatcher, McKnight, Baker, Arsal, & Roberts, 2011). The broadening of the use context also has brought about the study of a new set of risks (Brengman & Karimov, 2012; Yao, Liu, & Yuan, 2013). Researchers have examined trust and risk in online whistle-blowing systems (Lowry, Moody, Galletta, & Vance, 2013), in location personalization in mobile applications (Ho & Chau, 2013), and in social networking sites (Lankton et al., 2012). Studies in more specific or specialized contexts also have introduced new antecedents and moderating variables. For example, proximity and diffused responsibility have been shown to affect trust in technology (Lowry et al., 2013).

Emergent technologies and risks

Risk for a mature technology is determined not solely by the safeguards built into or around it, but also by the possibilities of widespread use. That is, a growing user base, interactions with other related new technologies, and changing goals in technology use would pose new risks that had not previously been contemplated. Organizations increasingly have adopted big data and algorithms enabled by technology. Some researchers argued that these algorithms do a better job at decision making than humans do. For instance, McAfee (2013) highlighted that human parole boards perform much worse than simple formulas at determining which prisoners should be released from prisons, and that highly trained pathologists do not do as good a job at diagnosing breast cancer as image analysis software. In fact, the IBM Watson's analytic capabilities were packaged into a decision-support tool that could scan medical images and monitor diagnoses that were missed by human experts. This capability also highlighted potential liability issues for pathologists and hospitals and had implications for trust in the technology. Some researchers such as McAfee (2013) believe that the role of human experts should be to provide input to the model, and not just to use the technology as an input to human decision making.

Contextualized models of trust

Context encompasses more than just technology (i.e., trust in technology); it also encompasses the use context (including tasks), the users, and other situational opportunities, constraints, and functional relationships (Hong, Chan, Thong, Chasalow, & Dhillon 2013). That is, the context often dictates the relevant elements in the interaction and the role that trust plays in modeling that interaction. The four models discussed – Mayer et al. (1995), Lewicki and Bunker (1995), Meyerson et al. (1996), and McKnight et al. (1998) – all consider context as a moderator. The context influences the perception of trust antecedents, as well as the consequences of trust, by influencing individual and collective motivations and risk in the present. Generally speaking, context refers to the interrelated conditions, situations, opportunities, and constraints in which something exists or occurs (Bamberger, 2008; Hong et al., 2013; Johns, 2006).

Research on trust in virtual environments has examined scenarios entailing a wide variety of trust forms. In their review on trust in online environments, Li et al. (2012) alone listed 19 different forms of trust. Many of these forms were highly sensitive to the context. For example, swift trust might be most relevant in scenarios with temporary systems. Addressing trust in organizational studies, McEvily (2011) argued that the caveat of all these varying contexts and differing forms of trust is the lack of convergence on a set of robust measures on trust that would facilitate the integration of results. McEvily's (2011) concern suggests that as researchers develop contextualized models of trust, they not only must seek the conceptualization of trust that makes sense given the context, but also must link to commonly used and theorized forms and measures of trust. Authors must situate their findings and attempt a convergence of models. The description of a context relevant to trust is contingent on the interacting entities and, more importantly, on the meaning that these entities assign to artifacts and to each other (Wright & Ehnert, 2010).

In information systems, the issue of context has been conceptualized in mobile and pervasive computing: where information is located, and how and with whom the technology is being used (Schilit, Adams, & Want, 1994). Tarasewich (2003) suggested a model of context, stemming from the work of Schmidt, Beigl, and Gellersen (1999), that acknowledged the participants in an information system, the social and physical aspects around an activity (environment), and users' behaviors and tasks (activities). Reyes (2002) suggested classifying context variables by unit of analysis, based on the typology created by Swanson (1987) of information system use determinants at the individual, organization, and market levels. As these efforts suggest, a careful definition of context can establish the level of analysis for the trust referents; the situational, physical, social, and institutional factors considered; and the activities included in the interaction. Contextual models also need to consider the different maturity stages of the technology, changing risk conditions, and user adaptations. As people become more comfortable with technology, factors other than trust become direct antecedents, having direct effects on use. Trust becomes a moderator of the relationship between antecedents and use.

As the focus shifts to continued use of technology, other more dominating processes are at play, and trust does not directly affect a behavior (Dirks & Ferrin, 2001; Rousseau et al., 1998). Trust becomes a moderating factor. For example, in the continued use of an e-commerce site, past use rather than trust has a direct effect; trust is a moderating variable that either strengthens or weakens the relationship. Trust affects how the past use is assessed, and this assessment affects continued use. In mobile banking (Chung & Kwon, 2009), institutional trust in the financial technology industry moderates the relationship between past use and continued use. Trust influences how people interpret or evaluate information, which affects their attitudes and behaviors (Jarvenpaa et al., 2004). In a study on organizational learning, Sanzo, Santos, García, & Trespalacios (2012) found that the organizational trust climate moderated the relationship between open communication and learning. The social interactions enabled by the organizational trust climate strengthened both knowledge creation and knowledge exchange. Brahm and Kunze (2012) studied trust climates as a moderator for virtual teams. They showed that the relationship between team goals and task cohesion was moderated by the trust climate. Fulmer and Gelfand (2012) noted in their review that when trust is a moderator, the trust construct is often organizational, and the relationship that is moderated is on the interpersonal or team level.

Future Research

Our review suggests questions that bear further study include: (1) How should we conceptualize trust in emergent technologies? (2) What balance of trust and distrust needs to be achieved to avoid negative consequences of trust? (3) How do we contextualize our understanding of trust and technology?

Conceptualization of trust in emergent technologies

How should we conceptualize trust in technology in the digital age? Technology has evolved in ways that reflect an attempt to make technology resemble humans, as current studies have identified (Deng, 2012; Komiak & Benbasat, 2008; Lankton et al., 2015). Processes, algorithms, and actions try to emulate what a human can do (even with errors and emotion) in order to motivate interaction between humans and technology (Anthes, 2014; Frick, 2015). Although technology has been established as an object of trust in the current literature, the appearance of robots, software, and artifacts that can make decisions that affect humans contests the traditional non-volitional view of technology and presents artifacts as potential trustors. Whether their decisions are determined by algorithms or human input, these technologies incorporate criteria that evaluate and affect humans. If a machine can learn, evaluate, decide, and act, is trust then a process that ought to be considered in design? If so, how would machines carry out trust-enhancing and distrust-reducing processes? Humans might trust technology algorithms in a limited decision scenario (e.g., organizing a work team or assigning tasks in a project). In addition, although emotional cues help humans interact with the technology, they trust technology the most when, for example, robots engage in analytical reasoning rather than emotional processing (Anthes, 2014; Frick, 2015). Humanness and adaptive algorithms might further incorporate learning processes, which would help robotic technology understand which situations can be tackled, and which ones require solutions outside technology's control. If the grounds for the user–robot interactions stem from an understanding that the robot is competent to undertake a given task, robots might possess significant decision capabilities. However, full decision capabilities require a certain mastery of emotion and instinct that has not yet been duplicated in robots.

Much research is needed to further our understanding of "learning machines": To what extent is trust and distrust in technology driven by the artifact's agency? How do decisions, actions, changes, errors and omissions influence a person's trust and distrust in the learning machines? Also important is the question of how these concrete actions are interpreted as trust or distrust behaviors by a human. This interpretive capacity is key in predicting the behavior of a person toward a machine. For example, will a surgery robot be expected to always perform a bariatric surgery in the same way? If so, does this limitation make a user trust or distrust the problem-solving capabilities of the robot? The identification of the actions or action possibilities that are relevant to the user and their effects on trust and distrust need to be tackled in future studies. Also, the study of behaviors in humans that stem from trust and distrust in the technology is valuable for designers who struggle with technology acceptance. Affordance, social appraisal, and cognitive appraisal theories can help analyze the action possibilities of a technology, as well as the human and non-human features that influence an interaction. If a technology can in fact constitute a trustor in the human–technology relationship, how does trust form in a relationship with marked asymmetries in capabilities, emotional grounds, information flows, and reciprocity?

What balance of trust and distrust needs to be achieved to avoid negative consequences of trust?

Jarvenpaa and Majchrzak (2010) argued that virtual environments are best conceptualized as environments in which relationships remain ambivalent and in which vigilance needs to be harnessed against the excesses of both trust and distrust. Virtual environments are inherently asymmetric in knowledge flows, are prone to intended or unintended deception and misrepresentation, and are novel because of the constant and unpredictable nature of interdependent changes in technology, actors, and institutional rules. Some scholars have raised concerns about the lack of awareness of risks in virtual environments (Komiak & Benbasat, 2008). Although technology maturity and familiarity might affect this awareness, much of the risk is likely to be irreducible. Trust without the presence of distrust leads to the use of oversimplified heuristics that disregard the unknown risks.

In addition, existing studies in trust and technology often focus on the antecedents of trust in technology. We propose that future research should integrate the trust and distrust dynamic that is highly relevant in algorithmic-based technology, such as self-driving cars. For instance, drivers of self-driving cars have to trust the algorithms to anticipate traffic conditions and respond to moves from other drivers accurately, and they have to entrust their lives to these algorithms to get them to their destination safely. Trust might be warranted when conditions remain stable, but self-driving cars need to harness distrust to keep the driver vigilant, particularly when driving conditions are changing quickly. Technology designers thus have a considerable responsibility to implement characteristics in the technology (e.g., through user experience) that cultivate both trust and distrust in the technology. These design decisions need to address what balance of trust and distrust is optimal to ensure that individuals use the technology in the intended manner. A theory about the interplay between trust and distrust in the use of these emergent technologies would provide guidance on design choices and mechanisms that facilitate vigilant use of them.

Although the case for the coexistence of trust and distrust in technology has been built, new research questions must address the behaviors that result from a relationship in which there is both trust and distrust in the artifact. Relevant questions include: When should a person trust or distrust a machine to recommend a course of action versus make a decision about it? How do self-regulation and conflicting motivations play a role? For example, an online shopper might be highly motivated to purchase clothes online and might willingly provide credit card information, but they might abstain from providing bio data and personal information on preferences. The behaviors that arise in trusting and distrusting relationships between people and machines are important matters to address in the introduction of emergent technologies. The findings from these studies can provide novel models that describe the interplay of trust and distrust in technology, that reveal the risk coping behaviors of the entities interacting (human or technology), and that shed light on affordances that should be discouraged by technology designers.

How can we contextualize our understanding of trust and technology?

A final direction for future empirical work is the explicit consideration of technology as a context, and with that understanding, testing propositions in which trust is seen as a moderator. Although different trust models depict trust as an independent variable, one can question whether a construction of such trust models often might be misguided. Trust certainly influences what our choice set consists of, but trust might be a less significant criterion in the decision-making process than what research suggests. Doney and Cannon (1997) suggest that trust appears "to operate as an 'order qualifier,' not an 'order winner'" (p. 47). Order qualifiers are "those criteria that a company must meet for a customer to even consider it as a possible supplier," whereas order winners are "those criteria that win the order" (Hill 2000, p, 37). Today's digital technology makes architectures that rely on self-selection and self-control increasingly possible. But in these scenarios, might trust be a context or a climate, rather than a primary effect or influence?

Contextualization of trust and technology can help determine the scenarios that are best suited for human–machine interactions. How does trust in technology moderate behaviors in high-risk and low-risk situations? How does trust and distrust in technology change in different risk scenarios and use cases? The identification of the context variables that influence trust in machine–human relationships, with an attempt at convergence, can provide a basis to analyze what situations and which actions are acceptable for a human interacting with a volitional technology. For instance, a human who initially distrusts a cooking robot will probably supervise it in the kitchen. With time, the person could grow to trust the robot and not even flinch when it holds a knife.

Finally, models need to take into account that technology can also render trust irrelevant. In a recent meta-analysis on virtual teams, Breuer et al. (2016) found that when team interactions are automatically documented by the technology, trust becomes largely irrelevant. We have only scratched the surface of research on trust in virtual environments, and refer readers to reviews by Gefen, Karahanna, & Straub (2003), Li et al. (2012), Luo and Najdawi (2004), McKnight et al. (2002), Tams (2012), Sherchan, Nepal, & Paris (2013), and Stenmark (2013). These works are complemented by many other reviews on trust in organizational studies and management that to varying extents also address new forms of organizing (Colquitt, Scott, & LePine, 2007; Fulmer & Gelfand, 2012; Korsgaard et al., 2014; Kramer, 1999; Schoorman et al., 2007).

Conclusion

This chapter has reviewed trust research in e-commerce and virtual work and advocated for deeper conceptualizations of technology and trust, as well as accounting for context. We have identified the key trust models and trust concepts used in virtual environments as well as opportunities for trust research with emergent technologies. The chapter has demonstrated the evolving role of trust in virtual environments, as contexts, technology, and use cases change. The insights into how the presence or absence of trust affects interpersonal and human–technology relationships highlight how the pervasiveness of technologies transforms communications, organization forms, and other elements that are key in human work behavior.

References

Adjei, J. K. (2013). Towards a trusted national identities framework. *Info*, *15*(1), 48–60.

Alsop, S. (1998). A look back from 2004: The e-corporation and me. *Fortune- European Edition*, *138*, 115–118.

Anthes, E. (2014, June 10). You should learn to trust robots. It's for your own good. Retrieved February 1, 2016, from www.wired.com/2014/06/trust-robots-2/

Ba, S., & Pavlou, P. A. (2002). Evidence of the effect of trust building technology in electronic markets: Price premiums and buyer behavior. *MIS Quarterly*, *26*(3), 243–268.

Badrinarayanan, V., Becerra, E. P., Kim, C. H., & Madhavaram, S. (2012). Transference and congruence effects on purchase intentions in online stores of multi-channel retailers: Initial evidence from the US and South Korea. *Journal of the Academy of Marketing Science*, *40*(4), 539–557.

Bakker, R. M. (2010). Taking stock of temporary organizational forms: A systematic review and research agenda. *International Journal of Management Reviews*, *12*(4), 466–486.

Balasubramanian, S., Konana, P., & Menon, N. M. (2003). Customer satisfaction in virtual environments: A study of online investing. *Management Science*, *49*(7), 871–889.

Bamberger, P. (2008). From the editors beyond contextualization: Using context theories to narrow the micro-macro gap in management research. *Academy of Management Journal*, *51*(5), 839–846.

Bansal, G., & Zahedi, F. M. (2015). Trust violation and repair: The information privacy perspective. *Decision Support Systems*, *71*, 62–77.

Bansal, G., Zahedi, F. M. & Gefen, D. (2010). The impact of personal dispositions on information sensitivity, privacy concern and trust in disclosing health information online. *Decision Support Systems, 49*(2), 138–150.

Ben-Shalom, U., Lehrer, Z., & Ben-Ari, E. (2005). Cohesion during military operations a field study on combat units in the Al-Aqsa Intifada. *Armed Forces & Society, 32*(1), 63–79.

Blomqvist, K. (2005). Trust in a dynamic environment-fast trust as a threshold condition for asymmetric technology partnership formation in the ICT Sector. In K. Bijlsma-Frankema, & R. K. Woolthuis (Eds.), *Trust in pressure: Empirical investigations of trust and trust building in uncertain circumstances* (pp. 127–147). Cheltenham, England: Edward Elgar Publishing.

Blomqvist, K., Hurmelinna-Laukkanen, P., Nummela, N., & Saarenketo, S. (2008). The role of trust and contracts in the internationalization of technology-intensive Born Globals. *Journal of Engineering and Technology Management, 25*(1), 123–135.

Bonabeau, E. (2003). Don't trust your gut. *Harvard Business Review, 81*(5), 116–123.

Botsman, R. (2015, October 7). The changing rules of trust in the digital age. Retrieved April 7, 2017, from www.druckerforum.org/blog/?p=1043

Brahm, T., & Kunze, F. (2012). The role of trust climate in virtual teams. *Journal of Managerial Psychology, 27*(6), 595–614.

Brengman, M., & Karimov, F. P. (2012). The effect of web communities on consumers' initial trust in B2C e-commerce websites. *Management Research Review, 35*(9), 791–817.

Breuer, C., Huffmeier, J., & Hertel, G. (2016). Does trust matter more in virtual teams? A meta-analysis of trust and team effectiveness considering virtuality and documentation as moderators. *Journal of Applied Psychology, 101*(8), 1151–1177.

Burgoon, J. K., Bonito, J. A., Ramirez, A., Dunbar, N. E., Kam, K., & Fischer, J. (2002). Testing the interactivity principle: Effects of mediation, propinquity, and verbal and nonverbal modalities in interpersonal interaction. *Journal of Communication, 52*(3), 657–677.

Charki, M. H., & Josserand, E. (2008). Online reverse auctions and the dynamics of trust. *Journal of Management Information Systems, 24*(4), 175–197.

Chen, Y. H., Wu, J. J., & Chang, H. T. (2013). Examining the mediating effect of positive moods on trust repair in e-commerce. *Internet Research, 23*(3), 355–371.

Chiu, Y. T., & Staples, D. S. (2013). Reducing faultlines in geographically dispersed teams self-disclosure and task elaboration. *Small Group Research, 44*(5), 498–531.

Choi, J., & Nazareth, D. L. (2014). Repairing trust in an e-commerce and security context: An agent-based modeling approach. *Information Management & Computer Security, 22*(5), 490–512.

Chung, N., & Kwon, S. J. (2009). Effect of trust level on mobile banking satisfaction: A multi-group analysis of information system success instruments. *Behaviour & Information Technology, 28*(6), 549–562.

Colesca, S. E. (2009). Increasing e-trust: A solution to minimize risk in e-government adoption. *Journal of Applied Quantitative Methods, 4*(1), 31–44.

Colquitt, J. A., Scott, B. A., & LePine, J. A. (2007). Trust, trustworthiness, and trust propensity: A meta-analytic test of their unique relationships with risk taking and job performance. *Journal of Applied Psychology, 92*(4), 909.

Corbitt, B. J., Thanasankit, T., & Yi, H. (2003). Trust and e-commerce: A study of consumer perceptions. *Electronic Commerce Research and Applications, 2*, 203–215.

Cramton, C. D., & Hinds, P. J. (2014). An embedded model of cultural adaptation in global teams. *Organization Science, 25*(4), 1056–1081

Crisp, C. B., & Jarvenpaa, S. L. (2013). Swift trust in global virtual teams. *Journal of Personnel Psychology, 12*, 45–56

Culnan, M. J., & Armstrong, P. K. (1999). Information privacy concerns, procedural fairness, and impersonal trust: An empirical investigation. *Organization Science, 10*(1), 104–115.

Cyr, D. (2008). Modeling web site design across cultures: Relationships to trust, satisfaction, and e-loyalty. *Journal of Management Information Systems, 24*(4), 47–72.

De Jong, B. A., Dirks, K. T. & Gillespie, N. (In press). Trust and team performance: A meta-analysis of main effects, moderators and covariates. *Journal of Applied Psychology.*

De Meyer, A. (1991). Tech talk: How managers are stimulating global R&D communication. *MIT Sloan Management Review, 32*(3), 49.

Deng, L. (2012). Discrete emotions and trust in self-service technology. *Proceedings of the Southern Association for Information Systems Conference (SAIS 2012)*, 61–65.

Dimoka, A. (2010). What does the brain tell us about trust and distrust? Evidence from a functional neuroimaging study. *MIS Quarterly, 34*(2), 373–396.

Dirks, K. T., & Ferrin, D. L. (2001). The role of trust in organizational settings. *Organization Science, 12*(4), 450–467.

Doney, P. M., & Cannon, J. P. (1997). An examination of the nature of trust in buyer-seller relationships. *Journal of Marketing, 61*(2), 35–51.

Dowling, G. R., & Staelin, R. (1994). A model of perceived risk and intended risk-handling activity. *Journal of Consumer Research, 21*, 119–134.

Driskell, J. E., Radtke, P. H., & Salas, E. (2003). Virtual teams: Effects of technological mediation on team performance. *Group Dynamics: Theory, Research, and Practice, 7*(4), 297.

Erez, M., Lisak, A., Harush, R., Glikson, E., Nouri, R., & Shokef, E. (2013). Going global: Developing management students' cultural intelligence and global identity in culturally diverse virtual teams. *Academy of Management Learning & Education, 12*(3), 330–355.

Fang, Y., Qureshi, I., Sun, H., McCole, P., Ramsey, E., & Lim, K. H. (2014). Trust, satisfaction, and online repurchase intention: The moderating role of perceived effectiveness of e-commerce institutional mechanisms. *MIS Quarterly, 38*(2), 407–427.

Forsythe, S. M., & Shi, B. (2003). Consumer patronage and risk perceptions in Internet shopping. *Journal of Business Research, 56*(11), 867–875.

Frick, W. (2015). When your boss wears metal pants. *Harvard Business Review, 93*(6), 84–89.

Friedman, B., Khan, P. H., Jr., & Howe, D. C. (2000). Trust online. *Communications of the ACM, 43*(12), 34–40.

Fulmer, C. A., & Gelfand, M. J. (2012). At what level (and in whom) we trust across multiple organizational levels. *Journal of Management, 38*(4), 1167–1230.

Gammelgaard, J. (2010). Knowledge retrieval through virtual communities of practice. *Behaviour & Information Technology, 29*(4), 349–362.

Gefen, D. (2002). Reflections on the dimensions of trust and trustworthiness among online consumers. *ACM Sigmis Database, 33*(3), 38–53.

Gefen, D., Karahanna, E., & Straub, D. W. (2003). Trust and TAM in online shopping: An integrated model. *MIS Quarterly, 27*(1), 51–90.

Grazioli, S., & Jarvenpaa, S. L. (2000). Perils of Internet fraud: An empirical investigation of deception and trust with experienced Internet consumers. Part A: Systems and Humans, *IEEE Transactions on Systems, Man and Cybernetics, 30*(4), 395–410.

Gutek, B. A. (1995). *The dynamics of service: Reflections on the changing nature of customer/provider interactions*. San Francisco, CA: Jossey-Bass.

Hackman, J. R. (1974). *Group tasks, group interaction process, and group performance effectiveness: A review and proposed integration*. In L. Berkowitz (Ed.), *Advances in experimental social psychology* (Vol. *8*). New York: Academic Press.

Handy, C. (1995). Trust and the virtual organization. *Harvard Business Review, 73*(3), 40.

Heider, F. (1958). *The psychology of interpersonal relations*. New York, NY: John Wiley & Sons, Inc.

Henttonen, K., & Blomqvist, K. (2005). Managing distance in a global virtual team: The evolution of trust through technology-mediated relational communication. *Strategic Change, 14*(2), 107–119.

Higgins, E. T., Bargh, J. A., & Lombardi, W. J. (1985). Nature of priming effects on categorization. *Journal of Experimental psychology: Learning, Memory, and Cognition, 11*(1), 59.

Hill, T. (Ed.) (2000). *Manufacturing strategy – text and cases*. New York, NY: Palgrave.

Hinds, P. J., & Mortensen, M. (2005). Understanding conflict in geographically distributed teams: The moderating effects of shared identity, shared context, and spontaneous communication. *Organization Science, 16*(3), 290–307.

Ho, S. Y., & Chau, P. Y. K. (2013). The Effects of Location Personalization on Integrity Trust and Integrity Distrust in Mobile Merchants. *International Journal of Electronic Commerce, 17*(4), 39–72.

Hong, W., Chan, F. K., Thong, J. Y., Chasalow, L. C., & Dhillon, G. (2013). A framework and guidelines for context-specific theorizing in information systems research. *Information Systems Research, 25*(1), 111–136.

Iacono, C. S., & Weisband, S. (1997). Developing trust in virtual teams. In *Proceedings of the Thirtieth Hawaii International Conference on System Sciences, 2*, 412–420. Washington DC: IEEE.

Jarvenpaa, S. L., & Ives, B. (1993). Organizing for global competition. *Decision Sciences, 24*(3), 547–580.

Jarvenpaa, S. L., & Ives, B. (1994). The global network organization of the future: Information management opportunities and challenges. *Journal of Management Information Systems, 10*(4), 25–57.

Jarvenpaa, S. L., Ives, B., & Pearlson, K. (1995). Global customer service for the computer and communications industry. *Journal of Management Information Systems.*

Jarvenpaa, S. L., & Leidner, D. E. (1998). Communication and trust in global virtual teams. *Journal of Computer-Mediated Communication, 3*(4).

Jarvenpaa, S. L., & Majchrzak, A. (2010). Research commentary – vigilant interaction in knowledge collaboration: Challenges of online user participation under ambivalence. *Information Systems Research, 21*(4), 773–784.

Jarvenpaa, S. L., Shaw, T. R., & Staples, D. S. (2004). Toward contextualized theories of trust: The role of trust in global virtual teams. *Information Systems Research, 15*(3), 250–267.

Jarvenpaa, S. L., & Tiller, E. H. (2001). Customer trust in virtual environments: A managerial perspective. *Boston University Law Review, 81*, 665.

Jarvenpaa, S. L., & Todd, P. A. (1996). Consumer reactions to electronic shopping on the World Wide Web. *International journal of electronic commerce, 1*(2), 59–88.

Jarvenpaa, S. L., Tractinsky, N., & Vitale, M. (2000). Consumer trust in an Internet store. *Information Technology and Management, 1*, 45–71.

Johns, G. (2006). The essential impact of context on organizational behavior. *Academy of Management Review, 31*(2), 386–408.

Kanawattanachai, P., & Yoo, Y. (2002). Dynamic nature of trust in virtual teams. *The Journal of Strategic Information Systems, 11*(3), 187–213.

Kanawattanachai, P., & Yoo, Y. (2007). The impact of knowledge coordination on virtual team performance over time. *MIS Quarterly*, 783–808.

Kiesler, S., Siegel, J., & McGuire, T. W. (1984). Social psychological aspects of computer-mediated communication. *American Psychologist, 39*(10), 1123–1134.

Kim, D. J. (2014). A study of the multilevel and dynamic nature of trust in e-commerce from a cross-stage perspective. *International Journal of Electronic Commerce, 19*(1), 11–64.

Kim, G., Shin, B., & Lee, H. G. (2009). Understanding dynamics between initial trust and usage intentions of mobile banking. *Information Systems Journal, 19*(3), 283–311.

Kirkman, B. L., Rosen, B., Tesluk, P. E., & Gibson, C. B. (2004). The impact of team empowerment on virtual team performance: The moderating role of face-to-face interaction. *Academy of Management Journal, 47*(2), 175–192.

Knoll, K., & Jarvenpaa, S. L. (1995). Learning virtual team collaboration. *Proceedings of the 28th Hawaii International Conference on Systems Sciences Proceedings, 4*, 92–101.

Komiak, S. Y. X., & Benbasat, I. (2008). A two-process view of trust and distrust building in recommendation agents: A process-tracing study. *Journal of the Association for Information Systems, 9*(12), 727–747.

Komiak, S. Y. X., Wang, W., & Benbasat, I. (2005). Trust building in virtual salespersons versus in human salespersons: Similarities and differences 1. *E-Service Journal, 3*(3), 49–64.

Korsgaard, M. A., Brower, H. H., & Lester, S. W. (2014). It isn't always mutual a critical review of dyadic trust. *Journal of Management, 41*(1), 47–70

Kramer, R. M. (1999). Trust and distrust in organizations: Emerging perspectives, enduring questions. *Annual Review of Psychology, 50*(1), 569–598.

Kramer, R. M., & Lewicki, R. J. (2010). Repairing and enhancing trust: Approaches to reducing organizational trust deficits. *Academy of Management Annals, 4*(1), 245–277.

Kuo, F. Y., & Yu, C. P. (2009). An exploratory study of trust dynamics in work-oriented virtual teams. *Journal of Computer-Mediated Communication, 14*(4), 823–854.

Lankton, N. K., & McKnight, D. H. (2008). Do people trust Facebook as a technology or as a "person"? distinguishing technology trust from interpersonal trust. *14th Americas Conference on Information Systems 2008, 4*, 2577–2585.

Lankton, N. K., Mcknight, D. H., & Thatcher, J. B. (2012). The moderating effects of privacy restrictiveness and experience on trusting beliefs and habit: An empirical test of intention to continue using a social networking website. *IEEE Transactions on Engineering Management, 59*(4), 654–665.

Lankton, N. K., Mcknight, D. H., & Tripp, J. (2015). Humanness, and trust: Rethinking trust in technology, *Journal of the Association for Information Technology, 16*(10), 880–918.

Lee, M. K. O., & Turban, E. (2001). A trust model for consumer internet shopping. *International Journal of Electronic Commerce, 6*(1), 75–91.

Lee, J., Kim, H. J., & Ahn, M. J. (2011). The willingness of e-Government service adoption by business users: The role of offline service quality and trust in technology. *Government Information Quarterly, 28*(2), 222–230.

Lewicki, R. J., & Bunker, B. B. (1995). *Trust in relationships: A model of development and decline.* San Francisco, CA: Jossey-Bass.

Lewicki, R. J., McAllister, D. J., & Bies, R. J. (1998). Trust and distrust: New relationships and realities. *Academy of Management Review, 23*(3), 438–458.

Lewicki, R. J., Stevenson, M. A., & Bunker, B. B. (1997). The three components of interpersonal trust: Instrument development and differences across relationships. Unpublished manuscript.

Lewicki, R. J., Tomlinson, E. C., & Gillespie, N. (2006). Models of interpersonal trust development: Theoretical approaches, empirical evidence, and future directions. *Journal of Management, 32*(6), 991–1022.

Lewis, J. D., & Weigert, A. J. (2012). The social dynamics of trust: Theoretical and empirical research, 1985–2012. *Social Forces, 91*(1), 25–31.

Li, F., Pieńkowski, D., Van Moorsel, A., & Smith, C. (2012). A holistic framework for trust in online transactions. *International Journal of Management Reviews, 14*(1), 85–103.

Liang, H., Laosethakul, K., Lloyd, S. J., & Xue, Y. (2005). Information systems and health care-I: Trust, uncertainty, and online prescription filling. *Communications of the Association for Information Systems, 15*(1), 2.

Lim, K., Sia, C., Lee, M., & Benbasat, I. (2006). Do I trust you online, and if so, will I buy? An empirical study of two trust-building strategies. *Journal of Management Information Systems, 23*(2), 233–266.

Lippert, S., & Swiercz, P. (2005). Human resource information systems (HRIS) and technology trust. *Journal of Information Science, 31*(5), 340.

Lount, R. B., Zhong, C. B., Sivanathan, N., & Murnighan, J. K. (2008). Getting off on the wrong foot: The timing of a breach and the restoration of trust. *Personality and Social Psychology Bulletin, 34*(12), 1601–1612.

Lowry, P. B., Moody, G. D., Galletta, D. F., & Vance, A. (2013). The drivers in the use of online whistle-blowing reporting systems. *Journal of Management Information Systems, 30*(1), 153–190.

Lowry, P. B., Zhang, D., Zhou, L., & Fu, X. (2010). Effects of culture, social presence, and group composition on trust in technology-supported decision-making groups. *Information Systems Journal, 20*(3), 297–315.

Luhmann, N. (2000). Familiarity, confidence, trust: Problems and alternatives. *Trust: Making and Breaking Cooperative Relations, 6*, 94–107.

Luo, W., & Najdawi, M. (2004). Trust-building measures: A review of consumer health portals. *Communications of the ACM, 47*(1), 108–113.

Majchrzak, A., Jarvenpaa, S. L., & Hollingshead, A. B. (2007). Coordinating expertise among emergent groups responding to disasters. *Organization Science, 18*(1), 147–161.

Markiewicz, P., & Adamus, M. (2013). Trust and implicit information asymmetry in repeated games. In *Active citizenship by knowledge management & innovation: Proceedings of the Management, Knowledge and Learning International Conference 2013* (pp. 739–746). Celje, Slovenia: ToKnowPress.

Martins, L. L., Gilson, L. L., & Maynard, M. T. (2004). Virtual teams: What do we know and where do we go from here? *Journal of management, 30*(6), 805–835.

Mayer, R. C., Davis, J. H., & Schoorman, F. D. (1995). An integrative model of organizational trust. *Academy of Management Review, 20*(3), 709–734.

Maznevski, M. L., & Chudoba, K. M. (2000). Bridging space over time: Global virtual team dynamics and effectiveness. *Organization Science, 11*(5), 473–492.

McAfee, A. (2013). Big data's biggest challenge? Convincing people NOT to trust their judgment. *Harvard Business Review, 9.*

McEvily, B. (2011). Reorganizing the boundaries of trust: From discrete alternatives to hybrid forms. *Organization Science, 22*(5), 1266–1276.

McKnight, D. H. (2005). Trust in information technology. *The Blackwell Encyclopedia of Management, 7*, 329–331.

Mcknight, D. H., Carter, M., Thatcher, J. B., & Clay, P. F. (2011). Trust in a specific technology: An investigation of its components. *ACM Transactions on Management Information Systems, 2*(2), 12–32.

McKnight, D. H., Cummings, L. L., & Chervany, N. L. (1998). Initial trust formation in new organizational relationships. *Academy of Management Review, 23*(3), 473–490.

McKnight, D. H., Choudhury, V., & Kacmar, C. (2002). The impact of initial consumer trust on intentions to transact with a web site: A trust building model. *Journal of Strategic Information Systems, 11*(3–4), 297–323.

McKnight, D. H., Kacmar, C. J., & Choudhury, V. (2004). Dispositional trust and distrust distinctions in predicting high-and low-risk internet expert advice site perceptions. *E-Service, 3*(2), 35–58.

Meyerson, D., Weick, K. E., & Kramer, R. M. (1996). Swift trust and temporary groups. In R. Kramer & T. Tyler (Eds.), *Trust in organizations: Frontiers of theory and research* (pp. 166–195). Thousand Oaks, CA: Sage.

Milberg, S. J., Smith, H. J., & Burke, S. J. (2000). Information privacy: Corporate management and national regulation. *Organization Science, 11*(1), 35–57.

Möllering, G. (2005). The trust/control duality an integrative perspective on positive expectations of others. *International Sociology, 20*(3), 283–305.

Montague, E. N. H., Winchester, W. W., & Kleiner, B. M. (2010). Trust in medical technology by patients and healthcare providers in obstetric work systems. *Behaviour & Information Technology, 29*(5), 541–554.

Morey, T., Forbath, T. T., & Schoop, A. (2015). Customer data: Designing for transparency and trust. *Harvard Business Review, 93*(5), 96.

Mowshowitz, A. (1997). Virtual organization. *Communications of the ACM, 40*(9), 30–37.

Nambisan, S. (2002). Designing virtual customer environments for new product development: Toward a theory. *Academy of Management Review, 27*(3), 392–413.

Nohria, N., & Eccles, R. G. (1992). Face-to-face: Making network organizations work. In N. Nohria & R. G. Eccles (Eds.), *Networks and organizations.* Boston, MA: Harvard Business School Press.

O'Hara-Devereaux, M., & Johansen, R. (1994). *Globalwork: Bridging distance, culture, and time.* San Francisco, CA: Jossey-Bass.

O'Leary, M. B., & Mortensen, M. (2010). Go (con) figure: Subgroups, imbalance, and isolates in geographically dispersed teams. *Organization Science, 21*(1), 115–131.

Orlikowski, W. J. (2015, August). *OCIS keynote presentation: Performing research researching performativity.* Paper presented at the Academy of Management Conference, Vancouver, BC, Canada.

Ou, C. X., Pavlou, P. A., & Davison, R. (2014). Swift guanxi in online marketplaces: The role of computer-mediated communication technologies. *MIS Quarterly, 38*(1), 209–230.

Pavlou, P. A. (2003). Consumer acceptance of electronic commerce: Integrating trust and risk with the technology acceptance model. *International Journal of Electronic Commerce, 7*(3), 101–134.

Pavlou, P. A., & Dimoka, A. (2006). The nature and role of feedback text comments in online marketplaces: Implications for trust building, price premiums, and seller differentiation. *Information Systems Research, 17*(4), 392–414.

Pavlou, P. A., & Gefen, D. (2005). Psychological contract violation in online marketplaces: Antecedents, consequences, and moderating role. *Information systems research, 16*(4), 372–399.

Peñarroja, V., Orengo, V., Zornoza, A., & Hernández, A. (2013). The effects of virtuality level on task-related collaborative behaviors: The mediating role of team trust. *Computers in Human Behavior, 29*(3), 967–974.

Pentina, I., Zhang, L., & Basmanova, O. (2013). Antecedents and consequences of trust in a social media brand: A cross-cultural study of Twitter. *Computers in Human Behavior, 29*(4), 1546–1555.

Peters, L., & Karren, R. J. (2009). An examination of the roles of trust and functional diversity on virtual team performance ratings. *Group & Organization Management.*

Polzer, J. T., Crisp, C. B., Jarvenpaa, S. L., & Kim, J. W. (2006). Extending the faultline model to geographically dispersed teams: How colocated subgroups can impair group functioning. *Academy of Management Journal, 49*(4), 679–692.

Ratnasingam, P. (2005). Trust in inter-organizational exchanges: A case study in business to business electronic commerce. *Decision Support Systems, 39*(3), 525–544.

Ratnasingam, P., & Pavlou, P. A. (2003). Technology trust in internet-based inter organizational electronic commerce. *Journal of Electronic Commerce in Organizations, 1*(1), 17–41.

Reichheld, F. F., & Schefter, P. (2000). E-loyalty. *Harvard business review,78*(4), 105–113.

Reyes, G. (2002). Classifying information systems context variables through a review of recent is research on how context affects performance. *ACIS 2002 Proceedings* (pp. 16.1–16.10). Mt Pleasant, MI: ACIS.

Ridings, C. M., Gefen, D., & Arinze, B. (2002). Some antecedents and effects of trust in virtual communities. *Journal of Strategic Information Systems, 11*(3), 271–295.

Rotter, J. B. (1967). A new scale for the measurement of interpersonal trust. *Journal of Personality, 35*(4), 651–665.

Rousseau, D. M., Sitkin, S. B., Burt, R. S., & Camerer, C. (1998). Not so different after all: A cross-discipline view of trust. *Academy of Management Review, 23*(3), 393–404.

Ruf, C., & Back, A. (2014). Design requirements for collaboration processes to increase customer trust in mobile banking platforms. In *27th Bled eConference. eEcosystems* (pp. 279–291). Atlanta, GA: AIS.

Sanzo, M. J., Santos, M. L., García, N., & Trespalacios, J. A. (2012). Trust as a moderator of the relationship between organizational learning and marketing capabilities: Evidence from Spanish SMEs. *International Small Business Journal, 30*(6), 700–726.

Sarker, S., Ahuja, M., Sarker, S., & Kirkeby, S. (2011). The role of communication and trust in global virtual teams: A social network perspective. *Journal of Management Information Systems, 28*(1), 273–310.

Schilit, B., Adams, N., & Want, R. (1994, December). Context-aware computing applications. In *Mobile Computing Systems and Applications, 1994.* WMCSA 1994. First Workshop on (pp. 85–90). Washington DC: IEEE Computer Society.

Schmidt, A., Beigl, M., & Gellersen, H. W. (1999). There is more to context than location. *Computers & Graphics, 23*(6), 893–901.

Schoorman, F. D., Mayer, R. C., & Davis, J. H. (2007). An integrative model of organizational trust: Past, present, and future. *Academy of Management Review, 32*(2), 344–354.

Sheppard, B. H., & Tuchinsky, M. (1996). Micro-OB and the network organization. In R. Kramer & T. Tyler (Eds.), *Trust in organizations: Frontiers of theory and research* (pp. 140–165). Thousand Oaks, CA: Sage.

Sherchan, W., Nepal, S., & Paris, C. (2013). A survey of trust in social networks. *ACM Computing Surveys, 45*(4), 47.

Shneiderman, B. (2000). Designing trust into online experiences. *Communications of the ACM, 43*(12), 57–59.

Singh, J., & Sirdeshmukh, D. (2000). Agency and trust mechanisms in consumer satisfaction and loyalty judgments. *Journal of the Academy of Marketing Science, 28*(1), 150–167.

Sitkin, S. B., & Roth, N. L. (1993). Explaining the limited effectiveness of legalistic "remedies" for trust/distrust. *Organization Science, 4*(3), 367–392.

Smith, H. J. (1994). *Managing privacy: Information technology and corporate America.* Chapel Hill, NC: UNC Press Books.

Smith, H. J., Dinev, T., & Xu, H. (2011). Information privacy research: An interdisciplinary review. *MIS Quarterly, 35*(4), 989–1016.

Steiner, P. (1993). On the Internet, nobody knows you're a dog. *The New Yorker 69*(20), 61.

Stenmark, D. (2013). Distrust in information systems research: A need for stronger theoretical contributions to our discipline. *2013 46th Hawaii International Conference on System Sciences* (pp. 4532–4540). Washington DC: IEEE Computer Society.

Stewart, K. J. (2003). Trust transfer on the world wide web. *Organization Science, 14*(1), 5–17.

Stewart, K. J. and Y. Zhang (2003) Effects of hyperlink texts on trust transfer. In N. Sadeh (Ed.), *Proceedings of the Fifth International Conference on Electronic Commerce* (pp. 235–239). Pittsburgh, PA: ACM.

Stone, D. L., Krueger, D. & Takach, S. (2017). Social issues associated with the Internet at work. In G. Hertel, D. L. Stone, R. D. Johnson, & J. Passmore (Eds.), *The Wiley Blackwell handbook of the psychology of the internet at work* (pp. 423–448). Chichester: John Wiley & Sons, Ltd.

Sullivan, T. (2015). Transparency, trust, and bitcoin. *Harvard Business Review, June 15*, 118–119.

Swanson, E. B. (1987). Information systems in organization theory: A review. In *Critical issues in information systems research* (pp. 181–204). New York, NY: John Wiley & Sons, Inc.

Tams, S. (2012). Toward holistic insights into trust in electronic markets: Examining the structure of the relationship between vendor trust and its antecedents. *Information Systems and e-Business Management, 10*(1), 149–160.

Tarasewich, P. (2003). Towards a comprehensive model of context for mobile and wireless computing. In *Americas Conference on Information Systems* (pp. 114–124). Seattle, WA: Curran Associates, Inc.

Teo, T. S. H., Srivastava, S. C., & Jiang, L. (2008). Trust and electronic government success: An empirical study. *Journal of Management Information Systems, 25*(3), 99–132.

Thatcher, J. B., Carter, M., Li, X., & Rong, G. (2013). A classification and investigation of trustees in B-to-C e-commerce: General vs. specific trust. *Communications of the Association for Information Systems, 32*(1), 107–134.

Thatcher, J. B., Mcknight, D. H., Baker, E. W., Arsal, R. E., & Roberts, N. H. (2011). The role of trust in postadoption IT exploration: An empirical examination of knowledge management systems. *IEEE Transactions on Engineering Management, 58*(1), 56–70.

The Economist. (1997, May 8). Electronic commerce: in search of the perfect market. Retrieved April 7, 2017, from www.economist.com/node/596262

Tomlinson, E. C., Dineen, B. R., & Lewicki, R. J. (2004). The road to reconciliation: Antecedents of victim willingness to reconcile following a broken promise. *Journal of Management, 30*(2), 165–187.

Valentine, M. A., Nembhard, I. M., & Edmondson, A. C. (2015). Measuring teamwork in health care settings: A review of survey instruments. *Medical Care, 53*(4), e16–e30.

Vance, A., Elie-Dit-Cosaque, C., & Straub, D. W. (2008). Examining trust in information technology artifacts: The effects of system quality and culture. *Journal of Management Information Systems, 24*(4), 73–100.

Wakefield, R. L., Stocks, M. H., & Wilder, W. M. (2004). The role of web site characteristics in initial trust formation. *Journal of Computer Information Systems, Fall*(32), 94–103.

Wallage, P., Christiaanse, E., & Noteberg, A. (2003). Consumer trust in electronic channels: The impact of electronic commerce assurance on consumers' purchasing likelihood and risk perceptions. *E-service Journal, 2*(2), 46–67.

Wang, N., Shen, X. L., & Sun, Y. (2013). Transition of electronic word-of-mouth services from web to mobile context: A trust transfer perspective. *Decision Support Systems, 54*(3), 1394–1403.

Warkentin, M., & Beranek, P. M. (1999). Training to improve virtual team communication. *Information Systems Journal, 9*(4), 271–289.

Webster, J., & Wong, W. K. P. (2008). Comparing traditional and virtual group forms: Identity, communication and trust in naturally occurring project teams. *International Journal of Human Resource Management, 19*(1), 41–62.

Whitener, E. M., Brodt, S. E., Korsgaard, M. A., & Werner, J. M. (1998). Managers as initiators of trust: An exchange relationship framework for understanding managerial trustworthy behavior. *Academy of Management Review, 23*(3), 513–530.

Wilson, J., Straus, S., and McEvily, B. (2006). All in due time: The development of trust in computer mediated and face-to-face teams. *Organizational Behavior & Human Decision Processes 99*, 16–33.

Worchel, P. (1979). Trust and distrust. *The Social Psychology of Intergroup Relations, 174*, 187.
Wright, A., & Ehnert, I. (2010). Making sense of trust across cultural contexts. In M. N. K. Saunders, D. Skinner, G. Dietz, N. Gillespie, & R. Lewicki (Eds.), *Organizational trust: A cultural perspective* (pp. 107–126). Cambridge: Cambridge University Press.
Yao, H., Liu, S., & Yuan, Y. (2013). A study of user adoption factors of mobile banking services based on the trust and distrust perspective. *International Business and Management, 6*(2), 9–14.
Ye, Y. (2010). Correlates of consumer trust in online health information: Findings from the health information national trends survey. *Journal of Health Communication, 16*(1), 34–49.
Yousafzai, S. Y., Pallister, J. G., & Foxall, G. R. (2003). A proposed model of e-trust for electronic banking. *Technovation, 23*(11), 847–860.
Zaheer, A., McEvily, B., & Perrone, V. (1998). Does trust matter? Exploring the effects of interorganizational and interpersonal trust on performance. *Organization Science, 9*(2), 141–159.
Zolin, R., Hinds, P. J., Fruchter, R., & Levitt, R. E. (2004). Interpersonal trust in cross-functional, geographically distributed work: A longitudinal study. *Information and Organization, 14*(1), 1–26.

7

Workplace Cyberdeviance

Steven D. Charlier, Gary W. Giumetti, Cody J. Reeves, and Lindsey M. Greco

Introduction

Technology can be used in various ways to enable individuals to engage in behaviors that detract from the achievement of organizational goals – these particular types of counterproductive work behaviors (or intentional, harmful behaviors enacted by members of an organization that go against an organization's legitimate interests; Sackett & DeVore, 2001) are collectively commonly known as "cyberdeviance." Cyberdeviance has emerged in recent years as a prominent area of focus for scholars in a number of fields, including management and information systems. This focus appears to be warranted, given the heavy reliance on the Internet and other communication technologies in today's workplace (Harter, Agrawal, & Sorenson, 2014), the high prevalence rates of various cyberdeviant behaviors (Garrett & Danziger, 2008), and the cost of such behaviors to organizations (Websense, 2006). Cyberdeviance can take many forms, ranging from relatively innocuous (e.g., cyberloafing, broadly defined in the literature as the use of technology at work for personal purposes; Lim, 2002) to more severe forms of deviance (e.g., hacking, software piracy, theft, and cyber-aggression). With regard to prevalence, over 80% of employed adults report engaging in cyberloafing (Garrett & Danziger, 2008; Lim & Teo, 2005; Vitak, Crouse, & LaRose, 2011) and a majority of working adults report experiencing cyber-aggression (i.e., the intentional use of information and communication technologies to behave aggressively; Ford, 2013). In turn, these cyberdeviant behaviors can be costly to organizations, as they may lead to increased employee withdrawal and reduced individual and organizational performance, estimated to cost US organizations up to US$200 billion annually (Websense, 2006).

Whereas counterproductive work behavior, as a general concept, has featured in the literature for some time (see Robinson & Bennett, 1995, for a foundational treatise on employee deviance), research specifically in the area of cyberdeviance is still in its relative infancy. Interestingly, many of the issues that faced employee deviance researchers in the mid-1990s are relevant to cyberdeviance researchers today: scattered research efforts,

The Wiley Blackwell Handbook of the Psychology of the Internet at Work,
First Edition. Edited by Guido Hertel, Dianna L. Stone, Richard D. Johnson, and Jonathan Passmore.

construct proliferation and misspecification, and the lack of a unifying framework. Nevertheless, it is clear from the existing body of cyberdeviance research that the cyberdeviance construct, while overlapping in some ways with counterproductive work behavior, encompasses a broader range of behaviors and focuses more explicitly on the role of technology in enabling deviant behavior. As such, the goal of this chapter is to propose a new typology and framework for viewing cyberdeviance and its associated behaviors, and review the existing literature within this new framework. Our typology of behaviors features three broad categories of cyberdeviance: intrapersonally focused behaviors (cyberloafing), interpersonally focused cyberdeviance (CD-I), and organizationally focused cyberdeviance (CD-O). This will serve as a road map for the literature review that follows. We conclude the chapter with our recommendations for future research, both from a methodological and a content-specific perspective.

Cyberdeviance: Definition and Typology

Existing research on cyberdeviant behaviors lacks consistency in defining and measuring the overall cyberdeviance construct and several of its underlying subcomponents. For example, Weatherbee (2010) defines cyberdeviance as "voluntary behaviour using information and communications systems which either threatens or results in harm to an organization, its members, or stakeholders" (p. 39). Others add that cyberdeviance violates formal or informal organizational norms (Mastrangelo, Everton, & Jolton, 2006; Venkatraman, 2008), in keeping with Robinson and Bennett's (1995) definition of workplace deviance. Still others have limited their focus of cyberdeviance to cyberloafing alone (Zoghbi Manrique de Lara, 2006). We propose a holistic definition of cyberdeviance, describing it as: behavior that takes place using information and communication technologies (ICTs, such as email, text messaging, or other computer technologies), which violates workplace norms, and has the potential to harm individual employees, the organization as a whole, or both (Robinson & Bennett, 1995; Venkatraman, 2008; Weatherbee, 2010).

Prior attempts at categorizing the cyberdeviance space have largely aligned with Robinson and Bennett's (1995) typology of general counterproductive work behavior, but other strategies have included the use of dichotomies (e.g., counterproductive/nonproductive), degrees on a continuum (e.g., impact on organization, reliance on a computer), or distinct groups of behaviors (e.g., entertainment/downloading, work–life balance/personal). Robinson and Bennett's (1995) typology utilizes a two-dimensional structure, with one dimension relating to the target of the deviance, and the other dimension dealing with the severity of the deviance, resulting in four categories of deviance (i.e., production, property, political, personal). Robinson and Bennett (1995) are widely cited in organizational research on general workplace deviance, although existing research has tended to focus primarily on the target of the behavior (interpersonal vs. organizational), and does not incorporate the severity of the behavior (minor vs. serious; see Berry, Ones, & Sackett, 2007).

The typology proposed in this review maintains the distinction regarding the target of the cyberdeviant behavior – cyberdeviance directed at individuals (CD-I) vs. cyberdeviance directed at the organization itself (CD-O). However, we hereby propose a third type of behavior, one that includes behaviors that are focused within an individual, or intrapersonally. In an effort to be consistent with the extant literature, we refer to these behaviors as *cyberloafing*. The important distinction here is that behaviors that fall under the category of cyberloafing may not always reach the threshold for "deviance." That is, they generally lack a clear external target, and have even been shown to be beneficial to the enactor and/or organization over time (see Anandarajan, Simmers, & D'Ovidio, 2011; Lim & Chen, 2012). Accordingly, we recognize three categories of cyberdeviant behaviors in this review: cyberloafing, CD-I, and CD-O. Within both the CD-I and the CD-O dimensions, we believe that further delineation can be made between behaviors that are largely focused on ***communication*** versus those that are focused on the ***acquisition*** of some "thing" (which generally refers to information) (see Figure 7.1).

As defined previously, cyberloafing behaviors do not negatively impact overall individual productivity, and when the intent of the behavior is not to "strike back" at the organization or a specific individual, the focus of the behavior can best be described as intrapersonal in nature. These behaviors can potentially provide long-term benefits to the individual (increased positive emotions, reduced stress) and the organization (increased overall productivity) that outweigh the short-term costs of time off task (Coker, 2013; Lim & Chen, 2012). But there are two important boundary conditions to these behaviors. First, the focus of the behaviors must remain introspective – if the motivation to enact one of these behaviors is to inflict harm on another person or the organization, it is no longer cyberloafing (i.e., intrapersonal). Second, if the behavior begins to impede the enactor's overall work productivity, then its categorization should be at the organizational level, akin to the "production deviance" area within the general counterproductive work behavior literature (Hollinger & Clark, 1982; Robinson & Bennett, 1995). These boundary conditions are represented by the arrows between the cyberloafing and CD-I/CD-O boxes in Figure 7.1.

In regards to interpersonally focused cyberdeviance, most of the CD-I behaviors included in our model are focused on communication with other individuals within the organization, for example cyberaggression, e-harassment (using digital resources to harass others),

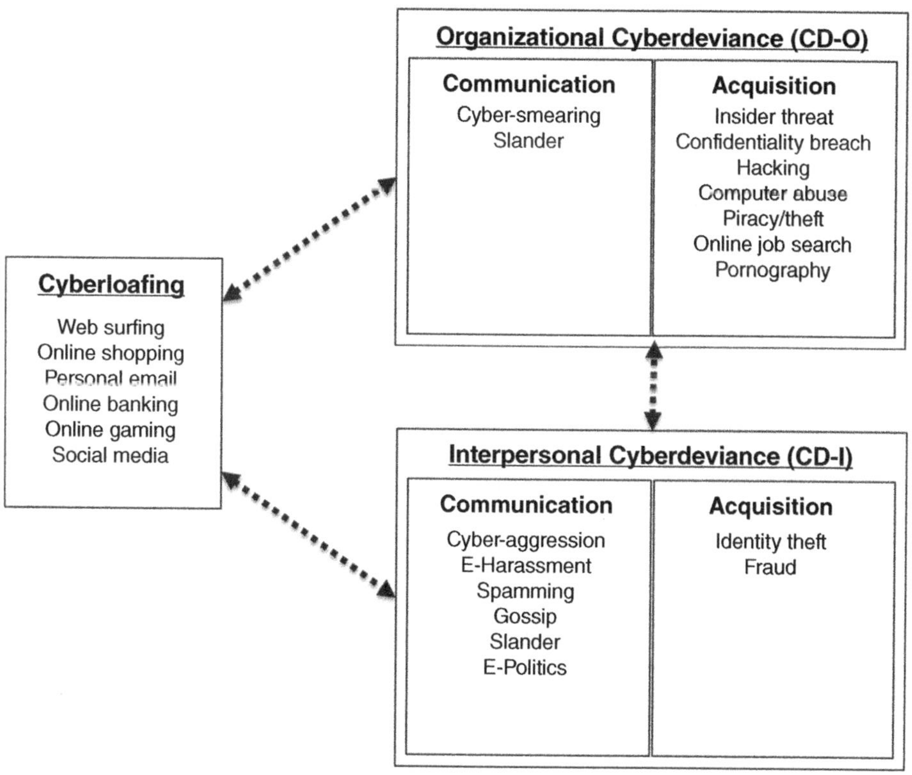

Figure 7.1 Typology of cyberdeviant behaviors.

spamming (the indiscriminate distribution of electronic messages without consideration of message appropriateness or receiver permission; Turban, Lee, King, & Chung, 2000), and gossip (via computer networks). A second category of behaviors where the target is another individual within the organization focuses on the acquisition of information or something of monetary value. For example, perpetrators of identity theft clearly are targeting a particular individual; however, the enactor of the cyberdeviance behavior is predominantly focused on acquiring information about the individual in the hopes of garnering financial benefits from the illegal use of that information. Likewise, those attempting to defraud others are also targeting individuals with their behavior, but again, the focus of the behavior is not simply to communicate with the target. Instead, the goal is the acquisition of information, goods/materials, or money in a surreptitious manner.

A similar subcategorization can be made within the CD-O category. Some cyberdeviance behavior that is targeted at organizations may focus primarily on communication, including cyber-smearing (cyber-libel) or slander. These behaviors generally refer to posting defamatory or disparaging comments about a company on message boards, in chat rooms, or on other sites. For these behaviors, the enactor of the behavior is primarily concerned with communicating a negative message about the organization to a larger audience. It is also possible for the target of this type of communication to be an individual; however, that would place the behavior within the CD-I dimension in our proposed framework. Nevertheless, the majority of cyberdeviance behaviors directed at an organization focus on the acquisition of information or digital resources (e.g., theft/fraud, confidentiality breach, and software piracy) as the end goal.

With this background, we review the current state of cyberdeviance research. We begin with our newly proposed category of intrapersonally focused cyberdeviance, or cyberloafing, followed by a review of the literature on CD-I and CD-O. Within each section, we review research within two groupings: antecedents of the cyberdeviant behavior and the outcomes.

Intrapersonal Cyberdeviance (Cyberloafing)

This category of studies focuses on behaviors that are not directed specifically towards an individual or the organization. These behaviors are referenced in the literature by a number of terms including cyberloafing, cyberslacking, nonwork-related computer use, personal web use, personal email use, and so forth. These behaviors appear to occur frequently in the workplace; one study (Lim & Chen, 2012) reported employees spend an average of 51 minutes per workday engaged in cyberloafing. In collecting the literature for our review, it was clear that this area of research had the most activity by a large margin within the overall cyberdeviance arena, as there are more than 50 empirical studies on this topic alone. We begin with a review of prior research that focuses on antecedents of cyberloafing.

Antecedents of cyberloafing

Individual differences Several studies have looked at the effects of demographic and personality-related differences on cyberloafing, and some tentative conclusions can be reached. First, the preponderance of evidence suggests that age and gender are both significant predictors of cyberloafing. Younger workers tend to be more predisposed to engaging in cyberloafing, personal Internet use at work, social media use, personal email, and other similar behaviors within this category (Askew et al., 2014; Everton, Mastrangelo, & Jolton, 2005; Gallagher, 2009; Garrett & Danziger, 2008; Henle, Kohut, & Booth, 2009;

Mastrangelo et al., 2006; Phillips & Reddie, 2007; Ugrin, Pearson, & Odom, 2007; Vitak et al., 2011); however, one study found a positive relationship between age and cyberloafing (Restubog et al., 2011). The results are equally consistent in regard to gender: males tend to engage in more cyberloafing than females (Andreassen, Torsheim, & Pallesen, 2014; Askew et al., 2014; Blanchard & Henle, 2008; Everton et al., 2005; Garrett & Danziger, 2008; Henle et al., 2009; Mastrangelo et al., 2006; Restubog et al., 2011; Vitak et al., 2011). But the reasons behind these findings are still highly speculative. For example, it is possible that younger workers are more comfortable with technology, and in turn, would be more likely than older workers to look toward technologically based alternatives for conducting deviant behavior. Although the results are rather consistent across studies, research on *why* these relationships exist for cyberloafing is still lacking.

Another individual difference domain that has received attention in the cyberdeviance literature is personality. In terms of the "big five" personality traits (conscientiousness, extraversion, agreeableness, openness to experience, and neuroticism; e.g., John & Srivastava, 1999), one finding that is consistent across multiple studies is that conscientiousness is negatively related to cyberloafing (Andreassen et al., 2014; Everton et al., 2005; Jia, Jia, & Karau, 2013; O'Neill, Hambley, & Bercovich, 2014). Some support exists for a positive relationship between cyberloafing and two other big five factors, neuroticism and extraversion (Andreassen et al., 2014; Jia et al., 2013). A fourth factor, agreeableness, may have a negative relationship with cyberloafing (Brock, Martin, & Buckley, 2013; O'Neill et al., 2014). Nevertheless, this line of research related to the big five is relatively limited.

Other research has focused on narrower personality traits as antecedents of cyberloafing. As would be expected, self-control has been found to be negatively related to cyberloafing (Gorsuch, 2014; Restubog et al., 2011). Relatedly, researchers have found significant positive relationships with cyberloafing and impulsiveness/sensation seeking (Everton et al., 2005), powerlessness/external locus of control (Blanchard & Henle, 2008; Blau, Yang, & Ward-Cook, 2006; Chen, Chen, & Yang, 2008), and Internet addiction/compulsive computer use (Chen et al., 2008; Mastrangelo et al., 2006). Together, these findings suggest that some employees may engage in cyberloafing behavior out of impulsivity or a sense of helplessness, rather than a purposeful or directed motivation.

Organizational policies/monitoring The situational variable that has received the most attention when studying cyberloafing appears to be organizational policies regarding appropriate computer use. Studies have focused primarily on the effects that these policies have on three outcomes: attitudes toward cyberloafing, cyberloafing behaviors, and justice perceptions. Multiple studies have examined the effectiveness of such organizational policies and found that they generally have a negative relationship with attitudes toward cyberloafing behaviors (Andreassen et al., 2014; D'Arcy & Devaraj, 2012; Henle et al., 2009; Strader, Fichtner, Clayton, & Simpson, 2011; Stratton, 2010). Training employees about appropriate use of computer resources also helps them to form more negative attitudes towards cyberloafing (Strader et al., 2011). Thus, the existing research provides evidence that a variety of organizational actions related to cyberloafing can be effective in influencing employee attitudes towards cyberloafing.

In turn, several studies have explored how organizational policies and monitoring efforts impact employee cyberloafing behavior directly. Multiple studies lend support to the notion that formal policies related to personal computer use at work (vs. a lack of policy) negatively impact employees' willingness to engage in cyberloafing behaviors (Andreassen et al., 2014; Henle et al., 2009; Jia et al., 2013). However, some specific policy characteristics appear to be more effective than others at deterring cyberloafing behaviors. Henle et al. (2009) found that periodic monitoring policies (vs. policies where monitoring occurs only for cause) are related to lower levels of cyberloafing. In a policy capturing study featuring students and working adults, Ugrin and Pearson (2013) found that for benign cyberdeviance behaviors such as checking personal email, a combination of policy decisions (including open acknowledgment of detection systems, public enforcement of policies, and the threat of termination) were necessary for reducing these types of behavior. For more severe behaviors such as accessing pornography, only detection systems seemed to influence employee self-reported behavior – additional safeguards and protocols did not have a significant impact. Indeed, the perceived risk of detection (Venkatraman, 2008) and the perceived consequences (Pee, Woon, & Kankanhalli, 2008) have both been supported as deterrents to cyberloafing behavior in other empirical studies.

Extant research also shows how organizational policies regarding appropriate computer use influence employees' perceptions regarding justice. Henle and colleagues (2009) found that zero-tolerance policies, progressive discipline, and the presence of appeals processes were all related to higher perceptions of policy fairness. However, organizations that use high levels of deterrence systems and procedures can create significant levels of mistrust and negative affect (i.e., feelings of unpleasantness and distress; Watson, Clark & Tellegen, 1988) among employees, as well as promote poor perceptions of organizational justice (see Zoghbi Manrique de Lara, 2006 and Ugrin & Pearson, 2013 for further discussions of these issues). Given the apparent trade-offs between increased policy and oversight and employee attitudes and perceptions, researchers may benefit from further exploration of the question "How much is too much?" with regard to different organizational policies. Indeed, this question applies to cyberloafing behavior in general.

Justice Beyond its relationship as an outcome to organizational policies on cyberloafing, organizational justice has also been an area of focus among researchers as an antecedent to cyberdeviance. Overall, the results remain somewhat mixed. Initial research suggested that cyberloafing is negatively related to perceptions of organizational justice (i.e., perceptions of justice related to process, procedures, and outcomes at an organizational level; Blau et al., 2006; Lim, 2002; Zoghbi Manrique de Lara, 2006) and interactional justice (i.e., perceptions of justice related to interpersonal interactions and communications from an authority figure within the organization; Venkatraman, 2008; Zoghbi Manrique de Lara, 2006). These findings were interpreted as evidence that employees use cyberdeviance as a means to retaliate against people or situations that are deemed unfair. However, several subsequent studies (Brock et al., 2013; Gallagher, 2009; Henle et al., 2009) found no relationship between organizational justice and cyberloafing behaviors. Restubog and colleagues (2011) showed significant negative relationships between various justice facets and cyberloafing, but these relationships were not statistically significant when controlling for the effects of age, gender, and time spent on the Internet for work purposes.

In their meta-analysis of counterproductive work behavior, Berry and colleagues (2007) found small to moderate relationships between counterproductive work behavior and justice – which seems to be the case for cyberloafing as well. Whereas some employees may view cyberloafing as a means to strike back at an individual or the organization for a perceived injustice, we suggest that perceived injustice plays a bigger role for more severe forms of cyberdeviance (CD-I and CD-O), and a relatively minor role for more benign behaviors such as cyberloafing.

Job and organizational characteristics Job- and organizational-level characteristics have been the focus of several studies in the cyberloafing literature. These studies have focused on three primary themes: socially constructed norms, stressors, and what researchers refer to as "facilitating conditions." First, socially constructed norms and roles appear to have a strong relationship with cyberloafing intent and behavior, in that the proclivity to engage in such behaviors is much higher when individuals perceive that these activities are accepted within the organizational culture or by their peers (Askew et al., 2014; Liberman, Seidman, Mckenna, & Buffardi, 2011; Moody & Siponen, 2013; Pee et al., 2008; Stratton, 2010). However, Askew et al. (2014) suggest that the type of norm being evaluated must be taken into account: descriptive norms (the extent to which others actually engage in cyberloafing activities) were related to cyberloafing activities, whereas prescriptive norms (the extent to which others would approve of cyberloafing activities) had no relation to cyberloafing.

Second, researchers have also investigated how job-related stressors can impact cyberloafing. Andreassen et al. (2014) found that challenge stressors (i.e., "obstacles to be overcome in order to learn and achieve"; LePine, Podsakoff, & LePine, 2005, p. 765) – operationalized in their study as positive work challenges and quantitative demands – were negatively related to social network use at work. Conversely, researchers have posited and found that hindrance stressors (i.e., demands that unnecessarily thwart performance and personal growth, including role overload, role ambiguity, and role conflict) are positively related to cyberloafing (Nixon & Spector, 2013; RuningSawitri, 2012). In the same vein, Ugrin et al. (2007) surveyed employees from multiple countries and discerned three different categories of worker based on their level of cyberloafing: "focused workers" (no cyberdeviance), "moderate emailers" (largely email and shopping), and "superslackers" (prominent levels of shopping, chatting, emailing, and/or media use). Among those who engaged in high levels of cyberloafing, the need for stress relief was cited as a motivating factor for cyberdeviance.

Ugrin et al. (2007) and Garrett and Danziger (2008) also found support for "facilitating conditions": high levels of autonomy (i.e., lack of oversight of computer activities) served as an enabling factor for cyberloafing. Indeed, those who occupy high-status and managerial positions with relatively more autonomy engage more significantly with cyberloafing activities at work than other groups (Andreassen et al., 2014; Garrett & Danziger, 2008) in spite of the fact that they tend to have more negative perceptions of cyberloafing acceptability (Andreassen et al., 2014; Strader et al., 2011). The term facilitating conditions can also refer to the work environment's propensity to facilitate employees' perceived ability to enact cyberloafing behaviors. Results in this area are somewhat mixed. Askew et al. (2014) found that the perceived "ability to hide" cyberloafing behavior was strongly related to the behavior itself. Pee et al. (2008) report that facilitating conditions (which they operationalized as two-dimensional: situational [e.g., conducive setting, access to resources] and internal [self-efficacy]) had a weak but positive relationship with both intent to engage in cyberloafing and actual cyberloafing. Finally, Moody and Siponen (2013) also investigated facilitating conditions (operationalized as the combination of organizational policies and "ready access" to the Internet) and found no effects on cyberloafing behavior or as a moderator of the relationship between intent and actual cyberloafing behavior. Overall, these results suggest that more research is needed to better determine the aspects of facilitating conditions that impact or do not impact behaviors.

Work–life balance An additional context in research within the cyberloafing domain relates to the juxtaposition of work and nonwork lives. The extant research supports a connection between the nonwork domain and cyberloafing behaviors via competing demands and cross-domain resource depletion. Anandarajan et al. (2011) interpreted one part of their findings regarding personal web usage at work as a desire to maintain work–life balance via the accomplishment of competing personal demands during work time (and using work resources). Along these lines, König and de la Guardia (2014) found that the level of an individual's "private demands" (i.e., nonwork obligations) were positively related to personal Internet use at work. Other researchers have viewed the relationship between work–life balance and cyberloafing from the perspective of resource depletion, in which demands from an individual's nonwork life interfere with his or her ability to self-regulate at work. Using this lens, Gorsuch (2014) found that work–family conflict was positively related to personal Internet use at work. Basing their work on the ego depletion model of self-regulation, Wagner, Barnes, Lim, and Ferris (2012) also found that episodes of sleep loss and sleep interruptions were positively related to more frequent and longer periods of cyberloafing at work. Although more research is needed to help clarify the relationships between cyberloafing and various aspects of work–life balance, as well as the directionality and mechanisms behind these relationships, the literature supports a link between the two.

Outcomes of cyberloafing

Surprisingly, research on the outcomes of cyberloafing is quite limited, but areas that have been empirically investigated to date include individual/organizational performance, work–life balance/conflict, affect, stress relief, and resource replenishment. From an individual performance perspective, a few studies show that performance rating by supervisors (Venkatraman, 2008) and self-report (O'Neill et al., 2014) is negatively related to cyberloafing. Although cyberloafing is purported to be a response to work–life imbalance, two studies (Gorsuch, 2014; König & de la Guardia, 2014) found that cyberloafing did not have a positive impact on work–life balance. In one of the few studies to empirically measure the effects of cyberloafing at the organizational level, Valli (2004) investigated nonbusiness-related Internet use in multiple sites. Internet log files from employee computers in three organizations were analyzed and all online activities were classified as business-related or nonbusiness-related, based on the type of website accessed. The amount of unacceptable Internet usage as a percentage of all Internet usage at these organizations ranged from 20.6% (medium-sized state government agency) to a staggering 74.6% (large university department). These results allowed the author to emphasize the large monetary impacts of unacceptable Internet usage.

Although it stands to reason that cyberloafing can have negative implications for individual and overall organizational productivity, empirical validation of this conclusion is lacking. A new stream of research – focused on the enactor of the behavior, as opposed to the organization – paints a rather different picture. This idea was first proposed by Mastrangelo et al. (2006), who developed a 41-item battery for "personal use of work computers," featuring a two-factor structure: counterproductive use and nonproductive use. The second category encompassed "behaviors that are not destructive, yet not directly productive" (Mastrangelo et al., 2006, p. 735). The authors clarify that although this type of deviant computer use is "usually a means of distraction" for some individuals, it may be "a means of destruction" (p. 735) for others, a notion that is reflected in our overall typology. That is, these behaviors can move from being rather innocuous (cyberloafing) to harmful (CD-I/CD-O) when enacted with a destructive purpose or performed with increased regularity (see Figure 7.1).

Lim and Chen (2012) investigated the effects of two specific nonwork-related behaviors (browsing and email communication), and found that browsing was positively related to positive affect (i.e., feelings of enthusiasm, activity, and alertness; Watson et al., 1988), whereas nonwork-related email communication was found to be positively related to negative affect. Thus, Internet browsing may give employees a needed "break" from work, while nonwork email may remind employees of additional responsibilities or stressors within their personal lives. Still, further research is needed to explore the causal pathways for these relationships.

Anandarajan et al. (2011) developed a battery of items for personal web usage at work; four clusters emerged from their data, depending on the level of perceived benefits to the individual or the organization. Most relevant to this discussion is the "work/family" cluster, defined as behaviors with direct benefits to the individual and potentially indirect benefits to the organization. Similar to previous scholars, the authors proposed that these behaviors can provide stress relief to individuals, promote psychological wellbeing, or be viewed as restorative behavior in maintaining a work–life balance or in response to working on personal time.

Relatedly, Coker (2013) experimentally manipulated nonwork-related web surfing and found that behaviors that may be traditionally classified as deviant may serve functionally to replenish individuals' attentional resources for subsequent work tasks. In an experiment, subjects were asked to perform several waves of a computerized vigilance task with 5-minute periods in between waves, during which participant activities varied based on their assigned experimental condition. Ultimately, task vigilance over time was found to be highest when participants took breaks to engage in workplace Internet leisure browsing (browsing Facebook) than when they engaged in a cognitively challenging online task (comparing health insurance plans) during breaks, took a break that did not involve using the Internet, or took no break at all. In total, we see an emerging picture of cyberloafing potentially leading to *positive* outcomes, both for the organization and the individual.

Interpersonal Cyberdeviance

We now turn to the second dimension of cyberdeviant behavior in our typology – CD-I. As noted previously, CD-I is behavior that takes place using ICTs, violates workplace norms, and either threatens to harm or results in harm to individual employees. CD-I behaviors can be categorized as involving communication (such as communicating mistreatment in the form of bullying, incivility, aggression, or sexual harassment) or the acquisition of information or objects (such as identity theft or fraud). We identified 21 empirical studies in the management or information technology literature focusing on communication-related CD-I, however, we were unable to locate any empirical studies that examined acquisition-based CD-I. This highlights an important gap in the existing research and our review focuses on communication-focused CD-I.

Because CD-I is technologically based, engaging in such behaviors involves written messages such as emails and texts, as opposed to verbal exchanges. Existing research on CD-I suggests a number of characteristics may be associated with perceiving a computer-mediated message as deviant. For example, Richard, Gacey, Young, and Lopez-Matos (2015) found that messages containing threatening language, profanity, use of all capital letters (indicating shouting), excessive punctuation, or unfair accusations were classified as aggressive by recipients and objective coders of the messages. Turnage (2008) adds that emails might be classified as deviant if they contain insults, sarcasm, or offensive statements, and Ritter (2014) notes that messages containing inappropriate jokes or sexually oriented comments are more likely to be perceived as deviant. Francis, Holmvall, and O'Brien (2015) suggest that an email may be perceived as CD-I if it violates norms for communication by being less polite, includes poor spelling or grammar, or is missing a greeting, subject, or salutation. Taken together, these studies underscore the defining features of CD-I: violation of norms and threats to harm employees at work.

Besides the characteristics of a message, certain features of the online environment may also make employees likely to either engage in CD-I or perceive a message as CD-I.

For example, email messages can easily be amplified from a one-on-one conversation to a large number of others by copying others or forwarding messages to one's entire address book (Flanagin, Pearce, & Bondad-Brown, 2009; Weatherbee & Kelloway, 2006). Moreover, the online disinhibition effect (Suler, 2004) suggests that individuals may be more likely to self-disclose inappropriate things or behave in ways that they would not usually behave in a face-to-face setting. This effect is attributed to the lack of media richness in some online communications (Ford, 2013). That is, low-richness online communications feature a greater perception of anonymity and a lack of nonverbal cues to guide behavior (Kiesler, Seigel, & McGuire, 1984; Ritter, 2014). Furthermore, these characteristics of online environments may make message recipients likely to interpret ambiguous messages as more negative than senders intended (Byron, 2008). Finally, a lack of organizational policies regarding online behavior and absence of monitoring may also contribute to the likelihood that individuals will engage in CD-I.

The prevalence rate of CD-I in the workplace appears to vary from one study to another, due perhaps to some degree to the form of interpersonal deviance being studied. For example, in a study of workplace cyberbullying (i.e., repeated intentional negative acts via technology that are directed at an organizational member who cannot easily defend themselves), Privitera and Campbell (2009) found that 10.7% of a sample of male manufacturing employees from Australia reported experiencing cyberbullying at work *in the past six months.* Another study of cyberbullying (Baruch, 2005) found a similar prevalence rate (9.2% *ever*) among a sample of 649 employees from the United Kingdom. However, Ford (2013) found a higher rate of cyberaggression in a sample of 492 US and Canadian employees obtained via StudyResponse, with 58% of the sample experiencing cyberaggression at least once *in the past year*. Rates of a more subtle form of cyberdeviance known as cyber-incivility (i.e., rude and discourteous behaviors that occur through information and communication technologies) also tend to be higher than rates for cyberbullying: Giumetti, McKibben, Hatfield, Schroeder, and Kowalski (2012) found that approximately 30% of their sample of 614 employees in southeastern United States reported experiencing at least one act of cyber-incivility *in the past six months.* Lim and Teo (2009) reported a much higher figure, 91% *in the past year*, among a sample of 192 financial employees from Singapore. Differences in reporting time frames, response scale options, and study locations make it hard to discern whether differences in response rates are due to the construct in question or to these other methodological issues. However, it is clear that various forms of CD-I are prevalent in many organizations today.

Existing face-to-face interpersonal deviance research has organized the various forms of deviance in terms of their level of intensity and intentionality (Andersson & Pearson, 1999; Hershcovis, 2011; Tepper & Henle, 2011). For example, the construct of incivility is defined as rude or discourteous behavior having low intensity and ambiguous intention to harm the target (Andersson & Pearson, 1999). In contrast, aggression and bullying are thought to have high intensity and unambiguous intent to harm others (Einarsen, 2000). Despite these differences in conceptualization, several researchers have argued that all of these different forms of deviant behavior can be classified as workplace aggression because they share common antecedents and outcomes, and suggest that doing so may aid future research in this area (Hershcovis, 2011). Therefore, in this chapter we review the various forms of aggressive CD-I behavior together rather than separately.

From a theoretical perspective, a large majority of this research is based in the stressor-strain model (e.g., see Bowling & Beehr, 2006), which argues that a number of individual and environmental factors serve as antecedents to workplace aggression (e.g., perceptions of injustice, organizational policies, and workplace norms). The experience of workplace aggression, moreover, is perceived as a stressor that is negatively related to a host of behavior- or attitude/perception-based outcomes (including burnout, counterproductive workplace behaviors, job satisfaction, and turnover intentions). Therefore, this review of CD-I is organized around two dimensions: (1) the antecedents versus outcomes of CD-I; and (2) whether the variables can be classified as behavior-based or attitude/perception-based.

Antecedents of CD-I

Perception-based antecedents Research in this area has tended to focus largely on attitudes/perceptions as antecedents of engaging in CD-I (i.e., aggression, bullying, incivility, or sexual harassment). For example, two studies examined perceptions of organizational justice as a predictor of engaging in CD-I (Mehdad & Nezhad, 2014; Weatherbee, 2007). In a Canadian government-based sample, Weatherbee (2007) found that perceptions of interactional injustice (or being treated impolitely or with disrespect by others at work) were not significantly related to engaging in cyber-aggression. In contrast, Mehdad and Nezhad (2014) found that, among a sample of Iranian employees in the education industry, both procedural injustice (i.e., perceiving that procedures in the organization are not being carried out fairly) and interactional injustice were significantly related to engaging in cyberbullying.

Weatherbee (2007) also tested two other perception-based variables as antecedents of cyber-aggression (organizational policies for email use and workplace norms), but found that neither variable was related to engaging in cyber-aggression. One final study, which utilized a sample of graduate and undergraduate students from southeastern United States, examined perceptions of the online environment as a predictor of engaging in CD-I (Ritter, 2014). Ritter found that participants who perceived the online environment as being a place where one can be anonymous, a place that is safe to communicate, and as a place where one is not being monitored by others were more likely to self-report engaging in cyber sexual harassment.

Behavior-based antecedents Other research has examined behaviors as potential antecedents of engaging in CD-I. For example, several studies have found that participants who reported experiencing cyber mistreatment (in the form of uncivil emails or other cyber-aggression) were more likely to perpetrate cyber-aggression in return (Francis et al., 2015; Martinell, 2010; Weatherbee, 2007). In addition to these behavioral antecedents, Francis et al. (2015) found that participants who were experiencing a higher level of workload also tended to engage in more cyber-aggressive behavior.

Demographics Several demographic variables have been examined as antecedents of perceiving messages as CD-I. For example, Turnage (2008) found that graduate students perceived aggressive emails as more intense than undergraduate students, but found no gender differences in these perceptions. Martinell (2010) also did not find differences between males and females in terms of their perceptions of cyber-aggression. However, both Ford (2013) and Giumetti et al. (2012) found that females reported experiencing greater levels of overall cyber-aggression than males. On the perpetration side, Lim and Teo (2009) found that male supervisors were more likely to perpetrate cyber-aggression than female supervisors, but there were no differences in rates of experiencing cyber-aggression between males and females. In total, generalizable relationships with regard to demographic variables and CD-I are yet to be determined.

Personality Personality variables have also been studied as individual difference factors in CD-I studies. For example, multiple studies have linked the trait of neuroticism (or the tendency to feel anxious, moody, or easily upset) to a greater likelihood of experiencing cyber-aggression (Giumetti et al., 2012; Weatherbee, 2007). Researchers have speculated that individuals with higher levels of neuroticism are more likely to perceive stressful experiences in their environment as harmful and respond more intensely to stressful stimuli than individuals with lower levels of neuroticism (Spector, Zapf, Chen, & Frese, 2000). Other personality variables that have been examined include positive affect and Machiavellianism. Hong et al. (2014) found that positive affect was not related to experiences of cyber-aggression among a sample of technology industry employees. With regard to Machiavellianism (i.e., a stable pattern of being deceitful and manipulative, Paulhus & Williams, 2002), Tripathi and Lim (2014) found that individuals with high levels of this trait tended to respond to experiences of cyber-aggression with direct and indirect revenge, whereas individuals low in this trait tended to respond with avoidance.

Outcomes of CD-I

The large majority of research in this area has focused on the outcomes of experiencing CD-I from the victim's perspective. The primary outcomes under study include employee attitudes or employee behaviors. From an attitudinal perspective, constructs that have been investigated in the CD-I literature include job satisfaction, organizational commitment, and engagement. Across multiple studies, participants who reported experiencing CD-I were likely to have lower levels of job satisfaction (Baruch, 2005; Giumetti, Saunders, Brunette, Graham, & DiFrancesco, 2016; Weatherbee, 2007), decreased organizational commitment (Lim & Teo, 2009), and decreased task engagement (Doll & Giumetti, 2015; Giumetti et al., 2013). From a theoretical perspective, affective events theory (Weiss & Cropanzano, 1996) can be used to understand the mechanism through which experiencing stressors, such as CD-I, can be linked with affective reactions and subsequent job attitudes and behaviors.

The range of behaviors that have been studied in association with CD-I includes task performance, counterproductive workplace behaviors, absenteeism, turnover intentions, and a variety of physical and mental health behaviors (Baruch, 2005; Doll & Giumetti, 2015; Ford, 2013; Giumetti et al., 2012; Giumetti et al., 2013; Hong, Chien-Hou, Hwang, Hu, & Chen, 2014; Lim & Teo, 2009; Park, Fritz, & Jex, 2015; Tripathi & Lim, 2014; Weatherbee, 2007; Young, Pita, White, Hickman, & Richard, 2012). For example, among a sample of employed adults from the United States and Canada, Ford (2013) found that cyber-aggression was a significant predictor of decreased psychological health, even after controlling for the media richness and perceptions of anonymity of the message. Along the same lines, Park et al. (2015) found that experiencing daily cyber-incivility was linked with affective and physical distress for individuals in the afternoon as well as the following morning. Additionally, Young et al. (2012) studied a sample of working adults and found that cyber-aggression predicted unique variance in interpersonal counterproductive workplace behaviors and strain over and above face-to-face workplace bullying.

Finally, in an experimental study of cyber-aggression among a sample of students from the southeastern United States, Giumetti et al. (2013) found that receiving a cyber-aggressive email was associated with lower performance on a simulated work task than receiving a cyber-supportive email, and this effect was mediated by the level of emotional energy participants reported. Taken together, these results suggest that experiencing CD-I in the workplace is linked with a host of negative outcomes for employees.

Organizational Cyberdeviance (CD-O)

The final dimension in our typology – cyberdeviance directed at the organization (CD-O) – includes a wide variety of behaviors, broadly categorized as unauthorized or illegitimate access to company resources or information using technological resources (Salem, Hershkop, & Stolfo, 2008). As depicted in Figure 7.1, the majority of CD-O behaviors concern acquisition of materials or goods. In this category, both insider threat (i.e., when individuals authorized to use information systems use their authorization to perpetrate harm; Randazzo, Keeney, Kowalski, Cappelli, & Moore, 2004) and computer abuse (i.e., unauthorized and deliberate misuse of assets of organizational information systems; Straub, 1990) have been used as blanket terms for behaviors such as confidentiality breach, computer hacking, software piracy/theft, online job searching on company time, and viewing pornographic material at work. In contrast, the communication-based CD-O behaviors have been largely neglected. We list two communication behaviors, cyber-smearing and slander, which refer to harassing or defamatory comments directed towards organizations online. However, studies on such behaviors to date (e.g., Slotkin & Vamosi, 2011; Workman, 2010, 2012) have focused solely on outsiders (i.e., those not employed at the organization) as perpetrators of the behavior rather than insiders (i.e., those employed at the organization), and these studies will not be reviewed here.

Antecedents of CD-O

Research on CD-O has focused on three broad categories of antecedents: individual difference variables, work attitudes (such as justice perceptions, trust, and commitment), and organizational monitoring and policies.

Individual differences A variety of individual demographic variables (e.g., age, gender), job related variables (e.g., job type, job level), and extra-organizational factors (e.g., financial difficulties, relationship problems) have been studied in relation to CD-O. Specific reports on gender as an antecedent to CD-O are mixed, as some studies report no gender differences (e.g., Mishra, Akman, & Yazici, 2006) whereas others found women were less likely to engage in computer abuse than their male counterparts (Lowry, Posey, Bennett, & Roberts, 2014). Age clearly has a significant impact on computer abuse, as younger people are consistently reported as engaging in more types of CD-O (e.g., Gan & Koh, 2006; Lowry et al., 2014; Mishra et al., 2006). In terms of job-related characteristics, software professionals are more likely to pirate software and access computing or telecommunication resources without permission than other professional workers (Oz, 2001). However, in a study of cases from the banking and finance sector, a major conclusion was that the perpetrators of CD-O did not share a common profile in terms of job title or manager/co-worker assessments (Randazzo et al., 2004). Similarly, Mishra and colleagues (2006) did not find any significant relationships between income, education, or job experience and the specific CD-O of pirating software.

It appears that extra-organizational factors such as marital and familial problems, personal finances, and substance addiction can also influence the intensity of disgruntlement and dissatisfaction with work, which may lead to cases of CD-O. Shropshire (2009) analyzed 62 intentional security breaches and categorized the breaches into two outcomes: acts of espionage (i.e., use of information systems and resources for unauthorized printing, copying, concealing, downloading, or saving) and acts of sabotage (i.e., disabling security systems, deleting data, destroying hardware or software, or corrupting organizational information systems). Espionage behavior was most highly correlated with personal financial changes and relationship changes (e.g., recent divorce or marriage, general relationship strains). Sabotage behavior, in contrast, was most highly correlated with job changes (i.e., recent firing, demotion, transfer, resignation, retirement, or lay-off), substance abuse, and relationship changes.

Attitudes Studies focusing on attitudes as antecedents to software piracy (i.e., unauthorized software copying) as a specific form of CD-O are most often grounded within ethical frameworks. For example, Taylor and Shim (1993) found that business executives and faculty members both had similar attitudes regarding the ethicality of pirating software, although faculty reported that they copied software more frequently than business executives. In a study of attitudes on software piracy, individuals who frequently pirated software generally possessed low ethical attitudes, had negative attitudes toward software publishers, had low regard for property rights, and were more likely to be students (Gan & Koh, 2006). Individuals with a technological background also reported being offered illegal copies of software at work more often than those with business backgrounds (Lending & Slaughter, 1999). Peace, Galletta, and Thong (2003) examined attitudes toward illegally copying software, although their focus was on the impact of the cost of the software and characteristics related to punishment. They found that software cost is positively related to attitudes toward piracy, meaning that for expensive software, the perceived benefits of piracy are increased by the high monetary value of the software itself.

Information on motivation and attitudes related to broader forms of CD-O rely on descriptive information drawn from industry reports. For example, the primary motivation for CD-O in a case analysis of the banking and finance sector appears to be financial gain (Randazzo et al., 2004). Beyond financial gain, about one-quarter (23%) of the cases studied were motivated by revenge, whereas lesser motives concerned dissatisfaction with the company management, culture, or policies (15%) and desire for respect (15%). In a separate report drawn from public records and Secret Service reports, 85% of the insiders committing CD-O experienced grievances or a negative work related-event before carrying out attacks. The most frequently reported motive for CD-O was revenge against the organization or members within the organization (Keeney et al., 2005). Behavioral models targeting prediction of insider threats suggest that the factors that are most likely to predict the risk of computer misuse are disgruntlement, disregard of rules, and disengagement (Greitzer, Kangas, Noonan, & Dalton, 2010).

Organizational trust and organizational commitment are frequently positioned as antecedents of CD-O. Studies have shown that lack of trust in one's organization leads to increased reports of CD-O (Lowry et al., 2014; Posey, Bennett, & Roberts, 2011). Moreover, the effects of distrust within organizations can be exacerbated by environmental instability. That is, CD-O was greatest when respondents reported both low stability in the organizational environment coupled with a lack of trust in the organization (Posey, Bennett, & Roberts, 2011). Trust has also been shown to mediate the relationship between explanation adequacy and CD-O such that adequate explanation of computer policies led to higher levels of trust, which was related to lower levels of CD-O (Lowry et al., 2014). Organizational bonds, such as attachment to the job, organizational commitment (Cheng, Li, Li, Holm, & Zhai, 2013), involvement, and loyalty (Lee, Lee, & Yoo, 2004) are all negatively associated with CD-O intentions.

Organizational policies/monitoring Studies focusing on deterrence examine whether organizational and legal polices have a positive or negative impact on CD-O. The main findings of studies related to deterrence are that the relationship between implementing computer monitoring systems and policies against computer misuse and CD-O is complicated at best. Computer monitoring has been shown to both increase and decrease CD-O (D'Arcy, Hovav, & Galletta, 2009; Posey, Bennett, Roberts, & Lowry, 2011) and the perceived severity and certainty of sanctions as a result of abusing computer-based resources appear to have inconsistent effects on enactment of CD-O. A notable challenge in reviewing this literature is the multiple operationalizations of both deterrence and CD-O. Deterrence safeguards ranged from detection and monitoring activities (e.g., Posey, Bennett, Roberts, & Lowry, 2011; Straub, 1990) to security awareness programs and codes of ethics (e.g., Harrington, 1996).

In terms of monitoring behaviors, Posey, Bennett, Roberts and Lowry (2011) found that higher levels of computer monitoring by the organization were positively related to reported CD-O and were influenced by feelings of privacy invasion and procedural injustice. In contrast, D'Arcy and colleagues (2009) found that computer monitoring, in conjunction with other policies, was negatively related to CD-O. Here the authors argued that computer monitoring was an effective deterrent because it increased the perceived consequences of misusing computer resources.

Regarding broader organizational policies, it appears that the awareness and presence of policies specifically prohibiting computer misuse does decrease CD-O. D'Arcy et al. (2009) found that user awareness of security policies and security education, training, and awareness decreased CD-O. Similarly, an information system-specific code of conduct decreased computer sabotage judgment and CD-O intentions (Harrington, 1996) and awareness of organizational rules against piracy had a negative impact on CD-O in both government and private sector organizations (Akman & Mishra, 2009). Straub (1990) found that information system security efforts, dissemination of information about penalties, guidelines for acceptable system use, and policies for system use all deterred CD-O. Only one study found contradictory results. In this study, both business executives and business professors were unaffected by the presence of a formal policy against software copying (Taylor & Shim, 1993).

A number of information science researchers have focused on deterring employee com puter crime using concepts from general deterrence theory. Central to this theory is the important role of *sanctions* associated with the crime; sanctions are assessed in terms of perceived *certainty* and perceived *severity*. The findings in regard to the role of sanctions in CD-O yield mixed results. Two studies have found that both the severity and certainty of sanctions deterred CD-O (Straub & Nance, 1990) and were related to unfavorable attitudes toward software piracy (Peace et al., 2003). In other studies assessing the impact of organizational control measures, only the perceived severity of sanctions, compared with certainty of sanctions, was related to CD-O intentions (Cheng et al., 2013; D'Arcy et al., 2009). In contrast, Lowry and colleagues (2014) found no effect for sanction severity and certainty on CD-O when the model included other individual and organizational factors.

There is also some evidence that the effect of deterrents in terms of perceived certainty and severity of punishment may be affected by national culture. Hovav and D'Arcy (2012) examined the different effects of procedural countermeasures (e.g., security policies and courses) and technical countermeasures (e.g., tracking data and security audits) on perceived sanctions. Each countermeasure was associated with increased perceptions of both the severity and the certainty of sanctions. However, in terms of outcomes, the *severity* of sanctions had a negative effect on CD-O intentions only for the US sample, while the *certainty* of sanctions acted as a deterrent for CD-O only in the Korean sample.

Deterrence theory also specifies an additional construct, *informal sanctions*, which refer to shame, morality, and societal norms, as forces which may discourage inappropriate behavior. However, fewer studies have examined the impact of informal sanctions as deterrents of CD-O. It appears that norms and social influences (Cheng et al, 2013; Lee et al., 2004; Li, Zhang, & Sarathy, 2010) are more effective deterrents than shame (Hu, Xu, Dinev, & Ling, 2011). For example, moral beliefs, or the beliefs that CD-O is morally unacceptable, are negatively associated with CD-O intentions (D'Arcy et al., 2009; Hovav & D'Arcy, 2012).

CD-O outcomes

Similarly to cyberloafing, research examining organizational or personal outcomes of CD-O is limited; in fact, we could not find empirical studies addressing organizational or personal outcomes of CD-O specifically from individuals acting from inside the organization. Some conclusions on organizational outcomes of CD-O can be drawn from a summary of cases identified through public media reports or criminal justice databases by Randazzo and colleagues (2004). The authors report that CD-O is most often associated with financial loss for the organization, a result of either direct theft or cost incurred in repairing damage done to online systems as a result of CD-O. Other reported harm to organizations included damages to business operations and damage to organizations' reputations.

It is important to note, however, that these cases are drawn from examples where individuals were caught and subsequently charged with a crime. There are undoubtedly many more instances of CD-O where perpetrators have not been caught or have not been formally charged with a crime. It is also possible that many employees who are caught engaging in CD-O experience a variety of organizational sanctions including warnings, probation, or termination, rather than facing criminal charges. Another key point to consider in studying CD-O outcomes is that many case studies are based on employees with a clear intent to engage in CD-O. There are likely many more sloppy or otherwise unmotivated employees who engage in CD-O (e.g., failure to maintain security or confidentiality) through accidental or otherwise careless actions. A clearer understanding of the outcomes of CD-O should include outcomes related to "careless" CD-O.

Future Research

We conclude with a brief discussion of topics for future research that we believe are fruitful avenues for continued exploration. Before delving into each of the cyberdeviance dimensions in our typology, one potential direction for future research that spans all three dimensions is to further elaborate on the underlying psychological processes that relate to workers engaging in various forms of cyberdeviance. For example, what motivational variables (such as goals or expectations) or individual difference variables (such as self-efficacy or self-regulation) might help to explain why an individual engages in cyberdeviant behavior at work? Generally speaking, the current literature has primarily focused on investigating constructs that predict the enactment of a particular cyberdeviant behavior, and to a lesser extent, what outcomes result from cyberdeviance. However, more complex theorizing and empirical testing of mediation/moderation models that dive deeper into how and why individuals exhibit cyberdeviance is needed across all of the dimensions. Studies that examine intraindividual differences in the willingness to engage in cyberdeviant behavior, including differences in the timing and rate of the behavior, might also prove particularly useful in the effort to better understand the underlying psychological processes that drive cyberdeviance.

Cyberloafing

There are a host of directions that future research might take specifically in relation to cyberloafing. One area might be to further examine the relationship of individual difference variables as antecedents of cyberloafing behavior. For example, whereas we know that younger and male employees tend to engage in more cyberloafing than older and female employees, less is known about why these differences exist. Qualitative research may help to uncover some of the reasons for these differences and could help to guide policy formation and intervention efforts. In addition to examining demographics, more focus is needed on the role of personality traits as antecedents of cyberloafing. Whereas we understand how certain big five personality traits and narrower traits (e.g., self-control, sensation seeking) might be related to cyberloafing behavior, less is known about other personality variables. Future research might examine how core self-evaluations (defined as a central assessment of one's worth, efficacy, and competence as a person; Judge, Erez, Bono, & Thoresen, 2003) relate to cyberloafing behaviors. Emerging research in the face-to-face deviance literature suggests that core self-evaluations is linked with both individual and organizational deviance through emotional exhaustion and trust in one's supervisor (Neves & Champion, 2015). Beyond core self-evaluations, future research should examine other possible personality variables that may play a role in cyberloafing behavior, such as narcissism (Judge, LePine, & Rich, 2006) and Machiavellianism (Zagenczyk, Restubog, Kiewitz, Kiazad, & Tang, 2014).

As more organizations put policies in place to deter employees from engaging in cyberdeviance, research may need to examine how such increased policy attention may impact employee attitudes and behaviors. Paschal, Stone, and Stone-Romero (2009) explored the role of email policy restrictiveness and policy justification on perceptions of privacy invasion and fairness. The authors found that having a restrictive email policy (i.e., employees may use email for business purposes only) was seen as both more invasive and less fair than a less restrictive email policy (i.e., employees may occasionally use email for personal use if it does not lead to increased costs for the company). However, providing justification for the policy increased fairness perceptions. Thus, preliminary evidence suggests that restrictive policies without justification are likely to be viewed negatively by employees. Further research is needed to examine other policy features and their impact on employee attitudes and behaviors related to cyberdeviance.

As noted earlier, a host of job characteristics have been identified that appear to be linked with increased cyberloafing behavior, including norms, stressors, and facilitating conditions. Research may be needed that incorporates cyberloafing within one of the theoretical stressor–strain models, such as the demand–control–person model (Rubino, Perry, Milam, Spitzmueller, & Zapf, 2012). In this model, increased job demands may be counteracted by having higher levels of job control (i.e., autonomy or decision latitude) as well as higher levels of emotional stability, thus lessening the strain outcomes for employees. When placed into this theoretical model, researchers may be able to explore the extent to which cyberloafing behaviors may help to mitigate the negative strain outcomes for employees experiencing high levels of job demands, and the role that autonomy and personality play in this relationship.

Another area for future research deals with the relationship between cyberloafing and work–life issues. One direction that researchers may take is to incorporate cyberloafing behavior within the framework of work–family boundary dynamics (Allen, Cho, & Meier, 2014). Boundary dynamics deal with the lines of separation between work and family roles and how individuals preserve and move across these lines. Increasing technology use by employees has made work–family boundaries more permeable, enabling employees to spend time cyberloafing at work and communicating with family members. Future research may consider exploring several individual difference variables that have been linked with work–family boundaries, such as boundary control (i.e., perceived control over work–nonwork boundaries; Kossek, Ruderman, Braddy, & Hannum, 2012) and integration/segmentation preferences (i.e., the degree to which employees desire to keep work and family separate; Kossek, Lautsch, & Eaton, 2006). Individuals with greater boundary control and segmentation preferences may be less likely to spend time cyberloafing and communicating with family (see also Rice, 2017, Chapter 9 in this volume, and Van Yperen & Wörtler, 2017, Chapter 8 in this volume).

One final and important avenue for future research on cyberloafing may be to understand the threshold at which cyberloafing stops being acceptable and starts being seen as counterproductive. At low levels, cyberloafing behavior may provide restorative benefits for employees (Coker, 2013) and may help employees to perform better on subsequent tasks. However, at higher levels, cyberloafing may be seen as destructive and counterproductive (Mastrangelo et al., 2006). The question remains – how much time spent cyberloafing is too much?

CD-I

The extant literature suggests that experiencing CD-I results in negative outcomes for individuals. However, our understanding of these relationships is largely limited to single time-point, self-report methodologies. This limits our ability to make causal inferences about whether experiencing CD-I does lead to negative outcomes, or whether there might be some third variable explanations or reverse-causation. Therefore, as is the case in much of the cyberdeviance literature, there is a need for research exploring how these relationships play out over time. More research is also needed on CD-I associated with acquisition, such as online identity theft or fraud within organizations. There are many questions that could be addressed, including determining the prevalence of these behaviors in the workplace, as well as understanding the environmental or individual difference factors that might predict acquisition-focused CD-I. Research on face-to-face employee theft suggests that employees with lower levels of moral development are more likely to engage in theft behaviors, as are employees who come from organizations without ethics policies (Greenberg, 2002). Employees may also be more likely to engage in identity theft or fraud if they perceive interactional injustice in their workplace environment (Skarlicki, van Jaarsveld, & Walker, 2008). Besides understanding the predictors of online identity theft or fraud in the workplace, we need to understand the consequences for victims and perpetrators. Lastly, a greater understanding of how organizations can prevent online identity theft and fraud among employees is needed. Research from the criminology literature suggests that protective software (such as anti-spyware and anti-virus programs) and individual identity protective behaviors can be effective at lessening the likelihood of identity theft (Holt & Turner, 2012; Lai, Li, & Hsieh, 2012).

Another avenue for future CD-I research is to more closely examine the role that individual difference variables (such as gender and personality) play in the experience of CD-I as both a victim and as a perpetrator. For example, existing research on face-to-face incivility suggests that individuals with certain personality traits are more likely to be targets of incivility, including those low in agreeableness and those high in neuroticism (Milam, Spitzmueller, & Penney, 2009). Therefore, researchers might attempt to answer the question of whether or not these or other personality traits may predispose employees to experiencing CD-I. Additionally, more research is needed to explore the theoretical mechanisms that might link antecedents (such as individual differences or situational characteristics) with outcomes of CD-I. Affective events theory is one model that might help to explain the link between CD-I as a stressor and subsequent emotional reactions, attitudes, and behaviors in the workplace. However, this mediational hypothesis remains largely untested in the empirical literature in a CD-I setting.

CD-O

There are many avenues for future research regarding attitudes toward CD-O. As is evident from our review, studies assessing CD-O attitudes have exclusively examined attitudes toward piracy while neglecting more serious forms of CD-O such as fraud and hacking. Work related to insider threats reveals that a majority of individuals committing CD-O are typical everyday computer users and not individuals heavily involved in information technologies or IT management (Gordon, Loeb, Lucyshyn, & Richardson, 2004). Accordingly, future work can be aimed at assessing the extent that lenient or positive attitudes toward abusing various types of computer-based resources at work may precede engagement in CD-O.

Few empirical studies have examined specific motivations for engaging in CD-O. The only information we have on motivations for CD-O comes from one empirical study and several industry reports. As such, we see this as one of the most promising areas for future research on cyberdeviance, as information on preventing CD-O must ultimately come from an understanding of the motives (e.g., retaliation, financial gain) that lead to such behavior. Whereas several pieces have established a theoretical basis for different motivations for CD-O (e.g., Willison, 2006, 2009), these tenets have yet to be tested.

Another area of future research for CD-O deals with measurement of the constructs that fall within this dimension, as there is a lack of valid data for different types of CD-O. By definition, CD-O encompasses many behaviors which are illegal and can result in organizational and/or criminal ramifications. Given that a large amount of information appears in reports sourced from criminal or media sources, one would assume that a large swath of CD-O behavior goes unreported and unanalyzed. In fact, most perpetrators of computer abuse are detected not by security measures, but by "accident" or "coincidence" by nonsecurity personnel (e.g., customers and/or managers; Keeney et al., 2005), again indicating that a large amount of CD-O goes undetected. And although most CD-O is carried out on the job, often the behaviors require little technical sophistication (Randazzo et al., 2004). Research is needed to assess the base rate of different types of CD-O that appear to be largely unreported at this point in time.

This brings us to a final problem related to the detection and reporting of CD-O: deviant behaviors are frequently interpreted as incompetence. That is, an employee who illegally copies software or uses privileges to access private information may often justify their actions by claiming it was accidental, or that they did not know it was against the rules (Pfleeger, Predd, Hunker, & Bulford, 2010). Given that most CD-O is discovered unintentionally, it may be easy for managers and others to write off less serious forms of CD-O as incompetence, especially when there is lack of direct proof to the contrary. Further exploration of the varying motives behind CD-O behaviors, particularly how often the behaviors are intentional versus unintentional, could provide valuable insights for how to best prevent CD-O in the workplace.

With regard to deterrence of CD-O behavior, there is a lack of consistency in findings, and this inconsistency offers many avenues for future research. Computer monitoring has been shown to both increase and decrease CD-O (D'Arcy et al., 2009; Posey, Bennett, Roberts, & Lowry, 2011) and the perceived severity and certainty of sanctions as a result of abusing computer-based resources appear to have inconsistent effects on enactment of CD-O. The effects of monitoring, policies, and policy enforcement have clear links to work on justice perceptions (e.g., Posey, Bennett, Roberts, & Lowry, 2011) and existing retaliatory frameworks for more general types of counterproductive work behavior (e.g., Fox, Spector, & Miles, 2001; Jones, 2009; Nixon & Spector, 2013). In addition, a notable problem in regard to this literature is the multiple operationalizations of both deterrence and CD-O. It is quite plausible that employees will respond differently to various deterrence initiatives (e.g., monitoring systems, codes of ethics, awareness programs), resulting in different rates of CD-O. Future research may help to tease apart these different responses and the resultant CD-O behavior.

There is also some evidence that the effect of deterrents in terms of perceived certainty and severity may be affected by culture (Hovav & D'Arcy 2012). However, existing research remains limited in the breadth of national cultures that have been investigated in regards to deterrence and CD-O. Another promising avenue for future research may be to examine the role of informal sanctions (e.g., shame, morality, societal norms) as forces which may discourage inappropriate behavior. Finally, a host of research options exist with regard to a better overall understanding of the outcomes of CD-O, as well as the communication-based areas of CD-O (e.g., slander and cyber-smearing). Both areas are virtually untapped in the current literature.

Conclusion

In closing, whereas cyberdeviance research is still an emerging field of study, it continues to grow in importance and focus within both the academic world and in practice as the rates of Internet use at work continue to rise. In this chapter, we have presented a review and new typology of cyberdeviance that we hope will provide greater clarity to this emerging aspect of current and future work life. It is clear that this area remains a fertile ground for continued research and exploration – one that promises to hold even greater importance as the unceasing innovation of new technology transforms our modern work environment.

References

Akman, I., & Mishra, A. (2009). Ethical behavior issues in software use: An analysis of public and private sectors. *Computers in Human Behavior*, *25*(6), 1251–1257.

Allen, T. D., Cho, E., & Meier, L. L. (2014). Work-family boundary dynamics. *Annual Review of Organizational Psychology and Organizational Behavior*, *1*, 99–121.

Anandarajan, M., Simmers, C., & D'Ovidio, R. (2011). Exploring the underlying structure of personal Web usage in the workplace. *Cyberpsychology, Behavior, and Social Networking*, *14*, 577–583.

Andersson, L. M., & Pearson, C. M. (1999). Tit for tat? The spiraling effect of incivility in the workplace. *Academy of Management Review*, *24*, 452–471.

Andreassen, C. S., Torsheim, T., & Pallesen, S. (2014). Predictors of use of social network sites at work - A specific type of cyberloafing. *Journal of Computer-Mediated Communication*, *19*(4), 906–921.

Askew, K., Buckner, J. E., Taing, M. U., Ilie, A., Bauer, J. A., & Coovert, M. D. (2014). Explaining cyberloafing: The role of the theory of planned behavior. *Computers in Human Behavior*, *36*, 510–519.

Baruch, Y. (2005). Bullying on the net: Adverse behavior on e-mail and its impact. *Information & Management*, *42*, 361–371.

Berry, C. M., Ones, D. S., & Sackett, P. R. (2007). Interpersonal deviance, organizational deviance, and their common correlates: A review and meta-analysis. *Journal of Applied Psychology*, *92*(2), 410–424.

Blanchard, A. L., & Henle, C. A. (2008). Correlates of different forms of cyberloafing: The role of norms and external locus of control. *Computers in Human Behavior*, *24*(3), 1067–1084.

Blau, G., Yang, Y., & Ward-Cook, K. (2006). Testing a measure of cyberloafing. *Journal of Allied Health, 35*(1), 9–17.
Bowling, N. A., & Beehr, T. A. (2006). Workplace harassment from the victim's perspective: A theoretical model and meta-analysis. *Journal of Applied Psychology, 91*(5), 998–1012.
Brock, M. E., Martin, L. E., & Buckley, M. R. (2013). Time theft in organizations: The development of the Time Banditry Questionnaire. *International Journal of Selection and Assessment, 21*(3), 309–321.
Byron, K. (2008). Carrying too heavy a load? The communication and miscommunication of emotion by email. *Academy of Management Review, 33*(2), 309–327.
Chen, J. V., Chen, C. C., & Yang, H. H. (2008). An empirical evaluation of key factors contributing to internet abuse in the workplace. *Industrial Management & Data Systems, 108*(1), 87–106.
Cheng, L., Li, Y., Li, W., Holm, E., & Zhai, Q. (2013). Understanding the violation of IS security policy in organizations: An integrated model based on social control and deterrence theory. *Computers & Security, 39*, 447–459.
Coker, B. S. (2013). Workplace internet leisure browsing. *Human Performance, 26*: 114–125.
D'Arcy, J., & Devaraj, S. (2012). Employee misuse of information technology resources: Testing a contemporary deterrence model. *Decision Sciences, 43*(6), 1091–1124.
D'Arcy, J., Hovav, A., & Galletta, D. (2009). User awareness of security countermeasures and its impact on information systems misuse: A deterrence approach. *Information Systems Research, 20*(1), 79–98.
Doll, J. L., & Giumetti, G. W. (2015, August). *Cyber-sexual harassment: Effects on affect, engagement, and performance.* Paper presented at the 75th annual meeting of the Academy of Management, Vancouver, BC, Canada.
Einarsen, S. (2000). Harassment and bullying at work: A review of the Scandinavian approach. *Aggression and Violent Behavior: A Review Journal, 5*, 371–401.
Everton, W. J., Mastrangelo, P. M., & Jolton, J. A. (2005). Personality correlates of employees' personal use of work computers. *CyberPsychology & Behavior, 8*(2), 143–153.
Flanagin, A. J., Pearce, K., & Bondad-Brown, B. A. (2009). The destructive potential of electronic communication technologies in organizations. In P. Lutgen-Sandvik, & B. Sypher (Eds.), *Destructive organizational communication: Processes, consequences, and constructive ways of organizing* (pp. 229–251). New York, NY: Routledge/Taylor & Francis Group.
Ford, D. P. (2013). Virtual harassment: Media characteristics' role in psychological health. *Journal of Managerial Psychology, 28*(4), 408–428.
Fox, S., Spector, P. E., & Miles, D. (2001). Counterproductive work behavior (CWB) in response to job stressors and organizational justice: Some mediator and moderator tests for autonomy and emotions. *Journal of Vocational Behavior, 59*(3), 291–309.
Francis, L., Holmvall, C. M., & O'Brien, L. E. (2015). The influence of workload and civility of treatment on the perpetration of email incivility. *Computers in Human Behavior, 46*, 191–201.
Gallagher, E. C. (2009). Narcissism and forgiveness as moderators of organizational justice and workplace counterproductive computer use. (Unpublished master's thesis.) University of Nebraska at Omaha. Retrieved April 7, 2017, from http://gradworks.umi.com/14/62/1462100.html
Gan, L. L., & Koh, H. C. (2006). An empirical study of software piracy among tertiary institutions in Singapore. *Information & Management, 43*(5), 640–649.
Garrett, R. K., & Danziger, J. N. (2008). On cyberslacking: Workplace status and personal internet use at work. *Cyberpsychology & Behavior, 11*(3), 287–292.
Giumetti, G. W., Hatfield, A. L., Scisco, J. L., Schroeder, A. N., Muth, E. R., & Kowalski, R. M. (2013). What a rude e-mail! Examining the differential effects of incivility versus support on mood, energy, engagement, and performance in an online context. *Journal of Occupational Health Psychology, 18*(3), 297–309.
Giumetti, G. W., McKibben, E. S., Hatfield, A. L., Schroeder, A. N., & Kowalski, R. M. (2012). Cyber-incivility @ work: The new age of interpersonal deviance. *Cyberpsychology, Behavior & Social Networking, 15*(3), 148–154.
Giumetti, G. W., Saunders, L. A., Brunette, J. P., Graham, P. G., & DiFrancesco, F. (2016). Linking cyber-incivility with job performance through job satisfaction: The buffering role of positive affect. *Psi Chi Journal of Psychological Research., 21*, 230–240.

Gordon, L. A., Loeb, M. P., Lucyshyn, W. & Richardson, R. (2004) *CSI/FBI computer crime and security survey*. Monroe, WA: Computer Security Institute.

Gorsuch, J. L. (2014). The effects of control and work/family centrality on the personal use of work computers. (Unpublished master's thesis). Cleveland State University. Retrieved April 7, 2017, from http://rave.ohiolink.edu/etdc/view?acc_num=csu1397838982

Greenberg, J. (2002). Who stole the money, and when? Individual and situational determinants of employee theft. *Organizational Behavior and Human Decision Processes*, *89*(1), 985–1003.

Greitzer, F. L., Kangas, L. J., Noonan, C. F. & Dalton, A. C. (2010). *Identifying at-risk employees: A behavioral model for predicting potential insider threats*. Richland, WA: Pacific Northwest National Laboratory.

Harrington, S. J. (1996). The effect of codes of ethics and personal denial of responsibility on computer abuse judgments and intentions. *MIS Quarterly*, 257–278.

Harter, J., Agrawal, S., & Sorenson, S. (2014, April). Most US workers see upside to staying connected to work. Retrieved April 7, 2017, from www.gallup.com/poll/168794/workers-upside-staying-connected-work.aspx

Henle, C. A., Kohut, G., & Booth, R. (2009). Designing electronic use policies to enhance employee perceptions of fairness and to reduce cyberloafing: An empirical test of justice theory. *Computers in Human Behavior*, *25*(4), 902–910.

Hershcovis, M. S. (2011). "Incivility, social undermining, bullying ... oh my!": A call to reconcile constructs within workplace aggression research. *Journal of Organizational Behavior*, *32*(3), 499–519.

Hollinger, R. C., & Clark, J. P. (1982). Formal and informal social controls of employee deviance. *The Sociological Quarterly*, *23*(3), 333–343.

Holt, T. J., & Turner, M. G. (2012). Examining risks and protective factors of on-line identity theft. *Deviant Behavior*, *33*(4), 308–323.

Hong, J., Chien-Hou, L., Hwang, M., Hu, R., & Chen, Y. (2014). Positive affect predicting worker psychological response to cyber-bullying in the high-tech industry in Northern Taiwan. *Computers in Human Behavior*, *30*, 307–314.

Hovav, A., & D'Arcy, J. (2012). Applying an extended model of deterrence across cultures: An investigation of information systems misuse in the US and South Korea. *Information & Management*, *49*(2), 99–110.

Hu, Q., Xu, Z., Dinev, T. & Ling, H. (2011) Does deterrence work in reducing information security policy abuse by employees? *Communications of the ACM*, *54*, 54–60.

Jia, H., Jia, R., & Karau, S. (2013). Cyberloafing and personality: The impact of the Big Five traits and workplace situational factors. *Journal of Leadership & Organizational Studies*, *20*(3), 358–365.

John, O. P., & Srivastava, S. (1999). The Big Five trait taxonomy: History, measurement, and theoretical perspectives. *Handbook of Personality: Theory and Research*, *2*, 102–138.

Jones, D. A. (2009). Getting even with one's supervisor and one's organization: relationships among types of injustice, desires for revenge, and counterproductive work behaviors. *Journal of Organizational Behavior*, *30*(4), 525–542.

Judge, T. A., Erez, A., Bono, J. E., & Thoresen, C. J. (2003). The Core Self-Evaluations Scale: Development of a measure. *Personnel Psychology*, *56*(2), 303–331.

Judge, T. A., LePine, J. A., & Rich, B. L. (2006). Loving yourself abundantly: Relationship of the narcissistic personality to self- and other perceptions of workplace deviance, leadership, and task and contextual performance. *Journal of Applied Psychology*, *91*(4), 762–776.

Keeney, M, Kowalski, E., Cappelli, D., Moore, A., Shimeall, T., & Rogers. S. (2005) *Insider threat study: Computer system sabotage in critical infrastructure sectors*. Pittsburgh, PA: US Secret Service and CERT Coordination Center/SEI, Carnegie-Mellon University.

Kiesler, S., Seigel, J., & McGuire, T. (1984). Social psychological aspects of computer-mediated communication. *American Psychologist*, *39*, 1123–1134.

König, C. J., & de la Guardia, M. E. C. (2014). Exploring the positive side of personal internet use at work: Does it help in managing the border between work and nonwork? *Computers in Human Behavior*, *30*, 355–360.

Kossek, E. E., Lautsch, B. A., & Eaton, S. C. (2006). Telecommuting, control, and boundary management: Correlates of policy use and practice, job control, and work-family effectiveness. *Journal of Vocational Behavior, 68*(2), 347–367.

Kossek, E. E., Ruderman, M. N., Braddy, P. W., & Hannum, K. M. (2012). Work–nonwork boundary management profiles: A person-centered approach. *Journal of Vocational Behavior, 81*(1), 112–128.

Lai, F., Li, D., & Hsieh, C. (2012). Fighting identity theft: The coping perspective. *Decision Support Systems, 52*(2), 353–363.

Lee, S. M., Lee, S. G., & Yoo, S. (2004). An integrative model of computer abuse based on social control and general deterrence theories. *Information & Management, 41*(6), 707–718.

Lending, D., & Slaughter, S. A. (1999, April). Understanding differences in ethical beliefs and behaviors toward software copying: The effects of organization culture. In *Proceedings of the 1999 ACM SIGCPR Conference on Computer Personnel Research* (pp. 253–260). New York, NY: ACM.

LePine, J. A., Podsakoff, N. P., & LePine, M. A. (2005). A meta-analytic test of the challenge stressor–hindrance stressor framework: An explanation for inconsistent relationships among stressors and performance. *Academy of Management Journal, 48*(5), 764–775.

Li, H., Zhang, J., & Sarathy, R. (2010). Understanding compliance with internet use policy from the perspective of rational choice theory. *Decision Support Systems, 48*(4), 635–645.

Liberman, B., Seidman, G., Mckenna, K. Y., & Buffardi, L. E. (2011). Employee job attitudes and organizational characteristics as predictors of cyberloafing. *Computers in Human Behavior, 27*(6), 2192–2199.

Lim, V. K. G. (2002). The IT way of loafing on the job: Cyberloafing, neutralizing and organizational justice. *Journal of Organizational Behavior, 23*(5), 675–694.

Lim, V. K. G., & Chen, D. J. (2012). Cyberloafing at the workplace: Gain or drain on work? *Behaviour & Information Technology, 31*(4), 343–353.

Lim, V. K. G., & Teo, T. S. H. (2005). Prevalence, perceived seriousness, justification and regulation of cyberloafing in Singapore: An exploratory study. *Information & Management, 42*: 1081–1093.

Lim, V. K. G., & Teo, T. S. H. (2009). Mind your E-manners: Impact of cyber incivility on employees' work attitude and behavior. *Information & Management, 46*(8), 419–425.

Lowry, P. B., Posey, C., Bennett, B., & Roberts, T. (2014). Leveraging fairness and reactance theories to deter reactive computer abuse following enhanced organisational information security policies: An empirical study of the influence of counterfactual reasoning and organisational trust. *Information Systems Journal, 25*, 193–273.

Martinell, J. (2010). The influence of sex on incivility in work-related email communication. (Unpublished master's thesis). Saint Mary's University, Halifax, Nova Scotia, Canada. Retrieved April 7, 2017, from http://library2.smu.ca/handle/01/23717#.VO9Jco4nm6E

Mastrangelo, P. M., Everton, W., & Jolton, J. A. (2006). Personal use of work computers: Distraction versus destruction. *Cyberpsychology & Behavior, 9*(6), 730–741.

Mehdad, A., & Nezhad, A. V. (2014). Relationship between perceived organizational injustice, perceived job stress, and cyber bullying: A test of mediation. *International Journal of Psychology and Behavioral Research, 3*(5), 374–385.

Milam, A. C., Spitzmueller, C., & Penney, L. M. (2009). Investigating individual differences among targets of workplace incivility. *Journal of Occupational Health Psychology, 14*(1), 58–69.

Mishra, A., Akman, I., & Yazici, A. (2006). Software piracy among IT professionals in organizations. *International Journal of Information Management, 26*(5), 401–413.

Moody, G. D., & Siponen, M. (2013). Using the theory of interpersonal behavior to explain non-work-related personal use of the internet at work. *Information & Management, 50*(6), 322–335.

Neves, P., & Champion, S. (2015). Core self-evaluations and workplace deviance: The role of resources and self-regulation. *European Management Journal*. Advance online publication.

Nixon, A. E., & Spector, P. E. (2013). The impact of technology on employee stress, health, and well-being. In M. D. Coovert & L. Foster-Thompson (Eds.), *The Psychology of workplace technology* (pp. 238–260). New York, NY: Routledge.

O'Neill, T. A., Hambley, L. A., & Bercovich, A. (2014). Prediction of cyberslacking when employees are working away from the office. *Computers in Human Behavior, 34*, 291–298.

Oz, E. (2001). Organizational commitment and ethical behavior: An empirical study of information system professionals. *Journal of Business Ethics, 34*(2), 137–142.

Paschal, J. L., Stone, D. L., & Stone-Romero, E. F. (2009). Effects of electronic mail policies on invasiveness and fairness. *Journal of Managerial Psychology, 24*(6), 502–525.

Park, Y., Fritz, C., & Jex, S. M. (2015). Daily cyber incivility and distress: The moderating roles of resources at work and home. *Journal of Management*. Advance online publication March 23, 2015. DOI:10.1177/0149206315576796

Paulhus, D. L., & Williams, K. M. (2002). The dark triad of personality: Narcissism, Machiavellianism and psychopathy. *Journal of Research in Personality, 36*(6), 556–563.

Peace, A. G., Galletta, D. F., & Thong, J. Y. (2003). Software piracy in the workplace: A model and empirical test. *Journal of Management Information Systems, 20*(1), 153–177.

Pee, L. G., Woon, I. M., & Kankanhalli, A. (2008). Explaining non-work-related computing in the workplace: A comparison of alternative models. *Information & Management, 45*(2), 120–130.

Pfleeger, S. L., Predd, J. B., Hunker, J., & Bulford, C. (2010). Insiders behaving badly: Addressing bad actors and their actions. *IEEE Transactions on Information Forensics and Security, 5*(1), 169–179.

Phillips, J. G., & Reddie, L. (2007). Decisional style and self-reported Email use in the workplace. *Computers in Human Behavior, 23*: 2414–2428.

Posey, C., Bennett, R. J., & Roberts, T. L. (2011). Understanding the mindset of the abusive insider: An examination of insiders' causal reasoning following internal security changes. *Computers & Security, 30*(6), 486–497.

Posey, C., Bennett, B., Roberts, T., & Lowry, P. B. (2011). When computer monitoring backfires: Invasion of privacy and organizational injustice as precursors to computer abuse. *Journal of Information System Security, 7*(1), 24–47.

Privitera, C., & Campbell, M. A. (2009). Cyberbullying: The new face of workplace bullying? *Cyberpsychology & Behavior, 12*(4), 395–400.

Randazzo, M. R., Keeney, M., Kowalski, E., Cappelli, D., & Moore, A. (2004). *Insider threat study: Illicit cyber activity in the banking and finance sector* (No. CMU/SEI-2004-TR-021). Pittsburgh, PA: Carnegie-Mellon University Software Engineering Institute.

Restubog, S. L. D., Garcia, P. R. J. M., Toledano, L. S., Amarnani, R. K., Tolentino, L. R., & Tang, R. L. (2011). Yielding to (cyber)-temptation: Exploring the buffering role of self-control in the relationship between organizational justice and cyberloafing behavior in the workplace. *Journal of Research in Personality, 45*(2), 247–251.

Richard, E., Gacey, H., Young, S., & Lopez-Matos, X. (2015, April). *Aggression in work-related email: A qualitative analysis.* Poster presented at the annual meeting of the Society for Industrial-Organizational Psychology, Philadelphia, PA.

Rice, R. E. (2017). Flexwork, work–family boundaries, and information and communication technologies. In G. Hertel, D. L. Stone, R. D. Johnson, & J. Passmore (Eds.), *The Wiley Blackwell handbook of the psychology of the Internet at work* (pp. 175–194). Chichester: John Wiley & Sons, Ltd.

Ritter, B. A. (2014). Deviant behavior in computer-mediated communication: Development and validation of a measure of cybersexual harassment. *Journal of Computer-Mediated Communication, 19*(2), 197–214.

Robinson, S. L., & Bennett, R. J. (1995). A typology of deviant workplace behaviors: A multidimensional scaling study. *Academy of Management Journal, 38*: 555–572.

Rubino, C., Perry, S. J., Milam, A. C., Spitzmueller, C., & Zapf, D. (2012). Demand–control–person: Integrating the demand–control and conservation of resources models to test an expanded stressor–strain model. *Journal of Occupational Health Psychology, 17*(4), 456–472.

RuningSawitri, H. S. (2012). Role of internet experience in moderating influence of work stressor on cyberloafing. *Procedia-Social and Behavioral Sciences, 57*, 320–324.

Sackett, P. R., & DeVore, C. J. (2001). Counterproductive behaviors at work. In N. Anderson, D. S. Ones, H. K. Sinangil,& C. Viswesvaran (Eds.), *Handbook of industrial, work and organizational psychology, Volume 1: Personnel psychology* (pp. 145–164). Thousand Oaks, CA: Sage Publications Ltd.

Salem, M. B., Hershkop, S., & Stolfo, S. J. (2008). A survey of insider attack detection research. In S. J. Stolfo, S. M. Bellovin, A. D. Keromytis, S. Hershkop, S. W. Smith, & S. Sinclair (Eds.), *Insider attack and cyber security. Beyond the hacker* (pp. 69–90). New York, NY: Springer US.

Shropshire, J. (2009). A canonical analysis of intentional information security breaches by insiders. *Information Management and Computer Security, 17*, 296–310.

Skarlicki, D. P., van Jaarsveld, D. D., & Walker, D. D. (2008). Getting even for customer mistreatment: The role of moral identity in the relationship between customer interpersonal injustice and employee sabotage. *Journal of Applied Psychology, 93*(6), 1335–1347.

Slotkin, M. H., & Vamosi, A. R. (2011). Erosion of a dental specialist's referral network: an economic loss template for cyber-defamation. *International Journal of Management and Decision Making, 11*(5), 344–357.

Spector, P. E., Zapf, D., Chen, P. Y., & Frese, M. (2000). Why negative affectivity should not be controlled in job stress research: Don't throw out the baby with the bath water. *Journal of Organizational Behavior, 21*(1), 79–95.

Strader, T. J., Fichtner, J. R., Clayton, S. R., & Simpson, L. A. (2011). The impact of context on employee perceptions of acceptable non-work related computing. In R. Luppicini (Ed.), *Moral, ethical, and social dilemmas in the age of technology: Theories and practice: theories and practice* (pp. 125–139). Hershey, PA: IGI Global.

Stratton, M. T. (2010). Uncovering a new guilty pleasure: A qualitative study of the emotions of personal web usage at work. *Journal of Leadership & Organizational Studies, 17*(4), 392–410.

Straub, D. W. (1990). Effective IS security: An empirical study. *Information Systems Research, 1*(3), 255–276.

Straub, D. W., & Nance, W. D. (1990). Discovering and disciplining computer abuse in organizations: a field study. *MIS Quarterly*, 45–60.

Suler, J. (2004). The online disinhibition effect. *Cyberpsychology & Behavior, 7*(3), 321–326.

Taylor, G. S., & Shim, J. P. (1993). A comparative examination of attitudes toward software piracy among business professors and executives. *Human Relations, 46*(4), 419–433.

Tepper, B. J., & Henle, C. A. (2011). A case for recognizing distinctions among constructs that capture interpersonal mistreatment in work organizations. *Journal of Organizational Behavior, 32*(3), 487–498.

Tripathi, N. & Lim, V. K. G. (2014, August). *Moderating effect of Machiavellianism and perceived similarity on cyber incivility & its outcomes.* Paper presented at the 74th Annual Meeting of the Academy of Management, Philadelphia, PA, USA.

Turban, E., Lee, J., King, D., & Chung, H. M. (2000). *Electronic commerce: A managerial perspective.* Englewood Cliffs, NJ: Prentice Hall.

Turnage, A. K. (2008). Email flaming behaviors and organizational conflict. *Journal of Computer-Mediated Communication, 13*(1), 43–59.

Ugrin, J. C., & Pearson, J. M. (2013). The effects of sanctions and stigmas on cyberloafing. *Computers in Human Behavior, 29*(3), 812–820.

Ugrin, J. C., Pearson, J. M., & Odom, M. D. (2007). Profiling cyber-slackers in the workplace: Demographic, cultural, and workplace factors. *Journal of Internet Commerce, 6*(3), 75–89.

Valli, C. (2004). Non-business use of the WWW in three Western Australian organisations. *Internet Research, 14*(5), 353–359.

Van Yperen, N. W. & Wörtler, B. (2017). Blended working. In G. Hertel, D. L. Stone, R. D. Johnson, & J. Passmore (Eds.), *The Wiley Blackwell Handbook of the psychology of the Internet at work* (pp. 157–174). Chichester: John Wiley & Sons, Ltd.

Venkatraman, S. (2008). The "darth" side of technology use: Cyberdeviant workplace behaviors. *Dissertation Abstracts International Section A, 69*, 4035.

Vitak, J., Crouse, J., & LaRose, R. (2011). Personal internet use at work: Understanding cyberslacking. *Computers in Human Behavior, 27*(5), 1751–1759. DOI:10.1016/j.chb.2011.03.002

Wagner, D. T., Barnes, C. M., Lim, V. K. G., & Ferris, D. L. (2012). Lost sleep and cyberloafing: Evidence from the laboratory and a daylight saving time quasi-experiment. *Journal of Applied Psychology, 97*(5), 1068–1076.

Watson, D., Clark, L. A., & Tellegen, A. (1988). Development and validation of brief measures of positive and negative affect: The PANAS scales. *Journal of Personality and Social Psychology, 54*(6), 1063–1070.

Weatherbee, T. G. (2007). Cyber aggression in the workplace: Construct development, operationalization, and measurement. (Unpublished doctoral dissertation). Saint Mary's University, Halifax, Nova Scotia, Canada. Retrieved April 7, 2017, from www.library2.smu.ca/bitstream/handle/01/9033/weatherbee_terrance_g_phd_2007.PDF?sequence=2

Weatherbee, T. G. (2010). Counterproductive use of technology at work: Information & communications technologies and cyberdeviancy. *Human Resource Management Review, 20*(1), 35–44. DOI:10.1016/j.hrmr.2009.03.012

Weatherbee, T. G., & Kelloway, E. K. (2006). A case of cyberdeviancy: Cyberaggression in the workplace. In E. K. Kelloway, J. Barling, & J. J. Hurrell, Jr. (Eds.), *Handbook of Workplace Violence* (pp. 445–487). Thousand Oaks, CA: Sage.

Websense, Inc. (2006). *Web@Work.* Retrieved April 7, 2017, from http://www.securitymanagement.com/archive/library/websense_technofile0906.pdf

Weiss, H. M. & Cropanzano, R. (1996). Affective events theory: A theoretical discussion of the structure, causes and consequences of affective experiences at work. *Research in Organizational Behavior, 18,* 1–74.

Willison, R. (2006). Understanding the perpetration of employee computer crime in the organisational context. *Information and Organization, 16*(4), 304–324.

Willison, R. (2009). *Motivations for employee computer crime: Understanding and addressing workplace disgruntlement through the application of organisational justice.* Technical Rpt. Working Paper No. 1. Copenhagen: Copenhagen Business School, Department of Informatics.

Workman, M. (2010). A behaviorist perspective on corporate harassment online: Validation of a theoretical model of psychological motives. *Computers & Security, 29*(8), 831–839.

Workman, M. (2012). Rash impulsivity, vengefulness, virtual-self and amplification of ethical relativism on cyber-smearing against corporations. *Computers in Human Behavior, 28*(1), 217–225.

Young, S., Pita, M., White, J., Hickman, A., & Richard, E. (2012). *%&$#?@!: A model of cyberaggression in email.* Poster presented at the meeting of Industrial Organizational and Organizational Behavior (IOOB), Orlando, FL, USA.

Zagenczyk, T. J., Restubog, S. D., Kiewitz, C., Kiazad, K., & Tang, R. L. (2014). Psychological contracts as a mediator between Machiavellianism and employee citizenship and deviant behaviors. *Journal of Management, 40*(4), 1098–1122.

Zoghbi Manrique de Lara, P. (2006). Fear in organizations: Does intimidation by formal punishment mediate the relationship between interactional justice and workplace Internet deviance? *Journal of Managerial Psychology, 21,* 580–592.

8

Blended Working

Nico W. Van Yperen and Burkhard Wörtler

Introduction

A pervasive and booming trend in modern societies is the increasing reliance on new and continuous improving information and communication technology (ICT) software, devices (computers, tablets, hand-held and wearable devices, etc.), and infrastructure (i.e., increased availability of high-speed broadband connections). This development provides flexibility in where, when, and how long people work, learn, and care, referred to as ***blended working*** (e.g., Van Yperen, Rietzschel, & De Jonge, 2014), ***blended learning*** (e.g., Hew & Cheung, 2014; Owston, 2013), and ***blended care*** (e.g., Wilhelmsen et al., 2013). This, in turn, creates opportunities for improving individual, organizational, and societal outcomes in ways that were not previously possible.

The central concept in this chapter is blended working, a term recently introduced by Van Yperen et al. (2014). Blended working combines on-site and off-site working, enabled by the utilization of ICTs that provide workers with almost constant access to job-relevant information and co-workers. Blended working refers to smooth and seamless time-independent working (flexibility in when and how long workers engage in work-related tasks) and location-independent working (flexibility in where work gets done). Working from the office, having a business meeting with colleagues in a restaurant, preparing a meeting in the train, online file sharing, and work-related use of tablets and smartphones are examples of blended working practices (De Jonge, Van Yperen, & Rietzschel, 2015). Often used related terms in the industrial and organizational psychological literature are teleworking and telecommuting, terms that suggest that workers are either remote workers (i.e., teleworkers) or office-based workers (Harris, 2003; Kurland & Bailey, 1999; Wilks & Billsberry, 2007). As noted by Hill, Ferris, and Märtinson (2003), the literature and language around telework is imprecise: "It is difficult to distinguish between the virtual office and varieties of telecommuting because terminology differs from study to study" (pp. 223–224). In contrast, the term blended working unambiguously refers to a work arrangement in which workers alter between traditional office working and working from home or another location at any time. Particularly knowledge and information workers, whose jobs mainly revolve around obtaining, analyzing, and sharing knowledge, typically *blend* on-site and off-site working and

The Wiley Blackwell Handbook of the Psychology of the Internet at Work,
First Edition. Edited by Guido Hertel, Dianna L. Stone, Richard D. Johnson, and Jonathan Passmore.

are flexible in when and how long they work (O'Neill, Hambley, Greidanus, MacDonnell, & Kline, 2009; Wheatley, 2012). That is, they may work during traditional office hours or choose to work in the evening, on weekends, or any combination of these. With regard to the location, workers may work at the office (which can include "hot desking," e.g., Millward, Haslam, & Postmes, 2007), at home, on the move, or at "neutral" workplaces that are shared, swapped, reserved, rented, or simply claimed for a time, etc. Similar to well-established terms as blended learning (e.g., Hew & Cheung, 2014; Owston, 2013) and blended care (e.g., Wilhelmsen et al., 2013), the term blended working (Van Yperen et al., 2014) probably best covers these flexible working practices and will be used throughout this chapter. More than similar terms such as flexible work (e.g., Rice, 2017, Chapter 9 in this volume), the term blended working emphasizes that it is just one of the manifestations of a more general trend in modern societies to combine multimedia and traditional face-to-face working, learning, and caring at any time and in any place.

Blended working has the potential to provide considerable benefits to both organizations and individual workers, but also evident are some potential negative consequences (see also Rice, 2017, Chapter 9 in this volume). We summarize and give an overview of the benefits and drawbacks of blended working, and show that blended working:

- may not work for everyone
- can be helpful to cope with high job demands
- may increase workers' job satisfaction
- may be particularly effective for specific groups (i.e., caregivers, older workers, and workers with a physical disability)
- may be most effectively implemented in companies characterized by trust and cohesion or in companies in which workers can be evaluated on the basis of their results.

We start by discussing possible consequences for the organization and the worker, before turning to the focus of this chapter, that is, a discussion of for whom and when blended working does (not) work. Finally, we discuss some promising new research directions.

Possible Consequences of Blended Working for Organizations

A general aim of blended working practices is to improve organizational performance, including productivity, customer service, workers' satisfaction and motivation, collaboration, and workplace utilization, and to reduce negative outcomes such as absenteeism, tardiness, turnover, errors, and time loss (*cf.*, Kurland & Bailey, 1999; Maruyama & Tietze, 2012; Patrickson, 2002; Van Yperen et al., 2014). In addition, blended working is a way for companies to reduce expenses, including real estate costs (Hill et al., 2003), to comply with government regulations regarding equal opportunities, and to demonstrate corporate social responsibility (Morgeson, Aguinis, Waldman, & Siegel, 2013). For example, due to reduced office space requirements, traffic, paper use, etc., blended working may have positive consequences for the environment, such as the reduction to sustainable levels of carbon dioxide emissions, deforestation, and greenhouse gases, and less need for additional office space and highway capacity (Lister & Harnish, 2011; McLennan, 2008).

Blended working may also have drawbacks for organizations. For example, in 2013, the chief executive officer (CEO) of Yahoo Marissa Mayer banned Yahoo employees from working from home because she felt that it undermined a collaborative, inventive environment (Allen, Golden, & Shockley, 2015; Raiborn & Butler, 2009). More generally, leaders may be worried that interpersonal relationships, and accordingly, the positive social and psychological climate, suffer. Indeed, in their meta-analysis, Gajendran and Harrison (2007) found that working remotely 2.5 or more days per week had a negative effect on co-worker relationship quality. However, there was no damaging effect on the relationship with the supervisor. Moreover, in a follow-up study, Gajendran, Harrison, and Delaney-Klinger (2015) found that the opportunity to work remotely from the office was associated with improved task and contextual performance rather than social costs.

Another potential undesired consequence for organizations may be productivity loss caused by cyberslacking, which occurs when workers use the Internet for nonwork purposes when they are on company time (Block, 2001; see also Charlier, Giumetti, Reeves, & Greco, 2017, Chapter 7 in this volume). Cyberslacking might be perceived as easier to engage in when working away from the office because it is more difficult to detect by supervisors and co-workers (O'Neill, Hambley, & Bercovich, 2014). In line with the contingency approach (e.g., Bailey & Kurland, 2002) and person–job fit theory (Kristof-Brown, Zimmerman, & Johnson, 2005), O'Neill et al. (2014) found preliminary evidence that individuals low in trait procrastination and high in agreeableness and honesty were most likely to avoid cyberslacking when working from home.

Possible Consequences of Blended Working for Workers

Blended working offers unprecedented opportunities for workers to decide when, where, and how to work (Gibson, Blackwell, Dominicis, & Demerath, 2002). Potential benefits include saving time (due to reduced commuting time) and freedom from workplace distractions and interruptions (Cutler, 2006; Van Yperen et al., 2014). Working connectedly may increase efficiency in information access and can provide workers with information and feedback that they would not have obtained as easily or quickly otherwise (Mazmanian, Orlikowski, & Yates, 2005). However, blended working may not be suitable for every worker. Person–job fit theory (Kristof-Brown et al., 2005) underlines that workers are most motivated and perform best when the requirements and affordances of the (physical and nonphysical) work environment are aligned with their own needs and abilities. In line with this approach, O'Neill et al. (2009) found that personality traits are differentially related to job effectiveness depending on whether one works in a traditional office-based setting, or in a setting which provides the opportunity to work remotely from the office. For example, sociability, that is, one's tendency to enjoy conversation, social interaction, and parties, may be an undesirable trait for blended workers because of relational and information impoverishment at work (Kurland & Bailey, 1999). On the other hand, working connectedly via online devices enables workers to maintain or even extend their contact with co-workers, and accordingly, may avoid social impoverishment and isolation when working off-site (Cutler, 2006; De Jonge, et al., 2015). In any case, to be successful, blended working practices should ensure opportunities for face-to-face communication, for developing supportive processes and structures, and for exchanging occupational knowledge (Maruyama & Tietze, 2012).

Other potential drawbacks of blended working for those working primarily away from the office are career stagnation ("out of sight, out of mind"), increased work–home interference, distraction and interruption by family members (particularly when there is no detached home office space), social loafing behavior in the (virtual) team, technostress, and procrastination (e.g., Allen, Johnson, Kiburz, & Shockley, 2013; Bartel, Wrzesniewski, & Wiesenfeld, 2012; Gerdenitsch, Kubicek, & Korunka, 2015; Greer & Payne, 2014; Kelliher & Anderson, 2010; Maruyama & Tietze, 2012; O'Neill et al., 2014; Perry, Lorinkova, Hunter, Hubbard, & McMahon, 2016; Salanova, Llorens, & Cifre, 2013; Wheatley, 2012).

Furthermore, being able to decide when, where, and how to work may come with the cost of increased complexity, and being constantly connected can induce feelings of external control, resulting from the pressure to be available anywhere, at any time. This presents the challenge to workers to manage the blurring of the home–work interface, which may put a strain on themselves and on their relations with partners, family members, and friends (see also Rice, 2017, Chapter 9 in this volume). Ambiguity about tasks and roles may arise from increased work–home interference, and because being continuously connected to co-workers makes it unclear whether, how, and when information will be pushed to one's workplace.

However, a large-scale meta-analysis by Gajendran and Harrison (2007) indicated that work arrangements allowing workers to perform their tasks while being remote from their office, had mainly beneficial effects on work–home balance, job satisfaction, and job performance, and generally had no detrimental effects on the quality of the relationship with the supervisor or on perceived career prospects. Other studies also support the idea that flexible work arrangements are likely to benefit both the employee and the employer in many ways (e.g., Maruyama & Tietze, 2012). The blurring of the home–work interface, or permeable borders between work and family, may even be a prerequisite for achieving work–life balance (Ashforth, Kreiner, & Fugate, 2000; Hill et al., 2003). Importantly, Gajendran and Harrison's (2007) meta-analysis also showed that the effects of flexible work arrangements were moderated by variables such as the intensity of, and experience with off-site working, and sex. Thus, the consequences of blended working may be more favorable (or unfavorable) for some people than for others, which raises the question for whom blended working is most likely to work (Matusik & Mickel, 2011; O'Neill et al., 2009). Van Yperen et al. (2014) recently demonstrated that the perceived personal effectiveness of blended working is contingent on the strength of the worker's psychological needs, that is, their need for autonomy, need for relatedness, need for competence, and need for structure (Gagné & Deci, 2005; Thompson, Naccarato, Parker, & Moskowitz, 2001).

Psychological Needs and the Perceived Personal Effectiveness of Blended working

Preeminent in today's theories of organizational behavior and work motivation is the concept of psychological needs (Gagné & Deci, 2005). Self-determination theory (SDT), one of the most influential psychological needs theories of the last three decades distinguishes three basic psychological needs (Deci & Ryan, 2002): (1) need for autonomy, that is, an individual's desire to feel volitional and to experience a sense of choice and psychological freedom; (2) need for competence, defined as an individual's need to feel competent and skilled; (3) need for relatedness, which denotes the need to feel connected to others. Although it has been argued that these psychological needs are intrinsically human (Deci & Ryan, 2002), in industrial and organizational psychology, psychological needs have typically been treated as individual difference variables, implying that people are viewed as differing in the *strength* of particular needs (Gagné & Deci, 2005; Hofer & Busch, 2011). For example, Van Yperen et al. (2014) hypothesized that a worker's perceived effectiveness of blended working was a function of the strength of the psychological needs for autonomy, competence, and relatedness. In addition, workers with a high need for a structured and predictable environment were expected to perceive blended working as personally ineffective because it fuels their aversion to ambiguity. Need for structure refers to a strong preference for structure and predictability and a low tolerance for ambiguity (Thompson et al., 2001). This need, which is completely neglected in SDT research, is positively related to feedback-seeking behaviors (Ashford & Cummings, 1985), managers' preference to guide their subordinates by planning and scheduling work tasks (Ehrhart & Klein, 2001), and a preference for a workplace that helps segment work and home domains (Van Yperen et al., 2014).

In line with their expectations, Van Yperen et al. (2014) found that the stronger the need for autonomy among workers, the more they felt that blended working works for them, because they have more discretion as to where and when to work. In contrast, workers' need for relatedness was negatively related to their perceived personal effectiveness of blended working. Indeed, blended working entails working away from the office and having flexible hours, which may cause workers to feel disconnected from others and socially isolated (Bartel et al., 2012; Cooper & Kurland, 2002; Maruyama & Tietze, 2012; O'Neill et al., 2009), which then obviously interferes with workers' need for relatedness. Furthermore, because blended working tends to create ambiguity (Slijkhuis, Rietzschel, & Van Yperen, 2013), workers' need for structure was inversely related to perceived personal effectiveness of time and location-independent working. Thus, the stronger their need for autonomy at work, and the weaker their needs for relatedness and structure at work, the more workers felt that blended working was effective for them.

Unexpectedly, Van Yperen et al. (2014) did not observe a link with workers' need for competence. On the one hand, this is remarkable because individuals' need for competence is positively associated with the wish to engage in challenging tasks, to acquire new skills, and to perform well (Van den Broeck, Vansteenkiste, De Witte, Soenens, & Lens, 2010). Blended working can be considered a challenging, innovative work arrangement that likely contributes to satisfying workers' need for competence. In line with this reasoning, Dikkers, Van Engen, and Vinkenburg (2010) demonstrated that career ambition was positively related to workers' utilization of work–home arrangements, including flexible hours and working from home. On the other hand, perceiving a job as new and challenging may be short-lived and primarily a function of work content rather than work arrangement. Furthermore, working away from the office may be associated with lower chances for promotion and expectations of less frequent and more ambiguous feedback, which undermines workers' desire to learn and to advance in their career (O'Neill et al., 2009). For examples, Busch (2008) noted that off-site working "can have a detrimental effect on junior employees, as they are unable to pick up many of the workplace cues they require for on the job success" (p. 39). If face-to-face contact is too limited, blended working may hamper the acquisition of tacit knowledge involving subjective insight and intuition. Particularly apprentices and junior workers may hugely benefit from learning by interaction with co-workers (i.e., learning by observation and knowledge sharing). Van Yperen et al. (2014) concluded that in terms of psychological needs, blended working seems to be most suitable for workers who are high in need for autonomy, low in need for relatedness, and low in need for structure (*cf.*, Dropkin, Moline, Kim, & Gold, 2016). Hence, blended working practices should not be seen as a 'one-size-fits-all' issue. Rather, to find the right solution for every worker, each individual's work-related psychological needs strengths should be considered to assess how well they fit with the core job characteristics (*cf.*, Hackman & Oldham, 1976; Kristof-Brown et al., 2005). For example, particularly for workers high in need for autonomy, the opportunity for blended working may be an effective resource to cope with high job demands.

Blended Working as Resource to Cope with High Job Demands

Quantitative job demands such as work overload, work pressure, or having too much work to do in too little time (Peeters, Montgomery, Bakker, & Schaufeli, 2005), are often perceived as a common contemporary cause of work-related stress (European Agency for Safety and Health at Work, 2013). According to the basic tenets of the job demand–control (JD-C) model (Karasek, 1979) and the job demands–resources (JD-R) model (Bakker & Demerouti, 2007), it is not high demands per se, but high demands in combination with a lack of job resources (e.g., job autonomy) that undermine desirable work outcomes such as intrinsic work motivation. Intrinsic motivation is the motivation to perform an activity for itself, in order to experience the pleasure and satisfaction inherent in the activity (Deci & Ryan, 2002).

Indeed, to keep workers motivated and satisfied in their jobs, and to keep them effective and productive, job autonomy is generally considered as a core job characteristic (Hackman & Oldham, 1976; Gagné & Bhave, 2011). O'Neill et al. (2009) demonstrated that relative to traditional office-based workers, workers who worked at least one day per month away from the office reported higher job autonomy. Hence, the perceived opportunity for blended working is likely to be accompanied by perceived job autonomy, and accordingly, potentially an effective resource for workers to cope with the increasing quantitative job demands typically observed in today's workplace (Van Yperen, Wörtler, & De Jonge, 2016). Specifically, both the JD-C model (Karasek, 1979) and the JD-R model (Bakker & Demerouti, 2007) predict that the arousal produced by high job demands will not undermine workers' intrinsic motivation when they perceive opportunities to manage and effectively cope with the high demands. In contrast, when such opportunities are absent (i.e., perceived job autonomy is low), the arousal associated with high job demands cannot be transformed into action and, consequently, will be directed internally and undermine intrinsic motivation (Van Yperen & Hagedoorn, 2003). Thus, the perceived opportunity for blended working may be particularly important for workers' intrinsic motivation when they find themselves in highly demanding jobs.

Indeed, a recent study by Van Yperen et al. (2016) suggests that the perceived opportunity for blended working is an effective, contemporary resource for workers to cope with increasing job demands. However, drawing on person–job fit theory (Kristof-Brown et al., 2005), they demonstrated that this expected pattern applied only to workers with a high need for autonomy. As shown in Figure 8.1, workers high in need for autonomy who perceived opportunities for blended working reported the highest levels of intrinsic work motivation as job demands increased. In contrast, workers high in need for autonomy who perceived little opportunity for blended working reported lower levels of intrinsic

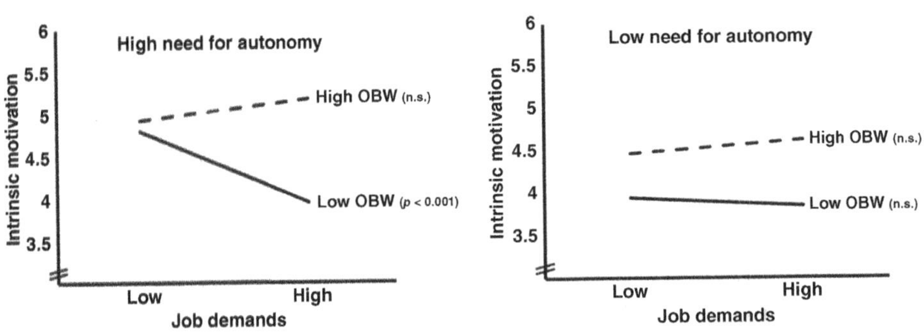

Figure 8.1 Interactive effect of job demands, perceived opportunity for blended working (OBW), and need for autonomy on intrinsic work motivation. "Low" and "high" reflect a value of 1 SD below and above the mean, respectively. n.s., not significant. *Source:* Adapted from Van Yperen, N. W., Wörtler, B., & De Jonge, K. M. M. (2016). Workers' intrinsic work motivation when job demands are high: The role of need for autonomy and perceived opportunity for blended working. *Computers in Human Behavior, 60*, 179–184.

motivation under conditions of high job demands. For workers low in need for autonomy, the perceived opportunity for blended working did not mitigate the negative effect of high job demands on intrinsic work motivation.

The findings presented in Figure 8.1 (Van Yperen et al., 2016) suggest that organizations and managers should create conditions at work that are conducive to workers' psychological needs (Williams, Halvari, Niemiec, Sørebø, Olafsen, & Westbye, 2014). Although it has been argued that the need for autonomy is intrinsically human (Deci & Ryan, 2002), workers who are relatively low in need for autonomy at work appear to profit less from the opportunity for blended working. Workplace interventions that provide structure and routine rather than job autonomy may work better for them, particularly when they have a high need for structure as well (Leone, Wallace, & Modglin, 1999; Van Yperen et al., 2014). Research has shown that managerial support for psychological needs is positively related to workers' psychological health, social wellness, work-related functioning, and job satisfaction (e.g., Gagné & Deci, 2005; Van den Broeck et al., 2010). For workers with a high need for autonomy, managers should provide opportunities for blended working, which may be particularly effective when job demands are high.

Blended Working and Job Satisfaction

In the field of industrial and organizational psychology, job satisfaction is a central variable of study. One reason is its positive relationships (albeit sometimes weak) with numerous desirable job outcomes, including job performance (Iaffaldano & Muchinsky, 1985), reduced turnover (Tett & Meyer, 1993), and organizational citizenship behavior (Organ & Ryan, 1995). Blended working is likely to pose a valuable means of increasing workers' job satisfaction and, accordingly, related desirable outcomes. Indeed, Gajendran and Harrison's (2007) meta-analysis revealed that working away from the office has modest but beneficial effects on workers' job satisfaction. Such a positive relationship is not limited to Western cultures. Bloom, Liang, Roberts, and Ying (2015) reported that in a nine-month experiment in China ($n = 249$ workers) that compared differences in outcomes such as job satisfaction between home-based working and traditional office-based working, job satisfaction was significantly higher for those working from home.

Other research suggests that the association between off-site working and job satisfaction is more complex, that is, contingent on its intensity. For example, Golden and Veiga (2005) found a curvilinear link between the extent of working outside the office and job satisfaction. At a certain extent of working remotely, the positive association with job satisfaction reached a peak and then started leveling off (*cf.*, Virick, DaSilva, & Arrington, 2010). In their sample, the threshold at which job satisfaction plateaued (and began to slightly decrease) was at around 15 hours of remote work per week. Moderators identified by Golden and Veiga (2005) were task interdependence (i.e., the degree to which organizational members must rely on one another to perform their tasks effectively given the nature of their job) and the extent to which workers had control over how an assigned task should be implemented. Workers occupying jobs which implied low levels of task interdependence and high levels of job control were inclined to experience comparatively higher levels of job satisfaction across all levels of remote work.

So far, it has been argued and demonstrated that blended working as a favorable work arrangement is contingent on both the person (e.g., psychological needs) and the context (e.g., low task interdependence). We now turn to specific subgroups of workers who may particularly benefit from blended working practices, namely workers who have dependent children, older workers, and workers with a physical disability.

Blended working and Specific Groups

Caregivers

The opportunity for blended working may be particularly effective for specific subgroups in organizations to keep them satisfied, motivated, and productive in their jobs (e.g., Maruyama & Tietze, 2012; O'Neill et al., 2009; Wheatley, 2012). For caregivers such as parents with dependent children, blended working may create opportunities for combining working life and childcare that were not previously available, and which are likely to improve and maintain work–life balance (Shockley & Allen, 2007). For example, Mustafa and Gold (2013) found that women with young children who worked from home tended to structure their temporal boundaries around their children's needs and school activities. O'Neill et al. (2009) showed that relative to traditional office-based workers, the workers who worked at least one day per month away from the office were more likely to have children under the age of five or between the ages of 5 and 17. In line with these findings, Van Yperen et al. (2014) found that number of children was positively related to the perceived effectiveness of time-independent working. Others have shown that the opportunity to work from home assists parents in managing both work and caring responsibilities (Maruyama & Tietze, 2012; Wheatley, 2012). For example, Dikkers et al. (2010) demonstrated that utilization of work–home arrangements (including flexible hours and working from home) by parents, and mothers in particular, was positively correlated with number of work hours and satisfaction with one's career.

Another reason blended working practices suit parents with dependent children is that it indirectly benefits the children's health. Working from home has been found to be associated with less frequent use of fast food for children's dinner (Allen, Shockley, & Poteat, 2008). This may also be regarded as an argument for blended working to pose a work arrangement with greater societal value. Note that organizational scientists have begun to consider children's health as even relevant to business concerns, because of its impact on outcomes of interest to organizations. Major, Cardenas, and Allard (2004) reviewed evidence that organizational outcomes such as healthcare expenses, lost time, and employer attractiveness, may be linked to child health issues. They argued that the opportunity to work from home allows parents to more easily and regularly gain preventive healthcare for their offspring, affords them the extra time they require to make satisfactory childcare arrangements in the event of an illness, and enables them to be the ones to care for their sick children. Such flexibility is likely to enhance the fulfillment of their job tasks and productivity.

Likewise, the opportunity for blended working can be particularly effective for workers with care responsibilities for family members with health problems or a disability. Specifically, a key worker benefit commonly associated with flexible working practices is being able to take dependable family members to doctor appointments, ensuring sick family members receive their regular medication, and generally being present for those in need (Major, Verive, & Joice, 2008).

Older workers

The workforce is aging more rapidly than ever before, as post-World War II cohorts reach ages 65 and over (Hedge & Borman, 2012). Many older workers are delaying their retirement not only as a result of, among other things, improved education and health, economic conditions, and flexible retirement options (Elias, Smith, & Barney, 2012; Hertel & Zacher, in press), but also to stay productive and mentally healthy (C. C. Lee,

Czaja, & Sharit, 2009). For organizations, it is important to retain older workers to avoid, or at least lower, the forecasted shortage of 20.8 million workers in the European Union by 2030 (Dropkin et al., 2016; Hertel & Zacher, in press; Sharit, Czaja, Hernandez, & Nair, 2009), and to keep workers with high levels of job expertise within the organization (Hedge & Borman, 2012). Relying on blended working practices may help retain older workers (De Jonge et al., 2015). Age is positively correlated with the need for autonomy (Kooij, de Lange, Jansen, Kanfer, & Dikkers, 2011) and the tendency to work from home (Bailey & Kurland, 2002; Gibson et al., 2002; Lister & Harnish, 2011), and negatively related to work–home segmentation preference, and the need for structure and relatedness (Van Yperen et al., 2014). Blended working fits these preferences and needs (Van Yperen et al., 2014), and accordingly, allows older workers to (re)design their jobs in a way that best suits their needs, including the need to avoid age-related stressors such as long commuting times, noise, and an overcrowded environment (Cutler 2006; Dropkin et al., 2016; Gibson et al., 2002; Hedge & Borman, 2012; Patrickson, 2002; Sharit et al., 2009), and their need for employment participation after their post-statutory retirement age (De Lange, Van Yperen, Van der Heijden, & Bal, 2010).

Other reasons blended working may work for older workers in particular is that they tend to thrive on their expertise and experience, they have had the time to earn the trust that is essential to remote working and to self-managing their work, and they may wish to continue working after retirement (Damman, 2016; De Lange et al., 2010; Dropkin et al., 2016) or move slowly towards retirement by shifting their emphasis more towards leisure time and nonwork activities (West & Anderson, 2005). Older workers and their partners are also more likely to face age-related health issues, and accordingly, mobility limitations (Dropkin et al., 2016; Lister & Harnish, 2011; Thompson & Mayhorn, 2012). Blended working practices allow older workers to balance their work and family responsibilities (Bailey & Kurland, 2002). Indeed, older ages have been found to be related to increased work–family balance, lower work–family conflict, greater job satisfaction and productivity, and lower absenteeism (Anderson, Bricout, & West, 2001; Hedge & Borman, 2012). Nevertheless, older workers may be ill-suited for using new Internet-related technologies due to lack of skill and experience with the computer technologies required for blended working, or lack of confidence or anxiety to use these technologies, which may be fueled by negative stereotypes about older workers (De Jonge et al., 2015; Elias et al. 2012; Hedge & Borman, 2012; Hertel & Zacher, in press; Raiborn & Butler, 2009; Van Dalen, Henkens, & Schippers, 2010). For example, people aged 50 years or more are less likely to own a computer or to use the Internet or computers in general (Cutler 2006; Hill et al., 2003). Only about 40% of those aged 65 years and over use the Internet (Charness, Fox, & Mitchum, 2010). These numbers are likely to increase rapidly in the (near) future because the next generations of older workers increasingly rely on ICTs (Thompson & Mayhorn, 2012). However, older workers often face perceptual, physical, and cognitive decline and may find it difficult to adapt rapidly to changing technological innovations (Thompson & Mayhorn, 2012; see also Burlacu, Truxillo, & Bauer, 2017, Chapter 20 in this volume).

Therefore, older workers in particular may require engaging in ICT training activities, which is advantageous for both the organization and themselves because training investments tend to pay off within a few years (De Jonge et al., 2015; Ng & Feldman, 2012; Patrickson, 2002). Older workers appear to be willing to learn the technological knowledge and skills required for their job, and their experience of success when using new technologies results in favorable attitudes towards it (Czaja & Moen, 2004; Cutler, 2006; Ng & Feldman, 2012; Sharit et al., 2009). To enable these positive outcomes, it is important to provide the right type of training (Cutler, 2006) and to include familiar tasks in the training program (Czaja & Moen, 2004). Possible physical and cognitive decline need to be taken into account, and the training program must be aligned with the needs of older workers (Dropkin et al., 2016; Thompson & Mayhorn, 2012; Sharit & Czaja, 2012). When older workers have successful experiences with computer technologies, they tend to experience these technologies as reducing the effort and time required to fulfill job tasks and as increasing their job performance, enabling them to keep working effectively and productive (Mitzner et al., 2010).

Workers with a Physical Disability

Among the working-age population in the European Union in 2002, 16.4%, reported that they had a long-term health problem or disability (Eurostat, 2003). Disability imposes severe challenges on the individuals affected, including transportation and lack of support services in the workplace (Schopp, 2004). More than traditional office-based work with a fixed work schedule, blended working poses a work arrangement that is conducive to (facilitated) employability of workers who have a physical disability. For instance, it provides a buffer against the stress of work–life disruptions for disabled people in particular because it allows them control over when and where to work. This is necessary for coping with issues such as fatigue, stamina, and pain, and it allows them to access their healthcare services during the workday (Linden, 2014). Besides facilitating work–life balance, blended working may also create work opportunities for individuals with disabilities who may not be otherwise employable (Hesse, 1996). On the downside, given that blended working for disabled workers may largely comprise home-based working, social and professional isolation (Cooper & Kurland, 2002) may become of concern for those with a high need for relatedness (Van Yperen et al., 2014). Overall, though, the benefits of blended working for disabled workers also points to the greater societal value inherent in such a work arrangement because it promotes equal opportunity. After all, the ability to work is a key component in achieving independence and inclusion in society.

Adoption of Blended Working in Organizations

Despite the advantages of blended working, its adoption has been slower than initially anticipated (Pyöriä, 2011). Management often has insufficient trust in workers working off-site, that is, they are apprehensive that workers will put in as much effort as possible in the absence of close supervision (e.g., Topi, 2004). Hence, it can be assumed that companies that offer the opportunity to work remotely are more likely to rely on incentive systems that compensate workers for results rather than time. Indeed, analyzing data from 122 CEOs in companies with an average number of 92 workers, Mayo, Pastor, Gomez-Mejia, and Cruz (2009) found that offering the opportunity to work away from the office for a certain number of hours per week was more likely among relatively young companies and companies with a high percentage of international workers, *but only* when the CEO reported high contingent reward leadership, that is, the provision of rewards in exchange for results.

Mayo et al. (2009) also found a stronger tendency to adopt flexible working practices in relatively small companies, probably due to their flexibility (or less bureaucracy) and the level of trust that may be higher as the company is smaller. That is, a smaller group size enhances group members' mutual awareness and the ability to monitor each other's behavior, and offers enough opportunities for face-to-face interactions. Research suggests that groupings of around 150 individuals (referred to as Dunbar's number) is the normative limit of the size of personal social networks among adults that can be held together through informal social control (Dunbar, 1993; Van Vugt & Kameda, 2012). Hence, an organizational climate of cohesion and trust might explain the higher prevalence of flexible working practices in smaller organizations which is, from a management perspective, associated with a higher willingness to take risk (Schoorman, Mayer, & Davis, 2007). In a management environment that trusts them, workers are more likely to organize themselves and their workload in such a way that does not undermine organizational goals. Thus, blended working practices may most effectively be implemented in companies characterized by trust and cohesion (which may be more prevalent in smaller companies), and in companies in which workers can be evaluated on the basis of their results.

Future Research

Blended working practices are associated with high perceived job autonomy (O'Neill et al., 2009; Van Yperen et al., 2014), which is generally considered as a core job characteristic (Gagné & Bhave, 2011; Hackman & Oldham, 1976). Similar to more traditional indices of job autonomy, our findings indicate that blended working is an effective resource for workers to cope with high job demands (Van Yperen et al., 2016). Future research may be directed at identifying implications of blended working for other domains in which job autonomy has been shown to be important, such as creativity (Hennessey & Amabile, 2010) and counterproductive work behaviors (Brink, Emerson, & Yang, 2016). In general, the impact of blended working on workers' *actual* job performance through job autonomy should be put to empirical test, which is likely to be a function of characteristics of both the worker (e.g., workers' psychological needs, personality factors) and the environment (e.g., type of work, leadership style, organizational culture).

Moreover, due to people's general preference for face-to-face interactions (e.g., Topi, 2004), future studies may be aimed at identifying the conditions under which supervision can be optimized in blended working practices. As emphasized by M. Lee (2014), the main challenge of leading geographically dispersed teams may be to overcome communication barriers and pitfalls that can trip up, including information and knowledge sharing, virtual etiquette (e.g., tone and style), and conflict management. One of the key contingencies in this regard may be the motivational climate within the team. Specifically, workers with different achievement goals differ in the way they develop and maintain relationships at work (Sijbom, Janssen, & Van Yperen, 2015) and in their willingness to share information and knowledge (Poortvliet, Janssen, Van Yperen, & Van de Vliert, 2007). Meta-analyses show that both other-referenced performance-approach (PAp) goals (aiming at performing better than others) and self-referenced mastery-approach (MAp) goals (aiming at performing better than one did before) positively affect performance at work and in other domains (Van Yperen, Blaga, & Postmes, 2014, 2015). However, PAp goals and MAp goals likely activate different action plan goals, which are defined as strategies or pathways for achieving desired goals (DeShon & Gillespie, 2005). PAp goal individuals tend to perceive *negative interdependence* with others because their goal can be reached only at the cost of others, that is, by outperforming others. For example, PAp goal workers are likely to perceive and approach their supervisor as someone who frustrates their goal of outperforming others, and to feel superior vis-à-vis others, including their supervisor (Van Yperen, in press). In contrast, MAp goal workers tend to perceive and approach supervisors as valuable sources of work-related knowledge, information, and experience that potentially serve their goal of learning, growth, and development. In turn, supervisors may provide these intrinsically motivated workers with support, decision latitude, and freedom so that they can initiate, control, and carry out their tasks more autonomously. Hence, MAp workers and their supervisors are likely to count on each other for support and loyalty, share important resources, and base their exchange relationship on mutual trust, respect, and obligation. For example, in a study of 170 workers from a Dutch energy supplier, Janssen and Van Yperen (2004) found that relative to PAp goal individuals, MAp goal individuals reported higher quality exchange relationships (i.e., more mutual trust and respect) with their supervisors.

Another interesting question is whether blended working can be learned (*cf.*, Greer & Payne, 2014), which might primarily be a function of workers' beliefs with regard to the malleability of relevant skills, attitudes, and psychological needs (Dweck, 1986). Most likely, workers can be trained to employ strategies for overcoming blended working challenges, including strategies to ensure accessibility of information and connectivity, to avoid distractions, to adopt a work-oriented mindset, and to plan and schedule one's working hours and tasks. On the other hand, our findings suggest that workers high in need for relatedness and structure, and low in need for autonomy perceive blended working as ineffective (Van Yperen et al., 2014). Hence, future studies may examine strategies that can render blended working practices suitable for those workers in particular.

Conclusion

The rapid advance of ever-improving ICT and computer technologies in modern societies provides a variety of opportunities to work, to learn, and to care for, among other things. These pervasive and booming trends in modern societies are referred to as *blended working* (e.g., Van Yperen et al., 2014), *blended learning* (e.g., Hew & Cheung, 2014; Owston, 2013), and *blended care* (e.g., Wilhelmsen et al., 2013). Blended working, the central concept in this chapter, combines on-site and off-site working at any time. We have shown that blended working has the potential to provide considerable benefits to both organizations and individual workers, but it is associated with serious drawbacks as well (see also Rice, 2017, Chapter 9 in this volume). Whether blended working works for workers seems to be a function of characteristics of both the worker (e.g., psychological needs, personality factors) and the environment (e.g., type of work, task interdependence, leadership style). By implication, blended working practices should not be seen as a "one-size-fits-all" issue (Van Yperen et al., 2014). For example, the perceived opportunity for blended working as an effective contemporary resource to cope with increasing job demands is relevant only to workers with a high need for autonomy at work (Van Yperen et al., 2016). Blended working seems particularly effective for specific groups such as caregivers, older workers, and workers with a physical disability (De Jonge et al., 2015; Van Yperen et al., 2014). Finally, based on Mayo et al. (2009), we have argued that blended working practices may be most effectively implemented in companies characterized by trust and cohesion, and in companies in which workers can be evaluated on the basis of their results. Given the rapid rate of improvements in ICT and computer technologies in modern societies provide opportunities for improving individual, organizational, and societal outcomes in ways that were not previously possible, both researchers and practitioners should develop and examine tailormade strategies that can render blended working practices suitable for all workers.

References

Allen, T. D., Golden, T., & Shockley, K. M. (2015). How effective is telecommuting? Assessing the status of our scientific findings. *Psychological Science in the Public Interest, 16*, 40–68. DOI:10.1177/1529100615593273

Allen, T. D., Johnson, R. C., Kiburz, K. M., & Shockley, K. M. (2013). Work-family conflict and flexible work arrangements: Deconstructing flexibility. *Personnel Psychology, 66*, 345–376. DOI:10.1111/peps.12012

Allen, T. D., Shockley, K. M., & Poteat, L. F. (2008). Workplace factors associated with family dinner behaviors. *Journal Of Vocational Behavior, 73*, 336–342. DOI:10.1016/j.jvb.2008.07.004

Anderson, J., Bricout, J. C., & West, M. D. (2001). Telecommuting: Meeting the needs of businesses and employees with disabilities. *Journal of Vocational Rehabilitation, 16*, 97–104.

Ashford, S. J., & Cummings, L. L. (1985). Proactive feedback seeking: The instrumental use of the information environment. *Journal of Occupational Psychology, 58*, 67–79. DOI: 10.1111/j.2044–8325.1985.tb00181.x

Ashforth, B. E., Kreiner, G. E., & Fugate, M. (2000). All in a day's work: Boundaries and micro role transitions. *Academy of Management Review, 25*, 472–491. DOI:10.5465/AMR.2000.3363315

Bailey, D. E., & Kurland, N. B. (2002). A review of telework research: Findings, new directions, and lessons for the study of modern work. *Journal of Organizational Behavior, 23*, 383–400. DOI:10.1002/job.144

Bakker, A. B., & Demerouti, E. (2007). The job demands-resources model: State of the art. *Journal of Managerial Psychology, 22*, 309–328. DOI:10.1108/02683940710733115

Bartel, C. A., Wrzesniewski, A., & Wiesenfeld, B. M. (2012). Knowing where you stand: Physical isolation, perceived respect, and organizational identification among virtual employees. *Organization Science, 23*, 743–757. DOI:10.1287/orsc.1110.0661

Block, W. (2001). Cyberslacking, business ethics and managerial economics. *Journal of Business Ethics, 33*, 225–231. DOI:10.1023/A:1012002902693

Bloom, N., Liang, J., Roberts, J., & Ying, Z. J. (2015). Does working from home work? Evidence from a Chinese experiment. *The Quarterly Journal of Economics, 130*, 165–218. DOI:10.1093/qje/qju032

Brink, A. G., Emerson, D. J., & Yang, L. (2016). Job autonomy and counterproductive behaviors in Chinese accountants: The role of job-related attitudes. *Journal of International Accounting Research, 15*, 115–131. DOI:10.2308/jiar-51194

Burlacu, G., Truxillo, D. M., & Bauer, T. N. (2017). Employee age differences in using Internet-based tools at work. In G. Hertel, D.L. Stone, R.D. Johnson, & J. Passmore (Eds.), *The Wiley Blackwell handbook of the psychology of the Internet at work* (pp. 449–480). Chichester: John Wiley & Sons, Ltd.

Busch, P. (2008). *Tacit knowledge in organizational learning*. Hershey, PA: IGI Publishing/IGI-Global. DOI:10.4018/978–1-59904–501–6

Charlier, S. D., Giumetti, G. W., Reeves, C. J., & Greco, L. S. (2017). Workplace cyberdeviance. In G. Hertel, D. L. Stone, R. D. Johnson, & J. Passmore (Eds.), *The Wiley Blackwell handbook of the psychology of the Internet at work* (pp. 130–156). Chichester: John Wiley & Sons, Ltd.

Charness, N., Fox, M. C., & Mitchum, A. L. (2010). Lifespan cognition and information technology. In K. Fingerman, C. Berg, T. Antonnuci, & J. Smith (Eds.), *Handbook of lifespan psychology*. New York, NY: Springer.

Cooper, C. D., & Kurland, N. B. (2002). Telecommuting, professional isolation, and employee development in public and private organizations. *Journal of Organizational Behavior, 23*, 511–532. DOI:10.1002/job.145

Cutler, S. J. (2006). Technological change and aging. In R. H. Binstock & L. K. Beorge (Eds.), *Handbook of aging and the social sciences* (6th ed) (pp. 257–276). Burlington, MA: Academic Press.

Czaja, S. J., & Moen, P. (2004). Technology and employment. In R. Pew & S. Van Hamel (Eds.), *Technology and adaptive aging* (pp. 150–178). Washington DC: National Research Council.

Damman, M. (2016). Blended work and employment participation of older workers: A further discussion. *Work, Aging and Retirement, 2*(4), 384–389. DOI:10.1093/workar/waw022

Deci, E. L., & Ryan, R. M. (2002). *Handbook of self-determination research*. Rochester, NY: University of Rochester Press.

De Jonge, K. M. M., Van Yperen, N. W., & Rietzschel, E. F. (2015). Age and blended working. In N. A. Pachana, & N. Thapa (Eds.), *Encyclopedia of geropsychology*. New York, NY: Springer.

De Lange, A. H., Van Yperen, N. W., Van der Heijden, B. I. J. M., & Bal, P. M. (2010). Dominant achievement goals of older workers and their relationship with motivation-related outcomes. *Journal of Vocational Behavior, 77*, 118–125. DOI:10.1016/j.jvb.2010.02.013

DeShon, R. P., & Gillespie, J. Z. (2005). A motivated action theory account of goal orientation. *Journal of Applied Psychology, 90*, 1096–1127. DOI:10.1037/0021-9010.90.6.1096

Dikkers, J., Van Engen, M., & Vinkenburg, C. (2010). Flexible work: ambitious parents' recipe for career success in The Netherlands. *Career Development International, 15*, 562–582. DOI:10.1108/13620431011084411

Dropkin, J., Moline, J., Kim, H., & Gold, J. E. (2016). Blended work as a bridge between traditional workplace employment and retirement: A conceptual review. *Work, Aging and Retirement, 4*, 373–383. DOI:10.1093/workar/waw017

Dunbar, R. I. M. (1993). Coevolution of neocortical size, group size, and language in humans. *Behavioral and Brain Sciences, 16*, 681–735. DOI: 10.1017/S0140525X00032325

Dweck, C. S. (1986). Motivational processes affecting learning. *American Psychologist, 41*, 1040–1048. DOI:10.1037/0003-066X.41.10.1040

Ehrhart, M. G., & Klein, K. J. (2001). Predicting followers' preferences for charismatic leadership: the influence of follower values and personality. *Leadership Quarterly, 12*, 153. DOI:10.1016/S1048-9843(01)00074-1

Elias, S. M., Smith, W. L., & Barney, C. E. (2012). Age as a moderator of attitude towards technology in the workplace: Work motivation and overall job satisfaction. *Behaviour & Information Technology, 31*, 453–467. DOI:10.1080/0144929X.2010.513419

European Agency for Safety and Health at Work. (2013). *European opinion poll on occupational safety and health*. Luxembourg: Publications Office of the European Union. Retrieved April 7, 2017, from https://osha.europa.eu/en/surveys-and-statistics-osh/european-opinion-polls-safety-and-health-work/european-opinion-poll-occupational-safety-and-health-2013

Eurostat. (2003). One in six of the EU working-age population report disability. News Release 142/2003, 5 December, 2003. Retrieved April 7, 2017, from http://ec.europa.eu/eurostat/web/products-press-releases/-/3-05122003-AP

Gagné, M., & Bhave, D. (2011). Autonomy in the workplace: An essential ingredient to employee engagement and well-being in every culture. In V. I. Chirkov, R. M. Ryan, & K. M. Sheldon (Eds.), *Human autonomy in cross-cultural context: Perspectives on the psychology of agency, freedom, and well-being* (pp. 163–187). Dordrecht, Netherlands: Springer.

Gagné, M., & Deci, E. L. (2005). Self-determination theory and work motivation. *Journal of Organizational Behavior, 26*, 331–362. DOI:10.1002/job.322

Gajendran, R. S., & Harrison, D. A. (2007). The good, the bad, and the unknown about telecommuting: Meta-analysis of psychological mediators and individual consequences. *Journal of Applied Psychology, 92*, 1524–1541. DOI:10.1037/0021-9010.92.6.1524

Gajendran, R. S., Harrison, D. A., & Delaney-Klinger, K. (2015). Are telecommuters remotely good citizens? Unpacking telecommuting's effects on performance via I-Deals and job resources. *Personnel Psychology, 68*, 353–393. DOI:10.1111/peps.12082

Gerdenitsch, C., Kubicek, B., & Korunka, C. (2015). Control in flexible working arrangements: When freedom becomes duty. *Journal of Personnel Psychology, 14*, 61–69. DOI:10.1027/1866-5888/a000121

Gibson, J. W., Blackwell, C. W., Dominicis, P., & Demerath, N. (2002). Telecommuting in the 21st century: Benefits, issues, and a leadership model which will work. *Journal of Leadership Studies, 8*, 75–86. DOI:10.1177/107179190200800407

Golden, T. D., & Veiga, J. F. (2005). The impact of extent of telecommuting on job satisfaction: Resolving inconsistent findings. *Journal of Management, 31*, 301–318. DOI:10.1177/0149206304271768

Greer, T. W., & Payne, S. C. (2014). Overcoming telework challenges: Outcomes of successful telework strategies. *The Psychologist-Manager Journal, 17*, 87–111. DOI:10.1037/mgr0000014

Hackman, J. R., & Oldham, G. R. (1976). Motivation through the design of work: Test of a theory. *Organizational Behavior and Human Performance, 16*, 250–279. DOI: 10.1016/0030-5073(76)90016-7

Harris, L. (2003). Home-based teleworking and the employment relationship: Managerial challenges and dilemmas. *Personnel Review, 32*, 422–437. DOI: 10.1108/00483480310477515

Hedge, J. W., & Borman, W. C. (2012). *The Oxford handbook of work and aging*. New York, NY: Oxford University Press. DOI:10.1093/oxfordhb/9780195385052.001.0001

Hennessey, B. A., & Amabile, T. M. (2010). Creativity. *Annual Review of Psychology, 61*, 569–598. DOI:10.1146/annurev.psych.093008.100416

Hertel, G., & Zacher, H. (In press). Managing the aging workforce. In C. Viswesvaran, N. Anderson, D. S. Ones, & H. K. Sinangil (Eds.), *The SAGE handbook of industrial, work, & organizational psychology* (2nd ed., Vol. *3*). New York: Sage.

Hesse, B. W. (1996). Using telework to accommodate the needs of employees with disabilities. *Journal Of Organizational Computing & Electronic Commerce, 6*, 327. DOI: 10.1080/10919399609540283

Hew, K. F. & Cheung, W. S. (2014). *Using blended learning*. Berlin: Springer.

Hill, E. J., Ferris, M., & Märtinson, V. (2003). Does it matter where you work? A comparison of how three work venues (traditional office, virtual office, and home office) influence aspects of work and personal/family life. *Journal of Vocational Behavior, 63*, 220–241. DOI:10.1016/S0001-8791(03)00042-3

Hofer, J., & Busch, H. (2011). Satisfying one's needs for competence and relatedness: Consequent domain-specific well-being depends on strength of implicit motives. *Personality and Social Psychology Bulletin, 37*, 1147–1158. DOI:10.1177/0146167211408329

Iaffaldano, M. T. & Muchinsky, P. M. (1985). Job satisfaction and job performance: A meta-analysis. *Psychological Bulletin, 97*, 251–273. DOI:10.1037/0033-2909.97.2.251

Janssen, O., & Van Yperen, N. W. (2004). Employees' goal orientations, the quality of leader-member exchange, and the outcomes of job performance and job satisfaction. *Academy of Management Journal, 47*, 368–384. DOI:10.2307/20159587

Karasek, R. A. (1979). Job demands, job decision latitude, and mental strain: Implications for job redesign. *Administrative Science Quarterly, 24*, 285–308. DOI:10.2307/2392498

Kelliher, C., & Anderson, D. (2010). Doing more with less? Flexible working practices and the intensification of work. *Human Relations, 63*, 83–106 24p. DOI:10.1177/0018726709349199

Kooij, D. M., de Lange, A. H., Jansen, P. W., Kanfer, R., & Dikkers, J. E. (2011). Age and work-related motives: Results of a meta-analysis. *Journal of Organizational Behavior, 32*, 197–225. DOI:10.1002/job.665

Kristof-Brown, A., Zimmerman, R., & Johnson, E. (2005). Consequences of individuals' fit at work: A meta-analysis of person-job, person-organization, person group, and person-supervisor fit. *Personnel Psychology, 58*, 285–342. DOI:10.1111/j.1744-6570.2005.00672.x

Kurland, N. B., & Bailey, D. E. (1999). Telework: The advantages and challenges of working here, there, anywhere, and anytime. *Organizational Dynamics, 28*, 53–67. DOI: 10.1016/S0090-2616(00)80016-9

Lee, C. C., Czaja, S. J., & Sharit, J. (2009). Training older workers for technology-based employment. *Educational Gerontology, 35*, 15–31. DOI:10.1080/03601270802300091

Lee, M. (2014). *Leading virtual project teams: Adapting leadership theories and communications techniques to 21st century organizations*. Boca Raton, FL: CRC Press.

Leone, C., Wallace, H. M., & Modglin, K. (1999). The need for closure and the need for structure: Interrelationships, correlates and outcomes. *Journal of Psychology, 133*, 553–562.

Linden, M. (2014). Telework research and practice: Impacts on people with disabilities. *Work: Journal of Prevention, Assessment & Rehabilitation, 48*, 65–67. DOI: 10.3233/WOR-141857

Lister, K., & Harnish, T. (2011). *The state of telework in the U.S.: How individuals, business, and government benefit*. San Diego, CA: Telework Research Network.

Major, D. A., Cardenas, R. A., & Allard, C. B. (2004). Child health: A legitimate business concern. *Journal of Occupational Health Psychology, 9*, 306–321. DOI:10.1037/1076-8998.9.4.306

Major, D. A., Verive, J. M., & Joice, W. (2008). Telework as a dependent care solution: Examining current practice to improve telework management strategies. *Psychologist-Manager Journal, 11*, 65–91. DOI:10.1080/10887150801967134

Maruyama, T., & Tietze, S. (2012). From anxiety to assurance: Concerns and outcomes of telework. *Personnel Review, 41*, 450–469. DOI:10.1108/00483481211229375

Matusik, S. F., & Mickel, A. E. (2011). Embracing or embattled by converged mobile devices? Users' experiences with a contemporary connectivity technology. *Human Relations, 64*, 1001–1030. DOI:10.1177/0018726711405552

Mayo, M., Pastor, J., Gomez-Mejia, L., & Cruz, C. (2009). Why some firms adopt telecommuting while others do not: A contingency perspective. *Human Resource Management, 48*, 917–939. DOI:10.1002/hrm.20322

Mazmanian, M. A., Orlikowski, W. J. & Yates, J. (2005). Crackberries: The social implications of ubiquitous wireless e-mail devices. In C. Sørensen, Y. Yoo, K. Lyytinen & J. I. DeGross (Eds.), *Designing ubiquitous information environments: Socio-technical issues and challenges* (pp. 337–344). New York: Springer.

McLennan, K. J. (2008). *The virtual world of work: How to gain competitive advantage through the virtual world place.* Charlotte, NC: Information Age Publishing.

Millward, L. J., Haslam, S. A., & Postmes, T. (2007). Putting employees in their place: The impact of hot desking on organizational and team identification. *Organization Science, 18*, 547–599. DOI:10.1287/orsc.1070.0265

Mitzner, T. L., Boron, J. B., Fausset, C. B., Adams, A. E., Charness, N., Czaja, S. J., ... Sharit, J. (2010). Older adults talk technology: Technology usage and attitudes. *Computers in Human Behavior, 26*, 1710–1721. DOI.10.1016/j.chb.2010.06.020

Morgeson, F. P., Aguinis, H., Waldman, D. A., & Siegel, D. S. (2013). Extending corporate social responsibility research to the human resource management and organizational behavior domains: A look to the future. *Personnel Psychology, 66*, 805–824. DOI:10.1111/peps.12055

Mustafa, M., & Gold, M. (2013). "Chained to my work?" Strategies to manage temporal and physical boundaries among self-employed teleworkers. *Human Resource Management Journal, 23*, 413–429. DOI:10.1111/1748-8583.1200

Ng, T. W. H., & Feldman, D. C. (2012). Evaluating six common stereotypes about older workers with meta-analytical data. *Personnel Psychology, 65*, 821–858. DOI:10.1111/peps.12003

O'Neill, T. A., Hambley, L. A., & Bercovich, A. (2014). Prediction of cyberslacking when employees are working away from the office. *Computers in Human Behavior, 34*, 291–298. DOI:10.1016/j.chb.2014.02.015

O'Neill, T. A., Hambley, L. A., Greidanus, N. S., MacDonnell, R., & Kline, T. J. B. (2009). Predicting teleworker success: An exploration of personality, motivational, situational, and job characteristics. *New Technology, Work & Employment, 24*, 144–162. DOI:10.1111/j.1468-005X.2009.00225.x

Organ, D. W & Ryan, K. (1995), A meta-analytic review of attitudinal and dispositional predictors of Organizational Citizenship Behavior. *Personnel Psychology, 48*, 775–802. DOI:10.1111/j.1744-6570.1995.tb01781.x

Owston, R. (2013). Blended learning policy and implementation: Introduction to the special issue. *The Internet and Higher Education, 18*, 1–3. DOI:10.1016/j.iheduc.2013.03.002

Patrickson, M. (2002). Teleworking: Potential employment opportunities for older workers? *International Journal of Manpower, 23*, 704–715. DOI:10.1108/01437720210453902

Peeters, M. C. W., Montgomery, A. J., Bakker, A. B., & Schaufeli, W. B. (2005). Balancing work and home: How job and home demands are related to burnout. *International Journal of Stress Management, 12*, 43–61. DOI:10.1037/1072-5245.12.1.43

Perry, S. J., Lorinkova, N. M., Hunter, E. M., Hubbard, A., & McMahon, J. T. (2016). When does virtuality really "work"? Examining the role of work-family and virtuality in social loafing. *Journal of Management, 42*, 449–479. DOI: 10.1177/0149206313475814

Poortvliet, P. M., Janssen, O., Van Yperen, N. W., & Van de Vliert, E. (2007). Achievement goals and interpersonal behavior: How mastery and performance goals shape information exchange. *Personality and Social Psychology Bulletin, 33*, 1435–1447. DOI:10.1177/0146167207305536

Pyöriä, P. (2011). Managing telework: risks, fears and rules. *Management Research Review, 34*, 386–399. DOI:10.1108/01409171111117843

Raiborn, C. & Butler, J. B. (2009). A new look at telecommuting and teleworking. *Journal of Corporate Accounting & Finance, 20*, 31–39. DOI:10.1002/jcaf.20511

Rice, R. E. (2017). Flexwork, work–family boundaries, and information and communication technologies. In: G. Hertel, D. L. Stone, R. D. Johnson, & J. Passmore (Eds.), *The Wiley Blackwell handbook of the psychology of the Internet at work* (pp. 175–194). Chichester: John Wiley & Sons, Ltd.

Salanova, M., Llorens, S., & Cifre, E. (2013). The dark side of technologies: Technostress among users of information and communication technologies. *International Journal of Psychology, 48*, 422–436. DOI:10.1080/00207594.2012.680460

Schoorman, F., Mayer, R. C., & Davis, J. H. (2007). An integrative model of organizational trust: Past, present, and future. *Academy of Management Review, 32*, 344–354. DOI:10.5465/AMR.2007.24348410

Schopp, L. H. (2004). Telework for persons with disabilities in the E.U. and the U.S.A.: What can we learn from each other? *Studies in Health Technology and Informatics, 106*, 47–51. DOI:10.3233/978-1-60750-948-6-47

Sharit, J., & Czaja, S. J. (2012). Job design and redesign for older workers. In J. W. Hedge, & W. C. Borman, (Eds.), *The Oxford handbook of work and aging* (pp. 454–482). New York, NY: Oxford University Press. DOI:10.1093/oxfordhb/9780195385052.013.0143

Sharit, J., Czaja, S. J., Hernandez, M. A., & Nair, S. N. (2009). The employability of older workers as teleworkers: An appraisal of issues and an empirical study. *Human Factors and Ergonomics in Manufacturing, 19*, 457–477. DOI:10.1002/hfm.20138

Shockley, K. M., & Allen, T. D. (2007). When flexibility helps: Another look at the availability of flexible work arrangements and work-family conflict. *Journal of Vocational Behavior, 71*, 479–493. DOI:10.1016/j.jvb.2007.08.006

Sijbom, R. B. L., Janssen, O., & Van Yperen, N. W. (2015). How to get radical creative ideas into a leader's mind? Leader's achievement goals and subordinates' voice of creative ideas. *European Journal of Work and Organizational Psychology, 24*, 279–296. DOI: 1080/1359432X.2014.892480

Slijkhuis, J. M., Rietzschel, E. F., & Van Yperen, N. W. (2013). How evaluation and need for structure affect motivation and creativity. *European Journal of Work & Organizational Psychology, 22*, 15–25. DOI:10.1080/1359432X.2011.626244

Tett, R. P. & Meyer, J. P. (1993). Job satisfaction, organizational commitment, turnover intention and turnover: Path analyses based meta-analytic findings. *Personnel Psychology, 46*, 259–294. DOI:10.1111/j.1744-6570.1993.tb00874.x

Thompson, L. F., & Mayhorn, C. B. (2012). Aging workers and technology. In J. W. Hedge, & W. C. Borman (Eds.), *Oxford handbook of work and aging* (pp. 341–361). New York, NY: Oxford University Press. DOI:10.1093/oxfordhb/9780195385052.013.0113

Thompson, M. M., Naccarato, M. E., Parker, K. C. H., & Moskowitz, G. (2001). The Personal Need for Structure (PNS) and Personal Fear of Invalidity (PFI) scales: Historical perspectives, present applications and future directions. In G. Moskowitz (Ed.), *Cognitive social psychology: The Princeton symposium on the legacy and future of social cognition* (pp. 19–39). Mahwah, NJ: Erlbaum.

Topi, H. (2004). Supporting telework: Obstacles and solutions. *Information Systems Management, 21*, 79–85. DOI:10.1201/1078/44432.21.3.20040601/82481.12

Van Dalen, H. P., Henkens, K., & Schippers, J. (2010). Productivity of older workers: Perceptions of employers and employees. *Population and Development Review, 36*, 309–330. DOI:10.1111/j.1728-4457.2010.00331.x

Van den Broeck, A., Vansteenkiste, M., De Witte, H., Soenens, B., & Lens, W. (2010). Capturing autonomy, competence, and relatedness at work: Construction and initial validation of the Work-Related Basic Need Satisfaction Scale. *Journal of Occupational & Organizational Psychology, 83*, 981–1002. DOI:10.1348/096317909X481382

Van Vugt, M., & Kameda, T. (2012). Evolution and groups. In J. Levine (Ed.), *Group processes* (pp. 297–332). New York: Psychology Press.

Van Yperen, N. W. (2017). Competence and the Workplace. In A. J. Elliot, C. S. Dweck, & D. S. Yeager (Eds.), Handbook of competence and motivation (2nd ed): theory and application. (pp. 635–654). New York, NY: Guilford Press.

Van Yperen, N. W., Blaga, M., & Postmes, T. (2014). A meta-analysis of self-reported achievement goals and nonself-report performance across three achievement domains (work, sports, and education). *PLoS One, 9*, e93594. DOI:10.1371/journal.pone.0093594

Van Yperen, N. W., Blaga, M., & Postmes, T. (2015). A meta-analysis of the impact of situationally induced achievement goals on task performance. *Human Performance, 28*, 165–182. DOI:10.1080/08959285.2015.1006772

Van Yperen, N. W., & Hagedoorn, M. (2003). Do high job demands increase intrinsic motivation or fatigue or both? The role of job control and job social support. *Academy of Management Journal*, *46*, 339–348. DOI:10.2307/30040627

Van Yperen, N. W., Rietzschel, E. F., & De Jonge, K. M. M. (2014). Blended working: For whom it may (not) work. *PLoS One*, *9*, 1–8. DOI:10.1371/journal.pone.0102921

Van Yperen, N. W., Wörtler, B., & De Jonge, K. M. M. (2016). Workers' intrinsic work motivation when job demands are high: The role of need for autonomy and perceived opportunity for blended working. *Computers in Human Behavior*, *60*, 179–184. DOI:10.1016/j.chb.2016.02.068

Virick, M., DaSilva, N., & Arrington, K. (2010). Moderators of the curvilinear relation between extent of telecommuting and job and life satisfaction: The role of performance outcome orientation and worker type. *Human Relations*, *63*, 137–154. DOI: 10.1177/0018726709349198

West, M. D., & Anderson, J. (2005). Telework and employees with disabilities: Accommodation and funding options. *Journal of Vocational Rehabilitation*, *23*, 115–122.

Wheatley, D. (2012). Good to be home? Time-use and satisfaction levels among home-based teleworkers. *New Technology, Work and Employment*, *27*, 224–241. DOI:10.1111/j.1468–005X.2012.00289.x

Wilhelmsen, M., Lillevoll, K., Risør, M. B., Høifødt, R., Johansen, M. L., Waterloo, K., ... Kolstrup, N. (2013). Motivation to persist with internet-based cognitive behavioural treatment using blended care: A qualitative study. *BMC Psychiatry*, *13*, 296. DOI:10.1186/1471–244X-13–296

Wilks, L., & Billsberry, J. (2007). Should we do away with teleworking? An examination of whether teleworking can be defined in the new world of work. *New Technology, Work & Employment*, *22*, 168–177. DOI:10.1111/j.1468–005X.2007.00191.x

Williams, G. C., Halvari, H., Niemiec, C. P., Sørebø, Ø, Olafsen, A. H., & Westbye, C. (2014). Managerial support for basic psychological needs, somatic symptom burden and work-related correlates: A self-determination theory perspective. *Work & Stress*, *28*, 404–419. DOI:10.1080/02678373.2014.971920

9

Flexwork, Work–Family Boundaries, and Information and Communication Technologies

Ronald E. Rice

Introduction

Flexwork involves workers having a choice in time, location, and duration of work-related tasks (Hill et al., 2008). Organizations have increasingly offered, and employees have increasingly used, flexwork options owing to organizational, technological, social, economic, and legislative forces (Kossek & Michel, 2010). The trend toward more flexible work arrangements involves transformations of work, office design, and work locations (Cowan & Hoffman, 2007; Felstead, Jewson, & Walters, 2005). Information and communication technologies, largely operating through the Internet and now wireless transmission, both facilitate and shape flexwork. Most flexwork would be impossible without the use of information and communication technologies (ICTs), as they allow communication, collaboration, and use of resources across time, space, participants, and work–family boundaries. With this increase in the use and forms of flexwork come change in, and concerns about, boundaries, balance, and conflict between work, life, and family domains. Flexwork may benefit one domain while harming another, or require reconceptualization of behaviors and norms in one or more domains. Another primary shaper of and contributor to the rise of flexwork and associated work–family boundary, balance and conflict is the increasingly pervasive use of ICTs at work, home, and other times and places. ICTs play a central role in these developments and implications, but have not been much considered in the field of flexwork research.

Organizational and social science research has responded to these developments, especially flexwork. Online searches (ProQuest social science and Google Scholar) using combinations of flexwork terms, work–family balance terms, and ICTs indicate limited coverage of the *intersection* among these topics – little between flexwork terms and ICTs, or work–family balance terms and ICTs, and essentially none among all three. This chapter

The Wiley Blackwell Handbook of the Psychology of the Internet at Work,
First Edition. Edited by Guido Hertel, Dianna L. Stone, Richard D. Johnson, and Jonathan Passmore.

summarizes conceptual definitions and distinctions, influences, and outcomes (positive, negative, conditional) associated with flexwork, work–family boundaries, and ICTs, and their interrelationships. The chapter also provides a model of the overall interrelationships of these important and growing aspects of organizational work and life, and suggests key research questions at the intersection of these three areas.

Overall Relationships

We begin with a general summary model of overall relationships among the central topics of flexwork, work–family balance, and ICTs. Figure 9.1 portrays how the relationships emerged from the review, but is used here to introduce the review topics and their relationships. Many factors – whether required or voluntary, facilitators or obstacles, legislation or norms, financial or demographic – influence organizational and individual awareness of and interest in adopting flexwork. Flexwork is a general term representing a portfolio of possible variations in amount, time, location, and control of work. There are many aspects of offering organizational flexwork, from formal, organization-wide policies to informal and unequal supervisory allowance and restrictions. Thus the level, types, and implications of individual flexwork use are diverse, both conceptually and empirically distinct from formal offerings. Use of flexwork is associated with a wide array of positive and negative outcomes, at organizational, individual, family, and social levels. Some of these are mediated by the characteristics of flexworkers' work–family boundaries, the crossing of those boundaries, and conflict arising in either work or family domains. Communication about work and family domains influence the interpretation, experience, and outcomes of those boundary characteristics and crossings, and their relation to outcomes. Many relationships between flexwork use, boundaries, and outcomes are moderated or mediated by a wide range of individual, organizational, and family characteristics. Some potentially

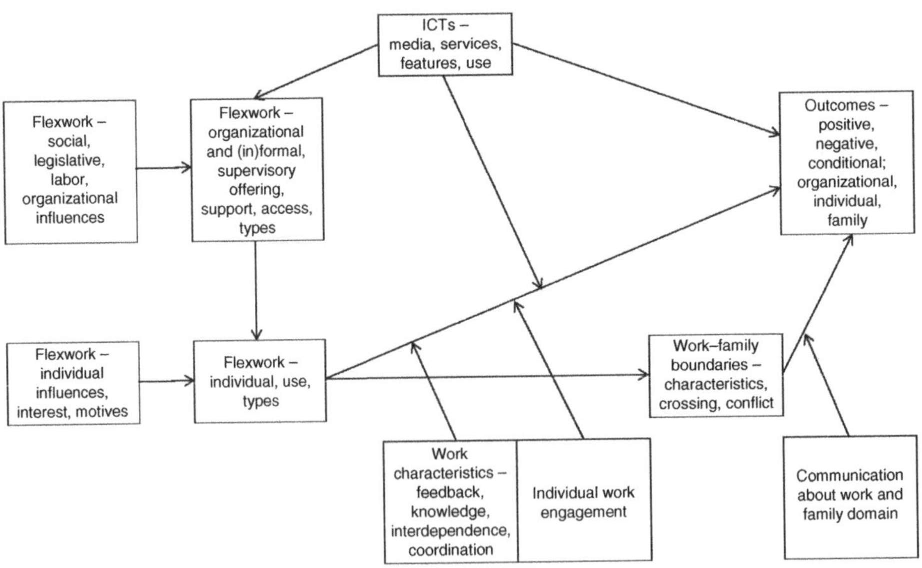

Figure 9.1 Overall model relating flexwork, work–family, boundaries, information and communication technologies, and outcomes.

relevant concepts affecting these relationships, such as work engagement, have not been much studied in the flexwork or boundary literature. Permeating all these concepts and relationships is the wide diversity, changing nature, and range of uses of ICTs – including infrastructure, devices, services, features, and extent of use. ICTs both enable, and affect the nature of, flexwork experience, work-family boundaries, and outcomes.

Flexwork

Flexwork or work flexibility includes flextime, part-time jobs, telecommuting/flexplace, job-sharing, compressed work weeks, unpaid personal leave, and sick leave for ill child care (Eaton, 2003). Related terms include, with variations in meaning, flexible work, flexibility in work, workplace flexibility, flexible work arrangements, or blended work. Van Yperen and Wörtler (2017, Chapter 8 in this volume), review related concepts, practices, and implications of blended working, involving the use of both multimedia and face-to-face communication and information sharing to engage in more or less seamless work across times and places. They discuss psychological and group needs, perceptions, and contexts, as well as organizational adoption of blended working.

Flexwork may also include telework or telecommuting, and thus is a more general concept. Telecommuting replaces at least some traveling to and from work, which occurs at a variety of home-based or nonhome-based contexts separate from the employee's supervisor (Mokhtarian, 1991). Telework may or not replace commuting; the US Telework Enhancement Act 2010 defines telework as:

> *a work flexibility arrangement under which an employee performs the duties and responsibilities of such employee's position, and other authorized activities, from an approved worksite other than the location from which the employee would otherwise work*
>
> US Office of Personnel Management
> (retrieved April 25, 2017 from www.opm.gov/FAQs, n.p.).

Kossek and Michel (2010) list four main categories of flexwork (with 14 more specific options): timing of work, location, amount and hours of work, and continuity/breaks. They also provide a comprehensive review of flexwork scheduling and associated outcomes. Matos and Galinsky (2014) list six general categories of flexwork (with 18 more specific options): flextime and place, choices in managing time, reduced time, caregiving leaves, time off, and flex careers. Their 2014 US National Study of Employers find that the options most commonly offered by organizations are: taking breaks (92% organizations offered this), time off for important family and personal needs (82%), and flextime (81%). These percentages are higher for smaller organizations (50–99 employees). Qualitative research by Cowan and Hoffman (2007) identified four valued categories of flexwork: time (e.g., scheduling, modifying work schedule to meet life demands), space (e.g., home, variation in preferences for permeability), evaluation (e.g., time at work, quality of results), and compensation (e.g., financial, time, overload).

Flexwork offering, use, and interest

The formal organizational offering of flexwork is just one component of the meaning of "flexwork" in practice. "Flexwork" includes what is formally provided or offered by organizations, what is offered to different employees, what is supported by managers/supervisors and organizational culture, and what is perceived and actually used by employees

(Eaton, 2003). The main distinction is between what is offered by the organization or supervisor and what is used by individuals.

A 2008 survey of US organizations with at least 50 employees found a range of flexible work arrangements, from flexible schedules (79%) to job-shares (29%) (Galinsky, Bond, & Sakai, 2008). The 2010 US Census Bureau's Survey of Income and Program Participation reported 6.6% employees worked exclusively from home on their primary job and 9.5% at least one day a week (Mateyka, Rapino, & Landivar, 2012). The Flexible Work Arrangements Factsheet (2010) reported that close to 29% of full-time employees used flexible work schedules, though only 38.8% of those participated in formal programs, meaning that a majority of those organizations offered flexwork informally or on an individual basis. Based on a 2014 web survey of 400 000 US federal employees (46.8% response) (US Office of Personnel Management, 2016), 29% of the sample participated in telework, although only 4% did so three or more days per week, 10% for one or two days per week, 4% no more than one or two days per month, and the remaining 11% did so infrequently or on an unscheduled basis. Beyond the 29% participating, 13% chose not to telework, and the remaining 57% did not telework due to barriers. Distinct from telework, alternative work schedules were used by a third, with 22% reporting that the option was not available to them.

Matos and Galinsky's (2014) report on a U.S. representative sample of 1,051 employers with 50 or more employees provides a recent and comprehensive summary of flexible work offerings, trends, and options. They describe increases in job flexibility in general, including options such as working remotely occasionally, increasing from 50% in 2008 to 67% in 2014, and control over overtime, up from 27% to 45%. On the other hand, provisions allowing extended time away from work largely saw declines, including practices such as job-sharing, down from 29% to 18%, and career breaks for personal and family responsibilities, down from 64% to 52%.

The Australian Public Service Commission (2013) update summarizes several government and industry reports. Overall, demand for more flexible work arrangements was growing in 2012. About 16% of respondents (working adults) to the fifth (2012) Australian Work and Life Index (Skinner, Hutchinson, & Pocock, 2012) worked at home on a regular basis; 44.2% worked at home sometimes (regularly or not). The eighth Household, Income and Labour Dynamics in Australia Survey (2010) indicated 22.8% worked at least some hours of their main job at home each week, with over 70% working fewer than 10 hours at home each work. Only 4.9% of all employed workers worked most of their time at home, and only 22.6% of those were employees as opposed to being self-employed. Hence current overall percentages of hours worked at home are low.

Flexwork is relevant to many employees. In 2010, an average of only 30.1% of employees in the European Union (EU) felt that their working hours fit their family/social commitments very well (Hoonakker, 2014). According to the Flexible Work Arrangements Factsheet (2010), 80% of US workers indicated preferences for more flexible work options, particularly if these options were not associated with negative implications at work. Of the 89% of workers who did not currently have any regular home workhours, 43% indicated they would like the option. In a 2000 EU survey, employee interest in telecommuting was also higher than employer interest (Peters, Tijdens, & Wetzels, 2004). And such programs are popular: of the 33% participating in them in the US Office of Personnel Management survey (2016), 89% reported they were satisfied with the alternative work schedules. A whitepaper on the Australian national broadband network documented that 60% of mature workers in that study felt they would move to telework and thus delay retirement, with 6% even willing to find such work in a new industry. Similarly, 62% of part-time and casual workers would accept telework if available (Australian Public Service Commission 2013).

Organizational and individual influences on flexwork

Influences on organizational offering of flexwork include characteristics of management, workforce, and organizations; organizational work–family culture; and supervisor support (McNamara, Pitt-Catsouphes, Brown, & Matz-Costa, 2012). Flexwork programs are often underused, partially due to organizational factors such as the nature of supervisory support, family-supportive culture, and reward systems (Batt & Valcour, 2003). Family support policy availability signals organizational concern, but actual use seems to be more influenced by personal interests and needs (Butts, Casper, & Yang, 2013).

Individual predictors of choice to use flexwork arrangements include longer tenure, supervisory responsibilities, and co-workers who used these arrangements (Lambert, Marler, & Gueutal, 2008). In the 2007–2008 Age and Generations Study, individual influences on access to and use of flexwork options (varying by program) included being a minority, female or disadvantaged individual, more education, and type of occupation (McNamara et al., 2012). For example, considering the temporal dimension of flexwork, on the one hand, workers in computer and math occupations were more likely to report using atypical schedules, compressed work weeks, or control when they took breaks. Management, on the other hand, was less likely to report occasional changes in when they started or quitted work.

Employees may be reluctant to use flexwork options out of concern for the effect on organizational support (Long, Kuang, & Buzzanell, 2013), careers (Hylmö & Buzzanell, 2002), performance evaluations and wages, and the "flexibility stigma" (Williams, Blair-Loy, & Berdahl, 2013). Some employees lack interest in or need for particular offerings (McNamara et al., 2012). Shockley and Allen (2010) concluded that need for role segmentation reduced flexwork use, while the relationship between need for achievement and flextime use was slightly strengthened by having greater family responsibility (i.e., children). Thus organizational options for and employees' actual use of flexwork are often weakly correlated.

Flexwork and telecommuting are not equally attractive to all job seekers. Rau and Hyland's (2002) survey showed that those with higher work–family role conflict were more attracted to organizations offering flexwork than to those providing telecommuting. But job seekers with lower work family role conflict were drawn more to organizations with telecommuting options, presumably because it involves lower transition costs and fewer out-of-role interruptions. A more subtle influence on flexwork offering and use involves discourses about and contradictions associated with "organizational policies and arrangements; workplace norms and practices; worker-supervisor relationships; and an individual's sense of agency" (Myers, Gailliard, & Putnam, 2012, p. 195). For example, mixed messages about flexwork, flexibility stigma, discrepancies between formal and informal access and use, wide variation in supervisory implementation, and diverse definitions of flexwork, all foster uncertainty and additional stress and conflict.

Outcomes associated with flexwork

Positive general outcomes associated with flexwork Flexwork can provide many benefits, including improved health and work–family balance, worker engagement, commitment, attachment, and lower absenteeism (see McNamara et al., 2012, for a review; for recent meta-analyses, see Gajendran & Harrison, 2007; Kossek, Pichler, Bodner, & Hammer, 2011). Flexwork is associated with increased (self-perceived) performance and decreased turnover, via increased affective commitment (loyalty and willingness to contribute) or citizenship behavior (Eaton, 2003). Work and family programs may help attract top workers, reduce work–family conflicts, increase peak productivity hours, increase commitment and motivation, more balanced use of organizational resources, decrease supervision/monitoring, and reduce absenteeism and turnover (Clifton & Shephard, 2004). Availability, and to less of an extent reported use, of work–family support policies are positively related to job satisfaction, affective commitment, and lower intentions to leave. These variables have a small influence on attitudes toward work – primarily as mediated by perceptions of those policies and by work–family conflict (Butts et al., 2013). Increased employer support of flexible and permeable work boundaries is associated with a variety of organizational benefits in terms of employee trust, commitment, satisfaction, attitudes, etc. (Scholarios & Marks, 2004).

Considering some studies on telework and telecommuting specifically, telework can increase job flexibility, allowing greater work–life balance, such as caring for dependents, balancing competing demands, improving relationships, and managing time (Gajendran & Harrison, 2007; Hilbrecht, Shaw, Johnson, & Andrey, 2008). Other potential benefits of telework include better work–family balance, lower travel and office costs, more opportunities for disabled workers, increased job performance, reduced sick days, and reduced air pollution (Wadsworth, Facer, & Arbon, 2010). Telework/telecommuting is positively associated with job satisfaction and performance (Bloom, Liang, Roberts, & Ying, 2013). One meta-analysis found a negative correlation between telecommuting and intention to leave (Gajendran & Harrison, 2007). However, a large-sample government survey of teleworking in US federal government agencies found no relationship between telecommuting and workers' turnover intentions, although employees working in agencies that denied this option reported higher turnover intention (Cailier, 2013).

Positive conditional outcomes associated with flexwork Leslie, Park, and Mehng (2012) discussed how the effects of flexwork on employee career success are influenced by the motivations managers attribute to the use of flexwork by employees: as a signal of commitment or of personal life management. If managers attribute use of flexwork as motivated by productivity, they perceived greater organizational commitment, generating a positive effect on employee rewards. However, the managerial attributions of employees' motivations were not correlated with the employees' own reported motivations. Thus both perceived and communicated motivations influence both flexwork offering and interpretations.

The type of flexwork location also matters. Morganson, Major, and Oborn (2010) discovered slightly greater work–family balance and job satisfaction among main office and home-based workers compared with those working from a satellite office or at a client's location, and also greater workplace inclusion in the main office. New work locations also involve issues of control and adaptation. In the context of open/office/open plan trends, managerial control of workspace affects employees' physical and psychological discomfort, and lowers organizational identification (because individuals' interests are constrained). Both these factors fully mediate the relationships between control and job satisfaction and wellbeing (Knight & Haslam, 2010).

Negative general outcomes associated with flexwork Flexwork programs raise issues of cost (insurance, productivity), labor force composition, equity for those without families, reduced salary compensation, uncertainty about actual organizational benefits, and intrusion into the worker's life (Clifton & Shephard, 2004). Flexwork options can increase isolation from the office environment and face-to-face interaction, thus lowering the valued job characteristic of task interdependence (Feldman & Gainey, 1997) as well as a sense of organizational affiliation (Morganson, Major, & Oborn, 2010). Although some meta-analyses demonstrate a relationship between flexwork and work–family conflict (Mesmer-Magnus & Viswesvaran, 2006), other studies do not (Gajendran & Harrison 2007; Golden, Veiga, & Simsek, 2006) (T. D. Golden, 2012).

One theoretical basis of flexwork is the need to balance energy and work exhaustion through resource recovery opportunities (Hobfoll, 1989). Applying this perspective, T. D. Golden (2012) found no direct effects of telework on exhaustion, but it did weakly increase the influence of time and strain-based work–family conflict, and strain-based family-to-work conflict, on exhaustion during regular and nonregular office hours. Another cost of telework includes stressful interruptions, which reduce organizational identification (Fonner & Roloff, 2012). Telework/telecommuting can generate role ambiguity, work–family spillover, lower control over planning and resources, isolation, and less feedback and support (Bloom et al., 2013).

Negative conditional outcomes associated with flexwork Rothbard, Phillips, and Dumas (2005) report that more access to integrating flexwork policies was negatively associated with, but more access to segmenting policies was positively related to, outcomes such as job satisfaction and organizational commitment, for those with a greater preference for role segmentation between work and family. Schieman and Glavin's (2008) analysis of the 2002–2003 US National Survey of the Changing Workforce (adult paid workers), showed that job attributes moderated the relationship between having work-related contact outside of working hours, and bringing work home, and work-to-home conflict. With low job autonomy, contact increased conflict; with more schedule control, bringing work home increased conflict. Thus autonomy and schedule control can be thought of as either resources or demands, affecting boundary blurring and conflict.

Flexworkers may have to continually reorder, perform, and reconceptualize their spatial and social relations both at work and at home. Richardson and McKenna (2014) describe how new flexworkers had to develop different within-home spatial and temporal boundaries. They needed to become better organized, and learn how to match tasks to places. They also became aware of the negative impact of distance on employee–manager trust, so had to display trustworthiness and make accomplishments more public. Relationships with work colleagues also changed. For example, flexworkers had to plan more intentionally to meet, interact with, and be seen by, co-workers in the physical office, to compensate for missing important and spontaneous interactions as well as to access needed information.

Summary

There are many forms of flexwork, grouped in various categories, but all fundamentally involve choices in time, duration, location, and continuity. They are also crucial distinctions between what organizations formally offer, what supervisors formally and informally offer or allow, what workers desire or prefer, and what workers actually use. And each of those forms of flexwork has both common and unique influences, at different levels (from legislative and occupation to supervisor assumptions, nature of the job, and employee concerns about career and work–family pressures). Flexwork may foster both positive and negative outcomes related to performance, hiring and retention, stress, and work–life balance. These relationships are typically moderated by other factors, such as motivations, preference, location, and organizational norms.

Work–Family Boundaries

During the industrial revolution, work became defined as organizational employment, and family life and work became segmented, with different cultures (Morf, 1989). However, today work and family activities increasingly cross the boundaries between each domain (Zedeck, 1992). This blurring of boundaries is due to many factors, including: policy and legislation, longer work hours, increased numbers of working women and single parents, an aging population (Kossek & Michel, 2010; Kossek & Ozeki, 1998; Martins, Eddleston, & Veiga, 2002), new organizational forms, such as increasing work–family role flexibility (Kossek & Michel, 2010); the use of multiple ICTs for work outside of the formal organization (Kossek & Lautsch, 2008) and part-time work; and increasing mobility as well as growing concern for nonwork life (Kossek & Michel, 2010).

Our focus here is on flexwork-related crossing and blurring of work–family boundaries. Managing this changing interrelationship between work and family is challenging for the organization, the employee, and the employee's family. Border and boundary theory in general (Michaelsen & Johnson, 1997; Nippert-Eng, 1996b) presume that institutions and individuals develop boundaries to simplify, organize, and give meaning to their domains. Boundaries have two distinct aspects: *flexibility* (ability to perform work outside of typical time and space constraints), and *permeability* (the extent one domain intrudes on the other, and how boundaries may expand or contract in response to these demands) (Hall & Richter, 1988). Flexibility and permeability not only make crossing boundaries easier but also blur boundaries, such as by bringing work home, or engaging in family communication at work. Organizational policies (e.g., flexwork) and climate affect boundary flexibility and permeability (Ashforth & Fried, 1988).

Many factors influence one's preference for and outcomes of work–family balance: individual characteristics, home and work social influence, meanings of work and family, organizational policies, work scheduling, support from other actors, and person–environment fit (affected by the nature of the boundary work) (Desrochers, Hilton, & Larwood, 2005). Boundary characteristics and individual preferences affect work–family boundary management styles (integrator, separator, alternating), and subsequent outcomes, such as inter-role conflict and stress (Kossek & Lautsch, 2012). Organizational context (more or less supportive) also moderates the relationship between boundary management style and work–family conflict.

Work–Family boundary crossing

According to work–family border theory (Clark, 2000), people daily cross borders between work, family and life domains, and interpret and shape those boundaries. Primary work-related transitions are between home and work, within work, and between work and "third places" (Hall & Richter, 1988). This spillover can be symmetric or asymmetric, sequential or simultaneous, creating both positive (job satisfaction) and negative (job stress) implications, or positive implications for one domain and negative implications for the other. Thus transitions across roles, ranging from highly segmented to highly integrated, can involve crossing boundaries with high or low flexibility and high or low permeability, and are associated with various benefits and costs (Ashforth, Kreiner, & Fugate, 2000).

This boundary-crossing affects work–family balance, and people may have different levels of support and communication about boundary-crossing issues. Because of daily boundary crossing, individuals need to manage the boundaries between and transitions across their domains and roles (e.g., home, work, school). This includes trying to balance each domain's demands, thereby reducing interference and conflict across the domains (Clark, 2000). Individuals may attempt to compensate in one domain for missing elements in the other (Staines, 1980). "Border crossers" proactively attempt to balance their work and family domains (Clark, 2000), across physical, temporal, and psychological boundaries, which vary in flexibility and permeability. Integrated role boundaries reduce the impact of interruptions, but require more boundary work (Rau & Hyland, 2002).

Individuals vary in their preferences for boundary crossing (flexibility, permeability, extent of symmetry in blurring; the fit between boundary strengths and preferences affects role conflict), and for role integration or role segmentation across either work-to-family or family-to-work boundaries (Nippert-Eng 1996a, 1996b). Individuals try to maintain order around a role by applying their own style of boundary management (Nippert-Eng, 1996a). Combinations of permeability and flexibility generate segmented or integrated styles (Bulger, Matthews, & Hoffman, 2007). Kossek and Lautsch (2008) introduced a third style, in which an individual alternates between the segmented and integrated styles. They also differ in the centrality of their work–family identities, and their experience of organizational support for work–family customization (Kossek & Lautsch, 2012; Kossek, Lautsch, & Eaton, 2005; Rothbard et al., 2005). In Hecht and Allen's (2009) study, greater home boundary strength was associated with role identification, while greater home as well as work boundary strengths was related to less work–family role conflict. High integration among roles and domains foster an encompassing identity, easier boundary transitions, and decreased negative affect related to interruptions. They also increase role blurring and diverse demands. Strong role identification may make role entry easier but it makes role exiting harder. Individuals also vary in their *interpretations* of their boundaries and the transitions. For example, a long commute may provide a valuable buffer between home and work roles for some people (Hall, 1990) but seriously degrade quality of life for others.

Employees attempt to manage their experienced boundary permeability and work-family balance through "enactment" of their environment, over which they have varying control and choice (Weick, 1979). Individuals accomplish these boundary characteristics through communication with supervisors, fellow workers, and family members to create the meaning of this enactment. Particularly important is "across-the-border" communication with members of one domain about their other domain (e.g., work–family, or family–work). Thus work–family boundaries are partially constituted through communication (Cowan & Hoffman, 2007). More supportive organizational culture, members' more positive perceptions of empowerment, and greater boundary permeability and flexibility, all increase cross-boundary communication.

This communication may be about *obligations* (e.g., scheduling, task requirements) or *understood meaningful* experience (e.g., the value of the work or family domain) (Clark, 2002). This communication is necessary to help others understand the boundary crosser's experience and meaning of the other domain (e.g., work-to-family and family-to-work) (Clark, 2002). In turn, this understanding can improve the negotiation of and adaptation to these crossings, and reduce role conflict and improve work–family balance. Clark's (2002) survey analysis concluded that such communication did vary by permeability of the boundaries, and did improve work satisfaction and functioning, and home and family activity satisfaction and functioning. In particular, interpersonal relationships were most important in cross-boundary communication, and communicating about domains as understood meaningful experience was influential, though communicating about domains as obligations was not.

Outcomes associated with work–family boundaries

Work–family integration can have positive or negative implications, depending on preference and context when not working in the office (Ahrentzen, 1990). For example, greater work–family integration and boundary blurring were related to more hours worked, distractions at home, and work–family conflict in Desrochers and colleagues' (2005) study. As a positive example, with some work at home, work–family conflict is lower with more bounded home work spaces (physical, behavioral, temporal, or social).

More directional boundary management can reduce directional interference, and thus lessen associated negative outcomes (Ashforth et al., 2000). The extent of formalization or rituals of transitions across roles and boundaries (such as explicit scheduling, or closing a home office door at the beginning of work) also moderates the relationship between some work at home and conflict.

Greater home boundary flexibility, and greater work boundary permeability, were both associated with more home communication about work in Clark's (2002) study. But greater work boundary permeability was negatively associated with work communication about family, and work boundary flexibility was not associated with communication across either domain. The least conflict arose from boundaries with high flexibility and low permeability (implying an underlying dimension of individual choice), but that relationship was moderated by work–family preferences.

Flexwork increases the likelihood of more frequent boundary crossing and more flexible and permeable boundaries (Clark, 2000). Some flexwork arrangements (in particular, flextime and compressed workweeks) have been associated with less work–family conflict, more work–family balance, less absenteeism, and better organizational outcomes (Baltes, Briggs, Huff, Wright, & Neuman, 1999). A survey of Hong Kong office workers showed that with greater work flexibility came less work spillover into home, home spillover into work, or job burnout. Work spillover into the home was a stronger significant influence on job burnout than was the significant effect of family spillover into work (Leung, 2011).

Flexwork and telecommuting offer different attributes relative to boundary management. Rau and Hyland (2002) found that flexwork allows less permeable boundaries, mid-range boundary flexibility, and greater role contrast. Telecommuting, however, supports more permeable and flexible boundaries, less role contrast, and easier role transitions.

ICTs, Flexwork, and Work–Family Boundaries

ICTs is a very general term. In management, information systems, and business research, it usually refers to more enterprise-level computer systems used to process, network, and distribute data, information, and communication content, both within and across people, units, and organizations. In research emphasizing the communication aspect ICTs can include any medium, from traditional radio and television through to cell phones and network infrastructure. Even more focused is the reference of ICTs to media, features, and services involving computers, digitization, and networking that support communication and knowledge sharing at various levels from dyadic to group and from within or across organizational boundaries. ICTs may be distinguished by their technological and software components, uses, and features, attributes or affordances (Faraj & Azad, 2012), such as searching, retrieving, and communicating. For example, ICTs now include "social media" (which are not a specific medium or device). The diffusion, capabilities, scale, scope, uses, roles and implications of ICTs in organizations are vast and continually evolving and being adapted (Rice & Leonardi, 2013). Thus there is no stable or comprehensive list of ICTs.

Our work, personal, and family lives and practices are embedded in, and mediated through, ICT infrastructures and use (Brynjolfsson & McAfee, 2011; Rainie, Wellman, & Chen, 2012). ICT use in general is positively associated with autonomy, skill variety, career opportunities, increased job mobility, and improved technology skills (Bloom et al., 2013). ICT use can provide more resources and communication for accomplishing work and improving the results from invested time and effort (Kubicek, Korunka, Paskvan, Prem, & Gerdenitsch, 2014). Furthermore, ICT use is constituted through social interaction in social contexts (Chesley, Siibak, & Wajcman, 2013). ICTs not only reshape the nature and networks of organizational communication, but also job conditions (Chesley, 2014).

Use of ICTs, flexwork, and work–family boundaries

Most flexwork would be impossible without the use of ICTs, as they allow communication, collaboration, and use of resources, across time, space, participants, and work–family boundaries. Influences on using ICTs after and outside of work include organizational provision of devices, and co-worker/organizational connectivity norms. Some other factors are how well media attributes fit with individual and work characteristics (Richardson & Benbunan-Fich, 2011). Because employees differ in their preferences for and experiences of boundary strength, their use of communication in general and ICTs in particular to manage these boundaries also likely varies (Park & Jex, 2011). Workers with greater affective attachment to their organization (related to organization identification, job involvement, and work engagement), are more likely to use media to extend their work after hours (Boswell & Olson-Buchanan, 2007).

Valcour and Hunter (2008) reviewed the influences and moderators (including macro and historical) associated with ICTs and work–family integration. Employees and their family members, through their work–family boundary-spanning practices and values, may use, adapt, and enact rules and resources for ICTs to reduce the asymmetry between work-to-family and family-to-work intrusions (Kennedy & Wellman, 2007). Such uses include contributing to organizational goals through home ICT use, and managing family interaction while at work (A. G. Golden, 2012). The initial organizational and family rules for ICTs use and resources (ICTs and knowledge about them) are adapted through interaction, communication, and practice within communities of users (A. G. Golden, 2012; Orlikowski, 2000). A. G. Golden's (2012) qualitative study identified three categories of ICT rules or structures in flexwork conditions: technology and tasks, spatial-social (where work is performed), and temporal-social (when work is performed, including whether family members are present or absent).

Outcomes associated with ICTs in flexwork and work–family boundaries

One of the paradoxes of ICTs for flexwork is that while the pervasiveness and use place more demands on workers and intensify the pace, stress, and even insecurity of work (Chesley, 2014), use of ICTs also reduces constraints on time, space, social interaction, and knowledge. Thus we see a variety of both positive and negative outcomes associated with ICTs and flexwork and work–family boundaries.

Positive outcomes associated with ICTs in flexwork ICTs can allow more flexibility at work or at home (Gajendran & Harrison, 2007). They can help integrate multiple work and nonwork roles by allowing balance, flexibility, and access, and by increasing autonomy and ability to accomplish tasks (Valcour & Hunter, 2008). ICT use can help balance work and home roles (Voydanoff, 2005) and facilitate greater work–family integration (Batt & Valcour, 2003). Positive aspects of household ICT use for work in Frissen's (2000) study included better organizing, increased flexibility and control, mobility, interactivity, time saving, ability to recover and improvise from changes, and being accessible to others. Considering telework and telecommuting specifically, benefits of teleworkers' use of ICTs include greater social presence and thus a greater sense of connectivity (Fonner & Roloff, 2012). Using cellular devices for telework helps workers manage integration and the work–family boundary (Kreiner, Hollensbe, & Sheep, 2009). A 2008 US national-level Pew survey showed that networked work (teamwork, telework, and other ICT use at work and outside of work) was associated with greater job decision latitude (skill development, job autonomy) (Chen & McDonald, 2015; Rainie et al., 2012).

Negative outcomes associated with ICTs in flexwork Disadvantages of using ICTs in flexwork include worker isolation, less organizational visibility, fewer rewards and face-to-face interactions, less observability by management, reduced control of important information, less teamwork, increased safety issues, and greater stress, burnout, and intention to leave (Ayyagari, Grover, & Purvis, 2011; Kaufman-Scarborough, 2006; Valcour & Hunter, 2008; Wadsworth et al., 2010). ICT use can foster greater workload, multitasking, poorer work environment, and stress (Bloom et al., 2013). Other negative aspects of ICTs include more demands on one's time, reduced flexibility and control, overload, invasion of home privacy, and pressure to always connect and respond (Frissen, 2000). Such pressures include excessive "workplace connectivity" (Schlosser, 2002), or a sense of always being "on call" (Tarafdar, Tu, Ragu-Nathan, & Ragu-Nathan, 2007).

ICT use can blur boundaries by presenting yet another job demand (Voydanoff, 2005), with intrusion of nonwork life by work life (Bloom et al., 2013), creating obligations of connectedness, reduced resources for home life, and increased work–family conflict (Valcour & Hunter, 2008). Several studies have considered the influence of mobile media on work–family boundary blurring, mostly finding negative effects (e.g., Chesley, 2005; Fenner & Renn, 2010; Kreiner et al., 2009). Hislop and Axtell (2011) noted that engineers who spent much time travelling to or working with external clients used their cell phones to manage work dependencies as well as the work/nonwork boundary. In the process, however, blurred and often unclear boundaries were created. Workers with greater job involvement and ambition are slightly more likely to use ICTs outside work hours, but that usage also creates slightly more work–family conflict, from both the employee's and the significant other's perspective (Boswell & Olson-Buchanan, 2007). About half of a US national sample agreed that using ICTs increased their stress and work–family boundary blurring and conflict (Fenner & Renn, 2010; Madden & Jones, 2008). Telework can harm work–family balance, by increasing work–family boundary permeability, total hours worked, and stress (Russell, O'Connell, & McGinnity, 2009).

Conditional outcomes associated with ICTs in flexwork and work–family boundaries ICT use is associated with both positive and negative outcomes in relation to work–family boundaries. These occur within varied contexts, moderators and mediators, such as managerial goals, labor power, organizational culture and supervisory support, institutional environment, occupations, need for coordination and interdependency, task requirements, job demands and resources, and individual and family characteristics (Valcour & Hunter, 2008). For example, employees with higher job satisfaction have indicated a positive assessment of use of Internet for accomplishing work-related tasks, and a more permeable family boundary and flexible work boundary (Leung, 2011). Those with lower job satisfaction reported greater work spillover and burnout. But higher family satisfaction was predicted by greater dependency on the Internet, less home permeability and more home flexibility. Thus to some extent job and family satisfaction compete with each other. Individuals with higher role identification (Ashforth et al., 2000) are more likely to broaden the boundary into or from the home domain, partially by reducing constraints on using ICTs for work purposes at home. Some of the telework/telecommuting outcomes vary by the type of work, with more positive implications for more professional workers (Bloom et al., 2013).

Park and Jex (2011) concluded that developing boundaries for using ICTs in cross-role activities mediated the influence of individual factors on work–family interference, differently for preference in each direction of boundary-crossing between work and family. And, those with higher work-role identification experienced more work-to-family interference. A study of employees from over 30 companies showed that while more use of work-related ICT outside of regular hours was associated with increased work–family conflict, more positive attitudes toward such ICT use decreased that conflict (Wright et al., 2014). Some of the negative implications of using ICTs to work at home can be moderated by time management strategies (Fenner & Renn, 2010). Thus, attitudes, type of flexwork, and the characteristics of boundaries interact in their influence on ICT outcomes.

Summary

Flexwork and ICTs are inextricably related. The term ICTs covers a wide range of systems, devices, and technologies. Of more conceptual importance are their capabilities, affordances, uses, roles, and implications. ICTs can provide job resources, fostering greater work–family flexibility and permeability, and allowing for more control over work and family demands. Yet they also can create more demands, fragmentation and interruptions, isolation, and work–family conflict. Some of these relationships are moderated by factors such as organizational and managerial goals, the nature of one's work, and boundary preferences.

Future Research

Each of the three main topics – flexwork, work–family boundaries, and ICTs – much less their intersections, could easily justify their own chapter or even book (as many of the references here reflect). Thus, due to space constraints, there are multiple limitations to the scope and depth of this review. Each of those limitations thereby provides opportunities for more detailed and nuanced reviews, research, and analyses.

Even though the initial search for the review included over 150 articles, because the literature in each of these areas is immense, the coverage here cannot be comprehensive. Researchers in each of these areas will easily identify other materials they might consider central. However, the goal was to identify and integrate concepts, theory and results representing the fundamental issues and relationships. For more detailed and comprehensive empirical results, see the referenced reviews and meta-analyses (e.g., Gajendran & Harrison, 2007; Kossek et al., 2011; McNamara et al., 2012).

This review has distinguished between the two basic contexts of flexwork – options offered by the organization, and those actually supported by supervisors and used by employees. Thus Figure 9.1 does not include a direct relationship between organizational offering and outcomes, because it is employee access and use that is associated with work–family boundary crossing, and most outcomes. However, while this review summarizes several typologies of flexwork, it does not segment results and relationships by any of the several general and detailed typologies of flexwork (e.g., Kossek & Michel, 2010; Matos & Galinsky, 2014). Given vagaries in what flexwork is, it would be a significant research challenge to develop a framework clearly linking relationships and outcomes to specific offered, supported, used, and interpreted flexwork types. A similar limitation and research opportunity applies to the access and use of types of ICTs. Different media, devices, features or affordances, and variations in use are likely to relate to different forms of flexwork, work–family boundary crossing, and outcomes. Some of the studies referenced here refer to specific devices (cell phones), services (email), or features (asynchronicity), while others refer to ICTs in general. The media affordance literature, in particular, argues against any particular technology having a particular effect. Instead, outcomes might be best associated with the opportunities for action that a (or category of) technology (texting, audio, group support) affords (e.g., pervasive awareness, searchability, mobility), within a given context (e.g., task, unit, organization, family; type of flexwork), in relation to agents' needs (knowing who knows what, obtaining answers to questions, minimizing interruptions) (Faraj, & Azad, 2012; Majchrzak & Markus, 2013). Also relevant is the portfolio of ICTs employees can use, including those provided or required by the organization, and those preferred and actually used by the employee. Thus, more specific syntheses would identify and interpret variations in relationships by ICT type, affordance, and use context.

Because there is little work on the role of communication on work–family boundary management and crossing (Clark, 2002), and related work–family conflict, this area provides good opportunities for assessing the role of ICTs. For example, does communicating about obligations or understood meaningful experiences, across work–family boundaries with varying permeability or flexibility, using different media (e.g., face-to-face, cell phone, social media) have different influences on outcomes such as work satisfaction or work–family conflict?

A rich opportunity for extending research into the relationships among flexwork, work–family boundaries, and ICTs would be including the work engagement concept and job resources and demands theory. Bakker, Schaufeli, Leiter and Taris (2008) define work engagement as a *motivational state* of work-related wellbeing consisting of vigor (energy), dedication (identification with work), and absorption (concentrated and positively immersed in work). "Engaged workers are more creative, more productive, and more willing to go the extra mile" (Bakker & Demerouti, 2008, p. 209). The primary work engagement antecedents are *job demands* (job features requiring costs, strain) and *resources* (job features functional to achieving work goals, reducing costs, allowing development) (Demerouti, Bakker, Nachreiner, & Schaufeli, 2001). Increasing one's work engagement facilitates the development and use of more task, social, and personal resources, which then enhance more work engagement, a gain spiral (Hobfoll, 2002). However, the opposite loss spiral may also occur. Thus greater work engagement may foster greater employee interest in flexwork, by fostering greater interest in and energy for working in new contexts. However, flexwork and associated ICT use, depending on whether they create more job resources (control over time, place, and amount of work) or job demands (interruptions, supervisory devaluation, work–family conflict), could strengthen or weaken work engagement. A study of health workers showed that greater work engagement was associated with less work-to-family conflict (Mauno, Kinnunen, & Ruokolainen, 2007). So far, however, little research has applied work engagement to flexwork and ICT contexts.

Conclusion

Work and family (more generally, nonwork life) provide the two major social contexts. Thus new developments in the nature of work (here, flexwork) and associated relationships between work and family (here, work–family boundaries) as well as in means for interacting and communicating within and across those domains (here, ICTs) are significant topics for individuals and organizations, for managers and researchers alike.

This literature review identifies several primary conclusions about research on, and across, these three topics. Flexwork is increasing, and includes a wide variety of options. Flexwork offering, meaning, use, and outcomes are influenced by legislative, organizational, supervisor, and individual factors. Flexwork may be associated with a variety of positive outcomes, such as better work–family balance, and performance. Flexwork can blur work–family boundaries, with both positive and negative consequences. Work and family boundaries vary in permeability and flexibility. But flexwork can also increase boundary crossing, and boundary flexibility and permeability, generating more work–family conflict. Communication about, and across, work-to-family and family-to-work domain boundaries both enacts these boundaries as well as influences work–family conflict. ICT use in flexwork has both positive and negative effects on boundaries, and those relationships are often moderated by work characteristics, and organizational and supervisory support for flexwork. ICT use and work characteristics may moderate some relationships between flexwork, boundary characteristics, and work–family conflict.

Clearly the interrelationships among flexwork, work–family boundaries, ICTs, and organizational and work–family outcomes provide not only a more integrated context for understanding the nature and outcomes of flexwork, but also provide useful, intriguing, and challenging opportunities for management, employees, consultants, and researchers.

References

Ahrentzen, S. B. (1990). Managing conflict by managing boundaries: How professional homeworkers cope with multiple roles at home. *Environment and Behavior, 22*(6), 723–752.

Ashforth, B. E., & Fried, Y. (1988). The mindlessness of organizational behaviors. *Human Relations, 41*(4), 305–329.

Ashforth, B. E., Kreiner, G. E., & Fugate, M. (2000) All in a day's work: Boundaries and micro role transitions. *Academy of Management Review, 25*(3), 472–491.

Australian Public Service Commission. (2013). Flexible work. *APS Human Capital Matters, Issue 7*. Retrieved April 25, 2017, from www.apsc.gov.au/__data/assets/word_doc/0005/57074/Human-Capital-Matters-Vol-7-2013-Flexible-Work.DOCX

Ayyagari, R., Grover, V., & Purvis, R. (2011). Technostress: Technological antecedents and implications. *MIS Quarterly, 35*(4), 831–858.

Bakker, A. B., & Demerouti, E. (2008). Towards a model of work engagement. *Career Development International, 13*(3), 209–223.

Bakker, A. B., Schaufeli, W. B., Leiter, M. P., & Taris, T. W. (2008). Work engagement: An emerging concept in occupational health psychology. *Work & Stress, 22*(3), 187–200.

Baltes, B. B., Briggs, T. E., Huff, J. W., Wright, J. A., & Neuman, G. A. (1999). Flexible and compressed workweek schedules: A meta-analysis of their effects on work-related criteria. *Journal of Applied Psychology, 84*(4), 496–513.

Batt, R., & Valcour, P. M. (2003). Human resources practices as predictors of work–family outcomes and employee turnover. *Industrial Relations, 42*(2), 189–220.

Bloom, N., Liang, J., Roberts, J., & Ying, Z. J. (2013). Does working from home work? Evidence from a Chinese experiment. National Bureau of Economic Research Working Paper No. 18871. Retrieved April 7, 2017, from www.nber.org/papers/w18871

Boswell, W. R., & Olson-Buchanan, J. B. (2007). The use of communication technologies after hours: The role of work attitudes and work-life conflict. *Journal of Management, 33*(4), 592–610.

Brynjolfsson, E., & McAfee, A. (2011). *Race against the machine*. Lexington, MA: Digital Frontier Press.

Bulger, C., Matthews, R., & Hoffman, M. E. (2007). Work and personal life boundary management: Boundary strength, work/personal life balance, and the segmentation–integration continuum. *Journal of Occupational Health Psychology, 12*(4), 365–375.

Butts, M. M., Casper, W. J., & Yang, T. S. (2013). How important are work-family support policies? A meta-analytic investigation of their effects on employee outcomes. *Journal of Applied Psychology, 98*(1), 1–25.

Cailier, J. G. (2013). Are teleworkers less likely to report leave intentions in the United States federal government than non-teleworkers are? *American Review of Public Administration, 43*(1), 72–88.

Chen, W., & McDonald, S. (2015). Do networked workers have more control? The implications of teamwork, telework, network, and ICTs for job decision latitude in American Workplace. *American Behavioral Scientist, 59*(4), 492–507.
Chesley, N. (2005). Blurring boundaries? Linking technology use, spillover, individual distress, and family satisfaction. *Journal of Marriage and the Family, 67*(5) 1237–1248.
Chesley, N. (2014). Information and communication technology use, work intensification and employee strain and distress. *Work, Employment and Society, 28*(4), 589–910.
Chesley, N., Siibak, A., & Wajcman, J. (2013). Information and communication technology use and work-life integration. In D. A. Major & R. Burke (Eds.), *Handbook of work-life integration of professionals: Challenges and opportunities* (pp. 245–268). Northampton, MA: Edward Elgar Publishing.
Clark, S. C. (2000). Work/family border theory: A new theory of work/family balance. *Human Relations, 53*(6), 747–770.
Clark, S. C. (2002). Communicating across the work/home border. *Community, Work, and Family, 5*(1), 23–48.
Clifton, T. J., & Shephard, E. (2004). Work and family programs and productivity: Estimates applying a production function model. *International Journal of Manpower, 25*(8), 714–728.
Cowan, R., & Hoffman, M. F. (2007). The flexible organization: How contemporary employees construct the work/life border. *Qualitative Research Reports in Communication, 8*(1), 37–44.
Demerouti, E., Bakker, A. B., Nachreiner, F., & Schaufeli, W. B. (2001). The job demands–resources model of burnout. *Journal of Applied Psychology, 86*(3), 499–512.
Desrochers, S., Hilton, J. M., & Larwood, L. (2005). Preliminary validation of the work-family integration-blurring scale. *Journal of Family Issues, 26*(4), 442–466.
Eaton, S. C. (2003). If you can use them: Flexibility policies, organisational commitment and perceived performance. *Industrial Relations, 42*(2), 145–167.
Faraj, S., & Azad, B. (2012). The materiality of technology: An affordance perspective. In P. Leonardi, B. Nardi, & J. Kallinikos (Eds.), *Materiality and organizing: Social interaction in a technological world* (pp. 237–258). Oxford: Oxford University Press.
Feldman, D. C., & Gainey, T. W. (1997). Patterns of telecommuting and their consequences: Framing the research agenda. *Human Resource Management Review, 7*(4), 369–388.
Felstead, A., Jewson, N., & Walters, S. (2005). The shifting locations of work: New statistical evidence on the spaces and places of employment. *Work, Employment and Society, 19*(2), 415–431.
Fenner, G. H., & Renn, R. W. (2010). Technology-assisted supplemental work and work-to-family conflict: The role of instrumentality beliefs, organizational expectations and time management. *Human Relations, 63*(1), 63–82.
Flexible Work Arrangements: The Fact Sheet. (2010). *Workplace flexibility 2010.* Washington DC: Georgetown University Law Center. Retrieved April 20, 2017, from workplaceflexibility2010.org/images/uploads/FWA_FactSheet.pdf
Fonner, K., & Roloff, M. (2012). Testing the connectivity paradox: Linking teleworkers' communication media use to social presence, stress from interruptions, and organizational identification. *Communication Monographs, 79*(2), 205–231.
Frissen, V. A. J. (2000). ICTs in the rush hour of life. *The Information Society, 16*(1), 65–75.
Gajendran, R. S., & Harrison, D. A. (2007). The good, the bad, and the unknown about telecommuting: Meta-analysis of psychological mediators and individual consequences. *Journal of Applied Psychology, 92*(6), 1524–1541.
Galinsky, E., Bond, J. T., & Sakai, K. (2008). *National study of employers.* New York, NY: Families and Work Institute.
Golden, A. G. (2012). The structuration of information and communication technologies and work–life interrelationships: Shared organizational and family rules and resources and implications for work in a high-technology organization. *Communication Monographs, 80*(1), 101–123.
Golden, T. D. (2012). Altering the effects of work and family conflict on exhaustion: Telework during traditional and nontraditional work hours. *Journal of Business Psychology, 27*(3), 255–269.
Golden, T. D., Veiga, J. F., & Simsek, Z. (2006). Telecommuting's differential impact on work-family conflict: Is there no place like home? *Journal of Applied Psychology, 91*(6), 1340–1350.

Hall, D. T. (1990). *Telecommuting and the management of work-home boundaries (No. 90)*. Boston, MA: Boston University, School of Management.

Hall, D. T., & Richter, J. (1988). Balancing work life and home life: What can organizations do to help? *Academy of Management Executive*, *2*(3), 213–223.

Hecht, T. D., & Allen, N. J. (2009). A longitudinal examination of the work–nonwork boundary strength construct. *Journal of Organizational Behavior*, *30*(7), 839–862.

Hilbrecht, M., Shaw, S. M., Johnson, L. C., & Andrey, J. (2008). "I'm home for the kids": Contradictory implications for work-life balance of teleworking mothers. *Gender, Work, & Organization*, *15*(5), 454–476.

Hill, J. E., Grzywacz, J., Allen, S., Blanchard, V., Matz-Costa, C., Shulkin, S., & Pitt-Catsouphes, M. (2008). Defining and conceptualizing workplace flexibility. *Community, Work and Family*, *11*(2), 149–163.

Hislop, D., & Axtell, C. (2011). Mobile phones during work and non-work time: A case study of mobile, non-managerial workers. *Information & Organization*, *21*(1), 41–56.

Hobfoll, S. E. (1989). Conservation of resources: A new approach at conceptualizing stress. *American Psychologist*, *44*(3), 513–524.

Hobfoll, S. E. (2002). Social and psychological resources and adaptation. *Review of General Psychology*, *6*(4), 307–324.

Hoonakker, P. (2014). Information and communication technology and quality of working life: Backgrounds, facts, and figures. In C. Korunka & P. Hoonakker (Eds.), *The impact of ICT on quality of working life* (pp. 9–24). Dordrecht: Springer.

Household, Income and Labour Dynamics in Australia Survey (2010). Melbourne, Australit: Melbourne Institute of Applied Economic and Social Research. Retrieved April 25, 2017, from http://melbourneinstitute.unimelb.edu.au/hilda

Hylmö, A., & Buzzanell, P. M. (2002). Telecommuting as viewed through cultural lenses: An empirical investigation of the discourses of utopia, identity, and mystery. *Communication Monographs*, *69*(4), 329–356. doi:10.1080/03637750216547

Kaufman-Scarborough, C. (2006). Time use and the impact of technology: Examining workspaces in the home. *Time & Society*, *15*(1), 57–80.

Kennedy, T. L. M., & Wellman, B. (2007). The networked household. *Information, Communication & Society*, *10*(5), 645–670.

Knight, C., & Haslam, S. A. (2010). Your place or mine? Organizational identification and comfort as mediators of relationships between the managerial control of workspace and employees' satisfaction and well-being. *British Journal of Management*, *21*(3), 717–735.

Kossek, E. E., & Lautsch, B. (2008). *CEO of me: Creating a life that works in the flexible job age*. Philadelphia, PA: Pearson/Wharton School Publishing.

Kossek, E. E., & Lautsch, B. A. (2012). Work–family boundary management styles in organizations: A cross-level model. *Organizational Psychology Review*, *2*(2) 152–171.

Kossek, E. E., Lautsch, B. A., & Eaton, S. C. (2005). Flexibility enactment theory: Implications of flexibility type, control, and boundary management for work family effectiveness. In E. E. Kossek & S. J. Lambert (Eds.), *Work and life integration: Organizational, cultural, and individual perspectives* (pp. 243–261). Mahwah, NJ: Lawrence Erlbaum.

Kossek, E. E., & Michel, J. S. (2010). Flexible work schedules. In S. Zedeck (Ed.), *APA handbook of industrial and organizational psychology* (Vol. 1, pp. 535–572). Washington DC: American Psychological Association.

Kossek, E. E., & Ozeki, C. (1998). Work–family conflict, policies, and the job–life satisfaction relationship: A review and directions for organizational behavior/human resources research. *Journal of Applied Psychology*, *83*(2), 139–149.

Kossek, E. E., Pichler, S., Bodner, T., & Hammer, L. B. (2011). Workplace social support and work–family conflict: A meta-analysis clarifying the influence of general and work–family-specific supervisor and organizational support. *Personnel Psychology*, *64*(2), 289–313.

Kreiner, G. E., Hollensbe, E. C., & Sheep, M. L. (2009). Balancing borders and bridges: Negotiating the work–home interface via boundary work tactics. *Academy of Management Journal*, *52*(4), 704–730.

Kubicek, B., Korunka, C., Paskvan, M., Prem, R., & Gerdenitsch, C. (2014). Changing working conditions at the onset of the twenty-first century: Facts from international datasets. In C. Korunka & P. Hoonakker (Eds.), *The impact of ICT on quality of working life* (pp. 25–42). Dordrecht: Springer.

Lambert, A. D., Marler, J. H., & Gueutal, H. G. (2008). Individual differences: Factors affecting employee utilization of flexible work arrangements. *Journal of Vocational Behavior*, *73*(1), 107–117.

Leslie, L. M., Park, T-Y., & Mehng, S. A. (2012). Flexible work practices: A source of career premiums or penalties? *Academy of Management Journal*, *55*(6), 1407–1428.

Leung, L. (2011). Effects of ICT connectedness, permeability, flexibility, and negative spillovers on burnout and job and family satisfaction. *Human Technology*, *7*(3), 250–267.

Long, Z., Kuang, K., & Buzzanell, P. M. (2013). Legitimizing and elevating telework: Chinese constructions of a nonstandard work arrangement. *Journal of Business and Technical Communication*, *27*(3), 243–262.

Madden, M., & Jones, S. (2008). *Networked workers*. Pew Research Center Publications, available at: http://www.pewresearch.org/pub/966/networked workers

Majchrzak, A., & Markus, M. L. (2013). Technology affordances and constraints in management information systems. In E. H. Kessler (Ed.), *Encyclopedia of management theory* (vol 2, pp. 832–835). Thousand Oaks, CA: Sage Publications.

Martins, L. L., Eddleston, K. A., & Veiga, J. F. (2002). Moderators of the relationship between work-family conflict and career satisfaction. *Academy of Management Journal*, *45*(2), 399–409.

Mateyka, P. J., Rapino, M. A., & Landivar, L. C. (2012). *Home-based workers in the United States: 2010. Household Economic Studies*. Current Population Reports. Washington DC: US Department of Commerce, US Census Bureau.

Matos, K., & Galinsky, E. (2014). *2014 National study of employers*. New York, NY: Families and Work Institute. http://www.familiesandwork.org/2014-national-study-of-employers/

Mauno, S., Kinnunen, U., & Ruokolainen, M. (2007). Job demands and resources as antecedents of work engagement: A longitudinal study. *Journal of Vocational Behavior*, *70*(1), 149–171.

McNamara, T. K., Pitt-Catsouphes, M., Brown, M., & Matz-Costa, C. (2012). Access to and utilization of flexible work options. *Industrial Relations*, *51*(4), 936–965.

Mesmer-Magnus, J. R., & Viswesvaran, C. (2006). How family-friendly work environments affect work/family conflict: A meta-analytic examination. *Journal of Labor Research*, *27*(4), 555–574.

Michaelsen, S., & Johnson, D. E. (Eds.) (1997). *Border theory: The limits of cultural politics*. Minneapolis, MN: University of Minnesota Press.

Mokhtarian, P. L. (1991). Defining telecommuting. Institute of Transportation Studies. Retrieved May 5, 2017, from http://escholarship.org/uc/item/35c4q71r.pdf

Morf, M. (1989). *The work/life dichotomy*. New York, NY: Quorum Books.

Morganson, V. J., Major, D. A., & Oborn, K. L. (2010). Comparing telework locations and traditional work arrangements: Differences in work-life balance support, job satisfaction, and inclusion. *Journal of Managerial Psychology*, *25*(6), 578–595.

Myers, K. K., Gailliard, B. M., & Putnam, L. L. (2012). Reconsidering the concept of workplace flexibility. In C. T. Salmon (Ed.), *Communication yearbook 36* (pp. 195–230). New York, NY: Routledge.

Nippert-Eng, C. E. (1996a). Calendars and keys: The classification of "home" and "work." *Sociological Forum*, *11*(3), 563–582.

Nippert-Eng, C. E. (1996b). *Home and work: Negotiating boundaries through everyday life*. Chicago: University of Chicago Press.

Orlikowski, W. J. (2000). Using technology and constituting structures: A practice lens for studying technology in organizations. *Organization Science*, *11*(4), 404–428.

Park, Y., & Jex, S. M. (2011). Work-home boundary management using communication and information technology. *International Journal of Stress Management*, *18*(2), 133–152.

Peters, P., Tijdens, K. G., & Wetzels, C. (2004). Employees' opportunities, preferences, and practices in telecommuting adoption. *Information & Management*, *41*(4), 469–482.

Rainie, L., Wellman, B., & Chen, W. (2012). Networked work. In H. Rainie, L. Rainie & B. Wellman (Eds.), *Networked: The new social operating system* (pp. 171–196). Cambridge, MA: MIT Press.

Rau, B. L., & Hyland, M. M. (2002). Role conflict and flexible work arrangements: The effects on applicant attraction. *Personnel Psychology, 55*(1), 111–136.
Rice, R. E., & Leonardi, P. M. (2013). Information and communication technology in organizations, 2000–2011. In L. Putnam & D. K. Mumby (Eds.), *Sage handbook of organizational communication* (3rd ed.) (pp. 425–448). Thousand Oaks, CA: Sage.
Richardson, J., & McKenna, S. (2014). Reordering spatial and social relations: A case study of professional and managerial flexworkers. *British Journal of Management, 25*(4), 724–736.
Richardson, K., & Benbunan-Fich, R. (2011). Examining the antecedents of work connectivity behavior during non-work time. *Information and Organization, 21*(3), 142–160.
Rothbard, N. P., Phillips, K. W., & Dumas, T. L. (2005). Managing multiple roles: Work-family policies and individuals' desires for segmentation. *Organization Science, 16*(3), 243–258.
Russell, H., O'Connell, P. J., & McGinnity, F. (2009). The impact of flexible working arrangements on work-life conflict and work pressure in Ireland. *Gender, Work & Organization, 16*(1), 73–97.
Schieman, S., & Glavin, P. (2008). Trouble at the border?: Gender, flexibility at work, and the work-home interface. *Social Problems, 55*(4), 590–611.
Schlosser, F. K. (2002). So, how do people really use their handheld devices? An interactive study of wireless technology use. *Journal of Organizational Behavior, 23*(4), 401–423.
Shockley, K. M., & Allen, T. D. (2010). Investigating the missing link in flexible work arrangement utilization: An individual difference perspective. *Journal of Vocational Behavior, 76*(1), 131–142.
Scholarios, D., & Marks, A. (2004). Work-life balance and the software worker. *Human Resource Management Journal, 14*(2), 54–74.
Skinner, N., Hutchinson, C., & Pocock, B. (2012). *The big squeeze: Work, home and care in 2012. Australian Work and Life Index.* Adelaide, Australit: Centre for Work Life, University of South Australia. Retrieved April 25, 2017, from www.wgea.gov.au/sites/default/files/AWALI2012-National%5B1%5D.pdf
Staines, G. L. (1980). Spillover versus compensation: A review of the literature on the relationship between work and non-work. *Human Relations, 33*(2), 111–129.
Tarafdar, M., Tu, Q., Ragu-Nathan, B., & Ragu-Nathan, T. S. (2007). The impact of technostress on role stress and productivity. *Journal of Management Information Systems, 24*(1), 301–328.
US Office of Personnel Management. (2016). Telework. Retrieved July 3, 2016, from https://www.opm.gov/FAQs/Search.aspx?q=telework
Valcour, P. M., & Hunter, L. W. (2008). Technology, organizations, and work-life integration. In E. E. Kossek & S. J. Lambert (Eds.), *Work and life integration: Organizational, cultural, and individual perspectives* (pp. 61–84). Mahwah, NJ: Lawrence Erlbaum.
Van Yperen, N. W., & Wörtler, B. (2017). Blended working In G. Hertel, D. L. Stone, R. D. Johnson, & J. Passmore (Eds.), *Handbook of the psychology of the Internet at work* (pp. 157–174). Chichester, UK: John Wiley & Sons, Ltd.
Voydanoff, P. (2005). Consequences of boundary-spanning demands and resources for work-to-family conflict and perceived stress. *Journal of Occupational Health Psychology, 10*(4), 491–503.
Wadsworth, L. L., Facer, R. L., & Arbon, C. A. (2010). Alternative work schedules in local government: Cui Bono? *Review of Public Personnel Administration, 30*(3), 322–340.
Weick, K. E. (1979). *The social psychology of organizing.* New York, NY: McGraw-Hill.
Wessel, D. (2012, December 19). For some, home=office. *The Wall Street Journal.* Retrieved April 7, 2017, from http://online.wsj.com/news/articles/SB10001424127887324677204578185180450152760
Williams, J. C., Blair-Loy, M., & Berdahl, J. L. (2013). Cultural schemas, social class, and the flexibility stigma. *Journal of Social Issues, 69*(2), 209–234.
Wright, K. B., Abendschein, B., Wombacher, K., O'Connor, M., Hoffman, M., Dempsey, M., ... Shelton, A. (2014). Work-related communication technology use outside of regular work hours and work life conflict: The influence of communication technologies on perceived work life conflict, burnout, job satisfaction, and turnover intentions. *Management Communication Quarterly, 28*(4), 507–530.
Zedeck, S. (1992). Introduction: Exploring the domain of work and family concerns. In S. Zedeck (Ed.), *Work, families, and organizations* (pp. 1–32). San Francisco, CA: Jossey-Bass.

10

Mobile Computing and Hand-Held Devices at Work

Humayun Zafar

Introduction

In 2010, *Fortune* magazine highlighted the stiff competition personal computers (PCs) were facing from mobile technologies (Weintraub, 2010). The upward trajectory and promise of mobile devices was clear. To some it looked like a replay of the 1980s and 1990s when the PC revolution freed organizations from centralized mainframes. To others it was a struggle to comprehend its impact and meaning. Mobile computing is defined as an ability to use computing technology while moving or roaming, while still maintain services and connectivity (Imielinski & Korth, 1996). Technologies that enable mobile computing include laptops, smart phones, tablets, and smart watches. Mobile computing has added a new dimension to how organizations and their employees work. As computers became increasingly affordable, mobile, and connected, more and more people began using home computers to work on after office hours. They also began to use their personal devices in the workplace. This led to the inevitable blurring of the lines between personal and work spaces, which unintentionally brought up concerns such as work–life balance, privacy, and security. It can also be seen as a new opportunity for user-driven innovation that nudges organizations forward, as long as appropriate policies and procedures are created and implemented (Scheepers & Middleton, 2013).

Consider, for example, the use of personal mobile devices at work. Gaff (2015) states that in addition to presence of policies that focus on device usage at the workplace, it is also important to address ownership issues. Although it is widely accepted that an employee owns the device and the personal information stored on it, the issue may arise if the device is lost. At that point it is important that a policy addresses this scenario, as it may help mitigate the likelihood of a data breach. If the lost device allows access to an organization's systems, the policy needs to allow for that organization to remotely wipe the device to protect its systems and materials. This also implies that the impacted employee loses personal information such as contacts, emails, and pictures. There may be reasons other than a device being lost for a wipe. For example, an employee may leave the organization, and as

The Wiley Blackwell Handbook of the Psychology of the Internet at Work,
First Edition. Edited by Guido Hertel, Dianna L. Stone, Richard D. Johnson, and Jonathan Passmore.

that personal device could have had sensitive organizational data on it, it may be incumbent on the organization to "sanitize" it before removing it from the corporate network.

Mobile computing has not been researched extensively, and due to its nature, it is multidisciplinary. Issues such as work–life balance, privacy and security, and innovation relate to fields such as psychology, information systems, management, and marketing. This makes framing the issue into a singular theory difficult.

The rest of the chapter is arranged as follows: the overview of mobile computing research in the next section is followed by a discussion of the current status of mobile computing at the workplace that focuses on the evolution of mobile computing trends, such as bring your own device (BYOD) and the Internet of Things (IoT). The chapter concludes after looking at future research in this area.

Mobile Computing Research

The idea of mobile computing has been around since the 1990s. Presently, mobile access is considered to be the new normal. This is reflected in research that has investigated the use of mobile devices and development of mobile services as essential components of any environment, and not just the workplace (Bang, Lee, Han, Hwang, & Ahn, 2013; Kietzmann et al., 2013; Sutanto, Palme, Tan, & Phang, 2013). However, mobile computing research, due to its nature, is interdisciplinary, which poses challenges to providing an integrated framework. According to Middleton, Scheepers, and Tuunainen (2014) there are two central tenets that frame mobile computing research: mobility and mobile use, and the overlapping mobile context.

Mobility and mobile use

This area looks at the evolution of the mobile artefact (e.g., a laptop, smartphone, tablet), along with development of mobile services, and characteristics of a mobile user. Due to advances in technology, modes of mobile communications have evolved from simple voice telephony and text messaging to email, instant messaging, and a host of social media services. Mobile devices have also become multipurpose because of the blurring of the lines between work and personal devices. Gerow, Ayyagari, Thatcher, and Roth (2013) state that organizational applications and services are productive by nature, while games and other entertainment options are hedonic. In between the extremes are services that offer a blend of both: usefulness and enjoyment. The positive correlation between productivity and efficiency when mobile technologies have been properly configured at the workplace is well established (Jeong, Lee, & Nagesvaran, 2016).

An important part of the mobile artefact is ubiquity, which is anytime, anywhere access (Balasubramanian, Peterson, & Jarvenpaa, 2002). In fact, it has also been argued that ubiquity is the most important attribute of mobile services (Watson, Pitt, Berthon, & Zinkhan, 2002). There is a level of expectation in regard to mobile computing availability at the workplace (Matusik & Mickel, 2011). Employees who are reliant on mobile computing appreciate the increased flexibility in where, when, and how they work (Golden & Geisler, 2007; Hislop & Axtell, 2011). However, such flexibility can be a double-edged sword. Murray and Rostis (2007) investigated the mobile device usage habits of virtual workers and found that as employers began to use the technologies to monitor employees, such workers were unable to reap the benefits of increased productivity expected from working virtually. Prasopoulou, Pouloudi, and Panteli's (2006) study of service firms had similar results. Workers frequently using cell phones for work communication during times usually devoted to private activities. Such temporal encroachment reflected these workers' positions as mid-level executives, jobs that required extensive and ongoing coordination among colleagues, clients, and associates.

Recent literature has also focused on services and affordances that offer value to organizations and individuals. In many ways the mobile artefact is also dissolving since computing power is now embedded in smaller devices. Gubbi, Buyya, Marusic, and Palaniswami (2013) state that the IoT will change the way we collect, analyze, and respond to data. Billions of sensors interact with each other, providing us with better technological interaction. This in turn leads to opportunities for organizations and individuals to create new business models and forms of interaction that take advantage of ubiquitous computing embedded in objects. Mobile computing adoption has also been investigated through the lens of a user's roles. For example, depending on the context, an individual can assume the role of an employee, a private person, entertainment seeker, a citizen, or an activist, each of which may be conflated (Duxbury, Higgins, Smart, & Stevenson, 2014; Middleton & Cukier, 2006; Scheepers, Scheepers, & Ngwenyama, 2006). Mobile device adoption can also be negatively impacted by factors such as device and user interface complexity (Islam & Fayad, 2003).

There is no doubt that mobility has rapidly evolved. For mobile computing to be effective in the workplace it has to be seamless and integrated into key business processes. Mobility also needs to improve employee productivity and support productive collaboration. As noted above though, only limited research has begun investigating mobility, and additional research is needed to fully understand both the positive and negative implications of mobility.

Overlapping mobile context

Mobile computing enables the interplay between the individual as a private person and their other assumed roles in a different context. Research has shown that in regard to user satisfaction there is an influence of professional contexts on mobile technology use in organizations. An individual's loyalty to their profession overrides their loyalty to a particular employer, and consequently such outer contexts significantly influence how individuals respond in inner contexts such as organizations (Scheepers et al., 2006). Scheepers et al. (2006) imply that as mobile technologies become pervasive in the organizational computing environment, it would be necessary to conceptualize user satisfaction more broadly. For example, traditional measures of satisfaction focus on factors such as information accuracy or completeness, and system flexibility and accessibility (Wixom & Todd, 2005). But in the mobile context, the very nature of the physical mobility of the technology and the new contexts of individual use, by way of mobility, require a more dynamic conceptualization of satisfaction.

Discourses in this context can be framed by considering the overlap between the work and private contexts with respect to specific boundaries (Ashforth, Kreiner, & Fugate, 2000; Clark, 2000). Mobile computing has changed attitudes toward work–life balance as it allows work flow across locations rather than being confined to a physical space. Due to its "always-on" nature, mobile computing encourages intermittent patterns of work, regardless of whether the employee is in the office or on vacation (Mazmanian, Orlikowski, & Yates, 2013; Middleton, 2008). Intertwined with the issue of "always-on" is the BYOD phenomenon. As more and more individuals bring their personal devices to work, they expect organizations to provide support to enable these technologies to interact with their work systems (Harris, Ives, & Junglas, 2012). This too plays a role in changing the context for a user. The private and work contexts are essentially swapped. Hence, the private use context becomes an outer context to the work environment.

There is an extensive array of research in the area of work–life balance (Guest, 2002; Jones, Burke, & Westman, 2013). A summary of this work is beyond the scope of this chapter, but we encourage interested readers to look at Rice (2017, Chapter 9 in this volume) and Van Yperen and Wörtler (2017, Chapter 8 in this volume), for more in-depth discussions on work–life balance and flexwork and blended work, respectively.

In sum, it is clear that mobile computing research needs to extend beyond the exploratory realm. Most of the studies presented here have focused on employees as users as opposed to potential impact on the enterprise. However, the blurring of the lines between personal and work lives makes it difficult to separate the intertwined and interdisciplinary nature of mobile technologies. The next section elaborates on this concept by presenting an overview of the evolution of mobile computing.

Evolution of Mobile Computing: From Personal Use to Bring Your Own Device

While the idea of mobile computing has been around since the 1990s, it has evolved from simple two-way radios to complex tablets that do almost everything that can be done by regular computers. Today it is common to see a mobile device (e.g. a laptop, tablet, smartphone) connected to a hotspot at a local coffee shop. Employees love their mobile devices for various reasons. The device can help an employee improve their productivity (Xu, Peak, & Prybutok, 2015) as well as manage work–life boundaries (Cousins & Robey, 2015). A fundamental benefit of such technologies is portability. They are easy to carry and work with while on the move. They invariably have attractive user interfaces and provide features such as wireless local area networks (LANs) that provide access to the Internet in any part of the world.

Based on the nature of such devices and their pervasiveness in the personal lives of individuals, it was only a matter of time before employees wanted to bring their own devices to work and expected the organizations to support them. Mobile computing has thus enabled a BYOD phenomenon. Although a recent phenomenon, BYOD is quickly becoming ingrained into organizational functioning. BYOD is defined as the practice of allowing employees to bring to the workplace their own mobile devices that are capable of connecting to the Internet (Madzima, Moyo, & Abdullah, 2014). The devices include laptops, smartphones, e-readers, and even gaming devices.

Advantages of bring your own device

With large numbers of employees already owning smart technology (e.g., smart phones and tablets), some organizations view this as an opportunity to implement new technology without having to pay for the devices themselves (French, Guo, & Shim, 2014). An employee is more likely to upgrade their devices to the newest features more frequently than what an organization can afford to budget for on an ongoing basis. Moreover, due to its popularity, BYOD is not just about using personal devices at work. It has also been associated with increase in employee productivity, and job satisfaction (Steinert-Threlkeld, 2011). A study by Osterman Research (Osterman, 2012) yielded similar results. This is a consequence of increased personal freedom since employees can use their preferred mobile devices in their favored locations and times.

Disadvantages of bring your own device

BYOD is also connected with certain risks. Some of these include higher levels of loss or theft of sensitive information, unauthorized access to information, and the loss of governance or control over sensitive business information. According to Citrix (2014), 16% of organizations have banned certain apps and personal email websites.

As BYOD continues to become more common in organizations, it is imperative that we understand not only potential disadvantages, but also future implications for productivity, organizational commitment, job satisfaction, work–life balance, and more. After all, it has been widely stated that BYOD offers cost efficiency for organizations as it does not provide mobile devices to employees (Eslahi, Naseri, Hashim, Tahir, & Saad, 2014). From a workforce point of view, organizations need to delve into this area because generation Y (commonly referred to as the millennials) will be nearly 50% of the US workforce in 2020, and 75% of the global workforce by 2030 (Citrix, 2014). The millennials represent the first generation that has grown up with the Internet as a way of life as opposed to being a novelty (Burlacu, Truxillo, & Bauer, 2017, Chapter 20 in this volume). Four out of five millennials in the United States have smartphones or tablets, and nearly one in five have shunned PCs in favor of using their mobile devices (Eslahi et al., 2014). In addition compared with either generation X or the baby boomers, millennials value leisure time and find work less central to their identity (Twenge, 2010). Thus they should value the use of tools which help them succeed at work while also allowing them the flexibility to pursue nonwork interests. However, although there are some obvious benefits to BYOD, it has also been attributed several challenges.

- *Device heterogeneity* (Guan, Ke, Song, & Song, 2011): Not all employees use the same device. From an information technology (IT) department point of view this presents challenges, because different types of device and operating system need to be supported. This can cause difficulties when an employee needs assistance.
- *Lack of control* (Barkhuus & Dey, 2003): Enforcing security is difficult because organizations generally do not have complete control over the types of applications put on the device. Although employees do not download games or other entertainment applications on traditional work computers, on their own devices that may not be the case.
- *Security* (Leavitt, 2013): There is no question that BYOD increases the risk of having a security breach. There are numerous reasons for this. When an employee leaves an organization, they do not have to return the device. It is likely that many company applications and other data may still be present on the device itself. This can lead to some organizational data being unsecure. Certain compliance regulations that a company may need to follow (e.g. the Health Insurance Portability and Accountability Act of 1996 [HIPPA] or the Gramm-Leach-Bliley Act [GLBA] in the United States) are difficult to enforce in such a case.
- *Infrastructure concerns* (Wolbach, Harkes, Chellappa, & Satyanarayanan, 2008): The types of device that may be present can vary. Each device may operate at a different speeds and with different operating systems. This may cause organizational issues in regard to setting up and maintaining an infrastructure. Also, if employees are able to bring their own devices, many more devices will be used than if the company was providing them. Employees might bring all of their phones, tablets, and computers to work. This may put additional strain on the company's network infrastructure.

Businesses cannot afford to ignore BYOD. Employees increasingly desire the ability to use their own devices at work. But, before embracing or rejecting BYOD, organizations need to be aware of the advantages and disadvantages of its adoption. As an enterprise-wide implementation, it is imperative that organizations take into consideration how employees perceive such ventures and the policies associated with BYOD.

Organizational responses to mobile and BYOD

Organizations have at least two approaches when considering how to respond to the BYOD phenomenon: passive or active. In the *passive* approach, employees are allowed to bring their personal devices to work and use them for work activities such as email and produce reports, but it does not do anything explicitly to support them. In the *active* approach organizations create an explicit BYOD policy and implement it in the work environment. This approach requires infrastructure to support and evaluate efficiency (Hockly, 2012). Both approaches lend themselves to subtle variations in how employees used their devices for nonwork (e.g., personal) and work while at work.

However, in the presence of either approach the organization will face challenge, especially if an employee's mobile device usage behavior is not monitored. Zhang, Oh, and Teo (2006) found that high levels of anonymity and a lack of BYOD policy are likely to encourage employees to engage in cyberloafing. Cyberloafing refers to employees using the Internet and other services provided by an organization (their employer) for nonwork purposes during work hours. Few studies have investigated the reasons behind cyberloafing (see Charlier, Giumetti, Reeves, & Greco, 2017, Chapter 7 in this volume). Some have indicated that it may be a response to mistreatment at the workplace. Lim, Teo, and Loo (2002) found that employees in a Singapore firm agreed that cyberloafing is justified when they put in extra effort to perform their jobs, work overtime without additional compensation, are asked to do excessive amounts of work, or are exposed to conflicting demands. Henle and Blanchard (2008) found that employees engage in cyberloafing more often if they perceive that their jobs are stressful. They concluded that cyberloafing helps employees cope with stressors in the workplace. But they also found cyberloafing was less likely to be used for coping with stress when employees perceived organizational sanctions against it (e.g., discipline).

Mobile computing meets cloud computing

Mobile computing can be linked to the growing use of cloud computing (Scarfo, 2012). Cloud computing is a set of network-enabled services providing scalable, personalized, inexpensive computing platforms on demand that can be accessed in a simple and pervasive way over the Internet (Bhardwaj, Jain, & Jain, 2010). It has become central to how organizations operate, and 80% of organizations would experience a severe impact if there was a cloud outage of just one day (Columbus, 2013). When employees routinely work outside an actual office, having easy access to data via mobile technologies such as smartphones or tablets is a huge advantage (Ankeny, 2011). According to Mahesh, Landry, Sridhar, and Walsh (2011) the need for employees to have remote access necessitates an effective cloud computing solution. Cloud computing in the context of mobility also helps in eliminating administrative overhead and permits access from any geographic location, and from any device (McAfee, 2011).

Mobile computing also encourages collaboration at the workplace (Krell, 2011). Cloud computing allows organizational stakeholders to share information and data, and store with and retrieve information from each other (Devaki, 2011; Jain, 2011; Sultan, 2011).

Document sharing and editing of a document by several people and collaboration is compelling users to adopt cloud computing in the context of mobile technologies (Marston, Li, Bandyopadhyay, Zhang, & Ghalsasi, 2011). However, with networked cloud-based technologies, security and privacy have been cited as a top concern for organizations (Payton, 2010). Cloud computing environments face many of the same threats as traditional corporate networks, but due to the vast amount of data stored on cloud servers and their centralized storage, cloud providers have become an attractive target. The severity of damage depends on the sensitivity of the data exposed. Common methods of reducing risks include authentication and encryption (Jain, 2011; Mahesh et al., 2011). Enhanced security is a possibility due to economies of scale as well as affordability of security experts (Neves, Marta, Correia, & Neto, 2011).

A global phenomenon

Although workplace BYOD adoption is a global phenomenon, Gartner (2013), US companies are twice as likely to allow BYOD as those in Europe, which historically has had low BYOD adoption rates. Gartner (2013) also found that employees in India, China, and Brazil were most likely to use a personal device at work, such as a smartphone. Bradley, Loucks, Macaulay, Medcalf, and Buckalew (2012) carried out an extensive study that focused on organizations of all sizes (small, medium, and large), and encompassed eight countries across three regions. They found that 60% of white-collar workers used a mobile device for work purposes; 47% of those workers were classified as mobile workers by their organizations. The study also found that in 2014, the average number of connected devices per white-collar worker was 2.8, up from an average of 2.3 in 2012, and that Brazil, Russia, and China led the surge. The study also indicated that India, China, and Mexico had the highest increase in the percentage of spending devoted to mobility in general. In relation to BYOD in particular, 42% of the workers owned the smartphones they used for work. IT leaders in the United States, Asia, and Latin America were extremely positive about BYOD, with up to 88% in both Mexico and China expressing optimism. In Europe, IT leaders were less certain, with just 51% overall saying it was a positive development (with a low of 40% in France). Finally, the study also found that employees liked the fact that BYOD gave them more control and enhanced their work experience, with some even indicating that it enabled bottom-up innovation.

There are a few reasons why Europe is not being enthused by BYOD. Much of BYOD involves corporate data being delivered to devices that are not managed by an organization's internal IT department (Kabanda & Brown, 2014). This has obvious security implications. In addition, BYOD could prove to be more expensive for organizations due to the difficulty of managing various platforms, which in turn may increase the security threats (Koh, Oh, & Im, 2014; Shim, Mittleman, Welke, French, & Guo, 2013). These concerns may be the reason why some organizations do not fully adopt the paradigm shift. For example, there was a 15% decrease in BYOD in western Europe in 2013 compared with 2012 (Van Leeuwen, 2014). Most western European organizations take a more guarded approach to BYOD implementations compared with the rest of the world (Infosecurity, 2012). In an Infosecurity (2012) survey, 92% of respondents in China stated that their employers allowed them to use mobile devices on their own networks, whereas only 37% from the UK were allowed to do the same. This may not entirely be a surprise to many, since Europe in general has had stringent regulations relating to information privacy and security in all realms.

There is no question that BYOD and mobility are widely adopted technology trends, with one being a growth engine of the other. The more we use our personal devices for work, the more we become reliant on mobile technologies. A survey carried out by the Ovum Institute found that in emerging markets (e.g. Brazil, India, Russia, and Malaysia), BYOD adoption rates were significantly higher than those in the more mature markets (Llogicalis, 2012). For instance, in Brazil, the adoption rate for BYOD was 65%. In fact, the average BYOD growth in emerging markets was greater than 74%. The report presented two reasons for the variation in growth rates between the two markets. First, emerging markets had a lower rate of organizations offering handsets and tablets to employees. Therefore, employees felt that in order to do their job well they had to use their own personal devices. Second, emerging markets had a high level of comfort with the blurring of the boundary between work and personal life. Employees in countries such as Brazil, South Africa, and Malaysia have more flexible attitudes to working hours and are more than happy to use their personal devices to complete their tasks (Stone, Krueger, & Takach, 2017).

The relatively high penetration of BYOD in emerging economies has been corroborated in more recent literature as well, with the main explanation being a high percentage of mobile subscriptions. Shakhnov, Zinchenko, Rezchikova, and Glushko (2015) stated that India, China, and South Africa show better BYOD penetration than the rest of the emerging markets. Almost 70% of Chinese, South African, and Indian workers subscribe to the BYOD concept, whereas in Russia, only 55% of the employees use mobile devices at the workplace.

This section has highlighted the evolution of the mobile computing space, and its affinity to cloud computing. Though challenges such as security and privacy will persist, it is important not to lose sight of its importance and relevance on the global front. For certain countries, mobile technologies offer an opportunity to provide communication and collaborative tools to areas that were previously inaccessible. However, mobile technologies have also begun to impact workforce trends.

Future Workforce Trends

The future of mobile will be driven, in part, by the demographic shifts in perceptions of and use of mobile technology. Millennials value the ability to work anywhere and at any time more than they do other benefits, including pay and promotion (Krautkremer, 2014). Working with researchers at the University of Southern California and London Business School, PricewaterhouseCoopers (2013) conducted a comprehensive two-year study of its global workforce. The results showed that 64% of millennials in their workforce would like to work from home, and 66% would like to shift their work hours if possible. Shades of findings that show how millennials value flexibility through mobility are brought up in varied contexts. Schawbel (2015) stated that millennials do not believe in a 9–5 workday, and in fact value the ability to work from home, or even a coffee shop.

In order to effectively utilize mobile technologies in the workplace, it is important to have a cloud network (Samaras, Daskapan, Ahmad, & Ray, 2014). A modern cloud network integrates reliability, and economics of the cloud with the flexibility of network virtualization. This is an entirely new networking paradigm that meets the expectations of the workforce. Organizations in highly regulated sectors such as healthcare, government, and finance will also need to focus more on providing secure, mobile solutions. Also, where a majority of the employees are accessing corporate Wi-Fi networks via their personal devices, there is a need to ensure that sufficient bandwidth is available and that the networks do not get clogged up.

Role of Training and Continuous Education

Though millennials are well versed in technology, and do not necessarily need to be extensively taught how to effectively use mobile technologies, a majority of the current workforce may not have the same advantage. According to the Pew Research Center, the millennials lead the way not only in terms of adopting mobile technologies such as smartphones, but also taking advantage of a wider range of functions that are made available through them (Zickuhr, 2011). This trend is not as strong for baby boomers. One way for an organization to succeed in an ever-changing market is to implement a strategy that maximizes the synergies between lifelong learning and workforce productivity. Ideally, all organizations should have a plan set out for an employee's career development path and the required training programs that will enable employees to develop necessary knowledge and skills.

The advent of mobile technologies has also allowed for the mobile learning sector to thrive. It has become a popular method of choice for training employees (Docebo, 2014). For example, as part of its BlueWork initiative, American Express incorporates training programs such as how to use new technology tools, tips and tricks on how to be a mobile employee, and how to lead in a mobile environment (Meister, 2013). This program allows the organization to connect various works styles with the appropriate workspaces. Consequently, employees are generally satisfied, and a high degree of efficiency and productivity is maintained.

The Internet of Things

Auto-ID Labs in the United States coined the term "Internet of Things" (Zorzi, Gluhak, Lange, & Bassi, 2010). Gubbi et al. (2013) provide a succinct definition of IoT: it is the "worldwide network of interconnected objects uniquely addressable based on standard communication protocols" (p. 1653) This is made possible by seamless ubiquitous sensing, data analytics, and information representation with cloud computing as the unifying framework. Since then there have been rapid developments in academia and industry, in the United States as well as in Europe and Asia. Countries or enterprises have put forward strategies and plans related to the IoT, such as Smart Planet, Sensing China, and u-Japan, while also incorporating new infrastructure standards (Kim et al., 2016; Ma & Zhang, 2011). In the workplace, IoT is not just about automating office temperature or turning on the office coffee pot. Corsello (2013) stated that IoT would allow employees to increase collaboration and productivity by helping them better manage aspects of their work lives. The following scenario, based on Zarrehparvar (2013), highlights the true capacity of IoT: I'm imagining I walk out of my front door with my device. My device locks the door behind me. It starts my car. It pays for my coffee at Starbucks. It knows that when I get in my car and I say I'm going to Starbucks, it has my order waiting for me when I get there. It recognizes that I'm late for a meeting and changes my meeting because it knows – by my location – that I'm not going to be at the office in time to be there for my videoconference meeting and it changes it to a voice call.

According to Conerstone's State of Workplace Productivity Report, almost 60% of the surveyed employees stated that they were willing wear devices as long as the device would help them do their job better (Cornerstone, 2013). There is also a move toward transforming workplace meetings, physically and virtually. Videoconferencing today is mostly relegated to use of a laptop, smartphone, or a monitor with a camera and a microphone.

There is not much freedom of movement as may be had in a meeting taking place physically. However, that is already about to change. A Cisco employee sets the stage with the following scenario:

> *Imagine you are an executive responsible for selecting a new logistics partner to ship your goods globally. You could send regional managers to visit facilities in each major city to verify their capabilities – or you could go yourself, virtually. You would simply tap an iPad screen to awaken the robot from its charging station in the vendor's facility, then tap where you want to go on a map. You would appear on the mobile TelePresence screen as the robot moves autonomously throughout the remote location – meeting with executives, talking with employees, and touring the facility. And you would do it all from your office, or home, or hotel room, or wherever you happen to be*
>
> Evans (2013, n. p.)

The rise of IoT during the same timeframe as BYOD has added a layer of complexity to organizational infrastructures. It is highly advisable that organizations thoroughly understand the risks involved in each scenario in which IoT is deployed and to communicate with employees, partners, and customers about security and privacy risks (Tankard, 2015). This is especially important if sensitive data are at risk.

Based on what has been specified in this section, it is entirely possible that one thinks about the relationship between IoT and mobility in terms of the many employee services being made available. But this is only a fraction of how mobile plays into, and with, IoT. Mobile computing actually makes IoT possible, and the presence of big data tools have made it more desirable. In many ways, it can be argued that IoT is not really about things, but about data. Organizations have to determine how to best manage mobile and its enablement of IoT.

Future Research

Mobile computing has opened up a number of research issues, and potential contradictions. The issues vary from organizational (technological, infrastructure based) to personal (bottom-up innovation, stress, work–life balance, etc.). Due to the blurring of the lines between personal and organizational lives, future research questions may need to delve into both areas to fully understand how mobile computing may impact work. There is the already stated element of organizational security risk due to security policies not being followed. Existing theories on information system (IS) security policies do not focus on how to deal with exceptional situations in which policies are in conflict with the business objectives of an organization. Siponen and Iivari (2006) have developed an IS security design theory framework that addresses this gap.

Innovation and adoption are key principles of mobile computing in the past, and will continue to play a role in the future. Focusing on cross-cultural issues with respect to mobile technology adoption (Loose, Weeger, & Gewald, 2013; Stone et al., 2017) through the lens of what Rogers (2010) specified in the diffusion of innovations theory can provide a more holistic picture of the phenomenon. Cross-cultural issues can be especially important in relation to the intrusive nature of organizational level mobile computing policies (e.g. wiping a device if an employee leaves). Another interesting future research direction may be the spectrum of social influences in mobile adoption. Eckhardt, Laumer, and Weitzel (2009) found that adoption can be impacted by different workplace referent groups, such as superiors and colleagues from the same department. Given the connected and social nature of mobile computing, organizational peers may have a large effect on mobile device usage and adoption. Further research could also assess presence of differences in the social influence of the various referent groups in private or workplace contexts (Weeger, Wang, & Gewald, 2016).

The role of organizational culture in BYOD adoption should also be investigated. For example, changes in organizations can be achieved in roughly two ways: top-down or bottom-up. Future work could focus on any variation that may exist when it comes to BYOD implementation in regard to implementing the mentioned change approaches and their impact on outcomes such as organizational processes, productivity, and employee satisfaction. Research is also needed on different types of BYOD program, and how employees react to them. For example, some organizations require employees to install corporate software that may control and/or monitor a device. This in turn may impact an employee's device usage behavior. Furthermore, alternative concepts and measurements of individual mobile device risk perception should be examined (Weeger & Gewald, 2014). Measurements of these perceptions can be important since one of the advantages of mobile computing has been increased efficiency and productivity. Herzberg's two-factor theory of motivation can potentially shed light on what needs to be done to motivate employees (House & Wigdor, 1967).

Due to the expansive nature of mobile technologies it is also imperative that researchers consider industry-specific effects of mobile computing. For example, the healthcare industry has kept up with these technologies since they are critical to the industry's shift to patient-centered and value-based care. But mobile technology in healthcare is often restricted to electronic data form completion (Faulds et al., 2016). Other uses in healthcare include the use of device global positioning technology to identify where and when data originates in epidemiological studies (Aanensen, Huntley, Feil, & Spratt, 2009). There is also potential to transfer in-built camera and sound recording data, hence extending the scope of future research in this direction. Needless to say, security and privacy concerns will need to be addressed alongside as well (Free, 2014). Some exciting developments have already taken place, which allow activity and physiological data from in-built sensors from hand-held and wearable devices to be collected using mobile technology (Jardine, Fisher, & Carrick, 2015).

Conclusion

The mobile ecosystem of devices, carriers, and app markets has continued to evolve rapidly. From a business perspective, the mobile technology trend has transformed work practices and workplaces. The workforce with its varying digital competencies will continue to evolve, providing organizations with a variety of opportunities to increase overall organizational and employee effectiveness.

The challenge is that unmanaged mobile technologies have the potential of creating security risks, which have implications for financial, legal, and reputational outcomes. To best address these risks, it is important to have a thorough understanding of employee behavior before creating specific policies. Although no mobile policy can ever be deemed complete, successful organizations will apply research-based findings to their policies in such a way that it supports employee productivity and flexibility regardless of time or location. Along with mobile computing, the IoT will only continue to grow. However, as noted in this chapter, security and privacy concerns need to be resolved. Until then organizations need to remain vigilant, ensuring that they compare the benefits gained from IoT against the security risks. Organizations also need to enact appropriate controls and policies, while closely monitoring what is connected to the corporate networks and how devices are performing.

Finally, researchers can play an important role in the advances of mobile computing. More work needs to be done to provide guidance about how best to utilize mobile technologies in the service of organizational goals (Colbert, Yee, & George, 2016). This is important because mobile technologies have also shaped individual behaviors such as creativity, information overload, and employee wellbeing (Van Knippenberg, Dahlander, Haas, & George, 2015), thus resulting in radical organizational changes.

References

Aanensen, D. M., Huntley, D. M., Feil, E. J., & Spratt, B. G. (2009). EpiCollect: linking smartphones to web applications for epidemiology, ecology and community data collection. *PloS one*, *4*(9), e6968.

Ankeny, J. (2011). Heads in the cloud. *Entrepreneur*, *39*(10), 50–51.

Ashforth, B. E., Kreiner, G. E., & Fugate, M. (2000). All in a day's work: Boundaries and micro role transitions. *Academy of Management Review*, *25*(3), 472–491.

Balasubramanian, S., Peterson, R. A., & Jarvenpaa, S. L. (2002). Exploring the implications of m-commerce for markets and marketing. *Journal of the Academy of Marketing Science*, *30*(4), 348–361.

Bang, Y., Lee, D.-J., Han, K., Hwang, M., & Ahn, J.-H. (2013). Channel capabilities, product characteristics, and the impacts of mobile channel introduction. *Journal of Management Information Systems*, *30*(2), 101–126.

Barkhuus, L., & Dey, A. (2003, October). *Is context-aware computing taking control away from the user? Three levels of interactivity examined.* Paper presented at the International Conference on Ubiquitous Computing, Seattle, WN, USA.

Bhardwaj, S., Jain, L., & Jain, S. (2010). Cloud computing: A study of infrastructure as a service (IAAS). *International Journal of Engineering and Information Technology*, *2*(1), 60–63.

Bradley, J., Loucks, J., Macaulay, J., Medcalf, R., & Buckalew, L. (2012). *BYOD: A global perspective. Survey Report.* San Jose, CA: Cisco IBSG Horizons.

Burlacu, G., Truxillo, D. M., & Bauer, T. N. (2017). Employee age differences in using Internet-based tools at work. In G. Hertel, D. L. Stone, R. D. Johnson, & J. Passmore (Eds.). *The Wiley Blackwell handbook of the psychology of the internet at work* (pp. 449–480). Chichester: Wiley-Blackwell.

Charlier, S. D., Giumetti, G. W., Reeves, C. J., & Greco, L. S. (2017). Workplace cyberdeviance. In G. Hertel, D. L. Stone, R. D. Johnson, & J. Passmore (Eds.). *The Wiley Blackwell handbook of the psychology of the internet at work* (pp. 131–156). Chichester: John Wiley & Sons, Ltd.

Citrix. (2014). *Enterprise mobility trends.* Retrieved January 2, 2016, from www.slideshare.net/citrix/citrix-xenmobile-enterprise-mobility-trends

Clark, S. C. (2000). Work/family border theory: A new theory of work/family balance. *Human Relations*, *53*(6), 747–770.

Colbert, A., Yee, N., & George, G. (2016). The digital workforce and the workplace of the future. *Academy of management Journal*, *59*(3), 731–739.

Columbus, L. (2013). *451 Research summit: Enterprises competing with digital infrastructure; cloud computing.* Retrieved January 27, 2016, from www.forbes.com/sites/louiscolumbus/2013/09/28/451-research-summit-competing-with-digital-infrastructure-cloud-computing/#1fd12eb43272

Cornerstone. (2013). *The state of workplace productivity report.* Retrieved January 30, 2016, from https://www.cornerstoneondemand.com/resources/research/state-of-workplace-productivity-2013

Corsello, J. (2013). *What the internet of things will bring to the workplace.* Retrieved February 11, 2016, from http://www.wired.com/insights/2013/11/what-the-internet-of-things-will-bring-to-the-workplace/

Cousins, K., & Robey, D. (2015). Managing work-life boundaries with mobile technologies: An interpretive study of mobile work practices. *Information Technology & People*, *28*(1), 34–71.

Devaki, S. (2011). File storage trends in cloud computing era. *Siliconindia*, *14*(8), 34–35.

Docebo. (2014). *E-Learning Market Trends and FOrecast 2014–2016 report*. Retrieved January 29, 2016, from www.wojde.org/FileUpload/bs295854/File/07r_41-.pdf

Duxbury, L., Higgins, C., Smart, R., & Stevenson, M. (2014). Mobile technology and boundary permeability. *British Journal of Management*, *25*(3), 570–588.

Eckhardt, A., Laumer, S., & Weitzel, T. (2009). Who influences whom? Analyzing workplace referents' social influence on IT adoption and non-adoption. *Journal of Information Technology*, *24*(1), 11–24.

Eslahi, M., Naseri, M. V., Hashim, H., Tahir, N. M., & Saad, E. H. M. (2014, April). *BYOD: current state and security challenges*. Paper presented at the IEEE Symposium on Computer Applications and Industrial Electronics (ISCAIE), Penang, Malaysia.

Evans, D. (2013). *New video collaboration robot: TelePresence gets moving*. Retrieved April 7, 2017, from http://blogs.cisco.com/news/new-video-collaboration-robot-telepresence-gets-moving

Faulds, M., Bauchmuller, K., Miller, D., Rosser, J., Shuker, K., Wrench, I., ... Mills, G. (2016). The feasibility of using "bring your own device" (BYOD) technology for electronic data capture in multicentre medical audit and research. *Anaesthesia*, *71*(1), 58–66.

Free, J. (2014). Real-world BYOD security BYOD security strategies from two distinct healthcare organizations. *Health Management Technology*, *35*(3), 14–17.

French, A. M., Guo, C., & Shim, J. (2014). Current status, issues, and future of bring your own device (BYOD). *Communications of the Association for Information Systems*, *35*(10), 191–197.

Gaff, B. M. (2015). BYOD? OMG! *Computer*, *48*(2), 10–11.

Gartner. (2013). *Gartner predicts by 2017, half of employers will require employees to supply their own device for work purposes*. Retrieved February 10, 2016, from www.gartner.com/newsroom/id/2466615

Gerow, J. E., Ayyagari, R., Thatcher, J. B., & Roth, P. L. (2013). Can we have fun @ work? The role of intrinsic motivation for utilitarian systems. *European Journal of Information Systems*, *22*(3), 360–380.

Golden, A. G., & Geisler, C. (2007). Work–life boundary management and the personal digital assistant. *Human Relations*, *60*(3), 519–551.

Guan, L., Ke, X., Song, M., & Song, J. (2011, May). *A survey of research on mobile cloud computing*. Paper presented at the 2011 10th IEEE/ACIS International Conference on Computer and Information Science, Sanya, China.

Gubbi, J., Buyya, R., Marusic, S., & Palaniswami, M. (2013). Internet of Things (IoT): A vision, architectural elements, and future directions. *Future Generation Computer Systems*, *29*(7), 1645–1660.

Guest, D. E. (2002). Perspectives on the study of work-life balance. *Social Science Information*, *41*(2), 255–279.

Harris, J., Ives, B., & Junglas, I. (2012). IT consumerization. When gadgets turn into enterprise IT tools. *MIS Quarterly Executive*, *11*(3), 99–112.

Henle, C. A., & Blanchard, A. L. (2008). The interaction of work stressors and organizational sanctions on cyberloafing. *Journal of Managerial Issues*, *20*(3), 383–400.

Hislop, D., & Axtell, C. (2011). Mobile phones during work and non-work time: A case study of mobile, non-managerial workers. *Information and Organization*, *21*(1), 41–56.

Hockly, N. (2012). Tech-savvy teaching: BYOD. *Modern English Teacher*, *21*(4), 44–45.

House, R. J., & Wigdor, L. A. (1967). Herzberg's dual-factor theory of job satisfaction and motivation: A review of the evidence and a criticism. *Personnel Psychology*, *20*(4), 369–390.

Imielinski, T., & Korth, H. F. (1996). Introduction to mobile computing. In T. Imielinski & H. F. Korth (Eds.), *Mobile computing* (pp. 1–43). Boston, MA: Springer.

Infosecurity. (2012). *Europeans exhibit more caution regarding BYOD*. Retrieved February 17, 2016, from www.infosecurity-magazine.com/news/europeans-exhibit-more-caution-regarding-byod/

Islam, N., & Fayad, M. (2003). Toward ubiquitous acceptance of ubiquitous computing. *Communications of the ACM*, *46*(2), 89–92.

Jain, V. (2011). How the cloud resonates with business today. *Siliconindia, 14*(10), 22–23.

Jardine, J., Fisher, J., & Carrick, B. (2015). Apple's ResearchKit: smart data collection for the smartphone era? *Journal of the Royal Society of Medicine, 108*(8), 294–296.

Jeong, M., Lee, M., & Nagesvaran, B. (2016). Employees' use of mobile devices and their perceived outcomes in the workplace: A case of luxury hotel. *International Journal of Hospitality Management, 57*, 40–51.

Jones, F., Burke, R. J., & Westman, M. (2013). *Work-life balance: A psychological perspective:* New York, NY: Psychology Press.

Kabanda, S., & Brown, I. (2014, December). *Bring-your-own-device (BYOD) practices in SMEs in developing countries – the case of Tanzania.* Paper presented at the 25th Australasian Conference on Information Systems, Auckland, New Zealand.

Kietzmann, J., Plangger, K., Eaton, B., Heilgenberg, K., Pitt, L., & Berthon, P. (2013). Mobility at work: A typology of mobile communities of practice and contextual ambidexterity. *The Journal of Strategic Information Systems, 22*(4), 282–297.

Kim, J., Yun, J., Choi, S.-C., Seed, D. N., Lu, G., Bauer, M., ... Song, J. (2016). Standard-based IoT platforms interworking: Implementation, experiences, and less ons learned. *IEEE Communications Magazine, 54*(7), 48–54.

Koh, E. B., Oh, J., & Im, C. (2014, March). *A study on security threats and dynamic access control technology for BYOD, smart-work environment.* Paper presented at the International MultiConference of Engineers and Computer Scientists, Hong Kong, China.

Krautkremer, T. (2014). *Beyond VPN: Supporting the mobile workforce.* Retrieved February 2, 2016, from www.networkcomputing.com/cloud-infrastructure/beyond-vpn-supporting-mobile-workforce/1367458936

Krell, E. (2011). The state of small business. *Baylor Business Review, 30*(1), 4–9.

Leavitt, N. (2013). Today's mobile security requires a new approach. *IEEE Computer, 46*(11), 16–19.

Lim, V. K., Teo, T. S., & Loo, G. L. (2002). How do I loaf here? Let me count the ways. *Communications of the ACM, 45*(1), 66–70.

Llogicalis. (2012). *BYOD: An Emerging Market Trend in More Ways Than One.* Retrieved January 10, 2016, from http://www.us.logicalis.com/globalassets/united-states/whitepapers/logicalisbyodwhitepaperovum.pdf

Loose, M., Weeger, A., & Gewald, H. (2013, August). *BYOD – the next big thing in recruiting? Examining the determinants of BYOD service adoption behavior from the perspective of future employees.* Paper presented at the 19th Americas Conference on Information Systems, Chicago, IL, USA.

Ma, T., & Zhang, C. (2011, December). *On the disruptive potentials in internet of things.* Paper presented at the IEEE International Conference on Parallel and Distributed Systems (ICPADS), Tainan, Taiwan.

Madzima, K., Moyo, M., & Abdullah, H. (2014, August). *Is bring your own device an institutional information security risk for small-scale business organisations?* Paper presented at the Information Security for South Africa (ISSA), Johannesburg, South Africa.

Mahesh, S., Landry, B. J., Sridhar, T., & Walsh, K. R. (2011). A decision table for the cloud computing decision in small business. *Information Resources Management Journal (IRMJ), 24*(3), 9–25.

Marston, S., Li, Z., Bandyopadhyay, S., Zhang, J., & Ghalsasi, A. (2011). Cloud computing – the business perspective. *Decision Support Systems, 51*(1), 176–189.

Matusik, S. F., & Mickel, A. E. (2011). Embracing or embattled by converged mobile devices? Users' experiences with a contemporary connectivity technology. *Human Relations, 64*(8), 1001–1030.

Mazmanian, M., Orlikowski, W. J., & Yates, J. (2013). The autonomy paradox: The implications of mobile email devices for knowledge professionals. *Organization Science, 24*(5), 1337–1357.

McAfee, A. (2011). What every CEO needs to know about the cloud. *Harvard Business Review, 89*(11), 124–132.

Meister, J. (2013). *Flexible workspaces: Employee perk or business tool to recruit top talent?* Retrieved November 20, 2015, from www.forbes.com/sites/jeannemeister/2013/04/01/

flexible-workspaces-another-workplace-perk-or-a-must-have-to-attract-top-talent/#1f38b9dd4382

Middleton, C., Scheepers, R., & Tuunainen, V. K. (2014). When mobile is the norm: researching mobile information systems and mobility as post-adoption phenomena. *European Journal of Information Systems, 23*(5), 503–512.

Middleton, C. A. (2008). Do mobile technologies enable work–life balance? In D. Hislop (Ed.), *Mobility and technology in the workplace* (*Vol. 9*, pp. 209–224). London: Routledge.

Middleton, C. A., & Cukier, W. (2006). Is mobile email functional or dysfunctional? Two perspectives on mobile email usage. *European Journal of Information Systems, 15*(3), 252–260.

Murray, W. C., & Rostis, A. (2007). Who's running the machine? A theoretical exploration of work stress and burnout of technologically tethered workers. *Journal of Individual Employment Rights, 12*(3), 249–263.

Neves, F. T., Marta, F. C., Correia, A. M. R., & Neto, M. d. C. (2011, October). *The adoption of cloud computing by SMEs: identifying and coping with external factors.* Paper presented at the 11ª Conferência da Associação Portuguesa de Sistemas de Informação (CAPSI 2011) – A Gestão de Informação na era da Cloud Computing, Lisboa, Portugal.

Osterman. (2012). *Putting IT back in control of BYOD.* An Osterman Research White Paper. Washinton DC: Osterman Research.

Payton, S. (2010). Fluffy logic. *Financial Management, 22*, 22–25.

Prasopoulou, E., Pouloudi, A., & Panteli, N. (2006). Enacting new temporal boundaries: The role of mobile phones. *European Journal of Information Systems, 15*(3), 277–284.

PricewaterhouseCoopers. (2013). Millennial workers want greater flexibility, work/life balance, global opportunities. Retrieved January 27, 2016, from http://press.pwc.com/global/nextgen-global-generational-study/s/376ce2a9–1769–46f2-a228–8b97d252f660

Rice, R. E. (2017). Flexwork, work-family boundaries, and information and communication technologies. In G. Hertel, D. L. Stone, R. D. Johnson, & J. Passmore (Eds.). *The Wiley Blackwell handbook of the psychology of the internet at work* (pp. 175–194). Chichester: John Wiley & Sons, Ltd.

Rogers, E. M. (2010). *Diffusion of innovations* (4th ed.). New York, NY: Simon and Schuster.

Samaras, V., Daskapan, S., Ahmad, R., & Ray, S. K. (2014, November). *An enterprise security architecture for accessing SaaS cloud services with BYOD.* Paper presented at the Australasian Telecommunication Networks and Applications Conference (ATNAC), Southbank, Victoria, Australia

Scarfo, A. (2012, November). *New security perspectives around BYOD.* Paper presented at the 7th International Conference on Broadband, Wireless Computing, Communication and Applications (BWCCA), Victoria, BC, Canada.

Schawbel, D. (2015). *The 4 things millennials want most from employers may surprise you.* Retrieved February 9, 2016, from www.businessinsider.com/what-millennials-want-most-from-employers-2015-4

Scheepers, R., & Middleton, C. (2013). Personal ICT ensembles and ubiquitous information systems environments: Key issues and research implications. *Communications of the Association for Information Systems, 33*(1), 381–392.

Scheepers, R., Scheepers, H., & Ngwenyama, O. K. (2006). Contextual influences on user satisfaction with mobile computing: Findings from two healthcare organizations. *European Journal of Information Systems, 15*(3), 261–268.

Shakhnov, V. A., Zinchenko, L. A., Rezchikova, E. V., & Glushko, A. A. (2015, September). *An Opportunity in Engineering Education: Russian BYOD Tendencies.* Paper presented at the International Conference on Interactive Collaborative Learning, Florence, Italy.

Shim, J., Mittleman, D., Welke, R., French, A. M., & Guo, J. C. (2013, August). *Bring your own device (BYOD): Current status, issues, and future directions.* Paper presented at the 19th Americas Conference on Information Systems, Chicago, IL, USA.

Siponen, M., & Iivari, J. (2006). Six design theories for IS security policies and guidelines. *Journal of the Association for Information Systems, 7*(1), 445–472.

Steinert-Threlkeld, T. (2011). BYOD: Save money, gain productivity. *Money Management Executive, 19*(46), 1–10.

Stone, D. L., Krueger, D., & Takach, S. (2017). Social issues associated with the Internet at work. In G. Hertel, D. L. Stone, R. D. Johnson, & J. Passmore (Eds.). *The Wiley Blackwell handbook of the psychology of the internet at work* (pp. 423–448). Chichester: John Wiley & Sons, Ltd.

Sultan, N. A. (2011). Reaching for the "cloud": How SMEs can manage. *International Journal of Information Management, 31*(3), 272–278.

Sutanto, J., Palme, E., Tan, C.-H., & Phang, C. W. (2013). Addressing the personalization-privacy paradox: An empirical assessment from a field experiment on smartphone users. *MIS Quarterly, 37*(4), 1141–1164.

Tankard, C. (2015). The security issues of the Internet of Things. *Computer Fraud & Security, 2015*(9), 11–14.

Twenge, J. M. (2010). A Review of the empirical evidence on generational differences in work attitudes. *Journal of Business and Psychology, 25*, 201–210.

Van Knippenberg, D., Dahlander, L., Haas, M. R., & George, G. (2015). Information, attention, and decision making. *Academy of Management Journal, 58*(3), 649–657.

Van Leeuwen, D. (2014). Bring your own software. *Network Security, 2014*(3), 12–13.

Van Yperen, N. W., & Wörtler, B. (2017). Blended working. In G. Hertel, D. L. Stone, R. D. Johnson, & J. Passmore (Eds.). *The Wiley Blackwell handbook of the psychology of the internet at work* (pp. 157–174). John Wiley & Sons, Ltd.

Watson, R. T., Pitt, L. F., Berthon, P., & Zinkhan, G. M. (2002). U-commerce: Expanding the universe of marketing. *Journal of the Academy of Marketing Science, 30*(4), 333–347.

Weeger, A., & Gewald, H. (2014, June). *Factors influencing future employees decision-making to participate in a BYOD program: Does risk matter?* Paper presented at the 22nd European Conference on Information Systems, Tel Aviv, Israel.

Weeger, A., Wang, X., & Gewald, H. (2016). It consumerization: Byod-program acceptance and its impact on employer attractiveness. *Journal of Computer Information Systems, 56*(1), 1–10.

Weintraub, S. (2010). *The numbers don't lie: Mobile devices overtaking PCs.* Retrieved July 15, 2015, from http://fortune.com/2010/08/11/the-numbers-dont-lie-mobile-devices-overtaking-pcs/

Wixom, B. H. & Todd, P. A. (2005), A theoretical integration of user satisfaction and technology acceptance. *Information Systems Research, 16*(1), 85–102.

Wolbach, A., Harkes, J., Chellappa, S., & Satyanarayanan, M. (2008, June). *Transient customization of mobile computing infrastructure.* Paper presented at the Proceedings of the First Workshop on Virtualization in Mobile Computing, Breckenridge, CO, USA.

Xu, C., Peak, D., & Prybutok, V. (2015). A customer value, satisfaction, and loyalty perspective of mobile application recommendations. *Decision Support Systems, 79*, 171–183.

Zarrehparvar, M. (2013). *Mobile is "the new IT": 5 reasons to spend more on enterprise mobility.* Retrieved February 11, 2016, from http://venturebeat.com/2013/07/27/mobile-is-the-new-it-5-reasons-to-spend-more-on-enterprise-mobility/

Zhang, D., Oh, L.-B., & Teo, H.-H. (2006, January). *An experimental study of the factors influencing non-work related use of IT resources at workplace.* Paper presented at the 39th Annual Hawaii International Conference on System Sciences, Kauai, Hawaii.

Zickuhr, K. (2011). Generations and their gadgets. *Pew Internet & American Life Project, 20.*

Zorzi, M., Gluhak, A., Lange, S., & Bassi, A. (2010). From today's intranet of things to a future internet of things: a wireless- and mobility-related view. *IEEE Wireless Communications, 17*(6), 44–51.

Part II
Organizational Perspectives

11

E-Recruiting: Using Technology to Attract Job Applicants

Derek S. Chapman and Anna F. Gödöllei

Introduction

The medium is the message.

Marshall McLuhan (1964, emphasis added)

Recruiting employees who add value to the organization is a key competitive advantage for organizations. Recruiting has undergone a revolution over the past 20 years such that both employer practices and applicant expectations have been radically altered by the influence of technology. In fact the use of communication technology in recruiting has largely supplanted traditional recruiting with what has become known as electronic recruiting or "e-recruiting." In this chapter, we examine the growing phenomenon of e-recruiting and review the relevant theory and empirical research examining this area of interest. We also propose a framework for studying this phenomenon, which we hope will guide future research in this exciting area.

We define e-recruiting as the use of communication technologies, such as websites and social media, to find and attract potential job applicants, to keep them interested in the organization during the selection processes, and to influence their job choice decisions. Although many practitioners perceive recruiting as consisting of both attraction and selection activities, most scholars perceive a benefit of maintaining a distinction between recruiting (attraction, maintenance, and job choice influence) and assessment and selection (Barber, Daly, Giannantonio, & Phillips, 1994; see also Dickter, Jockin, & Delany, 2017, Chapter 13 of this volume). However, as Rynes (1989) originally pointed out, our selection and recruiting systems are intertwined and interdependent. Applicants' reactions to selection procedures (such as online assessments) can also impact recruiting effectiveness and influence perceptions of the company (Gilliland, 1993; Ryan & Ployhart, 2000). Accordingly, we discuss both e-recruiting and applicant reactions to technology driven selection practices.

The Wiley Blackwell Handbook of the Psychology of the Internet at Work,
First Edition. Edited by Guido Hertel, Dianna L. Stone, Richard D. Johnson, and Jonathan Passmore.

First, we will review the history of e-recruiting and its emergence in the field. Next, we discuss the potential benefits of e-recruiting for organizations. In the third section we describe the current state of practices in e-recruiting; this is followed by a discussion of the limitations and dangers associated with e-recruiting in the fourth section. Finally, we introduce a model of e-recruiting and propose research directions for those interested in this topic.

A Brief History of E-Recruiting

It is exceedingly rare for industrial and organizational psychologists to witness the birth of a disruptive technology as it emerges in the field, so it is both exciting and a challenge to study a moving target such as e-recruiting. For many decades recruiting practices remained largely unchanged. Employers placed advertisements in newspapers or visited college campuses to meet and attract potential employees. Job advertisements were sent to employment centers and physically posted on large boards for job seekers to peruse in person. Information about the company and position were handed out in printed brochures. Recruiting interviews were conducted. However, in the 1990s, recruiting practices began to change dramatically. Beginning with widespread adoption of email in the early 1990s, recruiters were suddenly able to contact many potential applicants more directly than they had in the past. Obtaining lists of email addresses afforded recruiters fast and relatively effortless access to large numbers of potential applicants. Simultaneously, recruiters began creating online approaches to recruiting practices that were traditionally carried out using physical recruiting techniques.

Online job boards

One of the earliest attempts to translate the cumbersome traditional recruiting techniques to an e-recruiting approach was the online job board. Instead of posting job ads on job boards that required job seekers to physically travel to a location where the boards were set up, recruiters recognized that they could create an online virtual space, which potential job applicants could visit without leaving their computers. Creating virtual job ads for online job boards also made them easy to update, and the online boards were considerably cheaper than buying space in a newspaper.

The first of these online job boards, the Online Career Center, was introduced in 1992. As Dickey-Chasins (2012) notes, these early approaches were originally very broad in their targeting, in that they included jobs for many industries and for a variety of geographic locations. As they grew in popularity, they became more niche-targeted by geographic region and by industry. By the early 2000s, they had largely replaced newspaper-based recruiting as well as traditional job boards in employment centers. Online job boards such as Monster.com were visited by millions of job seekers per year and a majority of employers came to use these types of e-recruiting approaches over traditional methods. By 2007, US online recruiting revenues were predicted to grow to US$2.6 billion (Li, Charron, Roshan, & Flemming, 2002). In a matter of only 10 years, recruiting had been dramatically changed from an inefficient, print-based and face-to-face approach to more efficient and less expensive online technologies. The convenience of online job boards proved to be extremely popular with both employers and job seekers. As they evolved, options for using filters to search by region, industry, job title, and so forth have helped job seekers explore job opportunities in ways that would have been time consuming and expensive in the 1980s (Maurer & Liu, 2007). Other improvements included the opportunity for job seekers to post their résumés online to provide ready access to employers.

Job ad aggregators

Like most online technologies, e-recruiting accelerates the pace of innovation by opening up opportunities to capitalize on evolving technologies that influence practice. One such technology includes web-based aggregators which actively scan the Internet for information and compile it in a central location that is easily searched by users. For example, in the travel industry, aggregators such as Trivago search websites for hotel prices and aggregate the price information from many sites into one searchable location. This information can then be used by people to make informed choices and get the best deal for their vacation. In e-recruiting, aggregators took the form of websites such as Indeed and Eluta, which, beginning in the mid-2000s, started to take market share from the traditional job boards, eventually overtaking them for site traffic by 2011 (Dickey-Chasins, 2012). Employers could now bypass the relatively expensive and exclusive job boards by posting links to their job ads located on their own webpages on the aggregators. This exposed their job ads to an ever-increasing number of potential job applicants at substantially lower costs. Job seekers also had the advantage of increasing the number of employment opportunities available. However, like their job board predecessors, aggregators are limited in that they focus on one-way messaging. They introduce job seekers and employers to each other in ways that are fairly static. To learn more about the applicant (or the employer) required additional steps of formally applying for the job, meeting with recruiters, and so forth.

Employer websites

Employers have long used their own websites to provide information to potential applicants about the organization and opportunities for employment. Limited research has examined the features of employer websites that influence applicant perceptions of organizations. Parry and Tyson (2008) found that websites were more effective for larger UK companies, perhaps because larger and more well-known companies are more likely to attract general web traffic than smaller, lesser-known organizations. Cober and colleagues (Cober, Brown, Keeping, & Levy, 2004; Cober, Brown, & Levy, 2004; Cober, Brown, Levy, Cober, & Keeping, 2003) and more recently other researchers (De Goede, Van Vianen, & Klehe, 2011) have examined how website characteristics influence applicant perceptions and application behaviors. Four features of websites influence applicant perceptions.

Aesthetics. Cober et al. (2004) found that choices of colors, designs, layout, photographs, and graphics have a small but statistically significant impact on applicant perceptions of employers. Unfortunately, other than alerting web designers that they need to pay attention to this factor, the recruiting literature provides little guidance on the ingredients that will lead to aesthetically pleasing websites or whether there is agreement among job seekers regarding what is aesthetically appealing. On examining websites that were thought to be cutting edge and aesthetically appealing in the 1990s, it is easy to conclude that tastes and technologies change fairly often: what might be seen as pleasing today may be out of fashion in less than a year. It is also likely that employers' targeted demographics will influence the aesthetic choices they make. For example, Djamasbi, Siegel and Tullis (2010) found that there were significant differences between millennials and older generations in terms of what is aesthetically pleasing in websites. Their younger participants were more attracted to larger photographs and less attracted to text than older participants. Accordingly, it is unlikely that all demographics will respond similarly to web design characteristics. It is also unclear whether aesthetic preferences are context dependent. What might be perceived as stylish in one context (e.g., a retail website) might not be perceived favorably in a recruiting context. More research is needed on the influence of demographics on perceptions of aesthetics of recruiting website. We also suggest that employers are careful to test their designs with focus groups consisting of people from demographics similar to the intended target audience, and that they should periodically repeat this testing to ensure the site meets applicant expectations (see also Thielsch & Niesenhaus, 2017, Chapter 5 of this volume).

Navigability. Early research on website design for recruiting revealed that job seekers respond negatively when they experience difficulty finding desired information on employer websites (Cober et al., 2003; Lin, 2010; Selden & Orenstein, 2011). This issue may be somewhat mitigated by the proliferation of well-thought-out web design companies. These companies provide templates and web design tools that make it easy for even the smallest of organizations to design an easily navigable website. Nonetheless, employers need to ensure that there are links on their primary landing pages directing their website visitors to their career content directed at recruiting. Recent eye-tracking studies show that applicants spend much of their time looking for additional links when they visit websites (Allen, Biggane, Pitts, Otondo, & Van Scotter, 2013). It is essential to make these links obvious (e.g., buttons) and contrast them from the rest of the page.

Content. There is also evidence that the content of the recruiting messages contained in corporate websites influences applicants' perceptions (Allen, Mahto, & Otondo, 2007; Cober, Brown, & Levy, 2004; Van Birgelen, Wetzels, & van Dolen, 2008). Although the quality of the content is acknowledged as important, the literature is less informative about how to make content attractive to candidates. Kraichy and Chapman (2014) suggest that affect-based online messages are more persuasive than fact-based or cognitive messages. Schmidt, Chapman and Jones, (2015) found in a field experiment that fit information related to identifying what the company provides to applicants (needs–supplies fit) was more effective than fit information emphasizing what the company wants from applicants (demands–abilities fit). There are good reasons to expect that findings from the broader recruiting literature about what is attractive to applicants (see Chapman, Uggerslev, Carroll, Piasentin, & Jones, 2005) will translate to e-recruiting; however, researchers should be careful to test this assumption.

Interactivity. Websites have also evolved over time such that they now have the potential to bring employers and applicants together to communicate about employment opportunities in a two-way fashion. This interactivity leads to greater engagement with the employer and ultimately greater attraction to organizations. The nature of this interaction and whether certain forms of interaction are more influential than others is an area that needs exploring by e-recruiting researchers.

E-recruiting interviews

An early application of communication technologies in e-recruiting was the adoption of videoconference technology in employment and recruiting interviews (Chapman & Rowe, 2001). Although early videoconferencing technologies required multiple phone lines and large dedicated spaces, modern videoconferencing allows for inexpensive and reliable communication with candidates at a distance. There are benefits to videoconference interviewing at a distance, such as substantial savings on travel and accommodations, greater reach to distant applicants, and greater media richness than telephone interviews (Chapman, Uggerslev, & Webster, 2003). However, early research on the effectiveness of videoconference interviews showed that applicants tended to react less positively to these types of interview than face-to-face approaches (Chapman & Rowe, 2002; Chapman et al., 2003) particularly for unstructured interviews. Note, these studies consisted of field experiments in which real applicants were randomly assigned to videoconference or face-to-face interviews. More recently, it is apparent that improvements in videoconference technology and more widespread adoption of technologies such as Skype have not improved applicant reactions (Sears, Zhang, Wiesner, Hackett, & Yuan, 2013). Like many e-recruiting technologies, practitioners' perceptions of benefits (e.g., cost effectiveness, convenience) tend to outweigh most of the empirical research that examines the efficacy of various trendy approaches to attract applicants.

Social networking

The most recent phenomenon for e-recruiting is social networking. Few technologies have promised so much for recruiting as social media networks. Kohli, Suri, and Kapoor (2015) estimated that over 1.3 billion people use social media such as Twitter, Facebook, and Instagram. For some focused social media networks (e.g., LinkedIn), researchers have found that the majority of users are actively engaged in networking to either hire new colleagues or find a job for themselves (Stopfer & Gosling, 2013). Considering one of the primary goals of recruiting is to attract candidates to apply for your positions (Breaugh & Starke, 2000) this global recruiting technology has incredible potential. Human resources (HR) departments have faced enormous pressure to create an online presence in these social media spaces to attract applicants. Most HR managers have little or no training and limited experience with social media. As a result, the process of integrating social media with recruiting goals has been uneven with many failures along the way, although progress is clearly being made. Parry and Tyson (2008) for example, found that UK recruiter perceptions of e-recruiting effectiveness rose from only 48% in 2000 to 71% in 2006. As a field we are still learning how to effectively harness e-recruiting processes such as social networking to attract applicants (see also Lukaszewski & Johnson, 2017, Chapter 12 of this volume). We discuss the potential benefits for social networking in the next section.

Benefits of E-Recruiting

As we noted previously, the goals of e-recruiting are largely the same as traditional recruiting; that is, attracting job applicants, maintaining applicant interest and ultimately influencing their job choices. Despite ample positive anecdotal support for the benefits of e-recruiting, empirical research on the efficacy of e-recruiting for enhancing recruiting practice is remarkably scarce.

Volume management

One of the earliest concerns about the use of e-recruiting was the overwhelming volume of the resulting applicant pool and the costs associated with screening and assessing the potential of these applicants (Chapman & Webster, 2003; Parry & Tyson, 2008). In this respect, recruiting technology has become almost a two-edged sword: on the one hand, e-recruiting improves the potential to reach many applicants, on the other hand, it risks overwhelming the organization if there are no careful plans to deal with volume management. Early adopters of e-recruiting such as Procter & Gamble noted that failure to consider the need for effective screening is a key determinant of success with this strategy. Likewise, many recruiting technologies have made it easier for recruiters to track applicants and keep in touch with them at critical points. For example, most large enterprise software firms offer HR solutions that include tracking employees from potential applicants through hiring and ultimately into talent management.

Applicant quality

One of the key goals organizations report for their recruiting systems is to improve applicant quality (Breaugh & Starke, 2000). In addition, recruiting scholars have called for a focus on applicant quality in recruiting research (Carlson, Connerley, & Mecham, 2002). In its simplest form, attracting more applicants should result in better candidates. Assuming a random distribution of talent in the applicant pool, a gross increase in application volume should make it more likely to have enough high-quality applicants to fill positions (Boudreau & Rynes, 1985). Thus effective e-recruiting necessitates efficient, inexpensive technologies to screen and assess candidates. Few studies have explicitly examined how e-recruiting influences applicant quality. One field experiment of web-based recruiting ads demonstrated that online ads that focused on needs–supplies fit (i.e., emphasizing what the company can provide for the applicant) was significantly more effective at attracting a greater number of applicants than those that focused on other ways applicants might fit with the organization. Perhaps more importantly, the online ads focusing on needs–supplies fit were also more effective for attracting stronger applicants based on the résumés submitted and the likelihood with which applicants were selected to be interviewed by hiring managers (Schmidt et al., 2015).

E-recruiting also has the potential to target stronger applicants more precisely. Technological improvements allow employers to adjust and customize their website content to their visitors' characteristics. This is more typically used in marketing products where a visitor's marketing profiles from searches, social networking history, and so forth can influence what they see when visiting a website. Recruiting researchers have suggested that these marketing customizations can be used to tailor information to potential applicants based on their profiles, which can be derived from their online history, social networking information, or on the spot after completing on-site surveys (Dineen & Noe, 2009, Kraichy & Chapman, 2014). For example, Kraichy and Chapman (2014) demonstrated that tailoring online recruiting messages with emotion-based content was more influential in attracting potential applicants than cognitive/fact-based messages, particularly for those with low need for cognition. Dineen and Noe (2009) demonstrated that providing interactive feedback to online candidates about their fit with the organization had a significant influence on attraction. Thus, there is some promise that smarter, interactive online recruiting strategies can improve on static recruiting websites to attract stronger candidates. This approach to e-recruiting has significant potential, however, we also have to be cautious about not transgressing perceptions of procedural justice in the recruiting process (Ryan & Ployhart, 2000). Tailored or targeted e-recruiting messages may be effective but there is no information on how such approaches could affect applicant perceptions post-hire, or the complications that arise when managers deal with employees who were hired with different recruiting information emphasized. Additional care would be needed to ensure that protected groups were not adversely impacted by targeted recruiting information. While conceptually e-recruiting has the advantage of building in algorithms to target information more effectively to applicants, industry has been slow to adopt this approach and researchers have so far left this potential benefit largely unexplored.

Passive recruiting

Social media recruiting provides a forum for employers and potential job seekers to network passively in a manner that can facilitate attraction and hiring. As Sullivan (quoted in Imperato, 1998) pointed out, the best potential candidates are rarely conveniently unemployed at the moment employers are seeking to fill a position. In fact, the best applicants are likely happily employed with a competitor. Social networking allows passive job seekers to place their credentials online in ways that permit potential employers to find them. This is less labor intensive for the passive applicant than actively seeking employment and is thus a low-effort strategy for employed applicants in particular, to continue to market themselves in a less obvious manner. It also has the advantage of being potentially perceived as more benign by their employer than actively seeking employment. Nikolaou (2014) found that passive job seeking was most effective in online networking through social networking sites such as LinkedIn.

Signaling

Signaling theory has a long history in recruiting research. Spence (1973) argued that when making decisions about things such as job choice, people with limited information about the organization will extrapolate from the available information to make inferences about the organization as a whole. For example, if the only information one has about a company comes from a brief visit to their website, signaling theory suggests that you the individual will overweigh this information to make generalizations about the organization as a whole. Researchers have shown that website characteristics such as how aesthetically pleasing it is and how easy it is to navigate have a significant influence on how potential applicants perceive the organization and how likely they are to apply for a position with the organization (Cober et al., 2003). McLuhan (1964) famously stated that "the medium" *was* "the message." Researchers have proposed that recruitment media choices of employers impart a message about the type of company they are (Chapman et al., 2003, Cromheecke, Van Hoye, & Lievens, 2013; Sinar, Reynolds, & Paquet, 2003). Thus the potential for e-recruiting is not limited to transmitting intended recruiting messages more efficiently or more broadly but to signal to potential applicants that the organization is media savvy or trendy. More research is needed to identify the types of signals that e-recruiting sends to prospective applicants.

Critical contact

A related phenomenon, known as critical contact theory (Behling, Liebowitz & Gainer, 1968), also emphasizes the importance of first impressions on applicants. According to Behling et al. (1968), job seekers have difficulty differentiating among potential employers based on truly meaningful characteristics because they may not know enough about them (e.g., corporate culture) or because there is insufficient variance among alternatives to make a choice (e.g., when most companies offer similar compensation). They posited that this was a common phenomenon and it resulted in applicants relying too much on surface characteristics of the company based on early contact with agents of the organization (e.g., recruiters) or even such things as the physical surroundings of the company. Although critical contact was conceived with recruiting behavior and organizational physical characteristics in mind, the characteristics of e-recruiting interfaces with the potential applicant may also have a significant role. In some ways, poor website usability is not unlike a recruiter who is perceived as unknowledgeable or unfriendly. These pieces of information can be construed as evidence of incompetence or indifference – both of which are anathema to applicants. Likewise, an out-of-date and poorly maintained website may be viewed similarly to a dilapidated workplace. Accordingly, investment in an aesthetically pleasing, up-to-date, and easy-to-navigate website ought to increase recruiting effectiveness much in the way as investing in attractive office space and using competent, friendly recruiters.

Reach

The power of the Internet to reach large numbers of potential applicants is a significant advantage for e-recruiting (Gueutal, Stone, & Salas, 2005; Stone, Lukaszewski, & Isenhour, 2005). Employers are no longer constrained to local searches for employees, thereby facilitating national and global recruiting strategies. This can only result in a larger and stronger applicant pool. As noted earlier, larger applicant pools are needed for effective screening systems to work optimally (Boudreau & Rynes, 1985)

Networking

Networking is a key behavior in recruiting for both applicants and employers (Nikolaou, 2014; Van Hoye & Lievens, 2007). E-recruiting facilitates this networking process by providing easy access to potential candidates through platforms such as LinkedIn. Research is needed to identify the effectiveness of networking through these third-party websites relative to traditional networking (e.g., through associations, job fairs, and campus visits). The efficiency of networking through these sites has not been questioned. It is clear that many potential applicants can be contacted easily and cheaply relative to traditional networking. However, it is not known whether applicants contacted through networking sites such as LinkedIn are more or less likely to follow up contacts than those contacted through traditional networking.

Branding

Recruiting researchers have recognized that there are significant similarities between recruiting and marketing products and services. Both involve a *consumer* making a decision or choice among competing alternatives. In particular, researchers have drawn on the branding literature to explain recruiting processes (e.g., Cable & Yu, 2006). Although one would hope that important decisions such as employment choices would be immune to the cognitive shortcutting associated with branding, research suggests that branding can be effective and that simple familiarity with the organization influences applicant decisions (Chapman et al., 2005).

E-recruiting has enormous potential to create and reinforce employer brands. This can be accomplished through a combination of online advertising and a strong social media presence in sites that attract targeted applicants. Cable and Yu (2006) showed that media richness and source credibility were important determinants of effective recruiting images. However, employer image branding is most effective when potential applicants already have some familiarity with the organization and have a positive view of it.

Novelty

Recent recruiting research has found that potential applicants respond positively to the novelty of the recruitment medium (Cromheecke et al., 2013). In a given day, potential applicants are bombarded by messages urging them to buy products, try new services, visit websites featuring certain news stories, and so forth. Cutting through this clutter with a memorable recruiting message is a daunting task. E-recruiting offers the opportunity to be creative in presenting novel recruiting messages or enlisting novel recruiting media to highlight the company's recruiting message. An example of novelty in e-recruiting includes gamified recruiting approaches whereby employers such as Maersk, Marriot, L'Oreal, and the US Army use online games to attract potential applicants (Chow & Chapman, 2013).

Current Approaches to E-Recruiting

Online job boards and company websites

The most popular form of e-recruiting is online job boards and company websites. A survey of the German top 1000 companies revealed that companies primarily use their own websites (82%) and Internet job boards (60%) for recruitment (Laumer, Eckhardt & Weitzel, 2009) for recruitment. Applicants can search for numerous positions across multiple companies from wherever Internet access is available, and can filter their searches to better meet their needs. Large-scale collection of applicant information allows companies to search their own databases when future positions arise. E-recruiting can be effective; German top 1000 companies report that most of their new hires were generated from these two forms of e-recruitment: 35% recruited from the company website, and 33% recruited from online job boards.

Social media

The proliferation of social media websites such as Facebook. Twitter, and Google+, and specialized career networking social media sites such as LinkedIn and XING, have increased the ability of companies to identify and directly recruit individuals with a desired set of skills, even those currently employed with another company. The fastest growing demographic of social media users are middle aged (Meister, 2014). A survey of Belgian recruitment and selection specialists (Caers & Castelyns, 2010) showed that nearly half of the specialists had used LinkedIn to search for potential candidates to fill vacancies in their company. Specialists who were active users of the social media site were significantly more likely to use it for recruiting purposes.

Social media sites also offer an opportunity for employers to seek out additional information about applicants during the selection process, even before an interview. Seventy percent of specialists who were active LinkedIn users reported using the website to seek out information about applicants prior to an interview (Caers & Castelyns, 2010). Additionally, 43% of practitioners who were active users of Facebook used it to search additional information about applicants, raising concerns about the ethicality of personal social media websites influencing professional decisions. Although access to information prior to a more intensive selection process streamlines the hiring process, some information found on social media sites, such as the presence or lack of a picture of the applicant, introduces bias. Considering the consequences of a legally nondefensible selection practice that may introduce bias and may not be well received by applicants, companies may choose to evaluate whether "harmless" social media checks should be part of their regular prescreening processes (Smith & Kidder, 2010; see also Lukaszewski & Johnson, 2017, Chapter 12 of this volume).

Mobile recruiting

Rapid growth of technology has made the Internet accessible in virtually any location. Job searching is now done "on the go," with many applicants seeking information using their tablets and phones. In fact, 68% of job seekers reported using their mobile device to search for jobs at least once a week and 25% would not apply for a job at a company whose website was not mobile optimized (Glassdoor, 2013). Yet, only 20% of Fortune 500 companies have mobile-compatible websites (Meister, 2014). It is imperative for companies to adapt to the ever-changing technological landscape and ensure that company information is readily accessible on mobile devices. This includes ensuring that job application through mobile devices is easily accomplished, since 40% of current job seekers report that applying for a job on a mobile device is difficult or very difficult (Glassdoor, 2013). Some employers have philosophical or technical reasons for not permitting applications through cell phones. The quality of the website is necessarily compromised when optimizing for phone use, which could discourage applicants. Likewise, phones are likely to be used in distracting environments, making the recruiting message unclear. Employers want applicants to be focused on the recruiting message, which is less likely when using a phone. Finally, many selection assessments require a larger screen to adequately interface with the screening tool and to avoid issues with font size and readability, especially for items with multiple response options.

Gamification

Gaming elements may also be used to enhance a more traditional recruitment method (Chapman & Mayers 2015; Chow & Chapman, 2013). Gamification motivates behavior by fulfilling the inherent need for autonomy, competence, and relatedness (Przybylski, Rigby, & Ryan, 2010; Ryan, Rigby, & Przybylski, 2006). Thus well-constructed gaming elements can motivate desired behavior in both job seekers and employers. For example, companies may gamify referral programs by awarding points for employee referrals and bonuses for employees who refer an individual who is eventually hired. Job boards may be gamified through points and badges that reward job seekers for uploading useful information such as previous work history, referrals and résumés. Job application challenges might keep applicants engaged throughout the application process. On the other hand, employers could earn rewards for supplying all necessary information in the job postings or providing timely feedback to applicants. Leaderboards consisting of top-scoring companies could signal to applicants the companies that are the best to interact with during the selection process.

Gamification is a growing trend, with many companies employing gamified practices in HR management. In fact, Gartner predicted that by 2015 40% of Global 1000 organizations would be using gamification as a key tool for transforming business operations (Pettey, 2012). Research interest in the applications of gamification is also growing. The professional body, Society for Industrial and Organizational Psychology (SIOP), named "gamification" one of 2014's most popular trends (Munson, 2013) and called for research to understand whether gamification can be used in employee selection.

Many have supposed that employing gamified tools would benefit the recruitment process. Specifically, the embracement of a novel technology is presumed to have a positive effect on applicants' perception of a company and increase the likelihood of job acceptance. Research on these assumptions is scarce. Chow and Chapman (2013) showed that gamified recruitment tools enhance perceptions of the hiring company, although not any more than the company website. Gödöllei and Chapman (2016) also showed that applicant reactions and organizational attraction to companies using gamified selection measures were either worse or not significantly different from companies using a traditional selection measure. Thus it seems that anecdotal evidence regarding the positive effects of gamified recruitment is not supported by the scarce empirical evidence. However, this research is in its infancy. It is likely that the effectiveness of gamified recruitment is complex and depends on multiple variables related to specific gamified applications. For example, we are only beginning to understand whether the ease of use, complexity, fidelity, and playfulness of the gamified application affects its success. Future research is needed (see also Thielsch & Niesenhaus, Chapter 5 of this volume).

Limitations and Pitfalls of E-Recruiting

Enthusiasm for creating recruiting and selection technologies has had both positive and negative consequences. Many technology developers have a background in information technology and lack a basic understanding of HR processes, HR legal issues, psychometrics, and recruiting theory. As a result, the market is often inundated with apps and websites dedicated to helping organizations recruit and screen individuals with little or no regard for legal employment guidelines, in the USA, guidelines of the Equal Employment Opportunity Commission, SIOP selection practices, job analysis or competency modeling, predictive validity, applicant reactions, privacy, and so forth. As a result, marketing often trumps science in many of these ventures. There is a high risk that the e-recruiting industry will generate legal issues for ill-informed employers who are eager to adopt unproven technologies sold to them as cutting-edge solutions (Slovensky & Ross, 2012). Naturally we strongly encourage innovation in this space and make some suggestions in this regard. However, innovation needs to be tempered by careful and rigorous testing before bringing technologies to the market. Research lags industry adoption in all areas of e-recruiting and a greater focus on e-recruiting research is sorely needed.

Implementation issues

E-recruiting strategies have become popular despite many employers experiencing difficulties in implementing them. In 2002, Chapman and Webster (2003) surveyed North American HR vice-presidents, who were members of the Society for Human Resource Management (SHRM), about their use of technology in recruiting and selection. At that time many reported difficulties in implementing their e-recruiting strategies, such as technology issues, cost overruns, being inundated with unwanted applicants, and so forth. Similar problems were reported by a sample of UK HR practitioners surveyed in 2006 (Parry & Tyson, 2008). These surveys highlight the need for organizations to be thoughtful about their use of e-recruiting. A good place to start is delineating the recruitment goals of the e-recruiting strategy (see Breaugh & Starke, 2000). Contingency planning for outages, designing volume management strategies for screening large numbers of applicants, messages around privacy, and so forth are critical to achieving success.

Applicant reactions

How applicants respond to e-recruiting is critical to its success. The perceived fairness of recruiting and selection systems can have far-reaching implications for organizational attraction, application intentions, dropout rates, willingness to refer others, and even purchasing intentions (Gilliland, 1993; McCarthy et al., in press). Applicant reactions to e-recruiting and selection have been found to be predominantly unfavorable. Chapman et al. (2003) found applications reacted negatively to being interviewed using videoconference technologies. People also tend to react negatively to the use of social networking information in screening and selection (Nikolaou, 2014; Thielsch, Träumer, & Pytlik, 2012). Researchers need to identify what procedural justice rules (see Gilliland, 1993) are potentially violated by e-recruiting in order to recommend solutions to improve applicant reactions.

Ethics and privacy

The potential for abuse of online social networking information is troubling (Black, Stone, & Johnson, 2015). Scholars have questioned whether it is ethical for employers to look at social networking information at all (Clark & Roberts, 2010). As noted earlier, many e-recruiting applications are developed by individuals with little background in personnel selection or recruiting. Consequently, the absence of validity data underlying the use of these techniques is worrisome at best and legally irresponsible at worst. The impact of these approaches on generating diverse applicant pools is largely unknown. More research is needed to keep up with the escalation of practice in e-recruiting.

Organizational acceptance

In addition to worrying about applicant acceptance of e-recruiting practices, organizations may face resistance internally from stakeholders that could derail efforts to adopt e-recruiting practices. Stone, Stone-Romero, and Lukaszewski (2006) identified several potential threats to organizational stakeholders that could lead to resistance to adoption of e-recruiting, such as changes in information flows, social interactions and perceived control, and incongruence with company values. Failure to address these perceptions could result in internal stakeholders resisting the adoption of new e-recruitment approaches. For example, Chapman and Webster (2003) found some employers were worried that their image as "hands on" employers with a "personal touch" would be compromised by relying on e-recruiting solutions. Addressing these perceptions would be necessary to successfully adopt e-recruiting solutions.

HR Focus (2010) reported that companies fear losing control of their messaging in a social media context. Ostensibly, the more interactive and open social media networking is, the greater the likelihood of unintended messages or third-party messages influencing potential candidates. For example, Van Hoye and Lievens (2007) argue that unscripted third-party information about a company found in online posts (what they called "word of mouse") had a strong impact on applicant perceptions and attraction toward a company. This can certainly benefit the company if the information is positive. However, in the event that negative messages start appearing on social media websites, the company could be adversely impacted and have little opportunity to mitigate the bad press. Enterprising entrepreneurs have recognized the power of unscripted insider information in shaping potential applicants impressions of organizations.

Researchers have shown that perceptions of the credibility of recruiting information are important in determining attitudes (Mumford, 2012). Potential applicants can visit websites such as Glassdoor.com to find unsolicited opinions about companies from current and former employees. For example, at the time of writing this chapter, there were over 22,000 reviews on this single website of what it is like to work in IBM. These third-party websites can be highly useful for e-recruiting if the information is positive or highly detrimental to the organization's recruiting efforts if the information is negative. The potential for competitors to undermine an organization via anonymous negative postings is also a possibility, and this practice has been well documented in other contexts such as manipulating product or service reviews (e.g., Luca & Zervas, 2016). Ultimately, these technologies permit rapid and easy dissemination of recruiting information, which is not within the control of the organization. Researchers need to build on Van Hoye and Lievens' (2007) work to determine the impact of this third-party information on e-recruiting. Organizations would be wise to monitor these sites and attempt to address any negative perceptions of themselves by improving conditions. Some employers threaten their employees with termination if they discover that an employee is providing negative reviews of them online. While this may be an extreme reaction, both researchers and practitioners tend to view such negative pronouncements as counterproductive behavior.

Future Practice

Where is e-recruiting practice headed? The possibilities are endless. We speculate here on a few ideas which may or may not be already out there. Most organizations are reluctant to publicly reveal their proprietary approaches to business, including e-recruiting practices.

Search engine history

What individuals search for in Google or Bing or other search engines is accessible in aggregate to employers willing to pay for marketing information. This search history could reveal information about interests, values, occupations, demographics, purchase patterns, job-seeking activity, lifestyle, and a myriad other things that could potentially be used to tailor recruiting messages to potential applicants, including passive job seekers. One might ask questions such as "What are the most popular searches for accountants?", which could provide clues about recruiting message content appropriate for accountants. Search history can also reveal information that could identify the types of individual who are unhappy in their current positions and potentially seeking alternative employment. A candidate profile consisting of visits to LinkedIn twice a day, being a member of a job hunters group, and petroleum engineers in Houston group could identify a potential recruit for an oil and gas company.

Imagine a job seeker visiting an employer's career page, which identifies them and instantly accesses their search engine activity, which includes weekly shopping for pet food and frequent visits to a parenting advice column. Suddenly, next to the job description information they are looking at, they see pop ups with content related to these searches: picture of dog captioned "Did you know that X corp has a pet friendly policy and an annual 'bring your pet to work day'?" Underneath a photo of a young boy smiling and a caption "Thinking of moving to Calgary? Our public schools are among the best in the world according to X magazine!" The technology already exists to engage in this level of targeted recruiting but its use appears to be largely limited to selling products. However, it could be adapted to recruiting with little effort.

Location data

Most people are unaware that they are constantly beaming their location and identity to third parties. Marketing departments use location data to better understand their customers. For example, Wi-Fi location data can be used to track individuals in a mall and note which stores they visit and what order. More sophisticated systems can link to in-store sensors that tell how long a customer spent in each aisle and the product locations they stopped at. These data can be linked to other data about the customer and analyzed in ingenious ways. For example, a job seeker may have LinkedIn data regarding occupation linked to a particular phone location at a given time. By examining patterns of occupations, and attendance at certain sporting events at their local arena, a company could decide if it is a good idea to purchase advertising about its job vacancies at a certain sporting event. We are not aware of specific employers using this approach but it is technologically feasible.

It is clear that the potential for e-recruiting to target potential employees and customize recruiting content for them is nearly limitless. As we noted in the pitfalls section, practitioners need to be careful about transgressing privacy and ethical concerns. This is largely uncharted territory in the recruiting sphere and the potential negative repercussions may be greater in a recruiting context than for selling shampoo. It is often useful to test ideas on focus groups to gauge how they might be received by the public.

Future Research

Researchers interested in e-recruiting have an open field of unanswered, potential research questions. These questions range from practical issues such as efficaciousness of e-recruiting and development of e-recruiting methods to more theoretical underpinnings of e-recruiting processes and broader integration into theories of self, decision making, and so forth. Next we identify some research areas that seem central to moving our understandings of e-recruiting forward.

Efficacy of e-recruiting

A key area is the efficacy of emerging e-recruiting trends such as gamification, social media networking, electronic head hunting, online website design, and demographic interactions with e-recruiting content and design. We actively need to investigate these emerging and ongoing e-recruiting practices to determine their effectiveness and their relative efficacy compared with more traditional recruiting techniques. We also need to identify whether e-recruiting has the potential to create or remove demographic barriers in the recruiting process and to examine the potential for e-recruiting with various targeted demographic groups (see also Burlacu, Truxillo, & Bauer, 2017, Chapter 20 of this volume).

Theories to apply to e-recruiting

In addition to the efficaciousness of e-recruiting innovations, researchers need to identify and explore key e-recruiting constructs and determine the mechanisms affecting applicant perceptions of companies using e-recruiting. Several central theories around which some of this research could be based include critical contact theory (Behling et al., 1968) media richness theory, applicant reactions and justice theories (Gilliland, 1993), and signaling theory (Spence, 1973). Other relevant theories include social psychological theories of the self, decision-making theories, motivation theories and perception theories.

In addition, methodologies need to be developed to assist researchers in capturing e-recruiting processes. Some innovative approaches are: eye-tracking devices to establish what applicants are looking at when they visit websites (see Allen et al., 2013); click-through ratios that examine the efficacy of job ads unobtrusively by examining the number of ad views divided by the number of applicants (see Schmidt et al., 2015); longitudinal designs and diary approaches that follow applicants through their e-recruiting experiences; and qualitative interviews to obtain richer information on applicant perceptions of e-recruiting. Eye tracking could also reveal how candidates focus on e-recruiting messages, graphics, and so forth. Other physiological approaches ranging from skin galvanic response and heart rate to functional magnetic resonance imaging could prove useful in determining how e-recruiting information affects recruiting targets.

Conclusion

Despite the growing popularity and potential of e-recruiting, there is a dearth of research on the efficacy of these approaches for employee recruitment. Available research on applicant reactions gives us some information about the implementation of e-recruiting and particularly the use of online data and technologies for screening and selection. Better education for users of the technology and transparent, well-reasoned explanations for potential applicants are good starting points.

References

Allen, D. G., Biggane, J. E., Pitts, M., Otondo, R., & Van Scotter, J. (2013). Reactions to recruitment web sites: Visual and verbal attention, attraction, and intentions to pursue employment. *Journal of Business and Psychology, 28*(3), 263–285.

Allen, D. G., Mahto, R. V., & Otondo, R. F. (2007). Web-based recruitment: Effects of information, organizational brand, and attitudes toward a website on applicant attraction. *Journal of Applied Psychology, 92*, 1696–1708.

Allen, D. G., Van Scooter, J. R., & Otondo, R. F. (2004). Recruitment communication media: Impact on prehire outcomes. *Personnel Psychology, 57*, 143–171.

Barber, A. E., Daly, C. L., Giannantonio, C. M., & Phillips, J. M. (1994). Job search activities: An examination of changes over time. *Personnel Psychology, 47*(4), 739–766.

Behling, O., Labovitz, G., & Gainer, M. (1968). College recruiting: A theoretic base. *Personnel Journal, 47*(1), 13–19.

Black, S. L., Stone, D. L., & Johnson, A. F. (2015). Use of social networking Websites on applicants' privacy. *Employee Responsibilities and Rights Journal, 27*, 115–159.

Boudreau, J. W., & Rynes, S. L. (1985). Role of recruitment in staffing utility analysis. *Journal of Applied Psychology, 70*(2), 354–366.

Breaugh, J. A., & Starke, M. (2000). Research on employee recruitment: So many studies, so many remaining questions. *Journal of Management, 26*(3), 405–434.

Burlacu, G., Truxillo, D. M., & Bauer, T. N. (2017). Employee age differences in using Internet-based tools at work. In G. Hertel, D. L. Stone, R. D. Johnson, & J. Passmore (eds.). *The Wiley Blackwell handbook of the psychology of the internet at work* (pp. 449–480). Chichester: John Wiley & Sons, Ltd.

Caers, R., & Castelyns, V. (2010). LinkedIn and Facebook in Belgium: The influences and biases of social network sites in recruitment and selection procedures. *Social Science Computer Review, 29*, 437–448.

Carlson, K. D., Connerley, M. L. & Mecham, R. L., III. (2002). Recruitment evaluation: The case for assessing the quality of applicants attracted. *Personnel Psychology, 55*(2), 461–490.

Cable, D. M., & Yu, K. Y. T. (2006). Managing job seekers' organizational image beliefs: The role of media richness and media credibility. *Journal of Applied Psychology, 91*(4), 828–840.

Chapman, D. S., & Mayers, D. (2015). Recruitment processes and organizational attraction. In I. Nikolaou, & J. K. Oostrom (Eds.), *Employee Recruitment, Selection, and Assessment: Contemporary Issues for Theory and Practice* (pp. 27–42). New York, NY: Psychology Press.

Chapman, D. S., & Rowe, P. M. (2001). The impact of videoconference technology, interview structure, and interviewer gender on interviewer evaluations in the employment interview: A field experiment. *Journal of Occupational and Organizational Psychology, 74*(3), 279–298.

Chapman, D. S., & Rowe, P. M. (2002). The influence of videoconference technology and interview structure on the recruiting function of the employment interview: A field experiment. *International Journal of Selection and Assessment, 10*(3), 185–197.

Chapman, D. S., Uggerslev, K. L., Carroll, S. A., Piasentin, K. A., & Jones, D. A. (2005). Applicant attraction to organizations and job choice: A meta-analytic review of the correlates of recruiting outcomes. *Journal of Applied Psychology, 90*(5), 928.

Chapman, D. S., Uggerslev, K. L., & Webster, J. (2003). Applicant reactions to face-to-face and technology-mediated interviews: A field investigation. *Journal of Applied Psychology, 88*(5), 944.

Chapman, D. S., & Webster, J. (2003). The use of technologies in the recruiting, screening, and selection processes for job candidates. *International Journal of Selection and Assessment, 11*(2-3), 113–120.

Chow, S. & Chapman, D. S. (2013, June). Gamifying the employee recruitment process. Paper presented at the First International Conference of Gameful Design. Waterloo, ON, Canada.

Clark, L., & Roberts, S. (2010). Employer's use of social networking sites: A socially irresponsible practice. *Journal of Business Ethics, 95*, 507–525.

Cober, R. T., Brown, D. J., Keeping, L. M., & Levy, P. E. (2004). Recruitment on the Net: How do organizational Web site characteristics influence. *Journal of management, 30*(5), 623–646.

Cober, R. T., Brown, D. J., & Levy, P. E. (2004), Form, content, and function: An evaluative methodology for corporate employment web sites. *Human Resource Management, 43*, 201–218. DOI:10.1002/hrm.20015

Cober, R. T., Brown, D. J., Levy, P. E., Cober, A. B., & Keeping, L. M. (2003). Organizational web sites: Web site content and style as determinants of organizational attraction. *International Journal of Selection and Assessment, 11*(2-3), 158–169.

Cromheecke, S., Van Hoye, G., & Lievens, F. (2013). Changing things up in recruitment: Effects of a "strange" recruitment medium on applicant quantity and quality. *Journal of Occupational and Organizational Psychology, 86*, 410–416.

De Goede, M. E., Van Vianen, A. E., & Klehe, U. C. (2011). Attracting applicants on the Web: PO fit, industry culture stereotypes, and Website design. *International Journal of Selection and Assessment, 19*, 51–61.

Dickey-Chasins, J. (2012). Job board evolution. *HR Examiner*. Retrieved April 7, 2017, from www.hrexaminer.com/job-board-evolutio/

Dickter, D. N., Jockin, V., & Delany, T. (2017). The evolution of e-selection. In G. Hertel, D. L. Stone, R. D. Johnson, & J. Passmore (Eds.), *The Wiley Blackwell handbook of the psychology of the internet at work* (pp. 257–284). Chichester: John Wiley & Sons, Ltd.

Dineen, B. R., & Noe, R. A. (2009). Effects of customization on application decisions and applicant pool characteristics in a web-based recruitment context. *Journal of Applied Psychology, 94*(1), 224.

Djamasbi, S., Siegel, M., & Tullis, T. (2010). Generation Y, web design, and eye tracking. *International Journal of Human-Computer Studies, 68*(5), 307–323.

Gilliland, S. W. (1993). The perceived fairness of selection systems: An organizational justice perspective. *The Academy of Management Review, 18*(4), 694–734.

Glassdoor (2013). The rise of mobile job search. Retrieved April 7, 2017, from https://www.glassdoor.com/blog/infographic-rise-mobile-job-search/

Gödöllei, A. F., & Chapman, D. S. (2016). Gamified cognitive assessments in selection: Validity, discrimination and applicant reactions. *Academy of Management Proceedings 2016*(1), 17438.

Gueutal, D., Stone, L., & Salas, E. (2005). *The brave new world of e-HR: Human resources in the digital age*. San Francisco, CA: Pfeiffer.

HR Focus. (2010). Recruiting and marketing are top benefits of social media. Retrieved June 29, 2016, from http://hrpeople.monster.com/news/articles/3247-recruiting-and-marketing-are-top-benefits-of-social-media.

Imperato, G. (1998, December). How to hire the next Michael Jordan. *Fast Company*.

Kohli C, Suri R, & Kapoor A. (2015). Will social media kill branding? *Business Horizons, 58*, 35–44.

Kraichy, D., & Chapman, D. S. (2014). Tailoring web-based recruiting messages: Individual differences in the persuasiveness of affective and cognitive messages. *Journal of Business and Psychology, 29*(2), 253–268.

Laumer, S., Eckhardt, A., & Weitzel, T. (2009). Status quo and trends in e-recruiting – results from an empirical analysis. In *International Conference on Information Resources Management (CONF-IRM)*. Dubai, United Arab Emirates.

Li, C., Charron, C., Roshan, S., & Flemming, G. N. (2002). *Online recruitment grows up*. Cambridge, MA: Forrester Research.

Lin, H. F. (2010). Applicability of the extended theory of planned behavior in predicting job seeker intentions to use job-search websites. *International Journal of Selection and Assessment, 18*, 64–74

Luca, & Zervas, (2016). Fake it til you make it: Reputation and competition in Yelp review fraud. *Management Science*. http://dx.doi.org/10.1287/mnsc.2015.2304

Lukaszewski, K. M., & Johnson, A. F. (2017). Social networking systems, search engines, and the employment process. In G. Hertel, D. L. Stone, R. D. Johnson, & J. Passmore (Eds.), *The Wiley Blackwell handbook of the psychology of the internet at work* (pp. 231–256). Chichester: John Wiley & Sons, Ltd.

Maurer, S. D., & Liu, Y. (2007). Developing effective e-recruiting websites: Insights for managers from marketers. *Business horizons, 50*(4), 305–314.

McCarthy, J. M., Bauer, T. N., Truxillo, D. M., Anderson, N. R., Costa, A. C. & Ahmed, S. M. (In press). Applicant perspectives during selection: A review addressing "so what?," "what's new?," and "where to next?" *Journal of Management*.

McLuhan, M. (1964). *Understanding media: The extensions of man (pp. 23–25, 63–67)*. New York, NY: Signet.

Meister, J. (2014) The future of work: Why social HR matters. Retrieved April 7, 2017, from www.forbes.com/sites/jeannemeister/2014/01/06/2014-the-year-social-hr-matters/#2a16e2762dc1

Mumford, T. V. (2012). Whom to believe: Recruiting information source credibility and organizational attractiveness. *Business and Management Research, 1*(4), 63–80.

Munson, L. (2013). New year, new workplace! *SIOP Item of Interest*. Retrieved April 7, 2017, from www.siop.org/article_view.aspx?article=1203

Nikolaou, I. (2014). Social networking web sites in job search and employee recruitment. *International Journal of Selection and Assessment, 22*(2), 179–189.

Parry, E., & Tyson, S. (2008). An analysis of the use and success of online recruitment methods in the UK. *Human Resource Management Journal, 18*(3), 257–274.

Pettey, C. (2012) Gartner reveals top predictions for IT organization and users for 2013 and Beyond. *Gartner*. Retrieved April 7, 2017, from www.gartner.com/newsroom/id/2211115

Przybylski, A. K., Rigby, C. S., & Ryan, R. M. (2010). A motivational model of video game engagement. *Review of General Psychology, 14*(2), 154.

Ryan, A. M., & Ployhart, R. E. (2000). Applicants' perceptions of selection procedures and decisions: A critical review and agenda for the future. *Journal of Management, 26*(3), 565–606.

Ryan, R. M., Rigby, C. S., & Przybylski, A. (2006). The motivational pull of video games: A self-determination theory approach. *Motivation and Emotion, 30*(4), 344–360.

Rynes, S. L. (1989). The employment interview as a recruitment device. In R. W. Eder & G. R. Ferris (Eds.), *The employment interview: Theory, research and practice* (pp 127–141). Newbury Park, CA: Sage.

Schmidt, J. A., Chapman, D. S., & Jones, D. A. (2015). Does emphasizing different types of person–environment fit in online job ads influence application behavior and applicant quality? Evidence from a field experiment. *Journal of Business and Psychology, 30*(2), 267–282.

Sears, G., Zhang, H., Wiesner, W., Hackett, R., & Yuan, Y. (2013). A comparative assessment of videoconference and face-to-face employment interviews. *Management Decision, 51*(8), 1733–1752.

Selden, S., & Orenstein, J. (2011). Government e-recruiting web sites: The influence of e-recruitment content and usability on recruiting and hiring outcomes in US state governments. *International Journal of Selection and Assessment, 19*, 31–40.

Sinar, E. F., Reynolds, D. H., & Paquet, S. L. (2003). Nothing but net? Corporate image and Web-based testing. *International Journal of Selection and Assessment, 11*, 150–157.
Slovensky, R., & Ross, W. H. (2012). Should human resource managers use social media to screen job applicants? Managerial and legal issues in the USA. *Info, 14*, 55–69.
Smith, W. P., & Kidder, D. L. (2010). You've been tagged! (then again, maybe not): Employers and Facebook. *Business Horizons, 53*(5), 491–499.
Spence, M. (1973). Job market signaling. *Quarterly Journal of Economics*, 355–374.
Stone, D. L., Lukaszewski, K., & Isenhour, L. C. (2005). E-recruiting: Online strategies for attracting talent. In H. G. Gueutal, & D. L. Stone (Eds.), *The Brave New World of eHR* (pp. 22–53). San Francisco, CA: Jossey-Bass.
Stone, D. L., Stone-Romero, E. F., & Lukaszewski, K. (2006). Factors affecting the acceptance and effectiveness of electronic human resource systems. *Human Resource Management Review, 16*(2), 229–244.
Stopfer, J. M., & Gosling, S. D. (2013). Online social networks in the work context. In D. Derks & A. Bakker (Eds.), *The psychology of digital media at work* (pp. 39–59). London: Psychology Press.
Thielsch, M., T. & Niesenhaus, J. (2017). User experience, gamification, and performance. In G. Hertel, D. L. Stone, R. D. Johnson, & J. Passmore (Eds.), *The Wiley Blackwell handbook of the psychology of the internet at work* (pp. 79–102). Chichester: John Wiley & Sons, Ltd.
Thielsch, M. T., Träumer, L., & Pytlik, L. (2012). E-recruiting and fairness: The applicant's point of view. *Information Technology and Management, 13*(2), 59–67.
Van Birgelen, M. J., Wetzels, M. G., & van Dolen, W. M. (2008). Effectiveness of corporate employment web sites: How content and form influence intentions to apply. *International Journal of Manpower, 29*(8), 731–751.
Van Hoye, G., & Lievens, F. (2007). Investigating web-based recruitment sources: Employee testimonials vs word-of-mouse. *International Journal of Selection and Assessment, 15*(4), 372–382.

12

Social Networking Sites, Search Engines, and the Employment Process

Kimberly M. Lukaszewski and Andrew F. Johnson

Introduction

Managers around the world are responsible for making critical employment decisions (e.g., hiring, termination, promotion), and have to determine which individuals have the right knowledge, skills, and abilities to perform the job. They also have to ensure that employees are trustworthy, conscientious, and perform their roles in an ethical and moral manner. If managers do not assess the backgrounds of applicants, and these individuals violate laws as employees, organizations may be liable for negligent hiring (Bennett-Alexander & Hartman, 2015). Consequently, managers have been gathering and using information from social networking sites (SNS) and search engines to make employment decisions. For instance, one survey from the United States demonstrated that over 90% of recruiters and hiring managers visited a potential job applicant's SNS profile during the hiring process (Swallow, 2011), and they rejected nearly 70% of applicants based on the information acquired from them (Preston, 2011). Interestingly, SNS users typically construct and post personal profiles that are shared with friends and family, and they are also able to view and connect to the profiles of others (boyd & Ellison, 2007). When an employer uses a search engine, the search engine sends out programs (spiders) that access webpages and then uses an indexer to scan the results and create an overall index for the applicant (Webopedia.com, 2016). Popular SNS providers are Facebook and LinkedIn, and the most widely used search engines are Google and Baidu (often called China's Google) (Krawczyk, 2014).

In a recent survey by the Society for Human Resource Management (SHRM), 61% of global and national organizations reported scanning SNS and search engines to gain information about applicants that is not included on a résumé or application. Half (50%) also used SNS and search engines to verify the accuracy of information on an application

The Wiley Blackwell Handbook of the Psychology of the Internet at Work,
First Edition. Edited by Guido Hertel, Dianna L. Stone, Richard D. Johnson, and Jonathan Passmore.

or résumé, and 34% accessed information that is not readily available from background checks (SHRM, 2016). Although such searches often revealed an enormous amount of information, the data reported by these systems may or may not be relevant for employment decisions. For instance, these sites often contained information about aspects of a person's lifestyle, such as interests, family, disabilities, romantic relationships, sexual orientation, race, and religious or political views (Davison, Maraist, Hamilton, & Bing, 2012).

Survey findings also revealed that 39% of organizations used data from SNS, and 34% reviewed information obtained from search engines to make hiring decisions (SHRM, 2016). Apart from the United States, reliance on data from SNS and search engines seems to be somewhat consistent across different countries. A study conducted by the global consulting firm Development Dimensions International (Erker & Buczynski, 2009) found that companies in countries around the world review SNS or search engines: 46% in Germany reviewed SNS, 23% in France, 17% in Canada, 16% in Australia, and 12% in the United Kingdom. Fifty-two percent of employers in the United Kingdom who searched online profiles of applicants used the information in the hiring decisions (Peacock, 2009). In France, over 60% of interviewers used the information resulting from these searches to make a hiring decision (Erker & Buczynski, 2009). Despite this widespread use of SNS and search engines in the employment process, analysts have argued that many applicants and employees perceive such use of these sites as unfair and that it may have a negative impact on employment decisions (Rasmussen Reports, 2012). For instance, surveys revealed that approximately 70% of adults from the United States believe that employer access to their personal SNS is an unfair means of acquiring information (Rasmussen Reports, 2012). This is because the data posted in SNS are meant for friends and family, and individuals do not expect this private information to be used by employers. Another reason is that individuals perceive the review of SNS or search engine data are unfair because organizations do not always inform individuals that they will be reviewing the data in these sites to make employment decisions. Thus, individuals do not have the opportunity to make sensitive information private or to erase negative information. Furthermore, there appears to be some naivety on the side of job applicants – almost 44% of survey respondents in the United States did not see a connection between what they posted on SNS and their job-seeking outcomes (Erker & Buczynski, 2009). Interestingly, these findings are not just evident in the United States: surveys have shown that 25% of job applicants in Australia, France, and the United Kingdom did not assume a relation between their online posts and their job applications (Erker & Buczynski, 2009).

Given that many individuals believe that the use of SNS for employment decision-making is unfair, managers are often faced with an important dilemma. Should they use data acquired from SNS or a search engine to make employment decisions to avoid negligent hiring lawsuits or forgo a review of such data because they may lose out on talented applicants in the process? On the one hand, managers believe that the use of SNS and search engines is warranted to protect the organization, but on the other hand they are concerned that access to this information may violate individuals' perceptions of fairness or privacy and negatively affect job acceptance rates (Black, Stone, & Johnson, 2015). In view of the widespread use of SNS and search engines in the employment process, and applicants' concern about the extent to which this is unfair, the primary purposes of this chapter are to:

- consider the advantages and disadvantages of using SNS and search engine data to make employment decisions
- review the existing research on the use of SNS and search engines in the employment process

- apply a model of organizational justice to explain individuals' reactions to the use of this information in the employment context, and
- explore the avenues for future research and practice.

Advantages and Disadvantages of Using Social Networking Sites and Search Engine Data in Employment Decisions

Advantages

Using SNS and search engines has several advantages for organizations. For instance, SNS have become a source for talent acquisition (Wolk, 2004), and organizations now scan professional sites (e.g., LinkedIn) to identify individuals who are qualified for a job and contact them about job openings. This is especially helpful when trying to identify and attract passive job seekers. They also use search engines to do background checks on job applicants and current employees (Kluemper & Rosen, 2009). For example, organizations may scan SNS and search engines to determine whether current employees are involved in unethical behaviors, or are spreading negative information about their organization. As a result, the use of SNS and search engines helps organizations broaden applicant pools, and can help them attract and retain the best individuals for jobs. Furthermore, research has demonstrated that employment decision-makers can make fairly accurate inferences about applicants' personalities based on a review of these sites (Kluemper & Rosen, 2009; Kluemper, Rosen, & Mossholder, 2012). As a result, reviews of SNS and search engine information may help organizations streamline the initial screening process.

A recent survey by the Advisory, Conciliation, and Arbitration Services (ACAS) in the United Kingdom uncovered many other reasons why employment decision-makers used these sites (Broughton, Foley, Ledermaier, & Cox, 2013). The main two reasons offered were that using social media produced a wider net of applicants (65%), and that they were easier to use than some other methods (64%). In addition, 54% reported that the use of social media was less costly than other methods, 46% indicated that they were more likely to target unique applicants for jobs using these sites and 25% maintained that they could find information about applicants that other methods do not always provide (Broughton et al., 2013). Similarly, another survey revealed that 84% of global companies used SNS and search engines to recruit candidates (SHRM, 2016). Both professional and personal SNS provide employers with data that can be used to determine whether an applicant meets the basic qualifications of a job. Employers also use these sites to develop social connections with other professionals and receive recommendations from past employers or references.

In the past, SNS were built for niche markets such as racial/ethnic diversity groups or users interested in music or other hobbies (Edosomwan, Prakasan, Kouame, Watson, & Seymour, 2011). Thus organizations use these sites to help them identify minority candidates, and meet their diversity goals. In addition to identifying and targeting applicants, many organizations are using SNS as a recruitment tool because they provide a quick and effective means of disseminating information about job openings and about the company. These sites can make current employees aware of internal job openings, which encourage promotion from within their organizations. Thus, the use of SNS enables organizations to create a broad applicant pool, and is much less expensive than other recruiting methods (e.g., printed advertisements, search firms). Jobvite (2014) also found in their annual survey that 33% of recruiters did not spend any resources when they used social recruiting, and some organizations saw a large increase in the number of applicants (50% increase) when they used social media. Furthermore, in the United States, the cost for traditional job advertisements runs to almost US$3300, whereas web-based advertisements cost about US$375 (Searle, 2006 as cited in Broughton et al., 2013) In the United Kingdom, the cost of a traditional job advertisement runs to almost £7500, but social media adverts cost a little over £700. The use of SNS and search engine data can help organizations decrease the time needed to file job vacancies in executive, managerial, and nonmanagerial jobs (Jobvite, 2014; SHRM, 2016). This could be attributed to the ability to verify information quickly and obtain additional information about the candidate. About 44% of recruiters found an increase in the quality of candidates when they used SNS (Jobvite, 2014).

In sum, the advantages of using SNS and search engines in the employment process are that they decrease the cost and time associated with recruiting applicants, help organizations streamline the initial screening process, enable them to identify unique applicants and passive job seekers, and assist them with conducting online background checks to make sure that applicants meet role requirements.

Disadvantages

Although the use of SNS and search engines may benefit organizations in several ways, researchers argue there are also some disadvantages (Black et al., 2015). Many applicants and employees believe that the use of SNS and search engine data in the employment process is unfair and a potential invasion of privacy (e.g., Black et al., 2015; Rasmussen Reports, 2012). One survey found that 32% of job seekers are rather naïve and did not consider how employers might view information posted on SNS (Erker & Buczynski, 2009). As a result, information posted on these sites may unfairly stigmatize them in the employment process, and prevent them from being hired for jobs. For instance, a review of information acquired from SNS or search engines may enable employers to determine if individuals have an LGBT (lesbian, gay, bisexual, transgender) orientation, drink alcohol, smoke, or engage in risky hobbies (e.g., skydiving). Finally, if individuals perceive that employment processes are unfair they are less likely to accept jobs, more likely to have negative attitudes toward the organizations, and communicate these negative attitudes to other job applicants (Truxillo, Bauer, Campion, & Paronto, 2002). Thus, in the long run the use of SNS and search engines may tarnish the reputation of the organization and prevent them from hiring, promoting, or retaining the most talented employees.

Given the potential disadvantages associated with using SNS and search engines in the employment process, we use the model of organizational justice (Colquitt, Conlon, Wesson, Porter, & Ng, 2001; Greenberg, 1987; Leventhal, 1980) in this chapter to explain the factors that might affect individuals' reactions to the use of SNS or search engines in the employment process. Other research has examined how the use of SNS affects perceptions of invasion of privacy (Black et al., 2015), but to our knowledge no research has examined a model of the perceived fairness of using SNS or search engines to make employment decisions. Furthermore, relatively little research has assessed the factors that affect perceptions of fairness. Thus, in the next section we review the existing research on the use of SNS or search engines in the employment process. Then we apply the model of organizational justice (Colquitt et al., 2001; Greenberg, 1987), and use these factors to explain individuals' reactions to the use of SNS and search engines in employment decision-making.

Literature Review on the Use of Social Networking Sites and Search Engines in the Employment Process

To date, there has been relatively little empirical research on individuals' reactions to or the effectiveness of using SNS and search engines for employment decision-making (e.g., Kluemper et al., 2012). However, there have been several conceptual articles on the topic (e.g., Black & Johnson, 2012; Brown & Vaughn, 2011; Davison, Marasit, & Bing, 2011; Davison et al., 2012; McFarland & Ployhart, in press), which have identified several broad issues associated with the use of social media in the workplace. Most researchers offer an overview of the issues but they do not consider a theoretical basis for them. Since many of these articles present similar issues, we will highlight only a few of them here.

Davison et al. (2012) provided an overview of issues for employers to consider when using Internet screening (both search engines and SNS) in employment decisions. The researchers demonstrated there were opportunities to gather job-relevant information about candidates from SNS and search engines, such as knowledge, skills, and abilities (e.g., written communication, creativity). But they further indicated that much of the information available on SNS or search engines may not be relevant to the job (e.g., family photos, personal postings), and they questioned the reliability and validity of the information collected from such sources. They also highlighted the legal implications of Internet screening, including the violation of discrimination laws, privacy laws, and defamation laws, and noted that deviant or criminal behaviors may be revealed via Internet searches. However, they cautioned that this information may not be always accurate and that the use of such information may result in lawsuits and violations of employee rights. Brown and Vaughn (2011) outlined the risks associated with using SNS in the hiring process (e.g., invasion of privacy, lack of job relevance of information, lack of standardization and validity across applicants). They also advised employers that visual or textual information about protected groups (e.g., racial and ethnic minorities, older applicants) may be available on SNS and search engines, which could lead to biases in selection decisions. They concluded that the use of SNS and search engines in the hiring process may adversely impact applicants from protected groups.

Several other conceptual articles have examined SNS and search engines within a more specified context, and some of these provide a theoretical base for their argument (e.g., Abril-Sánchez, Levin, & Del Riego, 2012; Black et al., 2015; Brandenburg, 2008; Clark & Roberts, 2010; Drouin, O'Connor, Schmidt, & Miller, 2015; Genova, 2009; Myers, 2014). These articles will be discussed in the specific context they address (e.g., selection, laws). Some of the articles address the issue of fairness, and these will be discussed in the chapter where appropriate.

Privacy, SNS, and search engines

There is a great deal of sensitive information on the Internet and SNS about individuals (e.g., alcohol use, religious and political beliefs, dating behavior) (Brown & Vaughn, 2011; Davison et al., 2012). The use of SNS and search engines by individuals is often intended for family and friends and not employers, therefore, the use of such information in the employment process has raised concerns about invasion of privacy. Privacy is defined as the individual's ability to control information and the impressions that result from such information (Stone & Stone, 1990). Black et al. (2015) comprehensively reviewed the factors that influence perceptions of invasion of privacy when SNS are used in the hiring process. Drawing on the privacy model of Stone and Stone (1990), they considered the impact of information factors (e.g., type of information), procedural factors (e.g., permission for release of data), socio-cultural factors, and individual factors (e.g., gender, age, collective values) on invasion of privacy. In an empirical study, Stoughton, Thompson, and Meade (2015) found that pre-employment screening of SNS increased the perceptions of invasion of privacy, and decreased the perceptions of organizational justice, which led to decreased organizational attraction and increased intentions to sue the organization.

Ethical issues associated with using SNS and search engines

Clark and Roberts (2010) examined how using SNS and search engine information for character and background checks can be unethical and harmful for society because employers often conduct online background checks using SNS and search engines without applicants' knowledge. These online background checks have also led to termination of employment in a number of cases. For example, a major television network news producer was fired for blogging during his off-work hours, even though he never identified himself as an employee of the network (Wolgemuth, 2008). Clark and Roberts (2010) argued that most employers lack training in the use of SNS data in the termination process, and do not have policies governing the use of such information when making discharge decisions. They also indicated that the use of SNS in the employment process violates true reasons for online communications, invades personal privacy, and blurs the boundaries between work and personal life. Thus, they argued that employers should not use SNS online background checks and also offered some guidance for employers to become more socially responsible when using online background checks.

Legal issues and the use of SNS and search engines

As more employment decisions are being based on information collected from SNS or search engines, some research has reviewed court cases in the United States associated with using SNS and search engines in the hiring or termination process (e.g., Brandenburg, 2008). Other articles have addressed the legal implications of using SNS and search engines to make employment decisions (e.g., violations of civil rights and privacy laws) (Abril-Sánchez et al., 2012; Genova, 2009; Myers, 2014). The reader should note that these reviews are based on laws in the United State, but they also included findings from Canadian and American surveys on individuals' expectations regarding the use of SNS. For instance, Abril-Sánchez et al. (2012) reviewed the legal parameters and their ramifications in three specific areas: monitoring and surveillance of employee social media profiles, evaluating applicants' social media profiles, and limiting employees' off-duty online activities. They also included results of a survey of Canadian and American undergraduates regarding privacy expectations and SNS profiles, and the use of policies in the workplace. Results demonstrated that 92% indicated Facebook was their preferred SNS and only 2% had joined LinkedIn. The undergraduates often shared pictures of themselves (77%), their hometown (76%), and their real name (68%), posted their relationship status (62%), disclosed their dating interests (40%), and revealed their partner's name (25%). Another interesting finding was that 35% were concerned or very concerned if their employer accessed their information from SNS, and 49% found it somewhat or very inappropriate for employers to proactively search SNS to find future job applicants. Lastly, 56% found it very or somewhat inappropriate to use SNS to examine the character of a job applicant.

Other articles have also focused on specific contexts of legal issues. Drawing solely on laws of the state of California (United States), Genova (2009) examined privacy rights when using SNS to make decisions about discipline, termination, or denying employment based on individuals' off-duty behavior. Other researchers (e.g., Myers, 2014; Schmidt & O'Connor, 2015) examined specific court decisions where employees were fired based on their SNS posts. In several cases, employment was terminated because of SNS posts criticizing employers or customers (e.g., Facebook posts). For instance, an employee criticized the employer's strategy for withholding taxes on employees' paychecks. Another employee clicked "Like," indicating that he agreed with the criticism. Both employees were fired because the organization believed that the posted comments were maliciously untrue and revealed the employees' disloyalty toward the employer. However, the National Labor Relations Board (NLRB), the agency that rules on union–management disagreements in the United States, ruled that the employees' comments and "Like" postings on SNS were protected under the National Labor Relations Act of 1935 (NLRA), and the Facebook communications were not disloyal, defamatory, or maliciously untrue. That is, the NLRA provides American employees the right to engage in combined activity (two or more employees) to address work conditions and employment terms. Employers cannot interfere with this activity, which includes online social media (Schmidt & O'Connor, 2015). As a result, the employees were reinstated (Schmidt & O'Connor, 2015; Three D, LLC, d/b/a Triple Play Sports Bar & Grille v. NLRB, 2015). In another case, an American waitress was fired because she posted an angry note on Facebook about customers' tipping behaviors and used a derogatory name to refer to a customer. She also complained about lousy tippers, and wrote that "if you come into a restaurant and spend more than $50 you should be able to tip appropriately." A customer saw the post, printed a screenshot, and showed it to the restaurant's manager. The employer argued the waitress knew what she was doing when she posted the comment (Daily Mail, 2014). In a similar case in the United Kingdom, a guard at Buckingham Palace was fired after posting negative comments about Kate Middleton on Facebook. The posts were seen as violations of protocol or rules for guards. To our knowledge, these two terminations were not contested in the courts (Emerson, 2011).

Thus, it appears that employers are not only reviewing Facebook in order to hire people, but they are also looking at online posts to see if employees are criticizing or making derogatory comments about their employer or customers. Many employees believe that their Facebook posts are protected under freedom of speech laws, but this is not always the case. Freedom of speech laws protect individuals from governmental intrusions, but no such restrictions apply to private employers in the United States. Thus, employees should be aware that Facebook posts criticizing employers or make derogatory statements about customers may lead to termination, and they should avoid making negative comments about employers online, or posting comments on SNS or search engines (Guerin, 2016).

Relatively little research has examined the issue of fairness of using SNS or search engines in employment. There are two notable exceptions. Madera (2012) found that job applicants had lower perceptions of fairness and job pursuit intentions when an organization used SNS as a selection tool than when it did not use a SNS. Also, Stoughton et al. (2015) found that pre-employment screening of SNS increased the perceptions of invasion of privacy, and decreased the perceptions of organizational justice, which led to decreased organizational attraction and increased intent to sue. Taken together, research on individuals' reactions to the use of SNS revealed that more than half of survey respondents considered it inappropriate to assess an applicant's character using SNS, applicants felt that the selection process was more unfair when SNS were used in the process, and the use of SNS increased perceptions of invasion of privacy and decreased perceptions of fairness, which led to decreased organizational attraction. Thus, there is some evidence that the use of

SNS in the hiring process may influence perceptions of unfairness and may have a negative impact on individuals' attraction to organizations. Even though this research suggests that using SNS in the employment process may be perceived as unfair, the same research was not based in any theoretical model. However, we believe that the research would be more useful if we had a better understanding of why individuals perceive the use of these systems as unfair. Therefore, in this chapter we have applied the model of organizational justice (Colquitt et al., 2001; Greenberg, 1987; Leventhal, 1980) to understand the factors that affect individuals' reactions to the use of SNS and search engines in the employment process. We also use this model to offer directions for future research on reactions to the use of SNS or search engines.

Organizational Justice Theory

We argue, as do others (e.g., Madera, 2012; Stoughton et al., 2015), that organization justice theory can be used to explain individuals' reactions to the use of SNS and search engines in the employment process. We start by describing organizational justice theory and then consider each of the procedural dimensions noted by Leventhal (1980) to understand the reasons why individuals may perceive the use of SNS and search engines as unfair.

Organizational justice theory is based in social justice and equity models of motivation (Adams, 1965). It focuses on how individuals react to the fairness of the allocation of outcomes, and the fairness of organizational policies and procedures (Folger & Greenberg, 1985). These reactions are expressed as individuals' beliefs (e.g., perceptions of unfairness), subsequent job-related attitudes (e.g., job dissatisfaction), and behaviors (e.g., turnover, grievances) (Greenberg, 1987). The theory posits there are four dimensions of justice: distributive, procedural, interactional, and informational (Colquitt, 2001; Colquitt, Greenberg, & Zapata-Phelan, 2005). However, only two of these dimensions are relevant to this chapter, that is, distributive and procedural justice. Distributive justice refers to the perceived fairness associated with the decision regarding outcomes and the allocation of important outcomes (e.g., pay, promotion, layoffs) (Colquitt, 2001). Procedural justice is defined as the fairness of the processes that lead to outcomes (Colquitt, 2001). According to the model, when individuals perceive that the allocation of outcomes and processes are unfair they are likely to exhibit job dissatisfaction, refuse to accept job offers, display counterproductive work behaviors, or withdraw and leave the organization (Folger & Greenberg, 1985).

Furthermore, Leventhal (1980) developed six dimensions associated with procedural fairness: consistency, accuracy, correctability, representativeness, ethicality, and bias suppression. Leventhal argued that procedures:

- should be applied consistently across people and time
- should ensure that accurate information is collected and used in making decisions
- should have some mechanisms in place to correct flawed or inaccurate decisions
- should conform to personal or prevailing standards of ethics or morality as they pertain to the decision, and should ensure freedom from bias (e.g., ensure that a third party has no vested interest in a particular outcome).

Advanced notice is also a key determinant of fairness (Bies, 1987; Stone & Kotch, 1989). Given these potential determinants of procedural fairness, we now describe each of Leventhal's six dimensions and use this framework to explain why individuals may react negatively to the use of SNS and search engines in the employment process.

Consistency

An important determinant of procedural fairness is the notion that procedures should be applied consistently across people (Leventhal, 1980). As a result, if organizations do not review SNS or search engines consistently with regard to applicants or employees, then individuals are likely to believe that the process is unfair. When individuals are singled out for a review of their SNS or search engine information they may perceive that the organization does not trust them or is attempting to uncover negative information about them and not about others (e.g., alcohol or tobacco use, complaints about previous employers, pregnancy). Also, individuals may believe that employers are only reviewing SNS or search engines when they do not like an applicant or when applicants are members of minority groups. Therefore, this procedure may be perceived as a way of searching for negative information about some applicants and making sure that they are not hired. However, if employers review SNS and search engines consistently for all applicants, then individuals are unlikely to perceive that they are being singled out and treated unfairly.

We know of no specific research on this issue, but previous research on organizational justice supports the arguments about consistency of procedures in various employment contexts, such as performance evaluations (Greenberg, 1986), pay (Bartol, 1999; Ramaswami & Singh, 2003) accident compensation costs and employee sickness (Sashkin & Williams, 1990), and in promotion and selection decisions (Gilliland, 1993; Harris, Lievens, & Van Hoye, 2004). For example, Greenberg (1986) examined perceptions of fairness among 217 middle managers in three different industries (cable television, pharmaceuticals, and credit unions). The participants were asked to describe a time when they received an unfair or fair performance evaluation on the job. They were asked to write down one factor that made the evaluation fair or unfair, and one of these was the consistent application of procedures or standards. Ramaswami and Singh (2003) also found that fairness perceptions of salespeople were greater when there was consistent application of standards in regards to their merit pay. Sashkin and Williams (1990) examined data on accident compensation costs and levels of employee sickness obtained from retail store managers and department managers with regard to nine different dimensions of fairness (including consistency). Stores that were more consistent in their policies and practices had lower accident compensation costs and lower levels of employee sickness when compared with stores that were less consistent with their policies and practices. Furthermore, they found playing favorites was an action that employees noted when describing the unfair actions of their manager. Although research revealed that consistency of treatment is an important determinant of perceptions of fairness in multiple contexts, research is needed to assess the extent to which inconsistent use of SNS or search engines affects individuals' perceptions of fairness regarding their use.

Given the lack of research on the degree to which reviews of SNS or search engines are applied consistently, additional research is needed on the topic. In addition, we believe that if organizations review SNS or search engines for all applicants, then individuals are less likely to perceive that the use of these systems is unfair than when SNS or search engines are reviewed for only a subset of applicants. Furthermore, if information is only available for a subset of individuals then the information should not be used in employment decision-making. Organizations might also ask a third party to review the SNS or search engine information, so that protected information is not compromised. This procedure is similar to using background checks when screening applicants. Interestingly, in the United States, 59% of organizations do not have a formal policy regarding the use of SNS or search engines to screen job applicants (SHRM, 2016). On a global level, Proskauer (2014) found that 80% of companies have social media policies, but it was not clear which types of information were covered by these policies. As a result, there is no way to determine whether organizations review SNS or search engines for all job applicants or for just some of them. This process may put organizations at risk for civil rights lawsuits by applicants.

There are other ancillary issues to consider when incorporating the consistency rule into employment process. First, the review of SNS or search engine information may be problematic because only a subset of applicants have an SNS presence, and reviewing such information may give them an advantage or put them at a disadvantage in the hiring process. For example, it is not clear how organizations deal with applicants who do not have an SNS presence; thus it can be argued that employment decisions should be based on the same information across applicants or employees. Moreover, reactions to the lack of consistent application in policies is a problem across countries (Morris & Leung, 2000), and individuals in multiple countries (e.g., China, Japan, Korea, and the United States) have reacted negatively when organizations violate procedural justice issues (Kim & Leung, 2007).

Accuracy

Another procedural factor that affects perceptions of fairness is the accuracy of information (Leventhal, 1980). One reason why this may affect the fairness of reviewing SNS and search engine information is that individuals may be concerned that the data gathered from these systems are incorrect. For example, friends and family members may post information on these sites (e.g., person is a heavy drinker) that does not accurately reflect the person's background. A survey by SHRM (2016) showed a great deal of information posted on SNS is not correct. One case (Socorro *v.* IMI Data Search and Hilton Hotels, 02 C 8120 [United States District Court for the Northern District of Illinois, Eastern Division, 2003]) provides an excellent example of this problem. An executive at Hilton Hotels was fired because his online background check revealed he had been convicted of a crime and had served six months in jail. However, these data were incorrect, as he had never been convicted of a crime or served time in prison. Unfortunately, Hilton Hotels did not verify the results of the background check, and terminated his employment stating that he had lied on his application. Socorro was powerless to show them the data were incorrect or change their decision, and had numerous difficulties finding other jobs. Given that some data in an SNS or search engines may be inaccurate, individuals are likely to perceive that the use of information gathered from these systems is unfair.

A second reason that individuals are likely to believe that the use of data in an SNS or in search engines is unfair is that employers often make erroneous inferences based on information collected from these systems. As noted above, Hilton Hotels made the inference that their executive had lied on his application because the background check inaccurately showed he had been convicted of a crime and served time in prison. In another case, a young woman was fired from her student teaching job when the principal of the school saw a posting with a picture of her holding a cup with a caption that said "drunken pirate." Some of the young woman's friends had posted a picture of her wearing a pirate hat at a Halloween party, and holding up a cup as if for a toast. They also posted a caption stating she was a "drunken pirate" even though she was not drinking alcohol. Still, the school principal made the erroneous inference that the student was a heavy drinker, a bad role model for students, and discharged her from her job. Given that some information retrieved from SNS or a search engine may be inaccurate, and that employers often make erroneous inferences based on information collected from these systems, and thus individuals may experience negative job-related consequences (e.g., denial of job, promotion, or termination of employment), individuals are likely to perceive that the use of information gathered from these systems is unfair.

In support of this argument, several researchers have claimed that that the accuracy of information in SNS or search engines is a key determinant of individuals' reactions to the use of these systems in the employment context (e.g., Black et al., 2015; Brown & Vaughn, 2011; Davison et al., 2011; Kluemper, 2013, Kluemper & Rosen, 2009; Roth, Bobko, Van Iddekinge, & Thatcher, 2016; Van Iddekinge, Lanivich, Roth, & Junco, 2013). Parsons (2015) argued that some individuals post on SNS fake profiles and other false information about others competing for the same job, and these types of posts have a negative effect on the accuracy of data collected online. Other analysts maintain that information on SNS often contain exaggerations (Gilbert, 2013). Such posts are typically targeted at friends and family, and individuals may brag about their heavy drinking activities, upload risky pictures, or discuss their sexual prowess. These comments are often designed to create positive impressions among friends, and may not be accurate (Roth et al., 2016). As a result, employers should be skeptical when gathering and using information from these sites, and numerous sources should be utilized to verify the received information (Bohnert & Ross, 2010). Finally, overreliance on the information contained in these sites may cause employers to miss out on talented applicants.

Validity of information. Analysts have also voiced concern about the validity of job-related inferences made from data on SNS (Roth et al., 2016). Validity is the extent to which accurate or appropriate inferences are made from information. In a recent survey by SHRM (2016), 31% of the respondents believed that SNS profiles do not provide information on applicants' job potential or work performance, and should not be considered in employment decisions. Given that employers often try to determine a job applicant's suitability for the job or a current employee's promotion prospects, it is unclear whether the types of data found on SNS or through search engines are good sources of job-relevant information. Researchers have cautioned that the information collected from SNS and search engines may not be a valid predictor of job performance (e.g., Black et al., 2015; Brown & Vaughn, 2011; Kluemper, 2013; Roth et al., 2016; Van Iddekinge et al., 2013). Stone-Romero, Stone, and Hyatt (2003) suggested using alternative selection devices, for example, work samples or job performance data, as a better means of predicting job performance than the use of SNS or search engines. These researchers also found that the extent to which a selection technique predicts job performance is negatively related to perceptions of invasion of privacy.

There has been little research on the validity of inferences based on data acquired from SNS or search engines. Kluemper and Rosen (2009) showed that raters' inferences about applicants' personality based on a review of information on SNS may be positively related to job performance. Sixty-three undergraduate students rated personality characteristics and general mental ability from information gleaned from various Facebook pages. The personality ratings assigned by the students to the various viewed Facebook profiles were in agreement with the personality ratings given by those whose profile was viewed. However, Kluemper and Rosen (2009) warned about the preliminary nature of their results, and indicated that organizations should not generalize from these findings. In addition, they called for more research to establish the validity of information on SNS.

Bohnert and Ross (2010) found that raters' evaluation of a potential applicant is influenced by the applicant's résumé and SNS data. Raters who believe the SNS information was the most important factor for their ratings gave the lowest ratings to those with alcohol-related information on SNS and the highest ratings to those with family-related and profession-related information on SNS. Evaluators who found the résumé was the most important factor had less variability in their ratings, but exhibited a similar pattern of ratings. Overall, applicant SNS that featured a heavy emphasis on alcohol impacted their evaluation negatively.

Verification of data. Employers do not always verify the accuracy of information (e.g., pictures, personal information) posted on SNS or obtained from search engines. This has raised considerable concern and some organizations have stopped using SNS-derived information in the employment process. In a recent survey (SHRM, 2016), 47% of organizations noted having stopped reviewing SNS information because there was no way to verify confidently the accuracy of the information posted on these sites. Two considerations make verification an important issue when using SNS sites for information on applicants. The first is whether the inference being made on the basis of information obtained from SNS or search engines is valid and accurate. For example, a common problem associated with the use of these sites is that friends and family members may post pictures or comments that do not portray the individual in a positive light (e.g., a comment noting an individual was drunk when she was not). Many employers may make negative inferences about individuals based on pictures and postings, such as that the person is untrustworthy, unreliable, or has a propensity for taking risks (Lucero, Allen, & Elzweig, 2013). This may then lead to denial of employment or a promotion opportunity (Haefner, 2009). Inferences made from pictures posted on SNS or retrieved from search engines could also lead to termination of employment especially when the organization has a code of practice in regard to personal conduct.

Thus, verification of information obtained from SNS and search engines by employers cannot be overemphasized, especially before making critical employment decisions. Failure to verify such information could lead to poor decision-making (e.g., not hiring a qualified individual, wrongful termination) and possible violation of laws. For example, an individual denied employment or promotion, or fired because of an erroneous inference based on a photograph posted on a SNS may sue the organization based on civil rights, defamation, or other laws. Davison et al. (2009) raised an interesting issue about the use of photographs or postings to make hiring or firing decisions about individuals. They argued that in the United States, under the Americans with Disability Act (1990), an organization cannot fire an individual who was an alcoholic but is currently enrolled in a rehabilitation program, and there is no way to determine whether the individual is enrolled in rehabilitation based on information on an SNS or search engine. Therefore, employers may violate disability or civil rights laws if they make employment decisions based on information on an SNS or search engines without verifying this information.

A second problem with the use of SNS data by employers is that a person other than the applicant could post fake information to discredit the applicant. In the United States, only a few states have laws about Internet impersonation, thus there is a chance to discredit someone (e.g., ex-boyfriend, colleague, manager) through fraudulent SNS posts (Reznik, 2013). For example, a woman created a Facebook page in the name of her ex-boyfriend who was a narcotics detective. She noted that he frequently visited prostitutes and used drugs (Reznik, 2013). These postings were designed to damage his reputation, and if his employer had used this information, he could have been fired based on erroneous data. In another example, a football (soccer) player lost his 2015 contract just hours after signing with Barcelona B based on posts on a Twitter account from 2013 (Mendola, 2015). The posts in question criticized past players, management, and loss records of the team (Barcelona B). Further, he praised several competitors (i.e., other football teams) for their winning seasons (Mendola, 2015). Although the player maintained that a friend had posted the information on his account, the organization decided to void his contract (Mendola, 2015). Thus, employers need to be aware that third parties can use SNS and search engines to disseminate defamatory or inaccurate information about people, and should verify the information on a SNS or search engines before using it for decision-making. It merits noting that the narcotics detective did not lose his job and sued his ex-girlfriend for defamation. However, more research is needed to determine whether verifying information obtained from a SNS or search engines reduces perceptions of unfairness.

In view of the arguments noted above, we conclude that individuals may react negatively to the use of information on SNS or search engines in the employment context because the data gathered from these sources could contain inaccurate information. If an organization does not verify the accuracy of the information using multiple sources, it can lead to erroneous inferences about job performance. Previous research on fairness in the selection process supports these arguments: when information used to make selection decisions was accurate individuals were more likely to report that the process was fair than when the information was not accurate (Gilliland, 1993, 1995; Schmitt, Gilliland, Landis, & Devine, 1993; Truxillo et al., 2002).

Correctability

Another key determinant of fairness is based on the rule that there should be opportunity to reverse decisions made on the basis of questionable information (Leventhal, 1980). For example, individuals are likely to perceive that the process is unfair when employment decisions are made about them based on information retrieved from SNS or search engines, and they are not given an opportunity to contest or reverse negative decisions. In other words, when questionable information acquired from an SNS or using search engines is not verified, but the individual concerned is not given an opportunity to explain it, then they may believe that their rights have been violated. If the individual subsequently loses out on employment opportunities (i.e., being hired for a job, a promotion, a challenging job assignment), and there is no opportunity to appeal the decision then the individual may perceive the process is unfair or take legal action against the organization.

As noted previously, however, individuals may not be aware that organizations review information retrieved from these sites when making employment decisions. Often there is no recourse when employers make negative decisions based on this information, and individuals typically do not know why they were denied a job or promotion. In the United States, if a credit report is used as a factor in making an employment decision, and negative information is uncovered in the report, then the individual has a right to dispute the information under the Fair Credit Report Act (1970). Given that over 80% of credit reports contain errors (Dakss, 2004), this act allows individuals the opportunity to review and correct such errors. However, this is not the case for other types of information used in the hiring process in the United States. In other countries, such situations may be avoided because of laws that prevent or limit organizational access to online information about applicants and employees. For example, in the United Kingdom, the Data Protection Act (1998) requires employers to obtain consent before they can gather online data about an employee or applicant (Poerio & Bain, 2012).

Even though there is widespread use of SNS and search engines in the employment process, limited research has investigated an individual's ability to correct or contest decisions based on information derived from SNS or search engines. In a survey (SHRM, 2016), 47% of employers reported they did not allow the candidate to explain information obtained for the hiring process, 39% do not allow the individual to explain the information after the initial screening but prior to the decision to hire, and 14% do not allow them to explain information after the hiring decision. However, this survey did not assess applicants' perceptions of fairness about the ability to explain information obtained from SNS or search engines. Thus, research is needed to examine applicants' views about the fairness of the ability to explain information obtained for this process.

We know of no other research on this issue, but previous studies on organizational justice support arguments about the correctability rule in various employment contexts, such as performance evaluations (Barclay & Harland, 1995), in selection decisions (Barrett-Howard & Tyler, 1986; Gilliland, 1993), and in drug testing (Wagner & Moriarty, 2002). For example, Barclay and Harland (1995), drawing on a sample of 359 public health nurses, found that when peer performance evaluation systems allowed for correctability, the nurses were more likely to perceive that the process was fair than when there no opportunity for correctability. In addition, Wagner and Moriarty (2002) found that drug testing policies that did not allow for correctability were deemed as unfair.

Given the lack of research on the degree to which individuals can explain questionable information or appeal decisions based on the information gathered from SNS or search engines, additional research is needed to determine if the ability to explain or correct information retrieved from SNS or search engines enhances perceptions of fairness. When organizations decide to review information derived from SNS or search engines in the employment process, they should allow some kind of appeal process when questionable information is obtained. This appeal process should benefit both the organization and the individual: it should enable organizations to make better hiring decisions, and give applicants the opportunity to gain access to jobs.

Representativeness

This determinant of fairness is based on the rule that the basic concerns and values of society should being reflected in decision-making (Leventhal, 1980). For instance, research consistently shows that individuals in Western societies believe that employment decisions should be based on job-relevant information that has the ability to predict job performance (Heneman, Judge, & Kammeyer-Mueller, 2014; Stone & Stone-Romero, 1998). Therefore, information acquired from SNS or search engines should be clearly job relevant or should be able to predict job success. If information obtained from SNS or search engines is not considered job related (e.g., religious beliefs, disability, romantic relationships), then individuals are likely to perceive that the process of gathering information is unfair. Further, if the individual loses employment opportunities (i.e., access to jobs, promotions, or to job assignment opportunities) based on such nonrelevant information or using information with little or no ability to predict job performance, the individual will deem that the process is unfair.

One reason for this is that individuals are aware that a vast amount of information can be obtained from SNS or search engines, whether for professional purposes or for personal purposes with their family members and friends. They know that SNS or search engines often include photographs, and employers can determine their race, sex, age, or attractiveness based on them. SNS posts can also reveal information about personal interests, values, lifestyle, types of friends, children, etc. People often post information about their families, sexual orientation, drug/alcohol use, domestic partnerships, religious beliefs, and these types of data may be used to stigmatize them (Gross & Acquisti, 2005). Information about roomates or domestic partners may affect employers' inferences about the person's sexual orientation. Employers may also use information about alcohol use to make inferences about the applicant's dependability or trustworthiness (Bohnert & Ross, 2010). Thus, the kinds of information revealed on SNS or through search engines have the potential to embarrass or even discredit individuals if employers use such information when making employment decisions (Black et al., 2015). As a result, many individuals are likely to consider information contained in SNS, especially those used for personal purposes, as irrelevant for jobs and violation of their perceptions of fairness, and believe that they should not be used to make employment decisions (Grasz, 2009).

Although the collection of non-job-relevant information has long been considered an issue when employers use SNS or search engines information, there has been limited research on the topic. In one study, over 50% of hiring decision-makers reconsidered a candidate based on a review of information derived from SNS (Jobvite, 2014). Interestingly, over 60% of candidates being reconsidered on this basis were dropped. Negative items that were used as a basis for denial of employment included the use of profanity, spelling or grammatical mistakes, as well as references to illegal drug use, sexual activities, alcohol use, and attitudes toward guns (Jobvite, 2014). In another study, about 8% of employers admitted to firing an employee on the basis of their social media activities (Ostrow, 2009). Likewise, research consistently shows that the job relevance of information is an important predictor of perceptions of fairness and invasion of privacy (Barclay & Harland, 1995; Gilliland, 1993; Rousseau & Aquino, 1993; Sheppard & Lewicki, 1987; Stone-Romero et al., 2003). Thus, many researchers have cautioned employers about using SNS or search engines to obtain information about applicants or employees until the validity of the inferences made from such specific information can be determined (Brown & Vaughn, 2011). Some early research did suggest that employers can make accurate inferences about an applicant's personality based on a review of information derived from SNS, but the same research also cautioned employers about reading too much into this information (Back et al., 2010; Kluemper & Rosen, 2009). More research is needed to establish if the use of SNS-derived information can accurately predict job performance (e.g., trustworthiness, dependability, conscientiousness). Until then, the use of this information is likely to be considered unfair.

The issues noted here seem to be of lesser concern in countries other than the United States. In Germany, employment law only permits employers to ask applicants for their personal information if it is relevant to a specific job. As a result, it unlikely that an employer can legally obtain non–job-relevant information from search engines or SNS (Poerio & Bain, 2012). Although there are no specific employment law provisions in France that address retrieval of information from SNS and search engines, the general legal provisions prevent employers from asking about information that is unrelated to the job (Poerio & Bain, 2012).

Ethicality

This determinant of fairness is based on the rule that the procedures being used to make decisions must be compatible with the moral and ethical values of the individual (Leventhal, 1980). In other words, if an individual considers the acquisition and usage of information from SNS or search engines by employers is unethical or immoral, then its use in employment decisions (i.e., a new hire, a promotion, opportunity to lead a project, or even termination) will violate their sense of fairness.

There are many reasons why an individual may find the use of information retrieved from SNS and search engines by employers to be unethical or immoral. First, they might have an expectation of privacy, a right to privacy (Black et al., 2015), and the use of SNS or search engine information by employers may violate this perceived right. These beliefs about rights are often seen in countries that include explicit or implicit rights to privacy in their constitutions (Clark & Roberts, 2010). The extension of this right into the workplace may differ, depending on the country and its laws. However, individuals often believe this basic human right extends to the workplace, and view employers' use of

SNS or search engine information as unethical and unfair because they see posts on their personal SNS as private data that should not be viewed by current or potential employers. Second, there may be an imbalance between an individual's right to privacy and an organization's need to know information for employment decision-making. For example, privacy laws in some countries are geared toward the employer's need to know information contained in SNS or search engines. For example, Spain's Constitution (Article 18) grants a right to privacy (Poerio & Bain, 2012), and there is a belief that an employer's access to information derived from SNS of employees infringes their rights. However, this protection does not extend to publicly available data. Courts have also ruled that an employer has a right to fire an employee when they criticize or insult their employer on social media. For example, an employer in the United Kingdom fired an employee in 2013 on the basis of posts on Facebook made in 2011 about his colleague and drinking while on standby for work. Even though the employee was fired two years after he had made the posts, the termination was ruled as fair based on the Employment Rights Act (1996). Coverage of privacy rights may differ across countries, but in many cases individuals perceive that their rights are violated although court rulings tend to favor employers rather than employees (Poerio & Bain, 2012).

Although Clark and Roberts (2010) provide important insights regarding ethical issues associated with the use of SNS information by employers, little research has directly assessed this issue. Black et al. (2015) argued that a number of factors affect the degree to which individuals perceive that the use of SNS information invades their privacy, including procedural factors (e.g., permission for access to information, ability to correct information, verification of information) and individual/socio-cultural factors (e.g., collectivism, power distance, age, education). However, as the use of information retrieved from SNS and search engines continues to grow in employment decision-making, additional research is needed to determine the reasons that individuals may perceive the use of these systems as unethical or as an invasion of privacy.

Bias suppression

Another procedural factor that affects perceptions of fairness is bias suppression (Leventhal, 1980). Bias suppression occurs when personal self-interest and blind allegiance to rigid and strict preconceptions are avoided in the decision-making process. Bias suppression may affect the perceptions of fairness associated with the use of information derived from SNS and search engines in the employment process, because these sites often give employers access to information that may negatively affect their perceptions about individuals (e.g., pictures that show race, age, or posts that indicate religious and sexual orientations). This type of information typically elicits stereotypes or stigmatizations that have a negative impact on employment decisions. For example, when a prospective employer views a picture of an individual showing the person has a disability, this may lead them to use stereotypes about this subgroup to make decisions about the individual (e.g., person is incompetent, bitter, angry, emotional, or shy) (Stone & Colella, 1996). Another example is that posts on SNS may show that the person is in a same-sex relationship, and knowledge about this relationship may elicit stigma and an inaccurate inference that the person will not fit in the organization's culture (Badgett, Lau, Sears, & Ho, 2007).

When employers review SNS and search engines, they are often able to gain access to information that is not job-relevant or not included on resumes or job applications. Furthermore, there are rarely policies or practices that preclude decision-makers from using biases based on these data to make employment decisions. As a result, we believe that one reason individuals may perceive that the use of SNS or search engines as unfair is that the pictures and posts on these sites may actually increase biases, not suppress biases, in the employment decision-making process. For example, the use of these sites may give employers non–job-relevant information about age, race, religion, sexual orientation, gender, disability, and physical unattractiveness, and previous research has consistently revealed that these attributes often result in biases and unfair discrimination in decision-making (Dipboye & Colella, 2013; Shore et al., 2009). Furthermore, civil rights laws in the United States and Europe often preclude the use of this information in the employment process. However, there is no way of knowing the exact types of information collected from SNS or search engines by employers, and there are rarely procedures for curtailing employers' biases when they use SNS or search engines in employment process. Thus, it is not surprising that individuals are likely to perceive that the process is unfair when biases can creep into decision-making based on SNS or search engines. In the following sections, we consider three of the attributes that may result in biases if data are gathered from an SNS or a search engine (e.g., age, gender, race).

Age We know of no specific research that has examined whether the use of information derived from SNS and search engines invokes biases toward older workers or job applicants. However, older applicants are often stereotyped or stigmatized in the employment process (e.g., Hertel, van der Heijden, de Lange, & Deller, 2013; McGregor & Gray, 2002; Shore et al., 2009). For instance, a meta-analytic review conducted by Gordon and Arvey (2004) found that younger applicants and workers were evaluated more positively than their older counterparts. Shore et al.'s (2009) review demonstrated that older workers or applicants are more likely to experience unfair discrimination in the employment process than younger individuals. These results are consistent with prevailing negative stereotypes about older workers including the view that they are less flexible, less productive, less creative, and resistant to change, rigid, and uncomfortable with technology (Kulik, Perry, & Bourhis, 2000; Ringenbach & Jacobs, 1994; Rosen & Jerdee, 1976, 1977). Thus, when photos or posts on SNS or in search engines reveal the age of the applicant, the information may have a negative impact on employment decisions (e.g., hiring, promotion, performance appraisals). More research is needed, however, to determine if age-related information (e.g., marriage or graduation dates) on SNS or search engines biases employment decisions and to determine ways of controlling or suppressing biases when employers review information retrieved from SNS or search engines.

Gender There has been limited research regarding gender and the use of information obtained from SNS and search engines. However, Peluchette and Karl (2010) found males were more likely than females to post images on SNS that were sexually appealing, offensive, or wild. Males were also more likely to post problematic content on these sites (e.g., drug abuse, use of alcohol, sexual relations). Strano (2008) found that females were more likely than males to post pictures in which they were smiling, happy, and with friends. Van Iddekinge et al. (2013) found recruiters rated females higher than males on suitability for a job and overall skills' ratings, based on the individuals' SNS profiles. Females also received higher ratings on conscientiousness.

Surprisingly, research on gender biases based on SNS or search engine information appears to contradict research on gender biases in organizations. For instance, biases against women have been reported in terms of hiring and tenure decisions (Steinpreis, Anders, & Ritzke, 1999), performance ratings (Tsui & O'Reilly, 1989), and pay levels (Bielby & Baron, 1986). Steinpreis et al. (1999) found that both female and male academics were more likely to hire a male colleague than a female colleague even though they were equally qualified. Furthermore, when evaluating research, teaching, and service records of male and female job applicants, both male and female evaluators were more likely to provide positive evaluations of male applicants than of female applicants with an identical record.

In contrast, women may be rated higher than men when data are collected from SNS or search engines because women are less likely than men to provide posts or pictures of problematic behavior. Thus, it appears that SNS posts may be more problematic for men than women, and the content of the posts, rather than gender of the individual concerned, may be responsible for the ratings based on information obtained from SNS or search engines. However, it would still be considered unfair if employers draw on gender and online posts to make employment decisions because gender is not relevant to job decision-making (Brown & Vaughn, 2011). Furthermore, information on SNS or search engines may be posted by friends and family, and there is no way to determine if the data are accurate or just designed to make positive impressions on friends. Therefore, future research should assess the extent to which gender biases stem from posts on Internet sites or from gender information.

Race Although we know of no research on the degree to which racial information on SNS biases employment decisions, we do know from previous research that employment decisions about people from ethnic minorities are biased by their race or skin color (Stone, Stone, & Dipboye, 1992). For example, research reviews consistently show that people from racial minority groups are stereotyped more negatively than their White counterparts, and these stereotypes negatively influence employment decisions (e.g., hiring, promotion, job assignments, termination). When Van Iddekinge et al. (2013) asked recruiters to assess Facebook profiles of job seekers, they found White applicants received higher suitability ratings for jobs than both Black and Hispanic applicants. Thus, because images or posts on SNS or in search engines are likely to reveal a person's race or skin color, the use of information from these sites is likely to bias decisions about the person.

Recent research also shows that in the United States there are differences in biases toward particular ethnic minorities or immigrants from specific countries. For instance, Stone-Romero, Stone, and Hartman (2002) found that stereotypes were more negative for the following ethnic groups, in the following order: Mexicans, African Americans, Chinese, East Indians, and Native Americans. Thus, SNS or search engines may reveal information about an applicant's or worker's nation of origin (e.g., primary language, membership in ethnic organizations) that may also bias employment decisions. It would be interesting to know whether stereotypes of ethnic minorities in other nations also bias employment decisions about them (e.g., immigrants from Africa, Asia, or Middle-Eastern nations in Europe). Furthermore, research should examine the extent to which national origin information on SNS or search engines elicits stereotypes about applicants and negatively affects employment decisions.

Given that research confirms consistent biases against older workers and workers from racial minority groups, the increased use of SNS information in the employment processes may allow prejudice and negative biases to influence employment decisions. It is critical that researchers should assess the extent to which individual attributes such as age, gender, disability, and race influence decisions made on the basis of information derived from SNS or search engines as well as examine ways of suppressing these biases when employers gather information from SNS or search engines.

Advanced notice

Even though Leventhal (1980) did not specify that advanced notice was a determinant of procedural fairness, researchers suggest it should be added to Leventhal's list (Bies, Martin, & Brockner, 1993). This is because providing advanced notice to individuals about employment procedures (e.g., drug testing, terminations) enhances perceptions of procedural justice (Bies et al., 1993; Stone & Kotch, 1989). As most employers do not notify applicants or employees in advance that they will be reviewing information obtained from SNS or search engines, it is likely to have a negative impact on perceptions of fairness and perceptions of invasion of privacy (Black et al., 2015). We believe applicants and employees should provide advanced notice of the review for several reasons. First, when individuals know that employers will review their SNS data, they have an opportunity to edit the information posted, delete the SNS record, or change privacy settings to limit or deny access. Second, they could conduct a search on their own name in a search engine to become aware of the information that is associated with their name, including inaccurate information. They may have difficulties editing or deleting inaccurate information delivered by a search engine, but at least they could make the employer aware that such information, if obtained, will not be correct.

We know of no specific research on this issue, but previous research on organizational justice supports the use of advanced notice in various employment contexts such as such as drug testing (Cropanzano & Konovsky, 1996; Stone & Kotch, 1989), layoffs (Brockner et al., 1994), and terminations (Rousseau & Aquino, 1993). Stone and Kotch (1989) and Cropanzano and Konovsky (1996) found that advanced notice increased perceptions of fairness in drug-testing contexts. Similarly, Rousseau and Aquino (1993) reported advanced notice enhanced perceptions of fairness associated with terminations. Bies (1993) argued that advanced notice should be a central variable when examining the procedural fairness of decisions that may be negative, such as layoffs, disciplinary actions, and performance appraisals. Interestingly, Alder, Ambrose, and Noel (2006) examined the use of advanced notice on perception of fairness regarding Internet monitoring, but found that it did not have an impact on perceptions of fairness.

Given the lack of research on the use of advanced notice when using information from SNS or search engines, and that many individuals are unaware that employers are accessing their private data on SNS or via search engines (Vicknair, Elkersh, Yancey, & Budden, 2010), more research is needed on the topic. In addition, we maintain that advanced notice should be given to applicants and current employees about any review and use of information obtained from SNS or search engines in employment decisions. This gives individuals an opportunity to control the types of information posted, and the impressions presented to decision-makers. In some countries (e.g., France, Germany, and the United Kingdom) advanced notice may not be an issue because authorization is required when employers use such information for decision-making. However, to our knowledge, employers in the United States do not require permission from applicants or employees to review information posted on the Internet.

Future Research

With over 1.4 billion active monthly users of Facebook (Statistic Brain, 2015) and 400 million users of LinkedIn (Undercoverrecruiter, 2015) in 2015, it appears that the use of information derived from SNS in employment decisions is increasing. The vast amount of information contained in SNS and search engines means employers have access to data that can be used to enhance employment decisions because this information would often not be available otherwise or reflected in an applicant's résumé or a job application. However, the use of SNS and search engine data also has many negative consequences including applicants' and employees' perceptions of unfairness and decreased attraction to organizations that use these systems. In order to understand the factors that affect individuals' reactions to these systems, we used organizational justice theory and applied Leventhal's principles to the employment context. The application of this model open many new avenues for research on the use of SNS or search engines. First, future research should examine the degree to which applying the consistency rule affects applicants' or employees' reactions to these systems. As noted previously, many employers do not review information obtained from SNS or search engines for all employees but just a subset of them. Therefore, we maintain that establishing policies which require the consistent use of information obtained from SNS or search engines for all individuals may increase positive reactions to these practices. However, it is not clear how organizations apply this rule when individuals do not have SNS presence.

A second avenue for research is to examine the extent to which employers can make valid inferences about job performance based on SNS or search engine information. Preliminary research in this area by Kluemper and Rosen (2009) has found that employers' inferences about applicant personality predicted job performance, but much more research is needed to establish the validity of SNS or search engine information. Indeed, recruiters and employers are using SNS information for employment decisions without the benefit of evidence on its validity (Davison et al., 2012; Jobvite, 2014; SHRM, 2016). Research is also needed to examine the types of information employers are reviewing in these systems, and the degree to which this information predicts job performance.

A third avenue of research is to examine the different types of SNS and how they influence individuals' reactions to fairness. The purpose of a SNS may affect individuals' reactions to employers accessing it. For example, some SNS are designed for personal use, such as Facebook, and there are those that are designed for professional use, such as LinkedIn, which contain information about education and work experience rather than personal or sensitive information. Thus, when people post information on professional SNS they more likely expect that employers and other professionals will review their work-related information and we expect them to be less likely to view employers' review of these sites as unfair. When individuals post information on personal SNS they are unlikely to expect that employers will review the information because it is designed for friends and family and may be sensitive. Thus, we would expect individuals would be more likely to view employers' review of personal SNS as unfair. This could also be examined in the context of advanced notice. If the individual has time to edit or remove information on their SNS, especially information on their personal SNS, it could possibly increase the perceptions of fairness. Given the relevant information posted on professional SNS, the perceptions of fairness would probably not be affected by an advanced notice.

So far, most of the research on SNS and search engines in relation to employment has been conducted in the United States, and more research is needed in global environments. This is because there appear to be cross-cultural differences in reactions to the collection of employment information, and these cultural differences influence laws, social norms, and consequences for individuals (Black et al., 2015). For instance, there is some evidence that individuals in collectivist countries are less concerned about the privacy of their personal information than those in individualistic countries (Black et al., 2015). The rationale here is that those in collectivist cultures value personal relationships with others and believe that sharing information helps build personal relations (Triandis, 1994). Individualists are more concerned about privacy because they believe that personal data may be used against them (Triandis, 1994). For example, Marshall, Cardon, Norris, Goreva, and D'Souza (2008) found that students from individualistic nations (e.g., the United States) were less likely to disclose information about health, romances, school, or social events on SNS than those from collectivist cultures (e.g., India). Collectivists are also more likely to accept that organizations can intrude on their privacy than individualists (Milberg, Burke, Smith, & Kallman, 1995). In other research, individualistic nations such as Canada, the United Kingdom, Australia, and the United

States had the highest concerns for privacy and collectivistic nations such as Thailand and Japan had the lowest concerns for privacy (Milberg et al., 1995).

Given that privacy and fairness are related constructs (Eddy, Stone, & Stone-Romero, 1999), it may be the case that those from individualistic cultures (e.g., the United States, the United Kingdom, Australia, Canada) are more likely to view the collection of SNS or search engine data for employment as unfair than those from collective cultures (e.g., China, Japan, India, Colombia). However, research is needed to test this prediction. In addition, some countries have recently banned Facebook (e.g., China) and have created their own SNS, but it is not clear if these internal systems are being used to make employment decisions.

Conclusion

Organizations across the world are increasingly gathering information from SNS and search engines, and the data are being used to make employment decisions. This chapter discussed the positive and negative outcomes for employees and organizations of this practice. Advantages include lower recruitment costs, a potentially more robust recruitment field, and a wealth of information available to employers. Conversely, organizations face an uncertain legal and ethical landscapes in how employee and applicant information is obtained and for what purposes the data are used. In addition, analysts have argued that the use of SNS and search engine data are perceived as unfair by applicants and employees, and may negatively affect attraction to and retention in organizations.

We also reviewed the existing literature on the topic, which indicated that the issue of individuals' perceptions of fairness has not been adequately addressed in the literature. We found that many of the conceptual papers on the fairness of using SNS or search engines in employment lacked a theoretical foundation. This prompted us to use organizational justice theory as a framework to understand individuals' reactions to the use of SNS and search engines in employment decision making. As part of this theory, we used the following determinants of procedural justice based on Leventhal (1980): consistency, accuracy, correctability, representativeness, ethicality, and bias suppression. In addition, advanced notice was included.

Although there is widespread use of SNS and search engines in the employment process, there is little evidence that the inferences made from these data are valid. In addition, the legal framework and ethical considerations have not kept pace with the swift changes in technology underpinning these practices. As organizations throughout the world contend with the changing technology, legal and ethical landscapes surrounding the use of SNS and search engines, we hope that this chapter prompts increased research on the acceptance and effectiveness of using these new systems in the employment context.

References

Abril-Sánchez, P., Levin, A., & Del Riego, A. (2012). Blurred boundaries: Social media privacy and the twenty-first-century employee. *American Business Law Journal*, *49*(1), 63–124.

Adams, J. S. (1965). Inequity in social exchange. In L. Berkowitz (Ed.), *Advances in experimental social psychology* (Vol. 2, pp. 267–299). New York, NY: Academic Press.

Alder, G. S., Ambrose, M. L., & Noel, T. W. (2006). The effect of formal advance notice and justification on internet monitoring fairness: Much about nothing? *Journal of Leadership & Organizational Studies*, *13*(1), 93–108.

Americans with Disability Act (1990). *Americans with Disabilities Act*. Retrieved January 15, 2016, from https://www.ada.gov/pubs/ada.htm

Back, M. D., Stopfer, J. M., Vazire, S., Gaddis, S., Schmukle, S. C., Egloff, B., & Gosling, S. D. (2010). Facebook profiles reflect actual personality, not self-idealization. *Psychological Science, 20*(3), 372–374.

Badgett, M. V., Lau, H., Sears, B., & Ho, D. (2007). *Bias in the workplace: Consistent evidence of sexual orientation and gender identity discrimination.* The Williams Institute, UCLA School of Law.

Barclay, J. H., & Harland, L. K. (1995). Peer performance appraisals: The impact of rater competence, rater location, and rating correctability on fairness perceptions. *Group & Organization Management, 20*(1), 39–60.

Barrett-Howard, E., & Tyler, T. R. (1986). Procedural justice as a criterion in allocation decisions. *Journal of Personality and Social Psychology, 50*(2), 296–304.

Bartol, K. M. (1999). Reframing salesforce compensation systems: An agency theory-based performance management perspective. *Journal of Personal Selling & Sales Management, 19*, 1–16.

Bennett-Alexander, D., & Hartman, L. (2015). *Employment law for business* (8th ed). New York, NY: McGraw Hill.

Bielby, W. T., & Baron, J. N. (1986). Men and women at work: Sex segregation and statistical discrimination. *American Journal of Sociology, 91*(4), 759–799.

Bies, R. J. (1987). The predicament of injustice: The management of moral outrage. In L. L. Cummings & B. M. Shaw (Eds.), *Research in organizational behavior, (Vol 9*, pp. 289–319). Greenwich, CT: JAI Press.

Bies, R. J. (1993). Privacy and procedural justice in organizations. *Social Justice Research, 6*(1), 69–86.

Bies, R. J., Martin, C. L., & Brockner, J. (1993). Just laid off, but still a "good citizen?" Only if the process is fair. *Employee Responsibilities and Rights Journal, 6*(3), 227–238.

Black, S. L. & Johnson, A. F. (2012). Employers' use of social networking sites in the selection process. *Journal of Social Media in Society, 1*(1), 7–29.

Black, S. L., Stone, D. L., & Johnson, A. F. (2015). Use of social networking websites on applicants' privacy. *Employee Responsibilities and Rights Journal, 27*(2), 115–159.

Bohnert, D., & Ross, W. H. (2010). The influence of social networking web sites on the evaluation of job candidates. *Cyberpsychology, Behavior, and Social Networking, 13*(3), 341–347.

boyd, d. m., & Ellison, N. B. (2007). Social network sites: Definition, history, and scholarship. *Journal of Computer-Mediated Communication, 13*(1), 210–230.

Brandenburg, C. (2008). Newest way to screen job applicants: A social networker's nightmare. *Federal Communications Law Journal, 60*(3), 597–626.

Brockner, J., Konovsky, M., Cooper-Schneider, R., Folger, R., Martin, C., & Bies, R. J. (1994). Interactive effects of procedural justice and outcome negativity on victims and survivors of job loss. *Academy of Management Journal, 37*(2), 397–409.

Broughton, A., Foley, B., Ledermaier, S., & Cox, A. (2013). *The use of social media in the recruitment process.* Retrieved March 22, 2016, from http://apo.org.au/files/Resource/ACAS_social-media-in-the-recruitment-process_2013.pdf

Brown, V. R., & Vaughn, E. D. (2011). The writing on the (Facebook) wall: The use of social networking sites in hiring decisions. *Journal of Business and Psychology, 26*(2), 219–225.

Clark, L. A., & Roberts, S. J. (2010). Employer's use of social networking sites: A socially irresponsible practice. *Journal of Business Ethics, 95*(4), 507–525.

Colquitt, J. A. (2001). On the dimensionality of organizational justice: A contrast validation of a measure. *Journal of Applied Psychology, 86*(3), 386–411.

Colquitt, J. A., Conlon, D. E., Wesson, M. J., Porter, C. O., & Ng, K. Y. (2001). Justice at the millennium: A meta-analytic review of 25 years of organizational justice research. *Journal of Applied Psychology, 86*(3),425–445.

Colquitt, J. A., Greenberg, J., & Zapata-Phelan, C. P. (2005). What is organizational justice? A historical overview. In J. Greenberg & J. A. Colquitt (Eds.), *Handbook of organizational justice* (pp. 3–56). Mahwah, NJ: Lawrence Erlbaum Associates.

Cropanzano, R., & Konovsky, M. A. (1996). Resolving the justice dilemma by improving the outcomes: The case of employee drug screening. *Journal of Business and Psychology, 11*(2), 239–263.

Daily Mail (2014). *Waitress fired after she complains about a bad tipper on Facebook – not realizing the customer was her Facebook friend.* Retrieved April 10, 2016, from www.dailymail.co.uk/news/article-2678132/Waitress-fired-complains-bad-tipper-Facebook-not-realizing-customer-Facebook-friend.html#ixzz4F46WadCW

Dakss, B. (2004). *4 in 5 credit reports have errors.* Retrieved February 10, 2016, from www.cbsnews.com/news/4-in-5-credit-reports-have-errors/

Data Protection Act (1998). *Data Protection Act.* Retrieved March 15, 2016, from https://www.gov.uk/data-protection|/the-data-protection-act

Davison, H. K., Marasit, C., & Bing, M. N., (2011). Friend or foe? The promise and pitfalls of using networking sites for HR decisions. *Journal of Business and Psychology, 26*(2), 153–159.

Davison, H. K., Maraist, C. C., Bing, M. N., Papinchock, J., Southwell, M., & Tamanini, K. (2009, April). *The accidental résumé: Legal and practical issues of e-screening.* Paper presented at the Annual Meeting of Society for Industrial and Organizational Psychology, New Orleans, LA, USA.

Davison, H. K., Maraist, C. C., Hamilton, R. H., & Bing, M. N. (2012). To screen or not to screen? Using the internet for selection decisions. *Employee Responsibilities and Rights Journal, 24*(1), 1–21.

Dipboye, R. L., & Colella, A. (Eds.) (2013). *Discrimination at work: The psychological and organizational bases.* New York: Psychology Press.

Drouin, M., O'Connor, K. W., Schmidt, G. B., & Miller, D. A. (2015). Facebook fired: Legal perspectives and young adults' opinions on the use of social media in hiring and firing decisions. *Computers in Human Behavior, 46*, 123–128.

Eddy, E. R., Stone, D. L., & Stone-Romero, E. F. (1999). The effects of information management policies on reactions to human resource information systems: An integration of privacy and procedural justice perspectives. *Personnel Psychology, 52*(2), 335–358.

Edosomwan, S., Prakasan, S. K., Kouame, D., Watson, J., & Seymour, T. (2011). The history of social media and its impact on business. *Journal of Applied Management and Entrepreneurship, 16*(3), 79–91.

Emerson, R. (2011). *13 controversial Facebook firing palace guards, doctors teachers, and more.* Retrieved May 1, 2016, from www.huffingtonpost.com/2011/10/17/facebook-firings_n_1003789.html

Employment Rights Act (1996). *Employment Rights Act.* Retrieved April 23, 2016, from www.legislation.gov.uk/ukpga/1996/18/contents

Erker, S., & Buczynski, K. (2009). *Are you failing the interview? 2009 survey of global interviewing practices and perceptions.* Pittsburgh, PA: Development Dimensions International.

Fair Credit Report Act (1970). *Fair Credit Report Act.* Retrieved March 29, 2016, from www.ftc.gov/enforcement/rules/rulemaking-regulatory-reform-proceedings/fair-credit-reporting-act

Folger, R., & Greenberg, J. (1985). Procedural justice: An interpretative analysis of personnel systems. In K. M. Rowland & G. R. Ferris (Eds.), *Research in personnel and human resource management* (Vol. 3, pp. 141–183). Greenwich, CT: JAI Press.

Genova, G. L. (2009). No place to play: Current employee privacy rights in social networking sites. *Business Communications Quarterly, 72*(1), 97–101.

Gilbert, J. (2013). *New research shows people lie on social media to seem more interesting.* Retrieved April 5, 2016, from http://elitedaily.com/news/technology/new-research-shows-people-lie-on-social-media-to-seem-more-interesting/

Gilliland, S. W. (1993). The perceived fairness of selection systems: An organizational justice perspective. *Academy of Management Review, 18*(4), 694–734.

Gilliland, S. W., (1995). Fairness from the applicant's perspective: Reactions to employee selection procedures. *International Journal of Selection and Assessment, 3*(1), 11–18.

Gordon, R. A., & Arvey, R. D. (2004). Age bias in laboratory and field settings: A meta-analytic investigation. *Journal of Applied Social Psychology, 34*(3), 468–492.

Grasz, J. (2009). F*orty-five percent of employers use social networking sites to research job candidates, CareerBuilder survey finds.* Retrieved April 3, 2016, from www.careerbuilder.com/share/aboutus/pressreleasesdetail.aspx?id=pr519&sd=8/19/2009&ed=12/31/2009

Greenberg, J. (1986). Determinants of perceived fairness of performance evaluations. *Journal of Applied Psychology, 71*(2), 340–342.

Greenberg, J. (1987). A taxonomy of organizational justice theories. *Academy of Management Review, 12*(1), 9–22.

Gross, R., & Acquisti, A. (2005). Information revelation and privacy in online social networks. In *Proceedings of the 2005 ACM Workshop on Privacy in the Electronic Society* (pp. 71–80).

Guerin, L. (2016). *Can I be fired for something I posted on my Facebook page?* Retrieved March 5, 2016, from www.employmentlawfirms.com/resources/employment/employee-rights/can-i-be-fired-for-something-i-posted-my-facebook-page

Haefner, R. (2009). *More employers screening candidates via social networking sites*. Retrieved December 10, 2015, from http://jmadden.wikispaces.com/file/view/Employers+Screening+Social+Networking.pdf

Harris, M. M., Lievens, F., & Van Hoye, G. (2004). "I think they discriminated against me": Using prototype theory and organizational justice theory for understanding perceived discrimination in selection and promotion situations. *International Journal of Selection and Assessment*, *12*(1–2), 54–65.

Heneman, H. G. III, Judge, T. A., & Kammeyer-Mueller, J. (2014). *Staffing organizations* (6th ed.). New York, NY: McGraw Hill.

Hertel, G., van der Heijden, B. I. J. M., de Lange, A. H., & Deller, J. (2013). Facilitating age diversity in organizations – part I: Challenging popular misbeliefs. *Journal of Managerial Psychology*, *28*(7/8), 729–740.

Jobvite (2014). *Social recruiting survey*. Retrieved November 28, 2014, from http://web.jobvite.com/FY15_SEM_2014SocialRecruitingSurvey_LP.html?utm_source=bing&utm_medium=cpc&utm_term=social%20recruitment&utm_content=2014socialrecruitingsurvey&utm_campaign=search_bing_jobvitecorporate

Kim, T. Y. & Leung, K. (2007). Forming and reacting to overall fairness: A cross-cultural comparison. *Organizational Behavior and Human Decision Processes, 104*, 83–95.

Kluemper, D. H. (2013). Social network screening: Pitfalls, possibilities, and parallels in employment selection. In T. Bondarouk & M. Olivas-Lujan (Eds.), *Advanced series in management* (Vol. *12*, pp. 1–21) Bingley, UK: Emerald Group Publishing.

Kluemper, D. H., & Rosen, P. A. (2009). Future employment selection methods: Evaluating social networking web sites. *Journal of Managerial Psychology*, *24*(6), 567–580.

Kluemper, D. H., Rosen, P. A., & Mossholder, K. W. (2012). Social networking websites, personality ratings, and the organizational context: More than meets the eye? *Journal of Applied Social Psychology, 42*(5), 1143–1172.

Krawczyk, K. (2014). *Google is easily the most popular search engine, but have you heard who's in second?* Retrieved February 12, 2016, from www.digitaltrends.com/web/google-baidu-are-the-worlds-most-popular-search-engines/#:KgUb9pjPi-FIoA

Kulik, C. T., Perry, E. L., & Bourhis, A. C. (2000). Ironic evaluation processes: Effects of thought suppression on evaluations of older job applicants. *Journal of Organizational Behavior*, *21*(6), 689–711.

Leventhal, G. S. (1980). What should be done with equity theory? In K. J. Gergen, M. S. Greenberg, & R. H. Willis (Eds.), *Social exchange: Advances in theory and research* (pp. 27–55). New York, NY: Plenum.

Lucero, M. A., Allen, R. E., & Elzweig, B. (2013). Managing employee social networking: Evolving views from the national labor relations board. *Employee Responsibilities and Rights Journal*, *25*(3), 143–158.

Madera, J. M. (2012). Using social networking websites as a selection tool: The role of the selection process fairness and job pursuit intentions. *International Journal of Hospitality Management*, *31*, 1276–1282.

Marshall, B. A., Cardon, P. W., Norris, D. T., Goreva, N., & D'Souza, R. (2008). Social networking websites in India and the United States: A cross-national comparison of online privacy and communication. *Issues in Information Systems, IX, 2*, 87–94.

McFarland, L. A., Ployhart, R. E. (in press). Social media in organizations: A theoretical framework to guide research and practice. Journal of Applied Psychology.

McGregor, J., & Gray, L. (2002). Stereotypes and older workers: The New Zealand experience. Social Policy Journal of New Zealand, 18, 163–177.

Mendola, N. (2015). *Hours after signing, Barcelona player loses contract due to Twitter*. Retrieved January 10, 2016, from http://soccer.nbcsports.com/2015/12/29/hours-after-signing-barcelona-player-loses-contract-due-to-twitter/

Milberg, S. J., Burke, S. J., Smith, H. J., & Kallman, E. A. (1995). Values, personal information privacy, and regulatory approaches. *Communications of the ACM, 38*(12), 65–74.

Morris, M., & Leung, K. (2000). Justice for all? Progress in research on cultural variation in the psychology of distributive and procedural justice. *Applied Psychology, 49*(1), 100–132.

Myers, C. (2014). The new water cooler: Implications for practitioners concerning the NLRB's stance on social media and workers' rights. *Public Relations Review, 40,* 547–555.

National Labor Relations Act (1935). *National Labor Relations Act.* Retrieved March 27, 2016, from https://www.nlrb.gov/

Ostrow, A. (2009). *Facebook fired: 8% of US companies have sacked social media miscreants.* Retrieved January 27, 2016, from http://mashable.com/2009/08/10/social-media=misuse/

Parsons, J. (2015). *Facebook's war continues against fake profiles and bots.* Retrieved April 30, 2016, from www.huffingtonpost.com/james-parsons/facebooks-war-continues-against-fake-profiles-and-bots_b_6914282.html

Peacock, L. (2009). *Social networking sites used to check out job applicants.* Retrieved February 22, 2016, from www.personneltoday.com/hr/social-networking-sites-used-to-check-out-job-applicants/

Peluchette, J., & Karl, K. (2010). Examining students' intended image on Facebook: What were they thinking?! *Journal of Education for Business, 85,* 30–37.

Poerio, J. M., & Bain, L. E. (2012). Social media in the workplace: employer protections versus employee privacy. *ABA International Law News, 41*(4).

Preston, J. (2011). *Social media history becomes a new job hurdle.* Retrieved March 17, 2016, from http://www.nytimes.com/2011/07/21/technology/social-media-history-becomes-a-new-job-hurdle.html

Proskauer (2014). *Social media in the workplace around the world 3.0.* Retrieved January 27, 2016, from www.proskauer.com/files/uploads/social-media-in-the-workplace-2014.pdf

Ramaswami, S. N., & Singh, J. (2003). Antecedents and consequences of merit pay fairness for industrial salespeople. *Journal of Marketing, 67*(4), 46–66.

Rasmussen Reports (2012). 33% think employers have the right to fire those who post inappropriate Facebook content. Retrieved December 4, 2015, from www.rasmussenreports.com/public_content/business/jobs_employment/march_2012/33_think_employers_have_right_to_fire_those_who_post_inappropriate_facebook_content

Reznik, M. (2013). Identity theft on social networking sites: Developing issues of internet impersonation. *Touro Law Review, 29,* 455–483.

Ringenbach, K. L., & Jacobs, R. R. (1994, April). *Development of age stereotypes in the workplace scale.* Paper presented at Annual Meeting of Society of Industrial and Organizational Psychology, Nashville, TN, USA.

Rosen, B., & Jerdee, T. H. (1976). The influence of age stereotypes on managerial decisions. *Journal of Applied Psychology, 62*(4), 428–432.

Rosen, B., & Jerdee, T. H. (1977). Too old or not too old. *Harvard Business Review, 55,* 97–106.

Roth, P. L., Bobko, P., Van Iddekinge, C. H., & Thatcher, J. B. (2016). Social media in employee-selection-related decisions: A research agenda for uncharted territory. *Journal of Management, 42*(1), 269–298.

Rousseau, D. M., & Aquino, K. (1993). Fairness and implied contract obligations in job terminations: The role of remedies, social accounts, and the procedural justice. *Human Performance, 6*(2), 135–149.

Sashkin, M., & Williams, R. L. (1990). Does fairness make a difference?. *Organizational Dynamics, 19*(2), 56–71.

Schmidt, G. B. & O'Connor, K. W. (2015). Fired for Facebook: Using NLRB guidance to craft appropriate social media policies. *Business Horizons, 58*(5), 100–112.

Schmitt, N., Gilliland, S. W., Landis, R. S., & Devine, D. (1993). Computer-based testing applied to selection of secretarial applicants. *Personnel Psychology, 46*(1), 149–165.

Sheppard, B. H., & Lewicki, R. J. (1987). Toward general principles of managerial fairness. *Social Justice Research, 1*(2), 161–176.

Shore, L. M., Chung-Herrera, B. G., Dean, M. A., Ehrhart, K. H., Jung, D. I., Randel, A. E., & Singh, G. (2009). Diversity in organizations: Where are we now and where are we going? *Human Resource Management Review, 19*(2), 117–133.

SHRM (2016). *SHRM survey findings – Using social media for talent acquisition – Recruitment and selection.* Retrieved March 22, 2016, from www.shrm.org/research/surveyfindings/pages/social-media-recruiting-screening-2015.aspx

Statistic Brain (2015). Facebook statistics. Retrieved March 1, 2016, from www.statisticbrain.com/facebook-statistics/

Steinpreis, R. E., Anders, K. A., & Ritzke, D. (1999). The impact of gender on the review of the curricula vitae of job applicants and tenure candidates: A national empirical study. *Sex Roles, 41*(7–8), 509–528.

Stone, D. L., & Colella, A. (1996). A model of factors affecting the treatment of disabled individuals in organizations. *Academy of Management Review, 21*(2), 352–401.

Stone, D. L., & Kotch, D. A. (1989). Individuals' attitudes toward organizational drug testing policies and practices. *Journal of Applied Psychology, 74*(3), 518–537.

Stone, E. F., & Stone, D. L. (1990). Privacy in organizations: Theoretical issues, research findings, and protection strategies. In K. M. Rowland & G. R. Ferris, (Eds.), *Research in personnel and human resources management* (pp. 349–411). Greenwich, CT: JAI Press.

Stone, E. F., Stone, D. L., & Dipboye, R. L. (1992). Stigmas in organizations: Race, handicaps, and physical unattractiveness. In K. Kelley (Ed.), *Issues, theory, and research in industrial/organizational psychology* (pp. 42–70). New York: Elsevier Science.

Stone, D. L., & Stone-Romero, E. F. (1998). A multiple stakeholder model of privacy in organizations. In M. Schminke (Ed.), *Managerial ethics: Moral management of people and processes* (pp. 35–59). Mahwah, NJ: Erlbaum.

Stone-Romero, E. F., Stone, D. L., & Hartman, M. (2002, April). *Stereotype of ethnic groups: Own and other ratings.* Paper presented at the Annual Meeting at Society of Industrial and Organizational Psychology, Toronto, Canada.

Stone-Romero, E. F., Stone, D. L., & Hyatt, D. (2003). Personnel selection procedures and invasion of privacy. *Journal of Social Issues, 59*(2), 343–368.

Stoughton, J. W., Thompson, L. F., & Meade, A. W. (2015). Examining applicant reactions to the use of social networking websites in pre-employment screening. *Journal of Business and Psychology, 30*(1), 73–88.

Strano, M. M. (2008). User descriptions and interpretations of self-presentation through Facebook profile images. *Cyberpsychology: Journal of Psychosocial Research on Cyberspace, 2*(2), 1–11.

Swallow, E. (2011). *How recruiters use social networks to screen candidates.* Retrieved April 16, 2015, from http://news.yahoo.com/recruiters-social-networks-screen-candidates-infographic-175214003.html

Triandis, H. C. (1994). *Culture and social behavior.* New York: McGraw-Hill.

Truxillo, D. M., Bauer, T. N., Campion, M. A., & Paronto, M. E., (2002). Selection fairness information and applicant reactions: A longitudinal field study. *Journal of Applied Psychology, 87*(6), 1020 – 1056.

Tsui, A., & O'Reilly, C. A. (1989). Beyond simple demographic effects: the importance of relational demography in superior-subordinate dyads. *Academy of Management Journal, 32*(2), 402–423.

Undercoverrecruiter (2015). *LinkedIn user stats: By networks, groups, profiles, and applications.* Retrieved January 10, 2016, from http://theundercoverrecruiter.com/infographic-linkedin-user-statistics-network-profiles-groups-applications/

Van Iddekinge, C. H., Lanivich, S. E., Roth, P. L., & Junco, E. (2013). Social media for selection? Validity and adverse impact potential of a Facebook-based assessment. *Journal of Management, 39*, 120–134.

Vicknair, J., Elkersh, D., Yancey, K., & Budden, M. C. (2010). The use of social networking websites as a recruiting tool for employers. *American Journal of Business Education, 3*(11), 7–13.

Wagner, K., & Moriarty, L. J. (2002). Perceived fairness of drug testing policies: An application of Leventhal's principles of procedural justice. *American Journal of Criminal Justice, 26*(2), 219–233.

Webopedia.com (2016). *Definition of a search engine.* Retrieved January 22, 2016, from www.webopedia.com/TERM/S/search_engine.html

Wolgemuth, L. (2008). Stupid computer tricks. *U.S. News & World Report, 144*(16), 67

Wolk, D. (2004). Social-networking sites pique the interest of company recruiters. *Workforce Management, 83*(8), 70–73.

13

The Evolution of E-Selection

David N. Dickter, Victor Jockin, and Tanya Delany

Introduction

In the United States and for much of the global workforce, electronic selection technology (e-selection) is an essential aspect of human resource (HR) management across a spectrum of job families and a range of job levels within organizations (Gueutal & Stone, 2005; Stone, Deadrick, Lukaszewski, & Johnson, 2015; Tippins, 2011, 2015). Just as HR evolved during the past century from transactional to strategic functions, e-selection has grown and improved. In the past 15 years, the technology has expanded into an array of interconnected workflows, systems, and workforce analytics (Thite, Kavanaugh, & Johnson, 2012), and e-selection innovations have become available globally (Bartram & Tippins, 2017). However, e-selection remains an under-researched area of applied psychology whose domain of possible applications grows along with the invention and uptake of new advances in computing. That such computerized, digitized, and often Internet-enabled innovations outpace theory development and empirical research is perhaps not surprising, and a variety of publications have been offered in recent years to begin to fill the gaps in the literature (see Ryan & Ployhart, 2014). The importance of this effort cannot be understated: as with traditional "analog" selection systems, the practical utility of a selection process can be compromised when improperly applied in the name of progress, expediency, or convenience (Guion, 2010; Kehoe, Dickter, Russell, & Sacco, 2005).

Much of the available literature on e-selection consists of results-focused publications (Marler & Fisher, 2013; Stone, Lukaszewski, Stone-Romero, & Johnson, 2013) and practice guidelines (e.g., Reynolds, 2011; Reynolds & Rupp, 2010; Tippins & Adler, 2011) but relatively little psychological theory development (Stone & Dulebohn, 2013). Therefore, with the acknowledgment that theory development is among our recommendations for research (as well as the unabashed admission that as authors we are long-time practitioners ourselves), we forge ahead in discussing the accumulated wisdom on these practical issues.

We start by highlighting e-selection problems and solutions discussed in the literature, including deployment, data security, equivalence, and unproctored testing. The next section

The Wiley Blackwell Handbook of the Psychology of the Internet at Work,
First Edition. Edited by Guido Hertel, Dianna L. Stone, Richard D. Johnson, and Jonathan Passmore.

surveys the landscape of e-selection tests and assessments and some of the considerations involved in their development and use. We then describe considerations for implementation, including being mindful of the audience members or stakeholders for e-selection; working with international laws, policies and data privacy standards that apply to all candidates as well as protected groups; delivering tests under unproctored (unsupervised) conditions; and communicating and training. In the last section we offer areas of future research for this rapidly evolving field.

E-selection Problems and Solutions

E-selection involves a range of well-documented challenges, and a variety of fixes and/or coping strategies have been researched and applied. We highlight some of the important problems and solutions in this section and illustrate them further in the section on implementation considerations. Complex and/or large-scale project management may be required to develop, deploy, and maintain selection systems (e.g., Reynolds & Dickter, 2010). We touch upon general implementation issues, as well as the need for data security that underlies all of e-selection. We then turn to the issue of measurement equivalence, applicable when test data are derived from more than one type of response mode (e.g., paper and technology, or different types of technology). Also important are concerns about cheating and security when Internet tests are delivered unproctored, which have been addressed in several ways. Computerized adaptive testing using item response theory is one such strategy, as well as a measurement improvement in its own right.

General deployment considerations

Perhaps one of the more important lessons for industrial and organizational psychologists about creating, deploying and sustaining e-selection programs is the benefit, if not the requirement, of attaining competencies traditionally reserved for information technology experts, such as software and database managers. E-selection managers must familiarize themselves with the techniques and costs of software development, configuration, and integration between different systems, particularly applicant tracking and testing, and possibly others such as performance management. Prior to deployment it will be important to know how tests and data will be accessed, stored, and maintained (Reynolds & Dickter, 2010).

As just one example, many companies opt for a seamless integration between the e-selection platform and their applicant tracking system (if available), allowing administrators to manage the hiring process efficiently using one interface. The system might automatically send invitations to candidates who need to be tested, or make it easier for an administrator to register the candidate for assessment. However, if integration with the applicant tracking system is not possible, then to minimize cycle time and effort, organizations must determine the process and technology needs for separately delivering and scoring assessments, distributing results, and documenting decisions. Furthermore, integration may affect the ability to export data for research and improvements on the assessments. Those data can be rich and varied. For example, Box 13.1 illustrates the variety of data fields that might be included for research in a multinational selection program.

Box 13.1 A sample set of data variables an organization may export from an e-selection system for research and ongoing improvement of the system.

- Candidate first/last name
- Candidate email
- Applicant ID/ requisition ID
- Country applied to
- Candidate's residence country
- Candidate type
- Business unit
- Job code/job role
- Employment type – full time/ part time/ contractor
- Gender/race/ ethnicity/age
- Name of assessment
- Date of assessment
- Numeric score
- Theta score
- Percentile score
- T score
- Z score
- Language of assessment
- Unique test ID
- Final score
- Date of score
- Time to complete
- Subtest raw and final scores
- ID of item 1
- Response to item 1
- Response 1 correct or incorrect
- Complete time, item 1 (seconds)
- Theta after item 1
- Error after item 1
- PWD accommodation
- Score/result active/expired

Notes: Theta score is a score estimate computed using item response theory. PWD, people with disabilities; while it is useful to keep on record that an accommodation was given, organizations are also obligated to keep such personal information secure.

Data security

Here we provide an overview of the subject of security and direct readers to additional sources of information on this important subject, as the very nature of selection requires candidates or employees to sacrifice or place at risk some of their private information (Stone-Romero, Stone, & Hyatt, 2003). Industrial and organizational psychologists have an important role in understanding threats to security and privacy and shaping policies around e-selection processes and data use to safeguard individuals and organizations. Information security has been described in terms of three components: confidentiality, integrity (accuracy and completeness) and availability (see Wong & Thite, 2012, for a HR perspective on these components). Unfortunately, there are numerous possible sources of security threats, including human error, employee damage/sabotage, misuse, theft, computer fraud, viruses and other self-replicating programs, hackers, and natural disasters (Wong & Thite, 2012). Therefore, there is a need for more effective security policies regarding unauthorized access, disclosure, or loss of information (Zafar, 2013). There is also a need to update policies about privacy and data use as new sources begin to be mined (Zafar & Stone, 2015), such as social networking sites. Industrial and organizational psychologists must carefully consider how and what information to place online given the highly publicized hacking scandals that corporations and government organizations have experienced, such as the 22 million records and over 5 million sets of fingerprints stolen from the US Office of Personnel Management (Sanger, 2015). Given that any organization can become a target and that so much of business operations and commerce (including selection data and test information) are already stored online, all organizations and individuals are exposed to the threat of cybercrime. The picture becomes even more complicated with the use of cloud computing and software-as-a-service in HR, where e-selection data may be stored, accessed, distributed, and shared using outsourced firms (Wright, 2011).

Much of this chapter could easily have been devoted to the topic of the latest security methods. However, any detailed recommendations on such protocols are beyond the scope of this chapter and also begin to go out of date prior to publication. Recommended reading includes checklists from the Cybersecurity Unit, United States Department of Justice (2015) for procedures prior to and during a cyber "incident." Items on the preparedness checklist include identifying mission-critical data and using tiered levels of security for protection; 24/365 procedures for critical contact personnel and back-up personnel to follow up on emergencies; procedures for notifying law enforcement and other authorities; and procedures for preserving data related to intrusion to make them ready for forensic analysis. Also relevant are focal articles in the Society for Industrial and Organizational Psychology (SIOP) journal on big data, which include descriptions of security and privacy-preservation techniques such as data partitioning and encryption (see the big data discussion later in this chapter, and, for example, Guzzo, Fink, King, Tonidandel, & Landis, 2015). Industrial and organizational psychology would likely benefit by creating a collection of current online resources on the subject as well.

In addition to data security and the possible liability of the organization collecting data, at stake are applicant reactions to the selection experience, their attraction to and perceptions of the organization, and their perceptions of fairness and justice (Bauer et al., 2006; see also the section on the audience for selection). We discuss these in more detail in the section on implementation considerations for practitioners.

Equivalence considerations

The availability of multiple test administration modes (e.g., paper, computer, mobile device) brings into question whether measurement properties of a test might be different than intended. If a test was created on paper but delivered via computer, or if one candidate takes the test on a computer and another on a mobile device, how do we interpret the scores? Paper–computer equivalence and norming were early considerations in implementation. Mead and Drasgow (1993) found differences for speeded cognitive ability tests taken using paper-and-pencil versus computerized administration (see the section on cognitive ability tests). Highly speeded tests, for example, involve quickly answering short, low-difficulty items (e.g., comparing two strings of text). Test scores on such assessments are generally a function of how many items the candidate was able to complete within the allotted time, rather than of the difficulty of the items. For highly speeded tests, the differences in the required physical actions needed to fill in circles with a pencil and paper ad clicking on boxes or buttons with a mouse controller can become relevant. For example, responding via computer might well be easier and faster, and thus scores of the same assessment may be higher than paper-based for reasons unrelated to actual differences in the abilities the test is targeting. Problems arise if two modes of the assessment (e.g., paper-and-pencil and computerized) are delivered to the same applicant group, or if scores of the assessment delivered with one method are compared with score norms or cut scores that were based on candidates completing the assessment with the other method. In that case a score equating transformation would be necessary (American Psychological Association, 1986; Mead & Drasgow, 1993).

Researchers have examined a variety of other assessment types, from unspeeded ability testing to personality and biodata tests and situational judgment tests (SJTs) (see relevant sections in this chapter). Although there have been some inconsistent results, researchers have largely found score comparability for noncognitive and cognitive unspeeded tests (see review by Tippins, 2015). However, the line of inquiry has shifted to comparing ability score results for tests taken on different technologies that might affect test performance (e.g., screen sizes; Tippins, 2015). Current research focuses on how testing devices and Internet administration conditions may affect test results (e.g., scores and passing rates) in unproctored settings, as we will discuss next. Much work will be required to develop a model or taxonomy of factors that may affect scores, equivalence, and validity of assessments in different modes (Potosky, 2008; Stone et al., 2013).

Unproctored testing for computerized and mobile devices

The ability to test candidates under unsupervised, remote conditions brings the risk of cheating and compromised test security (e.g., Potosky, & Bobko, 2004; Tippins et al., 2006) and has generated much debate (Ryan & Ployhart, 2014; Tippins, 2009). Any lingering uncertainty about whether a test score represents an accurate measure of a candidate's attributes (including the true identity of the test taker) casts a shadow on the assessment's use. This should be less worrisome when a clearly correct answer is not evident (e.g., for personality and biodata assessments) or it is less obvious that the particular test represents a hurdle in the selection process, and more of a concern in a high-stakes situation that would motivate using outside sources or people to find correct answers (e.g., for cognitive ability and knowledge tests). Beaty et al. (2011) conducted a meta-analysis of studies with noncognitive assessments (for example, personality and biodata) in proctored and non-proctored settings that generally found consistent validities for proctored and unproctored tests. Arthur, Glaze, Villado, and Taylor (2010) reported similar scores when examinees took personality tests under high- and low-stakes conditions – as candidates and again as research participants – and scores that were somewhat higher for a highly speeded ability test, but not different enough to discount a practice effect. Karim, Kaminsky, and Behrend (2014) randomly assigned subjects (using Amazon's Mechanical Turk) to take cognitive tests either with or without webcam monitoring, using a within-subjects design (one test had searchable answers and the other did not). They found within-subject score differences suggestive of cheating in the latter group. Thus there is still room for caution (Ryan & Ployhart, 2014).

Despite wariness among many industrial and organizational psychologists, however, unproctored Internet testing, both cognitive and noncognitive, has become common (Burke, 2009; Lievens & Burke, 2011; Ryan & Ployhart, 2014). The international guidelines on computer-based and Internet-delivered testing (International Test Commission, 2006) recommend following unproctored testing with proctored on-site verification tests. It is unclear what percentage of organizations uses this protocol. Interestingly, Lievens and Burke (2011), in a study using verification for cognitive ability, reported little overall decrease in scores between unproctored and proctored conditions. Many practitioners are no longer dissuaded from unproctored testing given the practical benefits, and remote Internet testing appears to be here to stay. Yet more research must be done on the effectiveness of various techniques to improve unproctored Internet assessment and to deter cheating. Other techniques include identification verification using typing forensics, on-site score verification testing, and remote proctoring such as biometric verification, webcams, and screen-sharing (Karim et al., 2014).

While limited research has investigated the impact of smaller, mobile devices for completing assessments, the general trend again is a finding of no performance differences for noncognitive or personality-based assessments (Morelli, Mahan, & Illingworth, 2014; Lawrence, Wasko, Delgado, Kinney, & Wolf, 2013) and some impact for cognitive ability assessments (Doverspike et al., 2012; Impelman, 2013; Sanchez & Branaghan, 2011). Illingworth, Morelli, Scott, and Boyd (2015) examined a large archival database from 2010 of Internet-based, unproctored tests results for candidates using computers (approximately 98% of the sample) or smartphone devices. The tests were noncognitive assessments that included personality and biodata items (see the next section for a discussion of these item types). The researchers found comparable factor structures for the device groups, indicating they were measuring the same constructs, and slightly higher scores among the mobile group that were judged not to be of practical significance. However, Arthur, Doverspike, Muñoz, Taylor, and Carr (2014) analyzed personality and cognitive ability test data for another 3 million applicants taking remote and unproctored tests, including over 69 000 taking the test on mobile devices, and found that ability scores were substantially lower (total score standard mean difference of 0.90) for the mobile test takers. Personality test factor structures and scores were similar.

The literature on Internet testing still lacks clarity about the reasons for effects due to the monitoring (e.g., proctored vs. unproctored) and those due to technology devices (e.g., computer vs. smartphone), as well as environmental influences that might affect these, such as taking an unproctored test on a smartphone in an uncontrolled setting that may have noise or other distractions (Potosky, 2008). There is a need for more theoretical models and research on test modes, environments, and devices. We discuss devices further in the section on implementation considerations.

Computerized adaptive testing and item response theory

Computerized adaptive testing (CAT) leverages technology and item response theory to select items and compute a better estimate of a candidate's true characteristics on a construct, such as ability level, than is available using classical test theory (Embretson, 2000). Candidates receive items tailored to their strength or level on that construct, and need not answer questions that are less appropriate to their level (e.g., too difficult or too easy). This reduces overall item exposure and the required length of the test. One disadvantage is the required development of large item banks, which are expensive and could be vulnerable to cyber theft from hacking. Items that measure a construct at an ability level particularly well can also become over-exposed because they are often selected using the item response theory algorithm. Other item selection algorithms have been developed to reduce exposure, but many items may remain in common for individuals of the same ability level (Thissen & Mislevy, 2000).

CAT has spread from educational and professional licensure and certification testing into candidate e-selection. Industrial and organizational psychologists have reported on its use for ability testing (e.g., McCloy & Gibby, 2011) and personality testing (e.g., Stark, Chernyshenko, Drasgow, & Williams, 2006). Given the mainstream use of unproctored

Internet testing, it would make sense to use CAT for confirmation testing (Kantrowitz, Dawson, & Fetzer, 2011). Makransky and Glas (2011) developed such a method, using adaptive testing to create a shorter confirmation test, and reported success in detecting cheating in a simulation study. There is a lack of research on the success rate of such practices or an account of the extent to which they are in effect in organizations.

The Landscape of Tests and Assessments Used in E-Selection

The range of assessments used as part of employee selection systems has been altered and expanded with the advent of the Internet. In this section, we outline some of the major categories of personnel selection instruments that have been affected and continue to evolve as the result of computer and Internet technology.

Cognitive ability and knowledge testing

Though distinct, these types of measure are treated jointly here only because the manner in which they are assessed is so similar, specifically requiring test takers to identify one or more factually correct responses, typically from a list of provided options. Cognitive ability and knowledge tests are related largely because individual differences in cognitive ability can predict differences in the successful acquisition of domain-specific knowledge (Hunter, 1986). Correspondingly, the implications for computerization of cognitive ability and knowledge tests are relatively similar.

Cognitive ability tests seek to measure basic mental capabilities, such as reasoning, problem solving, verbal ability, and numerical ability (Ones, Dilchert, Viswesvaran, & Salgado, 2010; Salgado, Viswesvaran & Ones, 2005). Cognitive ability measures are generally among the best predictors of job performance, and the predictive value of these measures tends to increase as jobs become more complex, presumably because such jobs require the mastery of more complex principles and the solution of more complicated problems (Schmidt & Hunter, 1998). This has been found even though the test content of such fundamental ability measures often shows only an indirect resemblance to actual job tasks, or no resemblance at all. The term "knowledge test" is used to describe a different class of instruments, those that measure expertise required within specific content domains (Salgado et al., 2005) such as real estate principles or computer programming. These concepts are often relevant to only a specific group of settings or jobs (e.g., questions related to basic banking principles, for use in bank teller selection). Knowledge testing is also used widely in professional licensure (e.g., real estate broker, nursing) or certification testing, and used by some professional organizations to designate or credential those who have obtained a certain level of expertise and who are then preferred for entry into various positions (Browning, Bugbee, & Mullins, 1996).

The advent of computerized testing has created numerous advancements in the domain of cognitive ability and knowledge testing. One such benefit has been the ability to deploy different response modes not possible on paper assessments (e.g., see Silber & Foshay, 2010). For instance, drag-and-drop matching items can allow test takers to use a mouse or other pointing device to move response options into a set of predesignated locations. A test taker might be asked, for example, to match images of certain tools with images showing contexts in which using each tool might be appropriate. Hotspot items is another innovation that has become possible with the computerization of cognitive ability and knowledge assessments. The examinee might be presented with a photograph of a machine and asked to indicate which portion of it serves a particular stated function.

A click on that location of the photograph will be recorded. Answer choices are generally indicated by a marker image that appears at the point of the photo that is clicked. Such items might also require more than one response, requesting that multiple locations of the image be correctly identified.

Among the more important advances in the computerization of cognitive ability and knowledge testing is the introduction of audiovisual content, as this allows for a variety of new item stimuli. For example, a cognitive assessment may readily include spatial movement, as in a videogame experience (e.g., Ventura, Shute, Wright, & Zhao, 2013). For a knowledge test, a candidate may be presented with a video scenario showing a worker perform a particular task, and then be required to identify the next appropriate course of action.

One issue with these new technologies is that responding to test items now requires a different set of manual skills, and these could impact test scores for certain candidates. Drag-and-drop items, for example, require a level of manual dexterity that, while relatively low, exceeds what is required to respond to multiple choice questions. Applicants with disabilities might be disadvantaged by such test content, and may require special accommodations that would not have been necessary with traditional paper assessment. For that reason, some employers have discontinued the use of drag-and-drop matching items. We discuss testing for people with disabilities in more detail in the section on implementation considerations.

Some aspects of test computerization may change the difficulty of an assessment for nearly all test takers. These types of distributional differences are generally not observed for assessments where speed of responding is not critical to test scores. For example, cognitive ability and knowledge tests, where differences in test scores are driven by the inherent difficulty of the items (also called as power tests), rather than the rate at which the test taker is able to respond to them (speeded tests; Mead & Drasgow, 1993; see also Stone et al., 2013).

Personality, interests, and biodata surveys

As with cognitive ability and knowledge measures, we are treating personality, interest, and biodata measures jointly here because the manner in which they are assessed and thus the implications for their computerization are relatively similar. That is, these instruments are self-report survey assessments, typically consisting of questions that are answered using rating scales (e.g., a five-point scale from strongly agree to strongly disagree, or true/false) and sometimes multiple choice response formats.

The domain of personality encompasses behavioral dispositions on which people can vary, such as extraversion and conscientiousness. Though generally somewhat less predictive of job performance than measures of cognitive ability, personality assessments are widely used in personnel selection (Hough & Dilchert, 2010). Personality may be measured broadly, as in a five-factor model (Goldberg, 1993). There may be matching on particular profile scores (e.g., leadership), or alternatively, customized, job-specific instruments may be tailored to measure only the important dimensions of a job (e.g., customer service orientation). Industrial and organizational psychologists may use interest measures, such as scholastic interests or work preference (e.g., machines, data, or people), to assess candidates' potential fit with the required work activities. Biodata assessments consist of questions relating to the job applicant's actual work and life history, such as scholastic achievement, activities in past jobs, or activities and accomplishments outside of work or school (Stokes, Mumford, & Owens, 1994). Like personality measures, the items and dimensions assessed should be tailored to the job in question.

Unlike some of the other employment assessments discussed here, the impact of computerization on survey assessments has been more limited. As discussed above, because survey assessments are generally not timed, mode of responding (computers, mobile devices, paper) is generally not a concern as it is for some cognitive ability tests. One area where computerization has improved survey assessments, however, is the ability to enforce the completion of all items. In a traditional paper-and-pencil test, if an applicant leaves blank a personality question related to interpersonal skills, for example, there would be little grounds to infer anything about the level of interpersonal skills on the basis of that missing response. As practitioners we have observed that personality measures are typically scored by averaging rather than adding item scores for answered items, so that missing values are in effect ignored. Computerized administration, however, has created the opportunity to force completion of all survey items, eliminating the need to decide how missing responses should be scored. For example, job candidates can be presented with a pop-up message if they attempt to advance to a subsequent screen without having completed all items on that page. The result is a general increase in the reliability and construct validity of the resulting survey scales, as well as an increase in the comparability of results across job candidates.

Interviews

Interviews are widely used as part of employee selection for positions at all levels of the organization (Macan, 2009). Unstructured interviews, in which hiring managers or others ask unscripted questions and subjectively evaluate the effectiveness of the responses, are common, though research suggests they are less effective relative to more standardized assessments (Macan, 2009; Schmidt & Hunter, 1998). Structured interviews attempt to create such standardization by scripting questions and formalizing the evaluation of responses. They are generally developed to target particular competencies that have been linked to the job in question, for example via a job analysis. In addition, as part of their development, expert judges create and define the rules by which candidate responses are translated into scores. Structured interviews delve into some of the same territory as biodata assessments, in that they tend to explore particular behavioral examples from the candidate's past that are related to behaviors associated with performance in the job in question.

Technology has enabled construction of elaborate structured interviews. This extends beyond remote video (e.g., messaging services such as Skype, with bidirectional or one-way video) and recording interviews for ratings of performance later. Some newer products also include an ability to create programmed interviews, for which the hiring organization predefines written interview questions that are presented remotely in text format to the job candidate. The candidate then records video answers to each question. Programmed interviews, which do not require a representative from the hiring organization to participate at the time of the interview, can be scheduled or completed at a job applicant's convenience (Milne-Tyte, 2011). Because video responses are recorded, interviews may be stored for future reference or shared with other decision-makers involved in the hiring process. Some of these tools also allow multiple decision-makers to provide ratings on the candidate's responses, which can then be combined across raters and compared with ratings for other job applicants.

Despite these technological advances, it remains unclear whether the Internet interview provides better measurement than a similar interview conducted in person. While much more research is needed on the effectiveness of different types of technology-mediated interview (Stone et al., 2013), research on applicant reactions has been supportive of a variety of methods. For example, Chapman and Rowe (2002) found that candidates participating in technology-mediated interviews (teleconferences) were more attracted to the organization when the interview was structured, and that candidates participating in face-to-face interviews were more attracted to the organization when the interview was unstructured. Bauer, Truxillo, Paronto, Weekley, and Campion's (2004) research on interviewee perceptions of procedural justice provided support for the use of face-to-face, telephone, and interactive voice–response interviews.

Situational judgment tests

SJTs present a set of brief descriptions of work-related situations and respondents must evaluate a set of possible alternative responses to each (Corstjens, Lievens, & Krumm, 2017; Weekley & Ployhart, 2006). Respondents' answers can be scored based on different methods. For example, answers can be compared with those previously collected from a reference group (e.g., subject matter experts, large samples of laypeople, or a combination of these) with increased resemblance leading to higher scores, or answers can be evaluated against a chosen theory or framework (Corstjens et al., 2017). SJTs are not intrinsically measures of a particular candidate characteristic, but are rather a type of measurement method (Schmitt & Chan, 2006). When SJTs are used for pre-employment selection, the situations described are generally work-related and tailored to the selection scenario or target job. SJTs have become more popular as pre-employment assessments because they predict job performance across an array of contexts (McDaniel & Nguyen, 2001) while generally showing lower ethnic group differences than cognitive ability measures (McDaniel, Morgeson, Finnegan, Campion, & Braverman, 2001; Motowidlo, Dunnette, & Carter, 1990). SJT scores also tend to relate positively to job experience and job knowledge (Weekley & Jones, 1999). The fact that they tend to closely model actual job scenarios also generally increases acceptance of the assessment by test takers as related to the job.

By virtue of requiring reading, interpretation, and judgment, SJTs will generally call on some aspects of cognitive ability, and thus tend to correlate positively with cognitive ability measures (McDaniel et al., 2001). McDaniel and Nguyen (2001) and others have found that SJT scores also generally correlate positively with several of the big-five personality dimensions, most notably conscientiousness, emotional stability, and agreeableness. However, there is evidence that SJTs provide some incremental prediction of job performance beyond that provided by traditional psychological measures (Chan & Schmitt, 2002, Clevenger, Pereira, Wiechmann, Schmitt, & Harvey, 2001, Weekley & Jones, 1999).

Prior to the computerization of SJTs, scenarios were generally presented in written format. Job applicants would read a narrative description of scenarios, and then identify the effectiveness of a series of written responses. Computerization has allowed for the presentation of situations using visuals, audio, or videos. These presentation formats allow rendering situations more realistically, further improving on a principal advantages of SJTs, their apparent relationship to the job (i.e., face validity). Some research has also suggested that presenting SJT scenarios as audio or video rather than in written form reduces ethnic group score differences, and create additional improvements in candidate assessments of the relationship of the measure to the job (Chan & Schmitt, 1997; Konradt, Hertel, & Joder, 2003). Video scenarios can also convey important cues such as tone of voice, facial expressions, and body language, which could be relevant in many different workplace scenarios, such as sales or customer service situations.

The equivalence issue discussed earlier becomes relevant in that the effectiveness of the different response options could differ for the same responses when the scenario is presented in written format, where these cues were not available. Because of the additional visual and auditory information about scenarios that is presented when video is used, reference group answers should be collected on the basis of the videos rather than scripts. If the SJT is presented in both written and video format, studies may be undertaken to ensure the video and paper forms have similar score distributions and correlate with each other at or near their reliabilities (i.e., that they are parallel). The current trend, however, is to develop such instruments for computer administration only, thus standardizing the format of the instrument and side-stepping any issues related to the equivalence of paper and computer administrations.

Work simulations

Work simulations attempt to model one or more tasks required to perform a particular job (Fetzer & Tuzinski, 2013). These assessments may vary in the extent to which they simulate relatively general work tasks (e.g., keyboarding test) or assess domain-specific procedural job knowledge (e.g., machine operation or call-center work). Simulations may be relatively brief and focused, measuring only a narrow job task, or may be multifaceted and seek to simulate a broad range of job activities. Assessment centers, which are typically day-long sessions during which job candidates complete multiple assessments, often incorporate work simulations. Examples include inbox exercises, for which job candidates are presented with a set of correspondence and must formulate appropriate responses, and presentation exercises, for which job candidates are provided with a range of documents and background information and must deliver an oral presentation offering solutions and recommended courses of action. Electronic inbox exercises can now incorporate a simulated email interface and audio voicemail and record the job candidate's emailed replies, providing a highly realistic job simulation (McNelly, Ruggeberg, & Hall, 2011). As another example, call-center simulations can also be built to closely model the computer interface used by these workers, and incorporate audio recordings of incoming calls (e.g., Konradt et al., 2003). Candidates use the simulated interface to enter call information, research caller questions, and respond to customer inquiries. Even more advanced tasks, such as air traffic control, can now be more readily simulated because the work is increasingly computer-interfaced.

Perhaps more so than any other type of personnel selection instrument, work simulations have advanced and benefited from the advent of computerized test administration. As a result, their use in personnel selection has increased in recent years (Fetzer & Tuzinski, 2013). One reason that work simulations are so amenable to computerization is that work itself has increasingly become computer-based across a broad range of occupations and industries. For computer-based work, these simulations can therefore be realistic and comprehensive. Historically, basic paper-based psychological tests like visual scanning (comparing text strings for similarity) were used to select for positions involving transcription work. However today, the computerization of clerical work, a broad job category, has made it possible for work simulations to displace these older types of assessment. Another benefit of work simulations is that because they can often be made to closely resemble actual job tasks, they may be more likely to be accepted as fair by job applicants, and less likely to be successfully legally challenged.

Personality assessments or cognitive ability measures, which assess fundamental underlying psychological dimensions, should correlate with measures of job performance to be considered valid for selection into the jobs in question (criterion-related validity; e.g., Hough & Dilchert, 2010; Ones et al., 2010). Work simulations, however, may be considered valid for selection if a careful job analysis has been performed, establishing that the test adequately simulates a job task and that this task is an important part of the job (content validity; e.g., Guion, 1998). A computer-based simulation for a data entry role, for example, can be made to closely model actual data-entry work, with a similar entry interface and similar transcription materials, more fully modeling required job skills.

Game-like assessments

An emerging area in computerized instruments for personnel selection is game-like (or gamified) assessments (Fetzer & Tuzinski, 2013). Currently the term gamification appears to be used rather broadly. It can apply to the incorporation of game-like features, such as rules of play, rewards, or competition, or to the use technology from the entertainment industry that resembles aspects of videogames (e.g., see also O'Leary, Forsman, & Isaacson, 2017; Thielsch & Niesenhaus, 2017, Chapter 5 in this volume). The label gamification can also be applied to "serious games" designed to inform or teach (Michael & Chen, 2005). But the term sometimes simply refers to the addition of animated interactive activities into what are in effect work simulations. For example, Deloitte uses a 20-minute customized, gamified simulation that places candidates in real-life work situations. The aim of the game is to assess an applicant's problem-solving capabilities using videos and tasks from real Deloitte situations (Redrup, 2016). KPMG also uses games to assess candidates in Australia. The approach reportedly has provided a 79% reduction in applications needing to be screened and a 58% reduction in the numbers who get to the final interview (Woodward, 2015). PwC Hungary is also using this concept, administering a serious game for candidates, who are given tasks that assess business acumen, relational skills, and digital competency (Mak, 2015).

How far these kinds of assessments will make inroads into the domain of personnel selection remains to be seen. Outstanding questions include whether these instruments provide any useful incremental prediction of job performance relative to more traditional cognitive ability measures or other tests. Another challenge these assessments face is that familiarity and experience with gaming interfaces varies much more widely among typical samples of job applicants than does experience with more rudimentary aspects of computer use, such as the basic mouse and keyboard skills needed to complete traditional computerized assessments. This differential exposure could create score differences unrelated to differences in job performance. In addition, these assessments can sometimes require specialized development and deployment software. Finally, if gamified assessments do not clearly demonstrate or simulate some aspect of the job (content validity), then like traditional cognitive ability tests, it will be necessary to prove that the assessment actually predicts job performance (criterion-related validity).

Implementation Considerations for Practitioners

Organizations often use Internet-based e-selection systems for their scalability and efficiency relative to traditional assessments (Naglieri et al., 2004). However there are many factors to consider beyond efficiency when transitioning to an e-selection system, including access to tests, reactions of users, security and training (Box 13.2). In this section we highlight considerations for implementation by discussing each of these issues.

Box 13.2 Sample considerations when transitioning to an e-selection system

- Who is the **audience** and how will they interact with the e-selection tool?
- How will the organization manage **security** of assessment content?
- What **devices** will be supported by the e-selection tool?
- Will this be a **proctored** or **unproctored** assessment?
- How will access be ensured for **protected groups**, including **people with disabilities**?
- How will the organization **communicate** and **train** to support the assessment?

Note that with the increased use of technology in selection systems, the rollout process becomes more and more like a software implementation plan. Monitoring in the first days, weeks, and months will include collecting information about:

- Usage: for example, invitations sent, number of candidates who start and complete a test
- Troubleshooting: for example, help desk issues raised, defect rate),
- Data security: for example, content exposure,
- Maintenance: for example, keeping track of scores distributions and watching for score drift,
- User feedback: for example, monitoring social sites for candidate feedback/discussion as well as content exposure,
- New releases: for example, item/test updated, and new validation work.

The audience for e-selection

Researchers have described the importance of attending to the audience or stakeholders of selection systems (e.g., Gilliland & Cherry, 2000), and this is no less important when implementing an e-selection system (Kehoe et al., 2005). Three such stakeholders are the candidate, the organization's leadership (employer), and the administrator.

The candidate There are a number of convincing economic, legal, and psychological reasons for organizations to pay attention to how candidates react to selection systems (Hülsheger & Anderson, 2009). It has also been theorized that negative reactions to the selection process may influence the attitudes, performance, and work behaviors of candidates on the job once they are hired (Gilliland, 1993). To compound matters, negative reactions may persist into the work lifecycle because many organizations use standardized selection procedures for employee promotion (Ford, Truxillo, & Bauer, 2009; McCarthy, Hrabluik, & Jelley, 2009). This has direct implications for organizational productivity, morale, staff turnover/retention, counterproductive behaviors at work, and culture. Furthermore, inappropriate selection procedures may cause candidates to file complaints or take legal action that proves costly (Anderson, 2011). Criticisms of an organization's selection practices, traveling at the speed of social media, can also be damaging.

As part of the decision process, it is important to consider candidates' willingness to take part in e-selection and reactions to the experience (Ryan & Delany, 2010). For example, candidates who view the selection procedure as invasive may choose to withdraw from the selection pool (Macan, Avedon, Paese, & Smith, 1994), particularly when they are highly qualified and have competing offers. As a result, organizations may lose top talent to their competition. Disgruntled candidates also may develop a negative view of the organization and may communicate this perception to individuals in both their professional and social networks (Smither, Reilly, Millsap, Pearlman, & Stoffey, 1993). Social media sites, such as Glassdoor (glassdoor.com/reviews/index.htm), hold growing databases of company reviews including the assessment experience. It has been reported that on average, job seekers use 18 different sources when searching for a job — a steady increase from previous years that reinforces the idea of the consumer candidate (CareerBuilder, 2015). For example, sources may include "offline" interactions with employers and employees, and online networking with colleagues, exploration with search engines, and perusal of job boards and social media sites. This has given candidates a greater voice in shaping the practices a company adopts. Thus, evidence of a negative experience not only may have the detrimental effect of discouraging potentially strong candidates from applying or accepting an offer, but also may have direct implications for organizational image (Murphy, 1986), negatively impacting an employer's brand.

Some research has indicated that applicants react more positively to computerized or multimedia formats than to a paper format (Potosky & Bobko, 2004; Richman-Hirsch et al., 2000; Salgado & Moscoso, 2003). Potosky and Bobko (2004) reported that, on average, participants in their study enjoyed an Internet-based cognitive ability test more than a paper test. However, age was not well represented in this study and therefore the findings may not be inclusive of a wide range of candidates. In spite of possible differences among examinee groups in reactions to e-selection, its prevalence is part of the growing acceptance of technology in the end-to-end recruitment process. CareerBuilder (2014) reported that over 9 million workers conducted a job search using a mobile device in March 2013 alone, an increase from 2 million the prior year. Apart from age, it is also important to be aware of the how different cultures, minority groups and genders will use and react to the tools. In the United States, minority applicants may be more likely to use a mobile device for assessments (Pew Research, 2015; also see the Devices section). If taking a test using a mobile device is more difficult, this would represent a disadvantage to them (see also Stone, Krueger & Takach, 2017, Chapter 19 in this volume; and Burlacu, Truxillo & Bauer, 2017, Chapter 20 in this volume).

The employer The organization represents multiple levels of audiences or stakeholders. Certainly it is important to plan for how to obtain commitment by the organization's leadership to champion and help sustain the e-selection system beyond implementation. An enterprise-wide implementation will result in substantial out-year cost implications (e.g., personnel, licenses/hosting, help desk/desk side support, content updates, ongoing technology maintenance and updates) for which commitment should be obtained at the outset.

The overall organization – its brand and identity – also has a stake in e-selection, and there is a reciprocal relationship between branding and selection success. A clear and defined employer brand can improve application rates and provide an enviable, deep pool of talent. Therefore, branding can differentiate the organization in tight recruitment markets where there is competition for the best people. As the selection process is a prime opportunity to reinforce the employer brand, it is important for e-selection tools to align with the company's brand values. Without this alignment, desirable candidates may get an unintended impression, or leader support may be lost. For example, using gamification in a selection tool might appear to be a good strategy for engaging with potential candidates, but if this is a large deviation from the corporate culture and it does not resonate with senior executives, then the selection solution will not be sold in-house, regardless of the tool's merits in predicting success.

The line manager in the business unit being served by and using the e-selection process is another audience level in the organization. Sikora and Ferris (2014) argue that the organizational culture, climate, and politics surrounding line managers' work will drive the success or failure of HR implementations. For example, the organization's support of new HR practices, collaboration between HR and line management, and line managers' level of accountability in implementing HR tools and policies will all affect the success of an otherwise well-designed program with validated tools. If the organization rewards adoption of innovative programs, for example, then the use of the latest technology in a gamified assessment might drive adoption of the assessment and deliver results.

The administrator Selection system administrators directly drive the success of a program. It is important to be sure that they have ready access to the technology and that they are comfortable with using the type of technology in general (e.g., computers, operating systems, apps). If learning or using the system will increase their workload, then additional resources must be budgeted. Other questions for consideration are:

- How will technology affect time-to-hire?
- If it increases, will the recruitment team balk at using it?
- How will selection results be delivered to support hiring decisions?

International testing and data privacy

Over time countries have taken different approaches to regulating selection tools. For example, the United Kingdom, Norway, and Sweden have worked to set standards and qualifications in test use, and Italy and South Africa have moved to legally requiring that tests be used by psychologists only. As legislation continues to evolve and new precedents emerge, it is important to check and re-check local laws. Data privacy laws are also changing. The International Safe Harbor Privacy Principles have enabled compliant US companies to transfer personal data from the European Union to the United States but these are not applicable for transferring personal data from all countries. Challenges to these principles and new country laws require industrial and organizational psychologists who implement e-selection to keep their legal knowledge current. For instance, in 2014, a new personal data law emerged in Russia that includes a data localization requirement (the "Personal Data Law"), under which organizations using e-selection tools to assess Russian citizens or other individuals residing in Russia are required to process and store the data within the country. Another example, Singapore's Personal Data Protection Act (PDPA) added guidance in 2013 which, among other requirements, gave direction to organizations regarding notification requirements for the collection, use, or disclosure of personal data. This guidance even outlined the use of "cookies" for an Internet user's online activity, distinguishing in part between active consent on one the hand, and "the mere failure of an individual to actively manage his browser settings" on the other hand (see Singapore Personal Data Protection Commission, 2016, p. 53).

International/global solutions It is clear that selection managers deploying a global solution must be aware of local country laws, customs, and culture, and must align policies and processes accordingly. Programs can be tailored to particular countries or regions rather than adopting a "one-size-fits-all" approach. Some of this tailoring is by necessity, of course, as multinational corporations are subject to the legal requirements of the host country (e.g., US citizens protected by US employment laws). Sackett et al. (2010) provided an overview of the legal environment in over 20 countries on a range of issues; among others, these related to discrimination laws and disadvantaged groups, and making/refuting claims of discrimination. While few countries have case law around specific, *a priori* cause for discrimination (such as the four-fifths rule), each country has laws and practices preventing selection discrimination for a range of protected groups, and there is variation between countries that requires attention (Sackett et al., 2010).

Ryan and Tippins (2009) discussed various benefits and challenges of global selection. Those perhaps most relevant to e-selection are:

- identification of individuals who are eligible and effective in many settings and cultures,
- increase of efficiencies in hiring across geographies,
- ability to help establish a consistent global brand.

Challenges include varied levels of technology availability, security, and differences in workforce skills (e.g., varied levels of technology familiarity). Caliguiri and Paul (2010) described the tension in global organizations between centralization of control and responsiveness to local customs and culture. Attempts to standardize processes and systems globally will inevitably meet resistance locally. For example, while US tests might be available online in the United Kingdom, US-centric references or symbols (e.g., the dollar sign) might make them seem insufficiently customized. Similarly, a selection manager at an international organization might be pressured to begin testing rapidly (as soon as the technology can be accessed), prior to collecting proper norms for customer service orientation that vary across cultures. In our experience, candidates in different cultures could respond to tests in different ways, some of which might adversely affect measurement – for example, some candidates in collectivistic cultures might find it more natural than candidates from individualistic cultures to collaborate or share information on tests that are targeted for individual candidates.

Data privacy and security There are varying policies toward data privacy that restrict what candidate and employee information can be collected and maintained – the European Union, in particular, holds US laws insufficiently protective in this regard. However, in any country, overall data security and potential exposure to hacking are strong threats to industrial and organizational psychology and therefore for e-selection programs that contain sensitive personal records. Compliance with the data privacy laws of a particular country is required regardless of where a score or a report is being delivered, if the administrator will be accessing candidate results from a different country. And as described earlier, given the presence of HR functions online and that no organization's security protocols are unbreachable, much more work needs to be done on the best ways to secure and protect assessment records.

Proctored versus unproctored testing

Internet-based tests are typically unproctored and limit an organization's ability to verify the applicant's identity and deter cheating (Reynolds & Weiner, 2009; Ryan & Tippins, 2009). However, it is important to separate adoption of a particular e-selection tool from the mode of delivery, especially proctored versus unproctored. While unproctored Internet testing has its drawbacks, "brain dumps," test exposure, and test distribution happen in proctored environments as well (Drasgow, Nye, Guo, & Tay, 2009). The organization must be prepared to monitor the test regardless of the mode of administration. A large-scale international testing situation we know of at a multinational firm involved moving from paper testing to proctored computer adaptive tests to minimize side-by-side discussions that took place in large testing venues. The same e-selection system was delivered in other countries in an unproctored environment. The decision to use an e-selection system in a proctored or unproctored environment should be taken after considering many inputs from local HR and legal teams, as well as cultural factors. One approach is to group countries into high-risk, medium-risk or lower-risk categories. When possible, high-risk countries should adopt proctored testing environments. The International Test Commission (ITC) (2006) published its guidelines on computer-based and Internet-delivered testing, which provide recommendations for technology design, test security, administration policies and other tools and process that affect assessment quality. Open access, controlled delivery, supervised delivery, and managed delivery are defined and provide practitioners with greater insights into the options for e-selection systems.

Security Regardless of whether a proctored or unproctored delivery method is used, it is important that test content is monitored and evaluated on an ongoing basis to determine if cheating is impacting on test results. Techniques for complementing unproctored testing, such as remote proctoring (e.g., Karim et al., 2014) may be effective but require more research. Although one benefit of Internet-based testing is the quick and broad distribution of tests, a corresponding drawback is the rate at which test content becomes exposed. Overexposure may lead to an unfair advantage for those test takers with advance access to the test content.

Because applicants may use screen captures to distribute test question to other applicants, interrupt the test administration to search for the correct responses, or solicit assistance in completing the assessment, there is a potential for cheating to impact score results and disrupt the testing program on a large scale. Other techniques might be utilized – for example, timed assessments, random order of question delivery, or adaptive algorithms that typically deliver different test content to candidates – to help to minimize or mitigate the impact of cheating. Of course these are not, however, fool-proof. Therefore, just as with a paper-and-pencil assessment, introducing an e-selection tool will require ongoing monitoring and content development, especially as online content may become exposed at a faster rate than some traditional methods.

Devices Due to the broad adoption of technology-based assessment, the variety of computing devices or platforms available will keep increasing. Abasov (2013) wrote about the implications of 1.2 billion people accessing the web from a mobile device. With this proliferation, including smartphones and tablets, candidates are able to access e-selection tools on screen sizes and in locations that may not be conducive to giving their best performance. For example, completing a timed cognitive assessment while traveling on the subway to school might reasonably result in a lower test score than completing it on a larger tablet or laptop in a quiet environment. One important finding that needs more evaluation is that certain groups, including minorities and individuals in lower socio-economic strata, tend to invest more heavily in mobile devices compared with laptops or desktops. While the extent of the digital divide in the United States has decreased (64% of American adults own a smartphone), minority candidates are more likely to depend on their smartphone for Internet access (Pew Research Center, 2015). Therefore, diversity hiring efforts need to take into consideration device usage and its potential impact on hiring decisions.

Protected groups

As in a traditional assessment, e-selection systems must be designed to accommodate protected groups, including racial or ethnic minorities, older applicants, and people with disabilities. See Chapter 19 in this volume (Stone, Krueger, & Takach, 2017) for a detailed discussion of social and diversity issues, and Chapter 20 (Burlacu, Truxillo, & Bauer, 2017) for implications of the aging workforce on the Internet at work. Moreover, the digital divide exists in many societies, with variable access to computers or the Internet (Tippins, 2015), as well as lower computer skills and self-confidence. These differences may negatively affect scores on e-selection tools and result in loss of job opportunities. For example, in the Philippines it is a common practice to provide test centers to avoid candidates incurring prohibitive expenses on their smartphone personal data plans. Without this option, a socioeconomic divide would exist among the applicant pool. Thus, it would behoove employers to adopt practices such as using kiosks and keeping e-selection requirements minimal (through thoughtful design) so that minimal consumer data plans and Internet access can support testing.

Common practices for supporting people with disabilities in online assessments include additional time at the item level or test level, support with visual-assisted technology, and closed captions for hearing-impaired individuals (American Psychological Association, 2016). Not all assessment content can be technology-enabled and support people with disabilities, for example, diagrammatic reasoning or progressive matrix questions are particularly difficult, when technology enabled, to support visual assistive technology due to the presence of graphics or tabular representations. Even some of the new gamified assessments may be difficult to support as they frequently involve the use of colors. Most countries have compliance requirements that need to be adhered to in support of assessing people with disabilities. Resources on this topic include the Americans with Disabilities Section 508 of the Rehabilitation Act 1973, SIOP (2003) principles, the Standards for Educational and Psychological Testing (American Educational Research Association, American Psychological Association, National Council on Measurement in Education, Joint Committee on Standards for Educational, & Psychological Testing, 2014), the British Psychological Society testing guidelines (2016), and the Code of Practice from the Society for Industrial and Organisational Psychology of South Africa (2006). For requirements for a particular country, researchers should consult the American Psychology Association's directory of national associations of psychology (www.apa.org/international/networks/organizations/national-orgs.aspx).

Communication and training

Because of the potential complexities involved in using e-selection, program managers must carefully consider how and when to train the teams that will support the program (e.g., Kehoe et al., 2005; Reynolds & Dickter, 2010). Audiences to consider for training may include administrators, hiring managers, HR generalists, and help desk/desk-side support. Program managers should establish a cadence for receiving communications from selection program staff. Training may include an overview of the test and what it measures, the steps that must be taken to set up a test taker as well as policies (e.g., retesting situations, length of time a test score is valid), administration of the test for people with disabilities, and exceptions for the assessment. If the administrator is responsible for test interpretation, training should include how and when a compensatory strategy or a conjunctive strategy should be applied. The administrator may be the first point of contact for test taker questions. Therefore it is also important to identify and provide answers to common questions that test takers may ask. Most of these questions can be expected to be similar to questions about traditional tests, for example:

- What will my assessments be used for?
- What should I expect when taking the test?
- How long is the test?
- How can I prepare for the assessment?
- Will the results be shared with me?

Since desk-side support staff interact with candidates and represent the company, they will need to understand who these candidates may be (such as professionals, students, and degree of technical competence) and how the test should work. Having the trainees experience the test as a test taker to understand and anticipate user questions can be helpful.

Future Research

The pace of technological advancement has and will continue to change the workplace and HR management. In addition to big data and gamification there are a variety of sophisticated applications on the horizon, such as ubiquitous computing devices and Internet connectivity to measure candidate and employee capabilities and performance, and machine learning and artificial intelligence.

Big data

Frequently described in terms of their enormous volume, their broad of range of information elements, and their rapid speed of collection and accumulation (i.e., volume, variety and velocity; McAfee & Brynjolfsson, 2012), big data are recognized as valuable resources for decision-making in organizations. Guzzo et al. (2015) list a variety of examples of big data analytics in industrial and organizational psychology, such as analyzing Twitter feeds related to job attitudes, studying team interactions based on information from wearable sensors, and investigating employee "microbehaviors" such as pupil dilation. These data extend far beyond the localized datasets traditionally regarded as sufficient for selection and validation. This frontier in research is also characterized by complex, unstructured datasets and formats (e.g., free text, video) and analyzed *in toto* rather than in subsets, by, among others, computing experts and data scientists applying advanced statistical modeling (Ducey et al., 2015). Of concern is that these experts may have little understanding or regard for such psychological concepts as construct validity and theory-based approaches to selection. Nevertheless, it seems certain that big data will find its place in e-selection.

Privacy and ethics in use of the data are rightly of concern. Beyond the capacity for "linkage research" (e.g., efforts to connect and analyze selection with company performance measures within organizations), big data already include a broad range of linked records and may contain not only the data typically found in HR information systems, but also Internet data on individuals who are not aware of the current and potential implications of this tracking and collection. Another example is the use of social media information on prospective or current employees to support selection decisions. Organizations are already using social networking sites when making hiring decisions

(Brown & Vaughn, 2011), and there is evidence that analysis of text from social media can be used to predict personality traits (Park et al., 2015) (see also Lukaszewski and Johnson, 2017, Chapter 12 in this volume).

Guzzo et al. (2015) offer options for ensuring privacy and the proper use of personal information, and discuss challenges to obtaining informed consent. Possible strategies for data privacy protection include anonymization, aggregation, sharding (partitioning into small parts, each containing less of the private information), key coding of personal information, and using third parties for analyses, though each involves risks to security and/or information loss, and even de-identified or anonymized data can be "re-identified" by merging information with other datasets (Guzzo et al., 2015).

Simulations, virtual reality, and gamification

Technology has enabled ever-higher fidelity assessment centers to present test content with animation and voice, or to simulate work and gather candidate data using video and Internet presentation and recording. Virtual reality goggles or some other yet to be invented technology will create realistic, engaging experiences that should increase the likelihood that a candidate will respond more naturally or typically in performing the simulated task. Virtual reality representations also may increasingly correspond to the work environment – for example, virtual teams may routinely use goggles or other interfaces to interact as part of the normal work experience. Industrial and organizational psychologists could lead in understanding not only the implications for e-selection but also the changing nature of work itself.

Computing ubiquity and novel data collection

With the release of each new digital device (phone, tablet, watch, etc.) and software (operating system, app) arrives an opportunity for gathering new types of data (e.g., page views, speed of motion, biometrics). Gartner (2015) forecasts that there will be 25 billion Internet-connected things (from production-line machines to household refrigerators) by 2020 (see also Steil & Maier, 2017, Chapter 18 in this volume). The ubiquity of computing devices will allow for more data-points (e.g., more cameras, more in-room sensors) and new measures (e.g., heart rate, proximity/location within a room during a simulation) whose value in a selection context will need to be researched.

Machine learning and artificial intelligence

In order to process these big, novel and ubiquitous data, we will need to master advanced statistical methods and utilize ever greater computing resources. As noted by Zikopoulos, Eaton, deRoos, Deutsch, and Lapis (2012), 80% of the world's data are unstructured or semistructured. These include sources such as videos, pictures, audio files, free text fields, presentations, word-processing documents, email messages, and click streams. Industrial and organizational psychologists will need to partner with computer scientists, statisticians, data scientists, and others to learn from and apply methods to incorporate unstructured, imperfect data in organizational research (Ducey et al., 2015), using methods that accurately handle the ever-increasing velocity of talent information.

As technology is ever changing, e-selection will evolve also. Work is already underway to reimagine e-selection using machine learning. IBM industrial psychologists and technology researchers have collaborated on the Watson technology platform, which uses natural language processing and machine learning to reveal insights about candidates' personalities, technical/professional skills, and likelihood to fit within the IBM culture. Their goal is to train Watson to provide a consistent, fair, and engaging candidate experience. Anticipated benefits will include a reduction in resources and travel costs associated with in-person interviewing, elimination of unconscious bias, which can influence even the most experienced interviewer, and, most importantly, improvement in matching people to their next career role – a role in which they will be successful.

Research on the potential of the above applications and their impact on organizations and society might seem to always lag behind the rapid pace of technological change. However, having an understanding of how technology can be used to optimize talent management is a core competency for industrial and organizational psychologists and HR managers. Technology innovations represent an opportunity to add value to the measurement of employee capabilities (Reynolds & Rupp, 2010).

Conclusion

This chapter has provided an overview of e-selection problems and solutions, and the corresponding varieties of assessments currently in use. We emphasized that deploying these solutions requires industrial and organizational psychologists to become sufficiently conversant in information technology to take on big challenges such as data security, while being mindful of stakeholders, including the overall organization and its brand. To meet these challenges, a variety of new research areas must be explored. For example, researchers must develop theoretical models and research on test modes, environments and devices. More work is also needed on new constructs that might be assessed using simulations and gamification, as is work on the best methods for engaging the various audiences/stakeholders in all aspects of e-selection. Regarding security, industrial and organizational psychologists need to know more about the effectiveness of identity verification and cheating detection for unproctored Internet testing, including Computerized Adaptive Testing (CAT), remote proctoring, and forensic analysis. E-selection will begin to supply real-time, multi-point data collection using new Internet-connected devices. We need more research and more practical and ethical guidelines for utilization and statistical analysis of big data related to pre-employment selection, as well as methods for securing assessment-related records for datasets big and small.

References

Abasov, M. (2013). 13 stats to convince your boss to invest in mobile in 2013. Retrieved April 6, 2017, from www.mobify.com/blog/13-stats-to-convince-your-boss-to-invest-in-mobile-in-2013/

American Educational Research Association, American Psychological Association, National Council on Measurement in Education, Joint Committee on Standards for Educational, & Psychological Testing (2014). ***Standards for educational and psychological testing***. Washington DC: American Educational Research Association.

American Psychological Association (1986). ***Guidelines for computer-based tests and interpretations.*** Washington DC: American Psychological Association.

American Psychological Association (2016). ***Guidelines for assessment of and intervention with persons with disabilities.*** Retrieved April 6, 2017, from www.apa.org/pi/disability/resources/assessment-disabilities.aspx

Anderson, N. (2011). Perceived job discrimination: Toward a model of applicant propensity to case initiation in selection. Invited Distinguished Scholar Series keynote paper. ***International Journal of Selection and Assessment***, *19*, 229–244.

Arthur, W., Doverspike, D., Muñoz, G. J., Taylor, J. E., & Carr, A. E. (2014). The use of mobile devices in high-stakes remotely delivered assessments and testing. *International Journal of Selection and Assessment, 22*, 113–123.

Arthur, W., Glaze, R. M., Villado, A. J., & Taylor, J. E. (2010). The magnitude and extent of cheating and response distortion effects on unproctored Internet-based tests of cognitive ability and personality. *International Journal of Selection and Assessment, 18*, 1–16.

Bartram, D., & Tippins, N. (2017). The potential of online selection. In H. Goldstein, E. Pulakos, J. Passmore, & C. Semedo (Eds.), *Handbook of the psychology of recruitment, selection and retention* (pp. 271–292). Malden, MA: Wiley-Blackwell.

Bauer, T. N., Truxillo, D. M., Paronto, M. E., Weekley, J. A., & Campion, M. A. (2004). Applicant reactions to different selection technology: Face-to-face, interactive voice response, and computer-assisted telephone screening interviews. *International Journal of Selection and Assessment, 12*, 135–148.

Bauer, T. N., Truxillo, D. M., Tucker, J. S., Weathers, V., Bertolino, M., Erdogan, B., & Campion, M. A., (2006). Selection in the information age: The impact of privacy concerns and computer experience on applicant reactions. *Journal of Management, 32*(5), 601–621.

Beaty J. C., Nye, C. D., Borneman M. J., Kantrowitz, T. M., Drasgow, F., & Grauer, E. (2011). Proctored versus unproctored Internet tests: Are unproctored noncognitive tests as predictive of job performance? *International Journal of Selection and Assessment, 19*, 1–10.

British Psychological Society (2016). *The British Psychological Society Psychological testing centre.* Retrieved April 6, 2017, from https://ptc.bps.org.uk/ptc/guidelines-and-information

Brown, V., & Vaughn, E. (2011). The writing on the (Facebook) wall: The use of social networking sites in hiring decisions. *Journal of Business and Psychology, 10*, 11–21.

Browning, A. H., Bugbee, A. C., & Mullins, M. A. (Eds.) (1996). *Certification: A NOCA handbook.* Washington DC: National Organization for Competency Assurance.

Burke, E. (2009). Preserving the integrity of online testing. *Industrial and Organizational Psychology, 2*, 35–38.

Burlacu, G., Truxillo, D. M. & Bauer, T. N. (2017). Employee age differences in using Internet-based tools at work. In G. Hertel, D. L. Stone, R. D. Johnson, & J. Passmore (Eds.), *The Wiley Blackwell handbook of the psychology of the internet at work* (pp. 449–480). Chichester: John Wiley & Sons, Ltd.

Caliguiri, P., & Paul, K. B. (2010). Selection in multinational organizations. In J. L. Farr & N. T. Tippins (Eds.), *Handbook of employee selection* (pp. 781–799). New York, NY: Routledge.

CareerBuilder (2014). Trends in mobile adoption: An analysis of mobile recruitment efforts across industries. Retrieved April 6, 2017, from www.careerbuildercommunications.com/pdf/trends-in-mobile-adoption.pdf

CareerBuilder (2015). Nationwide study from CareerBuilder reveals six facts every employer should know about the candidate experience. Retrieved April 10, 2017, from www.careerbuilder.com/share/aboutus/pressreleasesdetail.aspx?sd=5%2F21%2F2015&id=pr894&ed=12%2F31%2F2015

Chan, D., & Schmitt, N. (1997). Video-based versus paper-and-pencil method of assessment in situational judgment tests: Subgroup differences in performance and face validity perceptions. *Journal of Applied Psychology, 82*, 143–159.

Chan, D., & Schmitt, N. (2002). Situational judgment and job performance, *Human Performance, 15*, 233–254.

Chapman, D. S., & Rowe, P. M. (2002). The influence of videoconference technology and interview structure on the recruiting function of the employment interview: A field experiment. *International Journal of Selection and Assessment, 10*, 185–197.

Clevenger, J., Pereira, G. M., Wiechmann, D., Schmitt, N., & Harvey, V. S. (2001). Incremental validity of situational judgment tests. *Journal of Applied Psychology, 86*, 410–417.

Corstjens, J., Lievens, F., & Krumm, S. (2017). Situational judgment tests for selection. In H. Goldstein, E. Pulakos, J. Passmore, & C. Semedo (Eds.), *The Wiley Blackwell handbook of the psychology of recruitment, selection and retention* (pp. 226–246). Malden, MA: Wiley-Blackwell.

Cybersecurity Unit, US Department of Justice. (2015). *Best practices for victim response and reporting of cyber incidents.* Retrieved April 6, 2017, from www.justice.gov/sites/default/files/criminal-ccips/legacy/2015/04/30/04272015reporting-cyber-incidents-final.pdf

Doverspike, D., Arthur, W., Taylor, J. E., & Carr, A. E. (2012, April). *Mobile mania: Impact of device type on remotely delivered assessments.* Paper presented at the 27th Annual Conference of the Society for Industrial and Organizational Psychology, San Diego, CA, USA.

Drasgow, F., Nye, C. D., Guo, J., & Tay, L. (2009). Cheating on proctored tests: The other side of the unproctored debate. *Industrial and Organizational Psychology, 2*, 46–48.

Ducey, A. J., Guenole, N., Weiner, S. P., Herleman, H. A., Gibby, R. A., & Delany, T. (2015). I-Os in the vanguard of big data analytics and privacy. *Industrial and Organizational Psychology: Perspectives on Science and Practice, 8*(4), 555–562.

Embretson, S. E. (2000). *Item response theory for psychologists.* Mahwah, NJ: Lawrence Erlbaum.

Fetzer, M., & Tuzinski, K. (Eds.), (2013). *Simulations for personnel selection.* New York, NY: Springer.

Ford, D. K., Truxillo, D. M., & Bauer, T. N. (2009). Rejected but still there: Shifting the focus in applicant reactions to the promotional context. *International Journal of Selection and Assessment, 17*, 402–416.

Gartner (2015). *The internet of things is a revolution waiting to happen.* Retrieved April 6, 2017, from www.gartner.com/smarterwithgartner/the-Internet-of-things-is-a-revolution-waiting-to-happen

Gilliland, S. (1993). The perceived fairness of selection systems: An organizational justice perspective. *Academy of Management Review, 18*(4), 694–734.

Gilliland, S., & Cherry, B. (2000). Customers of selection processes. In J. F. Kehoe (Ed.), *Managing selection strategies in changing organizations* (pp. 158–196). San Francisco, CA: Jossey-Bass.

Goldberg, L. R. (1993). The structure of phenotypic personality traits. *American Psychologist, 48*(1), 26–34.

Gueutal, H. G., & Stone, D. L. (Eds.) (2005). *The brave new world of eHR: Human resources in the digital age.* San Francisco, CA: Jossey-Bass.

Guion, R. M. (1998). *Assessment, measurement, and prediction for personnel decisions.* Mahwah, NJ: Lawrence Erlbaum.

Guion, R. M. (2010). Employee selection: Musings about its past, present, and future. In J. L. Farr & N. T. Tippins (Eds.), *Handbook of employee selection* (pp.943–958). New York: Routledge.

Guzzo, R. A., Fink, A. A., King, E., Tonidandel, S., & Landis, R. S. (2015). Big data recommendations for industrial-organizational psychology. *Industrial and Organizational Psychology: Perspectives on Science and Practice, 8*(4), 491–508.

Hough, L., & Dilchert, S. (2010). Personality: Its measurement and validity for employee selection. In J. L. Farr & N. T. Tippins (Eds.), *Handbook of employee selection* (pp.299–319). New York, NY: Routledge.

Hülsheger, U. R., & Anderson, N. (2009). Applicant perspectives in selection: Going beyond preference reactions. *International Journal of Selection and Assessment, 17*(4), 335–345.

Hunter, J. E. (1986). Cognitive ability, cognitive aptitudes, job knowledge, and job performance. *Journal of Vocational Behavior, 29*(3), 340–362.

Illingworth, A. J., Morelli, N. A., Scott, J. C., & Boyd, S. L. (2015). Internet-based, unproctored assessments on mobile and non-mobile devices: Usage, measurement equivalence, and outcomes. *Journal of Business Psychology, 30*, 325–343.

Impelman, K. (2013, April). *Mobile assessment: Who is doing it and how it impacts selection.* Paper presented at the 28th Annual Conference of the Society for Industrial and Organizational Psychology, Houston, TX, USA.

International Test Commission (2006). Guidelines for computer-based and Internet-delivered testing. *International Journal of Testing, 6*, 143–172.

Kantrowitz, T. M., Dawson, C. R., & Fetzer, M. S. (2011). Computerized adaptive testing (CAT): A faster, smarter, and more secure approach to pre-employment testing. *Journal of Business and Psychology, 26*, 227–232.

Karim, M. N., Kaminsky, S. E, & Behrend, T. S. (2014). Cheating, reactions, and performance in remotely proctored testing: An exploratory experimental study. *Journal of Business and Psychology, 29*, 555–572.

Kehoe, J. F., Dickter, D. N., Russell, D. P., & Sacco, J. M. (2005). e-Selection. In H. Gueutal, D. L. Stone, & E. Salas (Eds.), *The brave new world of eHR: Human resources in the digital age* (pp.54–103). Hoboken, NJ: John Wiley & Sons, Inc.

Konradt, U., Hertel, G., & Joder, K. (2003). Web-based assessment of call center agents: Development and validation of a computerized instrument. *International Journal of Selection and Assessment, 11*, 184–193.

Lawrence, A., Wasko, L., Delgado, K., Kinney, T., & Wolf, D. (2013, April). *Does mobile assessment administration impact psychology measurement?* Paper presented at the 28th Annual Conference of the Society for Industrial and Organizational Psychology, Houston, TX, USA.

Lievens, F., & Burke, E. (2011). Dealing with the threats inherent in unproctored Internet testing of cognitive ability: Results from a large-scale operational test program. *Journal of Occupational and Organizational Psychology, 84*, 817–824.

Lukaszewski, K. M., & Johnson, A. F. (2017). Social networking systems, search engines, and the employment process. In G. Hertel, D. L. Stone, R. D. Johnson, & J. Passmore (Eds.), *The Wiley Blackwell handbook of the psychology of the internet at work* (pp. 231–256). Chichester: John Wiley & Sons, Ltd.

Macan, T. (2009). The employment interview: A review of current studies and directions for future research. *Human Resources Management Review, 19*, 203–218.

Macan, T. H., Avedon, M. J., Paese, M., & Smith, D. E. (1994). The effects of applicants' reactions to cognitive ability tests and an assessment center. *Personnel Psychology, 47*(4), 715–738.

Mak, H. W. (2015). PwC's multipoly boosts employee recruitment and retention. Retrieved April 6, 2017, from www.gamification.co/2015/09/21/pwcs-multipoly-boosts-employee-recruitment-and-retention/

Makransky, G., & Glas, C. A. W. (2011). Unproctored Internet test verification: Using adaptive confirmation testing. *Organizational Research Methods, 1*, 608–630.

Marler, J. H., & Fisher, S. (2013). An evidence-based review of eHRM and strategic human resource management. *Human Resource Management Review, 23*(1), 18–36.

McAfee, A., & Brynjolfsson, E. (2012). Big data: The management revolution. *Harvard Business Review, 60*, 60–66.

McCarthy, J., Hrabluik, C., & Jelley, R. B. (2009). Progression through the ranks: Assessing employee reactions to high-stakes employment testing. *Personnel Psychology, 62*(4), 793–832.

McCloy, R. A., & Gibby, R. E. (2011). Computerized adaptive testing. In N. T. Tippins & S. Adler (Eds.) *Technology-enhanced assessment of talent* (pp. 153–189). San Francisco, CA: John Wiley & Sons, Inc.

McDaniel, M. A., Morgeson, F. P., Finnegan, E. B., Campion, M. A., & Braverman, E. P. (2001). Predicting job performance using situational judgment tests: A clarification of the literature. *Journal of Applied Psychology, 86*, 730–740.

McDaniel, M. A., & Nguyen, N. T. (2001). Situational judgment tests: A review of practice and constructs assessed. *International Journal of Selection and Assessment, 9*, 103–113.

McNelly T., Ruggeberg, B. J., & Hall, C. R. (2011). Web-based management simulations: Technology-enhanced assessment for executive-level selection and development. In N. T. Tippins & S. Adler (Eds.), *Technology-enhanced assessment of talent* (pp. 253–266). San Francisco, CA: John Wiley & Sons, Inc.

Mead, A. D., & Drasgow, F. (1993). Equivalence of computerized and paper-and-pencil cognitive ability tests: A meta-analysis. *Psychological Bulletin, 114*, 449–458.

Michael, D. R., & Chen, S. L. (2005). *Serious games: Games that education, train and inform.* Mason, OH: Course Technology.

Milne-Tyte, A. (2011, September 26). Seeking work? Ready your webcam. *The Wall Street Journal.*

Morelli, N. A., Mahan, R. P., & Illingworth, A. J. (2014). Establishing the measurement equivalence of online selection assessments delivered on mobile versus nonmobile devices. *International Journal of Selection and Assessment, 22*, 124–138.

Motowidlo, S. J., Dunnette, M. D., & Carter, G. W. (1990). An alternative selection procedure: The low fidelity simulation. *Journal of Applied Psychology, 75*, 640–647.

Murphy, K. (1986). When your top choice turns you down: Effects of rejected offers on selection test utility. *Psychological Bulletin, 99*, 133–138.

Naglieri, J. A., Drasgow, F., Schmitt, M., Handler, L., Prifitera, A., Margolis, A., & Velasquez, R. (2004). Psychological testing on the Internet: New problems, old issues. *American Psychologist, 59*, 150–162.

O'Leary, R. S., Forsman, J. W., & Isaacson, J. A. (2017). The role of simulation exercises in selection. In H. Goldstein, E. Pulakos, J. Passmore, & C. Semedo (Eds.), *The Wiley Blackwell handbook of the psychology of recruitment, selection and retention* (pp. 247–270). Malden, MA: Wiley-Blackwell.

Ones, D. S., Dilchert, S., Viswesvaran, C., & Salgado, J. F. (2010). Cognitive abilities. In J. L. Farr & N. T. Tippins (Eds.), *Handbook of employee selection* (pp. 255–275). New York, NY: Routledge.

Park, G., Schwartz, H. A., Eichstaedt, J. C., Kern, M. L., Kosinski, M., Stillwell, D. J., Ungar, L. H., & Seligman, M. E. P. (2015). Automatic Personality Assessment through Social Media Language. *Journal of Personality and Social Psychology, 108*, 934–952.

Pew Research Center (2015). U. S. smartphone use in 2015. Retrieved April 6, 2017, from www.pewInternet.org/2015/04/01/us-smartphone-use-in-2015/

Potosky, D. (2008). A conceptual framework for the role of administration medium in the personnel assessment process. *Academy of Management Review, 33*(3), 629–648.

Potosky, D., & Bobko, P. (2004). Selection testing via the Internet: Practical considerations and exploratory empirical findings. *Personnel Psychology, 57*, 1003–1034.

Redrup, Y. (2016). Deloitte creates customized game for recruiting graduates. Retrieved April 6, 2017, from www.afr.com/technology/deloitte-creates-customised-game-for-recruiting-graduates-20160227-gn5hyb

Rehabilitation Act 1973, as amended by § 508, 29 U. S. C. § 794(d) 3 Id. 4

Reynolds, D. H. (2011). Implementing assessment technologies. In N. T. Tippins & S. Adler (Eds.), *Technology-enhanced assessment of talent* (pp. 66–98). San Francisco, CA: John Wiley & Sons, Inc.

Reynolds, D., & Dickter, D. N. (2010). Technology and employee selection. In J. L. Farr & N. T. Tippins (Eds.), *Handbook of employee selection* (pp. 171–193). New York, NY: Routledge.

Reynolds, D. H., & Rupp, D. E. (2010). Advances in technology-facilitated assessment. In J. C. Scott & D. H. Reynolds (Eds.), *Handbook of workplace assessment: evidence-based practices for selecting and developing organizational talent* (pp. 609–641). San Francisco, CA: Jossey-Bass.

Reynolds, D. H., & Weiner, J. A. (2009). *Online recruiting and selection: Innovations in talent acquisition.* Malden, MA: John Wiley & Sons, Inc.

Richman-Hirsch, W. L., Olson-Buchanan, J. B., & Drasgow, F. (2000). Examining the impact of administration medium on examinee perceptions and attitudes. *Journal of Applied Psychology, 85*, 880–887.

Ryan, A. M., & Delany, T. (2010). Attracting job candidates to organizations. In J. L. Farr & N. T. Tippins (Eds.), *Handbook of employee selection* (pp. 127–150). New York, NY: Routledge.

Ryan, A. M., & Ployhart, R. E. (2014). A century of selection. *Annual Review of Psychology, 65*, 693–717.

Ryan, A. M., & Tippins, N. T. (2009). *Designing and implementing global selection systems.* Malden, MA: Wiley-Blackwell.

Sackett, P. R., Shen, W., Myors, B., Lievens, F., Schollaert, E., Van Hoye, G., ... Aguinis, H. (2010). Perspectives from twenty-two countries on the legal environment for selection. In J. L. Farr & N. T. Tippins (Eds.), *Handbook of employee selection* (pp. 651–676). New York, NY: Routledge.

Salgado, J. F., & Moscoso, S. (2003). Internet-based personality testing: Equivalence of measures and assesses' perceptions and reactions. *International Journal of Selection and Assessment, 11*, 194–205.

Salgado, J. F., Viswesvaran, C., & Ones, D. S. (2005). Predictors used for personnel selection: An overview of constructs, methods, and techniques. In N. Anderson, D. S. Ones, H. K., Sinangil, & C. Viswesvaran (Eds.), *Handbook of industrial, work, and organizational psychology volume 1: personnel psychology* (pp. 165–199). Thousand Oaks, CA: Sage.

Sanchez, C. A., & Branaghan, R. J. (2011). Turning to learn: Screen orientation and reasoning with small devices. *Computers in Human Behavior, 27*(2), 793–797.

Sanger, D. E. (2015, September). Hackers took fingerprints of 5.6 Million U. S. workers, government says. *The New York Times*, September 30, 2015.

Schmidt, F. L., & Hunter, J. E. (1998). The validity and utility of selection methods in personnel psychology: Practical implications and theoretical implications of 85 years of research findings. *Psychological Bulletin, 124*(2), 262–274.

Schmitt, N., & Chan, D. (2006). Situational judgment tests: method or construct? In J. A. Weekley & R. E. Ployhart (Eds.), *Situational judgment tests: Theory, measurement, and application* (pp. 135–155). Mahwah, NJ, US: Lawrence Erlbaum Associates.

Silber, K. H., & Foshay, W. R. (Eds.) (2010). *Handbook of improving performance in the workplace, volume 1: Instructional design and training delivery*. San Francisco, CA: Pfeiffer.

Sikora, D. M., & Ferris, G. R. (2014). Strategic human resource practice implementation: The critical role of line management. *Human Resource Management Review, 24*(3), 271–281.

Singapore Personal Data Protection Commission (2016). Legislation and guidelines. Retrieved April 6, 2017, from https://www.pdpc.gov.sg/legislation-and-guidelines/overview

Smither, J. W., Reilly, R. R., Millsap, R. E., Pearlman, K., & Stoffey, R. W. (1993). Applicant reactions to selection procedures. *Personnel Psychology, 46*(1), 49.

Society for Industrial and Organizational Psychology. (2003). *Principles for the validation and use of personnel selection procedures* (4th ed.). Bowling Green, OH: Society for Industrial and Organizational Psychology.

Society for Industrial and Organisational Psychology of South Africa (2006). *Code of practice for psychological and similar assessment in the workplace*. Garsfontein, South Africa: Society for Industrial and Organisational Psychology of South Africa.

Stark, S., Chernyshenko, O. S., Drasgow, F., & Williams, B. A. (2006). Examining assumptions about item responding in personality assessment: Should ideal point methods be considered for scale development and scoring? *Journal of Applied Psychology, 91*, 25–39.

Steil, J. J., & Maier, G. W. (2017). Robots in the digitalized workplace. In G. Hertel, D. L. Stone, R. D. Johnson, & J. Passmore (Eds.), *The Wiley Blackwell handbook of the psychology of the internet at work* (pp. 403–422). Chichester: John Wiley & Sons, Ltd.

Stokes, G. S., Mumford, M. D., & Owens, W. A. (Eds.) (1994). *Biodata handbook: Theory, research, and use of biographical information in selection and performance prediction*. Palo Alto, CA: CPP Books.

Stone, D. L., Deadrick, D. L., Lukaszewski, K. M., & Johnson, R. (2015). The influence of technology on the future of human resource management. *Human Resource Management Review, 25*(2), 216–231.

Stone, D. L., & Dulebohn, J. H. (2013). Emerging issues in theory and research on electronic human resource management (eHRM). *Human Resource Management Review, 23*(1), 1–5.

Stone, D. L., Krueger, D. & Takach, S. (2017). Social issues associated with the internet at work. In G. Hertel, D. L. Stone, R. D. Johnson, & J. Passmore (Eds.), *The Wiley Blackwell handbook of the psychology of the internet at work* (pp. 449–480). Chichester: John Wiley & Sons, Ltd.

Stone, D. L., Lukaszewski, K. M., Stone-Romero, E. F., & Johnson, T. L. (2013). Factors affecting the effectiveness and acceptance of electronic selection systems. *Human Resource Management Review, 23*(1), 50–70.

Stone-Romero, E. F., Stone, D., & Hyatt, D. (2003). Personnel selection procedures and invasion of privacy. *Journal of Social Issues, 59 (2)*, 343–368.

Thielsch, M. T. & Niesenhaus, J. (2017). User experience, gamification, and performance. In G. Hertel, D. L. Stone, R. D. Johnson, & J. Passmore (Eds.), *The Wiley Blackwell handbook of the psychology of the internet at work* (pp. 79–102). Chichester: John Wiley & Sons, Ltd.

Thissen, D., & Mislevy, R. J. (2000). Testing algorithms. In H. Wainer (Ed.), *Computerized adaptive testing: A primer*. Mahwah, NJ: Lawrence Erlbaum Associates.

Thite, M., Kavanaugh, M. J., & Johnson, R. D. (2012). Evolution of human resource management and human resources information systems: The role of information technology. In M. J. Kavanaugh, M. Thite, & R. D. Johnson (Eds.), *Human resource information systems* (pp. 2–34). Los Angeles, CA: Sage.

Tippins, N. T. (2009). Internet alternatives to traditional proctored testing: Where are we now? *Industrial and Organizational Psychology, 2*, 2–10.

Tippins, N. T. (2011). Overview of technology-enhanced assessments. In N. Tippins & S. Adler (Eds.), *Technology-enhanced assessment of talent* (pp. 1–18). San Francisco, CA: John Wiley & Sons, Inc.

Tippins, N. T. (2015). Technology and assessment in selection. *Annual Review of Organizational Behavior, 2*, 551–582.

Tippins, N. T., & Adler, S. (Eds.) (2011). *Technology-enhanced assessment of talent*. San Francisco, CA: John Wiley & Sons, Inc.

Tippins, N. T., Beaty, J., Drasgow, F., Gibson, W. M., Pearlman, K., Segall, D. O., & Shepherd, W. (2006). Unproctored Internet testing in employment settings. *Personnel Psychology, 59*, 189–225.

Ventura, M., Shute, V., Wright, T., & Zhao, W. (2013). An investigation of the validity of the virtual spatial navigation assessment. *Frontiers in Psychology, 4*, 1–7.

Weekley, J. A., & Jones, C. (1999). Further studies of situational tests. *Personnel Psychology, 52*, 679–700.

Weekley, J. A., & Ployhart, R. E. (Eds.) (2006). *Situational judgment tests: Theory, measurement, and application*. Mahwah: Lawrence Erlbaum Associates.

Wong, Y. K., & Thite, M. (2012). Information security and privacy in HRIS. In M. J. Kavanagh, M. Thite, & R. D. Johnson (Eds.), *Human resource information systems: Basics, applications and directions* (pp. 518–535). Thousand Oaks, CA: Sage.

Woodward, A. (2015). Game on: Using gamification for recruitment and engagement. Retrieved April 6, 2017, from www.hrmonline.com.au/section/featured/game-on-using-gamification-for-recruitment-and-engagement/

Wright, A. D. (2011). Cloud computing and security: How safe is HR data in the cloud? Retrieved April 6, 2017, from https://www.shrm.org/hrdisciplines/technology/articles/pages/cloudsecurity.aspx

Zafar, H. (2013). Human resource information systems: Information security concerns for organizations. *Human Resource Management Review, 23*(1), 105–113.

Zafar, H., & Stone, D. L. (2015). HRIS security and privacy. In M. Kavanagh, M. Thite, & R. D. Johnson (Eds.), *Human resource information systems: Basics, applications, and future directions* (3rd ed). Thousand Oaks, CA: Sage.

Zikopoulos, P. C., Eaton, C., deRoos, D., Deutsch, T., & Lapis, G. (2012). *Understanding big data: Analytics for enterprise class Hadoop and streaming data*. New York: McGraw-Hill.

14

E-Leadership

Surinder Kahai, Bruce J. Avolio, and John J. Sosik

Introduction

With the spread of information technology (IT), how we interact with, perceive, and influence others in organizations is changing. At the same time, IT is changing the nature and performance of work, including how work is completed and whether humans or IT complete it. Who has access to information, how information is acquired, stored, interpreted, and disseminated are also changing. Consequently, how leadership is both transmitted and received, and the context that embeds leadership are also not remaining the same (Avolio & Kahai, 2003; Avolio, Sosik, Kahai, & Baker, 2014; Kahai, 2012). We know that IT can be significantly shaped by the context and culture in which it is used, defined in part by the leadership within it (Avolio et al., 2014). What has been referred to as e-leadership focuses on this dynamic between IT and leadership. Specifically, it focuses on how IT changes leadership, how leadership changes IT, and how leadership must be practiced to adapt to the new socio-technological context. Accordingly, in line with this focus, Avolio et al. (2014) define e leadership as "a social influence process embedded in both proximal and distal contexts mediated by AIT [Advanced Information Technology] that can produce a change in attitudes, feelings, thinking, behavior, and performance" (p. 107).

This chapter discusses the various themes pertaining to e-leadership in the literature. It begins with a discussion of whether IT changes leadership and proceeds to discuss leadership in three IT-mediated contexts: same-place IT-supported meetings, virtual teams, and the age of social media. Subsequently, it discusses how leadership affects IT by influencing its adoption and assimilation and ends with topics for future research and conclusions.

Does Information Technology Change Leadership?

While it is clear that IT proliferation has changed the context in which leaders operate, it is not clear what those changes mean for leadership. For some scholars, the meaning of leadership has not changed. For instance, Sutton (2010) argues that like preindustrial leaders, leaders operating in the new IT dominated context still have to be competent,

---

The Wiley Blackwell Handbook of the Psychology of the Internet at Work,
First Edition. Edited by Guido Hertel, Dianna L. Stone, Richard D. Johnson, and Jonathan Passmore.

caring, and benevolent. In a similar vein, Champy (2010) argues that IT does not change the requirement that leaders relate to and personally engage others. For another set of scholars, the mediation of leadership by IT implies muting of certain leadership cues and behaviors. For example, Purvanova and Bono (2009) and Hoch and Kozlowski (2014) argue that when teams conduct their work via technology instead of meeting face to face, transformational leadership and leader–member exchange can suffer, because these rely on the transmission of nonverbal cues that communicate emotions. In the absence of visual and auditory cues, relevant emotional communication gets attenuated.

Den Hartog, Keegan, and Verburg (2007) refer to another characteristic of virtual work, or work that is made possible when technology brings together a set of dispersed individuals who would not connect otherwise, to question whether leadership behaviors would have any effect in virtual work settings. Specifically, they highlight the temporary, ambiguous, and frequently changing reporting relationships that characterize such work and ask whether leadership has any effect in such settings. Kahai, Sosik, and Avolio (2004) question whether participative leadership would be able to influence participation when a team uses a group support system (GSS). They argue that such a technology may substitute for a participative leader by providing opportunities for greater participation; it can also neutralize a directive leader by conflicting with the leader's style.

Avolio and Kahai (2003) present a different perspective. They suggest that advanced IT alters leadership systems in organizations by changing how information, an important basis of power, is acquired, stored, interpreted, and disseminated. By changing how people are influenced and how decisions are made, the authors argue that IT is changing expectations for effective leadership. For instance, stakeholders such as employees, customers, and investors, are much more powerful today due to unprecedented and easy access to information, media, and the ability to form influential collectives. The prominence of several moving parts is creating an unprecedented level of complexity and uncertainty for leaders and changing the meaning of leadership (Dinh et al., 2014). Effective leadership is likely to involve collaborative team processes and bottom-up follower-based processes in addition to hierarchical or top-down influences (Dinh et al., 2014). Only those leaders who are able to understand and harness these changes are likely to be more effective.

The work of some researchers provides an indication of what might constitute effective leadership in the new context dominated by IT. For instance, Wen, Yang, and Rosé (2015) demonstrate how the role of educators is changing. With the advent of massive open online courses or MOOCs, forms of social learning (Bandura, 1997) are emerging as ways to make online learning more engaging. Social learning is based on active and peer learning through team-based exercises, peer evaluation and feedback, and visible student work. Virtual teams are formed to enable such learning, and Wen et al. (2015) found that leadership behaviors in the form of team building, initiating structure, and collaboration are critical for effective team performance. While an educator still has to engage in top-down transmission of certain knowledge, a significant part of the educator's role is to set up conditions or processes that facilitate team building, initiation of structure, and collaboration.

In a similar vein, Mohammed and Nadkarni (2011) suggest a greater need for temporal leadership in virtual teams. Virtual teams, which are a result of increasing reliance on information technology for conducting work, often consist of individuals from different cultures (Maynard, Gilson, Young, Vartiainen, & Hakonen, 2017, Chapter 15 in this volume). Individuals from different cultures tend to have different visions of time (Saunders, Van Slyke, & Vogel, 2004). Those from Western cultures, such as Americans, for instance, possess a clock vision of time and see time as a scarce commodity that must be conserved. Those belonging to some of the Eastern cultures, such as Hindu or Buddhist cultures, on the other hand, adopt a timeless vision of time. They view the world as engaged in simultaneous destruction and creation. In such a view, the passage of time is not important because whatever goes, comes back again. Whatever is happening today will happen again in the future. Such and other time-related differences among team members from different cultures are likely to introduce temporal diversity within the team (Brislin & Kim, 2003). Mohammed and Nadkarni view temporal diversity as helping a team perform better provided it is managed appropriately. Their findings suggest that one way in which temporal diversity could be managed is via temporal leadership or leadership that focuses on scheduling deadlines, synchronizing team member behaviors, and allocating temporal resources (Mohammed & Nadkarni, 2011). Temporal leadership is becoming increasingly important due to the constant contact that comes with working in today's highly virtual and global organizations (Stokols, Mishra, Gould-Runnerstrom, & Hipp, 2009).

Some scholars highlight how increasing digitization might lead to new possibilities for leadership effectiveness. For example, Reeves, Malone, and Driscoll (2008) suggest that the technology employed by a virtual team can help someone emerge as a leader or become more effective as one, thereby challenging the idea that leadership is either inborn or developed via training. The authors studied leadership in virtual worlds, which are graphical three-dimensional (3D) simulations of physical spaces with inhabitants and found that technology may help a leader by keeping track of individual performance, implementing a nonmonetary reward system, and making team member capabilities, performance, and compensation transparent. The technology makes it possible for the leader to give rewards during or immediately after performance, thereby creating a perception of fairness. The leader is able to create or modify a strategy, select most suitable team members for a task, and decide their roles in real-time, because the technology provides information about the performance and capabilities of the team and its members on a continuous basis.

Reconciling different views of whether IT changes leadership

While the viewpoints described above may appear to somewhat contradictory, they are nevertheless valid and in need of reconciliation. One can view them as highlighting different but combinable aspects of leadership. Yammarino, Dansereau, and Kennedy's (2001) integrative model of leadership can help integrate the different accounts of how leadership is affected by increasing reliance on IT. According to this model (see Figure 14.1), leadership scholars have generally focused on five areas: fundamental human processes, leadership core processes, leadership outcomes, second-level leadership outcomes, and substitutes for leadership. Fundamental human processes refer to the psychological and related processes without which leadership would not be possible and they cover cognitive and emotional processes, attraction between individuals, communication, and the norms, values, and culture of collectives. Leadership core processes are the behaviors or qualities displayed to exercise leadership and they consist of charisma, transformational leadership, empowerment, providing task and relationship functions to groups, and supervision and management. Leadership outcomes cover the tactics (which is why we hereafter refer to this area as leadership tactics) of putting together leadership core processes. These tactics include team building, delegation, and participation in decision-making. The second-level leadership outcomes refer to the immediate outcomes of leadership core processes. These outcomes include performance, satisfaction, absenteeism, engagement of followers, and leader–follower relationships. Substitutes for leadership refer to leadership enhancers, neutralizers, and replacements.

Scholars who argue that leadership does not change in a digital world tend to focus on a leader's behaviors or qualities and second-level leadership outcomes in Yammarino et al.'s (2001) model. They are accurate in that leaders still need to display behaviors or qualities such as benevolence, competence, and caring (Sutton, 2010) and seek outcomes such as high level of follower engagement and high quality leader–follower relationships

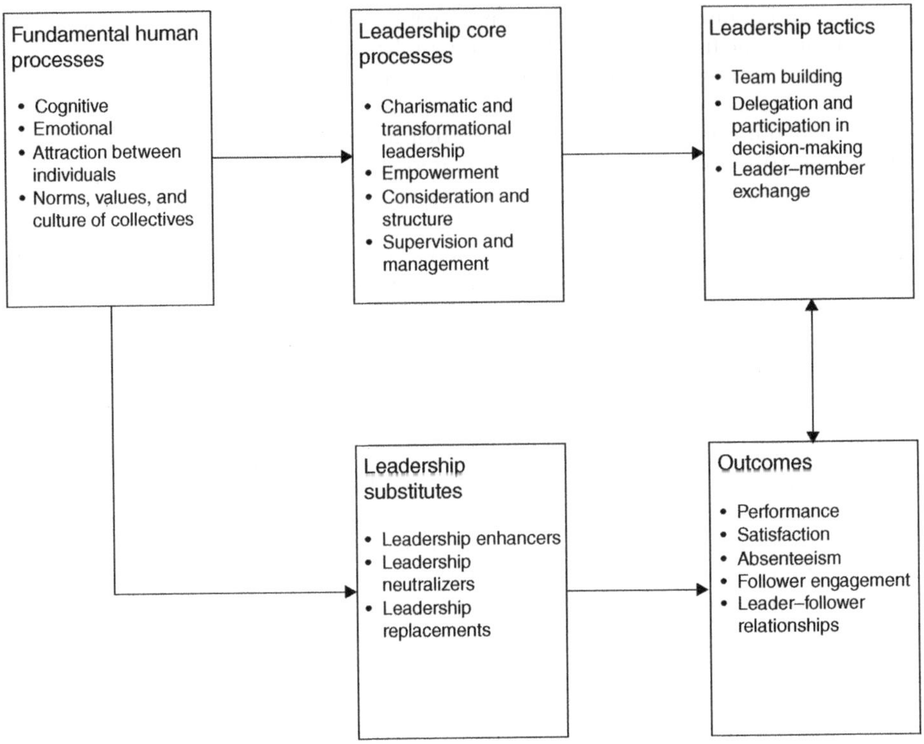

Figure 14.1 An integrative model of leadership. Adapted from Yammarino et al. (2001) with permission of American Psychological Association.

in a digital world (Champy, 2010) but they ignore other areas within Yammarino et al.'s (2001) model. They fail to acknowledge that when leadership is mediated by IT, the fundamental human processes (e.g., transmission of nonverbal cues) that enable leaders to display certain qualities and behaviors and relate to their followers are altered. They do not realize that leaders would need to make certain changes in order to overcome challenges and benefit from opportunities offered by new ways of connecting and accomplishing work in an IT-dominated context. Specifically, certain leadership behaviors or qualities (e.g., temporal leadership) and the tactics for displaying them may become even more critical and, therefore, would need to be scaled up, while others may need to be toned down (e.g., top-down leadership). Those who believe that leadership does not change in a digital world also miss the point that the locus of leadership may be shifting from individuals to emergent and dynamic interactions among various stakeholders (Uhl-Bien, Marion, & McKelvey, 2007) in a world that is being made increasingly complex by IT.

Scholars who argue that leadership and its development changes in an IT-dominated context are generally focusing on how IT is altering the fundamental human processes that underlie leadership. For them, IT is not an inert communication medium. Instead, it alters human cognition, emotions and communication as well as the norms, values, and culture of collectives and is, in turn, shaped by them (Avolio, Kahai, & Dodge, 2001). Scholars who question the relevance of leadership in an IT-dominated world are generally focusing on leadership neutralizers and substitutes. They believe that fundamental human processes are changing in a digital world and question whether leadership core processes, such as supervision, charisma, or empowerment, are likely to have any impact when leader–follower interactions are mediated by technology, which can potentially neutralize or substitute for leader behaviors. Scholars who point to the possibility that IT can help in the emergence or strengthening of leaders are focusing on IT as an enhancer of leadership.

Thus, IT is leading to changes in several aspects of leadership. The requirement for leader benevolence, competence, and caring is intact as is the need for leaders to seek outcomes such as high levels of follower engagement and high-quality leader–follower relationships. But the processes that the leaders need to engage in so that they are seen as displaying appropriate behaviors and qualities and are able to optimize outcomes are changing because IT is creating a new context for leadership. Technology is creating a smaller world and a leader is now able to reach and impact an international audience at the touch of a button. Leadership in such a context cannot remain the same as before. Beyond enabling a widespread, international audience with great efficiency, IT has not made it easier for a leader to be effective. An effective leader now needs to understand how national cultures may interact with technology to make a difference for leadership.

Leadership in Same-Place IT-Supported Meetings

The evolution of IT-supported work by groups or collectives has followed how IT networks have evolved over time (Kahai, 2012). Local area networks from the late 1980s and early 1990s, which spanned a limited physical distance (e.g., a room or a building floor), gave way to the larger Internet that spanned the whole world. The Internet also evolved from a network that largely maintained the social structure, with powerful entities such as businesses providing much of the content that we saw on the Internet, to one that enabled everyone to contribute and participate actively via tools such as blogs, wikis, media sharing sites, and social networking sites. Local area networks enabled the development of GSS or software to support same-place meetings and decision-making. The availability and connectivity of the Internet allowed organizations to conduct their work by tapping into human resources beyond their immediate confines and create virtual teams. When physical, social, and power distances reduced on the Internet during initial years of the twenty-first century, organizations turned their attention to online communities, which are allowing external stakeholders such as customers and others who share an interest with an organization to influence it and wield power in unprecedented ways. This section and the following two discuss leadership in these three situations.

GSS were designed to make same-place meetings and decision-making more effective. Meetings tend to be ineffective due to lack of structure, dominance by a few, inhibitions to participate, and lack of minutes. Decisions can be ineffective when groups place more emphasis on cohesiveness than decision quality, or are limited by inadequate information from a small number of sources. GSS can help overcome these issues by enabling anonymity, structure, parallel input, and a record of the interaction (Nunamaker, Dennis, Valacich, Vogel, & George, 1991). Studies on leadership in same-place IT-supported meetings have largely been laboratory studies and they can be divided into: studies that did not manipulate leadership; studies that manipulated leadership; and studies of GSS facilitation.

Studies that did not manipulate leadership behavior

These studies employed assigned or elected leaders without systematically manipulating leader behaviors. Though these studies did not manipulate leadership behavior, they made important contributions. First, they showed that leaders can influence others in computer-mediated groups. Harmon, Schneer, and Hoffman (1995) compared face-to-face and audio-conferencing groups and reported status differentiation and leader influence as remaining the same across both conditions. Lim, Raman, and Wei (1994) found that in leaderless groups, the GSS tended to promote equality of influence attempts, but in groups with an elected leader, the GSS did not stop the leader from exercising a greater degree of influence than others. In another paper (Lim, Raman, & Wei, 1990), the authors attribute this finding to the Singaporean culture within which this study was conducted. The Singaporean culture is characterized by high power distance and it recognizes and respects hierarchy; group members were willing to accept dominance by the elected leader despite the use of GSS which tends to discourage dominance. However, since the authors did not conduct a comparative study in which they contrasted leadership influence across cultures with different power distances, it is not clear whether the Singaporean culture accounted for leader dominance in GSS groups.

Indeed, George, Easton, Nunamaker, and Northcraft (1990) reported a similar finding with US groups: unequal participation was more likely in GSS groups with a leader and in leaderless, non-GSS groups. This finding lends further support to the idea that while GSS may equalize participation and influence attempts, it is likely to do so in leaderless groups; in groups with leaders, the leader can participate more and make greater attempts to influence the group's thinking and process. George et al. (1990) also reported anonymous GSS groups with leaders and non-anonymous leaderless GSS groups as being most satisfied. The authors argue that the direction provided by the leader may have made group members happier in the anonymous groups but it may have come in the way of the process in non-anonymous groups in which members knew who made which comment. Hiltz and colleagues reported positive effects of having a leader in GSS groups, apparently due to the direction or structure a leader may offer. They found that the presence of a leader was associated with higher decision quality in GSS groups (Hiltz, Johnson, & Turoff, 1991). Kim, Hiltz, and Turoff (2002) found a similar association; they also found that the presence of a leader was associated with greater levels of decision quality, communication effectiveness, and process satisfaction. The presence of a leader also helped reduce the effect of restrictive coordination structures on process satisfaction.

The presence of a leader in computer-mediated interaction can also be unfavorable. Barkhi, Jacob, Pipino, and Pirkul (1998) found that the presence of a leader among group members having different motives and working on a negotiation task lowered members' willingness to disclose information and increased their frustration with the process. Because of the mixed-motive nature of their task, it appears that group members viewed their leader as more likely to decide in a way that is harmful for them. These findings about a leader being able to influence the process and outcomes in a GSS setting are important because they suggest that the GSS, which is designed to create a level playing field and in, some cases, provide a structure to group tasks, does not substitute for a leader and render a leader's efforts as useless (Kahai et al., 2004).

Another contribution of studies that did not manipulate leadership comes from the finding that presence of a leader can interact with GSS features to influence group process and outcomes. Hiltz et al. (1991) found that assigned leadership reduced the level of agreement in groups when the statistical feedback feature of the GSS was turned on. Statistical feedback feature, it appears, may have served as a surrogate leader and conflicted with the designated leader by suggesting a course of action. Kim et al. (2002) reported that the nature of coordination supported by the GSS interacted with leader presence; sequential coordination groups reported greater process satisfaction in the presence of a leader. While the above studies make important contributions, they fail to provide practical recommendations and theoretical explanations for specific behaviors that leaders should display when leading in a GSS setting. Behavior reflects one mode of leadership transmission within social systems (Avolio et al., 2014).

Studies that manipulated leadership behavior

In studies manipulating leadership behavior, researchers trained confederates or a group member to display a certain leadership style (participative, directive, instrumental, transformational, or transactional) over a short period of time. These studies provide information about the effects of specific types of leadership behavior on leadership processes and outcomes, and enable the generation of practical recommendations. Overall, the studies show these leadership styles have a main effect on group process and outcomes and interact with contextual factors, which include nature of task, operating conditions, and anonymity, to influence process and outcome variables.

A salient set of findings deal with the interaction between transformational/transactional leadership styles and anonymity. Kahai, Sosik, and Avolio (2003) found that the effects of transformational versus transactional leadership on solution originality, group efficacy, and task satisfaction were enhanced in the presence of anonymity during a cognitive conflict task. Additionally, transformational leadership helped overcome social loafing that occurred when individual input was pooled for group rewards but was kept anonymous. It appears that anonymity makes it easier for a transformational leader to take attention away from individual identities and induce focus on the collective. Anonymity probably also makes a transactional leader's contingent rewarding less potent by making the recognition provided by the leader as impersonal, and preventing the recognition and rewarding of an individual's contributions. Anonymity's potential to enhance the effect of transformational leadership was also reported by Sosik, Kahai, and Avolio (1999),who found that transformational leadership's positive association with the psychological state of flow (i.e., intense concentration, enjoyment, and intrinsic motivation) was marginally stronger when followers participated anonymously than when they were identified.

The influence of anonymity on the effects of transformational leadership, however, may depend on the nature of task as suggested by Sosik, Avolio, and Kahai (1997). The authors found that transformational leadership was associated with higher levels of group potency than transactional leadership. However, this effect diminished in the presence of anonymity during idea generation, which involves low interdependence, but increased in the presence of anonymity during report writing, which involves high interdependence. The idea that the effect of transformational leadership may diminish in the presence of anonymity during a low interdependence task was supported by Sosik, Kahai, and Avolio's (1998) finding that transformational leadership promoted idea flexibility when individual input was identified but not when it was anonymous; anonymity may have substituted for leadership by encouraging flexibility of thinking. Taken together, these results appear to suggest that while anonymity substitutes for or diminishes the effect of transformational leadership during a task that does not require integration, such as idea generation, it enhances transformational leader's effect when a group is required to integrate its work or resolve conflicting viewpoints. The importance of anonymity in the role it has in the effects of leadership in GSS contexts is also borne out in Sosik et al. (1999). These authors found that flow mediated the effects of transactional and transformational leadership perceptions but only when followers participated anonymously. Many online interactions today are conducted under varying degrees of anonymity and thus these findings are relevant to the way leaders and followers interact with their organizational stakeholders and each other.

In addition to the above effects, transformational and transactional leadership also has other effects. Sosik et al. (1998) found higher levels of elaboration and originality during idea generation at higher levels of transformational leadership. Hoyt and Blascovich (2003) examined the effects of transactional and transformational leadership in both virtual and face-to-face settings and found no effects of setting. However, in comparison with the transactional leader condition, the transformational leader condition was associated with less output, the output was of higher quality and followers were more satisfied, felt more cohesive, trusted their leader more, and displayed greater value congruence. Additionally, trust fully mediated the effects of leadership on satisfaction and cohesiveness. When assigned an ethical scenario involving copying of "copyrighted" software in a GSS setting, participants in another study (Kahai & Avolio, 2006) were more likely to argue against illegal copying and also less likely to argue in favor of illegal copying when working with a transformational leader (confederate) than a transactional leader (confederate). Subsequent to group discussion, the authors measured intentions to copy the software and found that there was greater divergence in the intentions to copy among members of groups exposed to more arguments against illegal copying. They also found that members of groups exposed to fewer arguments in favor of copying displayed lower intentions to copy.

To develop specific behavioral prescriptions for individuals responsible for optimizing creativity processes in organizational groups assisted by GSS, Sosik, Avolio, and Kahai (1998) studied the effects of individual components of transactional and transformational leadership on group creativity. They found that the goal-setting and inspiring leadership components of these styles had positive effects on group creativity, whereas intellectual stimulation and individualized consideration, which are components of transformational leadership, had negative effects on group creativity. The authors offered several possible reasons for the negative effects. For instance, intellectual stimulation, which calls for questioning of assumptions, rethinking of ideas, and examination of ideas from different perspectives, may have promoted judicial thinking (i.e., questioning and evaluation of solutions) early on in the brainstorming process, thereby inhibiting the generation of a wider range of ideas. Also, since the groups were new, it is possible that group members did not have trust in their leader and saw the leader's consideration behavior as insincere, which may have then led to its negative effect on group creativity. Sosik (1997) demonstrated that leadership of GSS groups influences the nature of interaction within those groups. Regardless of whether the groups worked under identified or anonymous conditions, groups working under a leader displaying high level of transformational behavior generated more original solutions, supportive remarks, solution clarifications, and questions about solutions. These groups also reported higher levels of perceived performance, extra effort, and satisfaction with the leader across both contexts.

Research on leadership in GSS settings has also examined the effects of leadership behaviors other than transformational and transactional leadership behaviors. Arguing that participative leadership is more consistent with the participative spirit of GSS, Kahai, Sosik, and Avolio (1997) hypothesized that participative leadership is likely to lead to a greater number of solution proposals, critical remarks, and supportive remarks than directive leadership for a creative task in which followers provided input anonymously. The authors found support for their hypothesis in the case of supportive remarks but they found no difference in the frequency of critical remarks across the two leadership styles. Additionally, they found that participative leadership was associated with higher frequency of solution proposals than directive leadership for a semistructured problem whereas this pattern reversed for a structured problem. Probing deeper, Kahai et al. (2004) found that both perceptions of leader participativeness and directiveness were positively related to participation which, in turn, was positively related to group performance and negatively related to member satisfaction. Furthermore, the relationship of perceptions of leader participativeness with followers' participation strengthened with lack of structure whereas leader directiveness' relationship with participation strengthened with increasing structure. The authors concluded that it may not be accurate to assume that directive leadership is inconsistent with the participative spirit of GSS. In the presence of anonymity, participants probably ignored the controlling aspect of directive leadership and instead focused on the informational aspect, which may have increased their motivation to participate by reducing their role ambiguity.

The findings reported above allay concerns expressed by some researchers (e.g., Den Hartog et al., 2007; Purvanova & Bono, 2009) that transformational and other types of leaders who rely on emotional or long-term connection might be ineffective in computer-mediated settings. Despite employing leaders who merely facilitated the group process of a short duration (e.g., 20 minutes) and were not members of their groups, these studies showed that participative, directive, transformational, and transactional leadership can influence group process and outcome variables in computer-mediated settings.

Studies of GSS facilitation

A GSS facilitator's role has been identified as an important element of computer-mediated meetings (Dennis, Heminger, Nunamaker, & Vogel, 1990; Kolfschoten, Niederman, Briggs, & De Vreede, 2012; Niederman, Beise, & Beranek, 1996). Besides the role of a chauffeur who implements the GSS's features and enters or edits data when directed by the group, a GSS facilitator may also have directive and participative leadership roles. A facilitator may be directive when instructing a group on what features to use and how to use them (Clawson, Bostrom, & Anson, 1993; Dickson, Partridge, & Robinson, 1993) or when they providing expert advice (George, Dennis, & Nunamaker, 1992). A facilitator may be participative by creating and reinforcing an open and participative environment (Clawson et al., 1993).

It appears that process facilitation tends to reduce consensus. A facilitator tends to bring out a wide range of perspectives from group members, which can trigger disagreement in the group (Tan, Wei, & Lee-Partridge, 1999). George et al. (1992) studied GSS groups working on a task with creative and intellective components and found that directive process facilitation led to lower consensus after controlling for the number of alternatives generated. Dickson et al. (1993) found lower post-meeting consensus in facilitator-driven groups compared with chauffeur-driven groups working on a mixed-motive task, leading them to suggest that facilitation must be adaptive rather than restrictive. Tan et al. (1999) who studied the effects of both leadership and facilitation found that facilitation interacted with leadership to influence post-meeting consensus. While consensus decreased with facilitation in the absence of leadership, it increased with leadership in the absence of facilitation. The authors speculated that the leader may have encouraged respect for and consideration of all perspectives introduced by group members and promoted their constructive resolution.

Process facilitation's effects on other variables do not seem as unfavorable as for consensus. These are generally argued to occur due to structures provided by a facilitator that guide the group to follow a desired or a heuristic-driven process; such a process prevents premature task closure and enables careful consideration of relevant issues. George et al. (1992) reported that directive process facilitation led to higher decision quality after controlling for the number of alternatives generated. Wheeler and Valacich (1996) found that restrictive process facilitation led to a GSS being appropriated more faithfully, that is, used in a manner consistent with a decision-making heuristic during a decision-making task. Faithful appropriation, in turn, was associated with higher decision quality. Miranda and Bostrom (1999), who compared the effects of restrictive process and content facilitation in face-to-face and GSS group meetings for a task with creativity and decision-making components, found that process facilitation had a positive effect on meeting process which, in turn, was positively related to satisfaction. Adaptive process facilitation of GSS group meetings studied by Anson, Bostrom, and Wynne (1995) led to marginally significant improvement in interaction quality and cohesion.

A few studies have examined the interaction effects of facilitation and GSS use and their results parallel those of studies that highlight the potential of GSS features to substitute for leadership (Hiltz et al., 1991). Anson et al. (1995) found a weak significant interaction effect of adaptive process facilitation and GSS use on performance. Specifically, they found that compared with the increase in performance in non-GSS groups, performance in GSS groups dropped with the introduction of facilitation. Miranda and Bostrom (1999) found that process facilitation interacted with GSS use to influence satisfaction negatively. Studies examining content facilitation are rare. Miranda and Bostrom, who also studied content facilitation found that it had a negative effect on meeting process. Furthermore, content facilitation interacted with GSS use to have a positive effect on satisfaction.

In general, while GSS features may be neutralizing some aspects of facilitation of group meetings and GSS facilitation may also be reducing consensus, a GSS group may do better on other meeting outcomes with facilitation than without. However, facilitation services are costly and their use appears to be restricted to high-value tasks rather than routine collaborative tasks, despite the significant benefits for organizations if facilitation was provided for the latter tasks as well (Kolfschoten et al., 2012). Using their "expert" judgment and data from interviews of GSS facilitators, Kolfschoten et al. (2012) offer two roads to cost-effective and sustained GSS-based collaboration within organizations: (1) reducing or removing the need for a distinct facilitator, and (2) in situations where expert GSS facilitation cannot be avoided, organizations should consider setting up in-house facilitation practice or hiring external facilitators, depending on whether tasks requiring facilitation are recurring or ad-hoc and the variety of these tasks.

Summary

For a summary of the above discussion, one can view it through the lens of Yammarino et al.'s (2001) integrative model of leadership. We see that a leader's core processes and tactics are impactful in GSS settings. However, the technology may introduce a substitute in the form of anonymity. Thus, while anonymity may enhance certain effects of transformational leadership that rely on the leader making the collective to which a group member belongs more salient, it can replace transformational leadership for outcomes such as idea flexibility. In the presence of anonymity in a GSS setting, transactional leadership loses some of its potency and directive leadership may not be seen as controlling but as informational. Other technology's features, such as statistical feedback, may also conflict with the leader and, thus, serve as neutralizers. Certain fundamental human processes do not change in GSS settings. Specifically, a collective's culture is likely maintained. To the extent identification is enabled, how group members may see the leader does not change as seen in the maintenance of status difference with the leader.

To the best of our knowledge, there is a lack of GSS studies of leadership in an international context, despite the growing relevance of the international nature of those one leads today. However, a few studies (e.g., Watson, Ho, & Raman, 1994; Mejias, Shepherd, Vogel, & Lazaneo, 1996; Daily, Whatley, Ash, & Steiner, 1996) have tried to examine how culture may make a difference in the effects of GSS technology. It appears that GSS technology, which attempts to harness the individualistic culture of the place of its design – North America – may give greater voice to individual participants in collectivistic cultures and reduce pressures to conform (e.g., see Watson et al., 1994; Mejias et al., 1996). When participants in a collectivistic culture perceive that sufficient agreement exists within the group, they may be willing to let a dominant member prevail in order to maintain group harmony. GSS may also help a culturally diverse group harness its advantage in producing unique ideas even further by reducing barriers to participation that may be introduced by diversity (Daily et al., 1996). The effects likely to be seen when leadership is thrown into the mix is an open question. Just like certain culture-based assumptions underlie GSS and give rise to different effects when the GSS is employed in a culture different from that of its origin, certain leadership styles are more consistent with certain cultures and can give rise to different effects for a particular task. For instance, Jung and Avolio (1999) found that while collectivists generated more ideas under a transformational leader, individualists generated more ideas under a transactional leader, when performing a brainstorming task. How the congruence between collectivistic culture and a transformational leader's emphasis on building identification with a collective vision is affected within a GSS context that assumes individualistic tendencies, is an example of the many questions that future research needs to study for an increasingly international workforce that leaders have to lead today.

Leadership in Virtual Teams

The Internet enabled the formation of work teams whose members interacted via communication technology instead of meeting face-to-face. Such teams are able to overcome the constraints of geography and time to conduct their work and are known as virtual teams (Maynard et al., 2017). A team's virtuality falls on a continuum instead of being a discrete variable (Schweitzer & Duxbury, 2010). A variety of dimensions may contribute to a team's virtuality, including the use of communication technology, geographic separation of team members, temporary nature of the team, cultural differences within the team, and informational value of a team's interactions (Balthazard, Waldman, & Warren, 2009; Hoch & Kozlowski, 2014; Kirkman & Mathieu, 2005; O'Leary & Cummings, 2007). Virtual team use by organizations is increasing (Hoch & Kozlowski, 2014), motivated by access to relevant talent from a global workforce, the desire to boost collaboration within the organization, and cost-reduction pressures (Minton-Eversole, 2012). But virtual teams present several challenges, including those related to relationship building – low cohesion, lack of identification, and low trust among team members (Den Hartog et al., 2007; Wiesenfeld, Raghuram, & Garud, 2001) – as well as those that are task-related –role ambiguity, performance ambiguity, and task ambiguity and salience (Den Hartog et al., 2007; Huang, Kahai, & Jestice, 2010). These challenges make leading a virtual team more difficult (Minton-Eversole, 2012). The leader of a virtual team is now expected to play more robust process facilitation and team development roles than in face-to-face teams to overcome the team's challenges (Bell & Kozlowski, 2002; Huang et al., 2010).

Reduced impact of leadership due to virtuality

Some researchers have questioned the ability of leaders to have an impact in virtual teams due to relative lack of nonverbal cues, the challenge of creating identification and trust, and the temporariness of such teams. These aspects of virtuality may make the leader less motivated to create shared meaning and values (Den Hartog et al., 2007). There is some support for their position. Howell and Hall-Merenda (1999) found that while transformational leadership had a positive effect on the performance of proximate followers, its effect on distant followers was negative. Keegan and Den Hartog (2004) similarly found that while transformational leadership increased commitment and motivation and individualized consideration reduced stress in traditional project teams, there were no such effects in temporary project teams. Hoch and Kozlowski (2014) reported that the positive effect of hierarchical leadership, indicated by transformational leadership, leader–member exchange, and supervisory career mentoring, reduced as team virtuality increased.

In a study comparing leadership emergence in virtual and face-to-face teams, Balthazard et al. (2009) observed that two personality aspects, extraversion and emotional stability, predicted transformational leadership emergence in face-to-face teams but not in virtual teams. These findings are probably due to the virtual context curtailing the formation of transformational leadership perceptions that can emerge from one's personality-related behaviors. In virtual settings, the authors found a team member's emergence as a transformational leader is predicted by the following attributes of that member's written communication: number of ideas contributed, frequency of communication, and grammatical complexity. The importance attributed to written communication in virtual teams is also evident in studies on leadership emergence in virtual teams (see Misiolek & Heckman, 2005; Yoo & Alavi, 2004). Misiolek and Heckman (2005) found that frequency of communication initiation and receipt predicted leadership emergence. Yoo and Alavi (2004) observed that those who emerged as leaders sent more electronic messages than others and demographics did not seem to have a role in their emergence as leaders. They also found that emergent leaders sent more task-oriented messages than others and they enacted the roles of initiator, scheduler, and integrator.

The idea that virtuality curtails a leader's ability was supported by Serban et al. (2015), who found that cognitive ability and media type interacted to predict transformational leadership emergence but in a way opposite to what they had predicted. The authors had predicted that the uncertainty and ambiguity in a virtual context would render cognitive ability more important for leadership emergence in virtual than in a face-to-face teams. They tested their predictions using three types of study: simulation, quasi-experiment, and laboratory experiment. However, the simulation study produced the opposite result: cognitive ability predicted leadership emergence more strongly in face-to-face teams. Possible reasons according to the authors were: the virtual context inhibits the transmission of cues that lead to perceptions of intelligence and, hence, of leadership; or intelligence predicts leadership performance in low-stress rather than high-stress situations in which experience may be more important for leadership performance. Since a virtual team is likely to produce a higher level of stress, intelligence may not have a role in this context as it does in a face-to-face situation characterized by lower stress. Serban et al. (2015) also examined the effects of extraversion, conscientiousness, self-efficacy, and comfort with technology on leadership emergence. While extraversion and conscientiousness were found to predict leadership emergence, these personality variables did not interact with media type in any of the three types of study they conducted. The authors attribute the inconsistency between their results and Balthazard et al.'s (2009) findings, who found personality variables to interact with media type, to the difference in how leadership emergence was operationalized across the two studies. Specifically, while Balthazard et al. measured perceptions of leadership displayed by a team member on a continuum, Serban et al. (2015) measured whether someone emerged as a leader on a binary scale (0 or 1) in all three of their studies.

Increasing relevance of leadership due to virtuality

Some researchers have suggested that the challenges created by virtuality will make leadership even more relevant. Purvanova and Bono (2009) argue that virtual teams create conditions of uncertainty and ambiguity in which transformational leaders can be impactful by providing structure, social context, and predictability. The authors' empirical study supported their argument: transformational leadership had a stronger impact on performance in virtual teams than in face-to-face teams. Kayworth and Leidner (2002), who studied only virtual teams, however support the idea about increasing relevance of leadership behaviors that help teams overcome their challenges. They observed that highly effective leaders mentored other members and displayed understanding towards them. They communicated regularly and promptly with sufficient details in their messages. Such leaders also clarified the roles of team members.

Bell and Kozlowski (2002) argue that the lack of face-to-face contact in virtual teams severely restricts a leader's ability to monitor members' performances, implement solutions to problems, and perform typical mentoring and developmental functions. Therefore, in virtual teams, a leader must employ substitutes and distribute these functions to the team itself. According to the authors, a leader may need to create a system whereby team members are able to regulate their own performance as a team. Indeed, Geister, Konradt, and Hertel (2006) demonstrated that when virtual team members provide feedback about the team's process to the team, the motivation, satisfaction, and performance of members improves. Hill and Bartol (2016) make use of Bell and Kozlowski's (2002) ideas to hypothesize that distributed or shared leadership is likely to be more relevant as virtuality increases. They also state that since each team member faces challenges unique to her or his situation, empowering virtual team members to regulate themselves makes more sense. The authors found that a team member's virtual teamwork situational judgment, an indicator of any member's capability to operate effectively in a virtual team situation, positively impacts their virtual collaboration behaviors directly and their individual performance indirectly only under conditions of high empowering team leadership. Additionally, empowering leadership's positive effect on team virtual collaboration and on team performance (indirectly via team virtual collaboration) increased with team dispersion.

In a similar vein, Kirkman, Rosen, Tesluk, and Gibson (2004) observed that team empowerment had a positive relationship with virtual team performance and this relationship was stronger for teams that had fewer face-to-face meetings and, thus, were more virtual. Hoch and Kozlowski (2014), however, did not observe any variation in the importance of shared leadership as a function of a team's virtuality – shared leadership improved team performance in virtual teams regardless of the degree of virtuality. Another piece of evidence supporting the suggestion that shared leadership makes sense for virtual teams comes from a longitudinal study by Carte, Chidambaram, and Becker (2006) who found that higher performing teams displayed more shared leadership behavior, focusing on monitoring group work than lower performing teams. These authors also found that the higher-performing teams showed more concentrated leadership behaviors focused on production activities than lower performing teams. The distribution of leadership role among team members was also observed in a laboratory study by Wickham and Walther (2007), who found that more than one leader can emerge in a virtual team due to the different roles required. Those who engaged in frequent communication, created perceptions of being intelligent, and provided encouragement to the team were more likely to emerge as leaders. Appointing someone as a leader did not ensure that the individual would be able to stay in that position. The potential for virtuality to make leadership more relevant is also suggested by Huang et al. (2010), who found that while transactional leadership improved task cohesion directly in a virtual team, transformational leadership improved it indirectly by improving the team's cooperative climate. These effects, however, were contingent on media richness, defined as a medium's ability to enable shared understanding to develop within a given time period (Daft & Lengel, 1986), and they occurred only in low media richness conditions. Media richness can be associated with virtuality, with lower media richness indicating greater virtuality (Kirkman & Mathieu, 2005). Transactional and transformational leaders are helpful for building task cohesion and a cooperative climate when lower media richness challenges a virtual team, whereas these team variables may result without facilitation when media richness is high, thereby making these leadership styles less relevant.

Thus, these studies suggest that when leaders provide functions that address the shortcomings of virtuality, they become more effective. This functional perspective of leadership is further supported by literature that demonstrates the effectiveness of management strategies designed to address what tends to be missing in virtual teams. For instance, Hertel, Konradt, and Orlikowski's (2004) field study showed that quality of goal setting, building task interdependence, and the use of team-based rewards in virtual teams were associated with higher team effectiveness. These strategies help by compensating for challenges experienced by virtual teams because of low co-presence.

More powerful leadership mechanisms due to virtuality

Research on leadership in virtual teams also suggests that virtual team conditions may make the mechanisms by which leadership operates even more powerful. Joshi, Lazarova, and Liao (2009) employed a field study to examine how inspirational leadership and team member dispersion affected commitment, trust, and team performance in virtual teams. Inspirational leadership, defined as one that communicates a compelling vision, energizes the team, and expresses confidence in the team's ability to achieve that vision, was expected to enhance commitment and trust. Using ideas from the social identity model of deindividuation effects (SIDE) (Lea & Spears, 1991), they also expected dispersion of team members to enhance this effect of inspirational leadership. Specifically, team member dispersion and the accompanying lack of face-to-face contact in virtual teams reduces members attention to individual differences and increases their sensitivity to cues emphasizing the team and its social identity (Lea & Spears, 1991). Leadership behaviors such as inspirational leadership behaviors are likely to have a stronger impact under these conditions by highlighting the identity and the vision of the collective. The authors also argued that dispersion is likely to make interpersonal bonding more challenging, thereby providing more room for inspirational leadership to impact trust formation. Indeed, inspirational leadership enhanced commitment and trust, which in turn improved team performance and this effect of inspirational leadership became stronger under conditions of higher dispersion.

Kahai, Sosik, and Avolio (2013) conducted a field study to examine the effects of transformational leadership and the nature of media on collaboration quality and team performance in virtual teams at a large technology company. They conceptualized collaboration quality as being reflected in learning within the team, coordination of effort within the team, commitment of team members towards team goals, and positivity in team interaction. The authors found support for the hypothesis that transformational leadership and media interact to influence collaboration quality. In teams relying on text-based electronic channels (i.e., email, discussion board, and chat) to a greater extent, they argued, the effect of transformational leadership on collaboration quality would be greater because the use of text decreases attention to individual differences and team members become more sensitive to a transformational leader's messages about the team's vision, its identity, and its ability for collective work. The authors also found that higher quality collaboration was associated with higher levels of performance and group efficacy.

Kahai, Huang, and Jestice (2012) conducted a laboratory study to examine whether the use of a virtual world by virtual teams moderates the effectiveness of a transformational leader. The visual channel of a virtual world can dilute a transformational leader's collaboration-boosting efforts by increasing the salience of individual differences. Consistent with this argument, the authors found that leadership style (transactional vs. transformational) interacted with the communication medium (instant messaging vs. virtual world) to influence feedback positivity, an indicator of the quality of collaboration within a team. Transformational leadership promoted greater feedback positivity than transactional leadership, but only for teams using instant messaging. For teams using a virtual world, the leadership style did not seem to influence feedback positivity.

Hambley, O'Neill, and Kline (2007a) argued that the criticism perceived by some when a transformational leader tries to intellectually stimulate others would be reduced in more virtual settings due to perceived anonymity. In more virtual settings, such attempts would instead be seen as constructive. The contingent rewarding behavior of a transactional leader would be more powerful in less virtual settings because of lower perceived anonymity: rewards are more impactful when the person being rewarded can be identified. However, the authors did not find any effect of leadership style (transactional vs. transformational) on team interaction styles or team cohesion and performance in teams that communicated in one of the following ways: face-to-face, desktop videoconferencing, or text-based chat. They also did not find any interaction between leadership style and communication media. The authors attributed the lack of leadership effects to the short-term nature of their task and the scripted nature of leadership behaviors.

This review of leadership in virtual teams gives rise to an interesting question. Why is it that leadership, particularly transformational leadership, is challenged by the virtual context in some cases while in other cases, its impact is made stronger by the virtual context, either because it fills an important gap or because the mechanism by which it works becomes even more powerful? It is possible that the aspects of transformational leadership that become prominent in proximal and remote contexts are different. For instance, Purvanova and Bono (2009) found that while a leader displayed similar levels of overall transformational leadership across virtual and face-to-face settings, the frequencies and ratings of intellectual stimulation and individualized consideration displayed were higher in the face-to-face setting. Thus, it is likely that the leadership mechanisms causing variation in team performances in teams that meet face-to-face frequently are different from those operating in teams that do not meet face-to-face or do so infrequently. Maybe in the latter type of team, transformational leadership is able to take advantage of the absence of individuating cues and operates at the team level by highlighting the team, its identity, and vision. In teams that meet face to face, the leader may be operating via individualized consideration at the dyadic level because there are significant opportunities for them to have unplanned contact with others; unplanned communication can create perceptions of consideration by signaling to a follower that he or she is important (Kelley & Kelloway, 2012). Indeed, Humphreys (2002) observed that 89% of unique variance in the performance of followers in proximal sales environment was because of individualized consideration.

Thus, on the one hand, field studies that compare leadership in a virtual setting to that in a traditional setting are, perhaps, capturing the absence of individualized consideration and the accompanying drop in individual performance in the virtual condition. On the other hand, studies that examine the effects of transformational leadership at different levels of virtuality are already operating at low levels of unplanned communication that gives rise to individualized consideration, and the variation in the effect of transformational leadership is due to reduced salience of individuating cues and increased salience of the team. We should see this mechanism of team salience playing a more prominent role even in laboratory studies comparing face-to-face and virtual teams because unplanned communication is unlikely in face-to-face teams that have been put together temporarily for the purpose of the experiment.

Summary

The salient findings reviewed above can be summarized using Yammarino et al.'s (2001) integrative model of leadership. Some of the effects related to leadership in virtual teams have been attributed to changes in fundamental human processes caused by technology. Specifically, the lack of nonverbal cues due to virtuality have been proposed to be behind some of the results related to leadership in virtual teams, such as the inability of certain personality variables (extraversion and emotional stability) that predict transformational leadership in face-to-face teams to do the same in virtual teams. In the absence of nonverbal cues and, often, the visual and auditory channels as well, the human processes behind making sense of others and their situation come to rely on textual information and related meta-information (i.e., information about the textual information, such as when was it sent). Yet there are studies that suggest that virtuality may create a situation that is ripe for core leadership processes and tactics to have an even larger impact by compensating for losses in group processes introduced by virtuality. Additionally, certain leadership tactics, such as enabling shared leadership, gain more prominence in virtual teams and become helpful in overcoming the challenges posed by virtual teams. But for members to share leadership, the team leader need to engage in the process of empowering. There is also evidence that the mechanisms underlying core leadership processes may become more powerful in virtual team conditions. These conditions become enhancers of leadership effects such as when virtuality enhances the effects of inspirational leadership.

The virtual environment introduces both opportunities and challenges for leadership. Offering simple prescriptions beyond a few to a leader is not possible due to complex interactions between leadership and features of the virtual environment. Hambley, O'Neill, and Kline (2007b), for instance, learned from a field study that it is critical for virtual team leaders to establish regular team meetings and ensure that those meetings are well organized. They should personalize their relationships with team members and encourage them to do likewise. Additionally, leaders should make sure that they and the rest of the team can use different media effectively. Drawing on findings from the research literature, Roy (2012) argues that to be effective in a virtual environment, leaders need a variety of skills. They need to be able to establish trust, embrace diversity, foster a team spirit, motivate team members, and defuse the frustrations of team members as part of relationship-building skills. Leaders need the technical ability to use a variety of communication media, including videoconferencing, instant messaging, and email. Leadership skills are also needed; in addition to possessing emotional intelligence, leaders should be able to create an open and supportive environment and lead by example.

Since virtual teams are often formed to source the most appropriate talent for a task, irrespective of their location, they may include members from various national cultures. It is reasonable that, at the very least, leaders would need to be sensitive to the different cultures and adjust their leadership behaviors to avoid conflict or misunderstanding due to cultural differences. However, there is little help for leaders in this regard; despite the prevalence of international virtual teams, research on virtual teams, and not just that on leadership in such teams, lacks cross-cultural models (Wakefield, Leidner, & Garrison, 2008).

Wakefield et al.'s (2008) study on leadership and task conflict in virtual teams offers some findings on leadership and national culture even though that was not the study's purpose. The authors examined the effect of leadership roles on virtual team performance via team conflict and leader effectiveness. They employed virtual team members from both US and Korean firms as respondents and found that by exhibiting the roles of monitor, facilitator, and coordinator, virtual team leaders can reduce task conflict and can be seen as more effective. Furthermore, perceived leader effectiveness was positively related to team performance. The strength of the relationship between leader effectiveness and team performance was stronger for the Korean respondents than the US respondents. However, the relationship between leader roles and task conflict was weaker for Korean respondents than for US respondents. The authors attribute these effects to the individualistic and collectivistic orientations of US and Korean cultures. Due to emphasis of individuals from the United States on winning and individual rights, Americans may perceive a greater need for leader intervention to mitigate team conflict. On the other hand, Korean culture lays greater emphasis on cooperation, and individuals from such culture are more likely to defer to authority and believe that positive outcomes result from the collective effort of all members and attribute the team performance to the leader's actions.

Sarker and Schneider (2009) conducted a field experiment to examine factors that members of a virtual team working on an information systems development (ISD) task were likely to consider in thinking of a remote member as a leader. The authors found that team members from the US and the Scandinavian countries (Norway and Denmark) differed on factors they identified as important for leadership. The US team members considered the Scandinavian team members as leaders when the latter possessed ISD ability, exhibited high level of performance, and transferred knowledge to others in a cohesive team. However, none of these factors predicted the emergence of US team members as leaders for the Scandinavians. The authors attribute these results to the more individualistic culture of the United States, where leadership is likely to be seen as a property of the leader, and the more collectivistic culture of Scandinavian countries, where leadership is considered to be inseparable from the context and is likely to be assigned to those who are not different from others but are willing to equalize the status among all and create an informal, participative, and collaborative environment. With only six US team members being seen as leaders by the Scandinavian members in comparison with 23 Scandinavians being seen as leaders by the US team members, the authors conjectured that the Scandinavian preference for face-to-face communication may have been another reason for why the factors that predicted leadership for US team members did not do so for the Scandinavians.

Leadership in the Age of Social Media

The Internet makes it very easy for people and organizations to create social networks that facilitate communication, coordinated action, or association with a cause or interest across time and space. While several modern applications such as Facebook, Twitter, and LinkedIn are designed primarily to facilitate the formation of social networks, one can employ traditional tools such as email, discussion boards, instant messaging, and short message service (SMS) on cell phones, which are outside the category of modern social media, to form social networks easily (Kahai, 2012). While Kahai (2012) used anecdotes from the popular press to offer implications of modern social media applications for leadership, research on what social applications mean for leadership is largely restricted to traditional social applications. Kahai (2012) argued that social media make it possible for leaders to overcome hierarchical and geographic distances to connect authentically with followers far and wide. They allow leaders to monitor the participation of followers on these media and analyze this to gauge if their message is reaching them or to learn if some action is needed. Leaders can now employ social media to build social capital, an important leadership function from the social network perspective of leadership (Balkundi & Kilduff, 2006). They can employ social media to improve the quality of workplace interaction and enable workplace relationships that would not have existed otherwise; such interactions can help innovation by bringing together individuals with similar interests but from different knowledge domains. Leaders can also use social media tools to develop an accurate view of their social network, which can help them build successful coalitions in pursuit of organizational goals.

Kahai (2012) also argued that social media has its own set of dangers for leaders. Social media has made the Internet accessible on an unprecedented level to those not in positions of formal authority. Consequently, it is challenging for leaders in positions of authority to keep secrets for long. Today, armed with their cell phones equipped with capability to record the happenings around them and distribute them via the Internet to the rest of the world, the ordinary citizen has become a powerful news reporter with an eye and ear on everything. Social media can enable anyone with an idea that has mass appeal to stimulate the formation of a network that rises against the leader. The visibility into participation by others may tempt leaders to take unethical actions. The relative anonymity afforded by social media may also tempt a leader to behave inauthentically and take advantage of followers in several ways, including pretending to be one of them and using the information they share against them.

Empirical research on leadership in contexts supported by traditional social applications is largely restricted to online communities. Online communities are large collections of people who come together on the Internet because of a common interest (Preece, 2000). Members of online communities, who may span different nations, time zones, and organizations, pool their knowledge and resources to learn and/or create something. Members of a discussion forum geared towards improving photography and the developers of Linux Open Source Software (OSS) are examples of online communities. These communities tend to have fluid boundaries due to high turnover of members, and individuals gain influence on an emergent basis due to their expertise (Faraj, Jarvenpaa, & Majchrzak, 2011).

Huffaker (2010) studied the communication behaviors of online leaders, who were defined as those with the ability to stimulate: responses from others, conversations, and diffusion of their language. The author used automated text analysis to examine a large number of messages posted over two years on 16 Google groups to uncover the language and social behavior that predicted online leadership. The analysis revealed that high communication activity in the form of posts and replies, credibility (reflected in tenure), network centrality, and the use of messages that are affective in nature, longer, assertive, and possess language diversity predict online leadership. According to Huffaker (2010), online leaders seem to be relying on the potential for bonding or relationship development and maintenance in their behavior as evident in the leaders' tendency to reply to those who reply to them. Additionally, reply behavior of leaders outweighed their posting behavior. An interesting observation countered expectations. Specifically, brokering did not predict online leadership. The author argues that because all information is transparent and accessible to all in online discussion groups, everyone in the community could serve as a broker, thereby making it less likely for a single person to emerge as a leader. Johnson, Safadi, and Faraj (2015) report similar findings about the importance of network position and language in the emergence of online community leaders. The authors argue that traditional leadership theories cannot be employed to predict online community leaders. For instance, since communication in online communities in open and viewable to all, the ability for a leader to engage in differentiated leader–member exchange with someone is diminished and, therefore, the ability to directly apply leader–member exchange theory (Graen & Uhl-Bien, 1995) is constrained. It is the interpretation of text-based interactions observable to all that leads to someone being viewed as a leader. The authors used a combination of survey data (to identify leaders) and data collected from three discussion forums (to identify administrators, determine network position, and study the language used) to test their propositions. They found that beyond network factors in the form of formal role (as an administrator), centrality, membership in the core, and boundary-spanning role, language use factors predicted online community leadership. Specifically, those viewed as leaders by other participants post a large number of positive and concise posts in a simple language that is familiar to others.

The above reviewed studies examined leadership in discussion-oriented online communities in which members share existing knowledge. Online communities can also focus on creating an artifact such as a software, a knowledge base, or a movie. Researchers have examined leadership in such online communities. For example, Zhu, Kraut, and Kittur (2012) examined the community of editors on Wikipedia. They studied the messages that these editors left on each other's personal profile pages and classified them into four categories of leadership behavior: transactional, aversive, directive, and person-based. The authors found that one does not have to occupy a formal leadership role to influence others, where influence was indicated by motivating others to take an editing action; leadership behaviors by those in formal leadership positions as well as by others motivated action though formal leaders were more influential. Additionally, transactional, directive, and person-focused leadership behaviors promoted editing activity whereas aversive leadership reduced it. The authors also observed that editors who did not occupy formal leadership positions sent more than half of leadership messages, indicating the shared nature of leadership within the community of Wikipedia editors.

Eseryel and Eseryel (2013) looked at the online community of developers of Apache OSS. Using a grounded theory approach, the authors tried to answer the question of how leaders emerge and strategically influence the actions of others who form a part of self-managing and global software development team. They found that the leaders in their study exhibited action-oriented transformational leadership. Specifically, someone has to be actively contributing software first in order to emerge as a transformational leader unlike the traditional conceptualization of transformational leadership which emphasizes vision development and member management and facilitation. Those who are action oriented are seen as knowledgeable role models who guard the project. Such leaders provide vision for the project via their actions. They refrain from stating a grand vision and asking others to follow that vision. Instead, they role model the actions that they would like others to take and bring about transformational change in the structure and workings of the team's processes and product. Unlike the traditional transformational leadership identified in the literature, which separates the leader initiating the transformation from those who undergo transformation, the action-oriented transformational leader in the software development community is transforming the community's outcomes as well as being influenced by the transformation.

Luther and Bruckman (2008) studied leadership in three online communities whose members collaborate to create 'collabs' or computer-animated movies. While effective leadership is critical for the success of collabs, providing it is challenging because of the unique features of the production process and the lack of technological support available to the leader. The authors identified how the unique challenges provided by collabs make a difference for leaders of such projects relative to those who lead OSS development or open-content publishing projects such as Wikipedia. Specifically, these challenges have four themes: completion, originality, subjectivity, and ownership. Collabs are all-or-nothing projects and only completed work is released. On the other hand, OSS projects are characterized by frequent releases and Wikipedia articles are released continuously. The burden on a leader increases as releases become less frequent because the expectations for perfection increase and challenge the leader even more. In collab projects, originality is a sought-after goal whereas in OSS projects it is not attempted due to inherent difficulties and in Wikipedia it is prohibited (by way of original research). The goal of originality in collabs makes the work of the leader more challenging. Collabs are also open-ended because of the subjectivity involved in artistic expression, which leads to disagreement among artists and increases challenges for a leader. OSS, on the other hand, are more closed-ended since they try to replicate existing commercial products and Wikipedia projects have straightforward goals in the form of a familiar format and formal tone of a reference work. A collab leader cannot change any artist's work without the artist's permission whereas in OSS and Wikipedia projects, changes are often made to the code or to an article written by someone else. In sum, according to the authors, online creative collaboration projects differ based on their nature, thereby leading to a different set of challenges for leaders of such projects.

Summary

Social media make the tactic of monitoring by leaders for both positive and negative purposes easier, and certain social technologies may enhance this monitoring ability more than others. While social technologies can help a leader shape the fundamental cognitive and emotional processes of their followers and build social capital, they can also help individuals who have a more compelling message rise by neutralizing the existing leader's power. Indeed, empirical research suggests that leadership in the age of social media is likely to be emergent and informal. It is others in an online community who recognize someone to be a leader. Formal hierarchies are rare and governance tends to be bottom-up. When there are formal positions, those in such positions are not necessarily seen as leaders.

Unlike in face-to-face situations in which multiple communication channels are available to convey information that helps in leadership emergence, it is the quality and quantity of online communication that ends up predicting leadership emergence in online communities enabled by social media. A leader has to engage in core leadership processes and tactics (directive, transactional, and person-based) via the written message. Beyond the written message, the actions of a leader become the influence tactic. But for this action to be influential, its output has to be visible and thus, to the extent the output is visible, the effects of action may be enhanced or limited. Leadership in online communities may also be neutralized by the nature of the work of a community, such as that of a collab community. While the stream of empirical research on leadership in online communities recognizes that the mediated, text-based nature of interactions is likely to render leadership in online communities different from those in traditional face-to-face settings, there is little discussion of whether and how the online nature of communities is changing the fundamental way leadership should be displayed. Also missing is a discussion of how different national cultures make a difference for leadership emergence in online communities.

Effect of Leadership on Information Technology

The widespread deployment of IT is requiring that leaders play a more proactive role in the assimilation and management of information technology to ensure effective organizational performance (Ross & Weill, 2002). Specifically, IT implementation oftentimes requires a leader's attention because of changes required in leadership style, rewards, and location of decision-making authority (Hitt & Brynjolfsson, 1997). The absence of these changes is partly to blame for a lack of demonstrable relationship between IT spending and corporate performance (Brynjolfsson, 1993; Farrell, 2003; Grover, Jeong, & Segars, 1996). Our focus here is on leadership processes in a context in which work is supported to a significant extent by information systems. We begin by discussing leadership processes required to complement IT application and accompanying changes in work practices. We then discuss leadership processes that impact the assimilation and management of information systems.

Complementary Leadership Processes

When IT investments are made, successful IT assimilation requires complementary changes in several aspects of an organization, including leadership, rewards, and location of decision-making authority. Any misalignment between IT, leadership, rewards, and location of decision-making authority can lead to little or no gain from IT investments (Hitt & Brynjolfsson, 1997). We focus here on leadership requirements that accompany IT investments.

Hitt and Brynjolfsson (1997) found that greater levels of IT deployment are being accompanied by greater decentralization of decision-making, more team work, and more knowledge work. Knowledge work is difficult to measure, making it difficult to motivate performance by tying rewards to behavior. In such situations, team building and goal alignment are likely to be adopted by organizations as mechanisms for motivating performance beneficial to the organization. Team building, whereby organizational members collaborate on interdependent tasks, may be achieved by building a sense of collective confidence and fostering high levels of social exchanges among members of a team (Chun, Cho, & Sosik, in press). Goal alignment, whereby organizational members internalize the interests of the organization, may be achieved by creating a shared vision through inspirational leadership and high-quality goal-setting processes (Hertel et al., 2004; Hitt & Brynolfsson, 1997).

Jarvenpaa and Tanriverdi (2003) highlight the leadership challenges facing organizations as they transition towards virtual knowledge networks that consist of cooperative intraorganizational and interorganizational relationships. The nature of knowledge and knowledge work raises conflicts over the allocation of rewards from knowledge resources in virtual knowledge networks. In turn, these conflicts hurt knowledge creation. In such circumstances, leaders will need to focus on harnessing key social relationships and filling the "voids of trust" that exist in the network. To build trust, they will need to shift from being firm-centric to network-centric. When taking action, leaders will need to balance the interests of various stakeholders in the firm's value chain.

Additionally, with greater IT deployment, technical skills are emerging as an important leadership requirement. Leaders need to be comfortable using IT in order to interact with and monitor the performance of followers and organizational units (Kayworth & Leidner, 2002; Roy, 2012; Townsend, DeMarie, & Hendrickson, 1998). Bassellier, Benbasat, and Reich (2003) found that IT competence, made up of IT-related knowledge and experience of a business manager, accounted for 34% of the variance in managers' intentions to champion IT. Beath (1991) provides an additional perspective on IT champions, observing that to succeed, the champions need assistance from the IT function for: developing a persuasive case in support of their vision, obtaining flexible and relevant implementation resources, and gathering political support for their vision.

Assimilation and management of IT

It has long been recognized that senior leadership support is important for successful application of IT (Jarvenpaa & Ives, 1991; Oz & Sosik, 2000). Jarvenpaa and Ives (1991) identify two forms of senior leadership support: involvement and participation. Involvement is a psychological state that refers to leader's perceptions and attitudes concerning IT and reflects the degree of importance placed on IT by the leader. Involvement serves the purpose of sending the right supportive signals regarding IT to the organization.

Participation, on the other hand, refers to behaviors related to IT planning, development, and implementation (e.g., chairing an executive steering committee, requesting progress reports for important IT projects, approving IT projects). Such leadership support also prevents expensive long-term IT projects from being abandoned (Oz & Sosik, 2000).

According to Chatterjee, Grewal, and Sambamurthy (2002), senior management can affect assimilation of IT by manipulating institutional structures of signification, legitimization, and domination. These structures create norms and values regarding how managers should engage in structuring actions. Leaders can manipulate structures of signification, which give meaning and serve as cognitive guides, through inspirationally motivating behaviors that offer a vision for the organization and how IT fits into that vision. Leaders can influence the structures of legitimization, which endorses behavior, by discussing opportunities and risks with the application of IT. By believing in IT, participating in IT strategy and projects, and using IT, leaders can be role models and send signals that legitimize their followers' participation in IT projects and adoption of IT. Leaders can manipulate structures of domination, which regulate behavior, through mandates and policies regarding IT adoption and use.

With the use of modern social applications increasing in organizations, Richter and Wagner (2014) argue that leaders need to adapt to such applications because the influence processes that worked for them in offline settings or in settings dominated by older computer-mediated communication channels, such as email, are not going to be adequate. Modern social applications offer leaders the opportunity to increase their contact and be more social with various stakeholders both inside and outside the organization. They also create challenges. In interviews with 24 individuals responsible for the implementation of social media in their organizations, the authors found that special measures need to be taken to engage and develop leaders. Specifically, leaders need to be convinced about, sensitized to, and coached about the new social applications.

Even before the advent of social applications in organizations, which require leaders to understand the interaction between IT and social phenomena both inside and outside the organization, there has been need for leaders to understand that IT implementation is a complex social phenomenon enabled only by effective leadership. Sarker and Lee (2003) examined the roles of three social enablers – strong and committed leadership, open and honest communication, and a balanced and empowered implementation team – during enterprise resource planning (ERP) implementation at well-established company. They found that while all three social enablers may contribute to ERP success, only strong and committed leadership throughout the life of an ERP implementation project and existing at various levels in the organization (top, project management, and IT function) served as a necessary condition for success. Ke and Wei (2008) argue that successful ERP implementation requires creating the right culture in an organization and leadership is the most critical factor for creating the required culture. An effective leader cultivates a culture of learning and development, participative decision-making, support and collaboration, power sharing, communication, and tolerance for conflict and risk, all of which are required to succeed with the implementation of ERPs.

Some research has attempted to address the type of leadership relevant during IT implementation. Wang, Chou, and Jiang (2005) studied 300 of the top 500 corporations in Taiwan and found that charismatic behavior of the ERP project leader was important for the cohesiveness and overall performance of the project team. Based on data from 36 Australian firms, Thite (2000) observed that a combination of transformational and technical leadership augmented transactional leadership displayed by IT project leaders to influence project success. Neufeld, Dong, and Higgins (2007) found that charismatic project leadership during IT implementation predicted the many antecedents to the intention to use the IT being implemented. Specifically, it predicted performance expectancy (user's expectancy that IT would lead to better performance), effort expectancy (user's expectancy that the effort needed to use the IT being implemented is not beyond her or him), social influence (normative influence to use the IT being implemented), and facilitating conditions (the presence of conditions which make it easier for users to adopt the IT being implemented).

Due to IT criticality and the need for complementary changes in several aspects of an organization when IT is implemented, IT management cannot be completely delegated to the IT function (Ross & Weill, 2002). Non-IT leaders need to be involved in critical IT-related decisions, which requires them to acquire an understanding of technology and how it interacts with the organization's strategy, culture, and structure (Armstrong & Sambamurthy, 1999; Bassellier, Reich, & Benbasat, 2001; Ross & Weill, 2002). Eom, Kahai, and Yayla (2015) found support for a model according to which the leadership of the chief information officer (CIO) has an important role in creating a state in which non-IT leaders are engaged with IT and IT is well aligned with business. Specifically, the authors show that a transformational CIO promotes mechanisms that help IT and business units develop shared domain knowledge. These mechanisms, in turn, not only enable alignment, but they also predict mechanisms that allow IT and business units to integrate their specialized knowledge and produce the organization's outputs. Under certain conditions (specifically, when mechanisms to develop shared domain knowledge exist but occur with relatively low frequency) knowledge integration mechanisms lead to greater IT–business alignment.

Summary

How does the research on IT and leadership fit within the integrative model of leadership (Figure 14.1). Core leadership processes and leadership tactics may need to change with greater deployment of IT. For instance, team-building and goal alignment become more critical with IT deployment and the accompanying decentralization of decision-making. Leaders need to develop strong technical skills, work more closely with the IT function, offer a vision that incorporates IT, and be sensitive to and communicate opportunities and risks with the application of IT in order to be effective and enable successful assimilation of IT. Leaders also need to cultivate a culture of learning and development, participative decision-making, and collaboration to succeed in the implementation of complex IT such as ERPs. Leaders may need to display more transformational leadership, which is critical for IT implementation and its alignment with business. While it is recognized that differences in national cultures might have a role in IT adoption and implementation (Griffith, 1998; Veiga, Floyd, & Dechant, 2001), and IT project leadership has been studied in contexts other than the United States (e.g., Thite, 2000), research on the effect of national cultures on leadership's influence on IT is missing.

Future Research

E-leadership is a much neglected area in research. In addition to understanding how IT changes leadership and vice-versa, research is needed to understand and accelerate the development of leadership suitable for a world in which technology touches everything that we do. For same-place settings, future research could examine how the widespread use of smartphones is affecting leadership by enabling back-channel communication

(Kahai, 2012; Zafar, 2017, Chapter 10 in this volume). Back-channel communication is that which occurs simultaneously with the main communication but is secondary to it. It can give rise to several issues related to leadership, which can be examined by research (Kahai, 2012). For instance, back-channel communication can influence the emergence of a leader by connecting someone to outside sources of information and make that person appear very knowledgeable. In fact, the mere presence of a smartphone makes individuals appear smarter (Clayton, Leshner, & Almond, 2015). Another leadership-related issue is the emergence of cliques when individuals with similar agendas message each other to interpret and respond to the ongoing discussion. Such cliques can affect the balance of power between the leader and the rest of the group.

While research on virtual teams is paying increased attention to leadership issues, some topics have seen little or no leadership research. Gilson, Maynard, Young, Vartiainen, and Hakonen (2015) identify several themes that future research on virtual teams should address, of which a few can easily be associated with leadership. Specifically, how can leadership help virtual teams overcome the challenges due to and, possibly, take advantage of (a) generational gaps among team members – a phenomenon that is likely to become more acute as more individuals from the millennial generation enter the workforce, and (b) subgroups that arise within virtual teams. Virtual teams face many disruptions and interruptions that lead to the need for adaptation. Transitions are also likely to be common given that many virtual teams may be short-term project teams (Powell, Piccoli, & Ives, 2004). What roles do virtual team leaders have in making their teams more adaptive and capable of handling disruptions and transitions effectively? Finally, how can virtual team leaders contribute to team members' wellbeing?

Kahai (2012) argued that the rise of social networks due to proliferation of social media is making organizations resemble complex systems. Leadership in such systems is not located only in individuals or in formal positions but also in the interaction among the system's agents (Uhl-Bien et al., 2007). Uhl-Bien et al. (2007) suggest three types of leadership as relevant for complex systems: administrative, adaptive, and enabling. The existing empirical research on leadership in the age of social media has focused on administrative and enabling leadership types, which tend to rest in individuals. It has not examined adaptive leadership, which arises from the emergent and dynamic interplay among the system's heterogeneous agents; this interplay gives rise to new patterns of behavior and modes of operating which enable the system to adapt, learn, and innovate. Kahai (2012) argues that IT has the potential to influence adaptive leadership and be influenced by it. Systematic research on this interplay between IT and adaptive leadership in complex systems is currently lacking and future research should address this gap.

IT implementation may require complementary changes in leadership behaviors of existing leaders. For organizations to take advantage of the possibilities offered by new IT, their culture may need to change and leaders have to lead this change by modeling what they expect from others. While research points to leadership behaviors or styles that may be beneficial for IT implementation (and, thus, suggests how leaders ought to change), it has not documented whether and how leaders embrace change. Nor has research considered the role of leadership behaviors in deciding which technology to implement and whether it should be implemented. Future research should examine these issues. Results from such research may have important implications for whether organizations should develop their existing leaders to change or they should change their leaders in conjunction with the implementation of IT that requires complementary change in leadership.

Conclusions

IT continues its relentless advance, leaving little opportunity for leadership systems to evolve coherently with it. Meanwhile, benefits from IT continue to lag behind their potential, to a significant extent due to the absence of understanding on the part of leadership about its role in the implementation and assimilation of IT. The limited research and accounts that exist inform us that what we know about leadership is not going to be sufficient for the digital age, which, by some accounts, has just begun to unfold. Given that even a single feature of technology, such as the anonymity or the addition of a virtual 3D environment it may enable, can bring about significant changes in the effects of leadership, it would be fair to say that we have been largely flying blind as technology continues to become more powerful and the number of features and possibilities it creates keep increasing. Research has much to catch up with if we aspire to do better with technology in the future.

References

Anson, R., Bostrom, R., & Wynne, B. (1995). An experiment assessing group support system and facilitator effects on meeting outcomes. *Management Science, 41*(2), 189–208.

Armstrong, C. P., & Sambamurthy, V. (1999). Information technology assimilation in firms: The influence of senior leadership and IT infrastructures. *Information Systems Research, 10*(4), 304–327.

Avolio, B. J., & Kahai, S. S. (2003). Adding the" E" to e-leadership: How it may impact your leadership. *Organizational Dynamics, 31*(4), 325–338.

Avolio, B. J., Kahai, S., & Dodge, G. E. (2001). E-leadership: Implications for theory, research, and practice. *The Leadership Quarterly, 11*(4), 615–668.

Avolio, B. J., Sosik, J. J., Kahai, S. S., & Baker, B. (2014). E-leadership: Re-examining transformations in leadership source and transmission. *The Leadership Quarterly, 25*(1), 105–131.

Balkundi, P., & Kilduff, M. (2006). The ties that lead: A social network approach to leadership. *The Leadership Quarterly, 17*(4), 419–439.

Balthazard, P. A., Waldman, D. A., & Warren, J. E. (2009). Predictors of the emergence of transformational leadership in virtual decision teams. *The Leadership Quarterly, 20*(5), 651–663.

Bandura, A. (1997). *Self-efficacy: The exercise of control.* New York, NY: Freeman.

Barkhi, R., Jacob, V. S., Pipino, L., & Pirkul, H. (1998). A study of the effect of communication channel and authority on group decision processes and outcomes. *Decision Support Systems, 23*(3), 205–226.

Bassellier, G., Benbasat, I., & Reich, B. H. (2003). The influence of business managers' IT competence on championing IT. *Information Systems Research, 14*(4), 317–336.

Bassellier, G., & Reich, B. H., & Benbasat, I. (2001). Information technology competence of business managers: A definition and research model. *Journal of Management Information Systems, 17*(4), 159–182.

Beath, C. M. (1991). Supporting the information technology champion. *MIS Quarterly, 15*(3), 355–372.

Bell, B. S., & Kozlowski, S. W. (2002). A typology of virtual teams implications for effective leadership. *Group & Organization Management, 27*(1), 14–49.

Brislin, R. W., & Kim, E. S. (2003). Cultural diversity in people's understanding and uses of time. *Applied Psychology, 52*(3), 363–382.

Brynjolfsson, E. (1993). The productivity paradox of information technology. *Communications of the ACM, 36*(12), 66–77.

Carte, T. A., Chidambaram, L., & Becker, A. (2006). Emergent leadership in self-managed virtual teams. *Group Decision and Negotiation, 15*(4), 323–343.

Champy, J. (2010). Does leadership change in a web 2.0 world? *HBR Blog Network.* Retrieved April 6, 2017, from http://blogs.hbr.org/imagining-the-future-of-leadership/2010/05/does-leadership-change-in-a-we.html

Chatterjee, D., Grewal, R., & Sambamurthy, V. (2002). Shaping up for e-commerce: Institutional enablers of the organizational assimilation of web technologies. *MIS Quarterly, 12*(2), 65–89.

Chun, J. U., Cho, K., & Sosik, J. J. (in press). A multilevel study of group-focused and individual-focused transformational leadership, social exchange relationships, and performance in teams. *Journal of Organizational Behavior.*

Clawson, V. K., Bostrom, R. P., & Anson, R. (1993). The role of the facilitator in computer-supported meetings. *Small Group Research, 24*(4), 547–565.

Clayton, R. B., Leshner, G., & Almond, A. (2015). The extended iSelf: the impact of iPhone separation on cognition, emotion, and physiology. *Journal of Computer-Mediated Communication, 20*(2), 119–135.

Daft, R. L., & Lengel, R. H. (1986). Organizational information requirements, media richness and structural design. *Management Science, 32*(5), 554–571.

Daily, B., Whatley, A., Ash, S. R., & Steiner, R. L. (1996). The effects of a group decision support system on culturally diverse and culturally homogeneous group decision making. *Information & Management, 30*(6), 281–289.

Den Hartog, D. N., Keegan, A. E., & Verburg, R. M. (2007). Limits to leadership in virtual contexts. *The Electronic Journal for Virtual Organizations and Networks, 9*, 54–63.

Dennis, A. R., Heminger, A. R., Nunamaker, J. F., & Vogel, D. R. (1990). Bringing automated support to large groups: the Burr-Brown experience. *Information & Management, 18*(3), 111–121.

Dickson, G. W., Partridge, J. E. L., & Robinson, L. H. (1993). Exploring modes of facilitative support for GDSS technology. *MIS Quarterly, 17*(2), 173–194.

Dinh, J. E., Lord, R. G., Gardner, W. L., Meuser, J. D., Liden, R. C., & Hu, J. (2014). Leadership theory and research in the new millennium: Current theoretical trends and changing perspectives. *The Leadership Quarterly, 25*(1), 36–62.

Eom, M., Kahai, S., & Yayla, A. (2015). Investigation of how IT leadership impacts IT-business alignment through shared domain knowledge and knowledge integration. *AMCIS 2015 Proceedings.* Retrieved April 6, 2017, from http://aisel.aisnet.org/amcis2015/HumanCapital/GeneralPresentations/7/

Eseryel, U. Y., & Eseryel, D. (2013). Action-embedded transformational leadership in self-managing global information systems development teams. *Journal of Strategic Information Systems, 22*(2), 103–120.

Faraj, S., Jarvenpaa, S. L., & Majchrzak, A. (2011). Knowledge collaboration in online communities. *Organization Science, 22*(5), 1224–1239.

Farrell, D. (2003). The real new economy. *Harvard Business Review, 81*(10), 104–12.

Geister, S., Konradt, U. & Hertel, G. (2006). Effects of process feedback on motivation, satisfaction and performance in virtual teams. *Small Group Research, 37*(5), 459–489.

George, J. F., Dennis, A. R., & Nunamaker Jr, J. F. (1992). An experimental investigation of facilitation in an EMS decision room. *Group Decision and Negotiation, 1*(1), 57–70.

George, J. F., Easton, G. K., Nunamaker Jr, J. F., & Northcraft, G. B. (1990). A study of collaborative group work with and without computer-based support. *Information Systems Research, 1*(4), 394–415.

Gilson, L. L., Maynard, M. T., Young, N. C. J., Vartiainen, M., & Hakonen, M. (2015). Virtual teams research 10 years, 10 themes, and 10 opportunities. *Journal of Management, 41*(5), 1313–1337.

Graen, G. B., & Uhl-Bien, M. (1995). Relationship-based approach to leadership: Development of leader-member exchange (LMX) theory of leadership over 25 years: Applying a multi-level multi-domain perspective. *The Leadership Quarterly, 6*(2), 219–247.

Griffith, T. L. (1998). Cross-cultural and cognitive issues in the implementation of new technology: focus on group support systems and Bulgaria. *Interacting with Computers, 9*(4), 431–447.

Grover, V., Jeong, S. R., & Segars, A. H. (1996). Information systems effectiveness: The construct space and patters of application. *Information & Management, 31*(4), 177–191.

Hambley, L. A., O'Neill, T. A., & Kline, T. J. (2007a). Virtual team leadership: The effects of leadership style and communication medium on team interaction styles and outcomes. *Organizational Behavior and Human Decision Processes, 103*(1), 1–20.

Hambley, L. A., O'Neill, T. A., & Kline, T. J. (2007b). Virtual team leadership: Perspectives from the field. *International Journal of E-Collaboration, 3*(1), 40–64.

Harmon, J., Schneer, J. A., & Hoffman, L. R. (1995). Electronic meetings and established decision groups: Audio-conferencing effects on performance and structural stability. *Organizational Behavior and Human Decision Processes, 61*(2), 138–147.

Hertel, G., Konradt, U. & Orlikowski, B. (2004). Managing distance by interdependence: Goal setting, task interdependence, and team-based rewards in virtual teams. *European Journal of Work and Organizational Psychology, 13*(1), 1–28.

Hill, N. S., & Bartol, K. M. (2016). Empowering leadership and effective collaboration in geographically dispersed teams. *Personnel Psychology, 69*(1), 159–198.

Hiltz, S. R., Johnson, K., & Turoff, M. (1991). Group decision support: the effects of designated human leaders and statistical feedback in computerized conferences. *Journal of Management Information Systems*, 81–108.

Hitt, L. M., & Brynjolfsson, E. (1997). Information technology and internal firm organization: An exploratory analysis. *Journal of Management Information Systems, 14*(2), 81–101.

Hoch, J. E., & Kozlowski, S. W. (2014). Leading virtual teams: Hierarchical leadership, structural supports, and shared team leadership. *Journal of Applied Psychology, 99*(3), 390–403.

Howell, J. M., & Hall-Merenda, K. E. (1999). The ties that bind: The impact of leader-member exchange, transformational and transactional leadership, and distance on predicting follower performance. *Journal of Applied Psychology, 84*(5), 680–694.

Hoyt, C. L., & Blascovich, J. (2003). Transformational and transactional leadership in virtual and physical environments. *Small Group Research, 34*(6), 678–715.

Huang, R., Kahai, S., & Jestice, R. (2010). The contingent effects of leadership on team collaboration in virtual teams. *Computers in Human Behavior, 26*(5), 1098–1110.

Huffaker, D. (2010). Dimensions of leadership and social influence in online communities. *Human Communication Research, 36*(4), 593–617.

Humphreys, J. H. (2002). Transformational leader behavior, proximity and successful services marketing. *Journal of Services Marketing, 16*(6), 487–502.

Jarvenpaa, S. L., & Ives, B. (1991). Executive involvement and participation in the management of information technology. *MIS Quarterly, 15*(2), 205–227.

Jarvenpaa, S. L., & Tanrıverdi, H. (2003). Leading virtual knowledge networks. *Organizational Dynamics, 31*(4), 403–412.

Johnson, S. L., Safadi, H., & Faraj, S. (2015). The emergence of online community leadership. *Information Systems Research, 26*(1), 165–187.

Joshi, A., Lazarova, M. B., & Liao, H. (2009). Getting everyone on board: The role of inspirational leadership in geographically dispersed teams. *Organization Science, 20*(1), 240–252.

Jung, D. I., & Avolio, B. J. (1999). Effects of leadership style and followers' cultural orientation on performance in group and individual task conditions. *Academy of Management Journal, 42*(2), 208–218.

Kahai, S. S. (2012). Leading in a digital age: What's different, issues raised, and what we know. In M. C. Bligh, & R. E. Riggio (Eds.), *Exploring distance in leader-follower relationships: When near is far and far is near* (pp. 63–108). New York, NY: Taylor & Francis/Routledge Publishing.

Kahai, S. S. & Avolio, B. J. (2006). Leadership style, anonymity, and the discussion of an ethical issue in an electronic context. *International Journal of e-Collaboration, 2*(2), 1–26.

Kahai, S. S., Huang, R., & Jestice, R. J. (2012). Interaction effect of leadership and communication media on feedback positivity in virtual teams. *Group & Organization Management, 37*(6), 716–751.

Kahai, S. S., Sosik, J. J., & Avolio, B. J. (1997). Effects of leadership style and problem structure on work group process and outcomes in an electronic meeting system environment. *Personnel Psychology, 50*(1), 121–146.

Kahai, S. S., Sosik, J. J., & Avolio, B. J. (2003). Effects of leadership style, anonymity, and rewards in an electronic meeting system environment. *The Leadership Quarterly, 14*(4–5), 499–524.

Kahai, S. S., Sosik, J. J., & Avolio, B. J. (2004). Effects of participative and directive leadership in electronic groups. *Group & Organization Management, 29*(1), 67–105.

Kahai, S., Sosik, J., & Avolio, B. (2013, August). *Effects of transformational leadership and media on collaboration and performance in virtual teams.* Paper presented at the Academy of Management Conference, Orlando, FL, USA.

Kayworth, T. R., & Leidner, D. E. (2002). Leadership effectiveness in global virtual teams. *Journal of Management Information Systems, 18*(3), 7–40.

Ke, W., & Wei, K. K. (2008). Organizational culture and leadership in ERP implementation. *Decision Support Systems, 45*(2), 208–218.

Keegan, A. E., & Den Hartog, D. N. (2004). Transformational leadership in a project-based environment: a comparative study of the leadership styles of project managers and line managers. *International Journal of Project Management, 22*(8), 609–617.

Kelley, E., & Kelloway, E. K. (2012). Context matters : Testing a model of remote leadership. *Journal of Leadership & Organizational Studies, 19*(4), 437–449.

Kim, Y., Hiltz, S. R., & Turoff, M. (2002). Coordination structures and system restrictiveness in distributed group support systems. *Group Decision and Negotiation, 11*(5), 379–404.

Kirkman, B. L., & Mathieu, J. E. (2005). The dimensions and antecedents of team virtuality. *Journal of Management, 31*(5), 700–718.

Kirkman, B. L., Rosen, B., Tesluk, P. E., & Gibson, C. B. (2004). The impact of team empowerment on virtual team performance: The moderating role of face-to-face interaction. *Academy of Management Journal, 47*(2), 175–192.

Kolfschoten, G. L., Niederman, F., Briggs, R. O., & De Vreede, G. J. (2012). Facilitation roles and responsibilities for sustained collaboration support in organizations. *Journal of Management Information Systems, 28*(4), 129–162.

Lea, M., & Spears, R. (1991). Computer-mediated communication, de-individuation and group decision-making. *International Journal of Man-Machine Studies, 34*(2), 283–301.

Lim, L. H., Raman, K. S., & Wei, K. K. (1990). Does GDSS promote more democratic decision-making? The Singapore experiment. In *System Sciences, 1990: Proceedings of the 23rd Annual Hawaii International Conference on System Sciences, 3*, 59–68. Washington, DC: IEEE Computer Society.

Lim, L. H., Raman, K. S., & Wei, K. K. (1994). Interacting effects of GDSS and leadership. *Decision Support Systems, 12*(3), 199–211.

Luther, K., & Bruckman, A. (2008). Leadership in online creative collaboration. In *Proceedings of the 2008 ACM Conference on Computer Supported Cooperative Work* (pp. 343–352). New York, NY: ACM.

Maynard, T., Gilson, L. L., Young, N. J., Vartiainen, M., Hakonen, M. (2017). Virtual teams. In G. Hertel, D. L. Stone, R. D. Johnson, & J. Passmore (Eds.), *The Wiley Blackwell Handbook of the psychology of the internet at work* (pp. 315–346). Chichester: John Wiley & Sons, Ltd.

Mejias, R. J., Shepherd, M. M., Vogel, D. R., & Lazaneo, L. (1996). Consensus and perceived satisfaction levels: A cross-cultural comparison of GSS and non-GSS outcomes within and between the United States and Mexico. *Journal of Management Information Systems, 13*(3), 137–161.

Minton-Eversole, T. (2012, July 19). Virtual teams used most by global organizations, survey says. Retrieved January 31, 2016, from www.shrm.org/hrdisciplines/orgempdev/articles/pages /virtualteamsusedmostbyglobalorganizations,surveysays.aspx

Miranda, S. M., & Bostrom, R. P. (1999). Meeting facilitation: Process versus content interventions. *Journal of Management Information Systems, 15*(4), 89–114.

Misiolek, N. I., & Heckman, R. (2005). Patterns of emergent leadership in virtual teams. In *Proceedings of the 38th Annual Hawaii International Conference on System Sciences (HICSS '05)* (Track 1 – Volume 01) (p. 49.1). Washington, DC: IEEE Computer Society.

Mohammed, S., & Nadkarni, S. (2011). Temporal diversity and team performance: The moderating role of team temporal leadership. *Academy of Management Journal, 54*(3), 489–508.

Neufeld, D. J., Dong, L., & Higgins, C. (2007). Charismatic leadership and user acceptance of information technology. *European Journal of Information Systems, 16*(4), 494–510.

Niederman, F., Beise, C. M., & Beranek, P. M. (1996). Issues and concerns about computer-supported meetings: The facilitator's perspective. *MIS Quarterly, 20*(1), 1–22.

Nunamaker, J. F., Dennis, A. R., Valacich, J. S., Vogel, D., & George, J. F. (1991). Electronic meeting systems. *Communications of the ACM*, *34*(7), 40–61.

O'Leary, M. B., & Cummings, J. N. (2007). The spatial, temporal, and configurational characteristics of geographic dispersion in teams. *MIS Quarterly*, *31*(3), 433–452.

Oz, E., & Sosik, J. J. (2000). Why information systems projects are abandoned: A leadership and communication theory and exploratory study. *Journal of Computer Information Systems*, *41*(1), 66–78.

Powell, A., Piccoli, G., & Ives, B. (2004). Virtual teams: A review of current literature and directions for future research. *ACM SIGMIS Database*, *35*(1), 6–36.

Preece, J. (2000). *Online communities: Designing usability and supporting sociability*. New York, NY: John Wiley & Sons, Inc.

Purvanova, R. K., & Bono, J. E. (2009). Transformational leadership in context: Face-to-face and virtual teams. *The Leadership Quarterly*, *20*(3), 343–357.

Reeves, B., Malone, T. W., & O Driscoll, T. (2008). Leadership's online labs. *Harvard Business Review*, *86*(5), 58–66.

Richter, A., & Wagner, D. (2014). Leadership 2.0: Engaging and supporting leaders in the transition towards a networked organization. In *Proceedings of the 2014 47th Hawaii International Conference on System Sciences* (pp. 574–583). Washington, DC: IEEE Computer Society.

Ross, J. W., & Weill, P. (2002). Six IT decisions your IT people shouldn't make. *Harvard Business Review*, *80*(11), 84–95.

Roy, S. R. (2012). Digital mastery: The skills needed for effective virtual leadership. *International Journal of e-Collaboration*, *8*(3), 56–66.

Sarker, S., & Lee, A. S. (2003). Using a case study to test the role of three key social enablers in ERP implementation. *Information & Management*, *40*(8), 813–829.

Sarker, S., & Schneider, C. (2009). Seeing remote team members as leaders: A study of US-Scandinavian teams. *IEEE Transactions on Professional Communication*, *52*(1), 75–94.

Saunders, C., Van Slyke, C., & Vogel, D. R. (2004). My time or yours? Managing time visions in global virtual teams. *The Academy of Management Executive*, *18*(1), 19–37.

Serban, A., Yammarino, F. J., Dionne, S. D., Kahai, S. S., Hao, C., McHugh, K. A. . . . Peterson, D. R. (2015). Leadership emergence in face-to-face and virtual teams: A multi-level model with agent-based simulations, quasi-experimental and experimental tests. *The Leadership Quarterly*, *26*(3), 402–418.

Schweitzer, L., & Duxbury, L. (2010). Conceptualizing and measuring the virtuality of teams. *Information Systems Journal*, *20*(3), 267–295.

Sosik, J. J. (1997). Effects of transformational leadership and anonymity on idea generation in computer-mediated groups. *Group & Organization Management*, *22*(4), 460–487.

Sosik, J. J., Avolio, B. J., & Kahai, S. S. (1997). Effects of leadership style and anonymity on group potency and effectiveness in a group decision support system environment. *Journal of Applied Psychology*, *82*(1), 89.

Sosik, J. J., Avolio, B. J., & Kahai, S. S. (1998). Inspiring group creativity comparing anonymous and identified electronic brainstorming. *Small Group Research*, *29*(1), 3–31.

Sosik, J. J., Kahai, S. S., & Avolio, B. J. (1998). Transformational leadership and dimensions of creativity: Motivating idea generation in computer-mediated groups. *Creativity Research Journal*, *11*(2), 111–121.

Sosik, J. J., Kahai, S. S., & Avolio, B. J. (1999). Leadership style, anonymity, and creativity in group decision support systems: The mediating role of optimal flow. *Journal of Creative Behavior*, *33*(4), 227–256.

Stokols, D., Mishra, S., Gould-Runnerstrom, M., & Hipp, J. (2009). Psychology in an age of ecological crisis: From personal angst to collective action. *American Psychologist*, *64*(3), 181–193.

Sutton, R. (2010, June 9). What every new generation of bosses has to learn. *HBR Blog Network*. Retrieved April 6, 2017, from http://blogs.hbr.org/cs/2010/06/good_bosses_have_a_passion_for.html

Tan, B. C. Y., Wei, K. K., & Lee-Partridge, J. E. (1999). Effects of facilitation and leadership on meeting outcomes in a group support system environment. *European Journal of Information Systems*, *8*(4), 233–246.

Thite, M. (2000). Leadership styles in information technology projects. *International Journal of Project Management, 18*(4), 235–241.
Townsend, A. M., DeMarie, S. M., & Hendrickson, A. R. (1998). Virtual teams: Technology and the workplace of the future. *The Academy of Management Executive, 12*(3), 17–29.
Veiga, J. F., Floyd, S., & Dechant, K. (2001). Towards modelling the effects of national culture on IT implementation and acceptance. *Journal of Information Technology, 16*(3), 145–158.
Wakefield, R. L., Leidner, D. E., & Garrison, G. (2008). Research note – a model of conflict, leadership, and performance in virtual teams. *Information Systems Research, 19*(4), 434–455.
Wang, E., Chou, H. W., & Jiang, J. (2005). The impacts of charismatic leadership style on team cohesiveness and overall performance during ERP implementation. *International Journal of Project Management, 23*(3), 173–180.
Watson, R. T., Ho, T. H., & Raman, K. S. (1994). Culture: A fourth dimension of group support systems. *Communications of the ACM, 37*(10), 44–55.
Wen, M., Yang, D., & Rosé, C. P. (2015). Virtual teams in massive open online courses. In C. Conati, N. Heffernan, A. Mitrovic, & M. F. Verdejo (Eds.), *Artificial intelligence in education* (pp. 820–824). Cham, Switzerland: Springer International Publishing.
Wickham, K. R., & Walther, J. B. (2007). Perceived behaviors of emergent and assigned leaders in virtual groups. *International Journal of E-Collaboration, 3*(1), 1–17.
Wiesenfeld, B. M., Raghuram, S., & Garud, R. (2001). Organizational identification among virtual workers: The role of need for affiliation and perceived work-based social support. *Journal of Management, 27*(2), 213–229.
Wheeler, B. C., & Valacich, J. S. (1996). Facilitation, GSS, and training as sources of process restrictiveness and guidance for structured group decision making: An empirical assessment. *Information Systems Research, 7*(4), 429–450.
Uhl-Bien, M., Marion, R., & McKelvey, B. (2007). Complexity leadership theory: Shifting leadership from the industrial age to the knowledge era. *The Leadership Quarterly, 18*(4), 298–318.
Yammarino, F. J., Dansereau, F., & Kennedy, C. J. (2001). Viewing leadership through an elephant's eye: A multiple-level multidimensional approach to leadership. *Organizational Dynamics, 29*(3), 149–163.
Yoo, Y., & Alavi, M. (2004). Emergent leadership in virtual teams: What do emergent leaders do? *Information and Organization, 14*(1), 27–58.
Zafar, H. (2017). Mobile computing and hand-held devices at work. In G. Hertel, D. L. Stone, R. D. Johnson, & J. Passmore (Eds.), *The Wiley Blackwell handbook of the psychology of the internet at work* (pp. 195–210). Chichester: John Wiley & Sons, Ltd.
Zhu, H., Kraut, R., & Kittur, A. (2012). Effectiveness of shared leadership in online communities. In *Proceedings of the ACM 2012 Conference on Computer Supported Cooperative Work* (pp. 407–416). New York, NY: ACM.

15

Virtual Teams

M. Travis Maynard, Lucy L. Gilson, Nicole C. Jones Young, and Matti Vartiainen

Introduction

Digitalization of work along with globalization has contributed to an increased use of virtual teams (VTs). VTs have been defined as:

> *two or more persons who collaborate interactively to achieve common goals while at least one of the team members works at a different location, organization, or at a different time so that communication and coordination is predominantly based on electronic communication media.*
>
> Hertel, Geister, and Konradt (2005, p. 71)

Given the continued increase in technological advancements, which has led to employees desiring more flexibility in their work arrangements, VT usage will only continue to expand (Gilson, Maynard, Jones Young, Vartiainen, & Hakonen, 2015). Over the past decade, much has been written about VTs. This chapter provides an overview of this literature by highlighting areas that have received considerable attention, as well as areas where we believe the field has evolved over the past decade. Specifically, the traditional way to examine VTs has been to compare them with face-to-face (FtF) teams. While this is still the case in much of the extant research, the concept of team virtuality has started to gain traction as a way to more accurately consider today's teams that rarely interact exclusively via FtF means and likewise, are rarely purely virtual. Instead, most teams exist along a virtuality continuum. Given the significance of the development of the team virtuality construct, after we highlight various theoretical and methodological perspectives that have been used to study VTs, we present an overview of team virtuality. In particular, we describe how the term has been conceptualized as well as the approaches that researchers have utilized in operationalizing team virtuality. Following this, we move to our review section where we delineate VT research by leveraging a prominent team effectiveness framework (input-mediator-output-input). We conclude the chapter by highlighting areas that are still in need of additional research consideration and developing a series of questions that we hope will help guide the research agenda in this domain.

The Wiley Blackwell Handbook of the Psychology of the Internet at Work,
First Edition. Edited by Guido Hertel, Dianna L. Stone, Richard D. Johnson, and Jonathan Passmore.

Theory and Methods in Research on Virtual Teams

Theoretical considerations

With so much interest in VTs, it is hardly surprising that researchers have been actively engaged in trying to further our understanding of what makes them effective. In striving to understand the key ingredients for VT success, a plethora of theories have been leveraged. Listing each of these theories would be overwhelming and not particularly informative, therefore, in this chapter, we will discuss two of the more prominent examples. Similarly, we highlight a few theories that have started to receive some traction but as of yet have not been fully integrated into the VT domain.

The first theory that has received a great deal of consideration is *leadership*. Overwhelmingly, transformational leadership (see Bass, 1999) has received the most attention. In VTs, transformational leadership has been argued to arise from personality and communication factors (Balthazard, Waldman, & Warren, 2009), and to increase performance, satisfaction (Purvanova & Bono, 2009), and motivation (Andressen, Konradt, & Neck, 2012). The second theoretical lens to garner a great deal of attention is *trust*. Trust continues to be considered one of the most important drivers of VT success (e.g., Breuer, Huffmeier, & Hertel [2016]; see also Jarvenpaa, Cantu, & Lim [2017], Chapter 6 in this volume). Research here has leveraged well-established models and theories (e.g., Jarvenpaa & Leidner, 1999; Mayer, Davis, & Schoorman, 1995; McAllister, 1995) and has found that trust develops during early VT interactions and is often based team members' ability (Clark, Clark, & Crossley, 2010), and can have positive effects on collective awareness (Daassi, Jawadi, Favier, & Kalika, 2006) and performance (De Jong & Elfring, 2010).

Beyond these more popular theories, it is also interesting to examine which theories have not been considered extensively and therefore, represent opportunities for future research within the VT domain. In fact, in the great deal of VT research in the past decade, several mainstream team-level theories have only been sparingly utilized. For example, given the importance of fairness and the consistent effects of procedural justice within the organizational team literature (see Cohen-Charash & Spector, 2001), we were surprised to only find two studies that examine justice in VTs (see Hakonen & Lipponen, 2008; Tangirala & Alge, 2006). Interestingly, both studies used the uncertainty management theory of fairness (Van den Bos & Lind, 2002). We were also intrigued by the dearth of work examining identification within VTs. In line with this contention, in 2005, Fiol and O'Connor argued "we know very little about the interrelationships among individual, group, and situational factors in the development of identification in VTs" (p. 20). During the early years of VT research, there was work that considered social identity (Bouas & Arrow, 1996; Wiesenfeld, Raghurum, & Garud, 1999). Since that time, this stream of inquiry has been largely ignored. We continue to believe that shared social identity, as noted by Martins, Gilson, and Maynard (2004) and prior meta-analyses (e.g., Riketta, 2005) has the potential to affect member extra-role behaviors and turnover intentions.

Furthermore, the social identity approach (SIA; Haslam, 2001), which explains the way in which people define themselves as members of a social group, is potentially an overarching (meta) theory that could help explain a wide variety of intra- and inter-group phenomena relevant to VTs; from groupthink to discrimination. An example of the use of SIA is the study by O'Leary and Mortensen (2010), which found that the social identity dynamics of different subgroup configurations resulted in different processes and outcomes. Related to the discussion on meta-theories, theoretical parsimony is something that appears to be lacking within the VT research domain. In reviewing the extant literature, we found many studies that appear to be rather atheoretical in nature or follow a process of picking from theory A to explain phenomena A and theory B for phenomena B. Recently, Wilson, Crisp, and Mortensen (2013) posit that virtuality should not be examined as an objective state, but rather as a construal-level theory.

In sum, we have highlighted some "favorite" theories, which continue to receive a great deal of research attention as well as a few other established team and organizational theories that remain less well integrated within the VT literature. Accordingly, we suggest that future research involving VTs should start to more fully consider which of the existing theories are appropriate to help us understand this domain.

Methodological considerations

In terms of methodological trends with the VT literature over the past decade, we note the following main categories: laboratory studies and student samples; cross-sectional design; level of analysis; technology; and social network analysis.

Laboratory studies and student samples A key methodological issue noted by Martins and colleagues (2004) was the dominance of lab studies using student samples. While there are numerous arguments for and against these types of method, which we will not discuss here (see, e.g., Gilson, Maynard, & Bergeil 2013; Levitt & List, 2007; Shadish, Cook, & Campbell, 2001), a substantial number of VT studies still fall into this category. That being said, the numbers of field settings using "real-life" organizational VTs is growing, as is research employing several interesting hybrid or quasi-experimental designs: projects conducted with students from different universities (e.g., Berg, 2012; Boros, Meslec, Curseu, & Emons, 2010). While these samples do employ student teams, the subjects are members of short-term project teams, who come together to complete a task, and then disband without ever meeting FtF. Whether these studies should be categorized as lab or field is an interesting question (see Grant & Wall, 2009). Further questions remain as to whether results obtained from such hybrid designs are truly generalizable to organizational-based VTs. With that being said, we commend the trend of more research focused on organizational teams and hope it continues.

Cross-sectional design Our review of the extant literature suggests that the overwhelming majority of field-based empirical studies are quantitative in nature and rely on survey-based data collection methods (e.g., Altschuller & Benbunan-Fich, 2010). Most of this work is cross-sectional (e.g., Golden, & Fromen, 2011), and cross-sectional designs are subject to common method variance issues (Podsakoff, MacKenzie, & Podsakoff, 2012; Schwab, 2005). Thus, moving forward we implore VT researchers to rely more heavily on longitudinal studies that overcome some of the inherent limitations of cross-sectional designs. Importantly, this is not a problem unique to the VT literature, as Mathieu, Maynard, Rapp, and Gilson (2008) made a similar call to the broader organizational team research community and research in work and organizational psychology in general.

Within the VT context, longitudinal designs provide an opportunity to capture the development of complex team dynamics. For instance, Geister, Konradt, and Hertel (2006) studied 52 student VTs for five weeks and found that performance increased over time for those VTs that used an online feedback system. Moreover, longitudinal field studies can more robustly address dynamic relationships by employing structural equation modeling (Bollen, 1989). Lastly, longitudinal designs are better able to capture the essence of time that has long been heeded as critical to both VT development and performance. For example, Metiu's (2006) four-year ethnographic examination revealed that status and geographic boundaries shaped interactions between VTs working in the United States and India. This type of in-depth, qualitative case study design is an example of the types of question that can be addressed with longitudinal work.

Level of analysis The final methodological issue noted in a 2004 review by Martins et al. dealt with positioning, theorizing, and methodological level of analysis. Some of the articles from the past decade that we reviewed continue to describe team phenomena, but in fact conduct analysis at the individual-level (e.g., Bourgault, Drouin, & Hamel, 2008). On a positive note, this trend appears to be fading and we found multiple examples of research that aggregated individual data to the team level. While aggregation often results in better alignment of the underlying theory and methodology, it does not provide an opportunity to scrutinize variables at different levels of analysis, or employ multilevel modeling techniques (Luke, 2004). That said, however, Arling and Subramani (2011) surveyed several on-going work teams, and employed multilevel modeling techniques, which allowed them to examine the effects of both individual-level virtuality and team-level constructs (FtF and ICT centrality) on team performance. By adopting a multilevel lens, researchers have access to a broader nomological network of constructs that should more accurately reflect the nature of variables that impact VTs.

Technology Whether laboratory or field-based studies are employed, the norm within the VT research has been to examine the impact of rather conventional technologies, such as email, chat, and discussion boards. Consequently, new developments such as three-dimensional (3D) virtual environments (3DVEs) that have become potential collaboration platforms for VTs (see Bosch-Sijtsema & Sivunen, 2013) have received scarce consideration. 3DVEs can be defined as communication systems in which multiple participants share the same 3D digital space despite occupying remote physical locations; this technology allows participants to navigate, manipulate objects, and interact with one another via avatars (Sallnäs, 2005; Yee & Bailenson, 2007). A recent review of 3DVE studies (Sivunen & Hakonen, 2011) found only one that considered 3DVEs as a potential technology used in VTs. However, our review identified a handful of work that did in fact consider novel technologies. For example, within their study of 24 student teams, Kahai, Huang, and Jestice (2012) examined VT leadership styles and feedback comparing instant messaging and 3DVEs. Interestingly, they found no differences between the two technologies, but more work is obviously needed within this arena before more definitive conclusions can be reached.

Social network analysis A promising trend within the empirical studies included in this review is the increased use of social network analysis (e.g., Behrend & Erwee, 2009; Espinosa & Clark, 2014; Hsu & Chou, 2009; Sarker, Ahuja, Sarker, & Kirkeby, 2011; Suh, Shin, Ahuja, & Kim, 2011). Social network analysis allows for visual illustration of the underlying team structure, which provides a richer understanding of team dynamics and team member relationships. Likewise, the use of social network analysis can be valuable in discovering the presence and the impacts of subgroups on teams across the entire virtuality continuum. Moreover, social network analysis produces quantitative statistics (e.g., centrality) that can be combined with other numerical indicators (e.g., questionnaire data; see Scott, 1988). For example, Capece and Costa (2009) used social network analysis methodology to study social networks and knowledge creativity in VTs and revealed how team configuration was related to dimensions of knowledge creativity over time. Espinosa and Clark (2014) found team knowledge network structures (isolation, centrality) were associated with outcomes such as task coordination, strategy coordination, and the proportion of team knowledge cliques.

Other approaches We note two other approaches that have been adopted by VT researchers. First, there are several examples of in-depth case studies that range from the examination of a single team (e.g., McAvoy & Butler, 2006) to qualitative work on multiple teams (e.g., Matlay & Westhead, 2005), as well as work that has examined small networks of VTs (e.g., Panteli & Duncan, 2004) and online communities (e.g., Chen, Li, Clark & Dietrich, 2013). Second, our review of the VT literature evidenced several well-designed empirical studies that exclusively focused on VTs (e.g., Martins & Shalley, 2011; Rack, Ellwart, Hertel, & Konradt, 2011). This last category, which we refer to as VT studies, is work that does not contrast VTs to FtF teams or measure virtuality. Instead, such VT studies predominantly focus on issues of technology, communication, or the management of information systems, using student teams in lab-type settings and often include short-term projects (e.g., Leonard & Haines, 2007; Ocker, 2007; Turel & Connelly, 2012). For reasons of space and coherence, case studies and VT studies are not reviewed within this chapter.

In sum, many methodological advances have been made during the past 10 years. We note here the increase in field studies, longitudinal work, and the adoption of techniques such as social network analysis. That being said, a great deal of work still uses student samples and lean technologies (i.e., only email or text) rather than some of the newer more complex technologies Again, our goal here is not to criticize this work as much of it is excellent and has provided the necessary foundation for us to understand VTs. However, the time is right to move forward and take some risks. The rapidly evolving technological landscape affords us more flexibility, allowing us to utilize devices such as mobile apps, handheld devices, smart watches and so on, to collect data and conduct research (Kozlowski, 2015). In addition to the continuous nature of the data provided by such devices, they might allow VTs to be examined within context and collect data from team members regardless of their geographic location.

Team Virtuality

When reviewing research and trends in the VT literature over the past decade, we see a greater level of emphasis being placed on virtuality and a focus on understanding how virtuality impacts team dynamics, and ultimately performance. In fact, a recent review by Gilson and colleagues (2015) reports that approximately a third of all empirical VT studies since 2004 has considered virtuality as a dimension impacting team performance. Most definitions suggest that virtuality is multidimensional and exists along a continuum from "not at all virtual" to "highly virtual" (e.g., de Guinea, Webster, & Staples, 2012). As a result of this evolution within the VT literature, there are numerous definitions of virtuality (Table 15.1). Furthermore, as suggested in Table 15.2, there also are various views regarding how team virtuality should be conceptualized. Specifically, Griffith, Sawyer, and Neale (2003) early on proposed three dimensions – geographic dispersion, percentage of time spent apart, and the use of technology (and technology support) for communication. Around the same time, Cohen and Gibson (2003) suggested that virtuality consisted of only two dimensions, electronic dependence and geographic dispersion. Martins and colleagues (2004) defined VTs as "teams whose members use technology to varying degrees in working across locational, temporal, and relational boundaries to accomplish an

Table 15.1 Prominent definitions of team virtuality.

Authors	*Definition*
Bierly et al., 2009	Virtuality – "the degree to which members use technology to interact across geographic, organizational, or other boundaries ..." (p. 551)
Chudoba et al., 2005	Virtuality depends on discontinuities in geography, time zone, organization, national culture, work practices, and technology
R. De Jong et al., 2008	"the extent to which team members use communication media to coordinate their actions and execute their task, taking into account; the extent to which the communication media are synchronous; and the extent to which the communication media convey para-verbal and nonverbal aspects of communication" (p. 366)
Gibson & Gibbs, 2006	"We conceptualize virtuality here as a multifaceted higher-order construct comprising four independent defining characteristics identified in previous literature: geographic dispersion, electronic dependence, dynamic structural arrangements, and national diversity" (p. 455)
Khan et al., 2010	"Team virtuality includes the extent to which team members rely on virtual tools to coordinate and execute team processes; the amount of informational value provided by such tools, and the synchronicity of team member interaction" (p. 2744)
Kirkman & Mathieu, 2005	"we define team virtuality using three dimensions: (a) the extent to which team members use virtual tools to coordinate and execute team processes, (b) the amount of informational value provided by such tools, and (c) the synchronicity of team member virtual interaction" (p. 702)
Paul et al., 2004	Virtual teams are those that cross national, functional, and organizational boundaries and are connected by telecommunications and information technology
Schweitzer & Duxbury, 2010	"the extent that the team members do not work face-to-face, the distance between team members, their configuration or collocation patterns, and the extent that they work asynchronously" (p. 278)
Shin, 2004	The degree to which a group has temporal, cultural, spatial, and organizational dispersion and communicates through electronic means
Suh et al., 2011	Virtuality – "geographic and temporal dispersion, along with technological support ..." (p. 353)
Zimmermann et al., 2008	Virtual settings are defined as: "interactions between people who are working at different locations and often in different time zones. Communication is mainly computer-mediated, but face-to-face interactions are used in addition. Each of the team members as well as the leader is located in a different place" (p. 322)

interdependent task" (p. 808), and Shin (2004) added organizational dispersion to the mix. Subsequently, several researchers have emphasized the salience of work practices when conceptualizing virtuality (e.g., Chudoba, Wynn, Lu, & Watson-Manheim, 2005). Specifically, Harvey, Novicevic, and Garrison (2005) included consideration of the transitory nature of the team structure, while Gibson and Gibbs (2006) discussed that work teams often structurally change membership, roles, and relationships among team members. Lastly, Kirkman and Mathieu (2005) highlighted the need to more deeply examine the types and reasons for technology use. Specifically, they argued that the amount of informational value provided by communication tools and the synchronicity of team members' interaction needed to be considered.

Table 15.2 Prominent conceptualized dimensions of team virtuality.

Authors	*Spatial (geographic) dispersion*	*Extent of ftf meetings*	*Time dispersion*	*Diversity*[1]	*Organizational boundaries*	*Work practices*	*Technology usage*	*Type of technology used*[2]
Chudoba et al., 2005	X		X	X	X	X	X	
Cohen & Gibson, 2003	X						X	
Dixon & Panteli, 2010	X		X	X	X	X	X	X
Gibson & Gibbs, 2006	X			X		X	X	
Gibson et al., 2011	X						X	
Griffith et al., 2003	X	X					X	
Harvey et al., 2005	X		X	X	X	X	X	
Hertel et al., 2005	X				X		X	X
Kirkman & Mathieu, 2005							X	X
Martins et al., 2004	X		X	X			X	
Paul et al., 2004				X	X		X	
Rapp et al., 2010							X	
Saunders & Ahuja, 2006	X						X	X
Schweitzer & Duxbury, 2010	X							X
Shin, 2004	X		X	X	X			
Staples & Webster, 2008	X		X					
Suh et al., 2011	X		X				X	
Zimmermann et al., 2008	X		X				X	

FtF, face to face.
[1] Includes cultural and/or national diversity.
[2] Richness, synchronicity, and/or information value of technology.

Team virtuality studies

Team virtuality operationalization A number of dimensions have been theorized as key components of the team virtuality construct. However, when researchers actually measure virtuality, it is typically through the use of four dimensions: diversity, geographic dispersion, technology usage, and the characteristics of the technology used, with the latter three receiving the bulk of research attention (Table 15.3). Below, we briefly review each of these four dimensions before launching into our review of the VT research.

Diversity. While diversity has had a prominent role as an input variable in most organizational team effectiveness models, diversity and in particular cultural and national diversity also has been discussed as a dimension of team virtuality (e.g., Hoch & Kozlowski, 2014; Muethel, Gehrlein, & Hoegl 2012a; Muethel, Siebdrat, & Hoegl, 2012b). In part, the diversity of VT members in terms of the cultures and national background is relevant given that especially in global VTs, members are drawn from multiple countries. For example, research has examined how differing cultural backgrounds (e.g., Zhang, Lowry, Zhou, & Fu, 2007 – Chinese and US), as well as the number of nationalities present (e.g., Hoch & Kozlowski, 2014) affects team processes and performance. Additionally, Pinjani and Palvia (2013) went beyond cultural and national diversity considerations and examined the role of functional and deep-level diversity finding that deep-level diversity was more salient in its influence on trust and knowledge sharing.

Geographic dispersion. While Kirkman and Mathieu (2005) question the necessity of geographic dispersion stating that, "even co-located team members can communicate and coordinate in a highly virtual manner even though they are in close physical proximity" (p. 702), as detailed in Table 15.3, many researchers view geographic dispersion as a salient dimension of virtuality (e.g., Webster & Wong, 2008). For example, Cramton and Webber (2005) had team leaders categorize their teams as either consisting of collocated members or geographically dispersed (at least 30% of members working from separate locations). O'Leary and Cummings (2007) differentiated dispersion further by proposing three critical dimensions of geographic dispersion: spatial (i.e., distance among team members), temporal (i.e., extent to which team members have overlapping work hours), and configurational (i.e., number of sites at which team members are located, their isolation from other members, and their balance between subgroups of members across sites). Cummings and Haas (2012) had team members detail the extent to which they were physically separate from each team member and then calculated an average team separation. Lastly, O'Leary and Mortensen (2010) considered geographic dispersion in their examination of the impact of geographically defined subgroups – one of the few studies we found to consider subgroups within virtuality research.

Technology usage. We highlight two studies that operationalized virtuality by examining team technology usage. First, Bierly, Stark, and Kessler (2009) asked team members to agree or disagree with statements such as "we primarily interacted through computer and telecommunications technologies" (p. 558). In the second, Stark and Bierly (2009) asked team members three survey questions to assess the extent of technology use (e.g., "Most of the day-to-day communication between team members is through computer or telephone interaction," p. 466). Beyond these two studies, several others have included technology use within a composite measure of virtuality (e.g., Cummings, Espinosa, & Pickering, 2009; Hoch & Kozlowski, 2014). These multidimensional measures of virtuality will be discussed below.

Table 15.3 Operationalized dimensions of team virtuality.

Authors	*Spatial (geographic) dispersion*	*Extent of FtF meetings*	*Time dispersion*	*Diversity*[1]	*Organizational boundaries*	*Work practices*	*Technology usage or support*	*Type of technology used*[2]
Andressen et al., 2012	X							X
Bierly et al., 2009							X	
Cramton & Webber, 2005	X							
Cummings et al., 2009	X		X					X
Cummings & Haas, 2012	X							
R. De Jong et al., 2008								X
Ganesh & Gupta, 2010	X						X	X
Hakonen & Lipponen, 2008	X	X						
Han et al., 2011							X	X
Henderson, 2008	X						X	
Hoch & Kozlowski, 2014	X			X			X	
Joshi et al., 2009	X		X	X				
Kirkman et al., 2013								X
Maynard et al., 2012								X
O'Leary & Mortensen, 2010	X							
Pinjani & Palvia, 2013				X			X	
Rico & Cohen, 2005								X
Schweitzer & Duxbury, 2010	X	X						
Staples & Webster, 2008	X	X	X					
Stark & Bierly, 2009							X	
Suh et al., 2011	X		X				X	
Zhang et al., 2007				X			X	

FtF, face to face.
[1] Includes cultural and/or national diversity.
[2] Richness, synchronicity, and/or information value of technology.

Type of technology used. In contrast to the examples discussed above that have used a generalized measure to assess teams' reliance on technology to communicate, several studies have delved more deeply into the various types of communication technologies. These operationalizations have taken multiple forms; for example, Rico and Cohen (2005) employed a single survey item that asked team members the extent to which communication was asynchronous with "excessive pauses" (p. 267) or more synchronous and characterized as being smooth and occurring at the same time. In contrast, researchers also have manipulated the type of technology to assess the impact of text-based communication versus video-based communication channels (e.g., González-Navarro, Orengo, Zornoza, Ripoll, & Peiró, 2010; Han, Hiltz, Fjermestad, & Wang, 2011). An interesting operationalization was adopted by Suh and colleagues (2011) who looked at the usage of personalized (i.e., email and instant message) as compared with communal computer-mediated communication (CMC; i.e., group discussions, group calendars, audioconferences, and videoconferences). Finally, with regard to technology, several research teams ask VT members to delineate the percentage of their communication that occurs through a list of FtF and CMC tools (e.g., Maynard, Mathieu, Rapp, & Gilson, 2012; Rapp, Ahearne, Mathieu, & Rapp, 2010). In a similar fashion, De Jong, Schalk, and Curseu (2008) asked team members to allocate 100% of their communication within the team to a variety of communication channels. Based on this allocation they determined a "virtuality score" for each communication medium using Baltes, Dickson, Sherman, Bauer, and LaGanke's (2002) work to assess synchronization and the presence of nonverbal and para-verbal cues.

Multiple dimensions of virtuality. While many of the examples noted above demonstrate that at times researchers examine virtuality through the "lens" of a single dimension, there are also several instances where virtuality is operationalized as a multidimensional construct (see Table 15.3). For example, Andressen et al. (2012) measured virtuality based on the frequency of CMC and physical distance. Frequency of CMC was assessed by the use of different electronic media in relation to overall communication frequency within the team. In terms of physical distance, these authors leveraged the work of O'Leary and Cummings (2007) to assess distance measures such as site index, isolation index, spatial distance, and time zone index. Another interesting point regarding this work is that the virtuality of the team as a whole as well as the virtuality of the leader, compared with the rest of the team, were measured, and differential relationships were found for both measures of virtuality.

Finally, we found several studies where various dimensions of virtuality are combined into a composite measure (e.g., Joshi, Lazarova, & Liao, 2009). As a specific example, Cummings et al., (2009) examined the interaction between communication synchronicity, spatial dispersion, and temporal dispersion and found support for synchronous communication being more strongly associated with reductions in coordination delays when members cross spatial boundaries. Another compelling example of various dimensions of virtuality within a single empirical measure is work by Ganesh and Gupta (2010). In this study, team members were asked to rate what percentage of their communication was conducted using various CMC tools. Following this, each CMC tool was weighted based on its relative media richness and finally, virtuality was calculated in conjunction with the physical and temporal dispersion of individuals who were part of the communication exchange. Similarly, some studies with multidimensional operationalizations of virtuality have included considerations of the extent of FtF meetings (e.g., Hakonen & Lipponen, 2008; Schweitzer & Duxbury, 2010; Staples & Webster, 2008).

In sum, our review of the past decade of VT research highlights that there are multiple conceptualizations (see Table 15.2) and ways in which virtuality has been measured (see Table 15.3). While there is a fair amount of consistency regarding the use of geographic dispersion and technology as prominent dimensions, the trend remains to operationalize virtuality as a single dimension. That said; the "single dimension" approach typically centers around either geographic dispersion or technology usage. However as discussed, research is starting to include multiple dimensions of virtuality within a single operationalization. This is an important development and one we hope will become the norm in the future. Beyond being consistent with the definitions, including multiple dimensions allows for a more fine-grained examination of the effects of the various dimensions on team dynamics and performance.

Virtual Team Literature Review

Over the past decade more and more VT researchers have moved away from strict comparisons of virtual and FtF teams and have increasingly considered team virtuality (e.g., Martins et al., 2004). Again, such an approach is consistent with the reality of most organizational teams, which are often hybrid in nature, meaning that they vary in terms of their degree of virtuality (e.g., Griffith et al., 2003; Kirkman & Mathieu, 2005). However, this is not to say that the tradition of comparing purely VTs and FtF teams is over; over 20% of empirical VT work between 2004 and 2014 compared FtF and VTs. Here we review studies that compare FtF and VTs as well as those that examine team virtuality, that is, studies which actually measure the extent of virtuality. We believe this distinction is important because comparisons of FtF and VTs only provide a comparison between "no virtuality" and some level of virtuality but do not actually examine the extent of virtuality and how it may impact team functioning.

Our review is structured in keeping with an organizational team effectiveness framework. While the classic team effectiveness I-P-O model described inputs, processes, and outcomes (Hackman & Morris, 1975), more recent theoretical work has expanded on this framework to include emergent states (i.e., cognitive, motivational, or affective states within the team) as another type of mediator between inputs and processes (Marks, Mathieu, & Zaccaro, 2001). This revised framework, which is referred to as the input-mediator-output-input (IMOI) framework (Ilgen, Hollenbeck, Johnson, & Jundt, 2005) will be utilized within this review as it is in keeping with the current literature.

Within the IMOI model, ***inputs*** represent the team's set of resources, both external (e.g., rewards from the organization) and internal (e.g., team composition or task design). In turn, inputs influence ***mediators***, which consist of a set of psychosocial mechanisms (i.e., processes, emergent states, and mixed mediators; Mathieu et al., 2008) that permit team members to combine the available resources for performing the work. Within our discussion of team processes, we consider the three categories introduced by Marks and colleagues (2001) – transition, action, and interpersonal. ***Transition processes*** pertain to team planning, goal specification, and strategy formulation. Subsequently, teams focus on task accomplish via ***action processes*** such as coordination, communication, and monitoring progress. Finally, ***interpersonal processes*** involve conflict resolution, as well as affect management, and confidence building. Finally, the ***outputs*** considered both affective outcomes as well as team effectiveness (Mathieu & Gilson, 2012) and can be considered from both a multidimensional and multilevel perspective (Mathieu et al., 2008).

The reader should note that in the period 2005–2015, over 1000 articles were published on VTs. However, this number represents practitioner papers, training reviews, academic work that is theoretical in nature, in-depth case studies, as well as more traditional qualitative and quantitative research. In narrowing the number of papers to include in this chapter, we focused solely on studies that were published in peer-reviewed academic journals and academic conferences proceedings. Based on this elimination/inclusion process, we identified 52 studies that compared FtF and VTs and 71 papers that empirically assessed virtuality. As detailed in Table 15.4, VT studies conducted over the past decade have addressed almost every category within the IMOI framework. Below we detail some of the key findings.

Inputs

Within the VT literature, the primary input variables considered have been virtuality, technology (e.g., Cheshin, Kim, Bos Nathan, Ning, & Olson, 2013; Pridmore & Phillips-Wren, 2011; Triana, Kirkman, & Wagstaff, 2012), team composition such as demographic characteristics and team membership changes (e.g., Dineen, 2005; D'Souza & Colarelli, 2010), and leadership (e.g., Whitford & Moss, 2009).

Virtuality as an input In contrast to research that has compared virtual and FtF teams, the majority of research that has examined virtuality has viewed it as a moderating variable, a point we revisit below. However, a few studies have assessed the impact of virtuality as an input (e.g., González-Navarro et al., 2010). Kratzer, Leenders, and Van Engelen (2006) provide evidence for a positive relationship between team creative performance, the variability of team member geographic dispersion, and variety of CMC tools used. Additionally, Kock and Lynn (2012) measured the variety of CMCs used by the team to communicate and found that virtuality helped to facilitate action processes (coordination), which in turn enhanced team efficiency and effectiveness. Peñarroja, Orengo, Zornoza, and Hernández (2013) noted that trust partially mediated the relationship between virtuality and coordination and fully mediated the relationship between virtuality, cooperation, and information exchange. Finally, Zhang and colleagues (2007) examined majority influence during decision-making processes and demonstrated that CMC reduced the occurrence of such influence.

Technology When it comes to technology, members of VTs who employ mixed-media report levels of satisfaction similar to those of FtF teams; however, both are higher than purely VTs (Kennedy, Vozdolska, & McComb, 2010). The frequency of online interactions influences motivational factors that affect knowledge sharing between dispersed team members (Tangirala & Alge, 2006). Furthermore, the actual point in time when the communication medium was introduced influences both perceived inclusion and individual participation. Interestingly, results suggest that women perceive greater inclusion in teams where CMC is used prior to the first meeting as opposed to conducting an FtF meeting first (Triana et al., 2012).

Compositional factors Research examining composition within VTs suggests that many of the negative effects associated with demographic differences appear to dissipate. For instance, Pinjani and Palvia (2013) found that deep-level diversity was negatively related to knowledge sharing, but that this negative relationship was dampened in teams with high task interdependence and greater use of collaborative technologies. Likewise, in a

Table 15.4 Exemplar studies on virtual teams.

Citation	*Input*	*Team processes*			*Emergent states*	*Moderators*	*Outcomes*	
		Transition	*Actional*	*Interpersonal*			*Affective*	*Effectiveness/performance*
Andres, 2012			X		X		X	X
Andressen et al., 2012	X				X	X		X
Berg, 2012	X			X				X
Bierly et al., 2009	X	X	X	X	X	X		
Bourgault et al., 2008	X		X			X		X
Branson et al., 2010	X		X			X		X
Cheshin et al., 2013				X	X			X
Chidambaram & Tung, 2005	X		X		X	X	X	X
Cramton & Webber, 2005	X	X	X	X				X
B. A. De Jong & Dirks, 2012				X	X			X
R. De Jong et al., 2008				X		X		X
Dineen, 2005	X		X		X		X	X
D'Souza & Colarelli, 2010	X							X
Furumo et al., 2009	X				X		X	
Glückler & Schrott, 2007	X		X					X
Hakonen & Lipponen, 2008	X				X	X		
Henderson, 2008	X		X				X	X
Hill et al., 2009	X				X	X		
Hoch & Kozlowski, 2014	X					X		X
Kahai et al., 2012	X			X	X	X	X	X
Kennedy et al., 2010			X	X	X		X	X
Kock & Lynn, 2012	X		X					X
Kratzer et al., 2006	X		X					X
Krebs et al., 2006	X				X	X		X
Lira et al., 2008				X	X	X		
Lowry et al., 2010	X			X	X			
Martinez-Moreno et al., 2009					X	X		X
Maynard et al., 2012	X	X			X	X		X
Mesmer-Magnus et al., 2011	X		X			X	X	
Muethel et al., 2012a	X							X

(*Continued*)

Table 15.4 (Continued)

Citation	*Input*	*Team processes*			*Emergent states*	*Moderators*	*Outcomes*	
		Transition	*Actional*	*Interpersonal*			*Affective*	*Effectiveness/performance*
Muethel et al., 2012b					X	X		X
O'Leary & Mortensen, 2010	X		X	X	X			
Peñarroja et al., 2013	X		X		X			
Pinjani & Palvia, 2013	X		X		X	X	X	X
Powell et al., 2006			X		X		X	
Pridmore & Phillips-Wren, 2011			X					X
Purvanova & Bono, 2009	X					X	X	X
Rapp et al., 2010	X	X	X		X	X		X
Rico et al., 2009			X		X	X		
Rico et al., 2011	X			X		X		X
Rico & Cohen, 2005	X					X		X
Rockmann & Northcraft, 2008		X		X	X		X	
Schweitzer & Duxbury, 2010	X							X
Staples & Webster, 2008	X		X		X	X		X
Staples & Zhao, 2006	X			X	X	X	X	X
Stark & Bierly, 2009	X	X		X		X	X	
Tangirala & Alge, 2006					X		X	
Triana et al., 2012	X						X	X
van der Kleij et al., 2009	X			X		X	X	X
Webster & Wong, 2008	X		X		X		X	
Whitford & Moss, 2009	X					X	X	

quasi-experiment where all electronic communication (i.e., email, discussion board) was anonymous, suggestions from "different others" were more readily presented, allowing multicultural teams to experience increased success (Berg, 2012). In fact, when examining heterogeneous teams, performance was higher in VTs as compared with FtF teams (Staples & Zhao, 2006).

In many instances, the effects of composition factors in VT environments are not straightforward. For instance, although fluidity of team membership in VTs decreases cohesiveness, it also decreases social loafing (Dineen, 2005). Social loafing is a topic that has increasingly been considered within the VT literature, with Alnuaimi, Robert, and Maruping (2010) suggesting that it is more problematic in VTs. In contrast, however Perry, Lorinkova, Hunter, Hubbard, and McMahon (2016) find no significant relationship between virtuality and social loafing, and instead highlight the importance of work–family responsibilities. In a study on preferences for group work, Stark and Bierly (2009) found that team satisfaction was enhanced as virtuality increased. Muethel and colleagues (2012a) examined shared leadership and found that mean team age, and national diversity were positively related to shared leadership whereas the female/male ratio was negatively related to shared leadership, which in turn was positively related to team performance. Similarly in a study of sales teams, Rapp and colleagues (2010) examined the impact of working together on overall team performance. Lastly, research has begun to consider more complex compositional variables such as multiple team membership and the amount of time members allocate to a focal team (e.g., Cummings & Haas, 2012; Maynard et al., 2012). Taken together, the range of results suggest that compositional factors continue to present research opportunities moving forward.

Leadership Understanding the effects of leadership in VTs continues to be of great interest to researchers (e.g., Andressen et al., 2012; Glückler & Schrott, 2007). Here, the various types of leadership behaviors and traits including inspirational (e.g., Joshi et al., 2009), transformational, and transactional (e.g., Huang, Kahai, & Jestice, 2010) have all been examined. For example, Hoch and Kozlowski (2014) found that team virtuality dampened the relationship between hierarchical leadership and team performance, while enhancing the relationship between structural supports and performance. Additionally, transformational leaders have been found to be more effective in virtual as compared to FtF teams (Purvanova & Bono, 2009).

However, researchers have not solely been focused on the leadership styles and behaviors most apt to lead to VT performance enhancement. Specifically, Whitford and Moss (2009) studied the recognition provided by leaders finding that teams prefer leadership that emphasizes personal recognition and that visionary leadership was positively related to work attitudes. Furthermore, perceived managerial support and trust in managers was shown to be highest in FtF teams, followed by VTs comprising core or permanent team members, and lowest within VTs consisting of contract employees (Merriman, Schmidt, & Dunlap-Hinkler, 2007). Henderson (2008) examined project managers and found that team members were more satisfied with their team and the leader when the leader was geographically distant from the team. In addition, team members perceive their leader as being better able to decode messages when that leader was geographically dispersed (Henderson, 2008). Finally, Glückler and Schrott (2007) looked at the communication behaviors of team members to understand the factors resulting in certain members emerging as team leaders.

Mediators

Team processes Compared with the traditional organizational team research, VT studies have highlighted stark differences in team processes. Additionally, while some research has examined the impact of virtuality on composite measures of team processes (e.g., Cramton & Webber, 2005), the majority of research has tended to single out one team process subdimension (Marks et al., 2001): transition, action, interpersonal processes (see Table 15.4). As such, we highlight research as related to each process.

Transition processes. To date, VT research has not extensively examined transition processes. Some exceptions are studies related to the development of strategies to tackle decision-making tasks (Rockmann & Northcraft, 2008) or the nomological network of team transition processes (e.g., Stark & Bierly, 2009). For example, Bierly and colleagues (2009) found that virtuality dampened the positive relationship between goal clarity and trust in their examination of new product development teams. In contrast, Rapp and colleagues (2010) provided evidence that team virtuality moderated the relationship between team experience and planning processes such that higher degrees of virtuality enhanced the positive effect of experience on planning. Finally, Maynard and colleagues (2012) examined transition processes (e.g., preparation activities) and assessed the impact that such processes had on team effectiveness as mediated by transactive memory systems, or what Wegner, Erber, and Raymond (1991) define as a "shared system for encoding, storing, and retrieving information" (p. 923).

Action processes. In contrast to transition processes, action processes have received a great deal of consideration (e.g., Glückler & Schrott, 2007; Rapp et al., 2010; Zhang et al., 2007). In terms of general themes, work here has examined communication and team interaction (e.g., Altschuller & Benbunan-Fich, 2010; Glückler & Schrott, 2007; Guo, D'Ambra, Turner, & Zhang, 2009; Han et al., 2011; Henderson, 2008; Powell, Galvin, & Piccoli, 2006; van der Kleij, Schraagen, Werkhoven, & De Dreu, 2009; Webster & Wong, 2008); cooperation (e.g., Bierly et al., 2009); coordination (e.g., Cummings et al., 2009; Kock & Lynn, 2012; Kratzer et al., 2006; O'Leary & Mortensen, 2010; Peñarroja et al., 2013); extra-role performance (e.g., Ganesh & Gupta, 2010); and knowledge sharing (e.g., Andres, 2006; Hill, Bartol, Tesluk, & Langa, 2009; Staples & Webster, 2008).

In terms of communication, there is evidence that with increased levels of team virtuality, participation in decision-making declines (e.g., Kennedy et al., 2010). However, some research suggests that VTs are more likely to share unique information than FtF teams (e.g., Mesmer-Magnus, DeChurch, Jimenez-Rodriguez, Wildman, & Shuffler, 2011). Given these mixed results in terms of communication, it is important to consider task and temporal factors when examining action processes within VTs. Specifically, Bourgault and colleagues (2008) studied decision-making quality in distributed teams and found that task formalization improved the quality of decision-making, which in turn positively affected team effectiveness. Likewise, Rico, Alcover, Sanchez-Manzanares, and Gil (2009) provided a longitudinal examination that leveraged the punctuated equilibrium model (e.g., long periods of stability with intermittent short periods of change; Gersick, 1991) in examining various types of team communication and its relationship with trust. These authors found that task-oriented, enthusiastic, predictable, and substantive responses had differential relationships with team trust depending on the specific point in the team's lifecycle when such communication occurred (midpoint or end of the project) as well as on the level of task interdependence.

Interpersonal processes. The bulk of VT research addressing interpersonal processes has focused on conflict (e.g., Bierly et al., 2009; Hill et al., 2009). For example, De Jong and colleagues (2008) examined the task conflict–team performance relationship and whether it was moderated by virtuality. These authors found that task conflict has a positive influence on performance at high levels of virtuality, but at low levels of virtuality, this relationship turns negative. Stark and Bierly (2009), in a study of 178 new product development teams, described how relationship conflict's detrimental impact on team member satisfaction strengthens as team virtuality increases. Likewise, O'Leary and Mortensen (2010) studied teams with geographic subgroups and found that the presence of subgroups triggers more conflict. Similarly, while increased relationship conflict decreased group potency in both FtF and VTs, this negative effect was stronger for VTs, meaning that VT members were less likely to believe their team could be effective when faced with relationship conflict (Lira, Ripoll, Peiró, & Orengo, 2008).

In addition to considerations of conflict, other interpersonal processes have been examined such as impression management (Altschuller & Benbunan-Fich, 2010), positive feedback (e.g., Kahai et al., 2012), and group interaction styles (e.g., González-Navarro et al., 2010). van der Kleij and colleagues (2009) found that when VTs use videoconferencing tools, members were more polite, took fewer turns, and interrupted less than FtF team members. As such, these results suggest that the antecedents and implications of interpersonal processes may be altered when examined within VTs.

Emergent states Although VTs appear to lag behind FtF teams with regards to the development of emergent states, numerous studies of team virtuality have included considerations of emergent states. Topics such as shared mental models (e.g., Andres, 2012), trust (e.g., Merriman et al., 2007; Rockmann & Northcraft, 2008), cohesion (e.g., Dineen, 2005; Guo et al., 2009; Huang et al., 2010; Kahai et al., 2012; Staples & Zhao, 2006), team efficacy (Kahai et al., 2012), justice (Hakonen & Lipponen, 2008), psychological safety (e.g., Kirkman, Cordery, Mathieu, Rosen, & Kukenberger, 2013), and transactive memory systems (Maynard et al., 2012; O'Leary & Mortensen, 2010) have all been explored. Below, we provide a summary of work that has examined shared mental models, trust, cohesion, as well as team identification and commitment.

Shared mental models. Some of the variance between virtual and FtF teams in terms of emergent states may be attributed to the use of technology. As shown by Andres (2012), technology may create increased delays in communication of information, decreasing the ability to form shared mental models. This is particularly noteworthy because Guo and colleagues (2009) found that the use of videoconferencing (a richer media) by VTs that possessed shared mental models resulted in the same level of engagement as evidenced within FtF teams.

Trust. Trust continues to receive a great deal of attention in VT research (e.g., Bierly et al., 2009; Peñarroja et al., 2013; Pinjani & Palvia, 2013; Rico et al., 2009). In part, this interest may be attributed to research suggesting that trust is more important in shaping team effectiveness under conditions of geographic dispersion, CMC, and national diversity, which are all often present in VTs (Muethel et al., 2012b). In particular, work suggests that the type of technology used by the VT is important in the development of trust. Specifically, findings suggest that technology that enhances the perception within the team that members are consistently available to each other (virtual co-presence), is significantly related to intra-team trust (Altschuller & Benbunan-Fich, 2010). However, several other studies suggest that trust development may not be different within VTs. Namely, in a study comparing heterogeneous and homogeneous teams, no significant differences in trust formation were found between FtF and VTs (Lowry, Zhang, Zhou, & Fu, 2010). These findings align with Krebs, Hobman, and Bordia's (2006) study, which highlighted that trust formation occurred in all teams over time, even in the presence of demographic dissimilarity. However, when simply comparing virtual and FtF teams without considering demographic implications, VTs appear to have lower levels of trust and support (Merriman et al., 2007). As such, it remains unclear whether virtuality truly shapes the development of trust within VTs or whether the formation of trust in such teams is more reflective of who comprises the team.

Cohesion. Cohesion is most commonly explored in combination with other processes, emergent states, or team compositional factors. For instance, research has considered which dialogue technique is best suited for assisting the development of shared mental models, which in turn, result in increased cohesion (Guo et al., 2009). Team size also has been found to have a significant negative relationship with team cohesiveness in both FtF and VTs, while dispersion does not appear to be related to team cohesiveness (Chidambaram & Tung, 2005). Thus, VTs may experience similar levels of cohesion as do FtF teams and while stability decreases cohesiveness, this effect is similar regardless of team type (Dineen, 2005).

Team identification and commitment. Hakonen and Lipponen (2008) examined 39 VTs from a variety of Finnish-based organizations and found that procedural justice was positively associated with team identification and that this relationship was enhanced as virtuality (measured as the extent of FtF meetings and geographical dispersion) increased. Additionally, Webster and Wong (2008) assessed identity within FtF, hybrid, and virtual teams and provided compelling evidence that local members of the hybrid teams had higher levels of identity with their fellow local members as compared with remote team members. Finally, Joshi and colleagues (2009) considered both commitment and trust and found that inspirational leadership was positively associated with both types of emergent state and that both relationships were positively moderated by team virtuality, such that inspiration leadership was more apt to lead to enhanced levels of commitment and trust within more virtual teams.

Outputs

In keeping with a number of established team outcome frameworks (e.g., Mathieu & Gilson, 2012), research comparing virtual and FtF teams have considered both affective and effectiveness outcomes.

Affective outcomes As discussed by Mathieu and Gilson (2012), beyond merely considering the overall performance of organizational teams, researchers need to also consider affective outcomes or reactions of team members such as satisfaction and viability. Interestingly, our review of the VT literature revealed almost every study that has examined affective reactions has focused on some form of satisfaction (e.g., Henderson, 2008; Pinjani & Palvia, 2013; Stark & Bierly, 2009; Webster & Wong, 2008). For instance, González-Navarro and colleagues (2010) found that in teams that rely heavily on CMC, having a constructive group interaction style is important for team satisfaction.

Additionally, two studies (e.g., Huang et al., 2010; Kahai et al., 2012) examined discussion satisfaction as an outcome variable. Huang and colleagues (2010) found that cohesion (which was impacted by leadership behaviors and media richness) was positively associated with discussion satisfaction. Similarly, Kahai and colleagues (2012) found that teams communicating via instant messaging with transformational leadership had enhanced feedback positivity, which in turn had a positive impact on discussion satisfaction among members.

Team effectiveness As detailed in Table 15.4, the vast majority of VT studies conducted over the past decade have included a consideration of team performance. Unfortunately, given the myriad constructs included when one considers team performance, it is difficult to definitely state the impact of virtuality on effectiveness. For example, Chidambaram and Tung (2005) found that while co-located team members contributed more to decision-making tasks, besides the number of ideas generated, there were no significant differences in team performance. Maynard and colleagues (2012) suggested that virtuality is positively related to team effectiveness via transition processes and TMS. Similarly, enhanced levels of electronic communication and media variety have been shown to positively impact coordination activities, which in turn were positively associated with team efficiency and effectiveness (e.g., Kock & Lynn, 2012). Conversely, Cramton and Webber (2005) demonstrated a negative relationship between virtuality (geographic dispersion), processes, and team effectiveness. Likewise, Schweitzer and Duxbury (2010) found negative relationships between their multidimensional measure of virtuality and team members' perceptions of performance.

Beyond the interest in explaining the impact of virtuality on performance, research also has considered broader measures of effectiveness. For instance, studies have focused on project quality (e.g., Altschuller & Benbunan-Fich, 2010) and team creativity (Han et al., 2011) In addition, the source of effectiveness measures also have varied across studies. For instance, several studies have leveraged team members' perceptions of effectiveness (e.g., Cramton & Webber, 2005; Pinjani & Palvia, 2013), others have utilized outside ratings by leaders or managers (e.g., Andressen et al., 2012; Cummings & Haas, 2012), and yet others have leveraged objective measures of performance (e.g., Rapp et al., 2010; Rico & Cohen, 2005).

Moderators

Beyond considering the direct relationships between inputs, mediators, and outcomes, we also briefly discuss various moderators that have been considered within the VT literature. We contend that it is important to discuss moderators as they yield some particularly interesting findings that serve to clarify some of the relationships with outcomes. While numerous moderators have been considered, the most prominent include team virtuality and task interdependence.

Team virtuality With regard to virtuality as a moderator Andressen and colleagues (2012) examined the relationship between transformational leadership and self-leadership and showed that team leader virtuality (if the leader worked remotely from the team) dampened this relationship. This study also examined team member virtuality and found that when team member virtuality was higher, the relationship between self-leadership and motivation was enhanced. Likewise, Muethel and colleagues (2012b) discovered that three virtuality dimensions (geographic dispersion, use of CMC, and national diversity) enhanced the positive relationship between trust and team effectiveness.

Task interdependence Another moderator that has received considerable attention is task interdependence – the degree to which the task being worked on requires members to work closely together to complete the work rather than handing over work in a sequential manner. Here, Rico and Cohen (2005) examined graduate students who were placed on VTs, and found that performance was largely impacted by the VT's ability to appropriately match the synchronicity of the information and communication technologies used by the teams to communicate and the level of interdependence of the team's task. Likewise, Staples and Webster (2008) examined the relationship between trust and knowledge sharing across various virtuality conditions and showed that this relationship was stronger when task interdependence was low. As such, beyond the impact of team virtuality on the relationship between trust and knowledge sharing, how closely team members needed to coordinate their actions (i.e., task interdependence) enhanced the effect that trust had on team knowledge sharing.

Other moderators In addition to the above, the availability of mixed or an array of media through which to communicate serves as a critical moderator improving the relationship between self-esteem and demographic differences, and creativity (Thatcher & Brown, 2010). Likewise, time pressure moderates the relationship between over-rich media and satisfaction, such that increasing time pressure decreases the need for complex media in VTs (Caballer, Gracia, & Peiró, 2005).

Future Research

Our review of the past decade of VT research shows the field continues to advance and that the state of the science is in a good place. This is particularly important given the relevance and growth of VT usage within organizations (Society for Human Resource Management, 2012). However, even with the amount of progress that has been made, there remain numerous research opportunities or 'gaps' that still exist. In this final section, we highlight a few of these opportunities, again using the IMOI framework.

Inputs

Technology, now probably more than ever, is an input variable in need of more detailed consideration. In addition, while leadership has over the past decade received increasing attention, some of the more nuanced forms of leadership (i.e., shared leadership) that may be particularly critical in VT settings have yet to receive much attention. We highlight the opportunities that exist involving contextual and compositional factors.

Technology Today, a multitude of technologies are available to teams and how best to leverage them, and when to use which one, are critical questions that need to be considered moving forward. Teams no longer solely rely on one technology such as email to communicate; instead, teams today have at their disposal a range of options that all have benefits and trade-offs. Related to which, an interesting avenue for future research focuses on the importance of an initial FtF meeting that early research found to be critical (e.g., Geber, 1995). Recent work finds that when students are assigned to VTs, the first thing they do is look up other members using social media (Maynard et al., 2012). Given such behavior, research is needed to explore whether new media (i.e., social media) might now replace the initial FtF meeting. Likewise, what is the role of social media with regard to group functioning? Social media is further interesting because much of it is personal in nature. Although early VT research found that teams that communicate mainly via technology focus primarily on task accomplishment (e.g., Massey, Montoya-Weiss, & Hung, 2003), it remains unclear whether social media augments or detracts from this focus, not to mention its influence on team interpersonal processes, affective outcomes, and member liking.

While the broader role of technology has received attention in the extant literature and is acknowledged as important, there is a lack of research on some of the "newer" technologies. Despite the frequent acknowledgment that new technologies are common (Purvanova & Bono, 2009), most published work continues to examine instant messaging, phone, videoconferencing, or email platforms (e.g., Andres, 2006; Boh, Ren, Kiesler, & Bussjaeger, 2007; Caballer et al., 2005; Guo et al., 2009). We acknowledge that some of this might be attributable to a lag in publication time. That being said, from what we see in the literature, exploring more complex technological features such as mobile apps, avatars, or Second Life may be more realistic and answer interesting questions with regard to team processes. Additionally, as we consider technology, the age of team members should become a compositional factor that needs to be more fully integrated. To date, age has not received much attention in the VT literature, however, the millennial generation that is now entering the workforce has grown up using multiple complex technologies simultaneously. How then will technology and age interact to influence technology choice, types of communication, team processes, and subsequent team outcomes? The salience of age of personnel within organizations is discussed more broadly by Burlacu, Truxillo, and Bauer (2017; Chapter 20 in this volume).

Leadership While leadership has received a great deal of research attention over the past decade, one area that has received limited consideration, and might be particularly relevant is shared leadership (see Carte, Chidambaram, & Becker, 2006; Muethel et al., 2012a for exceptions). Traditional theories of leadership conceptualize leadership as a hierarchical, top-down, and formal position within an organization (Pearce, 2004). Given this, the leadership function is usually assigned to a single individual. However, as work becomes more complex and technologically enabled, it is harder to find an individual who has all the knowledge, skills, and abilities necessary to lead (Pearce & Manz, 2005). This might be particularly true in virtual contexts where individuals are often brought together because of their particular areas of task-relevant expertise (e.g., Townsend, DeMarie, & Hendrickson, 1998) and not because of their leadership expertise.

Shared leadership is characterized as mutual influence among team members in areas related to direction, motivation, and support (Carson, Tesluk, & Marrone, 2007). While shared leadership is beneficial for performance in FtF teams (Carson et al., 2007), Pearce, Yoo, and Alavi (2004) found that shared leadership was a better predictor of problem-solving quality than traditional vertical leadership in a sample of social workers working on educational programs in VTs. This finding was extended by Hoch and Kozlowski (2014), who found shared leadership to be positively linked to performance, regardless of the degree of virtuality. Given the promising results of these studies, more research is needed to fully understand the role of shared leadership within VTs and questions such as: When working in a VT, how do multiple leaders emerge? How are team functions, processes, or tasks shared among the said leaders? These questions would be valuable to consider within VT research over the coming decade. Finally, Kahai, Avolio, and Sosik (2017, Chapter 14 in this volume) offer a broader discussion of the leadership topic within this context.

Contextual factors In the IMOI model, organizational and team contexts play a crucial role as input factors. VTs are complex, adaptive, and dynamic systems that exist in the larger context of people, tasks, technologies, and settings (e.g., Martins et al., 2004). A unique feature of VTs has been that they allow members to work together regardless of time, space, and place. One of the outcomes of technological developments is that they afford each individual member the opportunity to work flexibly from multiple different places. Increasing mobile technologies and wireless connections in particular have made it possible for VT members to not only work in and from multiple places, but also move between locations. This mobility means that work can be done from changing environments. However, to date, research has yet to fully appreciate this fact and the influence it may have on the internal dynamics and the performance of VTs. As such, we would encourage future VT research to investigate the role of mobility and the use of multiple locations by a single member.

Compositional factors Another gap we want to call attention to is the need for attention to "new" composition factors such as: familiarity, subgroups, and multiple team membership. While a few studies have included member familiarity (e.g., Bierly et al., 2009; Stark & Bierly, 2009), this area is in need of more attention. For instance, Tangirala and Alge (2006) found that familiarity mediated the relationship between media and fairness perceptions. Here, future research could explore the role that familiarity plays as a mediator as well as moderator of relationships involving virtuality. Specifically, are members who are more familiar with one another better able to overcome certain technology or team member difficulties? Lastly, research has yet to explore the means by which familiarity develops within virtual contexts and to what extent virtuality affects team familiarity. Lastly, research is starting to explore the topics of subgroups (e.g., Fiol & O'Connor, 2005; Polzer, Crisp, Jarvenpaa, & Kim, 2006) along with multiple team membership and their influence on VT dynamics and performance (e.g., Maynard et al., 2012). That said, the attention given to these topics is limited and more work is certainly warranted. In particular, as demonstrated by Connaughton, Shuffler, and Goodwin (2011), multi-team system research and theories (e.g., Marks, DeChurch, Mathieu, Panzer, & Alonso, 2005) could be more fully leveraged to provide insight into how technology can facilitate or hinder multiple teams working together.

Team processes as mediators

While research has considered all three team processes, the level of attention has varied, with action processes receiving the most consideration. Within interpersonal processes, conflict management has received a fair amount of attention whereas motivation and affect management are much less studied. However, the area that we understand least well is transition processes, which is ironic given that VTs often come together to work on specific task, and hence the planning should be critical.

Transition processes While VTs such as distributed procurement teams within larger organizations may retain the same level of team virtuality throughout the team's lifecycle, others are more short term in nature as they come together to address a specific problem and disband once the task is complete (e.g., Powell, Piccoli, & Ives, 2004). For these types of team, planning should be paramount in determining how well the team members will work together. For example, it would be interesting to examine how team charters (Mathieu & Rapp, 2009) might play out in VT settings. Studies also could compare planning at different stages, the impact that planning may have when performed by individual members versus the team as a whole, and what technologies are best suited for planning processes. Future research in this area could leverage work on goal setting and management by objectives within the broader organizational team literature. These ideas serve only to highlight that there is much work that can, and should be done with regard to transition processes in VTs.

Emergent states

While trust, cohesion, commitment, identity, and more recently shared cognition (e.g., Chen et al., 2013) have all received attention, other emergent states such as group potency and team empowerment have not. While group potency has been demonstrated to be a powerful predictor of team performance (e.g., Gully, Incalcaterra, Joshi, & Beaubien, 2002), the VT literature remains silent here. Likewise, in a review of the team empowerment literature, Maynard, Gilson, and Mathieu (2012) found only one study that considered VTs (Kirkman, Rosen, Tesluk, & Gibson, 2004), which revealed that psychological empowerment was a stronger predictor of effectiveness when teams were not able to meet FtF (i.e., higher team virtuality). Empowerment initiatives have been found to enhance performance, well-being, and positive attitudes (e.g., Kirkman & Rosen, 1999), which raises numerous research questions within the VT area, such as: Does this same relationship hold when employees are members of VTs? More fundamentally, what does empowerment look like in a VT? And how does virtuality affect empowerment?

Outcomes

The preponderance of VT research has focused on team effectiveness and performance-type outcome measures. Moving forward, we urge researchers to keep going with this trajectory, but where possible to use objective measures of performance. For example, researchers could address questions such as how VTs perform on a specific task and if performance is improved when all of the best individual performers are on one team. Furthermore, understanding members' affective reactions is also relevant. Do members want to work together in the future? Are they satisfied with how the team performed as a unit? Research has started to address some of these questions, but much more depth is needed. Lastly, as Mathieu and Gilson (2012) outlined, what determines whether a team is effective is complex and can vary based on context. Accordingly, researchers could explore whether there are other measures of effectiveness that are more relevant to VTs and does this change as virtuality increases, or decreases?

Moderators

Time has been cited as being a critical yet understudied construct in organizational behavior research (Shipp, Edwards, & Lambert, 2009). Given that VTs are often formed to work on time-sensitive projects (e.g., Caballer et al., 2005), are there time implications with regard to team composition, technology, leadership, processes, and outcomes? In addition, what are the roles of various time conceptualizations on VT processes and outcomes (i.e., chronological time, time orientation, time pacing, time pressure)? The time literature argues that some individuals have a long-term orientation while others are more short-term focused (Shipp et al., 2009). When teams communicate via technology, do these orientations become more salient? For example, some team members might reply to an email as soon as they get it while others might wait 24 hours. How is this time lag perceived? These research questions reinforce our prior assertion that VT research needs to more fully adopt a longitudinal approach to better capture temporal considerations.

Conclusion

In conclusion, VT researchers have been extremely active over the past decade, and this chapter has provided a review and proposed avenues in need of additional consideration. A great deal of work has examined how VTs compare to their FtF counterparts, as well as the impact of virtuality on team dynamics and performance. Likewise, much of this work has been designed to leverage in-depth quantitative as well as qualitative analyses to understand these teams in context. Each of these trends demonstrates how VT research has grown and developed. However, as the use of VTs continues to rise, we urge researchers to continue and expand this line of inquiry. We look forward to the next decade of VT research.

References

Alnuaimi, O. A., Robert, L. P., & Maruping, L. M. (2010). Team size, dispersion, and social loafing in technology-supported teams: A perspective on the theory of moral disengagement. *Journal of Management Information Systems, 27*(1), 203–230.

Altschuller, S., & Benbunan-Fich, R. (2010). Trust, performance, and the communication process in ad hoc decision-making virtual teams. *Journal of Computer-Mediated Communication, 16*(1).

Andres, H. P. (2006). The impact of communication medium on virtual team group process. *Information Resources Management Journal, 19*(2), 1–17.

Andres, H. P. (2012). Technology-mediated collaboration, shared mental model and task performance. *Journal of Organizational & End User Computing, 24*(1), 64–81.

Andressen, P., Konradt, U., & Neck, C. P. (2012). The relation between self-leadership and transformational leadership: Competing models and the moderating role of virtuality. *Journal of Leadership & Organizational Studies, 19*(1), 68–82.

Arling, P. A., & Subramani, M. (2011). The effect of virtuality on individual network centrality and performance in on-going, distributed teams. *International Journal of Internet & Enterprise Management, 7*(4), 325–348.

Baltes, B. B., Dickson, M. W., Sherman, M. P., Bauer, C. C., & LaGanke, J. S. (2002). Computer-mediated communication and group decision making: A meta-analysis. *Organizational Behavior and Human Decision Processes, 87*, 156–179.

Balthazard, P. A., Waldman, D. A., & Warren, J. E. (2009). Predictors of the emergence of transformational leadership in virtual decision teams. *Leadership Quarterly, 20*(5), 651–663.

Bass, B. M. (1999). Two decades of research and development in transformational leadership. *European Journal of Work and Organizational Psychology, 8*(1), 9–32.

Behrend, F. D., & Erwee, R. (2009). Mapping knowledge flows in virtual teams with SNA. *Journal of Knowledge Management, 13*(4), 99–114.

Berg, R. W. (2012). The anonymity factor in making multicultural teams work: Virtual and real teams. *Business Communication Quarterly, 75*(4), 404–424.

Bierly, P. E., Stark, E. M., & Kessler, E. H. (2009). The moderating effects of virtuality on the antecedents and outcome of NPD team trust. *Journal of Product Innovation Management, 26*(5), 551–565.

Boh, W. F., Ren, Y., Kiesler, S., & Bussjaeger, R. (2007). Expertise and collaboration in the geographically dispersed organization. *Organization Science, 18*(4), 595–612.

Bollen, K. A. (1989). *Structural equations with latent variables.* New York: John Wiley & Sons, Inc.

Boros, S., Meslec, N., Curseu, P. L., & Emons, W. (2010). Struggles for cooperation: Conflict resolution strategies in multicultural groups. *Journal of Managerial Psychology, 25*(5), 539–554.
Bosch-Sijtsema, P. M., & Sivunen, A. (2013). Professional virtual worlds supporting computer-mediated communication, collaboration, and learning in geographically distributed contexts. *IEEE Transactions on Professional Communication, 56*(2), 160–175.
Bouas, K. S., & Arrow, H. (1996). The development of group identity in computer and face-to-face groups with membership change. *Computer Supported Cooperative Work (CSCW), 4*, 153–178.
Bourgault, M., Drouin, N., & Hamel, E. (2008). Decision making within distributed project teams: An exploration of formalization and autonomy as determinants of success. *Project Management Journal, 39*, S97–S110.
Branson, L., Steele, N. L., & Chung-Hsien, S. (2010). Differences in the decision making of constructively styled virtual and face-to-face teams regarding performance appraisal. *International Journal of Business & Public Administration, 7*(2), 11–22.
Breuer, C., Huffmeier, J., & Hertel, G. (2016). Does trust matter more in virtual teams? A meta-analysis of trust and team effectiveness considering virtuality and documentation as moderators. *Journal of Applied Psychology. 101*, 1151–1177.
Burlacu, G., Truxillo, D. M. & Bauer, T. N. (2017). Employee age differences in using Internet-based tools at work. In G. Hertel, D. L. Stone, R. D. Johnson, & J. Passmore (Eds.), *The Wiley Blackwell handbook of the psychology of the internet at work* (pp. 449–480). Chichester: John Wiley & Sons, Ltd.
Caballer, A., Gracia, F., & Peiró, J. (2005). Affective responses to work process and outcomes in virtual teams: Effects of communication media and time pressure. *Journal of Managerial Psychology, 20*(3/4), 245–260.
Capece, G., & Costa, R. (2009). Measuring knowledge creation in virtual teams through the social network analysis. *Knowledge Management Research & Practice, 7*(4), 329–338.
Carson, J. B., Tesluk, P. E., & Marrone, J. A. (2007). Shared leadership in teams: An investigation of antecedent conditions and performance. *Academy of Management Journal, 50*(5), 1217–1234.
Carte, T., Chidambaram, L., & Becker, A. (2006). Emergent leadership in self-managed virtual teams. *Group Decision & Negotiation, 15*(4), 323–343.
Chen, X. G., Li, X., Clark, J. G., & Dietrich, G. B. (2013). Knowledge sharing in open source software project teams: A transactive memory system perspective. *International Journal of Information Management, 33*(3), 553–563.
Cheshin, A., Kim, Y., Bos Nathan, D., Ning, N., & Olson, J. S. (2013). Emergence of differing electronic communication norms within partially distributed teams. *Journal of Personnel Psychology, 12*(1), 7–21.
Chidambaram, L., & Tung, L. L. (2005). Is out of sight, out of mind? An empirical study of social loafing in technology-supported groups. *Information Systems Research, 16*(2), 149–168.
Chudoba, K. M., Wynn, E., Lu, M., & Watson-Manheim, M. B. (2005). How virtual are we? Measuring virtuality and understanding its impact in a global organization. *Information Systems Journal, 15*(4), 279–306.
Clark, W. R., Clark, L. A., & Crossley, K. (2010). Developing multidimensional trust without touch in virtual teams. *Marketing Management Journal, 20*(1), 177–193.
Cohen, S. G., & Gibson, C. B. (2003). In the beginning: Introduction and framework In *Virtual teams that work: Creating conditions for virtual team effectiveness* (pp. 1–13). San Francisco, CA: Jossey-Bass.
Cohen-Charash, Y., & Spector, P. E. (2001). The role of justice in organizations: A meta analysis. *Organizational Behavior and Human Decision Processes, 86*, 278–321.
Connaughton, S., Shuffler, M., & Goodwin, G. F. (2011). Leading distributed teams: The communicative constitution of leadership. *Military Psychology, 23*(5), 502–527.
Cramton, C. D., & Webber, S. S. (2005). Relationships among geographic dispersion, team processes, and effectiveness in software development work teams. *Journal of Business Research, 58*(6), 758–765.
Cummings, J. N., Espinosa, J. A., & Pickering, C. K. (2009). Crossing spatial and temporal boundaries in globally distributed projects: A relational model of coordination delay. *Information Systems Research, 20*(3), 420–439.

Cummings, J. N., & Haas, M. R. (2012). So many teams, so little time: Time allocation matters in geographically dispersed teams. *Journal of Organizational Behavior, 33*(3), 316–341.

Daassi, M., Jawadi, N., Favier, M., & Kalika, M. (2006). An empirical investigation of trust's impact on collective awareness development in virtual teams. *International Journal of Networking & Virtual Organisations, 3*(4), 378–394.

De Guinea, A. O., Webster, J., & Staples, D. S. (2012). A meta-analysis of the consequences of virtualness on team functioning. *Information & Management, 49*(6), 301–308.

De Jong, B. A., & Dirks, K. T. (2012). Beyond shared perceptions of trust and monitoring in teams: Implications of asymmetry and dissensus. *Journal of Applied Psychology, 97*(2), 391–406.

De Jong, B. A., & Elfring, T. (2010). How does trust affect the performance of ongoing teams? The mediating role of reflexivity, monitoring and effort. *Academy of Management Journal, 53*(3), 535–549.

De Jong, R., Schalk, R., & Curseu, P. L. (2008). Virtual communicating, conflicts and performance in teams. *Team Performance Management, 14*(7/8), 364–380.

Dineen, B. R. (2005). TeamXchange: A team project experience involving virtual teams and fluid team membership. *Journal of Management Education, 29*(4), 593–616.

Dixon, K. R., & Panteli, N. (2010). From virtual teams to virtuality in teams. *Human Relations, 63*, 1177–1197.

D'Souza, G. C., & Colarelli, S. M. (2010). Team member selection decisions for virtual versus face-to-face teams. *Computers in Human Behavior, 26*(4), 630–635.

Espinosa, J. A., & Clark, M. A. (2014). Team knowledge representation: A network perspective. *Human Factors, 56*(2), 333–348.

Fiol, C. M., & O'Connor, E. J. (2005). Identification in face-to-face, hybrid, and pure virtual teams: Untangling the contradictions. *Organization Science, 16*(1), 19–32.

Furumo, K., de Pillis, E., & Green, D. (2009). Personality influences trust differently in virtual and face-to-face teams. *International Journal of Human Resources Development & Management, 9*(1), 36–58.

Ganesh, M. P., & Gupta, M. (2010). Impact of virtualness and task interdependence on extra-role performance in software development teams. *Team Performance Management, 16*(3/4), 169–186.

Geber, B. (1995). Virtual teams. *Training, 32*(4), 36–40.

Geister, S., Konradt, U., & Hertel, G. (2006). Effects of process feedback on motivation, satisfaction, and performance in virtual teams. *Small Group Research, 37*(5), 459–489.

Gersick, C. J. (1991). Revolutionary change theories: A multilevel exploration of the punctuated equilibrium paradigm. *Academy of Management Review, 16*(1), 10–36.

Gibson, C. B., & Gibbs, J. L. (2006). Unpacking the concept of virtuality: The effects of geographic dispersion, electronic dependence, dynamic structure, and national diversity on team innovation. *Administrative Science Quarterly, 51*(3), 451–495.

Gibson, C. B., Gibbs, J. L., Stanko, T. L., Tesluk, P., & Cohen, S. G. (2011). Including the "I" in virtuality and modern job design: Extending the job characteristics model to include the moderating effect of individual experiences of electronic dependence and copresence. *Organization Science, 22*, 1481–1499.

Gilson, L. L., Maynard, M. T., Jones Young, N., Vartiainen, M., & Hakonen, M. (2015). Virtual teams research: Ten years, ten themes, and ten opportunities. *Journal of Management, 41*, 1313–1337.

Gilson, L. L., Maynard, T., & Bergeil, E. B. (2013). Working in virtual teams: An experiential activity. *Small Group Research, 44*(4), 412–427.

Glückler, J., & Schrott, G. (2007). Leadership and performance in virtual teams: Exploring brokerage in electronic communication. *International Journal of e-Collaboration, 3*(3), 31–52.

Golden, T. D., & Fromen, A. (2011). Does it matter where your manager works? Comparing managerial work mode (traditional, telework, virtual) across subordinate work experiences and outcomes. *Human Relations, 64*(11), 1451–1475.

González-Navarro, P., Orengo, V., Zornoza, A., Ripoll, P., & Peiró, J. M. (2010). Group interaction styles in a virtual context: The effects on group outcomes. *Computers in Human Behavior, 26*(6), 1472–1480.

Grant, A. M., & Wall, T. D. (2009). The neglected science and art of quasi-experimentation: Why-to, when-to, and how-to advice for organizational researchers. *Organizational Research Methods*, *12*(4), 653–686.

Griffith, T. L., Sawyer, J. E., & Neale, M. A. (2003). Virtualness and knowledge in teams: Managing the love triangle of organizations, individuals, and information technology. *MIS Quarterly*, *27*, 265–287.

Gully, S. M., Incalcaterra, K. A., Joshi, A., & Beaubien, J. M. (2002). A meta-analysis of team-efficacy, potency, and performance: Interdependence and level of analysis as moderators of observed relationships. *Journal of Applied Psychology*, *87*(5), 819–832.

Guo, Z., D'Ambra, J., Turner, T., & Zhang, H. (2009). Improving the effectiveness of virtual teams: A comparison of video-conferencing and face-to-face communication in China. *IEEE Transactions on Professional Communication*, *52*(1), 1–16.

Hackman, J. R., & Morris, C. G. (1975). Group tasks, group interaction process, and group performance effectiveness: A review and proposed integration. In L. Berkowitz (Ed.), *Advances in Experimental Social Psychology* (Vol. *8*, pp. 45–99). New York, NY: Academic Press.

Hakonen, M., & Lipponen, J. (2008). Procedural justice and identification with virtual teams: The moderating role of face-to-face meetings and geographical dispersion. *Social Justice Research*, *21*(2), 164–178.

Han, H. J., Hiltz, S. R., Fjermestad, J., & Wang, Y. Q. (2011). Does medium matter? A comparison of initial meeting modes for virtual teams. *IEEE Transactions on Professional Communication*, *54*(4), 376–391.

Harvey, M., Novicevic, M. M., & Garrison, G. (2005). Global virtual teams: A human resource capital architecture. *International Journal of Human Resource Management*, *16*(9), 1583–1599.

Haslam, S. A. (2001). *Psychology in organizations: The social identity approach*. Newbury Park, CA: Sage.

Henderson, L. S. (2008). The impact of project managers' communication competencies: Validation and extension of a research model for virtuality, satisfaction, and productivity on project teams. *Project Management Journal*, *39*(2), 48–59.

Hertel, G., Geister, S., & Konradt, U. (2005). Managing virtual teams: A review of current empirical research. *Human Resource Management Review*, *15*(1), 69–95.

Hill, N. S., Bartol, K. M., Tesluk, P. E., & Langa, G. A. (2009). Organizational context and face-to-face interaction: Influences on the development of trust and collaborative behaviors in computer-mediated groups. *Organizational Behavior and Human Decision Processes*, *108*(2), 187–201.

Hoch, J. E., & Kozlowski, S. W. J. (2014). Leading virtual teams: Hierarchical leadership, structural supports, and shared team leadership. *Journal of Applied Psychology*, *99*(3), 390–403.

Hsu, J. L., & Chou, H. W. (2009). The effects of communicative genres on intra-group conflict in virtual student teams. *International Journal of Distance Education Technologies*, *7*(1), 1–22.

Huang, R., Kahai, S., & Jestice, R. (2010). The contingent effects of leadership on team collaboration in virtual teams. *Computers in Human Behavior*, *26*(5), 1098–1110.

Ilgen, D. R., Hollenbeck, J. R., Johnson, M., & Jundt, D. (2005). Teams in organizations: From input-process-output models to IMOI models. *Annual Review of Psychology*, *56*, 517–543.

Jarvenpaa, S. L., Cantu, C. & Lim, S. Y. (2017). Trust in virtual online environments. In G. Hertel, D. L. Stone, R. D. Johnson, & J. Passmore (Eds.), *The Wiley Blackwell handbook of the psychology of the internet at work* (pp. 103–130). Chichester: John Wiley & Sons, Ltd.

Jarvenpaa, S. L., & Leidner, D. E. (1999). Communication and trust in global virtual teams. *Organization Science*, *10*, 791–815.

Joshi, A., Lazarova, M. B., & Liao, H. (2009). Getting everyone on board: The role of inspirational leadership in geographically dispersed teams. *Organization Science*, *20*(1), 240–252.

Kahai, S., Avolio, B. J. & Sosik, J. (2017). E-leadership. In G. Hertel, D. L. Stone, R. D. Johnson, & J. Passmore (Eds.), *The Wiley Blackwell handbook of the psychology of the internet at work* (pp. 285–314). Chichester: John Wiley & Sons, Ltd.

Kahai, S. S., Huang, R., & Jestice, R. J. (2012). Interaction effect of leadership and communication media on feedback positivity in virtual teams. *Group & Organization Management*, *37*(6), 716–751.

Kennedy, D. M., Vozdolska, R. R., & McComb, S. A. (2010). Team decision making in computer-supported cooperative work: How initial computer-mediated or face-to-face meetings set the stage for later outcomes. *Decision Sciences, 41*(4), 933–954.

Khan, M. M., Lodhi, S. A., & Makki, M. A. M. (2010). Moderating role of team working environment between team implicit coordination and performance. *African Journal of Business Management, 4*, 2743–2752.

Kirkman, B. L., Cordery, J. L., Mathieu, J., Rosen, B., & Kukenberger, M. (2013). Global organizational communities of practice: The effects of nationality diversity, psychological safety, and media richness on community performance. *Human Relations, 66*(3), 333–362.

Kirkman, B. L., & Mathieu, J. E. (2005). The dimensions and antecedents of team virtuality. *Journal of Management, 31*(5), 700–718.

Kirkman, B. L., & Rosen, B. (1999). Beyond self-management: Antecedents and consequences of team empowerment. *Academy of Management Journal, 42*, 58–74.

Kirkman, B. L., Rosen, B., Tesluk, P. E., & Gibson, C. B. (2004). The impact of team empowerment on virtual team performance: The moderating role of face-to-face interaction. *Academy of Management Journal, 47*, 175–192.

Kock, N., & Lynn, G. S. (2012). Electronic media variety and virtual team performance: The mediating role of task complexity coping mechanisms. *IEEE Transactions on Professional Communication, 55*(3), 325–344.

Kozlowski, S. W. (2015). Advancing research on team process dynamics: Theoretical, methodological, and measurement considerations. *Organizational Psychology Review, 5*(4), 270–299.

Kratzer, J., Leenders, R., & Van Engelen, J. M. L. (2006). Managing creative team performance in virtual environments: An empirical study in 44 R&D teams. *Technovation, 26*(1), 42–49.

Krebs, S. A., Hobman, E. V., & Bordia, P. (2006). Virtual teams and group member dissimilarity–Consequences for the development of trust. *Small Group Research, 37*(6), 721–741.

Leonard, L. N. K., & Haines, R. (2007). Computer-mediated group influence on ethical behavior. *Computers in Human Behavior, 23*(5), 2302–2320.

Levitt, S. D., & List, J. A. (2007). What do laboratory experiments measuring social preferences reveal about the real world? *Journal of Economic Perspectives, 21*, 153–174.

Lira, E. M., Ripoll, P., Peiró, J. M., & Orengo, V. (2008). How do different types of intragroup conflict affect group potency in virtual compared with face-to-face teams? A longitudinal study. *Behaviour & Information Technology, 27*(2), 107–114.

Lowry, P. B., Zhang, D. S., Zhou, L. N., & Fu, X. L. (2010). Effects of culture, social presence, and group composition on trust in technology-supported decision-making groups. *Information Systems Journal, 20*(3), 297–315.

Luke, D. A. (2004). *Multilevel modeling*. London: Sage.

Marks, M. A., DeChurch, L. A., Mathieu, J. E., Panzer, F. J., & Alonso, A. (2005). Teamwork in multiteam systems. *Journal of Applied Psychology, 90*(5), 964–971.

Marks, M. A., Mathieu, J. E., & Zaccaro, S. J. (2001). A temporally based framework and taxonomy of team processes. *Academy of Management Review, 26*, 356–376.

Martinez-Moreno, E., González-Navarro, P., Zornoza, A., & Ripoll, P. (2009). Relationship, task and process conflicts and team performance: The moderating role of communication media. *International Journal of Conflict Management, 20*(3), 251–268.

Martins, L. L., Gilson, L. L., & Maynard, M. T. (2004). Virtual teams: What do we know and where do we go from here? *Journal of Management, 30*(6), 805–835.

Martins, L. L., & Shalley, C. E. (2011). Creativity in virtual work: Effects of demographic differences. *Small Group Research, 42*(5), 536–561.

Massey, A. P., Montoya-Weiss, M. M., & Hung, Y. T. (2003). Because time maters: Temporal coordination in global virtual project teams. *Journal of Management Information Systems, 19*(4), 129–156.

Mathieu, J. E., & Gilson, L. L. (2012). Critical issues and team effectiveness. In S. J. W. Kozlowski (Ed.), *Oxford handbook of industrial and organizational psychology* (pp. 910–930). New York, NY: Oxford University Press.

Mathieu, J. E., Maynard, M. T., Rapp, T., & Gilson, L. L. (2008). Team effectiveness 1997–2007: A review of recent advancements and a glimpse into the future. *Journal of Management, 34*(3), 410–476.

Mathieu, J. E., & Rapp, T. L. (2009). Laying the foundation for successful team performance trajectories: The roles of team charters and performance strategies. *Journal of Applied Psychology, 94*(1), 90–103.

Matlay, H., & Westhead, P. (2005). Virtual teams and the rise of e-entrepreneurship in Europe. *International Small Business Journal, 23*(3), 279–302.

Mayer, R. C., Davis, J. H., & Schoorman, F. D. (1995). An integrative model of organizational trust. *Academy of Management Review, 20*, 709–734.

Maynard, M. T., Gilson, L. L., & Mathieu, J. E. (2012). Empowerment – fad or fab? A multilevel review of the past two decades of research. *Journal of Management, 38*(4), 1231–1281.

Maynard, M. T., Mathieu, J. E., Rapp, T. L., & Gilson, L. L. (2012). Something(s) old and something(s) new: Modeling drivers of global virtual team effectiveness. *Journal of Organizational Behavior, 33*(3), 342–365.

McAllister, D. J. (1995). Affect- and cognition-based trust as foundations for interpersonal cooperation in organizations. *Academy of Management Journal, 38*, 24–59.

McAvoy, J., & Butler, T. (2006). A paradox of virtual teams and change: An implementation of the theory of competing commitments. *International Journal of e-Collaboration, 2*(3), 1–24.

Merriman, K. K., Schmidt, S. M., & Dunlap-Hinkler, D. (2007). Profiling virtual employees: The impact of managing virtually. *Journal of Leadership & Organizational Studies, 14*(1), 6–15.

Mesmer-Magnus, J. R., DeChurch, L. A., Jimenez-Rodriguez, M., Wildman, J., & Shuffler, M. (2011). A meta-analytic investigation of virtuality and information sharing in teams. *Organizational Behavior and Human Decision Processes, 115*(2), 214–225.

Metiu, A. (2006). Owning the code: Status closure in distributed groups. *Organization Science, 17*(4), 418–435.

Muethel, M., Gehrlein, S., & Hoegl, M. (2012a). Socio-demographic factors and shared leadership behaviors in dispersed teams: Implications for human resource management. *Human Resource Management, 51*(4), 525–548.

Muethel, M., Siebdrat, F., & Hoegl, M. (2012b). When do we really need interpersonal trust in globally dispersed new product development teams? *R & D Management, 42*(1), 31–46.

Ocker, R. J. (2007). A balancing act: The interplay of status effects on dominance in virtual teams. *IEEE Transactions on Professional Communication, 50*(3), 204–218.

O'Leary, M. B., & Cummings, J. N. (2007). The spatial, temporal, and configurational characteristics of geographic dispersion in teams. *MIS Quarterly, 31*(3), 433–452.

O'Leary, M. B., & Mortensen, M. (2010). Go (con)figure: Subgroups, imbalance, and isolates in geographically dispersed teams. *Organization Science, 21*(1), 115–131.

Panteli, N., & Duncan, E. (2004). Trust and temporary virtual teams: Alternative explanations and dramaturgical relationships. *Information Technology & People, 17*(4), 423–441.

Paul, S., Seetharaman, P., Samarah, I., & Mykytyn, P. P. (2004). Impact of heterogeneity and collaborative conflict management style on the performance of synchronous global virtual teams. *Information & Management, 41*(3), 303–321.

Pearce, C. L. (2004). The future of leadership: Combining vertical and shared leadership to transform knowledge work. *Academy of Management Executive, 18*(1), 47–57.

Pearce, C. L., & Manz, C. C. (2005). The new silver bullets of leadership: The importance of self- and shared leadership in knowledge work. *Management Department Publications. Paper 72.* Retrieved April 6, 2017, from http://digitalcommons.unl.edu/managementfacpub/72

Pearce, C. L., Yoo, Y., & Alavi, M. (2004). Leadership, social work and virtual teams: The relative influence of vertical vs. shared leadership in the nonprofit section. In R. E. Riggio & S. Smith-Orr (Eds.), *Improving leadership in nonprofit organizations* (pp. 180–203). San Francisco, CA: Jossey Bass.

Peñarroja, V., Orengo, V., Zornoza, A., & Hernández, A. (2013). The effects of virtuality level on task-related collaborative behaviors: The mediating role of team trust. *Computers in Human Behavior, 29*(3), 967–974.

Perry, S. J., Lorinkova, N. M., Hunter, E. M., Hubbard, A., & McMahon, J. T. (2016). When does virtuality really "work"? Examining the role of work-family and virtuality in social loafing. *Journal of Management, 42*(2), 449–479.

Pinjani, P., & Palvia, P. (2013). Trust and knowledge sharing in diverse global virtual teams. *Information & Management, 50*(4), 144–153.

Podsakoff, P. M., MacKenzie, S. B., & Podsakoff, N. P. (2012). Sources of method bias in social science research and recommendations on how to control it. *Annual Review of Psychology, 63*, 539–569.

Polzer, J. T., Crisp, C. B., Jarvenpaa, S. L., & Kim, J. W. (2006). Extending the faultline model to geographically dispersed teams: How colocated subgroups can impair group functioning. *Academy of Management Journal, 49*(4), 679–692.

Powell, A., Galvin, J., & Piccoli, G. (2006). Antecedents to team member commitment from near and far: A comparison between collocated and virtual teams. *Information Technology & People, 19*(4), 299–322.

Powell, A., Piccoli, G., & Ives, B. (2004). Virtual teams: A review of current literature and directions for future research. *ACM Sigmis Database, 35*(1), 6–36.

Pridmore, J., & Phillips-Wren, G. (2011). Assessing decision making quality in face-to-face teams versus virtual teams in a virtual world. *Journal of Decision Systems, 20*(3), 283–308.

Purvanova, R. K., & Bono, J. E. (2009). Transformational leadership in context: Face-to-face and virtual teams. *Leadership Quarterly, 20*(3), 343–357.

Rack, O., Ellwart, T., Hertel, G., & Konradt, U. (2011). Team-based rewards in computer-mediated groups. *Journal of Managerial Psychology, 26*(5), 419–438.

Rapp, A., Ahearne, M., Mathieu, J., & Rapp, T. (2010). Managing sales teams in a virtual environment. *International Journal of Research in Marketing, 27*(3), 213–224.

Rico, R., Alcover, C. M., Sanchez-Manzanares, M., & Gil, F. (2009). The joint relationships of communication behaviors and task interdependence on trust building and change in virtual project teams. *Social Science Information Sur Les Sciences Sociales, 48*(2), 229–255.

Rico, R., Bachrach, D. G., Sanchez-Manzanares, M., & Collins, B. J. (2011). The interactive effects of person-focused citizenship behaviour, task interdependence, and virtuality on team performance. *European Journal of Work and Organizational Psychology, 20*(5), 700–726.

Rico, R., & Cohen, S. G. (2005). Effects of task interdependence and type of communication on performance in virtual teams. *Journal of Managerial Psychology, 20*(3/4), 261–274.

Riketta, M. (2005). Organizational identification: A meta-analysis. *Journal of Vocational Behavior, 66*, 358–384.

Rockmann, K. W., & Northcraft, G. B. (2008). To be or not to be trusted: The influence of media richness on defection and deception. *Organizational Behavior and Human Decision Processes, 107*(2), 106–122.

Sallnäs, E.-L. (2005). Effects of communication mode on social presence, virtual presence, and performance in collaborative virtual environments. *Presence, 14*, 434–449.

Sarker, S., Ahuja, M., Sarker, S., & Kirkeby, S. (2011). The role of communication and trust in global virtual teams: A social network perspective. *Journal of Management Information System, 28*(1), 273–309

Saunders, C. S., & Ahuja, M. K. (2006). Are all distributed teams the same? Differentiating between temporary and ongoing distributed teams. *Small Group Research, 37*, 662–700.

Schwab, D. P. (2005). Research methods for organizational studies. Mahwah, NJ: Erlbaum.

Schweitzer, L., & Duxbury, L. (2010). Conceptualizing and measuring the virtuality of teams. *Information Systems Journal, 20*(3), 267–295.

Scott, J. (1988). Social network analysis. *Sociology, 22*(1), 109–127.

Shadish, W. R., Cook, T. D., & Campbell, D. T. (2001). *Experimental and quasi-experimental designs for generalized causal inference.* Boston, MA: Houghton Mifflin.

Shin, Y. (2004). A person-environment fit model for virtual organizations. *Journal of Management, 30*(5), 725–743.

Shipp, A. J., Edwards, J. R., & Lambert, L. S. (2009). Conceptualization and measurement of temporal focus: The subjective experience of the past, present, and future. *Organizational Behavior and Human Decision Processes, 110*, 1–22.

Sivunen, A., & Hakonen, M. (2011). Review of virtual environment studies on social and group phenomena. *Small Group Research, 42*(4), 405–457.

Society for Human Resource Management. (2012). *Virtual teams.* Retrieved January 15, 2014, from www.shrm.org/research/surveyfindings/articles/pages/virtualteams.aspx

Staples, D. S., & Webster, J. (2008). Exploring the effects of trust, task interdependence and virtualness on knowledge sharing in teams. *Information Systems Journal, 18*(6), 617–640.

Staples, D. S., & Zhao, L. (2006). The effects of cultural diversity in virtual teams versus face-to-face teams. *Group Decision & Negotiation, 15*(4), 389–406.

Stark, E. M., & Bierly, P. E. (2009). An analysis of predictors of team satisfaction in product development teams with differing levels of virtualness. *R&D Management, 39*(5), 461–472.

Suh, A., Shin, K. S., Ahuja, M., & Kim, M. S. (2011). The influence of virtuality on social networks within and across work groups: A multilevel approach. *Journal of Management Information Systems, 28*(1), 351–386.

Tangirala, S., & Alge, B. J. (2006). Reactions to unfair events in computer-mediated groups: A test of uncertainty management theory. *Organizational Behavior and Human Decision Processes, 100*(1), 1–20.

Thatcher, S. M. B., & Brown, S. A. (2010). Individual creativity in teams: The importance of communication media mix. *Decision Support Systems, 49*(3), 290–300.

Townsend, A. M., DeMarie, S., & Hendrickson, A. R. (1998). Virtual teams: Technology and the workplace of the future. *Academy of Management Executive, 12*(3), 17–29.

Triana, M. D., Kirkman, B. L., & Wagstaff, M. F. (2012). Does the order of face-to-face and computer-mediated communication matter in diverse project teams? An investigation of communication order effects on minority inclusion and participation. *Journal of Business and Psychology, 27*(1), 57–70.

Turel, O., & Connelly, C. (2012). Team spirit: The influence of psychological collectivism on the usage of e-collaboration tools. *Group Decision & Negotiation, 21*(5), 703–725.

Van den Bos, K., & Lind, E. A. (2002). Uncertainty management by means of fairness judgments. In M. P. Zanna (Ed.), *Advances in experimental social psychology* (Vol. *34*, pp. 1–60). San Diego, CA: Academic Press.

van der Kleij, R., Schraagen, J. M., Werkhoven, P., & De Dreu, C. K. W. (2009). How conversations change over time in face-to-face and video-mediated communication. *Small Group Research, 40*(4), 355–381.

Webster, J., & Wong, W. K. P. (2008). Comparing traditional and virtual group forms: Identity, communication and trust in naturally occurring project teams. *International Journal of Human Resource Management, 19*(1), 41–62.

Wegner, D. M., Erber, R., & Raymond, P. (1991). Transactive memory in close relationships. *Journal of Personality and Social Psychology, 61*, 923–929.

Whitford, T., & Moss, S. A. (2009). Transformational leadership in distributed work groups: The moderating role of follower regulatory focus and goal orientation. *Communication Research, 36*(6), 810–837.

Wiesenfeld, B. M., Raghurum, S., & Garud, R. (1999). Communication patterns as determinants of organizational identification in a virtual organization. *Organization Science, 10*, 777–790.

Wilson, J., Crisp, C. B., & Mortensen, M. (2013). Extending construal-level theory to distributed groups: Understanding the effects of virtuality. *Organization Science, 24*(2), 629–644.

Yee, N., & Bailenson, J. (2007). The Proteus Effect: The effect of transformed self representation on behavior. *Human Communication Research, 33*, 271–290.

Zhang, D. S., Lowry, P. B., Zhou, L. N., & Fu, X. L. (2007). The impact of individualism–Collectivism, social presence, and group diversity on group decision making under majority influence. *Journal of Management Information Systems, 23*(4), 53–80.

Zimmermann, P., Wit, A., & Gill, R. (2008). The relative importance of leadership behaviours in virtual and face-to-face communication settings. *Leadership, 4*(3), 321–337.

16

Online Employee Surveys and Online Feedback

Bernad Batinic and Carrie Kovacs

Introduction

The use of Internet-based technologies in employee surveys is widespread (Hossiep & Frieg, 2008; Mueller, Straatmann, Hattrup, & Jochum, 2014); new media are used not only for data collection, but also in survey planning, data analysis, communication of results, as well as in the implementation and evaluation of follow-up measures (Kraut, 2006). While these new technologies offer many advantages, such as ease and speed of use, they also present a number of new challenges (Kraut et al., 2004). In fact, the rapidity with which Internet-based and mobile technologies are evolving means that many of their uses and dangers have only begun to be explored (e.g., Alam & Hamida, 2014; Swan, 2013; Wells, Bailey, & Link, 2013).

The first part of our chapter elaborates on technical aspects of the uses of Internet-based technologies in organizations throughout the survey process, focusing on several examples of best practice. This includes project management in online employee surveys, content and design, as well as electronic and automatized reporting of the results. The second part of the chapter focuses on specific issues relevant to data quality, including mode effects, response rates, survey design, and benchmarks. In the third and last part of the chapter, we give an overview of recent developments in data collection and consider future developments. Our chapter concludes with a discussion of the risks as well as opportunities inherent to these developments.

Employee Surveys

Employee surveys are an important part of organizational development and numerous companies conduct them routinely (Borg & Mastrangelo, 2008; Kraut, 2006). Usually in such surveys, employees are asked about their attitudes and opinions with the help of a standardized questionnaire. Results are generally aggregated into single indicator values, which may be calculated both for the organization as a whole and for relevant

---

The Wiley Blackwell Handbook of the Psychology of the Internet at Work,
First Edition. Edited by Guido Hertel, Dianna L. Stone, Richard D. Johnson, and Jonathan Passmore.

subgroups (teams, regions, etc.). The main goal of this traditional form of employee survey is to determine the status quo; this information is then used both to guide strategic decisions for the future and to evaluate the effectiveness of decisions and policies made in the past (de Beuckelaer & Lievens, 2009). In this sense, employee surveys are a diagnostic instrument that shows the strengths and weaknesses of an organization.

Employee surveys are also a form of communication and informational exchange between employees and management. By giving employees the opportunity to participate in employee surveys, an organization signals that employees' feedback and participation in organizational development are welcomed. This also raises certain expectations (Sanchez, 2007). For instance, if employees are questioned about the quality of food in the company cafeteria, they may reasonably assume that their answers will have an impact on that food in the future. At the same time, asking them to complete this survey creates the impression that management is currently dealing with the topic "cafeteria" and may be planning changes in terms of food quality, cafeteria organization, or the like. Thus, employee surveys are not only a way for companies to gather information; they are already an intervention for organizational development in and of themselves. Thus, communication of results and the effectiveness of the follow-up process are central aspects of effective employee surveys (Borg, & Mastrangelo, 2008; Church et al., 2012; de Waal, 2014; Hyland & Dotan-Eliaz, 2012).

In addition to such traditional employee surveys, there are many other specialized forms of organizational surveys (e.g., 360-degree feedback). These specialized forms may complement or even replace traditional forms in a given context. Woods and West (2010) provided an example of the use of a traditional employee survey in their description of a large German bank, whose policy included deployment of a full-scale employee questionnaire at least once every three years (case study 4.1 [pp.133–139]). Besides the survey, the bank asked a sample of 4000 employees to complete an annual corporate identity survey (CIS), whose results were analyzed on a divisional/regional level, and 1000 employees were asked to complete pulse questionnaires (e.g., health check) up to three times per year.

Alongside data gathered through survey questions, the human resource department of an organization has access to countless other sources of nonreactive data, which it can use to better understand the organization, for instance, data about productivity, sick leave, and turnover. Yet in practice, these different types of data are rarely connected with each other, presumably due to the complexity of determining which of many possible sources of data are relevant, equivalent, and available, not to mention the considerable time, effort, and costs needed to answer these questions and handle multiple data sources (Podsakoff, MacKenzie, Lee, & Podsakoff, 2003; Podsakoff, MacKenzie, & Podsakoff, 2012). Concerns about data protection and effective data merging techniques add to these difficulties (Ford, Willey, White, & Domagalski, 2015).

Similar problems plague attempts to include qualitative data collection methods in employee surveys. Some organizations employ qualitative focus-group discussions with their employees as an independent source of employee feedback. In large organizations with thousands of employees, however, focus-group participants generally represent only a small segment of employees, and results can only be generalized to the rest of the organization in a limited way. Often, standardized employee questionnaires end with an open-ended question or text box where employees have the opportunity to make further comments or suggestions (Poncheri, Lindberg, Thompson, & Surface, 2008; Zuell, Menold, & Körber, 2015). When a large number of employees takes advantage of this opportunity, however, companies rarely have access to adequate methods for analyzing and interpreting the resulting slew of qualitative data (Borg & Zuell, 2012).

How the Internet Has Changed the Survey Process

In general, the delay between the collection of data and the communication of survey results and implementation of follow-up measures should be as short as possible. In the context of very large or complex surveys – for instance, the evaluation of large international organizations – this is difficult to achieve using paper-and-pencil questionnaires. Collecting the physical questionnaires, entering employees' answers into the computer, analyzing the data, and communicating results (along with relevant benchmarks) to individual teams costs time and money, and is prone to errors. As a consequence, more and more companies are using the Internet to implement their employee surveys (Mueller et al., 2014; Thompson, Surface, Martin, & Sanders, 2003).

The early years of this digital transition saw a focus mainly on the phase of data collection. Classic paper-and-pencil questionnaires were replaced with online questionnaires; organizations recruited employees by sending them a link to the webpage where the given questionnaire could be found. Data from this questionnaire was coded and saved automatically in a digital format, replacing the time-intensive and error-prone manual entry of data. Nevertheless, this early form of online employee survey had some problems. First, a simple webpage link allowed individuals to take part in the survey easily (even accidentally) on multiple occasions, potentially leading to unintended and undetected double entries within datasets. Second, it was difficult to identify employees who had already completed the questionnaire, which meant that reminder emails were often sent to all employees, including those who had already participated.

These two problems have led to the refinement of this method in the form of online participant management systems. Data from the organization, including employees' email addresses, as well as other relevant information such as employee team, language, or country, can now be entered into the participant databank in the survey management software. This allows the system to send out an individualized link to each employee. When an employee follows his or her individualized link, the management system registers this activity and connects the resulting survey responses to the individual participant (for perceived anonymity and response behavior, see Mueller et al., 2014).

Modern online survey systems contain at least two separate databases: one database contains participant email addresses and other meta-information (including whether a survey has been completed by a given participant); a second, separate database contains the survey data itself. Although the system can connect a participant's meta-data with his or her survey responses, this information is not available to the survey administrator. Removing any direct linkage of personal data to survey responses makes anonymous participation possible. At the same time, this technology allows survey administrators to send employees individualized invitations and appropriate reminder messages while preventing repeated survey participation. This can all be done without sacrificing meta-data such as team affiliation, enabling the calculation of team- or other group-level results. These basic functions of online surveys are the starting point for the use of new media in the full survey process. This survey process can be divided into four broad phases: planning, data collection, data analysis and communication of results, and evaluation and follow-up.

The survey process for online employee surveys: Phase 1, planning

Any employee survey begins with a planning phase to formulate the goals of the survey and decide on a form of execution, to determine the questions and any other data collection instruments, fix a time schedule, and choose a method for communicating survey results. Especially in the case of large organizations, the planning phase usually requires a steering committee responsible for the full survey process. In addition to human resource personnel, this committee may include managers, union representatives, consultants, experts, and other service providers. Steering committees of 20 or more people are not unusual.

Powerful Internet-based technologies make it possible for such large teams to work together on documents and projects using online editing systems, Internet-based conference systems, and other digital tools, without sharing the same physical location. Thus, the planning and development of the survey occurs largely online, saving substantial time and financial resources in this first phase of survey development. It also ensures that decisions made in this phase are electronically documented. In the course of planning an employee survey, before data collection, intra-organizational advertising is also important (Callegaro, Manfreda, & Vehovar, 2015). The survey is usually announced in advance, and company management provides information about the survey's aims and purpose in the company newsletter or through emails from management to all employees. The goal of this advertising is to emphasize the importance of the employee survey in the hopes of achieving higher participation rates (e.g., see Borg & Mastrangelo, 2008, pp. 191, 236).

Phase 2: Data collection

The distribution of email invitations (e.g., through the survey management system) marks the start of the actual survey. These invitation emails serve the function of informing participants about the purpose of the survey and providing an Internet address where the online questionnaire can be found (Mavletova, Deviatko, & Maloshonok, 2014). Online questionnaires offer a broad range of design and layout possibilities, such as the use of pictures, dynamically filtered questions, or multiple language versions of the survey (see Keusch, 2012; Keusch, & Zhang, 2015). Complex data analyses can be performed immediately following or even during survey participation. Current survey management systems offer opportunities for survey administrators to continuously track response rates. Thus, if participation in the survey is particularly low for a certain segment of the organization, additional advertising measures can be taken while the survey is still running. Almost all online employee surveys offer their participants survey support in the form of an email address and/or a telephone hotline, though this support is seldom utilized. It is also possible to use mixed-mode surveys, for example, combining paper-and-pencil and online questionnaires in the same survey (de Beuckelaer & Lievens, 2009). One practical reason for this method is limited Internet access among some employees, such as workers in a production line.

Phase 3: Data analysis and communication of results

Compared with paper-and-pencil surveys with manual data entry, online surveys provide the data immediately and with high accuracy. Most survey management systems offer methods for exporting survey responses in a variety of formats (e.g., as labeled data for statistical software programs). For a long time, it was common practice for organizations or survey service providers to analyze the exported data to create static survey result reports and to distribute these reports to interested parties (e.g., via email). If an organization is small or if the results are only presented to top management and individual organizational committees, the cost–benefit ratio of this procedure is acceptable. If a large number of reports are necessary, however, this procedure becomes too costly. For instance, in a company with 10 000 employees spanning three locations with different languages (e.g., the United States, Japan, Germany), sending both a full report and a team-specific report to each individual team would require thousands of different reports. The necessary resources and potential for errors are correspondingly high. Since about 2010, most companies create such reports digitally and automatically using survey management software (e.g., see Borg & Mastrangelo, 2008, p. 297). Prerequisite for such automated reports is careful consideration of report structure and content during the planning phase as well as the transfer of appropriate organizational information (e.g., team membership) into the survey management system before data collection.

Phase 4: Evaluation and follow-up

An important factor in determining the success of employee surveys is the quality and effectiveness of the follow-up process (de Waal, 2014). Ideally, this process draws upon the survey results in order to develop appropriate follow-up measures. The organization decides which areas will be subject to change, and what methods should be employed to respond to specific survey results. Of course, this can occur at the organizational level as well as at the level of individual departments or teams; some follow-up measures may be short-term policies, others may be long-term developments. In a large organization, thousands of individual follow-up measures may accumulate through a single survey. This also offers an opportunity for digital support, usually in the form of so-called action planners. These allow each team to specify its individual follow-up plan, to document its progress, to publish success stories and to access best practice examples from other departments. A meta-interface allows top management to see the whole organization's progress in implementing follow-up measures and to act quickly at the first sign of negative developments. The action planner allows employees to access materials from workshops or to sign up for moderated workshops on specialized topics from the employee survey. Such a company-wide administrative system makes it possible to standardize the follow-up process (Hyland & Dotan-Eliaz, 2012).

Data Quality in Online Employee Surveys

In a broad sense, the same factors that help ensure the quality of data in any survey also play a role for online employee surveys. Thus, just as in traditional surveys, the first step toward high-quality data is simply investing time in the *full* research process (e.g., Kothari, 2004), not just in its most visible or easily implemented aspects. Early and binding decisions about concrete aims and scope of the survey, adequate research and reflection on the benefits and costs of available instruments and methods for achieving these aims, clear communication with all stakeholders – including participants – before, during, and after the survey, as well as concrete *a priori* plans for how quantitative and qualitative data are to be analyzed, reported, and potentially integrated into a follow-up process are all prerequisites for high-quality surveys with positive organizational effects (e.g., see Sanchez, 2007). In the course of these general methodological considerations, however, some aspects seem to be particularly relevant and/or challenging for the specific case of online employee surveys. We now explore several of these aspects, i.e., online method effects and social desirability, response rates, survey design, and benchmarks.

Online method effects and social desirability

The Internet has had substantial impact on survey methodology. It has changed the way subjects are recruited, the type of data available for analysis, and even the meaning of key concepts such as anonymity and public versus private behavior (Kraut et al., 2004). Survey-makers' ability to guarantee confidentiality has become not only an ethical but also a potentially complex technical issue (e.g., Stanton & Rogelberg, 2001), leading some researchers to argue that the greatest risk for participants in online survey contexts stems from the possibility of unforeseen disclosure of personal data (Kraut et al., 2004). In this sense, the online survey context differs in fundamental ways from traditional paper-and-pencil surveys. But do these context factors actually change the way that participants respond to specific survey questions?

Survey results can be accurate only to the extent that participants are able and willing to provide accurate answers. One systematic threat to this response accuracy occurs when participants consciously or unconsciously skew their answers to better conform to perceived social norms (e.g., Krumpal, 2013). There is reason to suspect that the online survey mode may decrease this risk of social desirability bias in comparison to other presentation modes. Increased anonymity seems to lead to higher levels of self-disclosure; this finding has been used to explain higher self-disclosure rates found in computer-mediated versus face-to-face communication (Joinson, 2001). The same rationale holds in explaining why self-administered surveys (such as web surveys) tend to show higher reported levels of socially stigmatized characteristics and behavior than interview-administered surveys, which involve more social interaction (Krumpal, 2013).

There is also evidence that the online presentation mode alone – without any differences in personal interaction or nonverbal communication – may reduce social desirability, though these results are somewhat mixed. One meta-analysis comparing the online, offline (computer-administered), and paper survey formats found no differences in participants' responses to questions designed to detect social desirability bias (Dodou & de Winter, 2014). In contrast, a second meta-analysis found significantly higher reported incidences of socially undesirable behaviors in computerized rather than paper-and-pencil surveys, suggesting that the online presentation mode may produce less biased reports of actual behavior (Gnambs & Kaspar, 2015). These seemingly contradictory results might be because of problems with the validity of social desirability scales (e.g., Nederhof, 1985), though Dodou and de Winter (2014) showed that the level of participant anonymity *is* associated with differences in social desirability as measured by such scales. Alternatively, this might reflect different effects of survey mode on different types of social desirability. For instance, an increased disinhibition and sense of anonymity created by online surveys might decrease conscious manipulation of one's image for the benefit of others (impression management) and simultaneously increase self-absorption and the unconscious motivation to present oneself positively (self-deceptive enhancement; Booth-Kewley, Larson, & Miyoshi, 2007). However, Dodou and de Winter's meta-analysis – while showing slightly different effects of survey mode depending on which specific desirability scale was used – does not support this claim.

Finally, the different results might be a true indication that questions about sensitive behavior are generally more vulnerable to social desirability bias as compared with items of conventional social desirability scales. At first glance, this seems counterintuitive, since social desirability scales often do ask participants to report behaviors that are considered undesirable but common (Nederhof, 1985) – in other words, sensitive behaviors. However, the very fact that the behaviors included in these scales are "common" means that they are likely to be *less* socially sanctioned than the sensitive behaviors drawn on by survey mode research. Moreover, the aggregation of multiple behaviors in such scales may dilute effects of specific items; because survey mode studies aggregate answers across survey mode, they can focus on desirability effects for specific behaviors without confounding social desirability and true individual behavior. Thus, the idea that bias can be found more easily in behavioral reports than in social desirability scales – while highly speculative – is not implausible. All these interpretations, however, imply that online surveys are at least comparable with paper-and-pencil surveys in terms of minimizing social desirability bias as a whole, and that online surveys might be superior for gathering (sensitive) behavioral data.

Response rates of online employee surveys

In order to draw valid conclusions from employee feedback surveys, it is vital to collect valid data from an appropriate sample of employees. So what constitutes an appropriate sample? How many employees need to be included? What factors in the choice and recruitment of survey participants may endanger the validity of the collected data? The answers to these questions depend on a variety of factors.

One of the first decisions made about a survey in the planning phase is the formulation of survey goals. Assuming that organizations wish to fulfill these goals in an economical way, the next step is to reflect on *whose* survey participation is necessary to achieve these goals. Some survey purposes simply do not require information that is representative for all employees, so that it is not necessary to recruit all employees for survey participation (Borg & Mastrangelo, 2008). Instead, the collection of detailed input from a small but particularly informative sample of employees (e.g., a focus group, impacted teams) may be more efficient than a company-wide survey. Often, however, questions of prevalence *are* relevant. Evaluation of the relevance of previously identified problems within the organization or the acceptance and practicability of proposed solutions are implicitly questions about how the whole group of employees feels or can be expected to respond. The same is true for any survey intended simply to provide continual indicators of company effectiveness – such as employee satisfaction, work engagement, or organizational commitment – in a generalized preventive or evaluative sense. In these cases, information representative for *all* employees is necessary. Finally, any survey goal that includes a participative element, such as giving employees a voice in shaping company policy, also requires information – or at least the opportunity to provide information – for all employees. In these cases, full participation of all company employees is desirable.

Decisions about what constitutes a survey's target group are likely to be impacted by the way that online surveys have changed the structure of the cost–benefit calculations involved in sample recruitment. Highly automated and standardized processes – such as online surveys consisting largely of closed-answer questions – vastly decrease the cost of gathering and analyzing additional data. The highest investment of resources occurs at the level of content development (defining aims and choosing or developing appropriate survey questions), technical implementation (creating or choosing survey software, entering questions, planning and administering access methods, etc.), and analysis management (deciding on and carrying out appropriate calculations and presentation of results). Once the collection and analysis structures are in place, even substantial differences in the amount of gathered data have little impact on collection or processing time. Thus, particularly for online surveys, a full employee census is often a feasible and even cost-efficient aim. After all, when additional information incurs almost no cost, even minor benefits will easily outweigh that cost.

Online surveys have not, however, solved the problem of lack of participation in voluntary surveys. Estimates of the percentage of invited participants who complete a survey (i.e., the response rate) for online surveys differ by context and target participants.

For instance, overall response rates in online panels have been estimated to lie somewhere between 50% and 72% (Batinic & Moser, 2005; Göritz & Crutzen, 2012). However, since response rates in organizational research (53%, Baruch & Holtom, 2008, p. 1148) or specifically among managerial respondents (47%) and top executives (37%, Anseel, Lievens, Schollaert, & Choragwicka, 2010, p. 341) are considerably lower, online employee surveys may show substantial differences in comparison with overall online response rates.

Regardless of such inconsistencies, all these estimates are considerably less than the ideal rate of 100%. In fact, an employee survey's achieved response rate can be an important indicator of survey quality. This is especially true to the extent that employee surveys embody the principle of employee participation in organizational development. In the way of a political election process, high rates of survey participation can also act as a signal of support for and thus legitimation of company policy and/or leadership – independent of any methodological aspects.

On an individual employee level, survey nonresponse has been linked to low organizational commitment as well as to the belief that the organization does not take the survey seriously (Rogelberg, Luong, Sederburg, & Cristol, 2000). Thus, very low response rates in an organization can indicate an overall low commitment of employees; they might also signal that the organization has either not effectively utilized survey information in the past or has failed to effectively communicate this utilization to employees. Even simpler issues such as overly long, complex, or numerous surveys may also have a role in decreasing response rates (Baatard, 2012; Fan & Yan, 2010; Rogelberg & Stanton, 2007).

Determining the response rate for paper-and-pencil questionnaires is fairly straightforward (Rea & Parker, 2014): it is simply the ratio of the number of completed questionnaires to the total number of distributed questionnaires. The phase between distribution and data entry has no relevance for the calculation of the response rate; or at least, no data are available for this time period. In contrast, due to the greater transparency of the digital data collection process, the calculation of response rates for online surveys is more difficult. This difficulty can be illustrated with the following example: an organization sends email invitations to all of its 10 000 employees; 200 of these emails are answered with out-of-office messages (illness, vacation, business trip, etc.). Of the remaining 9800 employees, 7600 visit the survey welcome page, of which 7100 click on the button to begin the actual questionnaire. The survey is discontinued somewhere in the middle by 350 participants, and 6750 participants complete the full questionnaire. To describe the response rate, all these numbers need to be included in the survey results report; reducing the process to a 67.5% response rate discards important information.

Reporting of response rates for online questionnaires is generally less consistent than for traditional survey methods. In an experimental study, Saunders (2012) found that online questionnaires had significantly higher response rates than paper-and-pencil questionnaires, but that this difference mostly disappeared when incomplete questionnaires (more common in online surveys) were omitted from analysis. Thus, any comparison of online response rates for different surveys requires the evaluation – and transparent reporting – of equivalent response rate measures. Careful analysis of points of attrition within the survey (i.e. pages where large numbers of participants left the survey) may also help identify particularly problematic questions or suggest optimal questionnaire lengths for future surveys.

Of course, documenting the number of nonrespondents in a survey is necessary and informative. But what can be done to ensure that as many employees as possible *do* respond? A plausible – though somewhat draconian – approach to this problem is to make participation in employee surveys mandatory. Without discussing legal or ethical aspects of such a policy, one strong argument against it is simply pragmatic: while it is certainly possible to coerce unwilling employees into filling out a form, there is no way to guarantee honest answers under such conditions (according to Borg & Mastrangelo, 2008, p. 52, participation should be voluntary). Thus, one may have solved the issue of representative survey participants at the expense of representative survey answers while possibly also adversely affecting the organization–employee relationship.

A more moderate and vastly preferable strategy is to increase the likelihood that employees will voluntarily participate in the survey. One fairly straightforward way to increase participation is through communication (Callegaro, Manfreda, & Vehovar, 2015; Keusch, 2012). Transparent information about data privacy, including clear sample size limits for calculating and reporting results of subgroups, is important in guaranteeing anonymity and soothing fears of possible reprisals against employees who give honestly negative feedback (Borg & Mastrangelo, 2008). However, more empirical research on this topic is needed.

Conversely, motivation to participate can be increased by heightening its expected utility. This can be done by monetary or lottery rewards (e.g., Rose, Sidle, & Griffith, 2007; Singer & Ye, 2013; Donato & Link, 2013), but also through individualized feedback on survey results (Nestler, Thielsch, Vasilev, & Back, 2015). As already mentioned, the expectation (strengthened or weakened by past experience) that survey results will positively impact organizational policies may also have a role (Rogelberg et al., 2000). Importantly, only those individual incentives whose value is directly dependent on the true accuracy of survey data (e.g., individualized feedback, appropriate policy changes) are likely to motivate employees to participate as well as to give honest and thoughtful responses – though more research is needed to confirm this hypothesis. In contrast, any incentives linked directly to desired survey outcomes (e.g., rewards based on positive responses and "most motivated" teams) are likely to decrease data quality by pushing employees to mask problems and exaggerate positive developments. This assumption is supported by evidence that applicants can and sometimes do selectively distort their responses in a job selection context (Birkeland, Manson, Kisamore, Brannick, & Smith, 2006; Roulin, Krings, & Binggeli, 2016; Tett, Freund, Christiansen, Fox, & Coaster, 2012), which also links strong incentives to specific survey responses.

In addition to substantive incentives, subtle communicative incentives may also impact response rates. For instance, one way to increase participation rates is simply to send out survey invitations with a personal salutation (Sánchez-Fernández, Muñoz-Leiva, & Montoro-Ríos, 2012; Sauermann & Roach, 2013) – preferably from a high-status sender (Joinson & Reips, 2007). Similarly, sending reminder emails has a substantial effect on survey participation (Batinic, 2003; Sauermann & Roach, 2013), although this effect decreases over multiple reminder cycles. In terms of running time, online surveys tend to have shorter data collection phases than paper-and-pencil surveys.

In evaluating survey response rates, however, it is important to keep in mind that response rate is only one indicator of data quality – in fact, it is one indicator of how well the participant sample can be assumed to represent the complete staff of an organization. Another indicator of such representativeness is the extent to which individual socio-demographic and organizational characteristics (e.g., gender, type of job contract, and management level) are distributed similarly among survey participants and among all members of the organization. However, sample representativeness is important primarily in regard to the characteristics about which one intends to draw conclusions. Thus, a survey on employee satisfaction may recruit a sample that is perfectly representative of a company's demographic profile and still produce biased results due to the sample's lack of representativeness for the level of job satisfaction found in the company as a whole. If survey participation is voluntary, then the likelihood of participation will be impacted by individuals' level of motivation to participate. That motivation is likely to vary with job satisfaction. Other personal characteristics like conscientiousness or agreeableness may also impact willingness to participate (Nestler et al., 2015) and confuse the issue further, given that these traits are likewise linked to job-related outcomes like satisfaction (Judge, Heller, & Mount, 2002). With the possible exception of full census surveys, there is no failsafe way of preventing missing data from distorting survey results; however, Rogelberg and Stanton (2007) provide a more comprehensive list of strategies for identifying and dealing with nonresponse bias in organizational research.

In the early days of the Internet, similar considerations of systematic sampling biases led to criticisms of online surveys (e.g., Couper, 2000). If collecting data via new technologies skews accessibility towards more technically savvy participants and leads to overrepresentation of this group in any given online sample (Kwak & Radler, 2002), resulting samples will be nonrepresentative in regard to many characteristics under scrutiny. For instance, if technical interest, knowledge, or equipment are related to better or worse employee outcomes (satisfaction, engagement, commitment, etc.), then estimates of these outcomes based on the skewed sample will also, necessarily, be biased.

Over the years, some of these concerns have decreased. As Internet use has become more prevalent, the ability and opportunity to complete web surveys has become less and less a matter of specialized technical know-how or interest, making such differences less relevant to the likelihood of survey completion – although there are still systematic demographic differences between survey participants who chose web versus email formats (Kraut et al., 2004; Shin, Johnson, & Rao, 2012; see also Stone, Krueger, & Takach, 2017, Chapter 19 in this volume). In professions where computer and Internet use is required constantly, the online mode can no longer be seen as a substantial obstacle to survey completion. Yet for the wide range of professions that do not employ personal computers as a matter of course – whose tasks may even preclude the immediate presence of a computer – this concern continues to be quite relevant.

The use of mobile technologies is a promising way of decreasing this concern in many contexts, though it is not without its own challenges (see also Zafar, 2017, Chapter 10 in this volume). As access to and use of smartphones and tablets becomes more a matter of course, skewed sampling in mobile surveys due to technical knowledge or equipment becomes less likely. At the same time, the mobile nature of this form of Internet access makes it easier to integrate into a wider array of working contexts. Completing a survey via cell phone has become a simple matter in many situations where a personal computer or laptop would be impractical. In fact, web surveys may become multimode surveys independently of survey developers' intentions when participants simply access the given webpages on their phones or tablets (De Bruijne & Wijnant, 2014b). If ease of access can be ensured for a larger proportion of employees, then feedback results are less likely to be biased by survey accessibility and the sample is more likely to be representative for all a company's employees – not just those whose job makes it easiest to complete a survey during office hours.

Online survey layout and design

In addition to its potential to decrease survey dropout, attractive and functional survey design helps to ensure good understanding of question content and to decrease unintentional response errors and omissions. It is therefore a powerful tool in increasing (or decreasing) data quality. The dynamic nature of online questionnaires offers many opportunities above and beyond those provided by paper-and-pencil surveys – with an analogous increase in the number of possible problems, errors, or suboptimal implementation decision. But which aspects should be taken into account when designing online employee surveys? We present a selection of several design aspects that are likely to impact data quality (e.g., see Seckler, Heinz, Bargas-Avila, Opwis, & Tuch, 2014).

Dynamic filtering In comparison with paper-and-pencil questionnaires, which necessitate presentation of all questions to all participants who receive them, online surveys can dynamically adapt their content on the basis of individual participant answers. Most online survey tools offer simple methods to make the visibility of some questions dependent on specific responses to others, meaning that inapplicable content can be preemptively omitted. Since greater survey length has been found to lead to higher dropout rates (e.g., Yan, Conrad, Tourangeau, & Couper, 2010), this reduction is likely to have a positive impact on participant retention.

Three aspects should be noted, however. First, dynamic filtering runs the risk of failing to ask applicable questions if filters are set incorrectly or if only a subset of all possible participant responses (including inconsistent responses or nonresponse to relevant filter questions) are considered in the filter definitions. Second, most survey software makes it possible to filter not only on the basis of participant answers but also on the basis of meta-data stored in the participant database. While dynamic filtering based on such data must always be considered in terms of its impact on anonymity and data protection, data that can safely be combined without allowing personal identification may prove useful (e.g., department membership or leadership role in larger organizations). Specifically, an employee survey can avoid unnecessary length by making use of information that is already available in the organizational database, determining how this information impacts the applicability of further questions, and adapting the questionnaire accordingly (e.g., omitting questions about team climate for employees who work alone). The third consideration involves the technical implementation of dynamic question filtering. Through the use of JavaScript and similar technologies, it has become possible to hide or show questions on the current questionnaire page in real time as participants enter their answers. This means that participants see immediately how their answers impact further survey content. While this presumably reduces unintentional errors, it may also incentivize a strategic choice of those answers that minimize questionnaire length. Placing dynamically adapted content on separate questionnaire pages reduces the chances of such an effect. In this context, it is worth noting that there is little evidence for any general advantage or disadvantage to scrolling versus paging survey designs (Peytchev, Couper, McCabe, & Crawford, 2006). Nevertheless, most current survey management systems only save completion times and response data when participants navigate from one survey page to the next, so paging may be preferable until/unless multisection scrolling solutions or real-time response tracking in the database become more common.

Required fields In contrast to paper-and-pencil surveys, online surveys allow survey creators to make responses to individual (or even all) questions in the survey mandatory for survey completion. This strategy is sometimes recommended in order to increase the number of completed questionnaires in the final sample (e.g., Cantrell & Lupinacci, 2007). It does, in fact, have practical implications for dynamic filtering, where the possibility of missing responses adds a whole branch to the filtering tree and may make the addition of disclaimers to some questions necessary. However, the chances that required fields have a mostly negative impact on data quality are high. If participants are forced to choose between answering a question they would rather – for whatever reason – omit, or discontinuing the whole survey, many will choose the second option. Thus, a greater proportion of complete datasets are acquired at the cost of higher survey dropout. To the extent that full participation is an employee survey goal, the benefit of any additional information – even information with gaps – outweighs the cost of analyzing incomplete datasets. Of course some participants may choose to answer required questions upon prompting. However, a portion of these participants can be expected to respond dishonestly (if the desire to omit was due to unwillingness to disclose true information) or randomly (if the desire to omit was due to lack of understanding or inapplicability of the question). Thus, the benefit of gathering more data may be outweighed by the cost of having substantial reason to mistrust that data's accuracy and the usefulness or even appropriateness of required fields in employee surveys cannot be taken for granted. Nevertheless, in some cases the cost–benefit ratio of individual required fields may be impacted by a particularly cogent reason for requiring specific information (e.g., substantial costs in filtering) in conjunction with widespread agreement on the lack of sensitivity of the requested data. Research suggests that if required fields *are* implemented, they should be highlighted by both a contrasting color and an asterisk (Pauwels, Hübscher, Leuthold, Bargas-Avila, & Opwis, 2009), and grouped together separately from nonrequired fields; all these factors decrease completion times and clear highlighting may lead to lower error rates and higher survey satisfaction among participants (Bargas-Avila et al., 2010).

Responsive layout Online surveys were developed to be accessed via personal computers, and this continues to be the primary form of access for most participants. However, with the rise of tablets and smartphones, unintended mobile access to web surveys rapidly increased. In fact, one study showed an impressive increase of 3% to 11%/16% mobile web completion rates in two Dutch panels during the period from 2012 to 2013 alone (De Bruijne & Wijnant, 2014b, p. 732). Another study that explicitly gave participants the opportunity to complete a survey via smartphone found that 57% took advantage of this option (Toepoel & Lugtig, 2014). Many survey software providers have taken this development into account by providing responsive layouts that dynamically adapt to users' screen sizes. Nevertheless, user-determined question content – not only system-generated layout – substantially affects ease of use on mobile appliances. Horizontal answer option layouts, large numbers of answer options, and half-open questions (i.e. with text entry fields) have been shown to adversely impact survey usability on smartphones (De Bruijne & Wijnant, 2014a). This means that all surveys should be actively tested in the tablet and smartphone layouts, both to prevent biases due to unintended mobile use and to fully harness the potential mobile surveys have to reach employees working outside an office setting.

Benchmarks for online survey results

Employee surveys often include subjective ratings (level of agreement, satisfaction, etc.). The meaning of these ratings in a broader context is difficult to discern because most ratings have meaning primarily in relation to other ratings. Thus, an average rating of "satisfied" means something different if earlier surveys in the same organization produced a rating of "very satisfied" (internal norm) – or if the employees of other comparable organizations tend to rate themselves as "dissatisfied" (external norm). Using such reference norms – also known as benchmarks – in order to interpret observed results can vastly increase survey usefulness. In general, it is more difficult to ensure that external benchmarks are appropriate for judging internal results, but even internal benchmarks can be problematic if they are set under conditions of change in the organization or in the survey process. Even small differences in presentation or wording can drastically change participant responses (e.g., Schwarz & Strack, 1991), so that the only truly appropriate benchmarks are those collected from a comparable organization using a virtually identical questionnaire – or at least an identical subset of questions.

Published and validated questions (e.g., scales of work engagement, job satisfaction, and wellbeing) can fulfill this quality criterion. Though survey developers may be tempted to change the wording of individual questions or answer options in order to better fit the needs of the organization – and many private consulting firms earn their living by creating veritable thickets of "customized" survey solutions – any advantages of greater individualization must be weighed against the definite disadvantages of losing sources of comparison data. Using existing questionnaires with available benchmark data is by far the method of choice. If a thorough search through available, validated survey scales fails to reveal any questions that can appropriately further a specific survey aim, organizations might consider conducting their own external study in order to elicit accurate benchmarks.

Online panel providers, in addition to offering survey software solutions, are also able to quickly and effectively recruit participants with appropriate demographic characteristics (e.g., comparable ages, educational backgrounds, fields of employment) to participate in privately commissioned surveys. Many "customized" employee surveys ask questions that are relevant for all working people, or at least for all people working in the fields and industries of interest. If questions dealing with very specific company policies are excluded, all other questions of an employee survey – complete in their form of presentation, wording, answer options, number, and any other aspects that could impact responses – can be presented to this broader sample. Provided that this sample is representative for the professional field as a whole, the resulting benchmarks allow an organization to interpret its own survey results in relation this broader professional context. In this case, the benefits of interpretable results can be expected to outweigh the reasonable – although not always negligible – costs of commissioning an external provider. Further empirical research on this topic is needed.

Unfortunately, differences in the method of survey administration (e.g., online vs. paper and pencil) also constitute a change in presentation that may impact responses (Burkill et al., 2016; Dillman & Smyth, 2007; Kays, Gathercoal, & Buhrow, 2012), and thus the validity of given reference norms. Allowing an online questionnaire to be accessed in multiple ways (mobile phone, tablet, laptop, personal computer, etc.) increases the ease of use for participants and thus promotes higher rates of participation. At the same time, different modes of presentation also necessitate new reference norms – or at least exploration of the question whether survey presentation has impacted those norms.

Until a larger number of published scales have been explicitly normed using new technologies, the best way of navigating this problem may be simply to track the method in which participants have accessed a given survey. If these data are collected during survey administration, participant responses can be analyzed separately by survey mode and effects of mode can be explicitly tested. Of course, if such effects are found, their causes may be difficult to discover. For instance, higher levels of engagement among employees answering questions on their cell phones may be because the lowest category option was cut off in some cell-phone questionnaire views – or that engaged workers tend to use their cell phones more often.

In fact, there is some evidence that specific demographic characteristics (e.g., age; De Bruijne & Wijnant, 2014b) make individuals more likely to use smartphones to complete web surveys. Because there is no way to determine whether observed differences are artifacts of survey mode or whether the chosen survey mode is an artifact of real *a priori* differences, there is no way to decide whether benchmarks should be corrected for survey mode or not. However, failing to find differences in response behavior among participants using cell phones versus other types of access suggests that values gathered using one method can still be relevant benchmarks for results found using the other. Thus, in the worst case, tracking mode of survey access will provide early warning for probable limitations in interpreting survey results. In the best case, tracking mode of survey access can support a decision to ignore this factor in setting benchmarks. Either way, the potential problem of multiple modes of survey access should be considered in survey planning so that the necessary information can be gathered during the data collection phase.

Though multiple access modes for online surveys can be a problem for identifying valid reference norms, the online survey format itself offers a huge advantage in presenting survey results in relation to such norms. The continuous update of data offers the opportunity to provide continually updated benchmarks. Dynamic visualization of results makes it possible to adapt reference groups either to criteria entered by the user viewing the data or to characteristics of that user themselves. This, in turn, makes it easier for employees to understand their own answers in relation to relevant internal or external comparison groups while also facilitating the use of benchmarks by decision-makers and other stakeholders in a company. As an additional benefit, displaying internal benchmarks as they are successively gathered, may help increase employees' motivation to participate in the survey by providing immediate feedback. Ultimately, if benchmarks are to be used in interpreting survey result on a somewhat regular basis, there is every advantage to be gained in automating the visualization of results in comparison to the chosen benchmarks with a special focus on flexibility and regular updates.

The Future of Survey and Feedback Technologies in Organizations

While the arrival of the Internet has dramatically impacted all aspects of the employee survey process, more recent innovations in mobile and tracking technologies promise to have just as radical an impact on the way employees communicate with and evaluate their employees. Greater technical mobility and higher prevalence of relevant technologies have increased our ability to gather more and more diverse data at lower and lower costs. Organizations with the ability to effectively use these new sources of data may experience significant advantages in relation to organizations who continue to rely solely on increasingly outmoded and limiting techniques. Currently, two of the most visible developments in data collection technology with distinct organizational potential are wearable technologies and automated survey processes.

Wearable technologies

Wearable technologies refer to portable (e.g., cell phone, sociometric badge), accessory (e.g., watch, head- or wristband), garment (e.g., shirt, onesie), or implantable (e.g., tattoo, contact lens) devices that incorporate electronic technologies or computers (e.g., see Alrige & Chatterjee, 2015; Patel, Park, Bonato, Chan, & Rodgers, 2012). They consist of hardware and sensors (e.g., global positioning system [GPS], light, proximity, accelerometer, microphone, or camera) for data collection, hard- and software for data transmission, as well as applications for data processing, analysis, and display (e.g., apps, Cloud computing services). With their help, a variety of physiological (e.g., heart rate, respiratory functioning), behavioral (e.g., activity, voice, contact), and environmental (e.g., location, noise, light) parameters can be constantly and conveniently tracked in real-life situations. At the moment, wearables are primarily used by individual consumers to monitor fitness and health indicators (Cadmus-Bertram, Marcus, Patterson, Parker, & Morey, 2015; PricewaterhouseCoopers, 2014; Swan, 2013) and in healthcare settings (Patel et al., 2012; Piwek, Ellis, Andrews, & Joinson, 2016).

Wearable technologies can also, however, assist workers to increase productivity and workplace safety by providing real-time feedback of relevant parameters. Practical examples include monitoring of air quality in hazardous environments (Alam & Hamida, 2014), tracking attention while driving (Donmez, Boyle, & Lee, 2007), and measuring the quality of cardiopulmonary resuscitation rendered by medical staff (Hostler et al., 2011; Kramer-Johansen et al., 2006). Real-time feedback allows employees to adjust their behavior in real time, increasing the potential of evaluative technology to head off problems the moment they begin to occur, instead of simply reporting them after the fact.

By monitoring behavioral data, wearables also become a tool for gathering information that complements and deepens organizations' understanding of survey responses. Employee surveys are generally based on self-reports, though 360-degree feedback expands this to subjective reports made about other people. Subjective (self-)reports have utility and flaws (e.g., Donaldson & Grant-Vallone, 2002) – like data resulting from any single collection method (Podsakoff et al., 2003). Advantages of the survey method are its face validity and practicality (i.e., cheap and convenient to collect). Moreover, many questions (e.g., job satisfaction and organizational commitment) refer to internal states that are difficult or even impossible to ascertain with observational methods. However, such reports require individuals to think about their answers and to communicate them in the survey setting. Thus, subjective (self-)reports cannot capture experiences that people are not aware of, do not recognize, are not able to identify and express verbally, or do not want to disclose. While there are ethical and legal limits to employers' right to gather data that employees do not wish to disclose, wearables can be a valuable tool in overcoming some of the flaws of survey methods, especially problems related to discrepancies between reported attitudes and observed behavior. Access to selected anonymized physiological and activity measures may allow employers to craft better organizational health initiatives – for instance, by identifying organizational factors related to high stress or low activity levels. Employee wellbeing can also be monitored on a deeper, more holistic level, ideally allowing for early intervention or even prevention of problems such as burnout – although this presupposes automated feedback techniques that can generate and send personalized results without violating employee privacy. The research on this emerging field is still developing.

Automated employee survey processes

A second change in data collection that is arguably less innovative than wearable technologies but which has enormous practical relevance for organizational evaluation is the boom in automated computer-assisted surveys. When we go online to book a flight, buy a concert ticket, or order something from an online shop, our action initiates a specific, preprogrammed process. For instance, our travel and personal data is transmitted to the airline, the airport receives information about the total number of passengers, and catering firms find out which meals have been ordered. This process is fully automated; designated employees may be responsible for monitoring the system, but the individual steps in the process are all triggered automatically. An analogous process can be implemented in employee surveys by defining conditions that automatically trigger specific follow-up questionnaires.

This concept is already widely used in customer satisfaction monitoring. For instance, after completing our flight or receiving our product, we may receive a short questionnaire asking us to evaluate our experience and report any problems. The full survey process is handled autonomously by the software, including the timing of invitations and reminders, as well as the analysis and disclosure of results. Survey administrators can view these results on an interactive dashboard, allowing them to examine results grouped by a variety of different characteristics. In the case of some consumer surveys – such as touchscreens found in public restrooms asking patrons to evaluate the facilities – results are immediately communicated to employees with the ability to correct deficits at once. Here, customer feedback fulfills the same role as real-time physical and physiological feedback provided by wearable technologies. Customer ratings are less objective, but they are also able to respond to a wider range of problems – including those not anticipated by survey creators in advance. They also provide an opportunity for gathering specific follow-up information as needed (e.g., presenting unsatisfied customers with a list of options or an open text box in which they can specify reasons for their dissatisfaction).

In organizations, interactive dashboards are currently used to manage 360-degree feedback results, but the potential for expansion – particularly in the context of continual automated formative evaluation – is enormous. Just as organizations need to be aware of customer satisfaction, so do they need information about organizational climate and employee satisfaction. Regular invitations to evaluate office environment, cafeteria food, information technology services, or administrative procedures can help management improve organizational climate in the same way that customer evaluations help improve customer service: by aiding the identification and elimination of problems and daily hassles (for an example of a validated online feedback system for teams, see Geister, Konradt, & Hertel, 2006). An important aspect of this type of evaluation is the definition of conditions that trigger survey invitations. Examples for reasonable triggers are the passage of a predetermined amount of time, arrival at a given milestone, initiation of a specific administrative process, personnel changes, or even failure to meet given productivity criteria. There is also always the option of triggering surveys manually, for instance at the prompting of a supervisor or through employee initiative. The important aspect is that automation and standardization of the feedback process has the potential to, first, ensure continuous feedback for monitoring long-term developments, and second, make responses to feedback more rapid, visible, and consistent.

Future Research

Most of this chapter's conclusions about the benefits and risks of online employee surveys are drawn from a combination of organizational psychology research and research on human–computer interaction, particularly research on web survey and online panel methodology. Even though both these fields are quite prolific, there is surprisingly little research linking online survey methods to the specific – and unique – context of organization-initiated employee surveys. Yet many of the systematic differences that can be expected between general online research surveys and employee surveys – such as existing relationships between and among survey creators and participants, greater possible consequences of (non)participation, or easier access to long-term participant pools – can also be expected to have a substantial impact on several central findings, from average response rates through the effectiveness of specific incentives on participation rates and response quality.

In some cases, these differences are actually opportunities for more in-depth examination of the antecedents and effects of survey outcomes. For instance, existing demographic and performance data may help researchers find evidence or moderators of nonresponse bias, while long-term employment contracts might help in accessing participants for longitudinal research. Regardless of these additional opportunities, however, an important challenge faced by online survey research is simply the broader and more explicit exploration of ways in which survey contexts – specifically an organizational evaluative context – may moderate results.

While we are still attempting to understand the effects of technology we already have and use, we may also look forward to examining the properties of all the new forms of data that are becoming easily available to us via modern technologies (Boyd & Crawford, 2012; Swan, 2013). The ways in which mobile Internet use and social media may change responses to web surveys or even the type of information which web-based studies can access is only beginning to be explored (e.g., Park et al., 2015; Keusch, & Zhang, 2015; Struminskaya, Weyandt, & Bosnjak, 2015; Wells et al., 2013). Beyond mere mobility, easier access to behavioral data may also help researchers heed the call for multimethod data collection techniques (e.g., Eid & Diener, 2006). All these possible advances, however, are predicated on the reliability and validity of our new data sources, the criteria which have only begun to be examined. Can shorter, more frequent mobile surveys increase the external validity of questionnaire results and help track changes, or do increased repetition and situational distraction outweigh these benefits? How accurate are wearable activity-tracking technologies, and to what extent can the data they provide predict organizational outcomes? Will the spread of behavior-based productivity measures change our understanding of the determinants of job achievement, or will they simply echo existent rating-based findings? Future research will hopefully answer these questions by making more frequent use of mobile questionnaires and tracking both quantitative and qualitative participant responses, by continuing to evaluate physiological and behavioral data through multiple instruments instead of blindly trusting wearable data, and by juxtaposing the results of traditional measures with results provided by the plethora of newly available technologies in order to better understand the meaning and uses of the information they offer in relation to the knowledge we already have.

Conclusion

Overall, online employee surveys provide a cost-efficient and effective tool to assist organizational evaluation and development. The use of Internet technologies can aid organizations throughout the entire survey process. To the extent that software solutions – from participant management systems through automated feedback and action planners – force steering committees to standardize central aspects of the survey process, these solutions have the potential to promote survey economy, transparency, and consistency. This is particularly relevant for organizations that plan regular evaluations, since starting costs are substantially higher than the costs of sustaining an efficiently running survey system (especially when adequate time has been invested in survey planning). Nevertheless, the quality of "solutions" can only ever be discussed in relation to the specific problems which these solutions are intended to solve. Modern technology offers a vast array of time- and error-saving options; the challenge is determining which methods can be implemented in particular situations to best meet one's substantive goals.

References

Alam, M. M., & Hamida, E. B. (2014). Surveying wearable human assistive technology for life and safety critical applications: Standards, challenges and opportunities. *Sensors, 14*, 9153–9209.

Alrige, M., & Chatterjee, S. (2015). Toward a taxonomy of wearable technologies in healthcare. In B. Donnellan, M. Helfert, J. Kenneally, D. VanderMeer, M. Rothenberger, & R. Winter (Eds.), *New horizons in design science: Broadening the research agenda* (pp. 496–504). Heidelberg: Springer.

Anseel, F., Lievens, F., Schollaert, E., & Choragwicka, B. (2010). Response rates in organizational science, 1995–2008: A meta-analytic review and guidelines for survey researchers. *Journal of Business and Psychology, 25*, 335–349.

Baatard, G. (2012). A technical guide to effective and accessible web surveys. *Electronic Journal of Business Research Methods, 10*, 101–109.

Bargas-Avila, J. A., Brenzikofer, O., Roth, S. P., Tuch, A. N., Orsini, S., & Opwis, K. (2010). Simple but crucial user interfaces in the World Wide Web: Introducing 20 guidelines for usable web form design. In R. Mátrai (Ed.), *User interfaces*. Rijeka: InTech.

Baruch, Y., & Holtom, B. C. (2008). Survey response rate levels and trends in organizational research. *Human Relations, 61*, 1139–1160.

Batinic, B. (2003). Internetbasierte Befragungsverfahren [Internet-based survey methods]. *Österreichische Zeitschrift für Soziologie, 28*, 6–18.

Batinic, B., & Moser, K. (2005). Determinanten der Rücklaufquote in Online-Panels [Determinants of response rates in online panels]. *Zeitschrift Für Medienpsychologie, 17*, 64–74.

Birkeland, S. A., Manson, T. M., Kisamore, J. L., Brannick, M. T., & Smith, M. A. (2006). A meta-analytic investigation of job applicant faking on personality measures. *International Journal of Selection and Assessment, 14*, 317–335.

Booth-Kewley, S., Larson, G. E., & Miyoshi, D. K. (2007). Social desirability effects on computerized and paper-and-pencil questionnaires. *Computers in Human Behavior, 23*, 463–477.

Borg, I., & Mastrangelo, P. M. (2008). *Employee surveys in management: Theories, tools, and practical applications*. Göttingen: Hogrefe & Huber.

Borg, I., & Zuell, C. (2012). Write-in comments in employee surveys. *International Journal of Manpower, 33*, 206–220.

Boyd, D., & Crawford, K. (2012). Critical questions for big data. *Information, Communication & Society, 15*, 662–679.

Burkill, S., Copas, A., Couper, M. P., Clifton, S., Prah, P., Datta, J., ... Erens, B. (2016). Using the web to collect data on sensitive behaviours: A study looking at mode effects on the British National Survey of Sexual Attitudes and Lifestyles. *PLOS ONE, 11*, e0147983.

Cadmus-Bertram, L. A., Marcus, B. H., Patterson, R. E., Parker, B. A., & Morey, B. L. (2015). Randomized trial of a fitbit-based physical activity intervention for women. *American Journal of Preventive Medicine, 49*, 414–418.

Callegaro, M., Manfreda, K. L., & Vehovar, V. (2015). *Web survey methodology*. Los Angeles: Sage.

Cantrell, M. A., & Lupinacci, P. (2007). Methodological issues in online data collection. *Journal of Advanced Nursing, 60*, 544–549.

Church, A. H., Golay, L. M., Rotolo, C. T., Tuller, M. D., Shull, A. C., & Desrosiers, E. I. (2012). Without effort there can be no change: Reexamining the impact of survey feedback and action planning on employee attitudes. In A. B. Shani, W. A. Pasmore, & R. W. Woodman (Eds.), *Research in organizational change and development* (Vol. 20, pp. 223–264). Bingley, UK: Emerald Group Publishing.

Couper, M. P. (2000). Web surveys: A review of issues and approaches. *Public Opinion Quarterly, 64*, 464–494.

De Beuckelaer, A., & Lievens, F. (2009). Measurement equivalence of paper-and-pencil and internet organisational surveys: A large scale examination in 16 countries. *Applied Psychology: An International Review, 58*, 336–361.

De Bruijne, M., & Wijnant, A. (2014a). Improving response rates and questionnaire design for mobile web surveys. *Public Opinion Quarterly, 78*, 951–962.

De Bruijne, M., & Wijnant, A. (2014b). Mobile response in web panels. *Social Science Computer Review, 32*, 728–742.

De Waal, A. (2014). The employee survey: Benefits, problems in practice, and the relation with the high performance organization. *Strategic HR Review, 13*, 227–232.
Dillman, D. A., & Smyth, J. D. (2007). Design effects in the transition to web-based surveys. *American Journal of Preventive Medicine, 32*, S90–S96.
Dodou, D., & de Winter, J. C. F. (2014). Social desirability is the same in offline, online, and paper surveys: A meta-analysis. *Computers in Human Behavior, 36*, 487–495.
Donaldson, S. I., & Grant-Vallone, E. J. (2002). Understanding self-report bias in organizational behavior research. *Journal of Business and Psychology, 17*, 245–260.
Donato, P., & Link, M. W. (2013). The gamification of marketing research. *Marketing News, 47*, 38–42.
Donmez, B., Boyle, L. N., & Lee, J. D. (2007). Safety implications of providing real-time feedback to distracted drivers. *Accident Analysis & Prevention, 39*, 581–590.
Eid, M., & Diener, E. (2006). *Handbook of multimethod measurement in psychology*. Washington, DC: American Psychological Association.
Fan, W., & Yan, Z. (2010). Factors affecting response rates of the web survey: A systematic review. *Computers in Human Behavior, 26*, 132–139.
Ford, J., Willey, L., White, B. J., & Domagalski, T. (2015). New concerns in electronic employee monitoring: Have you checked your policies lately? *Journal of Legal, Ethical and Regulatory Issues, 18*, 51–70.
Geister, S., Konradt, U., & Hertel, G. (2006). Effects of process feedback on motivation, satisfaction and performance in virtual teams. *Small Group Research, 37*, 459–489.
Gnambs, T., & Kaspar, K. (2015). Disclosure of sensitive behaviors across self-administered survey modes: A meta-analysis. *Behavior Research Methods, 47*, 1237–1259.
Göritz, A. S., & Crutzen, R. (2012). Reminders in web-based data collection: Increasing response at the price of retention? *American Journal of Evaluation, 33*, 240–250.
Hossiep, R., & Frieg, P. (2008). Der Einsatz von Mitarbeiterbefragungen in Deutschland, Österreich und der Schweiz [The use of employee surveys in Germany, Austria, and Switzerland]. *Planung & Analyse, 6/2008*, 55–59.
Hostler, D., Everson-Stewart, S., Rea, T. D., Stiell, I. G., Callaway, C. W., Kudenchuk, P. J. ... Nichol, G. (2011). Effect of real-time feedback during cardiopulmonary resuscitation outside hospital: prospective, cluster-randomised trial. *BMJ, 342*, d512.
Hyland, P., & Dotan-Eliaz, O. (2012). Employee surveys as catalysts for change: Turning data into action. In L. Gideon (Ed.), *Handbook of survey methodology for the social sciences* (pp. 499–516). New York, NY: Springer.
Joinson, A. N. (2001). Self-disclosure in computer-mediated communication: The role of self-awareness and visual anonymity. *European Journal of Social Psychology, 31*, 177–192.
Joinson, A. N., & Reips, U.-D. (2007). Personalized salutation, power of sender and response rates to web-based surveys. *Computers in Human Behavior, 23*, 1372–1383.
Judge, T. A., Heller, D., & Mount, M. K. (2002). Five-factor model of personality and job satisfaction: A meta-analysis. *Journal of Applied Psychology, 87*, 530–541.
Kays, K., Gathercoal, K., & Buhrow, W. (2012). Does survey format influence self-disclosure on sensitive question items? *Computers in Human Behavior, 28*, 251–256.
Keusch, F. (2012). How to increase response rates in list-based web survey samples. *Social Science Computer Review, 30*, 380–388.
Keusch, F., & Zhang, C. (2015). A review of issues in gamified surveys. *Social Science Computer Review, 35*, 147–166.
Kothari, C. R. (2004). *Research methodology: Methods and techniques*. New Delhi: New Age International.
Kramer-Johansen, J., Myklebust, H., Wik, L., Fellows, B., Svensson, L., Sørebø, H., & Steen, P. A. (2006). Quality of out-of-hospital cardiopulmonary resuscitation with real time automated feedback: A prospective interventional study. *Resuscitation, 71*, 283–292.
Kraut, A. I. (2006). Moving the needle: Getting action after a survey. In A. I. Kraut (Ed.), *Getting action from organizational surveys: New concepts, technologies, and applications*. San Francisco, CA: Jossey-Bass.
Kraut, R., Olson, J., Banaji, M., Bruckman, A., Cohen, J., & Couper, M. (2004). Psychological research online: Report of board of scientific affairs' advisory group on the conduct of research on the Internet. *American Psychologist, 59*, 105–117.

Krumpal, I. (2013). Determinants of social desirability bias in sensitive surveys: A literature review. *Quality & Quantity, 47*, 2025–2047.

Kwak, N., & Radler, B. (2002). A comparison between mail and web surveys: Response pattern, respondent profile, and data quality. *Journal of Official Statistics, 18*, 257–273.

Mavletova, A., Deviatko, I., & Maloshonok, N. (2014). Invitation design elements in web surveys – can one ignore interactions? *Bulletin of Sociological Methodology, 123*, 68–79.

Mueller, K., Straatmann, T., Hattrup, K., & Jochum, M. (2014). Effects of personalized versus generic implementation of an intra-organizational online survey on psychological anonymity and response behavior: A field experiment. *Journal of Business Psychology, 29*, 169–181.

Nederhof, A. J. (1985). Methods of coping with social desirability bias: A review. *European Journal of Social Psychology, 15*, 263–280.

Nestler, S., Thielsch, M. T., Vasilev, E., & Back, M. D. (2015). Will they stay or will they go? Personality predictors of dropout in an online study. *International Journal of Internet Science, 10*, 37–48.

Park, G., Andrew, H., Eichstaedt, J. C., Kern, M. L., Kosinski, M., Stillwell, D. J., ... Seligman, M. E. P. (2015). Automatic personality assessment through social media language. *Journal of Personality and Social Psychology, 108*, 934–952.

Patel, S., Park, H., Bonato, P., Chan, L., & Rodgers, M. (2012). A review of wearable sensors and systems with application in rehabilitation. *Journal of NeuroEngineering and Rehabilitation, 9*, 21.

Pauwels, S. L., Hübscher, C., Leuthold, S., Bargas-Avila, J. A., & Opwis, K. (2009). Error prevention in online forms: Use color instead of asterisks to mark required-fields. *Interacting with Computers, 21*, 257–262.

Peytchev, A., Couper, M. P., McCabe, S. E., & Crawford, S. D. (2006). Web survey design: Paging versus scrolling. *Public Opinion Quarterly, 70*, 596–607.

Piwek, L., Ellis, D. A., Andrews, S., & Joinson, A. (2016). The rise of consumer health wearables: Promises and barriers. *PLoS Medicine, 13*.

Podsakoff, P. M., MacKenzie, S. B., Lee, J.-Y., & Podsakoff, N. P. (2003). Common method biases in behavioral research: A critical review of the literature and recommended remedies. *Journal of Applied Psychology, 88*, 879–903.

Podsakoff, P. M., MacKenzie, S. B., & Podsakoff, N. P. (2012). Sources of method bias in social science research and recommendations on how to control it. *Annual Review of Psychology, 63*, 539–569.

Poncheri, R. M., Lindberg, J. T., Thompson, L. F., & Surface, E. A. (2008). A comment on employee surveys: Negativity bias in open-ended responses. *Organizational Research Methods, 11*, 614–630.

PricewaterhouseCoopers (2014). *Consumer Intelligence Series: The wearable future*. Retrieved April 8, 2017, from www.pwc.com/us/en/industry/entertainment-media/publications/consumer-intelligence-series.html

Rea, L. M., & Parker, R. A. (2014). *Designing and conducting survey research: A comprehensive guide* (4th ed.). San Francisco, CA: Jossey-Bass.

Rogelberg, S. G., Luong, A., Sederburg, M. E., & Cristol, D. S. (2000). Employee attitude surveys: Examining the attitudes of noncompliant employees. *The Journal of Applied Psychology, 85*, 284–293.

Rogelberg, S. G., & Stanton, J. M. (2007). Introduction: Understanding and dealing with organizational survey nonresponse. *Organizational Research Methods, 10*, 195–209.

Rose, D. S., Sidle, S. D., & Griffith, K. H. (2007). A penny for your thoughts: Monetary incentives improve response rates for company-sponsored employee surveys. *Organizational Research Methods, 10*, 225–240.

Roulin, N., Krings, F., & Binggeli, S. (2016). A dynamic model of applicant faking. *Organizational Psychology Review, 6*, 145–170.

Sanchez, P. M. (2007). The employee survey: More than asking questions. *Journal of Business Strategy, 28*, 48–56.

Sánchez-Fernández, J., Muñoz-Leiva, F., & Montoro-Ríos, F. J. (2012). Improving retention rate and response quality in Web-based surveys. *Computers in Human Behavior, 28*, 507–514.

Sauermann, H., & Roach, M. (2013). Increasing web survey response rates in innovation research: An experimental study of static and dynamic contact design features. *Research Policy, 42*, 273–286.

Saunders, M. N. K. (2012). Web versus mail: The influence of survey distribution mode on employees' response. *Field Methods, 24*, 56–73.

Schwarz, N., & Strack, F. (1991). Context effects in attitude surveys: Applying cognitive theory to social research. *European Review of Social Psychology, 2*, 31–50.

Seckler, M., Heinz, S., Bargas-Avila, J. A., Opwis, K., & Tuch, A. N. (2014). Designing usable web forms – empirical evaluation of web form improvement guidelines. In *Proceedings of the SIGCHI Conference on Human Factors in Computing Systems (CHI '14)* (pp. 1275–1284). New York, NY: ACM. https://dx.doi.org/10.1145/2556288.2557265

Shin, E., Johnson, T. P., & Rao, K. (2012). Survey mode effects on data quality: Comparison of web and mail modes in a U.S. national panel survey. *Social Science Computer Review, 30*, 212–228.

Singer, E., & Ye, C. (2013). The use and effects of incentives in surveys. *Annals of the American Academy of Political and Social Science, 645*, 112–141.

Stanton, J. M., & Rogelberg, S. G. (2001). Using Internet/Intranet web pages to collect organizational research data. *Organizational Research Methods, 4*, 200–217.

Stone, D. L., Krueger, D., & Takach, S. (2017). Social issues associated with the Internet at work. In G. Hertel, D. L. Stone, R. D. Johnson, & J. Passmore (Eds.), *The Wiley Blackwell handbook of the psychology of the internet at work* (pp. 423–448). Chichester: John Wiley & Sons, Ltd.

Struminskaya, B., Weyandt, K., & Bosnjak, M. (2015). The effects of questionnaire completion using mobile devices on data quality: Evidence from a probability-based general population panel. *Methods, Data, Analyses, 9*, 261–292.

Swan, M. (2013). The quantified self: Fundamental disruption in big data science and biological discovery. *Big Data, 1*, 85–99.

Tett, R. P., Freund, K. A., Christiansen, N. D., Fox, K. E., & Coaster, J. (2012). Faking on self-report emotional intelligence and personality tests: Effects of faking opportunity, cognitive ability, and job type. *Personality and Individual Differences, 52*, 195–201.

Thompson, L. F., Surface, E. A., Martin, D. L., & Sanders, M. G. (2003). From paper to pixels: Moving personnel surveys to the web. *Personnel Psychology, 56*, 197–227.

Toepoel, V., & Lugtig, P. (2014). What happens if you offer a mobile option to your web panel? Evidence from a probability-based panel of internet users. *Social Science Computer Review, 32*, 544–560.

Wells, T., Bailey, J., & Link, M. (2013). Filling the void: Gaining a better understanding of tablet-based surveys. *Survey Practice*, 6. Retrieved April 8, 2017, from http://surveypractice.org/index.php/SurveyPractice/article/view/25

Woods, S. A., & West, M. A. (2010). *The psychology of work and organizations.* Andover, UK: Cengage Learning EMEA.

Yan, T., Conrad, F. G., Tourangeau, R., & Couper, M. P. (2010). Should I stay or should I go: The effects of progress feedback, promised task duration, and length of questionnaire on completing web surveys. *International Journal of Public Opinion Research, 23*, 131–147.

Zafar, H. (2017). Mobile computing and hand-held devices at work. In G. Hertel, D. L. Stone, R. D. Johnson, & J. Passmore (Eds.), *The Wiley Blackwell handbook of the psychology of the internet at work* (pp. 195–210). Chichester: John Wiley & Sons, Ltd.

Zuell, C., Menold, N., & Körber, S. (2015). The Influence of the answer box size on item nonresponse to open-ended questions in a web survey. *Social Science Computer Review, 33*, 115–122.

17

E-Learning

Richard D. Johnson and Kenneth G. Brown

Introduction

There is no doubt that technology is transforming both industry training and education delivery. Organizations are investing over US$50 billion annually in e-learning initiatives (Piccioli, 2014) and nearly 70% of universities are offering online courses (Allen & Seaman, 2011). Researchers are also interested e-learning, with scholars from several disciplines such as education, management, information systems, psychology, and sociology conducting research in this area. Why is there such an interest in e-learning? First, e-learning can reduce costs and learning time (DeRouin, Fritzsche, & Salas, 2004; Kulik & Kulik, 1991; Salas, DeRouin, & Littrell, 2005; Welsh, Wanberg, Brown, & Simmering, 2003). Some large organizations have reported savings of over US$400 million in training costs (Mullich, 2004) and training time reductions of 50–60% (Gill, 2000). Second, e-learning is argued to provide greater learner control, flexibility, and customization (Cascio & Arguinis, 2005; Kraiger, 2003; Long & Smith, 2004). With e-learning, learners are able to complete coursework at their convenience, and are able to control the pace and scope of learning. Not surprisingly, in a recent survey of over 4000 colleges and universities, 65% of university chief academic officers believe that e-learning is critical to their university's future (Allen & Seaman, 2011).

However, moving learning online is complex and fraught with potential risks. E-learning can be isolating (Garrison & Arbaugh, 2007; Stonebraker & Hazeltine, 2004; Welsh et al., 2003), leaving trainees feeling disconnected from peers (Flood, 2002). Trainees may also find online learning to be less engaging than face-to-face training (Salas et al., 2005), which contributes to dropout rates approaching 80% for certain courses (Flood, 2002; Long, Dubois, & Faley, 2009). Despite these acknowledged shortcomings, research indicates that well-designed e-learning courses can be as effective as face-to-face courses (Alavi, Wheeler, & Valacich, 1995; Alavi, Yoo, & Vogel, 1997; Hiltz, 1994b; Hiltz & Wellman, 1997; Piccoli, Ahmad, & Ives, 2001). Successful programs are about much more than just the technology; success occurs through a combination of factors including technology, training design, trainee motivation, trainee behaviors, and other nontechnologic factors (Arbaugh, 2000b, 2001; Sitzmann, Kraiger, Stewart, & Wisher, 2006). We therefore agree with Brown and Charlier (2013) who argue that "research comparing classroom versus technology-mediated training is

The Wiley Blackwell Handbook of the Psychology of the Internet at Work,
First Edition. Edited by Guido Hertel, Dianna L. Stone, Richard D. Johnson, and Jonathan Passmore.

less useful than research demonstrating how to ensure effectiveness of a particular medium once it is selected" (p. 39). Given the extensive and increasing use of digital technology to deliver learning events, research is needed on how to ensure these technologies support learning.

This chapter has several goals. First, it outlines and defines e-learning in its various contexts. Second, it reviews several e-learning frameworks and provides an integrative framework though which the research on e-learning is reviewed. Finally, building on the findings from current research, the chapter provides some directions for future research.

E-Learning

The first question that needs to be addressed in any review on e-learning is precisely what is meant by the term. Early papers in this area identified all the ways in which e-learning and other virtual environments are different than face-to-face learning environments. For example, Piccoli and colleagues (2001) noted that virtual learning environments are distinct from traditional face-to-face environments in terms of time, place, space, technology, interaction, and control. As technology has evolved and matured, the range of options available has increased and thus the possible forms of virtual, digital, or electronic learning have expanded. More recent efforts have attempted to classify the different forms of technology-delivered training, rather than differentiating technology-delivered from face-to-face training. Bostrom (2003) focused on the dimensions of time and place. From this perspective, e-learning can include both synchronous (same time) and asynchronous (different time), and both co-located (same place) and distant (different place) learning. In this way, Bostrom illuminated four cells with only the synchronous and co-located learning capturing the typical face-to-face classroom.

On the basis of a review of 112 articles, Brown, Charlier, and Pierotti (2012) noted that the term "e-learning" is used for different forms of delivery (i.e., Internet and standalone devices such as digital music and video players) and for different types of events (formal training and ongoing, informal learning). Drawing on these distinctions, the authors proposed two dimensions that underlie most definitions of e-learning: the level of interactivity and the purpose of the event. The first dimension deals with the degree to which the learning experience allows for sustained interaction among learners and other system users (such as an instructor or expert guide, if one is present), ranging from low to high. On the low end of the continuum, a learning event can be static and solitary, such as with a podcast. On the high end it can be dynamic and cooperative, such as with an online synchronous discussion group facilitated by an expert. The second dimension captures whether the purpose of the learning stimulus is to serve as instruction or as information. Instruction is designed to fulfill predefined objectives set a by a designer or expert. Information, on the other hand, is designed for easy access and retrieval by any user. Information resources in this typology are systems that provide information that will be used for learning new knowledge and skill. These two independent dimensions can be crossed to yield four types of e-learning environments: static information (low-interaction information resource, such as online help); static instruction (low-interaction instruction, such as standalone); collaborative information (high-interaction information resource, such as a corporate wiki); and collaborative instruction (high-interaction instruction, such as a learning-oriented multiplayer simulation).

The distinction between Bostrom (2003) and Brown et al.'s (2012) frameworks illustrates the evolution of e-learning definitions to the point where distinguishing them from "classrooms" (same time, same place) is no longer central. Instead, Brown et al. (2012) emphasize that technology-delivered events can be designed with different goals in mind and can take on different features. In this way, Bostrom and Brown et al.'s frameworks are not mutually exclusive; they simply emphasize different dimensions of many involved in differentiating an increasing complex array of learning options. For purposes of our review, we focus on the instructional side of Brown et al.'s (2012) taxonomy. Although we acknowledge that e-learning can have broader meanings, we define it here as planned events designed to foster learning and which use network technologies to deliver substantial elements of the learning experience.

Reviews of E-Learning

The depth and breadth of e-learning research has led to many reviews on the various technology, design, and use issues which arise from its implementation. For example, Sitzmann et al. (2006) reviewed 96 studies that compared the effectiveness of web-based instruction to traditional face-to-face (FtF) instruction. They found that web-based instruction can be as effective as FtF instruction, and that the differences in effectiveness are less about technology and more about sound design principles. Brown and Charlier (2013) reviewed the literature on the factors that affected trainees' decisions to utilize e-learning and developed a model to help organizations and scholars better understand how to increase utilization of e-learning. Arbaugh, Desai, Rau, and Sridhar (2010) reviewed the literature on online and blended learning in management. Their specific focus was on identifying the major topics investigated and the outlets chosen for these publications. In addition, Arbaugh (2010) reviews the role of e-learning in different business disciplines. Reviews have also focused on other issues, for example Salas et al. (2005) developed a set of guidelines for practitioners to make e-learning more effective. Towler and Mitchell (2015) reviewed the work on the role of the trainer/instructor in facilitating e-learning. Brown et al. (2012) reviewed research on e-learning in industrial and organizational psychology but which is more limited in scope than this review. Finally, Wan, Fang, and Neufeld (2007) and Gurtner (2015) provide limited and focused reviews of research focusing on technology mediated learning and virtual learning environments, respectively.

Models of E-learning Effectiveness

Researchers have developed a number of models to guide e-learning research. Each model has focused on a series of student, instructor, and design inputs that are thought to influence learning processes and outcomes. The earliest model, proposed by Alavi and Leidner (2001), focused on three key factors that affect e-learning outcomes: instructional design, technology, and psychological learning processes (Figure 17.1). Instructional design focuses on the pedagogical design of the course including decisions on how training will be arranged, sequenced, presented, and synthesized by trainees. Technology focuses on how courses utilize technology in the training process, from electronic repositories, to decision support, to completely mediated e-learning coursework. Both instructional design and technology are thought to affect psychological learning processes. Psychological learning processes in this context are trainee or learner states that include cognitive processing, motivation to learn, training interest, and other learning processes. Finally these learning processes affect learning outcomes.

Although Brown (2001) did not develop a model focusing on all factors affecting e-learning, his model provides insights that are different from those identified by Alavi and Leidner (2001). Specifically, drawing on prior models of the effects of trainee characteristics (Noe, 1986), Brown's model suggests that individual characteristics such as computer experience, goal

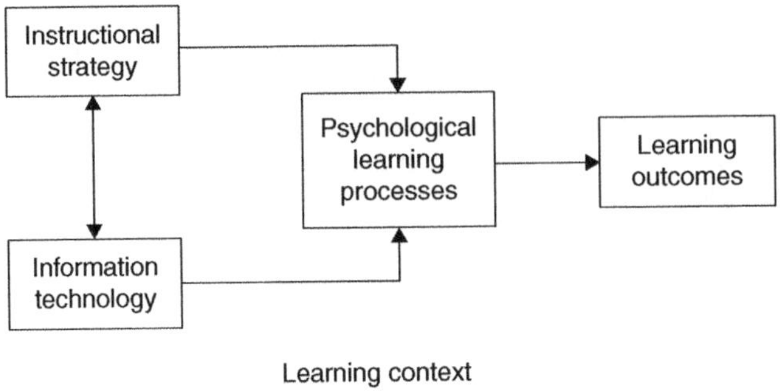

Figure 17.1 A framework for e-learning research. *Source:* Adapted from Alavi and Leidner (2001) with permission of Institute for Operations Research and the Management Sciences (INFORMS).

orientation, and age will affect a trainee's learning choices during training, such as practice time and off-task attention. Finally, these decisions are argued to affect learning outcomes. Thus, Brown's model (Figure 17.2) considers two new factors: first, this model includes a distinct focus on the characteristics the trainee brings to the training; and second, whereas Alavi and Leidner's model focuses on internal cognitive and motivation processes, Brown focuses on the actual learner behaviors during the training.

The final model discussed here was developed by Piccoli et al. (2001). This model focuses on two dimensions thought to affect outcomes: the human dimension and the design dimension (Figure 17.3). The human dimension focuses on student/trainee characteristics and behaviors as well as those of the trainer/instructor. This model expands on Brown's model in that it acknowledges the importance of the instructor in courses that are

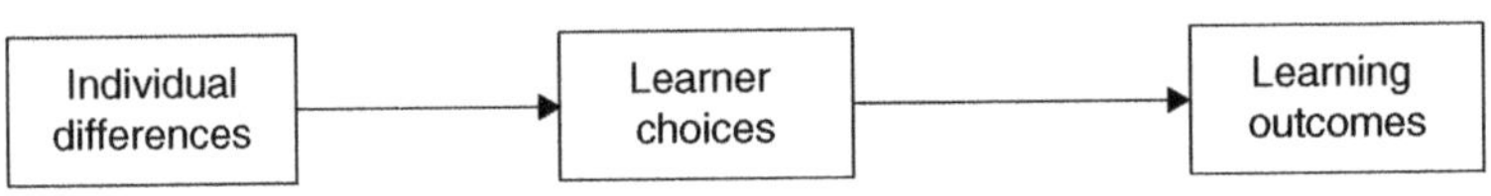

Figure 17.2 Individual differences and learner choices research model. *Source:* Adapted from Brown (2001), with permission from John Wiley & Sons.

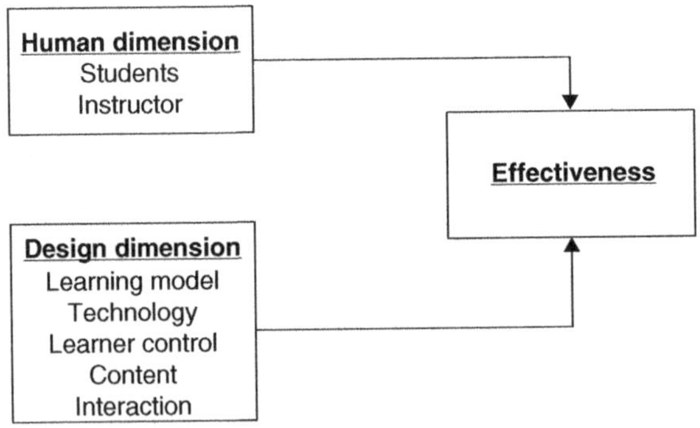

Figure 17.3 Dimensions and antecedents of e-learning effectiveness *Source:* Sufficiently adapted from Piccoli et al. (2001).

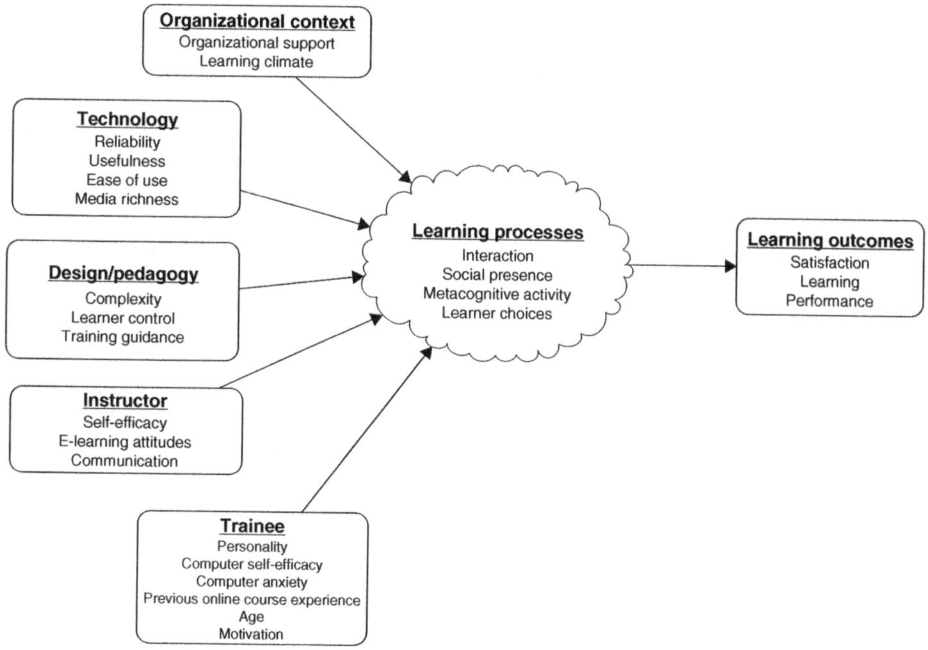

Figure 17.4 A model of e-learning effectiveness.

instructor managed. The design dimension focuses on course design considerations, such as the learning model utilized, learner control, technology, course content, and course interaction. The major shortcoming of this model is that it does not sufficiently recognize the importance of learning behaviors during training.

Based on these previous models and our review of e-learning research from education, management, information systems, psychology and sociology, it is clear that, as with training in general, e-learning outcomes are influenced by multiple inputs and processes. The model have we used to summarize the literature consists of five major inputs (Figure 17.4): organizational context, technology, design and pedagogy, instructor characteristics, and trainee characteristics. We start with the more distal, and nonmalleable factors, then turn to those factors over which designers may have more control. These inputs are then thought to affect learning processes such as peer and instructor interactions, social presence, meta cognitive activity, and the choices that learners make within the course. These processes influence learning outcomes. We place learning processes in a cloud to note that these are distinct, complex processes that should not be considered a single, higher-order construct. We now discuss the research findings behind each dimension of our model.

E-Learning Outcomes

When evaluating the effectiveness of e-learning, outcomes should parallel those with traditional training programs. The most studied outcomes of training are trainee reactions, learning, and on-the job performance at the individual and organizational levels. E-learning research has often focused on the reaction of overall trainee satisfaction as an outcome of interest (Johnson, Hornik, & Salas, 2008). Satisfaction may even more important in an e-learning setting due to the large dropout rates and the finding that individuals who are less satisfied with their experiences will be less likely to enroll in e-learning in the future (Carswell & Venkatesh, 2002; Lim, 2001).

In addition to satisfaction, researchers often have focused on the value or relevance of the training to the learner. Also called utility judgments, this reaction to training reflects the extent to which the leaners believe that the training will provide them with the knowledge and skills to improve their on-the-job performance or will contribute to a skill that will help them move forward with their education. Meta-analytic research has found that trainees' utility judgments are often a better predictor of training transfer than how well a learner performed during training (Alliger, Tannenbaum, Bennett Jr, Traver, & Shotland, 1997). More recent meta-analytic research demonstrates that overall satisfaction and utility judgments are correlated with similar relationships with learning outcomes (Sitzmann, Brown, Casper, Ely, & Zimmerman, 2008). Thus, rather than treating affective and utility reactions as separate, and moderately correlated concepts, they should be treated as lower-order factors of an overall satisfaction construct (Brown, 2005a). This model proposes that individuals may develop emotional reactions to training, and these emotional reactions may then affect (positively or negatively) perceptions in regards to the value of the training and even the usefulness of the technology. Although together the evidence seems to suggest that affective and utility reactions are best thought of as representing a single overall satisfaction construct, we discuss them separately because prior research has examined them separately.

In addition to research delving more deeply into reactions to training, there is research suggesting that the learning construct is more rich than originally conceptualized. An overall learning construct is too simplistic of a way of assessing learning. For example, Kraiger, Ford, and Salas (1993) argued that learning can reflect cognitive outcomes such as increased declarative knowledge, better mental models, and improved cognitive strategies. Learning can also improve skills. Skill-based outcomes capture the performance of tasks as they become faster, more fluid and of higher quality. Finally, there can be motivational and affective outcomes, such as improved self-efficacy, levels of goals set, organizational commitment, and commitment to diversity. The final two types of outcomes of interest to organizations are training transfer and organizational results. Training transfer captures both the maintenance of knowledge, skill, and attitude change back on the job, and the generalization of those changes to other stimuli and environments (Baldwin & Ford, 1988). Organization results reflect outcomes such as productivity increases, customer satisfaction, and profitability. These outcomes have rarely been studied within the e-learning context, and they reflect an overall training outcome, often influenced by external organizational events. Although relevant, they are beyond the scope of this review.

E-Learning Inputs

Organizational context

Organizational support Organizational support is a reflection of the extent to which an organization is committed to providing resources that support an employee's participation in e-learning. Employees are more likely to participate in e-learning when they are confident that the organization will provide them the necessary opportunities and resources (Maurer, Pierce, & Shore, 2002). Research has found that support in the form of tools, equipment, and financial support can affect employee participation in development activities (Hurtz & Williams, 2009; Noe & Schmitt, 1986). In addition, meta-analytic research has found that organizational support influences motivation to participate in training (Colquitt, LePine, & Noe, 2000).

When employees perceive stronger support for e-learning, they are more satisfied with e-learning and this relation is stronger when training complexity is higher (Sawang, Newton, & Jamieson, 2013). Organizational support is also positively related to utility judgments and perceptions of the quality of relationships in the environment (Byun & Mills, 2011; Cheng, Wang, Moormann, Olaniran, & Chen, 2012). Finally, Park, Sim, and Roh (2010) found that employees are more likely to transfer knowledge gained in e-learning when the company provides support. Together, these findings suggest that organizational support can improve learner motivation, satisfaction, and training transfer.

Learning climate Although scholars have long recognized the influence of climate and culture on work behavior, there is often confusion in terminology and inconsistency in measurement (Parker et al., 2003). For our purposes, we emphasize the importance of individual perceptions of the environment related to training and specifically the use of e-learning as a set of training tools. Parker et al. (2003) demonstrated that individual perceptions, labeled psychological climate, have significant relationships with employee attitudes, motivation, and performance. Prior research on training has demonstrated that training-related climate predicts learning and transfer (Holton, Bates, Seyler, & Carvalho, 1997; Rouiller & Goldstein, 1993; Tracey, Tannenbaum, & Kavanagh, 1995). To our knowledge research has not examined perceptions specific to the use of e-learning, but perceptions of the appropriateness of and support for those initiatives specifically should influence learners' beliefs about the value of e-learning within their context.

Technology

Reliability Given that content, communication, and interactions are mediated via technology, it is important that it is available when needed, provides accurate and timely information, and is flexible to meet the needs of the employee (Johnson, Gueutal, & Falbe, 2009). An important part of e-learning, any impacts of technology reliability on e-learning outcomes are mediated through its impact on learning processes (Alavi & Leidner, 2001), as instructors and learners fit the technology to the course design (Clark, 1994). When technology is reliable, students can spend less time focusing on navigating the environment, freeing cognitive resources to engage in deeper metacognitive and other learning processes. A number of studies have investigated the role of technology in e-learning, consistently finding that reliability was related to course attitudes (Webster & Hackley, 1997) and satisfaction (Islam, 2011; Johnson et al., 2009; Kim, Trimi, Park, & Rhee, 2012; Mohammadi, 2015). Research has also found that reliability was related to learner perceptions of course utility (Johnson et al., 2009; Liaw, 2008) and performance (Johnson et al., 2009).

Usefulness and ease of use One of the major themes of e-learning research by information systems is the adoption and continued use of these systems (Carswell & Venkatesh, 2002; Lee, Hsieh, & Chen, 2013; Ong, Lai, & Wang, 2004). Two factors in particular have been found to impact adoption decisions: perceived usefulness and ease of use. Perceived usefulness is the extent to which using a system would improve performance and perceived ease of use is defined as the extent to which using a system is free from effort (Davis, Bagozzi, & Warshaw, 1989). Research has also focused on how these factors affect e-learning process and outcomes.

Some studies have found that usefulness and ease of use are related to course satisfaction (Arbaugh, 2000b, 2005, 2014; Capece & Campisi, 2013; Johnson et al., 2008), but others report that only perceived usefulness was related to satisfaction (Arbaugh, 2000b, 2002). Still others have found no relation between perceived usefulness, ease of use, and course satisfaction (Arbaugh & Duray, 2002). Similarly some scholars have found statistically significant relations between perceived usefulness, ease of use and utility judgments (Arbaugh, 2014; Balakrishnan, 2014), although others have not (Arbaugh, 2002, 2005; Arbaugh & Duray, 2002; Johnson et al., 2008). Finally, the relations between usefulness, ease of use, and course performance are also equivocal, with one study finding that usefulness and ease of use were related to course grade (Galy, Downey, & Johnson, 2011), but other studies reporting no statistically significant relations (Arbaugh, 2014; Carswell & Venkatesh, 2002; Johnson et al., 2008).

What is interesting is that when only usefulness and ease of use are investigated, they are related to e-learning outcomes, but when they are entered with other variables such as interactions and social presence, their significance is reduced or removed. This suggests that e-learning systems that provide learners material and interaction capabilities in an easy to navigate and use manner lead them to interact more and increase peer connections. Thus, it is difficult to discern whether usefulness and ease of use have an important role or if they are a reflection of the organizational context and the design choices made by trainees and designers. Further research is needed to better understand the incremental validity and understanding of these constructs in light of context and design factors.

Media richness It has been argued that to maximize the effectiveness of e-learning environments, care should be taken to make the environment as rich as possible, supporting multiple communication and interaction opportunities, texts and video, and other tools (Salas et al., 2005). The ability of technology to support shared and coordinated behavior in groups has been described as its richness or synchronicity (Daft & Lengel, 1986; Dennis, Fuller, & Valacich, 2008). Media synchronicity theory posits that technologies differ in their ability to: provide immediate feedback, allow the trainee to rehearse or edit the message before sending, allow for communication parallelism (e.g., multiple simultaneous conversations), allow for message reprocessing (e.g., storing of information so that it can be recalled for future use), and support symbol variety (e.g., voice inflection, "body language") (Dennis et al., 2008). Previous research suggests that trainees will be more motivated in a richer communication environment, will communicate more, and should process course information more deeply. In turn, this should lead to higher e-learning outcomes. For example, Lan and Sie (2010) found that trainees' perceptions of the richness of email, short message service (SMS, e.g., text) and really simple syndication (RSS) differed depending on the dimension of richness being investigated (e.g., timeliness, clarity of message, and adaptability). In addition, e-learning systems that provide richer and more interactive tools have higher satisfaction and course performance (Hsieh & Cho, 2011; Johnson et al., 2009). Other research shows media synchronicity is positively related to satisfaction and utility judgments, but not training performance (Johnson et al., 2009). Interestingly, Alavi, Marakas, and Yoo (2002) found that a less rich medium led to better training outcomes than a more rich medium. We revisit this finding in the section on design fit.

Design and pedagogy

Complexity Training complexity reflects the extent to which learning tasks contain greater or lesser information load, information diversity, rate of information change, and interconnections between this information (Sweller, 1988; Wood, 1986). As learning tasks become more complex, the number of elements that must be processed in working memory increases (Van Merrienboer, Kester, & Paas, 2006). In addition, learners need to process more information concurrently, which creates an increased risk that they will not fully process all information. In addition, according to Bloom's taxonomy (Anderson & Krathwohl, 2001; Bloom & Krathwohl, 1956), as training material becomes more complex the focus shifts from the recall of factual information to the evaluation, application, and integration of knowledge to new contexts. Training research has found that both actual and perceived complexity are negatively related to learning outcomes (Argote, Insko, Yovetich, & Romero, 1995; Fisher & Ford, 1998; Warr, Allan, & Birdi, 1999).

Within the e-learning environment, the empirical investigation of complexity has been somewhat limited. But experts have argued that e-learning may be less appropriate for more complex, soft-skill types of training, such as leadership development and ethics, compared with more factual or hard-skilled training (Salas et al., 2005; Welsh et al., 2003). Research supports this argument, with Hornik, Saunders, Li, Moskal, and Dzuiban (2008) reporting that university e-learning outcomes (e.g., higher performance and lower course abandonment) were better for courses in disciplines that were more algorithmic in nature (e.g., chemistry, computer science). Empirical research has also found that more complex tasks are associated with lower e-learning performance (Granger & Levine, 2010; Yanson & Johnson, 2016), and that complexity and organizational support are interactively related with course satisfaction. Specifically, Sawang, Newton, and Jamieson (2013) found that the relation between complexity and satisfaction was highest when complexity and organizational support were both high and lowest when complexity was high but organizational support was low. Finally, Sawang et al. (2013) also found that complexity and organizational support interacted to predict performance such that performance was higher for complex tasks when the organizational support was higher.

Learner control It has been argued that one of the major benefits of e-learning is that it increases learner control (DeRouin et al., 2004). Learner control is the extent to which a trainee has control over instructional features such as pace, content, and structure of the training environment (Fisher, Wasserman, & Orvis, 2010). Learning environments with greater learner control are those where one or more learning activities are under the control of the trainee rather than the instructor or program. For example, with e-learning, trainees often can choose the time and place where to engage in training, how much to practice, and how many other activities to engage in while learning. But not everyone is prepared to effectively leverage this control. Granger and Levine (2013) argue that learner control adds an additional level of complexity to the training process that can negatively impact training outcomes. They noted when training was complex, students performed less well when they were given greater learner control. Research also suggests that shorter training programs with higher learner control leads to lower performance than those that are longer term because learners do not have the time to develop the skills that help them to leverage better the control available to them (DeRouin et al., 2004). Finally, Orvis, Brusso, Fisher, and Wasserman (2011) found that the relationship between learner control and performance was moderated by two personality traits, openness to experience and extraversion. Specifically, they found that individuals high in extraversion and openness to experience performed best in an environment with high learner control, but more introverted and less open individuals performed best in low learner control environments.

Higher levels of learner control can also improve learner satisfaction. Specifically, when trainees were able to have more control over the scheduling of learning activities, they had higher satisfaction than those who had lower control (Karim & Behrend, 2014). Providing learners with a choice of whether to engage in a higher or lower learner control environment also improves course satisfaction (Fisher et al., 2010; Orvis, Fisher, & Wasserman, 2009). Some studies report higher learner control to be associated with lower off-task attention (Orvis et al., 2009), but others have found that when individuals had higher control, as well as higher off-task attention, their performance was lower (Karim & Behrend, 2014). Thus, more research is needed to fully understand how learner control can affect learning behaviors, but it is clear that learner control affects them.

Overall these findings suggest that although learner control has the potential to enhance e-learning outcomes, the relationship between learner control and e-learning effectiveness may be complex and depends on factors such as personality, ability, experience, and motivation. It may be possible that learner control actually benefits those with higher motivation, ability, and experience to a greater extent than those who are lower in these areas. Therefore it is important to prepare those in e-learning to navigate an environment which is inherently flexible and has limited structure. Mechanisms to improve learning control, such as training in metacognitive behaviors can improve the effectiveness of learner control. We cover this aspect in greater detail in the section on metacognitive activity.

Training guidance Another way to improve e-learning outcomes is by providing guidance to trainees as to how best to utilize the tools and navigate the e-learning environment. In many ways, training guidance is closely related to learner choices and self-regulated behaviors in that much of the research on guidance has focused on helping trainees better employ self-regulated behaviors and to make improved learning choices. For example, Bell and Kozlowski (2002a) focused on the use of adaptive guidance to improve computer-based training outcomes. Adaptive guidance helps trainees focus on the material they need to learn and provides suggestion for how to improve based on their current performance. Bell and Kozlowski (2002a) found that adaptive guidance training led to trainees spending more time on the training task. In addition, adaptive guidance improved trainee knowledge and performance, and the effect was highest for more strategic and complex tasks. In another study, Bell and Kozlowski (2002b) found that framing training as exploratory and encouraging exploration as part of training led to higher metacognitive activity than when the training guidance focused on the training procedures/process. Furthermore, Santhanam, Sasidharan, and Webster (2008) found that trainees who were provided guidance in how to apply self-regulated learning strategies outperformed those who did not receive this guidance. In another study, when trainees were given feedback and prompts to encourage the use of self-regulated learning strategies, they outperformed those who did not receive this feedback (Sitzmann, Bell, Kraiger, & Kanar, 2009). This was true for both simple and complex training tasks, and the effects of feedback were strongest for those individuals with the higher cognitive ability and self-efficacy.

Finally, although self-regulated learning prompts and feedback will improve both time on task and performance, the relationship may be more complex than previously theorized (Sitzmann & Ely, 2010). Specifically Sitzmann and Ely (2010) found that providing continuous feedback and prompts leads to higher performance than prompting either early or late prompts. The effects of prompting do not produce immediate gains, but occur over time and are self-reinforcing. When trainees utilize self-regulated learning strategies, and perform well, they are likely to continue using these strategies in future phases of training. In addition, Sitzmann and Ely reported that when the prompts were removed, trainees stopped utilizing the self-regulated learning behaviors. Prompting trainees was associated with lower attrition rates, and that the effect was highest for those who performed well and received prompts.

Instructor

Instructor attitudes, e-learning confidence and behaviors can also affect e-learning processes and outcomes. E-learning is inherently more complex than FtF training due to the technology mediation of content, interactions, and learning processes (Hillman, Willis, & Gunawardena, 1994). Instructors can add value because they can help trainees to navigate and interact more effectively in the online environment. For example, trainees who have an instructor present during an e-learning course are more satisfied than those who received only computer simulations (Gupta & Bostrom, 2013). In addition, Arbaugh (2014) found that when an instructor had more e-learning teaching experience, students performed better, found the course to be more valuable, and had higher satisfaction than when he or she had less experience. But for instructors to maximize their value to trainees, they must find value in the e-learning process and be confident in utilizing the technology. Each has been investigated as contributors to better e-learning outcomes.

Self-efficacy Self-efficacy reflects an individual's belief in his or her ability to complete a specific task (Bandura, 1997). Self-efficacy and confidence to deploy technology in the classroom affects an instructor's use of technology in the classroom (Anderson, Groulx, & Maninger, 2011; Paraskeva, Bouta, & Papagianni, 2008). In the context of e-learning, the more confident the instructor is with technology, the more likely he or she will utilize the full set of resources available to them in the environment. In addition, when an instructor has higher efficacy, he or she should communicate more effectively with trainees, put more effort into managing the training environment, provide stronger support to trainees, and increase the social connections with students. The limited research that has focused on the role of self-efficacy in e-learning has found that instructors with higher efficacy are more likely to participate in e-learning (Ferdousi & Levy, 2010; Lee et al., 2013). In addition, instructor technical skills and confidence are reported to be related to e-learning effectiveness (Chien, 2012; Webster & Hackley, 1997).

E-learning attitudes Given that attitudes have long been of interest to training researchers, and have been shown to affect learning outcomes (Noe, 1986; Noe & Schmitt, 1986) it is not surprising that several studies have found that attitudes toward e-learning can affect instructor behaviors such as communication and system use. For example, instructor attitudes can affect trainee beliefs about the value of e-learning and their reactions to the training they receive (Chien, 2012; Webster & Hackley, 1997). In addition, research has found that positive instructor attitudes are associated with greater intention to use e-learning features by instructors (Liaw, Huang, & Chen, 2007). Instructors who embrace the e-learning tools should be able to utilize them to more effectively manage training, to communicate more with learners, to better manage a shared learning environment, and to provide more effective and timely feedback to trainees.

Communication Interactions with the instructor can have an important role in e-learning. Specifically, Arbaugh (2001) argues that instructor behaviors that reduce the social and psychological distance between learners, also called immediacy behaviors (Anderson &

Anderson, 1982), can positively impact the e-learning environment. Behaviors such as use of humor, referring to learners by name, and encouraging communication and feedback signal to learners that the instructor is available and willing to assist them, and that they are important to the instructor. A qualitative study (Inayat, Amin, Inayat, & Salim, 2013) found that students felt that instructor feedback was positively related to their course performance. Arbaugh and Rau (2007) observed that more interaction with the instructor was related to higher satisfaction and utility judgments. But the strongest and most consistent finding is that the more an instructor is actively engaged in interacting with learners (e.g., more immediacy), the more satisfied learners will be, the more valuable they will find the training, and the better they will perform (Arbaugh, 2014; Marks, Sibley, & Arbaugh, 2005; Sitzmann et al., 2008; Yang & Durrington, 2010). For a broader discussion of instructor facilitation in e-learning readers are encouraged to review the work of Towler and Mitchell (2015).

Trainee

In any learning environment, the trainee will bring with them certain characteristics which affect learning outcomes. For example, meta-analytic training research has found that individual characteristics such as cognitive ability, personality, anxiety, self-efficacy, age, and motivation to learn are related to learning outcomes (Colquitt et al., 2000). Each of these factors can also play a role in e-learning.

Personality Research linking personality to training outcomes has been widespread and has consistently found that personality can affect training outcomes. Two traits that have received strong attention in training research, and which have been found to affect e-learning outcomes, are locus of control and conscientiousness. Locus of control reflects the degree to which an individual believes that life events are either within or outside his or her control (Rotter, 1966). Locus of control is thought to affect how individuals approach new situations, their satisfaction, and performance in a variety of settings (Judge & Bono, 2001). In addition, meta-analytic training research has found that locus of control affects training motivation and outcomes (Colquitt et al., 2000). Furthermore, from a technological standpoint, the locus of control affects how individuals interact with and use technology (Hoffman, Novak, & Schlosser, 2003; Johnson, Veltri, & Thatcher, 2015).

Locus of control can be of particular importance for e-learning because of the inherent complexity of interacting in a technology-mediated learning environment. Individuals with a greater external locus of control may focus more on the shortcomings of the learning environment and be less likely to believe that they can effectively leverage the technology to learn compared with those with a more internal locus of control (Johnson et al., 2009). For this reason, trainees may place less effort into learning, interact less, and be more easily distracted by external factors during training. Existing research though is ambivalent on the how locus of control affects learning processes and outcomes in e-learning contexts. Yukselturk, Ozekes, and Turel (2014) found that this construct was a key differentiator between those who dropped out of e-learning and those who remained. Also, individuals with an internal locus had higher satisfaction than those with an external locus. But Johnson et al. (2009) did not find statistically significant relations between locus of control and utility judgements, satisfaction or course performance. Given both the limited number of studies in this area as well as the conflicting findings, more research is needed to better understand the role that locus of control plays in e-learning.

The second personality variable of interest to e-learning is conscientiousness. Individuals who are conscientious are thoughtful, responsible, dependable, and achievement oriented (Barrick, Mount, & Strauss, 1993). Meta-analytic research has found that conscientiousness is positively related to training outcomes (Colquitt et al., 2000). Within the e-learning environment, conscientiousness should also be important. As noted above, one the major advantages of e-learning is that learners can take advantage of learner control. Individuals who are more proactive, motivated, and diligent (e.g., conscientious) should communicate more, spend more time practicing and engaging in learning behaviors, and better take advantage of learner control capabilities. But research findings with respect to conscientiousness are also mixed. One study conducted in a longer-length training program (e.g., full semester) found that conscientiousness was related to learner effort and practice (assessed as number of times practice quizzes were taken) (Kleitman & Costa, 2014). Conversely, in a shorter-term training setting, research has not found statistically significant relations between conscientiousness and e-learning outcomes (Orvis et al., 2011). One potential explanation for the inconsistent findings is that length of training may matter. Because conscientiousness can drive pro-learning behaviors, its impact may be larger in longer-term training than in short-term training. Future research should investigate the role of conscientiousness in longer and shorter duration e-learning courses, given the contradictory findings to date.

Computer self-efficacy Computer self-efficacy (CSE) is an individual's belief in his or her ability to complete computer-related tasks (Marakas, Yi, & Johnson, 1998). Research has consistently found that individuals with higher CSE outperform those with lower computer self-efficacy (Compeau & Higgins, 1995; Johnson & Marakas, 2000). Within the e-learning environment CSE should affect the motivation to learn the material as well as how learners navigate the environment. With all interactions mediated via technology, trainees need to be confident in using technology to access course materials, to communicate with peers, to perform course tasks, and to take learning assessments. Individuals who are more efficacious should be better able to leverage the technology and achieve greater e-learning outcomes.

Johnson et al. (2008) argue that highly efficacious individuals will not only communicate more, but can also use the technology to maximize communication quality. Research has consistently supported these arguments with studies finding that CSE was related to metacognitive activity (Eom, 2011), course satisfaction and utility judgments (Eom, 2011; Johnson et al., 2008; Johnson et al., 2009; Liaw, 2008). Surprisingly though, CSE has not always been found to affect course performance. Johnson et al. (2008) reported a statistically significant relation between CSE and performance, but others report that, despite CSE and performance being significantly correlated, CSE was not related to course performance in the presence of other technological and learning process variables (Galy et al., 2011; Johnson et al., 2009; Sun, Tsai, Finger, Chen, & Yeh, 2008). One potential reason for this is that although CSE does affect learning outcomes, it does so by affecting more proximal learning processes, which in turn mediate the relationship between CSE and performance (Locke & Latham, 2004). In addition, successful performance in e-learning may depend not only on computer self-efficacy but also on learners' self-efficacy for the course topic (deNoyelles, Hornik, & Johnson, 2014).

Computer anxiety Computer anxiety is an effective state where individuals experience fear and apprehension about interacting with computers or technology (Thatcher & Perrewe, 2002). Computer anxiety is particularly important in e-learning because, as noted earlier, many (if not all) course interactions are mediated via technology. If individuals are apprehensive about using computers, they will be less likely to use the software tools available to them, and will likely spend less time learning, participate less, communicate less, and ultimately likely have lower outcomes. To date though, most research has not focused on the role of computer anxiety in e-learning processes or outcomes, but instead has looked at on how anxiety impacts decisions to use e-learning (Al-Busaidi, 2013; Jashapara & Tai, 2011; van Raaij & Schepers, 2008). Studies have reported though that higher computer anxiety during e-learning was related to lower course satisfaction (Saadé & Kira, 2009). Other research has found that learners with higher computer anxiety are less likely to participate and communicate via email and to have lower course performance than those with lower anxiety (Fuller, Vician, & Brown, 2006). Finally, one study found that in the presence of both perceived ease of use and usefulness, computer anxiety, although correlated with course grade, was not statistically significantly related to it in a regression analysis (Galy et al., 2011). Given the importance of anxiety in how people use technology and learn, we recommend more research in this area to assess its importance in e-learning.

Previous online course experience As students take more and more classes online, they should gain valuable experience, which will help them to better succeed in the environment. E-learning provides trainees with the flexibility and ability to arrange, design, and control their learning to a greater extent than is possible in a traditional classroom setting. But not all trainees are prepared to effectively leverage these learner control capabilities (Salas et al., 2005). Learners with more experience are less likely drop out of online courses (Dodd, Kirby, Seifert, & Sharpe, 2009) and more likely to participate, enjoy the course, and perform better than those with lower experience (Salyers, Carter, Myers, & Barrett, 2014). However, Yanson and Johnson (2016) found that although previous e-learning experience was correlated with performance outcomes, it was not statistically significantly related to outcomes in the presence of other factors such as socialization and training complexity. Arbaugh and colleagues have noted that although experience is related to course satisfaction and grades, it is not related to utility judgments (Arbaugh, 2002, 2005, 2008, 2014).

Together these findings suggest that although course experience is valuable, it may be more of a reflection of an increase in important in-training behaviors and skills. For example as experience is gained, trainees feel a greater sense of interaction with peers (Arbaugh, 2004). It is possible that experience provides trainees the skills to interact more effectively with others in the online environment. Thus, rather than simply being enrolled in multiple classes, it is a trainee's ability to interact that may most matter.

Age Another trainee characteristic that may affect e-learning outcomes is age. Previous research on training and development has argued that factors such as pending retirement, reduced access and support for training, increased anxiety, and lower self-efficacy may contribute to the decline in training and development as employees get older (Maurer, 2001; Maurer, Weiss, & Barbeite, 2003). It has also been argued that older trainees perform less well than younger employees because as employees age, they can experience reduced concentration, decreased memory capacity, and lower speed in performing tasks (Hertzog, Kramer, Wilson, & Lindenberger, 2008). However, meta-analytic research reveals that the relationship between age and training is complex. For example, Colquitt et al. (2000) found that age was positively related to training transfer, but negatively related to training motivation and performance. For e-learning, additional factors may cause age to affect learning processes and outcomes. Older learners may be less enthusiastic and comfortable with e-learning because they have less experience with computers, and are less comfortable with the computer replacing the trainer and classroom setting (Brown, 2001) Trainee expectations in regards to how best to deliver e-learning and what to include can also differ between older and younger generations (Becker, Fleming, & Keijsers, 2012).

Research on age has not borne out these perspectives. For example, Hashim, Ahmad, and Abdullah (2010) observed that older learners actually were more satisfied with e-learning than younger learners. Kim and Frick (2011) found that older learners were more motivated to participate in a self-paced, learner directed e-learning course than younger trainees. Gender may moderate the effect of age in e-learning such that older women have been found to have better learning performance than younger women (there were no differences between older and younger men) (Bausch, Michel, & Sonntag, 2014). Finally, other research has found no statistically significant relationship between age and e-learning outcomes (Brown, 2001; Chu, 2010). Together the findings suggest that older learners are not at a disadvantage in e-learning. For a fuller discussion of the role of age in the adoption and use of technology see Burlacu, Truxillo, and Bauer (2017, Chapter 20 in this volume).

Motivation E-learning research has focused on two types of motivation. The first is a general motivation to learn and the second is goal orientation. Motivation to learn reflects the desire of the trainee to learn the content of the training program. Theories of motivation suggest that the greater the desire to perform a task or behavior, the more likely the individual will be to engage in that behavior (Ajzen & Fishbein, 1980; Locke & Latham, 2002; Vroom, 1964). Meta-analytic research supports this argument, revealing that motivation to learn is positively related to learning outcomes such as satisfaction, skill development, and knowledge gain (Colquitt et al., 2000). Motivation to learn thus can influence the drive and effort that a trainee places into engaging in the behaviors and activities in the e-learning environment which can enhance opportunities for success. Although only a few studies have investigated motivation to learn in e-learning, they have consistently found that motivation to learn is related to learning behaviors and outcomes such as in class motivation (Kim & Frick, 2011), training effort (Brown, 2005b), and participation in e-learning (Garavan, Carbery, O'Malley, & O'Donnell, 2010). Given the findings from the broader training literature and the evidence of its importance in e-learning, more research is needed on trainee motivation as part of models of e-learning success. Research and resulting training practice may benefit from considering the potential multidimensionality of motivation to learn. Although many training studies have examined a single dimension of motivation with the motivation to learn construct, others have found unique effects for goals, goal commitment, and self-efficacy (Kim, Oh, Chiaburu, & Brown, 2012). As a result, researchers should examine different motivational theories and measures. Given the limited research in this area, and the need to more closely examine the unique role of different forms of motivation in e-learning, more research is also needed in this area.

The second type of motivation of interest to e-learning scholars is goal orientation. The setting of specific and challenging goals has long been shown to affect how well individuals perform (Locke & Latham, 2002). But beyond the proximal goal-setting process, researchers have theorized that individuals also have a broader general orientation toward goals and goal achievement (Dweck, 1986; Elliot & Dweck, 1988; Ford, Smith, Weissbein, Gully, & Salas, 1998). Two particular orientations are of interest in this study: performance and mastery orientation. Individuals with greater performance orientation seek opportunities to succeed and to demonstrate their skill proficiency. Ability is thus viewed as static and innate reflection of the individual. Individuals with higher performance orientation seek out situations where they can succeed and avoid situations where they may not (Ford et al., 1998). The second type of orientation is mastery orientation. Individuals with greater mastery orientation focus on learning as a way to improve competence, skill, and knowledge. Ability is viewed as cultivated over time through practice and experience. They seek out situations where they can improve skills, view errors as part of the learning process, and try to refine their skills through new pursuits.

Research has found that goal orientations can affect training processes and outcomes, such as metacognitive activity and performance (Ford et al., 1998; Johnson, 2005; Schmidt & Ford, 2003). E-learning research also suggests that goal orientations can affect e-learning processes and outcomes. For example, research has found that goal orientations are related to utility judgments, course satisfaction, and performance (Bell & Kozlowski, 2002b; Kickul & Kickul, 2006). But this relation is not as straightforward as theories have suggested. Bell and Kozlowski (2002b) found that there was an interaction between ability and mastery orientation such that having high mastery orientation did not improve performance in individuals with low ability but it improved performance in high-ability trainees. Wan, Compeau, and Haggerty (2012) found that higher performance orientation was negatively related to the use of self-regulated learning strategies, whereas a higher mastery orientation was positively related to the use of these strategies. In addition, learners with a higher performance orientation are more likely to engage in off-task behavior (Orvis et al., 2011), and less likely to practice (Brown, 2001; Orvis et al., 2011). Finally, conversely to the theorized relation, research has found that mastery orientation is negatively related to off-task attention, practice time, and post-test knowledge gain (Brown, 2001).

Learning Processes

Interaction

Interaction in e-learning has been defined as the "exchange of information between the various stakeholders in the course (e.g. peers, instructors, and other support staff)" (Johnson et al., 2008, p. 360), and has been argued to be the most effective and natural way to learn (Gunawardena, 1995; Hiltz, 1994). In fact researchers have argued that e-learning should be designed to support collaboration an interaction between peers (Benbunan-Fich & Arbaugh, 2006). The benefits of interaction include greater peer feedback, better ability to evaluate training progress, greater peer learning, and increased depth of information processing (Johnson et al., 2008; Piccoli et al., 2001). For example, in a qualitative study, Garrett (2011) observed that students felt that interactions allowed them to better know their peers, improve their work through peer feedback and evaluation, and to better learn the material. Interaction and collaborative oriented e-learning also improves trainee self-efficacy to a greater extent than individualized e-learning (Gupta & Bostrom, 2013), however, it is not only actual interactions that are important but also trainees' perceptions of how much they are interacting (Arbaugh, 2005).

Interaction has consistently been positively related to learning and performance (Alavi et al., 2002; Benbunan-Fich & Arbaugh, 2006; Gunawardena, Lowe, Constance, & Anderson, 1997; Schmidt & Ford, 2003). For example, research has found that more that learners participated in discussion boards, the better their course performance (Hwang & Arbaugh, 2006; Johnson et al., 2008; Johnson et al., 2009; Strang, 2011). In addition, using social network analysis, Lin, Huang and Chuang (2015) found that the more central a learner is to his or her communications network the higher his or her exam performance.

Interaction can also affect trainee reactions. When trainees interact more and process information more deeply, they are more likely to see the value in the training content. Furthermore, when they interact with peers, they should feel more connected and should be more positive and satisfied with their learning experiences. Research has found that courses designed to support collaboration and interaction lead to better trainee reactions (e.g., satisfaction and utility judgments) than those which were not. With only one exception (Arbaugh & Benbunan-Fich, 2007), the evidence suggests that the more a learner interacts with others in the environment, the more satisfied he or she will be with the training (Arbaugh, 2000a, 2002; Arbaugh & Hornik, 2006; Arbaugh & Rau, 2007; Johnson et al., 2008; Johnson et al., 2009). One mechanism through which interaction can affect satisfaction is through increased perceptions of peer connections and presence (Kim, Kwon, & Cho, 2011).

The results of research with respect to the relation between interaction and utility judgments are mixed. Some studies report a statistically significant relation between interaction and utility judgments (Arbaugh, 2000b, 2002; Arbaugh & Benbunan-Fich, 2007; Arbaugh & Rau, 2007; Yang & Durrington, 2010), but others have found either a marginally significant relation (Arbaugh & Hornik, 2006) or no relation between them (Johnson et al., 2008; Johnson et al., 2009).

Social presence

Social presence has been defined as the feeling, perception, and reaction of being connected to others when interacting, and the salience of others in interpersonal relationships (Short, Williams, & Christie, 1976); (Marks et al., 2005). Within the e-learning environment it has also been defined as the extent to which the technology enables students to create a warm, personal, sociable, and active environment and allows them to be connected in a shared learning space (Biocca, Harms, & Burgoon, 2003; Tu & McIsaac, 2002). Researchers have argued that social presence is a critical factor in the development of a shared learning environment and e-learning effectiveness (Gunawardena, 1995; Gunawardena & Zittle, 1997; Richardson & Swan, 2003). Also, women have a greater sense of social presence than men (Johnson, 2011) in e-learning.

When individuals perceive themselves to be part of a shared environment, they will exchange greater and more complex information (McGrath, Arrow, Grunfeld, Hollingshead, & O'Connor, 1993), may pay greater attention to messages produced by those with whom they are connected (Mackie, Worth, & Asuncion, 1990), and feel that they can better judge the quality of peer contributions (Lowry, Roberts, Romano, Cheney, & Hightower, 2006). By attending more greatly to messages made by others, learners should learn more than if they were not connected to others. What is interesting though is that empirical evidence does not support this claim. The three studies that have investigated social presence and performance did not find a statistically significant relationship between social presence and e-learning performance (Baturay, 2011; Johnson et al., 2008; Johnson et al., 2009). However, this does not mean that social presence is not important. Social presence helps learners engage more deeply with content and peers, and should increase the likelihood that the learners value the course. Empirical studies have also consistently found a positive relation between social presence and utility judgments (Arbaugh, 2001, 2002, 2008, 2014; Arbaugh & Hornik, 2006; Johnson et al., 2008; Johnson et al., 2009).

Finally, social presence can help trainees overcome feelings of isolation by enabling stronger peer connections, reducing feelings of isolation, strengthening feelings of community, and increasing course satisfaction. As with utility judgments, previous research has consistently found that social presence is positively related to satisfaction (Arbaugh, 2001, 2002, 2008, 2014; Arbaugh & Hornik, 2006; Gunawardena & Zittle, 1997; Johnson et al., 2008; Johnson et al., 2009). Only one study did not find a statistically significant relation (Kim et al., 2011). Overall, the evidence suggests that the importance of social presence is less about improving learning performance, and more about positively affecting the social context of learning. Helping trainees feel connected may not necessarily lead to better knowledge but will improve the learning environment. These findings are similar to the results of research on group support systems which found that the use of technology to support knowledge sharing within technology-mediated groups does not necessarily lead to better information sharing and decisions (Dennis, 1996).

Metacognitive activity

One of the hallmarks of most e-learning environments is the high levels of control placed in the hands of the trainees, but not all learners will naturally possess an awareness of their cognitive processes that can help them leverage this control to their advantage. Better utilization of self-regulated learning strategies via effective metacognitive activity can help. Metacognitive activity reflects an individual's awareness of and regulation of his or her cognitive processes (Flavell, 1979). When learners engage in more metacognitive activity, they can better monitor their learning, determine when problems are occurring, and make adjustments to their learning behavior (Ford et al., 1998). In the same vein, individuals who employ self-regulated learning strategies are more likely to be aware of their own strengths and weaknesses, set more realistic goals, better evaluate progress towards these goals, and better regulate learning behaviors (Bandura, 1991).

Those with greater metacognitive and self-regulation skills are essential in an e-learning environment because with them, learners can "make more informed decisions regarding what control strategies to utilize to progress in their learning" (Schmidt & Ford, 2003, p. 407). Both training research (Ford et al., 1998; Keith & Frese, 2005; Pintrich & DeGroot, 1990) and e-learning studies (Bell & Kozlowski, 2008; Gravill & Compeau, 2008; Lin et al., 2015l; Schmidt & Ford, 2003; Sitzmann & Ely, 2010; Wan et al., 2012) have consistently reported metacognitive and self-regulated strategies as positively related to learning performance and training transfer. However, it is important to note that one study did not find a positive relation between metacognitive activity and course performance (Johnson et al., 2009). In addition, trainees who can better monitor the environment and leverage it to their advantage should be able to more fully focus on the training material and adapt their behaviors within the environment to create a more positive learning experience. Research supports these arguments, with studies consistently finding a positive relation with utility judgments (Al-Samarraie, Teo, & Abbas, 2013; Johnson et al., 2009) and course satisfaction (Johnson et al., 2009; Wan et al., 2012).

Learner choices

One of the challenges in environments characterized by higher learner control is that trainees are more likely to abandon training before mastering the skills or material (Steinberg, 1989; Tennyson, 1980). One way to improve learning outcomes is to encourage trainees to engage in learning behaviors such as practice time and time on task, and to reduce off-task attention or behaviors. Research shows that to gain skill and expertise, an individual needs to attend to the task and practice (Ericsson, Krampe, & Tesch-Romer, 1993). In addition, those trainees whose focus on the task is greater may learn more than those who focus on non-task activities (Fisher & Ford, 1998; Kanfer & Ackerman, 1989). E-learning distractions can arise from phone calls, family interruptions, interface and technology issues, and more. Thus, the flexibility afforded by e-learning can actually be a doubled edged-sword, as these distractions may take the trainee's focus off of the training task at hand.

Several studies have investigated how learner choices affect e-learning outcomes. For example, Galy, Downey, and Johnson (2011) found that the ability to work independently was positively related to course grade. Several other studies found that practice time (Brown, 2001) and time on task (Brown, 2001; Sitzmann & Ely, 2010; Stark &

Krause 2009) were also positively related to course grade. Stewart, Stott, and Nuttall (2011) reported that students who spent more time in the online component of a blended learning course performed better than those who did not. With the exception of one study where off-task attention was not related to performance (Brown, 2001), research has also found that higher off-task attention is related to lower course performance (Gravill & Compeau, 2008; Karim & Behrend, 2014). Together the results suggest that trainee choices regarding attention, time on task, and practice all influence e-learning outcomes.

Future Research

As noted in our discussion, the literature on e-learning is both broad and deep. This particular review has focused on the organizational, technological, trainee/learner, instructor, course design, and learning process factors that affect e-learning outcomes such as satisfaction and learning. However e-learning continues to evolve, especially the technologies supporting e-learning. In this section, we discuss several of these changes as well as the research questions that organizations will face as they determine how to best leverage new technologies to improve e-learning outcomes.

Mobile learning

Mobile learning (mLearning) is the delivery of training or educational content through smaller, and portable, mobile devices such as tablets and cell phones. Just as with e-learning, one of the factors driving the use of mLearning is organizational demands for a highly skilled, flexible workforce that can gain knowledge and skills on an "as needed" basis rather than waiting for regularly scheduled training. The global market for mLearning is over US$8 billion and is expected to reach nearly US$15 billion by 2019 (Insight, 2015). The growth in mLearning is being fueled by the growth in employees who want to use their own devices for work; some estimates suggest that over 70% of employees want to bring their own device to work (Forrester, 2012). A full discussion of the use of mobile computing in organizations is beyond the scope of our discussion, but the interested reader is referred to Zafar (2017, Chapter 10 in this volume). MLearning provides additional flexibility over e-learning, because learners are able to access materials on their phone or tablet in places where traditional computer networks are unavailable.

A tacit assumption behind mLearning is that learners will be more motivated to attend training, because they can access these materials more conveniently. They can access documents at any time and in any place, can annotate and share files, and quickly search on keywords to find annotations or content (Chao & Chen, 2009; Zhang, Zhao, Zhou, & Nunamaker Jr, 2004). In addition, mobile devices can provide interactive multimedia experiences to learners on demand to reinforce training lessons. Although this notion seems similar to the arguments used to explain the use of e-learning, mLearning expands the context in which learning can occur. MLearning may encourage some individuals to learn in contexts that are less conducive to learning, depending on their motivation and experience. For example, it might make it more likely for those with low motivation, who also have trouble keeping attention on task, to engage in learning in contexts broader than either in-class learning or e-learning might allow. Although these arguments seem plausible, research is needed to determine how to best implement mLearning and how to leverage it within broader training programs. For example, Bargeron et al. (2002) found that the use of annotations can improve trainee engagement in asynchronous training groups, but they did not investigate performance improvements. In addition, Kaganer, Giordano, Brion, and Tortoriello's (2013) research on MBA students revealed that the students found a tablet to be more difficult to use than a computer when navigating documents with a lot of financial data. They were less satisfied with course materials, communicated less with peers, and had smaller increases in team cohesion than those students who did not use tablets.

Two additional, and related, technology issues with respect to the use of mobile devices are access and bandwidth. The move to mobile assumes that all employees will have access to mobile devices and that they have reliable networks with sufficient bandwidth over which to access training content. But this assumption may not be met. First, only about one-third of adults own a tablet and of these those with a college degree are more likely to own a tablet (49%) than those with no college degree (<30%). In addition, those aged over 50 years are much less likely to own a tablet as are those with lower income (Zickuhr, 2013). Furthermore, although recent research suggests that a similar percentage of White Americans have smartphones as Americans belonging to ethnic minority groups (e.g., Hispanic Americans and African Americans), there is still a gap in computer ownership and Internet use (Lopez, Gonzalez-Barrera, & Patten, 2013; Zickuhr, 2013). One of the risks of a mobile digital divide is that those with reduced access to mobile devices and networks may be less satisfied with mobile initiatives, may perform more poorly, and less likely to participate in development activities. They can then fall behind and not have the current skills that organizations seek when promoting employees to new positions or providing novel work opportunities. For more information on the impacts of the digital divide, see Stone, Krueger, and Takach (2017, Chapter 19 in this volume).

It is clear that organizations are embracing mLearning. Equally as clear is that researchers have not yet begun to rigorously assess how they can be utilized more effectively. For example, how and where can mobile devices be best utilized to deliver content, to increase learning motivation, and improve learning processes. As with web-based learning, it is not a question of whether mLearning is appropriate, but how it can be most effectively deployed. We believe that our model provides an excellent starting point for researchers to begin investigating this question.

Knowledge repositories

Knowledge repositories may also drive changes in e-learning. Employees can share knowledge on blogs, wikis, and other internal documents in knowledge repositories that can be accessed by other employees. Video repositories with annotation capabilities can be used to provide rich communications that extend learning beyond the formal training classroom (Steer, 2012). The challenge is that individuals are often hesitant to contribute to these repositories (Kankanhalli, Tan, & Wei, 2005). Although several studies have investigated how to improve knowledge sharing (Wasko & Faraj, 2000, 2005), research has yet to investigate their effectiveness at increasing employee knowledge and transfer. It could be argued that motivation to learn may be higher for those that utilize knowledge repositories, because they are only accessed when needed. Our model suggests that this would then improve learning outcomes such as learning and satisfaction. But the potential downsides to knowledge repositories must be considered. For example, as they are developed by others and not necessarily formally screened, they may contain incomplete or inaccurate information, it may be hard to search and find information, and they eliminate the development of a shared learning environment where learners can reinforce knowledge as they consume it together. Therefore research is needed to address how to integrate knowledge repositories most effectively into an organization's learning and development portfolio.

Virtual realities

Virtual reality is "a computer technology that enables users to view or 'immerse' themselves in an alternate [*sic*] world" (Aguinis, Henle, & Beaty Jr, 2001, p. 70). With virtual reality, the learner is immersed within a three-dimensional learning space where they can look at, move, and interact with objects within the environment as if they were real (Mujber, Szecsi, & Hashimi, 2004). For example, pilots have long trained in flight simulators which mimic the look and feel of a cockpit. Research on virtual simulations found that they improve training outcomes in fields such as medicine (Larsen, Oestergaard, Ottesen, & Soerensen, 2012), oil exploration (Brasil et al., 2011), and the military (Bowman & McMahan, 2007). The reduction in costs of virtual reality and the development of hand-held devices means that these devices are now available for a much wider set of training than was previously possible.

Similar to virtual reality, virtual worlds are increasingly being used in educational settings. Trainees in a virtual world are represented by avatars (or simulated bodies) that interact with the virtual environment, and can communicate with other avatars through text and voice. It has been argued that virtual worlds can increase learner engagement compared with traditional online courses, because learners interact in an environment which has richer visual, audio, and communication capabilities (Stone, Deadrick, Lukaszewski, & Johnson, 2015) and stimuli. Virtual worlds can improve learner engagement (Hearrington, 2010), interaction (Merchant et al., 2012), and learning outcomes (Hornik & Thornburg, 2010) in educational environments. In addition, research has found that factors such as computer self-efficacy are central to the development of skills in virtual worlds, just as they are in traditional online environments (deNoyelles et al., 2014). It must be noted that despite their promise, virtual worlds can be difficult in terms of communication and navigate (deNoyelles, 2012; Mennecke, Hassall, & Triplett, 2008). More research is needed to better understand how to deploy virtual reality and virtual worlds so that both learning processes and outcomes are enhanced. As we noted previously, investigating these technologies in light of the model outlined in this chapter can help scholars design studies to assess how to best improve both learning processes and both process and learning outcomes.

Gamification

In their review of the e-learning literature, Salas et al. (2005) recommended adding games into e-learning. They argued that the use of learning games would increase the attractiveness of e-learning, encourage practice, reduce the fear of testing, and encourage exploratory learning. The idea of including gaming elements in non-game contexts is known as gamification (Robson, Plangger, Kietzman, McCarthy, & Pitt, 2015). For example, traditional gaming elements can be included in the learning tasks themselves. In addition, achievement levels and badges, rewards, and leaderboards, can be used to identify the top performers and contributors in the learning environment. Gamification can improve motivation (Dominguez et al., 2013), participation (Snyder & Hartig, 2013) and learning (Foster, Sheridan, Irish, & Frost, 2012; McDaniel, Lindgren, & Friskics, 2012). For additional ideas on gamification we refer the reader to Thielsch and Niesenhaus (2017, Chapter 5 in this volume). For a fuller discussion of the use of simulation games in training, we encourage the researcher to refer to Grossman, Heyne, and Salas (2015).

Gamification is not without its risks though. For example, in some situations gamification has led to better hands-on task scores but lower written performance scores (Dominguez et al., 2013). In addition, some learners do not like the perceived competition inherent in the leaderboard process. Finally, poorly designed gamified activities can reduce learner engagement (Foster et al., 2012). To date, rather than identifying rich and theoretically driven models of the role that gamified activities can play in the learning process, much of the research has simply compared the effectiveness of gamified training to e-learning without gamification. Research is needed to understand more richly how gamification affects learning processes and outcomes.

Conclusion

Given the centrality of e-learning to both corporate and educational training and learning initiatives, it is important for organizations and educators to understand how best to leverage technology in the design of e-learning initiatives. To do this, they must have a better understanding of the various factors that can affect e-learning outcomes, such as training performance and satisfaction. This chapter sought to address these questions by reviewing a number of e-learning frameworks and integrating them into an overall framework. Specifically, this framework considers the role of the learner/trainee, the instructor, course design, technology, and the organization serve as inputs to the learning process. In turn, several motivational, behavioral and cognitive processes mediate these inputs and impact learning outcomes such as satisfaction and performance. Based on the review of this framework, we have provided a set of evidence-supported guidelines for research to build on and for designers to build more effective e-learning initiatives. Ultimately rather than chasing the latest technology to transform organizational and educational learning initiatives, a shifting target at best, scholars need to undertake studies that focus on how to best utilize technology, learners, instructors, and pedagogy to deliver e-learning that will most effectively meet the needs of all learners.

References

Ajzen, I., & Fishbein, M. (1980). *Understanding attitudes and predicting social behaviour:* Englewood Cliffs, NJ: Prentice Hall.

Al-Busaidi, K. A. (2013). An empirical investigation linking learners' adoption of blended learning to their intention of full e-learning. *Behaviour & Information Technology, 32*(11), 1168–1176.

Al-Samarraie, H., Teo, T., & Abbas, M. (2013). Can structured representation enhance students' thinking skills for better understanding of e-learning content. *Computers & Education, 69*, 463–473.

Alavi, M., & Leidner, D. E. (2001). Research commentary: Technology-mediated learning – A call for greater depth and breadth of research. *Information Systems Research, 12*(1), 1–10.

Alavi, M., Marakas, G. M., & Yoo, Y. (2002). A comparative study of distributed learning environments on learning outcomes. *Information Systems Research, 13*(4), 404–415.

Alavi, M., Wheeler, B. C., & Valacich, J. S. (1995). Using IT to reengineer business education: An exploratory investigation of collaborative telelearning. *MIS quarterly*, 293–312.

Alavi, M., Yoo, Y., & Vogel, D. R. (1997). Using information technology to add value to management education. *Academy of Management Journal, 40*(6), 1310–1333.

Allen, E., & Seaman, J. (2011). *Going the distance: online education in the United States, 2011.* Wellesley, MA: Babson Survey Research Group.

Alliger, G. M., Tannenbaum, S. I., Bennett Jr., W., Traver, H., & Shotland, A. (1997). A meta analysis of the relations among training criteria. *Personnel Psychology, 50*, 341–358.

Anderson, L. W., & Krathwohl, D. R. (2001). *A taxonomy for learning, teaching, and assessing: A revision of Bloom's taxonomy of educational objectives.* New York: Pearson.

Anderson, P., & Anderson, J. (1982). Nonverbal immediacy in instruction. In L. Barker (Ed.), *Communication in the classroom* (pp. 98–120). Englewood Cliffs, NJ: Prentice-Hall.

Anderson, S. E., Groulx, J. G., & Maninger, R. M. (2011). Relationships among preservice teachers' technology-related abilities, beliefs, and intentions to use technology in their future classroom. *Journal of Educational Computing Research, 45*(3), 321–338.

Aguinis, H., Henle, C. A., & Beaty Jr, J. C. (2001). Virtual reality technology: A new tool for personnel selection. *International Journal of Selection and Assessment, 9*(1–2), 70–83.

Arbaugh, J. B. (2000a). Virtual classroom characteristics and student satisfaction with internet-based MBA courses. *Journal of Management Education, 24*(1), 32–54.

Arbaugh, J. B. (2000b). Virtual classroom: An exploratory study of class discussion patterns and student learning in an asynchronous Internet-based MBA course. *Journal of Management Education, 24*(2), 213–233.

Arbaugh, J. B. (2001). How instructor immediacy behaviors affect student satisfaction and learning in web-based courses. *Business Communication Quarterly, 64*(4), 42–54.

Arbaugh, J. B. (2002). Managing the on-line classroom: A study of technological and behavioral characteristics of web-based MBA courses. *Journal of High Technology Management Research, 13*(2), 203–223.

Arbaugh, J. B. (2004). Learning to learn online: A study of perceptual changes between multiple online course experiences. *The Internet and Higher Education, 7*(3), 169–182.

Arbaugh, J. B. (2005). Is there optimal design for on-line MBA courses? *Academy of Management Learning & Education, 4*(2), 135–149.

Arbaugh, J. B. (2008). Does the community of inquiry framework predict outcomes in online MBA courses? *International Review of Research in Open and Distance Learning, 9*(2), 1–21.

Arbaugh, J. B. (2010). *Online and blended business education for the 21st century: Current research and future directions.* Oxford: Woodhead Publishing.

Arbaugh, J. B. (2014). System, scholar or students? Which most influences online MBA course effectiveness? *Journal of Computer Assisted Learning*, 351–362.

Arbaugh, J. B., & Benbunan-Fich, R. (2007). The importance of participant interaction in online environments. *Decision Support Systems, 43*(3), 853–865.

Arbaugh, J. B., Desai, A., Rau, B., & Sridhar, B. S. (2010). A review of research on online and blended learning in the management disciplines: 1994–2009. *Organization Management Journal, 7*(1), 39–55.

Arbaugh, J. B., & Duray, R. (2002). Technological and structural characteristics, student learning and satisfaction with web-based courses an exploratory study of two on line MBA programs. *Management Learning, 33*(3), 331–347.

Arbaugh, J. B., & Hornik, S. (2006). Do Chickering and Gamson's seven principles also apply to online MBAs. *The Journal of Educators Online, 3*(2), 1–18.

Arbaugh, J. B., & Rau, B. L. (2007). A study of disciplinary, structural, and behavioral effects on course outcomes in online MBA courses. *Decision Sciences Journal of Innovative Education, 5*(1), 65–95.

Argote, L., Insko, C. A., & Yovetich, N., & Romero, A. A. (1995). Group learning curves: The effects of turnover and task complexity on group performance. *Journal of Applied Psychology, 25*, 512–529.

Balakrishnan, V. (2014). Using social networks to enhance teaching and learning experiences in higher learning institutions. *Innovations in Education and Teaching International, 51*(6), 595–606.

Baldwin, T. T., & Ford, J. K. (1988). Transfer of training: A review and directions for future research. *Personnel psychology, 41*(1), 63–105.

Bandura, A. (1991). Social cognitive theory of self-regulation. *Organizational Behavior and Human Decision Processes, 50*, 248–287.

Bandura, A. (1997). *Self-efficacy: The exercise of control.* New York: W.H. Freeman and Company.

Bargeron, D., Grudin, J., Gupta, A., Sanocki, E., Li, F., & LeeTiernan, S. (2002). Asynchronous collaboration around multimedia applied to on-demand education. *Journal of Management Information Systems, 18*(4), 117–145.

Barrick, M. R., Mount, M. K., & Strauss, J. P. (1993). Conscientiousness and performance of sales representatives: Test of the mediating effects of goal setting. *Journal of Applied Psychology, 78*(5), 715–723.

Baturay, M. H. (2011). Relationships among sense of classroom community, perceived cognitive learning and satisfaction of students at an e-learning course. *Interactive Learning Environments, 19*(5), 563–575.

Bausch, S., Michel, A., & Sonntag, K. (2014). How gender influences the effect of age on self-efficacy and training success. *International Journal of Training and Development, 18*(3), 171–187.

Becker, K., Fleming, J., & Keijsers, W. (2012). E-learning: Ageing workforce versus technology-savvy generation. *Education & Training, 54*(5), 386–400.

Bell, B. S., & Kozlowski, S. W. J. (2002a). Adaptive guidance: Enhancing self-regulation, knowledge, and perception in technology-based training. *Personnel Psychology, 55*, 267–306.

Bell, B. S., & Kozlowski, S. W. J. (2002b). Goal orientation and ability: Interactive effects on self-efficacy, performance, and knowledge. *Journal of Applied Psychology, 87*(3), 497–505.

Bell, B. S., & Kozlowski, S. W. J. (2008). Active learning: Effects of core training design effects on self-regulatory processes, learning, and adaptability. *Journal of Applied Psychology, 95*(2), 296–316.

Benbunan-Fich, R., & Arbaugh, J. B. (2006). Separating the effects of knowledge construction and group collaboration in learning outcomes of web-based courses. *Information & Management, 43*(6), 778–793.

Biocca, F., Harms, C., & Burgoon, J. K. (2003). Toward a more robust theory and measure of social presence: review and suggested criteria. *Presence: Teleoperators and Virtual Environments, 12*(5), 456–480.

Bloom, B. S., & Krathwohl, D. R. (1956). *Taxonomy of educational objectives: The classification of educational goals. Handbook I: Cognitive domain.* UK: Longman Group.

Bostrom, R. (2003, August). *E-learning: Facilitating learning through technology.* Paper presented at the AMCIS 2003 Proceedings, Tampa, FL, USA.

Bowman, D. A., & McMahan, R. P. (2007). Virtual reality: how much immersion is enough? *IEEE Computer, 40*(7), 36–43.

Brasil, I. S., Neto, F. M. M., Chagas, J. F. S., de Lima, R. M., Souza, D. F. L., Bonates, M. F., & Dantas, A. (2011, May). *An intelligent agent-based virtual game for oil drilling operators training.* Paper presented at the Virtual Reality (SVR), 2011 XIII Symposium, Uberlandia, Brazil.

Brown, K. G. (2001). Using computers to deliver training: Which employees learn and why? *Personnel Psychology, 54*(2), 271–296.

Brown, K. G. (2005a). An examination of the structure and nomological network of trainee reactions: a closer look at "smile sheets." *Journal of Applied Psychology, 90*(5), 991–1001.

Brown, K. G. (2005b). A field study of employee e-learning activity and outcomes. *Human Resource Development Quarterly, 16*(4), 465–480.

Brown, K. G., & Charlier, S. D. (2013). An integrative model of e-learning use: Leveraging theory to understand and increase usage. *Human Resource Management Review, 23*(1), 37–49.

Brown, K. G., Charlier, S. D., & Pierotti, A. (2012). E-Learning at work: Contributions of past research and suggestions for the future. *International Review of Industrial and Organizational Psychology, 27*, 89–114.

Burlacu, G., Truxillo, D. M., & Bauer, T. N. (2017). Employee age differences in using internet-based tools at work. In G. Hertel, D. L. Stone, R. D. Johnson, & J. Passmore (Eds.), *Wiley Blackwell handbook of the psychology of the internet at work* (pp. 449–480). John Wiley & Sons, Ltd.

Byun, S., & Mills, P. J. (2011). Exploring the creation of learner-centered e-training environments among retail workers: A model development perspective. *Cyberpsychology, Behavior, and Social Networking, 14*(1–2), 66–69.

Capece, G., & Campisi, D. (2013). User satisfaction affecting the acceptance of an e-learning platform as a mean for the development of the human capital. *Behaviour & Information Technology, 32*(4), 335–343.

Carswell, A. D., & Venkatesh, V. (2002). Learner outcomes in an asynchronous distance education environment. *International Journal of Human-Computer Studies, 56*, 475–494.

Cascio, W. F., & Arguinis, H. (2005). *Applied psychology in human resource management* (6th ed.). Upper Saddle River, NJ: Pearson Prentice Hall.

Chao, P. Y., & Chen, G. D. (2009). Augmenting paper-based learning with mobile phones. *Interacting with Computers, 21*(3), 173–185.

Cheng, B., Wang, M., Moormann, J., Olaniran, B., & Chen, N. (2012). The effects of organizational learning environment factors on e-learning acceptance. *Computers & Education, 58*, 885–899.

Chien, T. (2012). Computer self-efficacy and factors influencing e-learning effectiveness. *European Journal of Training and Development, 36*(7), 670–686.

Chu, R. J. (2010). How family support and Internet self-efficacy influence the effects of e-learning among higher aged adults – analyses of gender and age differences. *Computers & Education, 55*, 255–264.

Clark, R. E. (1994). Media will never influence learning. *Educational Technology Research, 42*, 21–29.

Colquitt, J. A., LePine, J. A., & Noe, R. A. (2000). Toward an integrative theory of training motivation: A meta-analytic path analysis of 20 years of research. *Journal of Applied Psychology, 85*(5), 678–707.

Compeau, D. R., & Higgins, C. A. (1995). Application of social cognitive theory to training for computer skills. *Information Systems Research, 6*(2), 118–143.

Daft, R. L., & Lengel, R. H. (1986). Organizational information requirements, media richness and structural design. *Management Science, 32*(5), 554–571.

Davis, F. D., Bagozzi, R. P., & Warshaw, P. R. (1989). User acceptance of computer technology: A comparison of two theoretical models. *Management Science, 35*(8), 982–1003.

Dennis, A. R. (1996). Information exchange and use in group decision making: you can lead a group to information, but you can't make it think. *MIS Quarterly, 20*(4), 433–457.

Dennis, A. R., Fuller, R. M., & Valacich, J. S. (2008). Media, tasks, and communication processes: A theory of media synchronicity. *MIS Quarterly, 32*(3), 575–600.

DeNoyelles, A. (2012). Analysis in virtual worlds: The influence of learner characteristics on instructional design. In K. Seo (Ed.), *Using social media effectively in the classroom: Blogs, wikis, Twitter, and more* (pp. 2–18). New York, NY: Routledge.

DeNoyelles, A., Hornik, S. R., & Johnson, R. D. (2014). Exploring the dimensions of self-efficacy in virtual world learning: Environment, task, and content. *Journal of Online Learning and Teaching, 10*, 255–271.

DeRouin, R., Fritzsche, B., & Salas, E. (2004). Optimizing e-learning: Research-based guidelines for learner-controlled training. *Human Resource Management, 43*(2&3), 147–162.

Dodd, C., Kirby, D., Seifert, T., & Sharpe, D. (2009). The impact of high school distance e-learning experience on rural students' university achievement and persistence. *Online Journal of Distance Learning Administration, 12*(1).

Dominguez, A., Saenz-de-Navarrete, J., De-Marcos, L., Fernandez-Sanz, L., Pages, C., & Martinez-Herraiz, J. J. (2013). Gamifying learning experiences: Practical implications and outcomes. *Computers & Education, 63*, 380–392.

Dweck, C. S. (1986). Motivational processes affecting learning. *American Psychologist, 41*, 1040–1048.

Elliot, E. S., & Dweck, C. S. (1988). Goals: An approach to motivation and achievement. *Journal of Personality and Social Psychology, 73*, 171–185.

Eom, S. B. (2011). Relationships among e-learning systems and e-learning outcomes: A path analysis model. *Human Systems Management, 30*, 229–241.

Ericsson, K. A., Krampe, R. T., & Tesch-Romer, C. (1993). The role of deliberate practice in the acquisition of expert performance. *Psychological Review, 100*(3), 363–406.

Ferdousi, B., & Levy, Y. (2010). Development and validation of a model to investigate the impact of individual factors on instructor's intentions to use e-learning systems. *Interdisciplinary Journal of E-Learning and Learning Objects, 6*, 1–21.

Fisher, S. L., & Ford, J. K. (1998). Differential effects of learner effort and goal orientation on two learning outcomes. *Personnel psychology, 51*(2), 397–420.

Fisher, S. L., Wasserman, M., & Orvis, K. (2010). Trainee reactions to learner control: An important link in the e-learning equation. *International Journal of Training and Development, 14*(3), 198–210.

Flavell, J. H. (1979). Metacognition and cognitive monitoring: a new area of cognitive-developmental inquiry. *American Psychologist, 34*, 906–911.

Flood, J. (2002). Read all about it: On-line learning facing 80% attrition rates. *Turkish Online of Distance Education, 3.*

Ford, J. K., Smith, E. M., Weissbein, D. A., Gully, S. M., & Salas, E. (1998). Relationships of goal orientation, metacognitive activity, and practice strategies with learning outcomes and transfer. *Journal of Applied Psychology, 83*(2), 218–231.

Forrester Research (2012). The expanding role of mobility in the workplace. Retrieved April 7, 2017, from www.cisco.com/web/solutions/trends/unified_workspace/docs/Expanding_Role_of_Mobility_in_the_Workplace.pdf

Foster, J. A., Sheridan, P. K., Irish, R., & Frost, G. S. (2012, June). *Gamification as a strategy for promoting deeper investigation in a reverse engineering activity.* Paper presented at the 2012 American Society for Engineering Education Conference, San Antonio, TX, USA.

Fuller, R. M., Vician, C., & Brown, S. A. (2006). E-learning and individual characteristics: The role of computer anxiety and communication apprehension. *Journal of Computer Information Systems, 46*(4), 103–115.

Galy, E., Downey, C., & Johnson, J. (2011). The effect of using e-learning tools in online and campus-based classrooms on student performance. *Journal of Information Technology Education, 10,* 209–230.

Garavan, T. N., Carbery, R., O'Malley, G., & O'Donnell, D. (2010). Understanding participation in e-learning in organizations: A large scale empirical study of employees. *International Journal of Training and Development, 14*(3), 155–168.

Garrett, N. (2011). An e-portfolio design supporting ownership, social learning, and ease of use. *Educational Technology & Society, 14*(1), 187–202.

Garrison, D. R., & Arbaugh, J. B. (2007). Researching the community of inquiry framework: Review, issues, and future directions. *The Internet and Higher Education, 10*(3), 157–172.

Gill, M. (2000). E-learning technology and strategy for organisations. In K. Fry (Ed.), *The Business of e-learning: Bringing your organization in the knowledge economy.* Sydney: Australia University of Technology.

Granger, B., & Levine, E. (2010). The perplexing role of learner control in e-learning: Will learning and the transfer benefit suffer? *International Journal of Training and Development, 14*(3), 180–197.

Granger, B., & Levine, E. (2013). Complexity compromises learner-led e-learning, *T+D, 67*(4), 20.

Gravill, J., & Compeau, D. (2008). Self-regulated learning strategies and software training. *Information & Management, 45,* 288–296.

Grossman, R., Heyne, K., & Salas, E. (2015). Game- and simulation-based approaches to training. In K. Kraiger, J. Passmore, N. R. d. Santos, & S. Malvezzi (Eds.), *The Wiley Blackwell handbook of the psychology of training, development, and performance improvement* (pp. 202–222). Chichester: John Wiley & Sons, Ltd.

Gunawardena, C. N. (1995). Social presence theory and implications for interaction and collaborative learning in computer conferences. *International Journal of Educational Telecommunications, 1*(2), 147–166.

Gunawardena, C. N., Lowe, X., Constance, A., & Anderson, T. (1997). Analysis of a global debate and the development of an interaction analysis model for examining social construction of knowledge in computer conferences. *Journal of Educational Computer Research, 17*(4), 397–431.

Gunawardena, C. N., & Zittle, F. J. (1997). Social presence as a predictor of satisfaction within a computer-mediated conferencing environment. *American Journal of Distance Education, 11*(3), 8–26.

Gupta, S., & Bostrom, R. P. (2013). Research note – An investigation of the appropriation of technology-mediated training methods: Incorporating enactive and collaborative learning. *Information Systems Research, 24*(2), 454–469.

Gurtner, J.-L. (2015). Effective virtual learning environments. In K. Kraiger, J. Passmore, N. R. d. Santos, & S. Malvezzi (Eds.), *The Wiley Blackwell handbook of the psychology of training, development, and performance improvement* (pp. 188–204). Chichester: John Wiley & Sons, Ltd.

Hashim, R., Ahmad, H., & Abdullah, C. Z. (2010). Antecedents of ICT attitudes of distance education students. *Turkish Online Journal of Educational Technology, 9*(1), 28–36.

Hearrington, D. (2010). Evaluation of learning efficiency and efficacy in a multi-user virtual environment. *Journal of Digital Learning in Teacher Education, 27*(2), 65–75.

Hertzog, C., Kramer, A. F., Wilson, R. S., & Lindenberger, U. (2008). Enrichment effects on adult cognitive development: Can the functional capacity of older adults be preserved and enhanced? *Psychological Science in the Public Interest, 9*(1), 1–65.

Hillman, D. C., Willis, D. J., & Gunawardena, C. N. (1994). Learner-interface interaction in distance education: An extension of contemporary models and strategies for practitioners. *American Journal of Distance Education, 8*(2), 30–42.

Hiltz, S. R. (1994). *The virtual classroom: Learning without limits via computer networks.* Norwood, NJ: Intellect Books.

Hiltz, S. R., & Wellman, B. (1997). Asynchronous learning networks as a virtual classroom. *Communications of the ACM, 40*(9), 44–49.

Hoffman, D. L., Novak, T. P., & Schlosser, A. E. (2003). Locus of control, web use, and consumer attitudes toward internet regulation. *Journal of Public Policy & Marketing, 22*(1), 41–57.

Holton, E. F., Bates, R. A., Seyler, D. L., & Carvalho, M. B. (1997). Toward construct validation of a transfer climate instrument. *Human Resource Development Quarterly, 8*(2), 95–113.

Hornik, S., Saunders, C. S., Li, Y., Moskal, P. D., & Dzuiban, C. D. (2008). The impact of paradigm development and course level on performance in technology-mediated learning environments. *Informing Science: The International Journal of an Emerging Transdiscipline, 11*, 35–58.

Hornik, S., & Thornburg, S. (2010). Really engaging accounting: Second Life™ as a learning platform. *Issues in Accounting Education, 25*(3), 361–378.

Hsieh, P., & Cho, V. (2011). Comparing e-learning tools' success: The case of instructor-student interactive vs. self-paced tools. *Computers & Education, 57*, 2025–2038.

Hurtz, G. M., & Williams, K. J. (2009). Attitudinal and motivational antecedents of participation in voluntary employee development activities. *Journal of Applied Psychology, 94*(3), 635–653.

Hwang, A., & Arbaugh, J. B. (2006). Virtual and traditional feedback-seeking behaviors: Underlying competitive attitudes and consequent grade performance. *Decision Sciences Journal of Innovative Education, 4*(1), 1–28.

Inayat, I., Amin, R., Inayat, Z., & Salim, S. S. (2013). Effects of collaborative web based vocational education and training (VET) on learning outcomes. *Computers & Education, 68*, 153–166.

Ambient Insight (2015). The 2014–2019 Worldwide Mobile Learning Market. Retrieved April 24, 2017, from http://www.ambientinsight.com/Resources/Documents/AmbientInsight-2014-2019-Worldwide-Mobile-Learning-Market-Executive-Overview.pdf

Islam, N. (2011). The determinants of the post-adoption satisfaction of educators with an e-learning system. *Journal of Information Systems Education, 22*(4), 319–330.

Jashapara, A., & Tai, W. (2011). Knowledge mobilization through e-learning systems: Understanding the mediating roles of self-efficacy and anxiety on perceptions of ease of use. *Information Systems Management, 28*, 71–83.

Johnson, R. D. (2005). An empirical investigation of sources of application-specific computer-self-efficacy and mediators of the efficacy–performance relationship. *International Journal of Human-Computer Studies, 62*(6), 737–758.

Johnson, R. D. (2011). Gender differences in e-learning: Communication, social presence and learning outcomes. *Journal of Organizational and End User Computing, 23*(1), 79–94.

Johnson, R. D., Gueutal, H., & Falbe, C. M. (2009). Technology, trainees, metacognitive activity and e-learning effectiveness. *Journal of Managerial Psychology, 24*(6), 545–566.

Johnson, R. D., Hornik, S., & Salas, E. (2008). An empirical examination of factors contributing to the creation of successful e-learning environments. *International Journal of Human-Computer Studies, 66*(5), 356–369.

Johnson, R. D., & Marakas, G. M. (2000). Research report: the role of behavioral modeling in computer skills acquisition: toward refinement of the model. *Information Systems Research, 11*(4), 402–417.

Johnson, R. D., Veltri, N., & Thatcher, J. B. (2015). Beliefs and attributions toward computing technology: The moderating role of social cues in interfaces. *Journal of Organizational and End User Computing (JOEUC), 27*(3), 27–54.

Judge, T. A., & Bono, J. E. (2001). Relationship of core self-evaluations traits – self-esteem, generalized self-efficacy, locus of control, and emotional stability – with job satisfaction and job performance: A meta-analysis. *Journal of Applied Psychology, 86*(1), 80–92.

Kaganer, E., Giordano, G. A., Brion, S., & Tortoriello, M. (2013). Media tablets for mobile learning. *Communications of the ACM, 56*(11), 68–75.

Kanfer, R., & Ackerman, P. L. (1989). Motivation and cognitive abilities: An integrative/aptitude-treatment interaction approach to skill acquisition. *Journal of Applied Psychology, 74*(4), 657–690.

Kankanhalli, A., Tan, B. C., & Wei, K. K. (2005). Contributing knowledge to electronic knowledge repositories: An empirical investigation. *MIS Quarterly*, 113–143.

Karim, M., & Behrend, T. (2014). Reexamining the nature of learner control: Dimensionality and effects on learning and training reactions. *Journal of Business and Psychology, 29*, 87–99.

Keith, N., & Frese, M. (2005). Self-regulation in error management training: emotion control and metacognition as mediators of performance effects. *Journal of Applied Psychology, 90*(4), 677–691.

Kickul, G., & Kickul, J. (2006). Clossing the gap: Impact of student proactivity and learning goal orientation on e-learning outcomes. *International Journal on E-Learning, 5*(3), 361–372.

Kim, J., Kwon, Y., & Cho, D. (2011). Investigating factors that influence social presence and learning outcomes in distance higher learning. *Computers & Education, 57*, 1512–1520.

Kim, K., & Frick, T. W. (2011). Changes in student motivation during online learning. *Journal of Educational Computing Research, 44*(1), 1–23.

Kim, K., Oh, I. S., Chiaburu, D. S., & Brown, K. G. (2012). Does positive perception of oneself boost learning motivation and performance? *International Journal of Selection and Assessment, 20*(3), 257–271.

Kim, K., Trimi, S., Park, H., & Rhee, S. (2012). The impact of CMS quality on the outcomes of e-learning systems in education: An empirical study. *Decision Sciences Journal of Innovative Education, 10*(4), 575–587.

Kleitman, S., & Costa, D. S. J. (2014). The role of a novel formative assessment tool (Stats-mIQ) and individual differences in real-life academic performance. *Learning and Individual Differences, 29*, 150–161.

Kraiger, K. (2003). Perspectives on training and development. *Handbook of Psychology, 1*, 171–192.

Kraiger, K., Ford, J. K., & Salas, E. (1993). Application of cognitive, skill-based, and affective theories of learning outcomes to new methods of training evaluation. *Journal of Applied Psychology, 78*(2), 311–328.

Kulik, C. L. C., & Kulik, J. A. (1991). Effectiveness of computer-based instruction: An updated analysis. *Computers in Human Behavior, 7*(1), 75–94.

Lan, Y., & Sie, Y. (2010). Using RSS to support mobile learning based on media richness theory. *Computers & Education, 55*, 723–732.

Larsen, C. R., Oestergaard, J., Ottesen, B. S., & Soerensen, J. L. (2012). The efficacy of virtual reality simulation training in laparoscopy: A systematic review of randomized trials. *Acta Obstetricia et Gynecologica Scandinavica, 91*(9), 1015–1028.

Lee, Y., Hsieh, Y., & Chen, Y. (2013). An investigation of employees' use of e-learning systems: Applying the technology acceptance model. *Behaviour & Information Technology, 32*(2), 173–189.

Liaw, S. (2008). Investigating students' perceived satisfaction, behavioral intention, and effectiveness of e-learning: A case study of the Blackboard system. *Computers & Education, 51*, 864–873.

Liaw, S., Huang, H. M., & Chen, G. D. (2007). Surveying instructor and learner attitudes toward e-learning. *Computers & Education, 49*(4), 1066–1080.

Lim, C. K. (2001). Computer self-efficacy, academic self-concept, and other predictors of satisfaction and future participation of adult distance learners. *The American Journal of Distance Education, 15*(2), 41–51.

Lin, J., Huang, H., & Chuang, Y. (2015). The impacts of network centrality and self-regulation on an e-learning environment with the support of social network awareness. *British Journal of Educational Technology, 46*(1), 32–44.

Locke, E. A., & Latham, G. P. (2002). Building a practically useful theory of goal setting and task motivation: A 35-year odyssey. *American Psychologist, 57*(9), 705–717.

Locke, E. A., & Latham, G. P. (2004). What should we do about motivation theory? Six recommendations for the twenty-first century. *Academy of Management Review, 29*(3), 388–403.

Long, L. K., Dubois, C., & Faley, R. (2009). A case study analysis of factors that influence attrition rates in voluntary online training programs. *International Journal of E-learning, 8*(3), 347–359.

Long, L. K., & Smith, R. D. (2004). The role of web-based distance learning in HR development. *Journal of Management Development, 23*(3), 270–284.

Lopez, M. H., Gonzalez-Barrera, A., & Patten, E. (2013). *Closing the digital divide: Latinos and technology adoption*. Washington, DC: Pew Hispanic Center.

Lowry, P. B., Roberts, T. L., Romano, N. C. J., Cheney, P. D., & Hightower, R. T. (2006). The impact of group size and social presence on small group communication: Does computer-mediated communication make a difference. *Small Group Research, 37*, 631–661.

Mackie, D. M., Worth, L. T., & Asuncion, A. G. (1990). Processing of persuasive in-group messages. *Journal of Personality and Social Psychology, 58*, 812–822.

Marakas, G. M., Yi, M. Y., & Johnson, R. D. (1998). The multilevel and multifaceted character of computer self-efficacy: Toward clarification of the construct and an integrative framework for research. *Information Systems Research, 9*(2), 126–163.

Marks, R. B., Sibley, S. D., & Arbaugh, J. B. (2005). A structural equation model of predictors for effective online learning. *Journal of Management Education, 29*(4), 531–563.

Maurer, T. J. (2001). Career-relevant learning and development, worker age, and beliefs about self-efficacy for development. *Journal of Management, 27*(2), 123–140.

Maurer, T. J., Pierce, H., & Shore, L. (2002). Perceived beneficiary of employee development activity: A three-dimensional social exchange model. *Academy of Management Review, 27*(3), 432–444.

Maurer, T. J., Weiss, E. M., & Barbeite, F. G. (2003). A model of involvement in work-related learning and development activity: The effects of individual, situational, motivational, and age variables. *Journal of Applied Psychology, 88*(4), 707.

McDaniel, R., Lindgren, R., & Friskics, J. (2012, October). *Using badges for shaping interactions in online learning environments*. Paper presented at the 2012 IEEE International Professional Communication Conference, Orlando, FL, USA.

McGrath, J. E., Arrow, H., Grunfeld, D. H., Hollingshead, A. B., & O'Connor, K. M. (1993). Groups, tasks, and technology: The effects of experience and change. *Small Group Research, 24*, 406–420.

Mennecke, B., Hassall, L. M., & Triplett, J. (2008). The mean business of Second Life: Teaching entrepreneurship, technology and e-commerce in immersive environments. *MERLOT Journal of Online Learning and Teaching, 4*(3), 339–348.

Merchant, Z., Goetz, E. T., Keeney-Kennicutt, W., Kwok, O., Cifuentes, L., & Davis, T. J. (2012). The learner characteristics, features of desktop 3D virtual reality environments, and college chemistry instruction: A structural equation modeling analysis. *Computers & Education, 59*(2), 551–568.

Mohammadi, H. (2015). Investigating users' perspectives on e-learning: An integration of TAM and IS success model. *Computers in Human Behavior, 45*, 359–374.

Mujber, T. S., Szecsi, T., & Hashimi, M. S. (2004). Virtual reality applications in manufacturing process simulation. *Journal of Materials Processing Technology, 155*, 1834–1838.

Mullich, J. (2004). A second act for e-learning. *Workforce Management, 83*, 51–55.

Noe, R. A. (1986). Trainees' attributes and attitudes: Neglected influences on training effectiveness. *Academy of Management Review, 11*(4), 736–749.

Noe, R. A., & Schmitt, N. (1986). The influence of trainee attitudes on training effectiveness: Test of a model. *Personnel Psychology, 39*(3), 497–523.

Ong, C., Lai, J., & Wang, Y. (2004). Factors affecting engineers' acceptance of asynchronous e-learning systems in high-tech companies. *Information & Management, 41*, 795–804.

Orvis, K. A., Brusso, R. C., Fisher, S. L., & Wasserman, M. E. (2011). E-nabled for e-learning? The moderating role of personality in determining the optimal degree of learner control in an e-learning environment. *Human Performance, 24*, 60–78.

Paraskeva, F., Bouta, H., & Papagianni, A. (2008). Individual characteristics and computer self-efficacy in secondary education teachers to integrate technology in educational practice. *Computers & Education, 50*(3), 1084–1091.

Park, S., Sim, H., & Roh, H. (2010). The analysis of effectiveness on "transfer" through e-learning courses in industry and technology. *British Journal of Educational Technology, 41*(6), 32–34.

Parker, C. P., Baltes, B. B., Young, S. A., Huff, J. W., Altmann, R. A., LaCost, H. A., & Roberts, J. E. (2003). Relationships between psychological climate perceptions and work outcomes: A meta-analytic review. *Journal of Organizational Behavior, 24*(4), 389–416.

Piccoli, G., Ahmad, R., & Ives, B. (2001). Web-based virtual learning environments: A research framework and a preliminary assessment of effectiveness in basic IT skills training. *MIS Quarterly, 25*(4), 401–426.

Piccioli, V. (2014). *E-learning market trends & forecast 2014–2016 report.* Docebo. Retrieved April 24, 2017, from https://www.docebo.com/landing/contactform/elearning-market-trends-and-forecast-2014-2016-docebo-report.pdf

Pintrich, P. R., & DeGroot, E. V. (1990). Motivational and self regulated learning components of classroom academic performance. *Journal of Educational Psychology, 82*, 33–40.

Richardson, J. C., & Swan, K. (2003). Examining social presence in online courses in relation to students' perceived learning and satisfaction. *Journal of Asynchronous Learning Networks, 7*(1), 68–88.

Robson, K., Plangger, K., Kietzman, J. H., McCarthy, I., & Pitt, L. (2015). Is it all a game? Understanding the principles of gamification. *Business Horizons, 588*(4), 411–420.

Rotter, J. B. (1966). Generalized expectancies for internal versus external control of reinforcment. *Psychological Monographs: General and Applied, 80*(1), 1–28.

Rouiller, J. Z., & Goldstein, I. L. (1993). The relationship between organizational transfer climate and positive transfer of training *Human Resource Development Quarterly, 4*(4), 377–390.

Saadé, R., & Kira, D. (2009). Computer anxiety in e-learning: The effect of computer self-efficacy. *Journal of Information Technology Education, 8*, 178–191.

Salas, E., DeRouin, R., & Littrell, L. (2005). Research-based guidelines for designing distance learning: What we know so far. *The Brave New World of eHr*, 104–137.

Salyers, V., Carter, L., Myers, S., & Barrett, P. (2014). The search for meaningful e-learning at Canadian universities: A multi-institutional research study. *International Review of Research in Open and Distance Learning, 15*(6), 313–337.

Santhanam, R., Sasidharan, S., & Webster, J. (2008). Using self-regulatory learning to enhance e-learning-based information technology training. *Information Systems Research, 19*(1), 26–47.

Sawang, S., Newton, C., & Jamieson, K. (2013). Increasing learners' satisfaction/intention to adopt more e-learning. *Education & Training, 55*(1), 83–105.

Schmidt, A. M., & Ford, J. K. (2003). Learning within a learner control training environment: The interactive effects of goal orientation and metacognitive instruction on learning outcomes. *Personnel Psychology, 56*, 405–429.

Short, J., Williams, E., & Christie, B. (1976). *The social psychology of telecommunications.* New York, NY: John Wiley & Sons, Inc.

Sitzmann, T., Bell, B. S., Kraiger, K., & Kanar, A. M. (2009). A multilevel analysis of the effect of prompting self-regulation in technology-delivered instruction. *Personnel Psychology, 62*, 697–734.

Sitzmann, T., Brown, K. G., Casper, W. J., Ely, K., & Zimmerman, R. D. (2008). A review and meta-analysis of the nomological network of trainee reactions. *Journal of Applied Psychology, 93*(2), 280–295.

Sitzmann, T., & Ely, K. (2010). Sometimes you need a reminder: The effects of self-regulation on regulatory processes, learning and attrition. *Journal of Applied Psychology, 95*(1), 132–144.

Sitzmann, T., Kraiger, K., Stewart, D., & Wisher, R. (2006). The comparative effectiveness of web-based and classroom instruction: A meta-analysis. *Personnel Psychology, 59*(3), 623–664.

Snyder, E., & Hartig, J. R. (2013). Gamification of board review: A residency curricular innovation. *Medical Education, 47*(5), 524–525.

Stark, R., & Krause, U. (2009). Effects of reflection prompts on learning outcomes and learning behaviour in statistics education. *Learning Environments Research, 12*, 209–223.

Steer, D. (2012). Improve formal learning with social media. *T+D, 66*(12), 31–33.

Steinberg, E. R. (1989). Cognition and learner control: A literature review, 1977–1988. *Journal of Computer-Based Instruction, 16*(4), 117–121.

Stewart, M., Stott, T., & Nuttall, A. (2011). Student engagement patterns over the duration of level 1 and level 3 geography modules: Influences on student attendance, performance and use of online resources. *Journal of Geography in Higher Education, 35*(1), 47–65.

Stone, D. L., Deadrick, D. L., Lukaszewski, K. M., & Johnson, R. D. (2015). The influence of technology on the future of human resource management. *Human Resource Management Review, 25*(2), 216–231.

Stone, D. L., Krueger, D., & Takach, S. (2017). Social issues associated with the internet at work. In G. Hertel, D. L. Stone, R. D. Johnson, & J. Passmore (Eds.), *Wiley Blackwell handbook of the psychology of the internet at work* (pp. 423–448). Chichester: John Wiley & Sons, Ltd.

Stonebraker, P. W., & Hazeltine, J. E. (2004). Virtual learning effectiveness: An examination of the process. *The Learning Organization, 11*(3), 209–225.

Strang, K. D. (2011). How can discussion forum questions be effective in online MBA courses? *Campus-Wide Information Systems, 28*(2), 80–92.

Sun, P., Tsai, R. J., Finger, G., Chen, Y., & Yeh, D. (2008). What drives a successful e-learning? An empirical investigation of the critical factors influencing learner satisfaction. *Computers & Education, 50*, 1183–1202.

Sweller, J. (1988). Cognitive load during problem solving: Effects on learning. *Cognitive Science, 12*(257–285).

Tennyson, R. D. (1980). Instructional control strategies and content structure as design variables in concept acquisition using computer-based instruction. *Journal of Educational Psychology, 72*(4), 525–532.

Thatcher, J. B., & Perrewe, P. L. (2002). An empirical examination of individual traits as antecedents to computer anxiety and computer self-efficacy. *MIS Quarterly, 26*(4), 381–396.

Thielsch, M. T., & Niesenhaus, J. (2017). User experience, gamification, and performance. In G. Hertel, D. L. Stone, R. D. Johnson, & J. Passmore (Eds.), *Wiley Blackwell handbook of the psychology of the internet at work* (pp. 79–102). Chichester: John Wiley & Sons, Ltd.

Towler, A., & Mitchell, T. (2015). Facilitation in e-learning. In K. Kraiger, J. Passmore, N. R. d. Santos, & S. Malvezzi (Eds.), *The Wiley Blackwell handbook of the psychology of training, development, and performance improvement* (pp. XX–XX). Chichester: John Wiley & Sons, Ltd.

Tracey, J. B., Tannenbaum, S. I., & Kavanagh, M. J. (1995). Applying trained skills on the job: The importance of the work environment. *Journal of Applied Psychology, 80*(2), 239–252.

Tu, C. H., & McIsaac, M. (2002). The relationship of social presence and interaction in online classes. *American Journal of Distance Education, 16*(3), 131–150.

Van Merrienboer, J. J. G., Kester, L., & Paas, F. (2006). Teaching complex rather than simple tasks: Balancing intrinsic and germane load to enhance transfer of learning. *Applied Cognitive Psychology, 20*, 343–352.

van Raaij, E. M., & Schepers, J. J. L. (2008). The acceptance and use of a virtual learning environment in China. *Computers & Education, 50*, 838–852.

Vroom, V. H. (1964). *Work and motivation*. Chichester: John Wiley & Sons, Ltd.

Wan, Z., Compeau, D., & Haggerty, N. (2012). The effects of self-regulated learning processes on e-learning outcomes in organizational settings. *Journal of Management Information Systems, 29*(1), 307–339.

Wan, Z., Fang, Y., & Neufeld, D. J. (2007). The role of information technology in technology-mediated learning: A review of the past for the future. *Journal of Information Systems Education, 18*(2), 183–192.

Warr, P. B., Allan, C., & Birdi, K. (1999). Predicting three levels of learning outcomes. *Journal of Occupational and Organizational Psychology, 72*, 351–375.

Wasko, M. M., & Faraj, S. (2000). "It is what one does": Why people participate and help others in electronic communities of practice. *Journal of Strategic Information Systems, 9*(2), 155–173.

Wasko, M. M., & Faraj, S. (2005). Why should I share? Examining social capital and knowledge contribution in electronic networks of practice. *MIS Quarterly*, 35–57.

Webster, J., & Hackley, P. (1997). Teaching effectiveness in technology-mediated distance learning. *Academy of Management Journal, 40*(6), 1282–1309.

Welsh, E. T., Wanberg, C. R., Brown, K. G., & Simmering, M. J. (2003). E-learning: Emerging uses, emperical results and future directions. *International Journal of Training and Development, 7*(4), 245–258.

Wood, R. E. (1986). Task complexity: Definition of the construct. *Organizational Behavior and Human Decision Processes, 37*(1), 60–82.

Yang, Y., & Durrington, V. (2010). Investigation of students' perceptions of online course. *International Journal on E-Learning, 9*(3), 341–361.

Yanson, R., & Johnson, R. D. (2016). An empirical examination of e-learning design: The role of trainee socialization and complexity in short term training. *Computers & Education, 101*, 43–54.

Yukselturk, E. (2009). Do entry characteristics of online learners affect their satisfaction? *International Journal on E-Learning, 8*(2), 263–281.

Yukselturk, E., Ozekes, S., & Turel, Y. K. (2014). Predicting dropout student: An application of data mining methods in an online education program. *European Journal of Open, Distance and E-Learning, 17*(1), 119–133.

Zafar, H. (2017). Mobile computing and hand-held devices. In G. Hertel, R. D. Johnson, D. Stone, & J. Passmore (Eds.), *Wiley Blackwell handbook of the psychology of the internet at work* (pp. 195–210). Chichester: John Wiley & Sons, Ltd.

Zhang, D., Zhao, J. L., Zhou, L., & Nunamaker Jr, J. F. (2004). Can e-learning replace classroom learning? *Communications of the ACM, 47*(5), 75–79.

Zickuhr, K. (2013). *Tablet ownership in 2013.* Washington DC: Pew Research Center.

Part III

Societal and Cross-Sectorial Perspectives

18

Robots in the Digitalized Workplace

Jochen J. Steil and Günter W. Maier

Introduction

Flexible, intelligent, and interactive robots are regarded as a key technology of digitalization. Digitalization denotes the increasing pervasion of our daily lives by computerized devices and digital services that are ultimately offered through the Internet. New robotics applications strongly benefit from this process. It is highly likely that they will influence the ways we work and live at least as much as the invention of the automobile. Robotic assistants are being installed worldwide through programs of innovation in production. In Europe, this is happening within the conception of manufacturing 2.0, also known as "Industry 4.0," the fourth industrial revolution in Germany and beyond. In the United States, a similar agenda is being pursued in the context of cyber-physical systems and the Internet-of-Things (IoT) paradigms, and in China, in the context of a manufacturing innovation program reaching into the year 2025 under the conception "MiC2025" (Made in China 2025). The far-reaching consequences raise many questions in ethical, legal, and societal domains, as well as for the organization of work, which might change to a large extent. As opposed to traditional industrial robots, a new generation of robots has appeared that is lightweight, can safely operate in the same workspace as a human, and can even interact physically during the collaborative execution of tasks. Many of these robots are impedance controlled, that is, they can measure external forces and react to them with spring-like soft characteristics. This feature enables sophisticated interaction with the environment or with human co-workers.

Furthermore, increasingly, humanoid robots promise to truly implement the dream of a *living machine* – which has been a topic of science fiction for at least a century. But what types of robots will reach the goal of intuitive interaction with their human users? Although lightweight and versatile robot arms may provide assistance in many manipulation tasks, more humanoid robots can tap into the rich repertoire of human behavioral

The Wiley Blackwell Handbook of the Psychology of the Internet at Work,
First Edition. Edited by Guido Hertel, Dianna L. Stone, Richard D. Johnson, and Jonathan Passmore.

stereotypes, and thereby facilitate interaction and communication. Both types of robot, however, can make use of the ample resources available via the Internet by employing various digital tools. These reach from deep learning for improving the perception-for-action to Internet-based speech processing and dialog. Additionally, they may use data and knowledge bases and machine learning to support their world models, and for offline mining of interaction data. These robots are powerful tools that can be used far beyond mere automation. They pave the way for new processes and applications, make production systems flexible, and thereby open unprecedented opportunities and challenges for the novel organization of work.

However, so far, there is hardly any empirical evidence on the psychological consequences and implications of the deployment of such robot technology within the actual workplace. To date, new robotics has been more of a promise than an everyday phenomenon, and thus this chapter presents a mixture of related research from human–robot interaction to a description of likely implications. The chapter focuses on the sector of industrial production because we anticipate that it will be the first and, by far, the most relevant application field in the near future. Indeed, many companies have already started to create reference and test applications for assistive tasks in this field. We first introduce some aspects of the novel and advanced robot technology that are highly relevant to user interaction and the digitalization of production. We then proceed to analyze the implications from three different viewpoints. First, we take the user perspective, including deliberations on the relevance of personalization and connections to big-data processing. Second, we discuss the issue from the developer's and system designer's perspective. Along the way, implications for the future of work in the digitalized age as well as psychological implications are considered. Finally, we briefly discuss which findings may be transferred to more general domains beyond production, and what users may possibly expect from robotic agents in other sectors. A short discussion wraps up the main conclusions.

The "New Robotics": A Versatile and Powerful Technology

Since the introduction of the first freely programmable PUMA Unimation robot in 1978, industrial robots have been used mainly for the automation of flow production. They are employed to achieve short cycle times, submillimeter repeatability, and high availability. Worldwide, the majority of such robot installations are in the automobile industry, where the degree of automation is more than 90% in car body manufacturing. Industrial robots are potentially dangerous machines and are typically operated strictly separately from humans for safety reasons. In 2015, the International Federation of Industrial Robots (IFR, 2015) predicted a steady increase in installed units, reaching about 400 000 newly installed units per year worldwide by the year 2020, some of which will replace older installations. The total number of robots operating in production in 2016 is around 2 million units, which, after all, is not so impressive and strongly contrasts the common public perception that industrial production is highly dominated by robots. By and large, industrial mass production is still characterized by rigid automization through customized and highly specialized nonrobotic manufacturing technology, with the involvement of a moderate, but continuously increasing share of classical industrial robots. We will refer to these kinds of robots as *old robots* to distinguish them from the class of more versatile *new robots* described below. Furthermore, the degree of automation in sectors not devoted to highly standardized mass production was still relatively small in the year 2016. For instance, even in the pioneering automobile industry, the degree of automation in assembly is estimated to be only about 20%. It is exactly this situation that assumes a large application potential and raises high expectations for the introduction of new types of robots. To understand the envisaged change, it is worth considering the factors that contribute to the still relatively low degree of robotic automation today.

First, for many years, it has hardly been possible to allow a robot and a human to work in the same workspace for safety reasons. The respective International Organization for Standardization (ISO) standard (ISO 10218–2, version 2011) on collaboration in robotics was very restrictive in speed and force of a robot in interaction to a degree that true interaction was hardly possible. However, based on seminal work in safety analysis, new measurement devices, and novel control schemes, it has been shown in the years since 2005 that these restrictions could be overcome through better technology and, in particular, better control algorithms (Haddadin, Albu-Schäffer, & Hirzinger, 2009). Consequently, a substantial number of lightweight and smaller robots have been developed, which can measure forces in various ways in order to implement safe stopping on collision, and are thereby intrinsically safe. This has been acknowledged by the ISO, which published in the year 2016, an updated ISO/TS 15066 that defines anew under which conditions contact and collaboration is now allowed. It is worth noting, however, that different robots achieve this by means of different technological approaches and principles. These include sensing internal motor currents of the robot for detecting force and collision based on comparison with internal prediction (e.g., Universal Robots UR series, ABB YuMi), the use of a special sensor in the robot base (e.g., FANUC CR-35iA), the joint-wise direct measurement of force along the robot manipulator (e.g., Kuka iiwa, KBee Franka), the use of passive elastic elements to safely give way (e.g., Bionic Systems Biorob), or through the use of a kind of sensorized skin that detects collisions (e.g., Bosch APAS). This nonexhaustive list of available robot technology demonstrates that there is a large number of technological solutions, all with the same goal of safe interaction with humans. For a comparative overview on many of the products on the market in the year 2016 provided by a commercial reseller, see Robotiq (2016).

A second major obstacle for the application of robots beyond specialized mass production has been their lack of flexible integration into automation and production systems. Traditionally, each change in the production process has required a substantial programming and reconfiguration effort, together with an intricate risk and safety analysis, which is costly and can only be done by specialists. This sharply contrasts with a high and ever-increasing need for flexible automation to react to quickly changing customers' demands and agile development of new products (Gerber, Theorin, & Johnsson, 2014). The latter has created a large demand for more flexible hardware and software solutions that will become available only through an increasing degree of digitalization, for instance, in the form of a vision of a smart factory (Zuehlke, 2010). This demand is also known as *mass customization*, and creates market pressure on manufacturers to introduce innovations that facilitate and flexibilize the respective system's engineering. Mass customization, in turn, will open new markets for a new generation of flexible robots, although, to date, the particular impact of robotics in this context is hardly considered in the literature on production and automation systems.

Progress toward more flexible integration of robotics and standard automation technology is fueled by the general trend for digitalization in production. Novel smart factories are comprised of digitally interconnected cyber-physical devices that are embedded in more flexible modeling and programming infrastructures. Additionally, a much higher degree of modeling and vertical integration from the shop floor up to enterprise IT (Gerber et al., 2014; Slama, Puhlmann, Morrish, & Bhatnagar, 2015) is realized and often relies on Internet services, such as Cloud-based computing, service portals, and remote control. The respective software systems then provide opportunities to integrate flexible robots more seamlessly (Wrede, Beyer, Dreyer, Wojtynek, & Steil, 2016). In an analogy, this is reminiscent of the development from classical rigid and simple cell phones, with a few basic functions, to modern programmable smartphones, with their agile app-programming systems that allow for fast and distributed development of new applications. Production systems in the predigitalized age were rigid and nonflexible, with expert programming realized on the PLC (programmable logic controller), possibly including classical industrial robots. Digitalized production now allows one to connect to a multitude of sensors and possibly robots through the Internet-of-Things (IoT) network technology and flexible programming methods (Slama et al., 2015). Together, this provides the technological ecosystem for realization of more widespread collaborative robotics.

A third important factor for the still apparent lack of widespread practical application of the new generation of robots is a general uncertainty about their actual interaction capabilities. On the one hand, these depend on the robot technology itself. On the other hand, they are greatly enhanced by the ability of robots, like all computer-controlled devices, to access all kinds of digital resource. We discuss these two issues in turn. The physical hardware of the robots addressed in this chapter accommodates an increasing number of features and capabilities. Their increased functionality makes it difficult to judge their suitability for a particular application even for experts, let alone for the average user. The typical robot considered here is relatively small and lightweight (10–20 kg), with a range of 1.5–2 m. It allows for physical integration in any production or other technical system, and often can be mounted on a mobile platform, which itself can be considered as another robot. Due to its lightweight structure, it has a small payload of a few hundred grams up to 2 kg. Probably, the only common feature of all these new robots is that they measure external forces, either directly through sensors or indirectly through observing the respective internal disturbances, in order to implement safety features. They thereby allow for physical contact. The usage and implementation of this feature, however, already can be quite different. Some robots use the force measurements just for collision detection and safe stop in order to allow the robot to operate in the same workspace as the user. Others base their whole control on force measurements, which enable much more sophisticated interaction, because in contact situations, the robot can behave like a physical spring and apply forces in flexible ways (Villani & De Schutter, 2008). What this can mean for the users will be discussed in the next section. Conversely, seen as a software system, robots can be equipped with all the perceptual and knowledge-based artificial intelligence (AI) that is available anywhere in the digital world. Therefore, new robots can utilize the increasingly sophisticated tools from computer vision, if there is a camera installed somewhere in the application or on the robot itself (numerous respective works deal with robot vision). But robots also may speak and listen to the user, they may implement memory, use Internet-based resources, or use knowledge-based decision-making systems. Endowing robots with respective cognitive capabilities beyond vision alone, and integrating Internet resources, is therefore a growing and active field of research (Beetz et al., 2015; Tenorth & Beetz, 2013). Robots differ, however, from ordinary computers or other smart devices in the fact that they are physically embodied and actively controlled. They thus bridge the Internet-based digital world and the physical real world. Consequently, the question of how their physical embodiment interferes with or reinforces their Internet-based and digitally realized capabilities is an important matter of active research.

These obstacles have slowed the introduction of new robots into production and other applications, but will eventually be overcome through progressing digitalization, and especially when more experience with the new technology is available. This is because the major driver – customer demand – is very strong. Customers expect Internet-based configuration and customization of products, which, in turn, increases the need for flexibilization of production on the manufacturer's side. Furthermore, the technological advantages are apparent. Force measurements, beyond their usage for safety, also enable various novel teaching and programming by demonstration methods (for an overview, see Billard, Calinon, Dillmann, & Schaal, 2008). That is, physical guidance of the robot through a user, so-called kinesthetic teaching, is enabled and can even be extended into interfaces tailored to nonexpert users (Wrede et al., 2013). This can provide an effective means of interactively configuring applications and contributing to the reduction of programming and engineering costs. It shall not be omitted that these new interaction capabilities come at the cost that the control, trajectory planning, and task representation for such robots are much more difficult and not easy to automatize. Although well-designed interfaces may hide this complexity from the user, it inevitably leads to a high degree of sophistication in the robot control and the respective software engineering. Therefore, the trend toward collaborative robots that are easy to use increases the complexity of the enabling technologies.

In summary, robot technology that measures and controls forces is available. Applications for workspace sharing are being realized, mostly as pilot studies in the larger industries, and new programming by demonstration schemes facilitates the engineering. Initial software tools that automatically evaluate robot trajectory and speed for the design of safe applications are available. New applications for robots arise in flexible automation to react to quickly changing customer demands. They pave the way for the development of new products under the notion of mass customization. However, assistance for assembly under the concept of hybrid manufacturing and the robotization of logistics are new and important fields for the deployment of novel robotics.

An important distinction between classical industrial and novel robotics is that the latter is not per se automation prone. The new robots rather serve as versatile tools that can make workplaces and automation systems more adaptable, flexible, and configurable. Whether these capabilities will mainly be used to fully automize processes by replacing human work, or to assist humans through cooperation and humanization of work, is an open question and a matter of labor organization. Both kinds of utilization are feasible and reasonable. There is still much unhealthy or dangerous work that should be automated for the sake of workers. Moreover, there is a large demand for assistance to support workers rather than to replace them. This is reinforced by the demographic challenges across developed countries, but are felt earliest in (early-)industrialized countries, in Japan, and increasingly in China (Hertel & Zacher, in press). The new robot technology may support both strategies – full automation or assistance – and thereby creates a large design space for restructuring work. This may have major psychological implications. If it is true that robots create opportunities for a novel organization of work, then many fields of working life and work processes will be affected, starting with teamwork, leadership, work design, competencies, and human resource development, as will be proactive behavior and organizational citizenship behavior. For example, with the new robots it will be possible to create hybrid working teams, composed of humans and robots working together interactively. All that is known about psychological team processes has to be extended to this new situation, where some members of the team are not humans. How does team climate – comprising, for example, task orientation, participative safety, or support for innovation (Anderson & West, 1998) – develop in such a team? How do leaders distribute tasks between human and robot followers? How do leader–subordinate dyads develop if there are human and robot subordinates? What should be the leading principle for work allocation in such hybrid teams (e.g., maximize short-term performance, maximize skill variety of humans)? How much and how often should human co-workers be supported by their robot co-workers at the risk of demotivating the humans or forgetting skills? How can employees satisfy their basic needs for autonomy, competency, and relatedness (Deci, Olafsen & Ryan, 2017; Deci & Ryan, 2000) when they are faced in their work environment with highly competent, autonomous robots? How should team development be designed to introduce robots as co-workers? How can employees proactively act in a technical environment, which persistently monitors, assists, and corrects their behavior? Many of these questions are at the intersection between people, technology, and organization. Research in this field is still in its nascent stage (Cascio & Montealegre, 2016).

Finally, there are a number of unsolved technological challenges. True co-manipulation, where the robot actively regulates its impedance, is yet to be realized. The implicit or explicit negotiation between the robot and a user on how to split and share joint tasks, possibly in dynamical fashion, is left to future generations of robots that will have considerably more cognitive capabilities. However, initial studies have been conducted about decision-making in collaborating human–robot teams. Gombolay, Gutierrez, Clarke, Sturla, and Shah (2015) conducted a laboratory study in which the authority over how best to allocate work was shared between robots and human workers or could cede completely to the robot. They found that under some conditions, the robot could outperform a human worker with respect to maximizing human–robot team fluency and worker satisfaction. Furthermore, people preferred to cede their control authority to the robot while still valuing human co-workers higher than the machines. Interestingly, the authors recommended "that roboticists provide robots with as much autonomy as possible to support human–robot team coordination. Robot teammates with the ability to autonomously allocate and schedule tasks can improve both task completion time ..." (p. 310). This surely calls for much stronger cognitive capabilities on the side of the robot.

Some current research questions include: how to teach and represent impedance-based tasks that include forces; how to understand the semantics of forces in co-manipulation and to adapt to them; how to adapt in shared autonomy; and how to use variable impedance to adapt to the user for a much better personalization. Note that these research questions all refer to user interaction and thereby directly demand interdisciplinary collaboration between robotics and psychology for creation and proper evaluation of user interfaces and user studies. The realization of further capabilities, however, will not change the character of the new robot technology. It will rather increase the flexibility of the technical systems and thereby also increase the options for novel labor organization and possibly completely new and unforeseen application fields.

"New Robotics" From the User's Perspective

From the perspective of the user, interaction with versatile robots has a number of intricate and novel features. Consider the running example of a human-size, lightweight torso robot with two arms that have roughly the same length and movement range as in humans. Imagine the robot interacting for assembly of some workpiece. Such a robot has been commercially available since 2016, and such scenarios have been considered under the concept of hybrid assembly (Heinze et al., 2016; Wallhoff et al., 2010). In this example, the robot and the user share the same workspace, take turns in their actions, and need to be aware of each other to avoid collisions. All this is typically very much pre-programmed and the robot utilizes force measurements for collision detection and safe retreat in case the user performs some unexpected action. In the future, with more cognitive capabilities available in the robot, they might also have a shared plan that can even be adaptive, that is, the robot and the human might dynamically change their division of labor, adapting to the execution of subtasks, etc. Although at the time of writing, such sophisticated features as dynamically shared plans were not been realized in practice, there is active research in this field. We come back to this toward the end of this section. A number of questions arise, of which three are now addressed in the order of increasing complexity.

Human-like appearance and cognition?

First, there is an ongoing discussion about the kinds of robot that are needed. Shall they be human-like in their embodiment and kinematics? A typical argument is that robots that operate in an environment made for humans must have human-like form to be useful, which is possibly most apparent for the robotnaut robot in space (Diftler et al., 2011). A similar rationale holds in hybrid assembly, which assumes that robots should seamlessly integrate with their human co-workers and flexibly replace them for certain tasks. Then human-like size and form are important. The robot's appearance also influences and facilitates both the implicit and explicit communication between the robot and the human, which has been a strong motivation behind the development of humanoid robots for many years (e.g., Ishiguro et al., 2001). Interestingly, the robotics industry demonstrates the different poles of the discussion implicitly through their products. A number of companies offer human-like torsos, others stick to the classic design of robot arms. One major argument in favor of human-like bodies is that they enable easier feedback to the user and make the robot's movement more interpretable. A number of studies from human–robot interaction have shown that humans are sensitive to such nonverbal implicit channels of communication and themselves systematically modify their behavior accordingly (e.g., Vollmer et al., 2014). In simple terms: if the robot does not have a head with two cameras as eyes, it cannot nod or shake it to express feedback to a human action in the most intuitive way that humans understand, and it cannot "look" into the direction of its next move to make its behavior more predictable for others. In the hybrid assembly scenario, the robot may signal that it expects a handover of a workpiece merely through presenting it to the human with an arm forward. The humanoid form thereby enables intuitive interaction that taps into the human's natural understanding of movement and action.

The communication capabilities that come with the human-like appearance of a robot also have the potential to cause confusion and misunderstanding. They greatly increase the complexity of interaction and may introduce ambiguity, for instance, pointing with two arms to two different directions or positions. The very advantage of addressing the human tendency toward anthropomorphism of the robot through the human-like embodiment for communication is also its disadvantage. It evokes a number of other cognitive processes, which may be undesired. There is evidence that the degree of human-likeness of a robot influences the communication and behavior of an interacting human and biases the degree of attribution of mental states to the robot (Krach et al., 2008). Humans apply not only communicative stereotypes toward robots, but also other stereotypes, such as gender stereotypes (Eyssel & Hegel, 2012), occupancy stereotypes (Tay, Jung, & Park, 2014), or stereotypes toward children (Vollmer et al., 2009). This all may be rather undesired in a work environment. There seems to be a thin line between facilitating interaction and raising exaggerated expectations that actually lead to disappointment and successive dismissal of robots. What this means for the deployment of robots at work has not been fully explored. It seems that the appearance of robots influences what people expect from them, how they perceive them, and how they evaluate them (e.g., Goetz,

Kiesler, & Powers, 2003). Human-like appearance, however, is not always desirable. For example, Hinds, Roberts, and Jones (2004) demonstrated experimentally that people felt less responsible for their work if the collaborating robot was human-like compared with collaborations where robots were machine-like. This responsibility diffusion may arise from the expectation that human-like robots have human-like capabilities leading individuals to relinquish their responsibility toward their robot partners.

Another important consequence of robot-related factors may be trust in the robot collaborator. A meta-analysis showed that performance-related characteristics of robots (e.g., robot behavior, dependability, predictability) have an impact on trust in robotic systems (Hancock et al., 2011). Trust in the capabilities and reliability of robots is of utmost importance for increasing the use of robots in the workplace. Robots could be seen as a resource that helps to reduce stressors or facilitate handling of stressors. However, lacking trust in a technical member of the team enhances stress (e.g., proper working of the robot is unpredictable) and, in the long run, the human teammates may refuse using the robot, carrying out the robot's task on their own.

All in all, the psychological implications of human-like appearance are not as unambiguous as earlier speculations gave reason to expect. On the one hand, human-like appearance furthers identification as well as trust, yet on the other hand, it may undermine one's willingness to take over responsibilities.

Usability, User Experience, and Acceptance

The second set of questions revolves around usability and user experience. There is neither much research nor practical experience in this field in robotics applications, simply because in traditional industrial robotics, the goal was replacement of humans in the workplace rather than interaction with them. Although the focus of the new generation of robots is interaction, so far hardly any studies have considered interaction with naïve users from a feasibility or usability perspective. By far, the most considered application is in teaching the robot some target movement or target positions, namely kinesthetic teaching (Calinon & Billard, 2007; Kormushev, Calinon, & Caldwell, 2011). Consider again the example of the hybrid assembly: imagine you have to take the robot arm and guide it around to teach it how to pick up workpieces, how to hand over objects, how to stretch out to the user, and how to enable coordination with the other arm. This is a demanding task, although necessary to fit the robot to a new application, given that it is undesired to call in an expensive robotics expert for reprogramming. It turns out that naïve users do not have a good understanding how a robot moves. A key element is that moving a single joint at a time (e.g., through a keyboard or panel) is difficult to interpret because it may cause other joints to move as well to fulfill constraints, and it generally may move large parts of the robot (Kirstein, Fischer, & Solvason, 2014). Conversely, kinesthetically teaching the robot through physical guidance is not always easy. To enable such guidance, the robot typically is in a gravitation compensation mode and appears weightless and freely movable to the user. The user, however, must control all the joints along the full arm and has to care for the task at the same time. This is demanding, even for experts, and asks too much from a naïve user, who needs more sophisticated interfaces to perform this task. For instance, in one user study (Wrede et al., 2013), a rather sophisticated interaction design of several stages, including machine learning, was shown to work with factory floor workers.

As kinesthetic teaching is a new kind of shopfloor programming, many questions remain open from a psychological point of view: Which competencies will be helpful for the *trainers*? Who will become a successful programmer? To what extent does kinesthetic programming influence the acceptance of robotic colleagues or a fear of job loss?

Furthermore, new kinds of kinesthetic programming of robots allows for new approaches, explicating the knowledge structure of experts in skilled crafts and trades. Often the knowledge structure of experts consists of such highly automatized behavior that they are not aware of, or hardly can verbalize their knowledge (Hacker, 2003). Thus, kinesthetic teaching may offer a new way of making experts' knowledge explicit, a motivation that also has been stressed and explored from a robotics point of view (Steil et al., 2014).

Given that there is need to support users, this could possibly be realized through interfaces in other modalities. A common tool to support robot applications in the laboratory and during the development process are three-dimensional (3D) visualizations. It could be expected that naïve users would profit from a simultaneous display of the robot in the virtual world. Such an interface, however, in itself poses problems in matching the visualization to the real world. It also demands a close, sensor-based interconnection of the real application and the simulation. To realize this for the hybrid assembly scenario unfolded above would require monitoring the workpieces, to somehow track the human movements, and to synchronize the visual interface with the actual robot (e.g., Bannat et al., 2009). In a larger context, multimodal interaction with robots is a large research field, mostly for cognitive and humanoid research robots, which shall not be further reviewed here, but contributes to the development of interfaces for a more applied work domain. All additional multimodal interaction, however, requires further sensing and increases the system's complexity. It bears the risk that users will become confused if some components fail, for example, if an object is not properly recognized due to occlusion. The monitoring of the robot and additional visualization obviously also increase the cognitive load for the user, although currently no studies have quantified such effects. In summary, the usability of interactive robotics applications has not yet been evaluated sufficiently, but there is some evidence that a careful interaction design is of highest importance.

Similar to usability, the user experience has hardly been considered for such applications. As discussed earlier, humanoid form and communicative resources, which can be greatly enhanced by networked, Internet-connected digital interfaces (e.g., for speech or object recognition) may lift the interaction with a robot into a cognitive domain. Much can be gained in this respect from the rapid development of digital interfaces, for instance, on smartphones. It can be anticipated that users even expect intelligence (e.g., in visual recognition of faces or speech output) from any complex technical device in the near future, including robots. Everything known for these nonembodied interfaces carries over to robotics and has to be considered for successful interaction as well. But the embodiment of the robot adds a further, emotionally important channel, which is movement. The latter activates incredibly sophisticated perception and recognition mechanisms in the human, for instance, the capability to recognize emotions, or expertise in recognition of movement and postural features. In one study (Roether, Omlor, Christensen, & Giese, 2009), recordings of human emotional gaits were modeled through machine learning, and crucial movement features were identified and displayed through animated avatars. Subjects could then recognize the emotions from this avatar motion with high accuracy. Other studies based on point displays have also shown that minimal movement information in the form of moving dots on a screen is sufficient for emotion recognition (Atkinson, Dittrich, Gemmell, & Young, 2004). Certainly, these perceptual mechanisms also apply to robot movement, which will be interpreted by the human as meaningful and as an expression of an emotional state. To achieve a good level of user experience, it will thus be necessary to fulfill relatively high demands in interaction and strike the balance between naturalness, functionality, and easy-to-use interfaces. Altogether, both the physical embodiment and the increasing cognitive capabilities raise high expectations on the user's side. The user experience depends on the degree to which these expectations are matched, and the potential for disappointment is large. Fortunately, some evidence has suggested that users can also experience much forgiveness for mistakes on the robot side. Salem, Eyssel, Rohlfing, Kopp, and Joublin (2013) conducted a study where a humanoid robot assisted a user through a combination of verbal and gestural instruction in putting kitchen equipment into a cupboard. The gestures were manipulated to be consistent or sometimes contradictory with respect to verbal instruction. Surprisingly, although the robot's contradictory instructions "negatively affected the participants' task-related performance, they still rated the robot as being more likable, [and] reported greater shared reality with the robot, ..." (p. 321). It is not known whether this carries over to more applied scenarios at a workplace and with simpler robots. Nonetheless, the combination of human-like appearance and cognitive capabilities, together with the ability to implement some human-like traits, may determine the user experience with the co-working robot to a large extent. To err is human – but to do so may also assist in the interaction with robots.

User adaptation, competencies, and long-term effects

Up to now, we know very little about the long-term effects of technical assistance by robots. On the one hand, this assistance may provide support in coping with harmful physical or psychological stressors, such as heavy weight, one-sided physical strain, workload in general, or processing a lot of information simultaneously. On the other hand, the use of intelligent, adaptive robots may lead to such simplified tasks lacking task significance, identity, complexity, skill variety, and the affordance for problem solving – all characteristics which encompass a typical *good* job (e.g., Morgeson & Humphrey, 2008). Furthermore, following the disuse or "use-it-or-loose-it" (e.g., Bosma et al., 2003; Kohn & Schooler, 1983; Roberts, Fuhrer, Marmot, & Richards, 2011) and the deskilling (Geiselman, Johnson, & Buck, 2013) hypotheses, continuous support by intelligent technical systems might lead to a decline in cognitive functioning, procedural knowledge, or manual capabilities. Especially in the field of aviation, automation-related incidents and accidents are often discussed. For example, the National Transportation Safety Board stated that one cause of the crash of Asiana Flight 214 in 2013 was the airline policy recommendation for the use of full automation flights, potentially leading to skill loss for manual flight operations. A meta-analysis documented the trade-off between benefits and drawbacks of technical support of employees or automation (Onnasch, Wickens, Li, & Manzey, 2013): on the one hand, automation leads to improved system performance on routine tasks. On the other hand, manual performance is lower in high automation configurations where automation fails compared with manually working continuously. Therefore, we need to know more about the right short- and long-term balance between support and challenge for workers cooperating with intelligent systems in general and robots in particular.

Will users in the long run adapt and learn how the robots behave, and then accept them? Some pioneering studies on robot–child interaction have shown that companion robots may raise the user's curiosity at first, which then quickly decreases once the standard features are explored (Kanda, Hirano, Eaton, & Ishiguro, 2004). In a similar vein, in a work environment, it seems that employees undergo a process of becoming familiar with robots at the workplace (Wurhofer, Meneweger, Fuchsberger, & Tscheligi, 2015), starting with ambivalent expectations and feelings (e.g., uncertainty, novelty) to adaption and habituation of worker's behavior to the robot. However, habituation to robots' motions is one of the greatest judgment errors in human–robot interaction, resulting in injuries from industrial robots (Murashov, Hearl, & Howard, 2016).

Most likely, the picture will become even more complex with increasing cognitive capabilities of the robot. Assume that the robot will not just execute a predefined sequence of motions, but rather will be adaptive and, for instance, implicitly negotiate the division of labor with the user, possibly based on learning on the robot side (Mason & Lopes, 2011). Then, much of the discussed questions recur at a higher level, including how feedback to the users shall be organized, how expectations can be managed, how users will interpret the robot's movements, and which interfaces are necessary. However, users also prescribe intentions to the robots and may perceive strategic behavior of a robotic agent beyond movement. For example, Kulms, Mattar, and Kopp (2016) performed a study in which a human and a virtual agent solve a puzzle cooperatively, where the strategies for the agent can be manipulated easily to be more compliant to the human or to be more or less skilled. They found that "perceived trustworthiness and competence – attributions that are crucial for social interactions – were affected both by an agent's compliance and competence. This shows that in cooperation, humans are very sensitive ..."(p. 1342). Perception of critical robot cooperation strategies may then lead to acceptance issues because they, for instance, will not violate the user's sense of justice about workload sharing. Following this idea, Töniges, Ötting, Wrede, Maier, and Sagerer (2017) suggested that system design should implement ways to inform human co-workers in a fair manner about decisions of the technical system. More investigations in this direction are needed. Finally, it is again an open question as to how long-term learning effects will play out once robots are more widespread and their capabilities and limitations are better known. It might be the case that the usage of robots develops similarly to, for example, usage of car navigation systems, which, after an initial adaptation and learning phase, have become an integral part of daily life, including their well-known limitations.

Personalization and big data

One of the big promises of assistive robotics is the use of their flexibility for personalization. Every user or worker could receive exactly the amount of assistance that he or she needs, including optimal ergonomics (Busch, Thomas, Deuse, & Kuhlenkötter, 2012). To realize this promise, the robot or the system in which it is integrated needs to collect data about the user. This has so far been considered again mainly for humanoid and cognitive research robots under the paradigm of adapting for shared plans (Mason & Lopes, 2011). However, with the rise of machine learning and big data, and the practically unlimited amount of computation available through the Internet, the question arises as to what the robot can learn, not only about the task, but also about a single user, based on personal interaction data. In some sense, the configuration of a robot (or any flexible digitalized production system) that adapts to a user can be interpreted as a particular form of personalized data. Assume that in the hybrid assembly scenario, the robot and the user jointly fit some part into a frame through co-manipulation, that is, both parties grasp and move the part jointly. To enable this and, of course, for safety reasons, the robot will measure the forces which the user applies, and then adapt its own impedance. This is important not only for task fulfillment, but also for personalization, which may include the optimization of ergonomic factors for the user.

The flipside of the coin is that the robot will also measure when the user is shaky, whether he or she becomes stronger or weaker systematically, or whether there are particular days or times when the user is more or less effective. Similar to popular fitness trackers or in combination with them, such data reveal much about the worker and may be stored and processed offline, for example, in combination with camera images and quality assurance reports. In short, the robot will deliver a comprehensive assessment of health and wellbeing of the interacting person. This is thus highly sensitive and may be compared with and treated like other medical data. Robots hereby can function as highly efficient means for thorough monitoring of a worker, given that data can be processed and analyzed through respective big data methods, as they are increasingly available on the Internet and by specific service providers.

A particular problem of assistive robotics in the workplace is that the measurement of data cannot simply be abandoned or limited. It is essential both for safety in interaction and for execution of the robot function, which cannot be carried out in a collaborative manner without measuring the external forces caused by the human user. A typical approach of respective regulation is to discard such data for further processing. This option, however, is also problematic, because the data might open opportunities, for example, for timely prediction of serious diseases. In such cases, the employer may even have the obligation to use data for the benefit of the worker. Personalization of a workplace can undoubtedly be an important means to make the work healthier, to keep workers in process for longer times, and to optimize performance. These considerations show that robots may serve not only as flexible tools in the workplace, but in connection with data processing and the respective digital resources of the Internet, they also qualify for the purpose of surveillance. Consequently, the respective robot applications inherit all concerns about data security that are inherent to all digital applications that record human data.

Integration and Application Development

A largely underestimated topic is the complexity of the integration of robots in full systems, be it for more flexible automation systems or other applications. In traditional robotics, an industrial robot comes with its own controller box and can be programmed independently of any other devices or systems, which are rather considered as peripheral to the robot. In the smart factory, new robots are connected to other devices, networks, or Internet resources, which is known as the *cyber–physical systems approach* (Lee, Bagheri, & Kao, 2015). Furthermore, if used as a component technology to make automation systems or workplaces more flexible and adaptive to the user, robots have to fit to the digital infrastructure and modeling approaches of such larger systems. In this sense, they are considered as a component of a system rather than the system's master (Wrede et al., 2016). This leads to a number of perspective shifts on the part of developers, system designers, and production engineers.

First, the existing separation of rigid automation technology and industrial robotics slowly vanishes. Novel production systems that are increasingly digitalized can more easily accommodate robots everywhere, from flexible material handling to human–robot collaboration. Thus, on the one hand, system engineers have an unprecedented freedom for creating novel workflows and completely restructured production systems. This freedom of choice, on the other hand, implies an increased complexity, which arises not only from the robot and its versatile movements, but also from the induced flexibility in the overall system. Consider again the example of hybrid assembly. The robot can grasp its parts from various places, which implies that there does not need to be a fixed workplace arrangement. This, in turn, provides the option to have another robot put parts on the table, whenever and where ever there is space. This consequently opens the possibility for intralogistics to have a mobile supply robot wandering around and delivering parts just in time, all without a fully preprogrammed schedule and plan. To create such systems, digital engineering and simulation must be employed, and the engineering work becomes much more interdisciplinary, in particular, toward software engineering (incorporating, e.g., simulations of human–robot interaction). Second, psychological expertise is required in different ways. On the one hand, it is needed to support the interaction of system designers with the technical system. This requires highly qualified work on the meta-level of interaction with the system, that is, developer-interface engineering, in analogy to the more established user-interface engineering. Here, the developer is the user and he or she interacts with the development process. On the other hand, the user perspective needs to be considered in the systems design. The collaborative and interactive element, which is strongly focused through robotics, requires changing the traditional practice of ***waterfall engineering*** for production systems. The latter refers to a development process that starts from a well thought out blueprint and then builds the system accordingly, which is reasonable if technical specifications can be made in detail and in advance. Typically, interaction with users is foreseen neither in this process, nor in the resulting system. This traditional separation of user and production system is rescinded in the digitalized and robotized world, and therefore usability, user experience, and ergonomic issues have to be taken into account in development of production systems and, more generally, digitalized workplaces (Heinze et al., 2016). It, however, also means following strategies of a more user-centered design, which has been studied intensively and has been subject to various norms and best practices for a long time (Bevan, Carter, Earthy, Geis, & Harker, 2016; International Organization for Standardization, 2010), but mostly for product development and human–computer interfaces and not for production engineering. A respective change of culture in the latter toward much more agile and iterative development seems inevitable, which in turn requires interdisciplinary collaboration, this time including experts on usability and human-centered design.

Third, the overall systems complexity, including new robots, not only increases from the point of view of system design, but also with respect to system safety. Although the new robots themselves are safe in the sense that they employ safe operation and safe stops based on force measurements, the overall system of which they are a part is not automatically also safe. Consider that in the hybrid assembly case, the robot takes a pointed instrument, for instance, a screw driver, to fix something on the frame. This immediately endangers the human as the robot in principle could poke her in the eye with this instrument. Obviously, this must not be possible an intrinsically safe application, which may require that the human leaves the joint workspace or that there is a reorganization of the process, such that the human performs this potentially dangerous action. In this scenario, the robot again is a powerful tool that enables the assembly, but also creates new risks if used in improper ways. In classical engineering, these risks have to be assessed through a well-defined process of risk analysis (e.g., DIN EN ISO 12100). However, already in traditional automation systems, only experts are capable of conducting a proper risk analysis. All the more, this holds if new robots are involved. It can be expected that for a long time there will be uncertainties about best practices in risk analysis, and possibly specialized services will have to be created for this purpose, either by robot manufacturers, specialized integrators, or other specialized robot development companies.

These more general considerations of how robots influence production systems and workplace development raise the expectation that new high-tech jobs will be created in substantial numbers at the intersection of systems engineering, artificial intelligence, software, and robotics. Interdisciplinary collaboration will be essential and basic software engineering is always required. Such versatile workers are scarce among engineers, let alone among less technology-oriented experts, both in computer and software engineering. Moreover, the engineering and software costs exceed, to a large extent, the initial investment in hardware and robots in a frequently changing and flexible production system for mass customization. Therefore, finding and educating the right people and creating services that cut across the issues of usability, interface design, and safety will be essential in making widespread employment of robots economically feasible.

Robots in Other Application Domains

What can be gained from the evaluation of industrial application of novel robots and transferred to more general domains? Technology-wise the vision of the universal assistive humanoid robot in homes or care is still far away because both the physical capabilities and the complexity of cognitive architectures do not easily scale to complex environments such as households, for example. However, three directions of development shall be briefly discussed and put into the context discussed in this chapter.

One direction that social robots appear to be taking is as embodied communication agents for special purposes, which may relatively quickly enter daily life. They can be used in teaching, surveillance, or entertainment functions, based on the described Internet resources in knowledge engineering and AI, whereas the embodiment adds emotional factors in comparison with plain computer interfaces, and fosters engagement of the users. Typical research projects that are implementing field studies are using robots as motivational trainers for children with diabetes (Coninx et al., 2016), for exploring assistive functions in homes for the elderly (Gross et al., 2015), or using similar applications that do not require sophisticated mechanical or physical capabilities. Similar arguments about the role of appearance and human-likeness apply in this domain, in fact, most of the cited studies originate in this domain of social robots. From a psychological point of view, it will be important to design respective interfaces, to evaluate interaction quality, and to finally assess long-term effects. In particular, the feasibility and efficiency of robots for supporting motivational and learning factors should be investigated, which is also a major finding of studies on child–robot interaction (see, e.g., Coninx et al., 2016). In the wider perspective of a digitalized world with quasi-unlimited access to Internet resources, again, such social robots will inherit all the problems and benefits of general AI. Their particular distinguishing feature is their embodiment.

A second direction is robotic assistant systems which can be used for very specific tasks, and in particular teleoperated robots which have more character of an advanced manual tool. Examples are the da Vinci robot for surgery or remote-controlled robots for inspection or bomb disarming. These robots do not move autonomously and typically do not interact with humans, they are operated by humans. A crucial factor for their success, and this has also been discussed above, is the usability of the respective operator interfaces. These can have high complexity, multimodal feedback channels combining sound, vision, and touch, and need to be carefully designed, as prominently manifested in the DARPA Robotics Challenge, the biggest robotics competition worldwide, which featured competitive tasks for remote-controlled humanoid robots in a disaster-response scenario. Summarizing lessons learned, the team winning the pretrial in 2013 stresses that "The operator is a key part of the system. ... This means that the operator must have an appropriate level of skill, must be provided an effective interface to leverage that skill, and must have adequate training" (Johnson et al., 2015, pp. 203–204).

Third, the rapid progress of robotic technology has the potential for disruptive innovation. Completely new fields of application may emerge. A good example is quadcopters, which spread at breath-taking speed. Existing at best as research prototypes, a few years ago, they have created many new services that extend from advertising holiday homes to inspection of industrial facilities to delivery of goods. Such disruptive innovation caused by robots may affect whole sectors and, although technologic, can quickly lead to the emergence of a new organization of work, complete with new qualifications and entirely new job profiles. When realizing services for humans, such robots again will be accepted more easily and have a larger chance of market success if they are endowed with intuitive and usable interfaces. However, the rapidity with which, for example, new touch interfaces have been accepted indicates that humans may adapt to new forms of interaction and applications very quickly if a substantial advantage is apparent. From a psychological perspective, the main questions for disruptive innovation will be rather classically socio-psychological in nature, similar to previous steps in industrial automation: What qualifications will be needed in such new sectors? Will robots rather replace or enhance humans? Which kinds of job will be lost and what will happen to the respective workforce?

Future Research

In summary, we see five major directions of research in the future. First, there is considerable need for the development and proper evaluation of user interfaces. Generally there is both a lack of effort and lack of experts in user-interface design in robotics, which has even been noted in the robotics community itself (Murphy, 2015). Novel interfaces that combine physical, verbal, and multimodal channels in general can be designed and technically realized. This, in turn, demands interdisciplinary collaboration between roboticists, computer scientists, and psychologists for the proper evaluation of user interfaces and the creation of meaningful user studies. Field tests and studies with nonexpert users outside the lab are rare because they are difficult to set up, that is, companies have to be convinced to evaluate different robot designs in their real work environment. Nevertheless, it is necessary to make an effort to connect with practitioners in the field to conduct such studies, as these yield the most meaningful results. Not everything in the lab works in the field, certainly not in interaction with complex and increasingly cognitively capable robots.

Second, it is important to study the effects of learning on the user side, especially if robots become more adaptive. Until 2016, the overwhelming majority of human–robot studies were conducted in the lab, on very short timescales in fixed scenarios. It is, however, highly likely that users will quickly adapt and learn about the strength and weaknesses of robots once they are introduced in everyday interaction. Thus it will be important to validate the degree to which results of ad hoc studies on acceptance of service robots, for instance, persevere in the longer term. Subsequently, it will be relevant to investigate whether users also learn to interpret and adapt to learning on the robot side. Vollmer et al. (2014) have already shown that humans modify their teaching behavior when the robot apparently learns. But the conditions for successful interaction with such adaptive robots remain largely unknown.

Third, the social dynamics of human–robot teams require greater attention. Robots will become increasingly "intelligent," and they will be able to process more and more complex information around them. To allow more sophisticated interactions with human co-workers, it will be necessary to understand the kinds of emotional and social intelligence that are essential for robots to be accepted and trusted as full members of a team. Principles of anthropomorphism may be used initially to embed the robots into social entities in the workplace (*cf.* Epley, Waytz, & Cacioppo, 2007), such as enhance perceived similarity (e.g., humanoid behavior, appearance). In this early phase of human–robot interaction in the workplace, anthropomorphism may be needed to reduce uncertainty among human co-workers. Later, when co-workers have had more, and positive, experiences, investigating how interaction dynamics unfold within human–robot teams will be a key consideration.

Fourth, we need to know more about how tasks should be allocated between robots and their human co-workers, and how the tasks should be designed. Work tasks should be designed such that health is preserved and employees have the opportunity to satisfy their basic psychological needs (i.e., their need for autonomy, competence, and relatedness). From a health-oriented point of view, it might not just be about robots supporting their human co-workers by taking over dangerous tasks (e.g., defusing bombs, fire fighting in the first row) or carrying heavy loads. We need to understand more about adaptive, health-oriented, holistic task allocation between robots and humans: sometimes it may be helpful for robots to encourage their human colleagues to do manual work to strengthen their physical health. We also need to understand how more and more perfect technical systems will allow humans to experience autonomy and competence at work. Will robots at least provide some opportunities to improve relatedness at work and how should such robots be designed?

Finally, since the initial work by Organ (1988), we have learned how important organizational citizenship behavior is for the successful functioning of organizations. In the long run, however, will such proactive behavior still be relevant if technical systems and robots monitor and support their human co-workers? Perhaps we need new kinds of encouragement for human co-workers, enacted by technical systems, in order for them to still show responsibility.

Conclusion

Collaborative, interactive robots are an important, flexible, and versatile technology. They have the potential for disruptive innovation and novel organization of work on different scales, from new automation or hybrid assembly to systems design for the development of smart factories. Access to the Internet makes robots even more powerful and adds capabilities such as big data processing and employment of knowledge bases, while their physical embodiment renders them acceptable as partners for the human worker. Assistive robotics promise to empower users, although the flipside will be an unprecedented capacity for surveillance as interacting robots necessarily measure high-quality user data. From a psychological point of view, several research directions are possible. The user perspective on novel interaction and interfaces needs to be explored and evaluated with respect to usability, user experience, and long-term adaptation effects. The latter also concerns the possibility that continued assistance may lead to a decline in cognitive functioning when, for instance, only trivial tasks are left to the user. It will thus be important to strike the right balance between support and challenge among workers cooperating with intelligent robots. On a systems level, a more user-centered design methodology must also be introduced for production systems, whereas interdisciplinary collaboration between engineers, computer scientists, and psychologists will be required. Finally, classical problems related to innovation and technological development on a larger scale will need to be addressed. Whether the new robots will primarily be used to replace or to assist humans may turn out to be the most important question for their impact on society. As a technology, they have the potential for both.

References

Anderson, N. R., & West, M. A. (1998). Measuring climate for work group innovation: Development and validation of the Team Climate Inventory. *Journal of Organizational Behavior, 19*(3), 235–258.

Atkinson, A. P., Dittrich, W. H., Gemmell, A. J., & Young, A. W. (2004). Emotion perception from dynamic and static body expressions in point-light and full-light displays. *Perception, 33*, 717–746.

Bannat, A., Gast, J., Rehrl, T., Rösel, W., Rigoll, G., & Wallhoff, F. (2009). A multimodal human-robot-interaction scenario: Working together with an industrial robot. In J. A. Jacko (Ed.), *Human-computer interaction. Novel interaction methods and techniques* (pp. 303–311). Berlin: Springer.

Beetz, M., Bartels, G., Albu-Schäffer, A., Bálint-Benczédi, F., Belder, R., Beßler, D., ... Worch, J.-H. (2015). Robotic agents capable of natural and safe physical interaction with human co-workers. In *2015 IEEE/RSJ International Conference on Intelligent Robots and Systems (IROS)* (pp. 6528–6535). New York, NY: IEEE.

Bevan, N., Carter, J., Earthy, J., Geis, T., & Harker, S. (2016). New ISO standards for usability, usability reports and usability measures. In M. Kurosu (Ed.), *Human-computer interaction. Theory, design, development and practice* (pp. 268–278). Cham, Switzerland: Springer International Publishing.

Billard, A., Calinon, S., Dillmann, R., & Schaal, S. (2008). Robot programming by demonstration. In B. Siciliano & O. Khatib (Eds.), *Springer handbook of robotics* (pp. 1371–1394). Berlin: Springer.

Bosma, H., van Boxtel, M. P. J., Ponds, R. W. H. M., Houx, P. J., Burdorf, A., & Jolles, J. (2003). Mental work demands protect against cognitive impairment: MAAS prospective cohort study. *Experimental Aging Research*, 29, 33–45.

Busch, F., Thomas, C., Deuse, J., & Kuhlenkötter, B. (2012). *A* hybrid human-robot assistance system for welding operations – methods to ensure process quality and forecast ergonomic conditions. In S. J. Hu (Ed.), *Technologies and systems for assembly quality, productivity and customization–proceedings of 4th CIRP Conference on Assembly Technologies and Systems (CATS)* (pp. 151–154).

Cascio, W. F., & Montealegre, R. (2016). How technology is changing work and organizations. *Annual Review of Organizational Psychology and Organizational Behavior, 3*, 349–375.

Calinon, S., & Billard, A. (2007). Active teaching in robot programming by demonstration. In *RO-MAN 2007 – The 16th IEEE International Symposium on Robot and Human Interactive Communication* (pp. 702–707). New York, NY: IEEE.

Coninx, A., Baxter, P., Oleari, E., Bellini, S., Bierman, B., Henkemans, O. B., ... Belpaeme, T. (2016). Towards long-term social child-robot interaction: Using multi-activity switching to engage young users. *Journal of Human-Robot Interaction, 5*, 32–67.

Deci, E. L., Olafsen, A. H., & Ryan, R. M. (2017). Self-determination theory in work organizations: The state of a science. *Annual Review of Organizational Psychology and Organizational Behavior, 4*, 19–43.

Deci, E. L., & Ryan, R. M. (2000). The "what" and "why" of goal pursuits: Human needs and the self-determination of behavior. *Psychological Inquiry, 11*, 227–268.

Diftler, M. A., Mehling, J. S., Abdallah, M. E., Radford, N. A., Bridgwater, L. B., Sanders, A. M., ... Ambrose, R. O. (2011). Robonaut 2 – the first humanoid robot in space. In *2011 IEEE International Conference on Robotics and Automation (ICRA)* (pp. 2178–2183). New York, NY: IEEE.

Epley, N., Waytz, A., & Cacioppo, J. T. (2007). On seeing human: A three-factor theory of anthropomorphism. *Psychological Review, 114*, 864–886.

Eyssel, F., & Hegel, F. (2012). (S)he's got the look: Gender stereotyping of robots. *Journal of Applied Social Psychology, 42*, 2213–2230.

Geiselman, E. E., Johnson, C. M., & Buck, D. R. (2013). Flight deck automation: Invaluable collaborator or insidious enabler? *Ergonomics in Design: The Quarterly of Human Factors Applications, 21*(3), 22–26.

Gerber, T., Theorin, A., & Johnsson, C. (2014). Towards a seamless integration between process modeling descriptions at business and production levels: Work in progress. *Journal of Intelligent Manufacturing, 25*, 1089–1099.

Goetz, J., Kiesler, S., & Powers, A. (2003). Matching robot appearance and behavior to tasks to improve human-robot cooperation. In *Proceedings of the 2003 IEEE International Workshop on Robot and Human Interactive* (pp. 55–60). IEEE.

Gombolay, M. C., Gutierrez, R. A., Clarke, S. G., Sturla, G. F., & Shah, R. A. (2015). Decision-making authority, team efficiency and human worker satisfaction in mixed human–robot teams. *Autonomous Robots, 39*, 293–312.

Gross, H. M., Mueller, S., Schroeter, C., Volkhardt, M., Scheidig, A., Debes, K., ... Doering, N. (2015). Robot companion for domestic health assistance: Implementation, test and case study under everyday conditions in private apartments. In *2015 IEEE/RSJ International Conference on Intelligent Robots and Systems (IROS)* (pp. 5992–5999). IEEE.

Hacker, W. (2003). Action Regulation Theory: A practical tool for the design of modern work processes? *European Journal of Work and Organizational Psychology, 12*, 105–130.

Haddadin, S., Albu-Schäffer, A., & Hirzinger, G. (2009). Requirements for safe robots: Measurements, analysis and new insights. *The International Journal of Robotics Research, 28*, 1507–1527.

Hancock, P. A., Billings, D. R., Schaefer, K. E., Chen, J. Y. C., de Visser, E. J., & Parasuraman, R. (2011). A meta-analysis of factors affecting trust in human-robot interaction. *Human Factors, 53*, 517–527.

Heinze, F., Klöckner, M., Wantia, N., Rossmann, J., Kuhlenkötter, B., & Deuse, J. (2016). Combining planning and simulation to create human robot cooperative processes with industrial service robots. *Applied Mechanics & Materials, 840*, 91–98.

Hertel, G., & Zacher, H. (in press). Managing an aging workforce. In C. Viswesvaran, N. Anderson, D. S. Ones, & H. K. Sinangil (Eds.), *The SAGE handbook of industrial, work, and organizational psychology (2nd ed.)*. Thousand Oakes, CA: Sage.

Hinds, P. J., Roberts, T. L., & Jones, H. (2004). Whose job is it anyway? A study of human-robot interaction in a collaborative task. *Human-Computer Interaction, 19*, 151–181.

International Federation of Robotics. (2015). *World Robotics 2015 – Industrial Robots, International Federation of Robotics*. Frankfurt am Main: IFR.

Ishiguro, H., Ono, T., Imai, M., Maeda, T., Kanda, T., & Nakatsu, R. (2001). Robovie: An interactive humanoid robot. *Industrial Robot: An International Journal, 28*, 498–504.

ISO (2010). *ISO 9241-210: Ergonomics of human-centred system interaction – Part 210: Human-centred design for interactive systems*. Geneva: International Organization for Standardization.

Johnson, M., Shrewsbury, B., Bertrand, S., Wu, T., Duran, D., Floyd, M., ... Carff, J. (2015). Team IHMC's lessons learned from the DARPA robotics challenge trials. *Journal of Field Robotics, 32*(2) 192–208.

Kanda, T., Hirano, T., Eaton, D., & Ishiguro, H. (2004). Interactive robots as social partners and peer tutors for children: A field trial. *Human-Computer Interaction, 19*, 61–84.

Kirstein, F., Fischer, K., & Solvason, D. (2014). Human embodiment creates problems for robot learning by demonstration using a control panel. In *Proceedings of the 2014 ACM/IEEE International Conference on Human-Robot Interaction* (pp. 212–213). New York: ACM.

Kohn, M. L., & Schooler, C. (1983). *Work and personality: An inquiry into the impact of social stratification*. London: Erlbaum.

Kormushev, P., Calinon, S., & Caldwell, D. G. (2011). Imitation learning of positional and force skills demonstrated via kinesthetic teaching and haptic input. *Advanced Robotics, 25*, 581–603.

Krach, S., Hegel, F., Wrede, B., Sagerer, G., Binkofski, F., & Kircher, T. (2008). Can machines think? Interaction and perspective taking with robots investigated via fMRI. *PLoS ONE, 3*(7), e2597.

Kulms, P., Mattar, N., & Kopp, S. (2016). Can't do or won't do? Perceived competence, warmth, and cooperativeness in human-agent cooperation. In J. Thangarajah, K. Tuyls, C. Jonker, & S. Marsella (Eds.), *Proceedings of the 15th International Conference on Autonomous Agents and Multiagent Systems (AAMAS 2016)* (pp. 1341–1342). Richmond, SC: International Foundation for Autonomous Agents and Multiagent Systems.

Lee, J., Bagheri, B., & Kao, H.-A. (2015). A cyber-physical systems architecture for industry 4.0-based manufacturing systems. *Manufacturing Letters, 3*, 18–23.

Mason, M., & Lopes, M. (2011). Robot self-initiative and personalization by learning through repeated interactions. In *Proceedings of the 6th ACM/IEEE International Conference on Human-Robot Interaction* (pp. 433–440). New York, NY: ACM.

Morgeson, F. P., & Humphrey, S. E. (2008). Job and team design: Toward a more integrative conceptualization of work design. In J. J. Martocchio (Ed.), *Research in personnel and human resource management* (pp. 39–91). Bingley, UK: Emerald Group Publishing.

Murashov, V., Hearl, F., & Howard, J. (2016). Working safely with robot workers: Recommendations for the new workplace. *Journal of Occupational and Environmental Hygiene, 13*, D61–D71.

Murphy, R. R. (2015). Meta-analysis of autonomy at the DARPA Robotics Challenge trials. *Journal of Field Robotics, 32*, 189–191.

Onnasch, L., Wickens, C. D., Li, H., & Manzey, D. (2014). Human performance consequences of stages and levels of automation: An integrated meta-analysis. *Human Factors, 56*, 476–488.

Organ, D. W. (1988). *Organizational citizenship behavior: The good soldier syndrome*. Lexington, MA: Lexington Books.

Roberts, B. A., Fuhrer, R., Marmot, M., & Richards, M. (2011). Does retirement influence cognitive performance? The Whitehall II study. *Journal of Epidemiological Community Health, 65*, 958–963.

Robotiq (2016). *Collaborative robot ebook*. Retrieved June 30, 2016, from http://blog.robotiq.com/collaborative-robot-ebook

Roether, C. L., Omlor, L., Christensen, A., & Giese, M. A. (2009). Critical features for the perception of emotion from gait. *Journal of Vision, 9*(6), 1–32.

Salem, M., Eyssel, F., Rohlfing, K. J., Kopp, S., & Joublin, F. (2013). To err is human(-like): Effects of robot gesture on perceived anthropomorphism and likability. *International Journal of Social Robotics, 5*, 313–323.

Slama, D., Puhlmann, F., Morrish, J., & Bhatnagar, R. M. (2015). *Enterprise IoT: Strategies and best practices for connected products and services*. Sebastopol, CA: O'Reilly.

Steil, J. J., Emmerich, C., Swadzba, A., Grünberg, R., Nordmann, A., & Wrede, S. (2014). Kinesthetic teaching using assisted gravity compensation for model-free trajectory generation in confined spaces. In F. Röhrbein, G. Veiga, & C. Natale (Eds.), *Gearing up and accelerating cross-fertilization between academic and industrial robotics research in Europe* (pp. 107–127). Cham: Springer International Publishing.

Tay, B., Jung, Y., & Park, T. (2014). When stereotypes meet robots: The double-edge sword of robot gender and personality in human–robot interaction. *Computers in Human Behavior, 38*, 75–84.

Tenorth, M., & Beetz, M. (2013). KnowRob: A knowledge processing infrastructure for cognition-enabled robots. *The International Journal of Robotics Research, 32*, 566–590.

Töniges, T., Ötting, S., Wrede, B., Maier, G. W., & Sagerer, G. (2017). An emerging decision authority: Adaptive cyber-physical system design for fair human-machine interaction and decision processes. In H. Song, D. B. Rawat, S. Jeschke, & C. Brecher (Eds.), *Intelligent data-centric systems: Sensor collected intelligence. Cyber-physical systems: Foundations, principles, and applications* (pp. 419–430). Amsterdam: Elsevier BV.

Villani, L., & De Schutter, J. (2008). Force control. In B. Siciliano & O. Khatib (Eds.), *Springer handbook of robotics* (pp. 161–185). Berlin: Springer.

Vollmer, A.-L., Lohan, K. S., Fischer, K., Nagai, Y., Pitsch, K., Fritsch, J., ... Wrede, B. (2009). People modify their tutoring behavior in robot-directed interaction for action learning. In *2009 IEEE 8th International Conference on Development and Learning* (pp. 1–6). IEEE.

Vollmer, A.-L., Mühlig, M., Steil, J. J., Pitsch, K., Fritsch, J., Rohlfing, K. J., & Wrede, B. (2014). Robots show us how to teach them: Feedback from robots shapes tutoring behavior during action learning. *PloS one, 9*(3), e91349.

Wallhoff, F., Blume, J., Bannat, A., Rösel, W., Lenz, C., & Knoll, A. (2010). A skill-based approach towards hybrid assembly. *Advanced Engineering Informatics, 24*, 329–339.

Wrede, S., Beyer, O., Dreyer, C., Wojtynek, M., & Steil. J. J. (2016, June). *Vertical integration and service orchestration for modular production systems using business process models.* Paper presented at the 3rd International Conference on System-Integrated Intelligence: New Challenges for Product and Production Engineering, Paderborn, Germany.

Wrede, S., Emmerich, C., Grünberg, R., Nordmann, A., Swadzba, A., & Steil, J. (2013). A user study on kinesthetic teaching of redundant robots in task and configuration space. *Journal of Human-Robot Interaction, 2*, 56–81.

Wurhofer, D., Meneweger, T., Fuchsberger, V., & Tscheligi, M. (2015). Deploying robots in a production environment: A study on temporal transitions of workers' experiences. In J. Abascal, S. Barborsa, M. Fetter, T. Gross, P. Palanque, & M. Winkler (Eds.), *Human-Computer Interaction – INTERACT 2015 (Part III)* pp. 203–220). Heidelberg: Springer.

Zuehlke, D. (2010). SmartFactory – towards a factory-of-things. *Annual Reviews in Control, 34*, 129–138.

19

Social Issues Associated with the Internet at Work

Dianna L. Stone, Dianna Krueger, and Stephen Takach

Introduction

The Internet has had a profound effect on almost every aspect of our lives, and changed the way that we purchase products, apply for jobs, access healthcare, and communicate with others (Stone, Deadrick, Lukaszewski, & Johnson, 2015). In addition, the Internet now mediates the relationship between individuals and organizations (e.g., supervisors and subordinates), and reduces distances so that many employees can work from remote locations and virtual teams can operate worldwide (e.g., Hertel, Geister, & Konradt, 2005).

Research on the use of the Internet in work organizations reveals several important benefits, including decreased transaction times, increased efficiency, and enhanced communication (Kavanagh, Thite, & Johnson, 2015). The Internet has also had some unintended consequences that affect individuals and organizations. For instance, the use of the Internet has evoked concerns about the digital divide, adverse impact against women and racial/ethnic minorities, and violations of individuals' rights to privacy. These concerns have prompted the passage of many laws designed to balance the rights of individuals and organizations (e.g., the European Directive, 1995), but analysts are still concerned that the Internet may exacerbate many employment problems (e.g., high unemployment rates, low salary levels). As a result, in this chapter we consider two key social issues associated with the widespread use of Internet-based work: the digital divide, and the extent to which it invades personal privacy. We also discuss the results of research on these issues, and offer suggestions for future research and practice on the topics. It merits emphasis that we do not discuss all social issues associated with Internet use (e.g., age issues, bullying, and counterproductive behavior) as these topics are examined in other chapters in this volume (e.g., see Burlacu, Truxillo, & Bauer, 2017, Chapter 20, and Charlier, Giumetti, Reeves, & Greco, 2017, Chapter 7). In addition, our review is not meant to be a comprehensive evaluation of the literature on the digital divide or perceptions of invasion of privacy in work-related contexts.

The Wiley Blackwell Handbook of the Psychology of the Internet at Work,
First Edition. Edited by Guido Hertel, Dianna L. Stone, Richard D. Johnson, and Jonathan Passmore.

The Digital Divide

One important social issue associated with the use of the Internet at work is that there is a digital divide, or gap between those who have access to computers and the ability to use them, and those who do not (Hoffman & Novak, 1998). For instance, there are racial, ethnic, gender, educational, socioeconomic status, age, and national differences in access to and the ability to use the Internet (i.e., Hoffman & Novak, 1998; Jackson, Ervin, Gardner, & Schmitt, 2001; Van Dijk & Hacker, 2003). Given that the Internet is now used for job applications, those individuals without Internet access are less likely to gain employment opportunities than those who have access to it. For instance, surveys indicated that there is a worldwide digital divide, and Asian, European, and North American nations account for the majority (89%) of the world's Internet users, with African nations having low usage rates (9.8%) (Internetlivetstats, 2015) that likely limit employment prospects. It merits emphasis that we do not mean to imply that Internet use is a panacea, but note that the Internet now controls access to many important work outcomes (e.g., jobs, promotions), and is required for numerous jobs. As a result, the lack of access to and use of the Internet could magnify many work-related problems (e.g., adverse impact against women and minority group members). In the United States, there is also a digital divide between majority group members (European-Americans) and some ethnic/racial minorities, and those with low rather than high levels of education or socioeconomic status (Pew Research Center, 2016). In particular, European-Americans are more likely to have access to and the ability to use the Internet (74%) than African-Americans (64%), Hispanic-Americans (53%), and Native Americans (10–15%) (Pew Research Center, 2016). However, this difference is not consistent across all technological platforms. For example, members of all subgroups appear to have greater access to smartphones than the Internet, but it is often difficult to complete job applications using smartphones (Pew Research Center, 2016). Furthermore, there are socioeconomic discrepancies in access to and use of the Internet. A recent Pew Internet survey (Pew Research Center, 2016) found that 97% of those with incomes above US$100,000 have access to the Internet, but only 42% of people with incomes less than US$10,000 have the same access. Apart from the ethnic and socioeconomic disparities in Internet access, individuals' educational level also affects their use of the Internet. Not surprisingly, those with higher educational levels are more likely to utilize the Internet than those with lower educational levels. In the United States, 89% of those with a college education use the Internet, but only 57% with a high school education and 37% with no high school education use it (Pew Research Center, 2016).

Theoretical Bases for the Digital Divide

Although most research on the digital divide has been somewhat atheoretical, social cognitive theory (Bandura, 1986) can be used to explain gender and racial/ethnic differences in Internet and computer use (hereinafter these terms are used interchangeably). As a result, we use a social cognitive framework to explain gender and racial/ethnic differences in attitudes toward and use of computers. For example, access to computers gives individuals the opportunity to master computer-related tasks, and this mastery has a positive impact on computer self-efficacy (CSE) levels which, in turn, enhances computer use and access to job opportunities. Consistent with these arguments, social cognitive theory maintains that individuals' cognitions, especially self-efficacy beliefs, influence their motivation and behaviors. Self-efficacy refers to the strength of one's beliefs in one's own ability to successfully complete tasks or reach goals (Bandura, 1986). These beliefs determine: how much effort people expend on tasks, the extent to which they engage in particular behaviors, and how long they will persevere when faced with obstacles. Quite simply, the theory predicts that individuals with high self-efficacy are more likely to expend effort to complete a task, and to persist longer in those efforts than those with low self-efficacy (Bandura, 1986). However, low self-efficacy can lead people to believe tasks are harder than they actually are, which results in decreased confidence and performance and increased stress levels. Even though numerous studies have found that high self-efficacy is positively related to performance, some recent research indicated that high self-efficacy led to over-confidence in one's abilities and decreased effort expended on tasks which was even less than in low self-efficacy (Bandura & Jourden, 1991; Stone, 1994; Vancouver, Thompson, Tischner, & Putka, 2002). In particular, two studies by Vancouver et al. (2002) revealed that high self-efficacy was negatively related to task performance. One reason for this is that "high self-efficacy leads to complacent self-assurance which creates little incentive to expend the effort needed to attain high levels of performance" (Bandura & Jourden, 1991, p. 949). Despite these results, we argue that CSE levels still have important implications for understanding gender and racial/ethnic differences in computer usage rates.

Social cognitive theory also predicts that self-efficacy beliefs are influenced by several factors: one factor is successful experience or mastery at a task. Success at a task enhances self-efficacy, whereas failure lowers it. Another factor affecting self-efficacy beliefs is modeling or vicarious experience. For instance, when a similar person can perform a task then an individual is likely to believe that they can succeed at the task too. Stated differently, when we see someone who is similar to ourselves succeed then our own self-efficacy will increase. However, when similar others fail then our self-efficacy will decrease. Thus, successful or unsuccessful role models have a major impact on self-efficacy beliefs.

There are several reasons for gender and racial/ethnic differences in computer use: differences in access to computers, CSE levels, and motivation to use computers. It can also be argued that differences in the ability to use computers has a negative impact on many job-related outcomes (e.g., access to jobs, assignment to challenging jobs requiring computer use, promotions, pay raises, and training opportunities) (hereinafter these job outcomes are termed job opportunities). It merits emphasis that 62% of jobs in the United States require Internet use, and these jobs typically have higher pay levels than others. In some societies, men gain computer experience playing male oriented videogames, but until recently few women played these games because they were designed primarily for men. As a result, women often have lower levels of CSE than Caucasian men (Levine & Donitsa-Schmidt, 1998). (Note: Throughout this chapter we use the terms women and men to refer to Caucasian women and men. We use the term racial/ethnic minorities to refer to both women and men who are members of these groups.)

Likewise, racial/ethnic minorities and Caucasians who live in low socioeconomic status (SES) areas have little access to computers and fewer opportunities to master computer-related tasks than those who live in high SES areas. As a result, the lack of successful experiences with computers is likely to have a detrimental impact on their CSE levels and job prospects. These arguments are supported by considerable research in management information systems (MIS) (Compeau & Higgins, 1995; Marakas, Yi, & Johnson, 1998) showing that CSE beliefs are positively related to computer attitudes and usage rates.

Social cognitive theory also suggests a second reason for gender and racial/ethnic differences in computer usage rates, that is, women and racial/ethnic minorities have fewer successful role models who use computers than among Caucasian men. For instance, some reports indicated that 16% of those in high-tech jobs are women, and only 6% are Hispanic Americans or African Americans (Cheng, 2015). As a result, the lack of successful role models is likely to have a negative influence on their CSE beliefs, and motivation to apply for high-tech jobs or those jobs requiring computer use.

Given this theoretical structure, we maintain that women's and racial/ethnic minorities' access to computers affects their opportunities to succeed on computer-related tasks, and these success experiences, or the vicarious success of similar role models, influence their self-efficacy beliefs, attitudes toward computers, motivational levels, and access to job opportunities. In view of these arguments, we provide a brief review of the research related to gender and racial/ethnic differences in the elements of this model. It merits noting that most of this research was conducted in the United States, but we specify the names of countries when the research was conducted in other contexts. We also consider the limited research on national differences in the digital divide.

Access to technology

Several researchers have argued that a key factor that accounts for the digital divide is access to computers and/or the Internet (Bertot, 2003; Van Dijk & Hacker, 2003). For instance, Bertot (2003) maintained that many factors affect the digital divide, such as access to technology including the Internet and computers, and access to broadband (i.e., a high-capacity transmission technique using a wide array of frequencies which enables large numbers of simultaneous communications). More specifically, Bertot indicated that access to technology often varies with type of employment, access to libraries, and location. Bertot also reported that access to broadband services is often inadequate and extremely costly in many rural areas, and these communities do not always have the funds needed to create a broadband infrastructure. Thus, free information in these areas is often based on printed documents requiring a physical presence. A more recent report indicated that access to broadband is still a problem in rural areas of the United States (Beede & Neville, 2015). For instance, 76% of urban US residents have access to broadband whereas 65% of rural US residents have the same access, and only 18% of rural US residents have high-speed broadband compared with 65% of urban residents (Beede & Neville, 2015). However, Germany has fairly equal access to broadband in urban and rural areas (Initiative D21 e.V., & TNS Infratest, 2015).

Finally, Bertot argued that individuals need to know how to use the technology to locate and determine what information is useful. In addition, Bertot (2003) and Beede and Neville (2015) claimed that differences in access to computers and broadband are likely to perpetuate the digital divide, and people from racial and ethnic minority groups have less computer access because they live and work in low SES areas. As a result, the racial digital divide is exacerbated by low income levels which affect access to technology, and lack of access to technology may increase unemployment rates in these areas. In support of this argument, Van Dijk and Hacker's (2003) research among Dutch and German Internet users, and Chinn and Fairlie (2006) found that income was the most important factor affecting digital skills. Chinn and Fairlie (2006) used data from the International Telecommunications Union (ITU) World Telecommunication Indictor databases. ITU surveyed 30 countries in seven geographic areas (East Asia and the Pacific; Europe and Central Asia; Latin America and the Caribbean; Middle East and North Africa; North America; South Asia; and Sub-Saharan Africa). Computer and Internet penetration rates were analyzed per 100 people.

Apart from access to computers, a survey by Van Dijk and Hacker (2003) found that subjective factors had a negative effect on digital skill levels, attitudes toward technology and motivation, whereas increased opportunities to practice had a positive influence on these issues. This survey also revealed that women in Europe have less access to the Internet and lower digital skills than those in the United States. Similarly, Durndell and Haag (2002) found that compared with men, women in Romanian universities spent less time on computers and the Internet, had more negative attitudes toward computers, greater computer anxiety, and lower levels of CSE. Consistent with these findings, Stanley (2003) maintained that access is not the only factor affecting computer usage rates, but that psychosocial factors (e.g., CSE) are also key determinants of the digital divide. Taken together these results suggested that although access to the Internet or computers may affect individuals' digital skills and CSE, other factors may also influence rates of computer usage (e.g., stereotypes, social roles). Furthermore, research has revealed that there may be demographic differences (e.g., gender, race, ethnic) in digital skills, CSE, and motivation levels.

Demographic factors

In the following sections we provide a brief review of the research on gender and racial/ethnic differences. For age differences, see Burlacu et al. (2017, Chapter 20 in this volume).

Gender differences A number of studies have examined the degree of gender differences in digital skills, CSE, and Internet usage rates (e.g., Bimber, 2000; Cooper, 2006; Durndell & Haag, 2002). Most of this research has found that women's digital skills, CSE, and computer anxiety are lower than those of men, and are socially rather than biologically based (Levine & Donitsa-Schmidt, 1998). One important implication of these findings is that if women have low levels of digital skills they may have fewer job opportunities than men and may be relegated to low-paying jobs. For example, women are employed in only 17% of high-tech jobs in the United States, and most of these jobs have relatively high salaries compared with other positions (Cheng, 2015). However, other studies have not found any gender differences in computer skills, self-efficacy levels, or Internet usage rates among undergraduate college students (Sam, Othman, & Nordin, 2005).

Despite these inconsistent findings, studies indicate that women's beliefs about their computer-related abilities are negatively related to their computer attitudes and rates of computer usage (Cooper, 2006). For example, Bimber (2000) found that women were less likely to use the Internet on a daily basis than men, but the gaps in usage rates were a function of SES, not gender itself. Consistent with these arguments, Cooper (2006) noted that women were less likely to use the Internet because of computer anxiety, and that this anxiety may stem from their lack of experience with computers. Cooper (2006) also reported that women's computer anxiety was affected by negative stereotypes about their lack of digital skills. Although these results are interesting, it should be noted that they cannot identify the direction of causality, and there may be reverse causality associated with some of the findings (e.g., lack of computer use may have a negative impact on beliefs about computer abilities).

Another study conducted in Eastern Europe (Durndell & Haag, 2002) found that women spent less time on the Internet than men because they had lower levels of CSE, higher levels of computer anxiety, and more negative attitudes toward the Internet than men. Still other studies report women are less likely than men to use the Internet at home than men, but this may be a function of their household and childcaring responsibilities rather than negative attitudes toward the Internet (Ono & Zavodny, 2003). Furthermore, Weiser (2000) observed gender differences in usage styles. For example, women use the Internet for interpersonal communication and educational assistance (e.g., chatting online, using email) more than men, but men use it more than women for entertainment and leisure (Cooper, 2006; Jones, Johnson-Yale, Millermaier, & Perez, 2009; Ono & Zavodny, 2003; Weiser, 2000). Apart from the differences in usage styles, several researchers have argued that social context factors, including differences in socialization, gender-role stereotypes, and stereotype threat are key determinants of the digital divide (e.g., Cooper, 2006; Koch, Müller & Sieverding, 2008; Lily, 1994; Whitley, 1997. For instance, Bimber (2000) found that women's use of technology was negatively affected by unfounded stereotypes about them (e.g., women are less technical and have lower computer skills than men). Likewise, Lily (1994) reported that gender-role stereotypes held by parents had a negative impact on female students' attitudes toward computers. Stereotype threat refers to the risk of confirming a negative stereotype about one's group (e.g., women have poor computer skills, Steele & Aronson, 1995). For instance, stereotype threat theory predicts that when women are faced with the stereotype that they will underperform on computer tasks relative to men they become anxious and their behavior will be consistent with the stereotype. In Germany, Koch, Müller, and Sieverding (2008) also found that stereotype threat influenced women's attributions regarding failure on a computer task. In particular, when women experienced stereotype threat they attributed failure to internal causes (e.g., lack of computer ability) but men attributed their failure to external factors (e.g., faulty technical equipment). Koch et al. (2008) argued that these attributions should impact women's CSE and motivation to use computers. In contrast, Whitley (1997) found that gender-role stereotypes negatively influenced CSE and attitudes toward computers among females. However, in this latter study, the gender differences in computer-related behaviors were very small. Cooper (2006) showed that in addition to stereotype threat, gender-role stereotypes also led to computer anxiety and poor computer performance. Group composition further influenced women's computer anxiety and CSE levels (Cooper, 2006). In particular, in mixed male and female groups, computer anxiety among females increased and their performance was hindered by the laughter and criticism of male group members. Interestingly, in mixed-gender groups male criticism and taunting may be reasons that women do not pursue careers in science, technology, and engineering (Johnson, Stone, & Phillips, 2008).

In sum, research on gender differences in the digital divide indicates that access to computers, CSE, computer anxiety, and gender-role stereotypes are negatively related to women's attitudes and computer use (Bimber, 2000; Durndell & Haag, 2002; Cooper, 2006). One work-related implication of these findings is that if women have lower levels of digital skills they may have less access to jobs, or more likely than men to be segregated into low-paying dead-end jobs. However, research also suggests that other factors influence these differences. For instance, stereotypes about women's computer abilities, criticism from men and parents, and stereotype threat may be important determinants of gender differences in computer use (e.g., Cooper, 2006; Lily, 1994). Although several factors have been noted to affect gender differences in computer usage rates, additional research is needed to identify strategies (e.g., use of supportive feedback or successful female role-models) for decreasing the digital divide.

Racial and ethnic differences Even though some researchers argued that the digital divide between racial/ethnic minorities and majority group members is narrowing, some studies have demonstrated persistence of the digital divide based on race or ethnicity (e.g., Cooper, 2006; Jackson et al., 2001; Mullis, Mullis, & Cornille, 2007). For instance, Pew Research Center (2016) surveys reveal that African Americans, Mexican Americans, and other Latin Americans still have less access to computers than their European American counterparts. Asians have slightly lower access rates to home computers than European Americans, whereas Native Americans have a lower probability of owning a computer than all other groups (Pew Research Center, 2016).

In other research when minority and majority group members had equal access to computers, the European American undergraduate students were still slightly more likely to use the Internet and email than the African American students, but there were no differences in their use of the World Wide Web (Jackson et al., 2001). (Note: We differentiate the term World Wide Web from the Internet because the Internet refers to the massive network of networks or a networking infrastructure, but the World Wide Web refers to the way of accessing the information over the Internet [Beal, 2016]). Interestingly, Jackson et al. (2001) also found that African American students were less likely than European Americans to trust computers, and more likely to believe that computers were taking control of their lives. In contrast, European American students were more likely to report that computers could help them gain control over their lives. One reason that African Americans were more likely to distrust computers than European Americans is that they perceived that government officials could track their web-based activities and messages (Jackson et al., 2001). Although not reported in the study, we believe that African Americans are concerned about access to their messages because they fear it will lead to unfair discrimination in the employment process. These students also reported more negative attitudes toward computers because they noted that their use exacerbates loneliness, depression, and social isolation (Jackson et al., 2001).

Given these findings, the racial digital divide may be due to a myriad of factors (see, e.g., Fairlie, 2004; Jackson et al., 2001; Slate, Manuel, & Brinson, 2002). For instance, Fairlie (2004) noted that racial or ethnic differences in computer usage rates may be because of low income levels, lack of education, and segregation into low-level jobs (Fairlie, 2004). In addition, language barriers may affect Mexican Americans' access to and use of computers (Fairlie, 2004). In support of this argument, Slate, Manuel, and Brinson's (2002) study of Hispanic college students revealed that when the primary language spoken at home was Spanish, the students had much less access to the Internet. When English was the primary language spoken at home, access increased and the students had lower levels of computer anxiety and observed that the Internet was more useful than those whose home language was Spanish (Slate et al., 2002). In all, this research suggests that the racial digital divide is in part due to differences in language, trust levels, and beliefs about the benefits of using computers (Jackson et al., 2001; Slate et al., 2002).

Other research argues that racial and ethnic differences are less pronounced than in previous years (Cotten & Jelenewicz, 2006; Jackson et al., 2001). For example, Cotten and Jelenewicz (2006) found that neither race nor Internet experience was related to undergraduate students' use of the Internet. Two other studies found African American undergraduate students reported higher levels of CSE than European American students (e.g., Johnson et al., 2008; Lent, Lopez, Lopez, & Sheu, 2008). Another a study of students at predominately African American colleges also reported higher levels of CSE than those at traditional European American schools (Lent et al., 2008). Johnson et al. (2008) also found that European American women had lower levels of CSE than all other groups (e.g., European American men, African American men and women). Yet the African American students were less likely to pursue careers in information technology (IT) than European-American males because of unfair discrimination in the IT field, as well as concern that they would be stereotyped as nerds if they pursued careers in the field.

As evident from our review, little research has examined the differences in CSE levels and computer use between Hispanic Americans and European Americans. One notable exception is the work of Johnson, Stone, and Navas (2011), who found that Hispanic adults were less likely to use computers than European American adults. However, the former were more likely to use cell phones than computers, but this usage may be problematic because it is difficult to apply for jobs or perform work-related tasks using cell phones. Interestingly, relative to the European Americans, the Hispanic Americans were more likely to endorse collectivist cultural values that include strong personal networks and preference for face-to-face rather than computerized communication channels. Furthermore, Hispanic Americans who endorsed high levels of collectivism were less likely to use computers and more likely to prefer face-to-face interaction than those who valued low levels of collectivism. Still, cultural values may be an important determinant of the digital divide among Hispanic Americans. These findings may help explain national differences in computer usage, but we know of no other research that has examined the extent to which differences in cultural values influence computer usage rates.

In sum, the research on racial and ethnic differences in computer use reveals that low income, low educational levels, and employment in low-level jobs were negatively related to access to and usage rates of computers and CSE (e.g., Fairlie, 2004). Apart from these factors, the racial digital divide may be, in part, due to differences in language, distrust of computers, and beliefs that computers contribute to social isolation (e.g., Jackson et al., 2001; Slate et al., 2002). Given these findings, the lack of computer skills among US racial and ethnic minorities may have a negative impact on their employment rates and salary levels because the majority of large companies use web-based recruiting and application systems (86%, Sierra-Cedar, 2015–2016).

A few studies have addressed factors underpinning cross-national differences in the digital divide and we discuss this research next.

Cross-national differences Even though there are still cross-national differences in access to computers and the Internet, some researchers argued that the digital divide is narrowing (Chen & Wellman, 2004). For instance, developing countries are rapidly catching up with the United States and other Western nations, and South Korea is the world leader in broadband use (Chen & Wellman, 2004). However, other research also indicates that some nations are still lagging behind others in Internet usage (e.g., Chen & Wellman, 2004; Corrocher & Ordanini, 2002): South Korea and the United States rank number one followed by the Scandinavian countries, the United Kingdom, and Japan. Germany ranks in the middle, but France, Italy, and Spain lag behind other Western nations in terms of geographic distribution of the Internet (Corrocher & Ordanini, 2002). Interestingly, gender differences in the digital divide are particularly problematic in Germany and Italy, but the gap in Internet usage is greatest between older women with low education and SES levels and all the other groups (Chen & Wellman, 2004). Thus, Chen and Wellman (2004) argued that the following factors continue to affect Internet usage rates: SES, gender, age, education, and geographic location (e.g., Internet penetration). Barriers to computer and Internet use in developing countries include the high costs of computers and broadband, lack of technical support, lack of relevant contexts, and unfamiliarity with English, the dominant language on the Internet. Similarly, Fuchs and Horak (2008) found that even though Africa comprises 14.1% of the world's population only 2.6% of Africans use computers, which may impact unemployment rates in this region. Structural inequalities in global access to the Internet in Africa compared to other countries include digital apartheid, and poor digital skills, and these are determinants of the digital divide in African nations.

In sum, most research on cross-national differences in the digital divide has focused on developing countries, and emphasized objective factors such as access to computers and the Internet. However, research has also revealed differences in Internet use among developed countries with fairly equal access to the Internet (e.g., France, Italy, and Spain have been slower to adopt computers and the Internet than other European nations, e.g., Germany). As a result, additional research is needed to examine the extent to which cultural factors such as collectivist versus individualistic social systems and so on affect computer and Internet usage rates. For instance, research showed that Italy and Spain have a more collectivist culture than Germany or the United Kingdom (Hofstede, 1980), and individuals in these nations may prefer face-to-face than computer communication.

Violations of Individual Privacy

Another key social issue associated with the increased use of the Internet is that it has the potential to violate actual or perceived rights to privacy. The Internet is now used for a multitude of work-related purposes (e.g., employee record-keeping, e-recruiting, e-selection), and a great deal of personal information (e.g., social security numbers, background check data) is shared and stored on it. Thus, individuals are justifiably concerned that first, the collection and storage of employment data on the Internet may prevent them from managing positive impressions, stigmatize them, or have a negative impact on job opportunities. Second, personal information stored on the Internet may be accessed by third parties and result in embarrassment (e.g., presence of a disability, same-sex domestic partner benefits) or identity theft. Identity theft is the deliberate use of someone else's identity, usually as a method to gain financial advantage in the other person's name, which may compromise the other person if they are held responsible for the perpetrator's actions (US Federal Trade Commission, 2016). A third reason that the Internet may violate personal privacy is that data stored on it may be inaccurate and erroneously discredit individuals. For example, in a survey by the Society for Human Resource Management (SHRM) and West Group (2000), 73% of participants reported errors in their background data that had resulted in the loss of job opportunities. Interestingly, privacy theorists have long argued that the storage of sensitive information on databases can lead to permanent marks that are extremely difficult to purge from these systems (Wasserstrom, 1978). For instance, a single poor performance appraisal can become a problem for an employee throughout his or her career. Given these concerns, it is clear that the rise in the use of the Internet-based human resource systems may have a negative impact on an employee's ability to control personal information and gain access to job opportunities.

Some recent cases illustrate how individuals' privacy concerns about the Internet may be warranted. For instance, Google and Facebook may violate privacy by tracking individuals' online activities and personal conversations, and then selling the data to employers and other organizations (USAToday, 2015). In response to these invasive actions, a Belgian court fined Facebook US$269,000 per day for tracking individuals for up to two years after they had visited the social media site even though they were not a member of the site (USAToday, 2015).

In another case, the US Office of Personnel Management (OPM, 2015) was the target of a data breach, and as many as 21.5 million records of current, former, or prospective employees were stolen. These records included employees' names, addresses, social security numbers, fingerprints, passwords, and detailed security-clearance background information. In addition, personal information on 10000 current and past employees of Seagate Technology, a large US data storage company, was breached (Privacy Rights Clearinghouse, 2016). The files contained employees' W-2 tax forms, social security numbers, and other sensitive data. The data were then used by the thieves to file false tax refund requests with the US Internal Revenue Service. Interestingly, some estimates indicate that over 184 million records have been breached in the United States since 2005 (Privacy Rights Clearinghouse, 2016).

Given the growing problems associated with using the Internet to store and transmit employee data, the primary purposes of this section of the chapter are to:

- use a psychological model to explain the factors that affect employees' perceptions of privacy invasion
- review the existing research on these factors
- consider some global privacy laws
- suggest directions for future research and practice on the issues.

Definition of privacy and privacy model

Information privacy (hereinafter referred to as privacy) has been defined as "an individual's ability to control personal information and the impressions garnered from it" (Stone & Stone, 1990, p. 354). Even though there are models of privacy in management information systems (MIS) (Culnan, 1993; Smith, 2004), and political science (Westin, 1967), we use the model developed by Stone and Stone (1990) to explain the factors associated with perceptions of invasion of privacy. The primary reason for this is that it is grounded in psychology and explains many of the underlying processes associated with employees' perceptions of invasion of privacy. It is based in the expectancy theory of motivation (Porter & Lawler, 1968), and suggests that applicants' and employees' motivation to control personal information is influenced by first, their values regarding control over personal information, second, beliefs that if the they exert effort they can control personal information, and, third, beliefs that if they can control information they can gain valuable job-related outcomes. The model also posits that individuals' motivation influences privacy-related attitudes and behaviors. In particular, employees are often motivated to control information to manage positive impressions and reduce stigmatization in the employment process. However, when they do not perceive that they have the ability to control the information, and they expect that lack of control leads to negative job outcomes they will react negatively. These negative reactions are similar to psychological reactance (Brehm, 1966) which argues that if a person's behavioral freedom is threatened they will become aroused, and then motivated to re-establish threatened freedoms. Thus, the privacy model argues that individuals' negative reactions to the lack of control over information will be expressed as perceptions of invasions of privacy, and individuals will be motivated to regain control by engaging in privacy protection strategies and behaviors (e.g., withholding information, disclosing inaccurate information, taking legal actions).

The model predicts that four kinds of factors affect privacy-related values, beliefs, and attitudes: informational factors (e.g., type of information collected and stored), procedural factors (e.g., ability to authorize access to and release of information), sociocultural factors (e.g., norms and cultural values), and individual factors (e.g., values, demographics, personality). For instance, the model suggests that individual's value controlling some types of personal information more than others. In particular, if the type of information requested by employers leads to negative job-related impressions (e.g., applicant has been laid off multiple times) then individuals are likely to have a greater desire for privacy than when the data do not lead to negative impressions (e.g., training history). In addition, the model predicts that some procedures used for managing employee data are likely to have a negative impact on individuals' expectancies about their ability to control personal information, and their beliefs that lack of control leads to negative job outcomes. For instance, when organizations give supervising managers access to employee information (e.g., disability) without gaining the individual's permission, employees' are likely to expect that they have no control over personal information. Given the lack of control, they may also perceive that they will experience negative job outcomes. Furthermore, the model predicts that sociocultural and individual factors may influence privacy-related values, beliefs, and attitudes. Specifically, the model argues that there are individual differences in the extent to which employees' value controlling personal information. For example, women and racial minorities may value controlling personal information more than majority group members because they are more likely to experience unfair discrimination in employment process than their counterparts. A simplified version of the privacy model is presented in Figure 19.1.

In the following sections we provide a brief review of the research on the factors that affect individuals' perceptions of invasion of privacy in the employment process. Note that

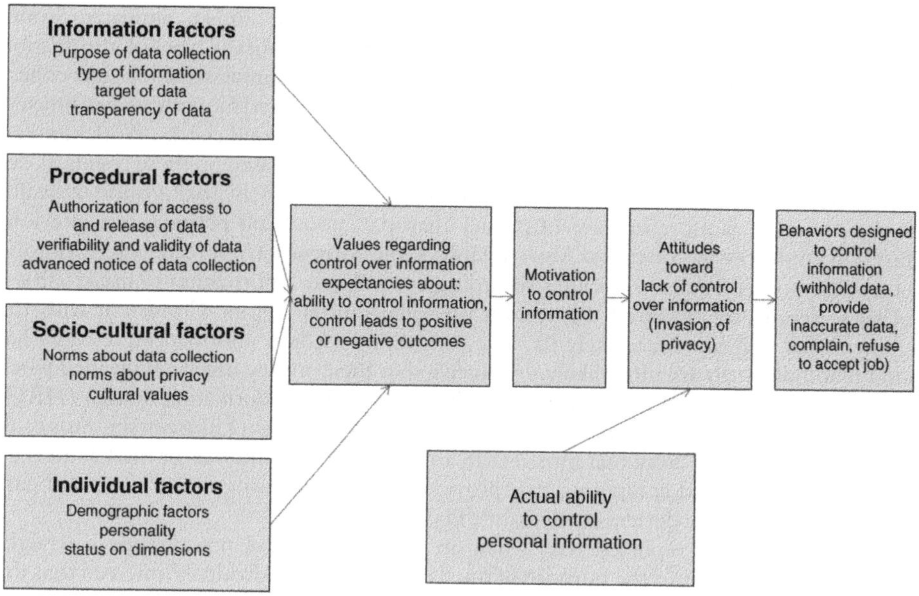

Figure 19.1 Modified privacy model. *Source:* Modified from Stone and Stone (1990).

in this context the term individuals refer to applicants and employees. We also consider the research on informational factors, procedural factors, sociocultural factors, and individual factors that may influence employees' reactions to the collection and storage of personal data on the Internet. Prior to discussing these factors we want to note that much of the research on privacy has focused on legal reviews (Westin, 1967), the dimensions and construct of privacy (Zwick & Dholakia, 2004), survey data (SHRM and West Group, 2000; Zureik, 2004), privacy and information systems (Bélanger & Crossler, 2011), and the contexts of e-commerce, consumer behavior, and data mining (Smith, Dinev, & Xu, 2011). Much less empirical research has examined the factors thought to affect employees' perceptions of invasion of privacy.

Information factors The privacy model suggests that a number of informational factors influence privacy-related values, beliefs, attitudes, and behaviors (Stone & Stone, 1990). For example, the model predicts that the type of information, purpose of the data collection, and the transparency of data collected and stored on the Internet should affect individuals' privacy perceptions. In particular, individuals typically value controlling information that may result in negative job impressions or stigmatization. They also value controlling information when it is collected for one purpose (e.g., recordkeeping) and then used for another purpose (e.g., disciplinary action). Similarly, people value controlling information when the purpose of the data collection is not transparent or obvious (Awad & Krishnan, 2006).

In support of these predictions, a survey revealed that individuals valued control over some types of data stored on the Internet more than others (Zureik, 2004). For instance, respondents indicated that they valued control over their social security numbers, health information, income, credit history, and phone number more than all other types of information. About half of them reported that they valued controlling data about their names and postal addresses on the Internet, but few valued controlling information about personal interests (e.g., hobbies) or email addresses (Zureik, 2004). Similarly, Stone-Romero, Stone, and Hyatt (2003) provided indirect evidence about the types of information that individuals value controlling. Their scaling study of 12 personnel selection data collection practices revealed that the following techniques were considered most invasive of privacy: lie detector, drug tests, genetic tests, medical exams, and background checks. Work samples, application blanks, and interviews were viewed as least invasive. Based on these results, it can be argued that information about criminal offenses, drug use, genetic background or health may be perceived as more invasive of privacy than data about job performance, previous education, and personal interests. Also, notably, some surveys have revealed cross-cultural or cross-national differences in values regarding control over information (Zureik, 2004), and these results are reviewed the section on sociocultural factors. Consistent with the argument that individuals are likely to value controlling medical information, research has found that individuals are more likely to perceive that their privacy had been invaded when medical data were collected and stored in a human resource information system (HRIS) than when only employment data were stored in these systems (Lukaszewski, Stone, & Stone-Romero, 2008). Note that most HRIS are stored on the Internet or the Cloud (i.e., a kind of Internet-based computing that provides shared processing of resources and data to computers and other devices on demand (Hassan, 2011).

Zureik (2004) also reported that when an organization did not provide a straight forward explanation about the purpose of the data collection, individuals' inferred that the organization had ulterior or harmful motives. Likewise, other research showed that when the purpose of collecting data was only beneficial to the organizations, individuals were more likely to perceive that their privacy had been invaded than when the purpose was beneficial to both individuals and organizations (Lukaszewski, Stone, & Johnson, 2014). Still other studies have found that when the purpose of the data collection was not transparent, individuals were less likely to disclose information on websites (Awad & Krishnan, 2006). See also Batinic and Kovacs (2017, Chapter 16 in this volume) for more details about the transparency of the purpose of data collection.

Procedural factors Stone and Stone's (1990) model predicts that several procedural factors are likely to affect privacy-related values, beliefs, and motivation to control information. For instance, many procedural factors may reduce perceived violations of privacy, including the ability to authorize access or release of personal information and the ability to check the validity or accuracy of employment data stored on the Internet (Stone & Stone, 1990).

Authorization for access to information The privacy model posits one reason that individuals are concerned about the collection and storage of their data on the Internet is fear the systems may allow unauthorized access to their private information (Black, Stone, & Johnson, 2015; Stone, Stone-Romero, & Lukaszewski, 2003). For example, individuals may perceive that if Internet users have access to their social security numbers or financial data they will experience identity theft. Recall that identity theft has been defined as the deliberate use of someone else's identity as a means of gaining financial advantage (Federal Trade Commission, 2016). In support of this argument, some reports indicate that identity theft is one of the primary consequences of breach of employee data (Privacy Rights Clearinghouse, 2010). Similarly, if unauthorized users have access to medical or sexual-orientation data, individuals may believe that they will experience embarrassment or loss of job opportunities. Interestingly, some survey research shows that these concerns may be well founded (SHRM & West Group, 2000). In the SHRM & West Group (2000) survey, 34% of the participating

HR professionals collected and stored medical and prescription drug information about employees. This employee information was often accessed by insurance companies and future employers without employees' permission. For example, prospective employers may review job applicant information on search engines or social media sites without informing job applicants (Black et al., 2015).

Furthermore, even though some US laws restrict the use of health data in the employment process (e.g., Americans with Disabilities Act [ADA] 1990; Health Insurance Portability and Accountability Act [HIPAA] 1996) employees have been fired when employers discover they are using prescription drugs for hypertension, diabetes, or pain control (Personnel Policy Service, 2013). In one case an organization established a policy requiring that employees report all drugs present in his or her body and prohibited the use of legal prescription drugs unless approved by a supervisor (e.g., Roe *v.* Cheyenne Mt Conference Resort, 172 F.3d 879 (10th Cir., 1997). Not surprisingly, the court ruled that requiring employees to report their legal prescription drug use was an invasion of privacy and violation of the ADA. Employees were also more likely to perceive a HRIS was invasive of privacy when personal information was accessed by users outside the organization than those inside it (Eddy, Stone, & Stone-Romero, 1999). Others have shown that the use of a HRIS was perceived as more of an invasion of privacy when supervisors were able to access data in employee records than when only HR professionals could access the data (Lukaszewski et al., 2014). It merits emphasis that the US health-related laws require storing medical data separately from other employment data, but some HRIS do not separate medical and employment records. For instance, US health laws allow a great deal of medical information to be stored in employee records (e.g., data from drug tests, family leave certifications, safety reports, workers' compensation). Similarly, affirmative action data in the United States (e.g., race, ethnicity, gender, age, disabilities), which are collected for government reporting, are not always separated from employment data in these systems. As a result, individuals may be concerned that sensitive data about their health, race, or disability status will be accessed by decision-makers, and used unfairly against them in the employment process. At present there are few legal restrictions on access to private sector data in the United States. Therefore, organizations concerned with protecting privacy may want to take steps to limit the degree to which unauthorized individuals have access to employee data. Note that privacy laws in Europe and other nations include restrictions on access to personal data.

Authorization for release of personal information Another concern about the collection and storage of personal data on the Internet is that individuals may perceive that their personal information will be released to others without their permission (Stone & Stone, 1990). For example, in a study by Linowes and Spencer (1998), 70% of employers disclosed employment data without the individuals' permission to creditors, 47% gave information to landlords, and 19% disclosed data to charitable organizations. Other reports indicate that organizations regularly sell data collected on recruiting websites (Stone, Lukaszewski, & Isenhour, 2005), and most organizations (60%) do not inform applicants or employees when they release personal information outside the organization (SHRM & West Group, 2000). Tolchinsky et al. (1981) and Fusilier and Hoyer (1980) support these arguments, indicating that employees were more likely to perceive that their privacy has been invaded when they did not give permission for the release of personal information then when they did give it. These studies were conducted many years ago, however, a recent survey by Pew Research Center (2016) revealed that individuals still prefer that organizations ask their permission before releasing personal data about them.

Ability to check validity or accuracy of data The privacy model also asserts that the lack of validity of data or the inaccuracy of information on the Internet may affect perceptions of invasion. Interestingly, survey research shows that data from background checks, credit checks, or social media are often incorrect (SHRM & West Group, 2000). Furthermore, Stone-Romero et al. (2003) revealed that the degree to which personnel selection information is used to erroneously discredit a person is the most important predictor of invasion of privacy. Still other research indicates that employees' perceptions of invasion of privacy were greater when they could not check and correct data in a HRIS than when they were given the opportunity to check and amend it (Lukaszewski et al., 2014). The use of employee self-service systems may alleviate some of the concerns about the degree to which inaccurate data are stored in HRIS. This is because with employee self-service systems, employees can access their records and check and make changes to those records (e.g., change address, marital status). However, these systems only give employees access to some of their employment data (e.g., personal information); they may not allow them to check the accuracy of all information in their files (e.g., background check data).

Inaccurate data in HRIS are especially problematic because they may stigmatize individuals unfairly, and result in denial of job opportunities. One case provides a vivid example of the concerns associated with inaccurate data (Socorro *v.* IMI Data Search and Hilton Hotels, 02 C 8120 [United States District Court for the Northern District of Illinois, Eastern Division, 2003]). An executive at Hilton Hotels was fired shortly after he was hired when data in his background check incorrectly noted that he had been convicted of a misdemeanor and he had served six months in jail. Hilton hired a firm, IMI Data, to conduct the background check, but did not check the accuracy of the findings. Socorro was not informed of the background check, but his managing director asked if he had ever been convicted of a crime or had spent six months in jail. Socorro replied truthfully that he had not. Although the data about Socorro's conviction and jail sentence were incorrect, Hilton terminated Socorro's employment for falsifying information on his job application. Hilton then told third parties that Socorro was fired because he lied on his application, and that he was a convict who had spent six months in jail. Subsequently, Socorro had a great deal of difficulty securing new employment because of the false and defamatory statements made by Hilton. Although he finally secured new employment, it was at a substantially lower rate of compensation than the Hilton position. It should be clear from this example that the storage and use of inaccurate data on the Internet may have a negative effect on organizations as well as individuals. For instance, when organizations make decisions based on erroneous data then they may fail to hire talented employees, and qualified applicants may be denied job opportunities. Interestingly, most individuals are not aware that the data stored on the Internet or in a HRIS are inaccurate, and in the United States individuals generally have little recourse when their data are erroneous (Karafin, 1998).

Sociocultural factors The model of privacy also indicates that a number of sociocultural factors (e.g., societal norms, cultural values) affect privacy-related values, beliefs, and attitudes. For instance, individuals learn norms, values, and societal expectations through the socialization process, and these factors differ across cultures (Laufer & Wolfe 1977). As Laufer and Wolfe (1977) noted "the mores of a community transmitted through language, tradition, and values constitute boundaries of consciousness about privacy" (p. 28). Thus, cultural norms and values should influence the degree to which employees' value controlling information, and their expectations of privacy. For instance, US cultural norms (Trice & Beyer, 1993) dictate the types of information that should and should not be revealed to others (e.g., data about personal hygiene should remain private), and the extent to which individuals perceive that they have rights to privacy. Thus we consider cultural differences in privacy values and rights below.

Cultural and national differences in privacy values Research consistently reveals that there are cultural and national differences in privacy values (Neuliep, 2014; Zureik, 2004). For instance, Zureik (2004) found that survey respondents in the United States, Mexico, and Japan were more likely to be concerned with the release of personal data to third parties and when information was stolen from databases than those in the United Kingdom. In addition, Dinev et al. (2006) found that Italian participants were less concerned about privacy but more concerned about government intrusions than those in the United States. Furthermore, Chinese people have a different conception of privacy than those in Western countries. In particular, they are less concerned than Westerners with regard to privacy of individuals and more concerned with privacy of the family unit (Neuliep, 2014). Individuals in South Africa are more concerned about privacy than those in the United Kingdom (Ball, Daniel, & Stride, 2012). The reasons for these differences are not clear, but South Africa's experience with apartheid may explain why they emphasize privacy more than other nations. Indirect evidence about cross-cultural or cross-national differences in privacy values comes from differences in the passage of legislation across nations. For instance, Sweden was the first country to enact comprehensive privacy legislation (Swedish Data Act 1973), which suggests that Swedish individuals place great emphasis on privacy values. Likewise, the European Union has long had strict privacy laws (Data Protection Directive, 1995) and the right to privacy is a highly developed area of law in Europe. In contrast, the United States has no comprehensive privacy law, but instead has developed privacy legislation on a state-by-state basis. Given these differences in legislation, it can be inferred that people in Europe should be more likely to stress privacy values than those in the United States. These differences may be particularly problematic when organizations work in a global context because there are unique laws in different nations, and multinational organizations may have to comply with multiple laws.

Cross-cultural theorists (Stone-Romero, Stone, & Salas, 2003) have also argued that people from individualistic cultures should be more likely to value controlling personal information than members of collectivist cultures. The primary reason for this is that individualists often value independence, freedom, autonomy, and self-reliance, and control over personal information should enable them to maintain their independence and freedom (Markus & Kitayama, 1991; Stone-Romero et al., 2003). For example, individualistic nations often support the view that each person "is an entity separate from every other and from the group, and is endowed with natural rights," including the right to privacy (Spence, 1985, p. 1288). In contrast, when people emphasize collectivism they typically use the group as the unit of analysis in social relationships, and value interdependence, security, and personalized relationships (Markus & Kitayama, 1991; Triandis, 1994). As a result, they may be less concerned about sharing personal information with others than individualists because it helps them cultivate personal relationships (Triandis, 1994). These arguments are supported by research indicating that individualists were more concerned about privacy than collectivists (Hann, Hui, Lee, & Png, 2002; Lowry, Cao, & Everard, 2011). Marshall et al. (2008) found that students from individualistic nations (e.g., the United States) were less likely to disclose information about health, romances, school, or social events than those from collective nations (e.g., India). Collectivists are also more likely to accept that organizations can encroach on their privacy than individualists (Milberg, Burke, Smith, & Kallman, 1995). In Milberg et al.'s (1995) research, nations scoring high on individualism, such as Canada, the United States, the United Kingdom, and Australia, had the highest concerns for privacy, and collectivistic nations, such as Thailand and Japan, had the lowest concerns.

Although there is some evidence that those from individualistic cultures are more likely to value control over information than those from collectivist cultures (e.g., Hann et al., 2002; Marshall et al., 2008), research results on this issue are mixed. As noted above, survey participants in the United Kingdom (Zureik, 2004) were less concerned with the release of data to third parties and the theft of data from databases than those from the United States, Mexico, and Japan. Given that Mexico and Japan are considered collectivistic nations, and the United Kingdom and the United States are viewed as individualistic countries (Hofstede, 1980), these results contradict those reported in other studies (e.g., Hann et al., 2002; Milberg et al., 1995). One possible reason is within-nation differences in cultural values (e.g., those with high SES are generally more individualistic even in collectivist nations than those with low SES) (Triandis, 1994). Thus, additional research is needed to determine the reasons for cross-cultural and cross-national differences in privacy values and beliefs.

Individual factors According to the privacy model, individual factors (e.g., gender, race/ethnicity, age, personality) affect privacy values, beliefs, and attitudes. Given space limitations, we only consider three of these factors (e.g., gender, race, personality) in the section that follows. It merits noting, however, that the research on demographic differences in privacy values and beliefs has been inconsistent, hence it is difficult to make valid generalizations (Smith et al., 2011).

Gender differences Several studies show that there are gender differences in privacy concerns (e.g., Ball, Daniel, & Stride, 2012; Friedman et al., 2006; Milne & Rohm, 2000). For instance, early research on privacy (Rosenbaum, 1973) found that women were more concerned about privacy in the selection process than men, and results of surveys reported the same findings (Culnan & Armstrong, 1999). Likewise, research in consumer behavior (Milne & Rohm, 2000) and human–computer interaction (Friedman et al., 2006) reports that women valued privacy more than men, and research in employment contexts also notes gender differences in concern for privacy (Ball et al., 2012). Although researchers do not provide explicit explanations for these differences, it may be that women express greater concern for privacy because they are more likely to experience unfair discrimination. However, more research is needed to uncover the reasons for gender differences in privacy values and beliefs.

Racial/ethnic differences Although there has been little research on racial or ethnic differences in concern for privacy, in some surveys, African-Americans and Hispanic Americans were more likely to value privacy than European-Americans (Culnan & Armstrong, 1999; UPI-Zogby, 2007). Similarly, African-Americans express greater concern for Internet privacy than their European-Americans counterparts (O'Neil, 2001). Furthermore, a study by Gandy (1993) reported in Neuliep (2014) found that African-Americans had different views on privacy than European-Americans. For example, educated African-Americans did not view the collection of data by businesses as invasive of privacy, but were concerned about privacy when they had lost control over their personal information. A possible reason is that loss of privacy makes African-Americans more susceptible to unfair discrimination in the employment process. Finally, compared with European-Americans, African-Americans were more concerned about privacy as it relates to relationships between individuals and the government (Neuliep, 2014).

Personality differences A few studies have examined the relationship between personality factors and concern for privacy or perceptions of invasion of privacy (e.g., Dinev & Hart, 2004; Lu, Tan, & Hui, 2004; Stone, 1986). Two studies found that those high in introversion were more likely to value privacy than those high in extraversion (Lu et al., 2004; Stone, 1986). This is because introverts are typically shy and retiring, and more likely to believe than extraverts that the disclosure of personal information will result in negative evaluations from others. Similarly, Dinev and Hart (2004) indicated that those individuals high in social awareness were more concerned about privacy than those who were low in that dimension.

Global privacy laws Over 40 years ago, the Swedish Data Act (1973) was the first privacy law to be passed, and since then over 109 countries have adopted privacy laws or directives (Greenleaf, 2015).

Table 19.1 is an abbreviated version of the table reported in Greenleaf (2015), and provides a list of 11 countries that have passed privacy laws, the title of the law, and the year in which the law was adopted. A complete list of global privacy laws can be found at Greenleaf (2015). The passage of these laws provides indirect evidence about the degree to which members of these nations and cultures value privacy.

The European Union established its European Data Directive in 1995, and has always been the primary leader in privacy legislation. Over 50% of the countries who have passed privacy laws are European states. However, it is interesting to note that many Asian nations have now passed privacy laws, even though privacy has not always been a key part of their cultural norms. For instance: Hong Kong, Japan, Macao, and Malaysia. In addition, Australia, eight Latin American, eight African, five Middle Eastern, two North American, and four Caribbean nations have adopted privacy laws. Based on this review, it can be argued that privacy is now a widespread global value, and we expect more nations will pass laws in the near future.

In sum, we presented E. F. Stone and D. L. Stone's (1990) privacy model to explain the factors that may affect individuals' privacy values, beliefs, and attitudes toward the collection and storage of employment data on the Internet. Our brief review revealed that

Table 19.1 Global privacy laws (Greenleaf, 2015).

Country	*Law*	*Date passed*	*Date of latest version*
Argentina	Personal Data Protection Act	2000	2000
Australia	Privacy Act of 1988	1988	2001
Bulgaria	Law of Protection of Personal Data	2002	2007
Canada	Personal Information Protection & Electronic Document Act	1983	2002
Chile	Privacy law	1999	1999
Colombia	Data Protection law	2008	2012
Denmark	Act on Processing Personal Data	1978	2000
Germany	Federal Data Protection Act	1977	2009
Hong Kong	Personal Data (Privacy) Ordinance	1995	1995
Japan	Act on the Protection of Personal Information	2003	2007
Malaysia	Personal Data Protection Act	2010	2013

Extract from table in Greenleaf (2015). Reproduced with permission of Elsevier.

the type of information collected and stored on the Internet, the purpose and the transparency of information, the ability to limit access to and release of personal data, the capacity to correct inaccurate data, and individual differences in cultural values, personality, and demographic background are likely to affect employees' perceptions of invasion of privacy. However, much of the existing evidence is based on survey studies. Other types of research are needed to examine the extent to which these factors affect perceptions of privacy. We now consider directions for research on privacy.

Future Research

Our review indicates that although access to computers and the Internet affects women's and racial/ethnic minorities' mastery of computer tasks and self-efficacy levels, we believe that access may become less important as communities provide Internet access in public libraries, Internet cafés, and schools. Some US foundations (e.g., the Bill & Melinda Gates Foundation) are providing computers to individuals in low SES areas of the United States and around the world (e.g., Africa). In other countries (e.g., Spain, Denmark, Finland) there is free public access to Wi-Fi at airports, in libraries, bookstores, and coffee shops, and individuals often share Wi-Fi connections in cities (German-way, 2016). Thus, we hope that increasing access to computers and the Internet will help alleviate some of the problems associated with the worldwide digital divide. However, research is needed to examine the effectiveness of these community-based strategies for decreasing the digital divide. Again, although increasing access to computers may help alleviate the digital divide, our review shows that the lack of computer ability and self-efficacy among women and racial/ethnic minorities still remain key determinants of the problem. However, some research points out that the lack of computer ability may not be the only cause of this problem (e.g., Jackson et al., 2001; Johnson et al., 2008). For instance, stereotyping, public criticism, and stereotype threat may be related to women's and minorities' reluctance to use computers or pursue careers in IT. Thus, future research should focus on the socio-psychological bases of the digital divide (e.g., stereotype threat, stereotypes), and examine strategies that can be used to overcome these challenges.

One strategy that might be used is to employ similar role models to show women and racial/ethnic minorities that people like them can successfully perform computer-related tasks, and these skills have a positive impact on job opportunities. The use of role models has been used to enhance motivation through vicarious learning and vicarious reward systems in multiple contexts (Bandura, 1986). Another strategy that can be used to increase CSE and computer use is to provide supportive feedback to those individuals learning computer-related tasks. If research shows that these strategies are effective then organizations might use them to ensure that all employees can use computers or the Internet. In addition, they might provide specific computer training to new employees to guarantee that they all have digital skills. Furthermore, given that the use of web-based recruiting systems may put women and racial/ethnic minorities at a disadvantage in the employment process, organizations might use multiple strategies to advertise jobs rather than limiting them to web-based systems. When applicants do not have the ability to complete online applications, organizations should provide special assistance so that all applicants can apply for jobs.

Research also indicates that the social context may influence gender differences in computer use. Thus, social role theory (Eagly & Wood, 2012) may provide a more specific explanation of gender differences in computer use than social cognitive theory

(Bandura, 1986). Social role theory predicts that gender differences represent gender role beliefs in a society (Eagly & Wood, 2012). These beliefs arise because people observe female and male behavior, and make inferences that they each possess dispositions and attributes for gender-typical roles. Societies prepare men and women for their roles though extensive socialization that promotes specific skills and behaviors. Thus, gender differences in computer use may be more of a function of socially created roles rather than dispositional skills and abilities. However, research is needed to assess the extent to which social role theory (Eagly & Wood, 2012) predicts gender differences in computer use.

In addition, equal employment opportunities for men and women are important goals in many societies, and women have entered male-dominated roles (e.g., IT, engineering, science). As women assume non-traditional roles, we hope that the gender role stereotypes and expectations for women and computer use will change over time (Eagly & Wood, 2012). However, research is needed to examine strategies for overcoming existing expectations about women's use of computers and their entry into computer-related fields (e.g., computer science or software engineering). As noted above, we believe that the use of similar role models who are successful in computer fields and supportive feedback may change women's beliefs about their CSE.

Suggestions for future research on Internet privacy

Our review indicates that the widespread use of the Internet to store applicant and employee information has prompted growing concerns about employee privacy. As a result, several countries have passed laws designed to protect individual privacy (Greenleaf, 2015). However, sociologists (Etzioni, 1998) have long argued that people and organizations do not always heed laws and may not obey them (e.g., people do not always obey speed limit laws). Therefore, research is needed to determine the types of employment data that people would like to control. For example, it may be the case that employees value controlling data about their health, disabilities, same-sex domestic partnerships, and performance evaluations more than other types of data. One reason is that the release of these types of data may have greater negative impact on individuals' ability to manage positive impressions, and may lead to greater stigmatization. Thus, organizations may want to limit the types of personal data that are collected and stored on Internet-based HR systems. In addition, organizations may give employees or HR professionals the ability to erase data that are out of date or no longer needed for decision-making. For example, a 30-year-old performance evaluation or disciplinary report may no longer be relevant for employment decisions.

Given that many organizations are operating in multinational environments, they may also want to consider cross-cultural differences in privacy-related values. For example, employees in individualistic cultures may be more concerned with protecting the privacy of their personal information than those in collectivist cultures. However, research is needed to determine if this cultural value and others (e.g., power distance, autonomy-oriented values) are associated with privacy-related values and beliefs. Research should also identify the degree to which culture influences the types of data that employees would like to protect. Similarly, research is needed to identify individual differences (e.g., gender, age, personality) in privacy values and beliefs. For example, some research on gender differences notes that women were often more concerned about privacy than men (e.g., Culnan & Armstrong, 1999; Friedman et al., 2006). Likewise, research shows introverts are more concerned with privacy than extraverts (e.g., Lu et al., 2004; Stone, 1986). However, these results should be replicated before generalizations can be made, and other individual differences factors should be examined in future research (e.g., generational differences, need for autonomy).

In addition, research should assess the acceptance and the effectiveness of privacy protection strategies that provide individuals with greater control over their personal data. For example, if there is evidence that employment-related information stored on the Internet is inaccurate then one strategy that might be used to protect privacy is to give employees the ability to check and correct inaccurate data stored in their files. Organizations may also want to use multiple sources to check the accuracy of background data about job applicants to ensure that the results of these checks are accurate and establish policies that restrict access to and the release of individual data on internet-based HR systems. Organizations often release employee data to creditors, landlords, or charitable organizations without gaining their permission (Linowes & Spencer, 1998). In addition, organizations are increasingly reviewing data on social media and Internet sites in the hiring process, and in most cases they do not ask applicants for permission to review this information (Black et al., 2015; see also Lukaszewski & Johnson, 2017, Chapter 12 in this volume for additional information on the use of social media or Internet sites in the employment process). As a result, organizations may want to establish policies that require employees or applicants to give permission before personal data are accessed or released to third parties. These practices should be highlighted in the organizational privacy policy that is posted and communicated to all applicants and employees.

Even though we have identified a few privacy protection strategies, it merits noting that technology is constantly changing so it may be necessary to monitor individuals' privacy concerns over time and develop new strategies as problems arise. Thus, we suggest that organizations should develop new policies and practices as new threats to personal privacy emerge.

Conclusion

Although the use of the Internet in the employment process is beneficial in many ways, we argued in this chapter that it also has some unintended consequences that negatively affect job applicants, employees, and organizations. For instance, it has created a digital divide, or gap, between those who have access to and the ability to use it and those who do not. This gap means that some individuals (e.g., women, racial/ethnic minorities) may be less likely to have access to job opportunities offered on the Internet, and this divide may prevent organizations from employing highly talented employees. We reviewed the existing theory to identify potential determinants of the digital divide, and hope that our evaluation will prompt future research and changes in practice on the issue. We also expect that our review and subsequent research can be used to ameliorate many of the employment problems caused by the digital divide.

Our review also revealed that there are growing concerns about the extent to which the increased use of the Internet violates applicants' and employees' perceived rights to privacy. We reviewed the existing theory and research on privacy to discover the potential factors that affect perceptions of invasion of privacy. We believe that identifying and researching these factors will help organizations develop sound privacy protection policies that can be used to give applicants and employees greater control over their personal information. These practices should also help create a balance between an organization's need for information and an individual's desires for privacy.

References

Americans with Disabilities Act (ADA) (1990). Pub. L. 101–326, 104 stat, 327, enacted July 26, 1990, codified at 42 U.S.C., 12101.

Awad, N., & Krishnan, M. (2006). The personalization privacy paradox: An empirical evaluation of information transparency and the willingness to be profiled online for personalization. *MIS Quarterly, 30*(1), 13–28.

Ball, K., Daniel, E., & Stride, C. (2012). Dimensions of employee privacy: An empirical study. *Information Technology & People, 25*(4), 376–394.

Bandura, A. (1986). *Social foundations of thought and action: A social cognitive theory*. Upper Saddle River, NJ: Prentice-Hall.

Bandura, A., & Jourden, F. J. (1991). Self-regulatory mechanisms governing the impact of social comparison on complex decision making. *Journal of Personality and Social Psychology, 60*, 941–951.

Batinic, B., & Kovacs, C. (2017). Online employee surveys and online feedback. In G. Hertel, D. L. Stone, R. D. Johnson, & J. Passmore (Eds.), *The Wiley Blackwell handbook of the psychology of the internet at work* (pp. 347–368). Chichester: John Wiley & Sons, Ltd.

Beal, V. (2016). The difference between the Internet and the world wide web. Webopedia. Retrieved April 25, 2017, from www.webopedia.com/Did you know/Internet/Web_vs_Internet.asp

Beede, D., & Neville, A. (2015). Broadband availability beyond the rural/urban divide. *Journal of Current Issues in Media and Telecommunications, 7*(4), 423–434.

Bélanger, F., & Crossler, R. (2011). Privacy in the digital age: A review of information privacy research in information systems. *MIS Quarterly, 35*(4), 1017–1042.

Bertot, J. C. (2003). The multiple dimensions of the digital divide: More than the technology "haves" and "have nots." *Government Information Quarterly, 20*(2), 185–191.

Bimber, B. (2000). Measuring the gender gap on the Internet. *Social Science Quarterly, 81*, 868–876.

Black, S., Stone, D. L., & Johnson, A. (2015). Use of social media websites on applicants' privacy. *Employee Responsibilities and Rights Journal, 27*, 115–159.

Brehm, J. W. (1966). *A theory of psychological reactance*. Cambridge, MA: Academic Press.

Burlacu, G., Truxillo, D. M., & Bauer, T. N. (2017). Employee age differences in using internet-based tools at work. In G. Hertel, D. L. Stone, R. D. Johnson, & J. Passmore (Eds.), *The Wiley Blackwell handbook of the psychology of the internet at work* (pp. 449–479). Chichester: John Wiley & Sons, Ltd.

Charlier, S. D., Giumetti, G. W., Reeves, C. J., & Greco, L. S. (2017). Workplace cyberdeviance. In G. Hertel, D. L. Stone, R. D. Johnson, & J. Passmore (Eds.), *The Wiley Blackwell handbook of the psychology of the internet at work* (pp. 131–156). Chichester: John Wiley & Sons, Ltd.

Chen, W., & Wellman, B. (2004). The global digital divide – within and between countries. *IT & Society, 1*(7), 39–45.

Cheng, R. (2015). *Women in tech: The numbers don't add up*. Retrieved April 24, 2017, from https://www.cnet.com.news/women-in-tech-the-numbers-dont-add-up.

Chinn, M. D., & Fairlie, R. W. (2006). ICT use in the developing world: an analysis of differences in computer and internet penetration. *Review of International Economics, 18*(1), 153–167.

Compeau, D. R., & Higgins, C. A. (1995). Computer self-efficacy: Development of a measure and initial test. *MIS Quarterly*, 189–211.

Cooper, J. (2006). The digital divide: The special case of gender. *Journal of Computer Assisted Learning, 22*(5), 320–334.

Corrocher, N., & Ordanini, A. (2002). Measuring the digital divide: A framework for the analysis of cross-country differences. *Journal of Information Technology, 17*(1), 9–19.

Cotten, S. R., & Jelenewicz, S. M. (2006). A disappearing digital divide among college students? Peeling away the layers of the digital divide. *Social Science Computer Review, 24*(4), 497–506.

Culnan, M. (1993). "How did they get my name?": An exploratory investigation of consumer attitudes toward secondary information use. *MIS Quarterly, 17*(3), 341–361.

Culnan, M., & Armstrong, P. (1999). Information privacy concerns, procedural fairness, and impersonal trust: *An empirical investigation Organization Science, 10*(1), 104–115.

Dinev, T., Bellotto, M., Hart, P., Russo, V., & Serra, I. (2006). Internet users' privacy concerns and beliefs about government surveillance: An exploratory study of differences between Italy and the United States. *Journal of Global Information Management, 14*(4), 57–93.

Dinev, T., & Hart, P. (2004). Internet privacy concerns and their antecedents: Measurement vailidity and a regression model. *Behaviour & Information Technology, 23*(6), 413–423.

Durndell, A., & Haag, Z. (2002). Computer self efficacy, computer anxiety, attitudes towards the Internet and reported experience with the Internet, by gender, in an East European sample. *Computers in Human Behavior, 18*(5), 521–535.

Eagly, A. H., & Wood, W. (2012). Social role theory. *Handbook of Theories of Social Psychology, 2*, 458–478.

Eddy, E., Stone, D., & Stone-Romero, E. (1999). The effects of information management policies on reactions to human resource information systems: An integration of privacy and procedural justice perspectives. *Personnel Psychology, 52*(2), 335–358.

European Directive 95/46/EC on individuals with regard to the processing of personal data and the free movement of such data [1995] OJ L281/31.

Etzioni, A. (1998). *The new golden rule: Community and morality in a democratic society.* New York, NY: Basic Books.

Fairlie, R. W. (2004). Race and the digital divide. *Contributions in Economic Analysis & Policy, 3*(1).

Federal Trade Commission (2016). Identity theft definition. Retrieved March 3, 2016, from https://www.consumer.ftc.gov/features/feature-0014-identity-theft.

Friedman, B., Kahn Jr, P., Hagman, J., Severson, R., & Gill, B. (2006). The watcher and the watched: Social judgments about privacy in a public place. *Human-Computer Interaction, 21*(2), 235–272.

Fuchs, C., & Horak, E. (2008). Africa and the digital divide. *Telematics and Informatics, 25*(2), 99–116.

Fusilier, M., & Hoyer, W. (1980). Short note: Variables affecting perceptions of invasion of privacy in a personnel selection situation. *Journal of Applied Psychology, 65*(5), 623–626.

Gandy, O. H. (1993). African-Americans and privacy: Understanding the Black perspective in the emerging policy debate. *Journal of Black Studies, 42*(2), 178–195.

German-way (2016). *Internet and computers in Germany.* Retrieved April 5, 2016, from www.german-way.com/travel-and-tourism/telephone-tips-for-germany/internet-computers-in-germany

Greenleaf, G. (2015). Global tables of data privacy laws and bills. *Privacy Laws & Business International Report*, 18–28.

Hann, I., Hui, K., Lee, T., & Png, I. (2002, December). *Online information privacy: Measuring the cost-benefit trade-off.* Paper presented at the ICIS 2002 Proceedings, Barcelona, Spain.

Hassan, Q. (2011). Demystifying cloud computing. *The Journal of Defense Software Engineering, 24*(1), 16–21.

Health Insurance Portability and Accountability Act of 1996, 42 U.S.C. § 201 (1996).

Hertel, G., Geister, S., & Konradt, U. (2005). Managing virtual teams: A review of current empirical research. *Human Resource Management Review, 15*(1), 69–95.

Hoffman, D. L., & Novak, T. P. (1998), Bridging the racial divide on the Internet, *Science, 280*, 390–391.

Hofstede, G. (1980). *Culture's consequences, international differences in work-related values.* Beverly Hills: Sage Publications.

Initiative D21 e.V., & TNS Infratest (2015). D21-Digital-Index 2015. Die Gesellschaft in der digitalen Transformation [D21 digital index 2015. Society in digital transformation]. Berlin. Retrieved April 7, 2017, from www.initiatived21.de/wp-content/uploads/2015/10/D21_Digital-Index2015_WEB.pdf

Internetlivestats (2015). Retrieved March 6, 2016, from www.internetlivestats.com/google-search-statistics/

Jackson, L. A., Ervin, K. S., Gardner, P. D., & Schmitt, N. (2001). The racial digital divide: motivational, affective, and cognitive correlates of internet use 1. *Journal of Applied Social Psychology, 31*(10), 2019–2046.

Johnson, R., Stone, D. L., & Navas, D. S. (2011, August). *Factors related to the digital divide: Hispanics' use of computers.* Paper presented at the national meeting of the Academy of Management, San Antonio, TX, USA.

Johnson, R. D., Stone, D. L., & Phillips, T. N. (2008). Relations among ethnicity, gender, beliefs, attitudes, and intention to pursue a career in information technology. *Journal of Applied Social Psychology, 38*(4), 999–1022.
Jones, S., Johnson-Yale, C., Millermaier, S., & Seoane Perez, F. (2009). Everyday life, online: U.S. college students' use of the Internet. *First Monday, 14*(10).
Karafin, C. T. (1998). Don't ask, don't tell: A discussion of employee privacy in cyberspace in light of McVeigh v. Cohen, et al. *Virginia Journal of Law and Technology, 3*, 1.
Kavanagh, M. J., Thite, M., & Johnson, R. D. (2015). *Human resource information systems: Basics, applications, and future directions.* Thousand Oaks, CA: Sage.
Koch, S. C., Müller, S. M., & Sieverding, M. (2008). Women and computers. Effects of stereotype threat on attribution of failure. *Computers & Education, 51*(4), 1795–1803.
Laufer, R., & Wolfe, M. (1977). Privacy as a concept and a social issue: A multidimensional developmental theory. *Journal of Social Issues, 33*(3), 22–42.
Lent, R. W., Lopez, A. M., Lopez, F. G., & Sheu, H. B. (2008). Social cognitive career theory and the prediction of interests and choice goals in the computing disciplines. *Journal of Vocational Behavior, 73*(1), 52–62.
Levine, T., & Donitsa-Schmidt, S. (1998). Computer use, confidence, attitudes, and knowledge: A causal analysis. *Computers in Human Behavior, 14*(1), 125–146.
Lily, S. (1994). Socioconomic status, parents' gender-role stereotypes, and the gender gap in computing. *Journal of Research on Computing in Education, 26*(4), 433–451.
Linowes, D. F., & Spencer, R. C. (1998). How employers handle employees' personal information: report of a recent survey. *Employee Rights & Employment Policy Journal, 1*, 153–293.
Lowry, P., Cao, J., & Everard, A. (2011). Privacy concerns versus desire for interpersonal awareness in driving the use of self-disclosure technologies: The case of instant messaging in two cultures. *Journal of Management Information Systems, 27*(4), 163–200.
Lu, Y., Tan, B., & Hui, K. (2004, December). Inducing customers to disclose personal information to internet businesses with social adjustment benefits. Paper presented at the ICIS 2004 Proceedings, Washington DC, USA.
Lukaszewski, K., & Johnson, A. (2017). Social networking systems, search engines, and the employment process. In G. Hertel, D. L. Stone, R. D. Johnson, & J. Passmore (Eds.), *The Wiley Blackwell handbook of the psychology of the internet at work* (pp. 231–256). Chichester: John Wiley & Sons, Ltd.
Lukaszewski, K. M., Stone, D. L., & Johnson, R. D. (2014, August). *Factors affecting individuals' reactions to the collection of data for a human resource information system.* Paper presented at the meeting of the National Academy of Management, Philadelphia, PA, USA.
Lukaszewski, K. M., Stone, D. L., & Stone-Romero, E. F. (2008). The effects of the ability to choose the type of human resources system on perceptions of invasion of privacy and system satisfaction. *Journal of Business and Psychology, 23*(3–4), 73–86.
Marakas, G. M., Yi, M. Y., & Johnson, R. D. (1998). The multilevel and multifaceted character of computer self-efficacy: Toward clarification of the construct and an integrative framework for *research. Information systems research, 9*(2), 126–163.
Markus, H., & Kitayama, S. (1991). Culture and the self: Implications for cognition, emotion, and motivation. *Psychological Review, 98*(2), 224–253.
Marshall, B., Cardon, P., Norris, D., Goreva, N., & D'Souza, R. (2008). Social networking websites in India and the United States: A cross-national comparison of online privacy and communication. *Issues in Information Systems, IX*(2), 87–94.
Milberg, S., Burke, S., Smith, H., & Kallman, E. (1995). Values, personal information privacy, and regulatory approaches. *Communications of the ACM, 38*(12), 65–74.
Milne, G., & Rohm, A. (2000). Consumer privacy and name removal across direct marketing channels: Exploring opt-in and opt-out alternatives. *Journal of Public Policy & Marketing, 19*(2), 238–249.
Mullis, R. L., Mullis, A. K., & Cornille, T. A. (2007). Relationships between identity formation and computer use among black and white emerging adult females. *Computers in Human Behavior, 23*(1), 415–423.
Neuliep, J. (2014). *Intercultural communication: A contextual approach* (6th ed.). Los Angeles, CA: Sage Publications.

O'Neil, D. (2001). Analysis of internet users' level of online privacy concerns. *Social Science Computer Review, 19*(1), 17–31.

Ono, H., & Zavodny, M. (2003). Gender and the Internet. *Social Science Quarterly, 84*(1), 111–121.

Personnel Policy Service (2013). Employees terminated for using prescription drugs. Retrieved March 3, 2017, from www.ppspublishers.com/biz

Pew Research Center. (2016). Internet, science, and technology. Retrieved March 3, 2016, from www.pewinternet.org/2016/01/14/privacy-and-information-sharing/

Porter, L., & Lawler, E. (1968). *Managerial attitudes and performance.* Homewood, IL: Irwin-Dorsey.

Privacy Rights Clearinghouse (2016). Retrieved March 6, 2016, from https://www.privacyrights.org

Rosenbaum, B. (1973). Attitude toward invasion of privacy in the personnel selection process and job applicant demographic and personality correlates. *Journal of Applied Psychology, 58*, 333–338.

Sam, H. K., Othman, A. E. A., & Nordin, Z. S. (2005). Computer self-efficacy, computer anxiety, and toward the Internet: A study among undergraduates in Unimas. *Educational Technology & Society, 8*(4), 205–219.

Sierra-Cedar (2015–2016). *Human resources system survey, 18th annual edition: Innovation, insights, and strategies.* Alpharetta, GA: Sierra-Cedar.

Slate, J. R., Manuel, M., & Jr, Brinson, K. H. (2002). The "Digital Divide": Hispanic college students' views of educational uses of the Internet. *Assessment & Evaluation in Higher Education, 27*(1), 75–93.

Smith, H. (2004). Information privacy and its management. *MIS Quarterly Executive, 3*(4), 201–213.

Smith, J., Dinev, T., & Xu, H. (2011). Information privacy research: An interdisciplinary review. *MIS Quarterly, 35*(4), 989–1015.

Society for Human Resource Management and West Group (2000). *2000 Workplace Privacy Survey.* Retrieved April 7, 2017, from www.shrm.org/research/surveyfindings/documents/workplace privacy survey.pdf

Socorro *v.* IMI Data Search and Hilton Hotels, 02 C 8120 (United States District Court for the Northern District of Illinois, Eastern Division, 2003).

Spence, J. (1985). Achievement American style: The rewards and costs of individualism. *American Psychologist, 40*(12), 1285–1295.

Stanley, L. D. (2003). Beyond access: psychosocial barriers to computer literacy special issue: ICTs and community networking. *The Information Society, 19*(5), 407–416.

Steele, C. M., & Aronson, J. (1995). Stereotype threat and the intellectual test performance of African Americans. *Journal of Personality and Social Psychology, 69*(5), 797–811.

Stone, D. L. (1986). Relationship between introversion/extraversion, values regarding control over information, and perceptions of invasion of privacy. *Perceptual and Motor Skills, 62*(2), 371–376.

Stone, D. L., Deadrick, D. L., Lukaszewski, K. M., & Johnson, R. (2015). The influence of technology on the future of human resource management. *Human Resource Management Review, 25*(2), 216–231.

Stone, D. L., Lukaszewski, K., & Isenhour, L. (2005). E-recruiting: Online strategies for attracting talent. In H. G. Gueutal, & D. L. Stone (Eds.), *The Brave New World of eHR* (pp. 22–53). San Francisco, CA: Jossey-Bass.

Stone, D. L., Stone-Romero, E. F., & Lukaszewski, K. M. (2003). The functional and dysfunctional consequences of using technology to achieve human resource goals. In D. L. Stone (Ed.), *Advances in human resources and cognitive engineering research,* Vol. *3,* (pp. 37–68). London: Elsevier.

Stone, D. N. (1994). Overconfidence in initial self-efficacy judgments: Effects on decision processes and performance. *Organizational Behavior and Human Decision Processes, 59*(3), 452–474.

Stone, E. F., & Stone, D. L. (1990). Privacy in organizations: Theoretical issues, research findings, and protection mechanisms. *Research in Personnel and Human Resource Management, 8*, 349–411.

Stone-Romero, E., Stone, D., & Hyatt, D. (2003). Personnel selection procedures and invasion of privacy. *Journal of Social Issues, 59*(2), 343–368.

Stone-Romero, E., Stone, D., & Salas, E. (2003). The influence of culture on role conceptions and role behavior in organisations. *Applied Psychology, 52*(3), 328–362.

Swedish Data Act (1973). The Data Act: SFS 1973:289 (in Swedish) Data.

Tolchinsky, P., McCuddy, M., Ganster, D., Adams, J., Woodman, R., & Fromkin, H. (1981). Employee perceptions of invasion of privacy: A field simulation experiment. *Journal of Applied Psychology, 66*(3), 308–313.

Triandis, H. (1994). *Culture and social behavior*. New York, NY: Mcgraw-Hill Book Company.

Trice, H., & Beyer, J. (1993). *The cultures of work organizations*. Englewood Cliffs, NJ: Prentice Hall.

UPI-Zogby. (2007). *UPI poll: Concern on health privacy*. Retrieved March 1, 2016, from www.upi.com/Top_News/2007/02/21/UPI-Poll-Concern-on-health-privacy/39291172098800

US Office of Personnel Management (2015). Cybersecurity incidents in Office of Personnel Management. Retrieved April 7, 2017, from https://www.opm.gov/cybersecurity/cybersecurity-incidents/

USAToday (2015). Facebook faces daily fine in Belgium over privacy. Retrieved March 1, 2016, from www.usatoday.com/story/tech/2015/11/10/facebook-faces-steep-fines-belgium-over-privacy/75500860/

Van Dijk, J., & Hacker, K. (2003). The digital divide as a complex and dynamic phenomenon. *The Information Society, 19*(4), 315–326.

Vancouver, J. B., Thompson, C. M. Tischner, E. C., & Putka, D. J. (2002). Two studies examining the negative effect of self-efficacy on performance. *Journal of Applied Psychology, 87*, 506–516.

Wasserstrom, R. (1978). Privacy: Some arguments and assumptions. *Philosophical Law*, 148–166.

Weiser, E. B. (2000). Gender differences in Internet use patterns and Internet application preferences: A two-sample comparison. *CyberPsychology and Behavior, 3*(2), 167–178

Westin, A. (1967). *Privacy and freedom*. New York, NY: Antheneum.

Whitley, B. E. (1997). Gender differences in computer-related attitudes and behavior: A meta-analysis. *Computers in Human Behavior, 13*(1), 1–22.

Zureik, E. (2004). Appendix A: Overview of public opinion research regarding privacy. The Surveillance Project, Public Opinion Workshop, March 3, 2004, Kingston, ON, Canada.

Zwick, D., & Dholakia, N. (2004). Whose identity is it anyway? Consumer representation in the age of database marketing. *Journal of Macromarketing, 24*(1), 31–43.

20

Employee Age Differences in Using Internet-Based Tools at Work

Gabriela Burlacu, Donald M. Truxillo, and Talya N. Bauer

Introduction

As the nature of work becomes more global and complex, Internet-based tools are increasingly being used to help organizations recruit, manage, and retain today's employees on a large scale (Bhatt, 2015; Marler & Fisher, 2013; Stone, Deadrick, Lukaszewski, & Johnson, 2015; Stone & Dulebohn, 2013). These tools provide organizations with numerous benefits including more efficient, streamlined processes and reduced cost compared with traditional, paper-and-pencil based human resources management (HRM) activities (Bauer, Truxillo, Paronto, Weekley, & Campion, 2004; Bhatt, 2015; Davoudi & Fartash, 2012; Parida & Nayak, 2012; Strohmeier, 2009). Although much emphasis has been given to the organization-level benefits of using these tools, relatively little attention has been paid to how these tools impact individuals' productivity and wellbeing.

One characteristic influencing the use of technology-based tools within and outside of work is employee age (e.g., Chang, McAllister, & McCaslin, 2015; Mitzner et al., 2010; Wolfson, Cavanagh, & Kraiger, 2014). Research examining age and technology has revealed a complex relationship in which individuals of different ages may have different needs, preferences, and requirements when it comes to technology design and use (e.g., Wolfson, Cavanagh, & Kraiger, 2014). Relatively little research has been conducted on age and the use of Internet-based work tools specifically, however, within both developmental and industrial and organizational psychology literatures chronological age has been extensively researched in other ways, which may provide insight into how employees of different ages might approach these tools. Given that the global workforce is aging and is concurrently becoming more diverse with regards to age (Hedge, Borman, & Lammlein, 2006; Leibold & Voelpel, 2006; Truxillo, Cadiz, & Hammer, 2015), we propose that the

The Wiley Blackwell Handbook of the Psychology of the Internet at Work,
First Edition. Edited by Guido Hertel, Dianna L. Stone, Richard D. Johnson, and Jonathan Passmore.

relationship between employee age and the increasingly popular use of Internet-based work tools needs to be better understood.

This chapter provides an overview of the current landscape of the aging workforce, as well as the kinds of tools that organizations are increasingly adopting to support their HRM processes. We then describe the lifespan changes in cognition, personality, motivation, and societal context that influence the way adults perceive and approach work. We apply these lifespan development theories and research findings toward exploring how employees of different ages will be impacted by use of Internet-based work tools, proposing further research that needs to be conducted in order to determine how these tools can be designed and deployed to drive effective use across the increasingly age-diverse workforce.

The Aging Workforce

Demographic analysis demonstrates that the average age of the industrialized workforce around the world is increasing and will continue to do so at least through the middle of the twenty-first century (Hertel & Zacher, in press; Leibold & Voelpel, 2006; Truxillo et al., 2015). This trend is due to a convergence of demographic, economic, and sociological changes. First, relatively low birth rates in developed nations mean fewer younger workers are entering the workforce. Second, recent economic downturns have influenced people to work later in life, resulting in older employees in the workplace than we have seen in the past. Third, medical advances have resulted in better overall health for older individuals as they progress into their senior years, and many of these healthy individuals choose to continue working, as work represents an important aspect of their mental and social wellbeing (Kim & Feldman, 2000). This longer life expectancy and better health in later years will also force retirement systems to raise the retirement age in order to stay solvent (Coile, Milligan, & Wise, 2016). It is estimated that by the year 2020, 25% of the workforce in the United States will be over the age of 55 years (Toossi, 2012). In the European Union nations, the rate of employment among people aged 55–64 years has increased from 39.9% in 2003 to 50.1% in 2013 and will likely continue increasing due to initiatives to enhance the workforce through 2020 (European Agency for Safety and Health at Work, 2014). This trend is occurring in Asia as well. For instance, Singapore's Ministry of Manpower reported that over 40% of people aged 65–69 years were still in the workforce in 2015, up from 24% in 2006 (Khalik & Wei, 2016). Finally, older individuals are also increasingly taking up bridge employment post-retirement, which allows them to continue working or return to work in a job that may be less demanding than a full-fledged career (Kim & Feldman, 2000; M. Wang, Adams, Beehr, & Schultz, 2009). As such, older workers are not found just at the top levels of organizations – there may be an increasing number of older workers in all job positions and at all organizational levels (Leibold & Voelpel, 2006; Truxillo et al., 2015).

As the proportion of older workers increases, the younger generation of workers, termed "millennials" (Myers & Sadaghiani, 2010; PricewaterhouseCoopers, 2011), is rising in numbers as well. It is estimated that by the year 2020, 45–50% of the developed workforce will consist of millennials (PricewaterhouseCoopers, 2011). These demographic trends are coinciding such that workers of all ages are working side by side as never before, creating much greater age diversity in the workplace. Furthermore, because of factors such as bridge employment, older workers may find themselves working *for* younger managers. Indeed, research is showing we have greater age diversity in today's organizations than we have in the past, and this increased diversity occurs at the workforce, organization, and even team level (Boehm & Kunze, 2015; Truxillo et al., 2015). In short, researchers and practitioners cannot afford to focus on just engaging millennials or on just engaging older workers – they need to consider holistically how these different age groups work together and what impact this has on individuals and the organizations in which they work. But because Internet-based tools are typically deployed in a one-size-fits-all fashion (Stone et al., 2015), little groundwork has been done to understand age differences in how these tools will be adopted and how they will impact attitudes and behaviors across the entire base of employees.

There is reason to believe that employees of different ages approach elements of work differently. Extensive research has been conducted to understand the attitudes (e.g., Ng & Feldman, 2010), motivations (e.g., Inceoglu, Segers, & Bartram, 2012; Kanfer & Ackerman, 2004; Kooij, de Lange, Jansen, Kanfer, & Dikkers, 2011), and behaviors (e.g., Ng & Feldman, 2008) of different age groups at work, even though relatively little research has looked at these relationships in the context of Internet-based work tools. We begin by defining what is meant by an older or younger worker and then review the literature on age and Internet use in general. We will then present developmental and sociological theories we believe are relevant to the study of age differences in this area. Very little research has been conducted to link these theories to the adoption of Internet-based tools that support a variety of HRM processes, but in subsequent sections we apply what little is known to discuss how age may impact experience with these tools, laying the groundwork for future research that needs to be conducted in this area by presenting a series of research questions (summarized in Box 20.1). It is critical for the field of industrial and organizational psychology to address these questions to gain a better understanding of the impact these tools have on employees of different ages as their prevalence in managing an increasingly age-diverse workforce grows.

Who is an older worker? Who is a younger worker?

Although age is typically studied as a continuous variable (e.g., Bertolino, Truxillo, & Fraccaroli, 2013; Carstensen, 1992; Ng & Feldman, 2008; 2013; Shore, Cleveland, & Goldberg, 2003), it has historically been difficult for researchers to ascertain when an individual becomes an "older worker." This categorization varies across cultures, industries, and national and other contexts. Stereotypes regarding age may influence when an employee can be considered "old," and these stereotypes differ depending on job and industry characteristics (Ng & Feldman, 2012; Truxillo et al., 2015). Legal aspects of age such as workplace protection and retirement age vary across countries. For instance, in the United States, once an individual turns 40, he or she becomes part of a legally protected class, and the typical retirement age is around 65. However, in China, the retirement age tends to occur earlier and thus their conceptualization of what constitutes an "older worker" may differ (Giles, Wang, & Cai, 2011). As people live longer and work later in life, traditional retirement ages in many nations are changing (Coile et al., 2016), and conceptualizations about older/younger workers are gradually changing as well (Truxillo et al., 2015). But as the global workforce also increases in age diversity, it becomes critical for researchers to better understand changes that occur within individuals throughout the lifespan and how these influence attitudes and behaviors regarding work and the tools employees are increasingly being called on to use in their jobs.

Familiarity with technology such as the Internet has often been studied as a generational issue, with older generations being further removed from and thus less familiar with Internet-based tools (Tacken, Marcellini, Mollenkopf, Ruoppila, & Szeman, 2005). However, there are several reasons why employee age, rather than generation, is critical to study with regards to attitudes toward and behaviors around use of Internet-based work tools. First, although generational differences have received much attention in the popular press, research has found few consistent differences across generations at work

(Cadiz, Truxillo, & Fraccaroli, 2015; Costanza & Finkelstein, 2015). There is considerable variability within generations in attitudes, motivations, and behaviors (Costanza & Finkelstein, 2015; Giancola, 2006; Parry & Urwin, 2011; Twenge, 2010), and there is reason to believe that the limited generation-based variations that have been found can be explained by the chronological aging process (Benson & Brown, 2011; Cadiz et al., 2015; Smola & Sutton, 2002). Second, generations tend to be relevant only in the cultural context in which they were created (Egri & Ralston, 2004; Yu & Miller, 2005). Generations, defined as groups of people that share birth years, birth location, and experiences of significant life events at key developmental stages (Kupperschmidt, 2000) may look different from one country or geographic region to the next. As we seek to develop theory and practice that is relevant worldwide, we need to determine how people differ along dimensions that are relevant across cultural contexts. Finally, the rise of Internet-based work tools means that generational category is no longer a valid reason for lack of familiarity with these tools. Employees have been increasingly called on to learn and use these tools, regardless of the generational category they belong to. Our task then is to identify how this call will impact the attitudes and behaviors of employees of all ages and how we can optimize these tools for all age groups, rather than identifying generations for whom these tools may or may not work.

Internet-based work tools

The use of Internet-based tools have enabled organizations to more effectively and efficiently manage their people, with a variety of HR practices now supported by online applications for many of today's companies (Bhatt, 2015; Cascio, 2000; Stone et al., 2015). These tools provide numerous benefits to organizations in the areas of recruitment, selection, training, and development, as well as for facilitating virtual team and e-leader communications and tasks. When it comes to recruiting new talent, Internet-based recruiting and marketing tools allow organizations to cast a wider hiring net than ever before, reaching qualified job candidates from all over the world (Chapman & Gödöllei, 2017; see Chapter 11 of this volume; Maurer & Liu, 2007; Venkatesh, 2013). Companies are increasingly using tools such as job-posting sites to more simply and inexpensively advertise and fill open job positions (Lee, 2005). Furthermore, first, because Internet-based tools have a broad reach, companies are able to increase their pool of qualified candidates and be more selective, increasing the chances that they will find the right candidates to fit with their culture, vision, and roles (Maurer & Liu, 2007; Singh & Finn, 2003). Second, Internet-based assessment and selection measures save companies time and money, enabling them to process a greater number of job applicants than was previously possible (Anderson, 2003; Bauer et al., 2004, 2006).

The application of Internet-based tools for employee training, learning and development has led to huge advancements in how organizations provide training content to employees. E-learning allows organizations to quickly deliver onboarding, training, and employee development content at a greater scale than ever before. It allows companies to cut costs associated with training instructors, pencil-and-paper based instruction, and securing physical locations for training courses (Cascio, 2000; Derouin, Fritzsche, & Salas, 2005; Johnson & Brown, 2017, see Chapter 17 in this volume; Wang, Wang, & Shee, 2007). Use of Internet-based tools has also enabled new communication channels in an increasingly virtual world. As organizations expand and adopt increasingly advanced technology, the concept of a "work team" is no longer limited to people who work closely together in the same location. Virtual teams are an increasingly common organizational structure in which people in different locations work together remotely (Cascio, 2000; Martins,

Gilson, & Maynard, 2004; Maynard, Gilson, Young, & Vartiainen, 2017, see Chapter 15 of this volume; Truxillo, Cadiz, & Rineer, 2014), sometimes in vastly different countries or cultures (Kraimer, Takeuchi, & Frese, 2014). Virtual teams function as regular work teams, but they have added challenges in effective communication and collaboration via use of Internet-based communication tools (Shachaf, 2008).

With the increase in prevalence of virtual teams, some attention has been given to the people who manage those teams. Virtual leadership, or e-leadership, not only carries the traditional challenges of managing people, but also requires leaders to manage effectively and communicate strategy via virtual communication channels. Furthermore, an e-leader may experience challenges in managing subordinate performance in a world where they rarely see that performance first-hand (Cascio, 2000; Kahai, Avolio, & Sosik, 2017, see Chapter 14 of this volume; Zigurs, 2003). Organizations are increasingly supporting managers in this task by using electronic and Cloud-based technology to track, measure, and document employee performance, particularly with the advent of "big data" (Stone et al., 2015). Some technology is also focused on assisting managers in coaching and providing effective feedback (London, 2003).

While the benefits of Internet-based technology are well documented, the success of any workplace technology intervention is heavily dependent on whether employees actually use it (Venkatesh & Bala, 2008). The extent to which employees adopt technology depends on many factors, including social norms and influence (Fishbein, 1979; Liker & Sindi, 1997), perceptions around usefulness and ease of use (Davis, 1989), self-efficacy in one's ability to effectively use the technology (Tarhini, Hone, & Liu, 2014), prior experience and proper training (Venkatesh, Morris, Davis, & Davis, 2003), and enjoyment derived from using the technology (Venkatesh, Thong, & Xu, 2012). This suggests that there are ways to design and position Internet-based work tools in a way that will foster employee adoption of those tools. However, as we discuss in this chapter, there are unique needs and preferences that employees of different ages may have regarding these tools as a result of cognitive, personality, motivational, and societal changes over their lifespan. These preferences need to be better understood for organizations to ensure Internet-based work tools are adopted by and effective for all employees.

Age and internet use

Although stereotypes suggest that older individuals are resistant to using Internet technology (Czaja et al., 2015; Mitzner et al., 2010), in 2000, Trocchia and Janda's US-based research showed that individuals over 50 years were an active and growing group when it came to regular use of the Internet, with peer comparisons, ease of use, and physical dexterity influencing the regularity with which these people used the Internet. Since then, more research has shown that a majority of older individuals use the Internet in the United States, with 88% of those aged 50–64 years reporting regular or daily use, as well as 57% of those over 65 years (Pew Research Center, 2014). In 2014, 18% of all Internet users in Europe were over the age of 55 years (Statista Dossier, 2014). As a group, older adults in developed nations have had the fastest growth in Internet usage over the second decade in 2000 (Chang et al., 2015). Occupational Internet use among older workers has also increased, particularly given the increased access to Internet-based tools, which eliminates a barrier that hinders some older adults from using technology recreationally (e.g., Czaja & Lee, 2003; Porter & Donthu, 2006).

Although limited research has focused on age and use of Internet-based work tools specifically, some research has looked at older adults' attitudes toward and adoption of technology both within and outside of work, revealing that older adults are a varied group when it comes to technology use. For instance, Czaja and Lee (2003) found that older adults tend to experience greater anxiety about computers and feel more negatively about the effort required to utilize technology. However, some evidence shows these effects disappear over time (Adams, Stubbs, & Woods, 2005; Pew Research Center, 2014; Weatherall, 2000). Tarhini and colleagues (2014) found that factors such as perceived ease of use, perceived usefulness of the technology, and self-efficacy are key determinants of intention to adopt new technology for all individuals, but age moderates this relationship, such that for older adults these determinants are particularly important. Rizzuto (2011) suggested that workers' acceptance of IT innovations at work depends in part on department-level climate and attitude toward these innovations. He found that older workers have more positive perceptions of IT innovations than younger workers when department climate around the innovations is perceived as positive. Furthermore, older adults' motivation to learn about and use technology may look different from that of their younger counterparts. Morris and Venkatesh (2000) found that while younger adults were more likely to adopt technology at work in a manner that was consistent with their attitudes toward the technology, older adults had other influences, such as perceived social norms around technology adoption and behavioral control. However, this work has not yet been expanded to determine how Internet-based HRM tools can be designed to foster perceptions of positive department climate, social norms and behavioral control.

By and large, researchers agree that older adults' relationship to technology is complex and needs to be properly addressed to level the playing field for employees of all ages in the increasingly technological workplace (Abbasi, Tarhini, Hassouna, & Shah, 2015; Charness, 2006). As such, age-related changes in cognition, personality, and motivation, as well as the broader societal context of aging at work, need to be taken into account when designing and deploying technological workplace interventions (Czaja et al., 2015; Wolfson et al., 2014; Truxillo et al., 2015). We now briefly review the changes occurring within individuals over the course of their lifespan that impact their attitudes, behaviors, and motivations at work. We also summarize research that has been conducted on the societal context within which older workers operate. We then link these age-related workplace factors to the study of attitudes toward and adoption of Internet-based work tools.

Within-Person Changes Over Lifespan

Age and cognition

Extensive research shows that as people age, they experience changes in memory and cognition. Although there is wide variance within age groups in terms of cognitive ability (Curtis, Windsor, & Soubelet, 2015), older adults tend to experience gains in some types of cognitive ability while simultaneously experiencing loss in others. Specifically, crystallized intelligence, or learned and practiced knowledge which includes job knowledge, tends to increase for older adults, while fluid intelligence, which includes working memory and the ability to quickly solve novel or unfamiliar problems, tends to decrease (Salthouse, 2011), beginning from the early twenties. This decrease impacts daily cognitive functioning in a variety of ways. For instance, older adults tend to have reduced processing speed (Salthouse, 1995), working memory capacity, and cognitive control processes (e.g., a decline in the ability to coordinate and integrate information; Salthouse, 2011). Verhaegen and Salthouse (1997) showed that the negative relationship between age and fluid intelligence is mediated by cognitive processing speed, such that older adults tend to process new information more slowly and therefore less effectively. In addition, decline in working memory impacts new skill acquisition (Kennedy, Patridge, & Raz, 2008) and search task performance (Sharit, Hernandez, Czaja, & Pirolli, 2008). Because working memory acts as "a system for the temporary maintenance and manipulation of information" (Baddeley, 1992, p. 281), this puts older adults at a disadvantage for quickly learning and comprehending new, complex information.

Although people tend to experience cognitive losses with age, crystallized intelligence, which continues to increase in the fifth and sixth decades of life, can largely offset these losses, especially through the adult working lifespan (Truxillo et al., 2015). In his review of age-related cognitive decline, Salthouse (2011) suggested that losses in fluid intelligence and reduced working memory capacity have limited or nonsignificant consequences for day-to-day functioning. He also suggested that contexts such as the workplace rarely require maximal working memory capacity to succeed, and that older adults can rely on accumulated knowledge, accommodations, and other determinants of success in life (for instance, conscientiousness) to offset cognitive losses. Basic cognitive training can significantly reduce the losses older adults may face: sustained participation in cognitively demanding, novel activities has been shown to have positive effects on older adults' memory (Park et al., 2014), as does training specifically targeted toward improving working memory (Zinke et al., 2014). Furthermore, both on-site and at-home instruction positively impact numerous types of cognitive functioning in adults over 50 years and over 65 years (Wolinsky et al., 2013). As such, although older adults tend to experience cognitive declines as a result of age, these declines are reversible and can be minimized in a context such as the workplace. To an extent, these cognitive changes have been applied toward the study of older adults and technology (e.g., Wolfson et al., 2014). In their review of technology-based training requirements for older workers, Wolfson and colleagues (2014) described the cognitive load theory (CLT) and the cognitive theory of multimedia learning (CTML). CLT suggests that humans have a limited working memory capacity in general, which can be overwhelmed during learning (Sweller, van Merrienboer, & Paas, 1998). CTML suggests that technology-based instruction methods that have multiple sensory demands (e.g., requiring learners to listen to a speaker while also viewing presentation slides) have an even greater likelihood of overwhelming working memory demands (Mayer, 2005). Because these theories posit that people have limited working memory in general, older adults' continued declines in working memory capacity may be particularly detrimental in how well they can learn from and use novel Internet-based HRM tools without proper training and guidance.

Age and personality

Personality traits tend to stay relatively stable throughout the lifespan, but certain personality characteristics differ between older and younger adults (Heggestad & Andrew, 2012; Roberts & DelVecchio, 2000). Older adults tend to exhibit more conscientiousness (Warr, Miles, & Platts, 2001), more agreeableness (Bliedorn et al., 2013), and less neuroticism (Roberts, Walton, & Viechtbauer, 2006), and some of these changes commence as individuals transition into early adulthood. However, older adults also tend to exhibit slightly reduced extraversion and openness to experience (Costa Jr, et al., 1984), although certain facets of extraversion, such as independence and self-confidence, increase with age (Roberts et al., 2006; Soto, 2015). These trends are robust across different cultures (McCrae et al., 1999). There is significant empirical evidence that personality traits shift to some extent with age, influenced by external factors such as physical and mental health (Chow & Roberts, 2014; Magidson, Roberts, Collado-Rodriguez, & Lejuez, 2014; Soto, 2015). Despite these overall trends, older adults are a varied group when it comes to personality, and it has been suggested that differences in cognitive ability may account for some of these variations. Specifically, researchers have suggested a link between conscientiousness, openness, and slower rates of cognitive decline (Curtis et al., 2014).

While personality is not necessarily indicative of how and when an individual will adopt Internet-based tools at work, differences in conscientiousness, openness to experience, and general willingness to share knowledge and experiences will likely predict performance in general (Barrick & Mount, 1991; Hurtz & Donovan, 2000) and possibly willingness to take on and adopt new tools at work. Major, Turner, and Fletcher (2006) found that proactive personality, openness, extraversion, and conscientiousness predict motivation to learn and develop. However, this work has not been extended to explain whether older adults will be more or less likely to adopt Internet-based work tools given their personality tendencies. Furthermore, it has not been explored how lifespan changes in extraversion may influence the use of communication-focused Internet work tools.

Age and work motivation

There is considerable evidence that motivation changes with age, which ultimately impacts how individuals approach work over their lifespan. An important note here is that motivation does not *decline* with age, but rather *shifts* to focus on other priorities (Kanfer & Ackerman, 2004). Several lifespan development theories have been proposed to explain why and how this happens. The theory of selection, optimization, and compensation (SOC theory; Baltes, 1987; Baltes & Baltes, 1990) suggests that as older adults experience losses in specific capabilities (e.g., in fluid intelligence), they *compensate* by *optimizing* the strengths they have remaining (e.g., performing tasks that require greater use of their crystallized intelligence), and *select* goals that allow them to do so. Research has shown that with increasing age, adults tend to exhibit more SOC behaviors (Freund & Baltes, 1998, 2002), and effective use of SOC strategies has been linked to healthy aging, both in general (Charles & Carstensen, 2009; Ouwehand, de Ridder, & Bensing, 2007) and in the context of the workplace (Kooij, 2015; Muller, Heiden, Herbig, Poppe, & Angerer, 2015; Weise, Freund, & Baltes, 2002; Zacher, 2015). Socioemotional selectivity theory (SST; Carstensen, 1991, 1992) also suggests that the kinds of activity and goals pursued are impacted by age, such that older adults are more focused on short-term rewards, positive emotional regulation, and strengthening social relationships due to their perception that the time they have is limited. In contrast, SST posits that younger adults, who perceive time as being relatively unlimited, are more motivated by opportunities for development and advancement. Research evidence supports SST in the workplace, with older employees exhibiting more organizational citizenship behaviors (Ng & Feldman, 2008), better attitudes toward work and co-workers (Ng & Feldman, 2010), and greater intrinsic work motivation (Inceoglu et al., 2012; Kooij et al., 2011). In addition, Kanfer and Ackerman (2004) described generativity, or concern for the success and wellbeing of future generations (McAdams, de St Aubin, & Logan, 1993; McAdams, 2001), as a motivation that emerges with age.

It is unclear how these age-related changes in motivation influence the adoption of Internet-based tools at work. However, it is possible that these tools can be presented in a manner consistent with the goals older individuals tend to pursue, driving adoption for this group. For instance, SST suggests that if Internet-based tools are positioned in a way that optimizes positive affect and meaningful social interaction, they will be more appealing to older adults. We will explore the implications of lifespan development and changes in motivation in more detail, but in general, more research is needed to determine how age and motivation interact to influence adoption of and experience with Internet-based work tools.

Older workers and the societal context

Although older workers are legally protected in many industrialized nations, they face stereotypes regarding their ability to perform and learn at work (Posthuma & Campion, 2009). These stereotypes exist in many countries and across different cultures (Shiu, Hassan, & Parry, 2015) and are prevalent across industries and job types in varying degrees. We focus here on stereotypes that negatively impact older workers, but age-related stereotypes exist for younger workers as well (Bertolino et al., 2013; Crumpacker & Crumpacker, 2007; Twenge, 2010; Truxillo, McCune, Bertolino, & Fraccaroli, 2012; Weiss & Maurer, 2004). These stereotypes can sometimes influence whether an individual is hired (Bal, Reiss, Rudolph, & Baltes, 2011), how performance is measured (Cleveland & Shore, 1992; Gordon & Arvey, 2004; Truxillo et al., 2014), and the types of opportunities an employee receives (Goldberg, Finkelstein, Perry, & Konrad, 2004; Maurer, Shultz, & Adams, 2007; Shiu et al., 2015). Furthermore, some research has examined the concept of meta-stereotypes, or beliefs that different groups may have about how *others* perceive them and how this influences behavior (e.g., Finkelstein, King, & Voyles, 2015). Such research builds on the stereotype threat literature, which suggests that older adults may perform in ways that match expectations of them, particularly when they are aware of those expectations (e.g., Hughes, Geraci, & De Forrest, 2013). Both age-related stereotypes and older adults' meta-stereotypes may influence the opportunities older workers receive with regard to learning and adopting technology at work (Morris & Venkatesh, 2000). However, more research is needed to determine the impact that these contextual elements may have on older adults' adoption of and experience with Internet-based tools that support HR processes.

Age and Technology as Applied to Different Human Resources Functions

We now discuss Internet-based tools that support a range of HR functions and the likely consequences for workers of different ages, particularly older workers. Note that in certain areas there is woefully little research. As such, we apply what we know about age-related changes in cognition, personality, and motivation, as well as stereotyping and meta-stereotyping, in discerning how age might affect the use of Internet-based tools at work. We also propose future research questions that will help us better understand how to optimize the use of Internet-based work tools for employees of all ages.

Age and e-recruiting

While little research has investigated the extent to which employees of different ages are effectively recruited using Internet methods, some research has explored age differences in the recruiting process in general. Such research has focused on the discrimination older adults face during recruitment, which may negatively impact their decision to move forward with an organization (Doverspike, Taylor, Shultz, & McKay, 2000; Hedge, 2008; Riach, 2007). Building on research that suggests that certain recruiting messages are more effective in attracting minority populations (e.g., Avery, 2003; Gully, Phillips, & Kim, 2014), different recruiting methods and messages may be warranted when attempting to recruit workers of different ages. For instance, Doverspike and colleagues (2000) suggest using images of older workers in job advertisements, removing age discrimination from all parts of the process, and considering and highlighting job design factors such as flexible work hours, which will be more attractive to older workers. Indeed, Rau and

Adams (2005) found that recruiting messages communicating schedule flexibility and equal opportunity employment were successful in attracting older workers seeking bridge employment. By contrast, younger workers are more attracted to messages focused on opportunities for development (a preference that is compatible with SST [Carstensen, 1991, 1992]), as well as a direct discussion of the reasons a particular company would be a desirable place to work relative to competitors.

Older-worker–friendly messaging in the recruiting process does more than simply allow older adults to feel represented – it provides an avenue for communicating a company's age-diversity climate, or the extent to which the organization demonstrates support for workers of all ages. Goldberg, Perry, Finkelstein, and Shull (2013) found that positive organizational diversity climate predicts the extent to which organizations target older workers in recruitment. Because organizations that develop practices to support and value older employees tend to be more successful in attracting and retaining older workers (Armstrong-Stassen & Schlosser, 2011), older job applicants are likely to have positive views of companies that specifically target them in their recruiting efforts. However, further research is needed to directly examine how companies can effectively tailor their recruiting messages around age diversity using Internet-based delivery methods. There is some evidence that online recruiting tool design features directly influence perceptions of an organization (e.g., Allen, Mahto, & Otondo, 2007; Cober, Brown, Keeping, & Levy, 2004; Maurer & Liu, 2007; Stone et al., 2015), but age differences in this area are not well understood and are a critical avenue for future research.

Researchers have also examined age differences in how employees seek out jobs in general, particularly as the number of older adults returning to work after retirement is rising (Wang et al., 2009). Findings in this area have been somewhat discouraging for older adults. For instance, Wanberg, Watt, and Rumsey (1996) showed that, on average, older adults tend to engage in job-seeking behaviors less frequently than their younger counterparts, and even when they do match younger adults' job-seeking frequency they are less likely to find work. Zacher (2013) showed that age is negatively related to job search intensity, although this is moderated somewhat by a proactive personality and perceptions of time and opportunities remaining in one's work life. Chan and Stevens (2001) found that adults over 50 years who had lost their jobs involuntarily were far less likely to be employed four years later. Kanfer, Wanberg and Kantrowitz's (2001) meta-analysis revealed that certain personality traits, such as openness to experience and extraversion, and other personal factors such as job search self-efficacy, were related to unemployment duration. This suggests that age-related within-person changes and other contextual influences work against older adults in the job search process, although older adults can be taught skills and capabilities that allow them to build self-efficacy in this area and appear more attractive to potential employers (Liu, Huang, & Wang, 2014; Taha, Czaja, & Sharit, 2015; Salmela-Aro, Mutanen, & Vuori, 2012; Vuori, Toppinen-Tanner, & Mutanen, 2012). Still, major research is needed to determine whether Internet-based recruiting tools in particular help older workers in their job search efforts by providing greater access to possible opportunities, or hurt older workers by introducing a potentially unfamiliar technology into an already complex process.

Limited research has investigated age differences in recruiting effectiveness, but there is some evidence that employees of different ages have different desires regarding and expectations of work. Older adults have greater intrinsic motivation (Inceoglu et al., 2012) and generativity motives (McAdams et al., 1993), which suggests that they are more attracted to job opportunities that emphasize these elements. Indeed, the job design literature suggests that older workers thrive in jobs that allow greater autonomy (Hertel et al., 2013; Ng & Feldman, 2014; Truxillo et al., 2012) and application of accumulated skills and knowledge (Kanfer & Ackerman, 2004; Zaniboni, Truxillo, & Fraccaroli, 2013). Kanfer and Ackerman (2004) also suggested that older adults are attracted by opportunities to exercise their generativity motives and provide mentorship and guidance to others. However, the jobs that attract employees of different ages do not depend just on within-person changes over the lifespan, but also on societal context: Truxillo and colleagues (2014) observed that older and younger workers may desire different things from work because of different career expectations, norms, and nonwork responsibilities (e.g., family duties) at different life stages. Further research is needed to determine how organizations can effectively communicate, using Internet-based delivery methods, the job aspects that may be more attractive to older workers.

While researchers suggest that organizations should consider the needs of job applicants of all ages in the recruiting process, little attention has been paid to the application of Internet-based tools for supporting these processes and how these tools impact the attitudes and experiences of older and younger workers. There is considerable opportunity to apply existing research on age-related changes in cognition, personality, motivation, and societal context toward understanding how Internet-based recruiting tools can be designed to effectively recruit workers of all ages: for instance, to determine how older job applicants utilize SOC strategies in a recruiting setting (selecting job postings designed to match their cognitive needs and so on), and how these strategies can be optimized through e-recruiting methods. Additionally, there is opportunity to study more deeply the tendency among older adults toward positive emotional regulation as posited by SST (Carstensen, 1991) and how this can be optimized via use of Internet-based recruiting tools. Finally, given the importance of perceived social norms and behavioral control in older adults' use of technology (Morris & Venkatesh, 2000), future researchers should address how e-recruiting tools can be designed to enhance these perceptions, in order to ensure older adults are motivated to learn and utilize the technology.

Age and e-selection

Some research has been conducted on older adults in the job selection process, and it has been shown that older job applicants are less likely to be hired (Bal et al., 2011), which may be due to age stereotyping (Perry, Kulik, & Bourhis, 1996; Posthuma & Campion, 2009; Richardson, Webb, Webber, & Smith, 2013; Truxillo et al., 2014) and possibly even due to perceptions of candidate over-qualification (Erdogan, Bauer, Peiro, & Truxillo, 2011). However, some evidence also exists that widely used selection measures adversely impact older adults. For instance, because of within-person changes across the lifespan in cognition and personality, traditional cognitive ability and personality measures may capture different data from employees of different ages (Truxillo et al., 2014). Klein, Dilchert, Ones, and Dages (2015) found that older job applicants tended to score lower on tests of general mental ability, particularly on tests assessing fluid intelligence capabilities such as inductive reasoning. Because older adults tend to engage in compensation behaviors that may offset cognitive declines (e.g., Kanfer & Ackerman, 2004), lower scores on certain cognitive ability tests may not be indicative of how an older employee will perform on the job. Furthermore, while measures of proactive personality predict performance in a training context (Major et al., 2006), Bertolino, Truxillo, and Fraccaroli (2011) found that this was only true for younger workers. These findings indicate that older adults may be at a disadvantage not just in the traditional recruiting process, but also in the job assessment and selection process.

Much of the research in this area focuses on the content within selection measures and how this impacts the hiring process for job applicants of different ages. While provision of these measures via use of Internet-based tools will likely not change the content (e.g., Bauer et al., 2006; Salgado & Moscoso, 2003), it is possible that the online nature of these tests will result in a different experience for job applicants of different ages. For instance, discomfort or unfamiliarity with technology can impact motivation to take tests and intentions toward an organization in an online selection context, as can concerns around information privacy (Bauer et al., 2006). These negative applicant reactions may influence test performance (Borman, Hanson, & Hedge, 1997). Older adults in particular display discomfort toward technology when it is unfamiliar or poses a possible information security risk (e.g., Mitzner et al., 2010). Furthermore, the decline in fluid intelligence and working memory capabilities with age (Salthouse, 2011) may be especially detrimental in an online measurement format in which applicants have to complete a test in limited time. However, the extent to which Internet-based selection test formats directly impact the performance of older job applicants relative to more traditional methods has not been studied.

In general, more research is needed to determine whether use of Internet-based selection tools negatively impact older workers, and whether these tools can be designed and positioned in a way that maximizes older workers' strengths while minimizing their limitations. Little is known about how lifespan changes in cognition, personality, and motivation can be addressed when designing online job selection tools. For instance, research has not determined whether there are design elements of e-selection measures and methods that can minimize age-related cognitive deficits in learning speed and capacity and maximize gains in crystallized intelligence, although providing unlimited time to complete an online selection test might help. Additional opportunities include exploring how older adults' tendencies toward optimization of current strengths (Baltes & Baltes, 1990) and positive emotional regulation (Carstensen, 1991, 1992) can be addressed in an Internet-based job selection process. Researchers should also explore the design elements of e-selection measures that foster job applicant perceptions of social norms and behavioral control (Morris & Venkatesh, 2000), given the importance of these perceptions for older adults. Finally, existing research around age-related bias in the job selection context (e.g., Richardson et al., 2013) has not been expanded to incorporate selection measures that are completed via the Internet. More research is needed to determine whether online selection measures exacerbate possible bias against older adults in the hiring process, or whether they can be designed to reduce age disparities in hiring.

Age and e-learning

Research shows that employees of different ages have different needs and requirements when it comes to training, regardless of the delivery medium. Kubeck, Delp, Haslett, and McDaniel (1996) found that older workers tend to take longer during training and they also tend to perform worse in a training context than their younger counterparts. Relatedly, stereotypes regarding the capabilities of older workers and relative lack of support may contribute to reduced self-efficacy and confidence in a training setting, inflating these negative age-related effects (Maurer, 2001). However, extensive research shows that age-related deficits in training outcomes disappear when the training is designed to meets the unique needs of older workers (e.g., Beier, 2008; Colquitt, LePine, & Noe, 2000; Kanfer, 1990; Maurer, Weiss, & Barbeitte, 2003). In her comprehensive review of training requirements for older workers, Beier (2008) described the distal traits and proximal processes that need to be taken into account when designing effective training for older workers. She suggested that factors such as employee cognitive ability, personality, self-efficacy, and goal preferences (all posited to change with age) greatly impact the way training is perceived and completed.

Although Ng and Feldman (2012) dispelled many commonly held stereotypes regarding older workers' performance and motivation, they found support for the idea that older workers may be less motivated to learn – which makes sense in terms of SST and the value of investing in development when time is perceived as being relatively limited. This may also be partially explained by the stereotype and meta-stereotype literatures (e.g., Finkelstein et al., 2015; Hughes et al., 2013). However, as with training performance, other research shows that when the training is designed and positioned in certain ways, the negative relationship between age and training motivation diminishes (e.g., Beier, Teachout, & Cox, 2012). For instance, older workers are just as motivated and capable in a training context as younger workers when they receive organizational support (Beier, 2014), when task self-efficacy is high (McCausland & King, 2014), and when the training is collaborative (Wolfson & Kraiger, 2014a). Aside from variations in training motivation, individual differences may also contribute to older employees' performance in learning and training exercises. Beier and Ackerman (2005) suggested that prior knowledge and ability is critical for older adults' acquisition of new knowledge. They found that levels of crystallized intelligence (posited to increase with age) predicted the extent to which older adults learned from an instructional video. Similarly, Carter and Beier (2010) reported that error management training, or training in which errors are framed as learning opportunities, was beneficial for learning among older adults, particularly when their pretraining cognitive ability was high. Self-pacing (Callahan, Kiker, & Cross, 2003) and generally allowing more time for completion of learning tasks (Beier, 2008) can greatly reduce age-related deficits in learning.

In relation to technology-based training in particular, similar requirements have been noted for older workers. Wolfson and colleagues (2014) proposed technology-based instruction should be highly structured, incorporate feedback and adaptive guidance, include metacognitive prompts, incorporate principles derived from CLT and CTML, and utilize a simple and consistent user interface to account for age-related cognitive changes. However, Goldstein and Ford (2002) argued many of these recommendations reflect best practices in training for all trainees. On applying some of these recommendations to a computer-based training course, Wolfson and Kraiger (2014b) observed that instructional coherence (or removing all irrelevant information from the course) improved training performance across all age groups, and that using advance organizers (e.g., bullet lists of the information that would be presented in the course) improved training performance among older adults. Furthermore, self-pacing is considered particularly beneficial for the learning of older adults and an advantage of offering training courses in a technology-based format (Derouin et al., 2005). However, as Beier (2008) posited, with traditional training design, simply crafting courses to promote older workers' training motivation and performance is not sufficient to eliminate age-related learning deficits. Contextual factors such as organizational support, trainee self-efficacy, and training content also need to be taken into account to ensure training effectiveness across all ages (Liu, Courtenay, & Valentine, 2011; Maurer et al., 2003; van Rooij, 2012). Although Wolfson and colleagues' (2014) recommendations give us an elementary understanding of how Internet-based training courses can be designed to support the needs of older workers effectively, further research is needed to determine how these courses and the processes surrounding them can be designed to promote self-efficacy and communicate organizational support.

Taken together, findings regarding age and learning suggest that older adults can be competent in a training context when the right contextual and course design elements are included. Due to limited research in the area to date, it is unclear how Internet-based training in particular will impact training motivation and performance among employees of different ages, and how these courses can be designed to best support the needs of all workers. We have speculated on how organizations might achieve this based on prior research findings around age and learning, but further research is needed to determine how age-related changes over the course of the lifespan influence learning and self-efficacy in an Internet-based training context. For instance, while some research has investigated the role of age-related cognitive changes in computer-based instruction (e.g., Wolfson et al., 2014), further research is needed to determine how e-learning in particular can be designed to offset cognitive declines and optimize cognitive strengths. Furthermore, little is currently known about how e-learning can be designed to maximize older adults' motivation to learn, account for older adults' tendency toward positive emotional regulation (Carstensen, 1991, 1992), and enhance perceptions of social norms and behavioral control (Morris & Venkatesh, 2000).

Career development Although it goes beyond traditional training, career development is also increasingly being supported in organizations using Internet-based tools (Crocitto, Sullivan, & Carraher, 2005; Hendrickson, 2003). Age differences in this area have been relatively understudied, but there is some evidence that older adults may not have the same developmental opportunities as their younger counterparts. For instance, Finkelstein, Allen, and Rhoton (2003) showed that older adults receive less mentoring. Older workers are also less likely to receive any form of career counseling from their supervisors (Cleveland & Shore, 1992), they tend to receive lower ratings of promotabilty (Shore et al., 2003), and they have a lower likelihood of actually receiving promotions (Goldberg et al., 2004). SST posits that older adults are less motivated by career advancement (Carstensen, 1991, 1992). However, Zacher and Frese (2011) found that effective use of SOC strategies predicts greater interest in furthering one's career. Age-related stereotyping may also lead to disparate treatment of older adults around career development. T. J. Maurer and colleagues (2007) suggested that employee development is influenced by negative stereotypes, discrimination, and differences in access to developmental opportunities among employees in different age groups. Further research is needed to determine whether supporting career development via use of Internet-based tools increases the detrimental effect of aging on development opportunities, or whether these tools can be leveraged to provide more equitable access to career advancement programs and training for workers of all ages.

Age and virtual teams

As virtual teams are rapidly gaining popularity, considerable research has investigated the antecedents of and barriers to virtual team effectiveness. This research has revealed that the more successful virtual teams are those in which team members have some common ground, loosely coupled work tasks, and readiness for collaboration and adopting collaboration technology (Olson & Olson, 2000). Furthermore, when virtual team members' collaboration frequency matches the style of work (e.g., more interrelated tasks requiring more frequent, scheduled communication), the teams will have higher performance (Rico & Cohen, 2005). It seems that although factors typically related to team performance, such as trust, are important, they are even more important in driving team outcomes when members do not meet face to face (Zigurs, 2003). However, the age composition of virtual teams and the impact this has on team member attitudes and performance remains unresearched (Gilson, Maynard, Young, Vartiainen, & Hakonen, 2015). This area seems particularly ripe for study, given possible differences in comfort levels, especially between age groups, in using social media to manage relationships.

Researchers examining traditional teams have given us some understanding about how age may impact team functioning, especially age disparity or diversity within the team: teams composed of members of varied ages may not perform as effectively as those composed of members closer in age (e.g., Kunze, Boehm, & Bruch, 2011; Liebermann, Wegge, Jungmann, & Schmidt, 2013; Molleman, 2005). Several theories have been applied toward explaining this phenomenon. Social identity theory (SIT; Tajfel, 1974) suggests that people categorize others as being part of their "in-group" or "out-group." Although a number of factors can cause these categorizations, demographic characteristics have been found to be particularly robust. The concept of relational demography suggests that individuals categorize others who are demographically similar to themselves as part of their "in-group" (Riordan, 2000), and display more positive attitudes towards them. Field studies of demographic differences in the workplace, however, have shown mixed results. With regards to age diversity in particular, some studies have reported negative outcomes (Kearney & Gebert, 2009; Kunze et al., 2011; Liebermann et al., 2013; Timmerman, 2000), whereas others show no effect (Horwitz & Horwitz, 2007; Webber & Donahue, 2001) or even positive outcomes (Boehm, Kunze, & Bruch, 2014; Kearney, Gebert, & Voelpel, 2009; Wegge, Roth, Neubach, Schmidt, & Kanfer, 2008). However, no research to date has examined outcomes for age-diverse teams that operate and communicate virtually.

Positive relationships have been reported between age and some components of job performance (Ng & Feldman, 2008) and workplace attitudes (Ng & Feldman, 2010). Together these findings indicate that older workers may be at an advantage in building relationships within a virtual team, as their performance and attitudes could favorably impact other team members. Indeed, SST suggests that older adults gravitate toward forming meaningful social relationships, which may be beneficial in building and maintaining relationships among virtual team members (Carstensen, 1992), and older adults' increased generativity motives (e.g., Kanfer & Ackerman, 2004) may help build effective team relationships in a virtual context. However, we base these suggestions on prior research that has examined age, job performance, and teams – researchers need to consider how these research streams can be combined to deepen our understanding of how age manifests itself in a virtual team setting. In addition, age-related lifespan changes in particular have not yet been applied toward understanding the experience of workers of different ages in virtual teams. For instance, little is known about how virtual team communication tools and processes can be designed with respect to age-related cognitive changes, and whether these tools can optimize older workers' cognitive strengths. The extent to which SOC strategies are used in virtual team settings, and how virtual team communication tools and methods can foster these strategies also needs to be better understood. Finally, how virtual team communication tools and methods can enhance older workers' perceptions of social norms and behavioral control, which influence their attitudes toward and adoption of these tools, also needs further attention.

Age and e-leadership

Virtual teams have been shown to rely more heavily on team member relationships than on the influence of one particular leader (Hoch & Kozlowski, 2014). However, as with all teams, a leader is critical for directing and driving virtual team performance (e.g., Cascio, 2000; Kearney & Gebert, 2009), and effective communication is an important component of this process. Although leaders may be challenged to use Internet-based communication channels effectively (Cascio, 2000; Walvoord, Redden, Elliott, & Coovert, 2008), leadership styles are evident even in virtual communications. Kelloway,

Barling, Kelley, Comtois, and Gatien (2003) conducted a study in which they sent emails to students from a "manager" that were written in such a way as to capture different management styles. The email recipients could distinguish between the different management styles in the emails, and emails drawing on a transformational leadership style in particular led to better perceived justice and greater satisfaction with the manager. Other research on e-leader effectiveness also suggests transformational leadership styles are associated with the greatest success (Grant, 2012; Judge & Piccolo, 2004), and that transformational leadership in particular predicts more successful outcomes for e-leaders than in typical face-to-face management (Purvanova & Bono, 2009; Zigurs, 2003). There is some evidence that transformational leadership behaviors increase with age (Barbuto, Fritz, Matkin, & Marx, 2007), which suggests that older leaders may be particularly effective in a virtual setting. However, little is known about how age-related changes in cognition, for both leaders and followers, impact experiences with and effectiveness of e-leader communication tools and processes. Further research is needed to determine how lifespan changes impact e-leader effectiveness, as well as how e-leader communication tools can be designed to enhance perceptions of social norms and behavioral control for both older leaders and followers.

A well-researched predictor of leadership effectiveness is leader–member exchange (LMX), or the quality of the relationship between a leader and his or her followers (e.g., Bauer & Erdogan, 2016). Little research has directly examined the impact of the age of the leader and follower(s) on LMX (Truxillo & Burlacu, 2016). Although followers' *expectations* of leaders and how this influences perceptions of their effectiveness has been examined, this has rarely been investigated in an e-leadership setting. For instance, although we know followers have more positive impressions of older leaders who display generativity and legacy beliefs, that is, beliefs that their work will be remembered and remain impactful after they are gone (Zacher, Rosing, Henning, & Frese, 2011; Zacher, Rosing, & Frese, 2011), no research has examined the extent to which these characteristics are evident in a virtual setting. Furthermore, age-related stereotypes in the context of e-leadership need to be better understood, particularly given e-leaders' reliance on technology and possible attitudes regarding older adults' capabilities in learning and using technology (e.g., Czaja et al., 2015).

Performance management Employees' performance management is one of the critical HRM responsibilities of a leader. The role of age in using and adopting Internet-based performance management tools has not been researched. However, employee and manager age have been studied in the context of traditional performance measurement and management. Such research has shown that age-related stereotypes influence expectations of how employees will perform, as well as ratings of that performance (Landy & Farr, 1980; Murphy & Cleveland, 1995). Older workers are generally expected to have lower job performance than younger workers (Gordon & Arvey, 2004; Hedge et al., 2006), possibly due to assumptions that older adults have lower ability and motivation (Kite, Stockdale, Whitley, & Johnson, 2005). It is worth noting, however, that studies examining job performance (Ng & Feldman, 2008) and motivation (Inceoglu et al., 2012) do not support these claims. Employee age has also been found to interact with supervisor age, such that younger supervisors tend to give older subordinates lower ratings of promotability (Shore et al., 2003). Further research is needed to determine whether use of Internet-based tools that allow companies to collect and report performance data can alleviate some of these stereotype-driven outcomes.

Although performance feedback continues to be an important component of performance management in organizational practice (Alvero, Bucklin, & Austin, 2001; Wang, Burlacu, Truxillo, James, & Yao, 2015), it is not always effective in eliciting improved performance. Kluger and DeNisi (1996) showed that 40% of feedback events lead to performance decline, which may be due in part to how the feedback is perceived by the receiving employee. Feedback that is perceived as being more credible and useful tends to elicit better outcomes (Brett & Atwater, 2001). Little research has examined feedback reactions and how they differ with employee age, but employees of different ages may desire different things from their feedback experiences. For instance, SST suggests that younger adults will be more concerned with growth and development, while older adults will be more concerned with positive emotions and meaningful relationships. This argument is supported by the findings of a feedback study which showed that older workers had feedback orientations focused on social awareness, while younger workers had feedback orientations focused on utility, or usefulness of the information contained in the feedback (Wang et al., 2015). Furthermore, they found that older workers had stronger positive reactions to feedback that was favorable and delivered in a polite, forthcoming manner, while younger adults had stronger positive reactions to feedback that contained high-quality information. Wang et al.'s (2015) findings imply that age differences in motivation may extend to the performance feedback experience, particularly influencing reactions to the delivery medium. For instance, older workers may find technology-delivered performance appraisal and Internet-based performance metrics less palatable, although this still needs to be empirically examined.

In general, further research is needed to determine the role that cognitive, motivational, and developmental lifespan changes play in the adoption of and experience with e-performance management tools. It is not yet understood how these tools can be developed and positioned to optimize older workers' motivation, or their perceptions of social norms and behavioral control around using these tools. Furthermore, because performance management impacts both leaders and followers, it is important to investigate the age-related factors that influence the acceptance, adoption, and optimization of this technology from the perspective of both groups.

Future Research

As noted throughout this chapter, little research has been conducted on the impact that use of Internet-based work tools has on the motivation, learning, and performance of employees of different ages. Yet the limited research on age-mindful HRM processes that accommodate, build, and develop workers of all ages (e.g., Boehm et al., 2014; Kooij, Jansen, Dikkers, & de Lange, 2014; Kunze et al., 2011; Truxillo et al., 2014; Truxillo et al., 2015), combined with the extensive research on lifespan and societal factors that influence age-related changes in cognition, personality, motivation, and self-efficacy and ability to use technology, provides a framework for research questions that industrial and organizational psychologists should address in the future and which are outlined in Box 20.1. We start by providing examples of broad research questions that may be addressed by combining the existing literatures on age and Internet-based HRM tools, which has not previously been done. Such an exercise will greatly advance our understanding of how these tools can be developed and positioned to meet the needs of employees of all ages, with particular attention paid to the cognitive, motivational, and contextual needs of older workers. We then provide additional research questions that build on existing findings around age and HR processes, which provide a fruitful

Box 20.1 Suggested research questions regarding age-related changes and Internet-based workplace technology use.

Age and e-recruiting
How can age-related changes in cognition, personality, motivation, and lifespan development be taken into account when designing e-recruiting tools and processes?

Research questions examining lifespan development influences on e-recruiting:

- How do older adults utilize SOC strategies in the context of e-recruiting, and how can recruiting tools be designed to maximize the use of these strategies?
- How can organizations factor in older adults' disposition toward positive emotional regulation in an e-recruiting context?
- Given findings around age and technology use, how can organizations design e-recruiting tools and methods that enhance perceptions of social norms and behavioral control?

Research questions applying existing findings around age and recruiting toward the study of Internet-based recruiting tools:

- What are the aspects of e-recruiting processes that minimize perceptions of age discrimination and maximize older adult representation?
- Can elements of e-recruiting be altered to reduce perceived discrimination and effectively communicate positive age diversity climate?
- How do the design features of e-recruiting tool influence older and younger adults' perceptions of an organization?
- How do Internet-based recruiting methods impact older adults' job search experience?
- How can e-recruiting tools and methods be designed to effectively communicate job aspects that are likely to appeal to older adults?

Age and e-selection
How can age-related changes in cognition, personality, motivation, and lifespan development be taken into account when designing e-selection measures and processes?

Research questions examining lifespan development influences on e-selection:

- How can e-selection measures compensate for age differences in learning speed and capacity?
- Are there specific elements that maximize older adults' cognitive strengths while minimizing their deficits?
- How can organizations factor in older adults' disposition toward positive emotional regulation and SOC strategies in e-selection?
- Given findings around age and technology, how can organizations design e-selection measures and methods that enhance perceptions of social norms and behavioral control?

(*Continued*)

Box 20.1 (*Continued*)

Research questions applying existing findings around age and selection toward the study of Internet-based selection tools:

- Do online selection assessments negatively impact the performance of older job applicants relative to traditional paper-and-pencil testing methods?
- Do online selection assessments exacerbate bias against older workers in the hiring process? How can these tools be designed to reduce bias-driven age disparities in hiring?

Age, e-learning, and career development
How can age-related changes in cognition, personality, motivation, and lifespan development be taken into account when designing e-learning courses and processes?

Research questions examining lifespan development influences on e-learning:

- How can e-learning methods compensate for age differences in learning speed and capacity?
- Are there specific elements that maximize older adults' cognitive strengths while minimizing their deficits?
- How can e-learning courses be designed and positioned to maximize older worker training motivation?
- How can organizations factor in older adults' disposition toward positive emotional regulation in e-learning?
- Given findings around age and technology, how can organizations design e-training courses and methods that enhance perceptions of social norms and behavioral control?

Research questions applying existing findings around age and learning toward the study of Internet-based training and development tools:

- How can organizational support be developed and communicated in an e-learning context to ensure older workers feel self-efficacious, both around technology use and around training capability?
- How do Internet-based career development tools impact the development experience of older workers?

Age and virtual teams
How can age-related changes in cognition, personality, motivation, and lifespan development be taken into account when designing virtual team communication tools and processes?

Research questions examining lifespan development influences on virtual teams:

- How do the motivations and preferences of older adults (e.g., forming meaningful social relationships, generativity) manifest themselves in a virtual team setting?
- Given age-related changes in cognition, what Internet-based team communication tools and processes meet the needs of employees of all ages?
- Are there elements of virtual team communication tools and processes that leverage older workers' cognitive strengths?

(*Continued*)

Box 20.1 (*Continued*)

- How do older workers utilize SOC strategies within virtual teams?
- Given the positive outcomes that result from greater use of SOC strategies, are there design elements of virtual team communication tools that will promote the use of these strategies for older adults?
- Given findings around age and technology, how can organizations design virtual team communication tools and processes that enhance perceptions of social norms and behavioral control?

Research questions applying existing findings around age and teams toward the study of virtual team functioning and communication:

- How does age diversity impact virtual team functioning?

Age, e-leadership, and performance management
How can age-related changes in cognition, personality, motivation, and lifespan development be taken into account when designing e-leader communication and e-performance management tools and processes?

Research questions examining lifespan development influences on e-leadership and e-performance management:

- Given age-related changes in cognition, how can Internet-based communication tools and processes be designed to meet the needs of leaders and followers of all ages?
- Given findings around age and technology, how can organizations design communication tools and methods that enhance perceptions of social norms and behavioral control, from the perspectives of both e-leaders and e-followers?
- How can e-performance management be designed and positioned to maximize older worker motivation?
- Given findings around age and technology, how can organizations design e-performance management tools and processes that enhance perceptions of social norms and behavioral control, both for leaders and followers?

Research questions applying existing findings around age and leadership toward the study of e-leader effectiveness, communication, and e-performance management capabilities:

- How can older leaders effectively communicate aspects of their leadership style, such as generativity and legacy beliefs, via virtual communication channels?
- How do age-related stereotypes manifest themselves in the context of e-leadership? What role does communication technology play in these stereotypes?
- How do Internet-based performance measurement tools impact performance ratings that may be age stereotype-driven?
- Given older adults' preferences regarding meaningful, positive social interactions and the fact that Internet-based performance management may be relatively impersonal, what are the design components of an e-performance management system that can optimize older adults' motivation to attend to and incorporate feedback?

framework for understanding how age manifests itself in the workplace but which have not been extended to the use of Internet-based tools that support organizations in recruiting, training, and managing employees.

Conclusion

The use of Internet-based tools supporting HR processes is solving numerous organizational challenges around managing costs, building efficiencies, and scaling effective workforce management. The tools have rapidly gained popularity (Bhatt, 2015; Cascio, 2000; Marler & Fisher, 2013) and will likely see increased use as the nature of work becomes more global, complex, and technologically based. Although these tools carry the expected challenges of end-user training and change management as with any new HRM process (Caldwell, 2001; De Leede & Looise, 2005; Voermans & van Veldhoven, 2007), there is reason to believe employees of different ages (particularly those who are older) may be differentially impacted by the use of these tools. Changes in cognition, personality, motivation, and socio-emotional preferences may all influence how these tools are perceived and to what extent they are effectively adopted. Societal context and age-related expectations and stereotypes may have a role as well. Given the changing demographic trends in industrialized nations with growing prevalence of older workers at all organizational levels, these impacts need to be better understood. Industrial and organizational and work psychologists are well positioned to lead future research in this area, providing organizations with guidance around how to use cost-saving yet effective technologies to manage their demographically diverse workforces.

References

Abbasi, M. S., Tarhini, A., Hassouna, M., & Shah, F. (2015). Social, organizational, demography and individuals' technology acceptance behavior: A conceptual model. *European Scientific Journal, 11*, 48–76.

Adams, N., Stubbs, D., & Woods, V. (2005). Psychological barriers to Internet usage among older adults in the UK. *Medical Informatics and the Internet in Medicine, 30*, 3–17.

Allen, D. G., Mahto, R. V., & Otondo, R. F. (2007). Web-based recruitment: Effects of information, organizational brand, and attitudes toward a Web site on applicant attraction. *Journal of Applied Psychology, 92*, 1696–1708.

Alvero, A. M., Bucklin, B. R., & Austin, J. (2001). An objective review of the effectiveness and essential characteristics of performance feedback in organizational settings (1985–1998). *Journal of Organizational Behavior Management, 21*, 3–29.

Anderson, N. (2003). Applicant and recruiter reactions to new technology in selection: A critical review and agenda for future research. *International Journal of Selection and Assessment, 11*, 121–136.

Armstrong-Stassen, M., & Schlosser, F. (2011). Perceived organizational membership and the retention of older workers. *Journal of Organizational Behavior, 32*, 319–344.

Avery, D. R. (2003). Reactions to diversity in recruitment advertising – are differences black and white? *Journal of Applied Psychology, 88*, 672–679.

Baddeley, A. (1992). Working memory: The interface between memory and cognition. *Journal of Cognitive Neuroscience, 4*, 281–288.

Bal, A. C., Reiss, A. E. B., Rudolph, C. W., & Baltes, B. B. (2011). Examining positive and negative perceptions of older workers: A meta-analysis. *Journals of Gerontology, Series B: Psychological Sciences and Social Sciences, 66*, 687–698.

Baltes, P. B. (1987). Theoretical propositions of life-span developmental psychology: On the dynamics between growth and decline. *Developmental Psychology, 23*, 611–626.

Baltes, P. B., & Baltes, M. M. (1990). Psychological perspectives on successful aging: The model of selective optimization with compensation. In P. B. Baltes & M. M. Baltes (Eds.), *Successful aging: Perspectives from the behavioral sciences* (pp. 1–34), New York, NY: Cambridge University Press.

Barbuto Jr, J. E., Fritz, S. M., Matkin, G. S., & Marx, D. B. (2007). Effects of gender, education, and age upon leaders' use of influence tactics and full range leadership behaviors. *Sex Roles, 56*, 71–83.

Barrick, M. R., & Mount, M. K. (1991). The big five personality dimensions and job performance: A meta-analysis. *Personnel Psychology, 44*, 1–26.

Bauer, T. N., & Erdogan, B. (2016). *The Oxford handbook of leader-member exchange.* New York, NY: Oxford University Press.

Bauer, T. N., Truxillo, D. M., Paronto, M., Weekley, J., & Campion, M. A. (2004). Applicant reactions to different selection technology: Face-to-face, interactive voice response, and computer-assisted telephone screening interviews. *International Journal of Selection and Assessment, 12*, 135–148.

Bauer, T. N., Truxillo, D. M., Tucker, J. S., Weathers, V., Bertolino, M., Erdogan, B., & Campion, M. A. (2006). Selection in the information age: The impact of privacy concerns and computer experience on applicant reactions. *Journal of Management, 32*, 601–621.

Beier, M. E. (2008). Age and learning in organizations. In G. P. Hodgkinson & J. K. Ford (Eds.), *International Review of Industrial and Organizational Psychology* (pp. 83–105). Chichester: John Wiley & Sons, Ltd.

Beier, M. E. (2014, May). *Motivation for training and development: Considerations for an aging workforce.* Paper presented at the Annual Conference of the Society for Industrial and Organizational Psychology, Honolulu, HI, USA.

Beier, M. E., & Ackerman, P. L. (2005). Age, ability, and the role of knowledge on the acquisition of new domain knowledge: Promising results in a real-world learning environment. *Psychology and Aging, 20*, 341–355.

Beier, M. E., Teachout, M. S., & Cox, C. B. (2012). The training and development of an aging workforce. In J. W. Hedge & W. C. Borman (Eds.), *The Oxford handbook of work and aging* (pp. 436–453), New York, NY: Oxford University Press.

Benson, J., & Brown, M. (2011). Generations at work: Are there differences and do they matter? *International Journal of Human Resource Management, 22*, 1843–1865.

Bertolino, M., Truxillo, D. M., & Fraccaroli, F. (2011). Age as moderator of the relationship of proactive personality with training motivation, perceived development from training, and training behavioral intentions. *Journal of Organizational Behavior, 32*, 248–263.

Bertolino, M., Truxillo, D. M., & Fraccaroli, F. (2013). Age effects on perceived personality and job performance. *Journal of Managerial Psychology, 7/8*, 867–885.

Bhatt, R. (2015). E-HRM review and implications. *Journal of Advances in Business Management, 1*, 387–389.

Bliedorn, W., Klimstra, T. A., Dennissen, J. J., Rentfrow, P. J., Potter, J., & Gosling, S. D. (2013). Personality maturation around the world: A cross-cultural examination of social-investment theory. *Psychological Science, 24*, 30–40.

Boehm, S. A., & Kunze, F. (2015). Age diversity and age climate in the workplace. In P. M. Bal, D. T. A. M. Kooij, & D. M. Rousseau (Eds.), *Aging workers and the employee-employer relationship* (pp. 33–55). Switzerland: Springer International Publishing.

Boehm, S. A., Kunze, F., & Bruch, H. (2014). Spotlight on age-diversity climate: The impact of age-inclusive HR practices on firm-level outcomes. *Personnel Psychology, 67*, 667–704.

Borman, W. C., Hanson, M. A., & Hedge, J. W. (1997). Personnel selection. *Annual Review of Psychology, 48*, 299–337.

Brett, J. F., & Atwater, L. E. (2001). 360 feedback: Accuracy, reactions, and perceptions of usefulness. *Journal of Applied Psychology, 86*, 930–942.

Cadiz, D. M., Truxillo, D. M., & Fraccaroli, F. (2015). What are the benefits of focusing on generation-based differences and at what cost? *Industrial and Organizational Psychology, 8*, 356–362.

Caldwell, R. (2001). Champions, adapters, consultants and synergists: The new change agents in HRM. *Human Resource Management Journal, 11*, 39–52.

Callahan, J. S., Kiker, D. S., & Cross, T. (2003). Does age matter? A meta-analysis of the effects of training method on older learner training performance. *Journal of Management, 29*, 663–680.

Carstensen, L. L. (1991). Socioemotional selectivity theory: Social activity in life-span context. *Annual Review of Gerontology and Geriatrics, 17*, 195–217.

Carstensen, L. L. (1992). Social and emotional patterns in adulthood: Support for socioemotional selectivity theory. *Psychology and Aging, 7*, 331–338.

Carter, M., & Beier, M. E. 2010. The effectiveness of error management training with working aged adults. *Personnel Psychology, 63*, 641–675.

Cascio, W. F. (2000). Managing a virtual workplace. *Academy of Management Executive, 14*, 81–90.

Chan, S., & Stevens, A. H. (2001). Job loss and employment patterns of older workers. *Journal of Labor Economics, 19*, 484–521.

Chang, J., McAllister, C., & McCaslin, R. (2015). Correlates of, and barriers to, internet use among older adults. *Journal of Gerontological Social Work, 58*, 66–85.

Chapman, D., & Gödöllei, A. F. (2017). E-recruiting: Using technology to attract job applicants. In G. Hertel, D. L. Stone, R. D. Johnson, & J. Passmore (Eds.). *The Wiley Blackwell handbook of the psychology of the internet at work* (pp. 213–230). Chichester: John Wiley & Sons, Ltd.

Charles, S. T., & Carstensen, L. L. (2009). Socioemotional selectivity theory. In H. Reis & S. Sprecher (Eds.), *Encyclopedia of human relationships* (pp. 1578–1581), Thousand Oaks, CA: SAGE Publications.

Charness, N. (2006). Work, older workers, and technology. *Generations, 2*, 25–30.

Chow, P. I., & Roberts, B. W. (2014). Examining the relationship between changes in personality and changes in depression. *Journal of Research in Psychology, 51*, 38–46.

Cleveland, J. N., & Shore, L. M. (1992). Self- and supervisory perspectives on age and work attitudes and performance. *Journal of Applied Psychology, 77*, 469–484.

Cober, R. T., Brown, D. J., Keeping, L. M., & Levy, P. E. (2004). Recruitment on the Net: How do organizational web site characteristics influence applicant attraction? *Journal of Management, 30*, 623–646.

Coile, C., Milligan, K. S., & Wise, D. A. (2016). Social security and retirement programs around the world: The capacity to work at older ages – introduction and summary. *NBER Working Paper No. 21939.* Retrieved April 7, 2017, from www.nber.org/papers/w21939

Colquitt, J. A., LePine, J. A., & Noe, R. A. (2000). Toward an integrative theory of training motivation: A meta-analytic path analysis of 20 years of research. *Journal of Applied Psychology, 85*, 678–707.

Costa Jr, P. T., McCrae, R. R., Zonderman, A. B., Barbano, H. E., Lebowitz, B., & Larson, D. M. (1986). Cross-sectional studies of personality in a national sample: II. Stability in neuroticism, extraversion, and openness. *Psychology and Aging, 1*, 144–149.

Costanza, D. P., & Finkelstein, L. M. (2015). Generationally based differences in the workplace: Is there a there there? *Industrial and Organizational Psychology, 8*, 308–323.

Crocitto, M. M., Sullivan, S. E., & Carraher, S. M. (2005). Global mentoring as a means of career development and knowledge creation: A learning-based framework and agenda for future research. *Career Development International, 10*, 522–535.

Crumpacker, M., & Crumpacker, J. M. (2007). Succession planning and generational stereotypes: Should HR consider age-based values and attitudes as a relevant factor or a passing fad? *Personnel Management, 36*, 349–369.

Curtis, R. G., Windsor, T. D., & Soubelet, A. (2015). The relationship between big-5 personality traits and cognitive ability in older adults – a review. *Aging, Neuropsychology, and Cognition: A Journal on Normal and Dysfunctional Development, 22*, 42–71.

Czaja, S. J., & Lee, C. C. (2003). The impact of the Internet on older adults. In N. Charness & K. W. Schaie (Eds.), *Impact of technology on successful aging* (pp. 113–133). New York, NY: Springer.

Czaja, S. J., Sharit, J., Charness, N., & Schmidt, A. C. (2015). The implications of changes in job demands for the continued and future employment of older workers. In L. M. Finkelstein, D. M. Truxillo, F. Fraccaroli, & R. Kanfer (Eds.), *Facing the challenges of a multi-age workforce: A use-inspired approach* (pp. 159–179). New York, NY: Taylor & Francis.

Davis, F. D. (1989). Perceived usefulness, perceived ease of use, and user acceptance of information technology. *MIS Quarterly, 13*, 319–339.

Davoudi, S. M., & Fartash, K. (2012). Electronic human resource management: New avenues which leads to organizational success. *Spectrum: A Journal of Multidisciplinary Research, 1*, 75–87.

De Leede, J., & Looise, J. K. (2005). Innovation and HRM: Towards an integrated framework. *Creativity and Innovation Management, 14*, 108–117.

Derouin, R. E., Fritzsche, B. A., & Salas, E. (2005). E-learning in organizations. *Journal of Management, 31*, 920–940.

Doverspike, D., Taylor, M. A., Shultz, K. S., & McKay, P. F. (2000). Responding to the challenge of a changing workforce: Recruiting nontraditional demographic groups. *Public Personnel Management, 29*, 445–459.

Egri, C. P., & Ralston, D. A. (2004). Generation cohorts and personal values: A comparison of China and the United States. *Organization Science, 15*, 210–220.

Erdogan, B., Bauer, T. N., Peiro, J. M., & Truxillo, D. M. (2011). Overqualified employees: Making the best of a potentially bad situation for individuals and organizations. *Industrial and Organizational Psychology, 4*, 215–232.

European Agency for Safety and Health at Work (2014). OSH in the context of an ageing workforce. Retrieved April 7, 2017, from https://osha.europa.eu/en/themes/osh-management-context-ageing-workforce

Finkelstein, L. M., Allen, T. D., & Rhoton, L. A. (2003). An examination of the role of age in mentoring relationships. *Group and Organization Management, 28*, 249–281.

Finkelstein, L. M., King, E. B., & Voyles, E. C. (2015). Age metastereotyping and cross-age workplace interactions: A meta view of age stereotypes at work. *Work, Aging and Retirement, 1*, 26–40.

Fishbein, M. (1979). A theory of reasoned action: Some applications and implications. *Nebraska Symposium on Motivation, 27*, 65–116.

Freund, A. M., & Baltes, P. B. (1998). Selection, optimization, and compensation as strategies of life management: Correlations with subjective indicators of successful aging. *Psychology and Aging, 13*, 531–543.

Freund, A. M., & Baltes, P. B. (2002). Life-management strategies of selection, optimization, and compensation: Measurement by self-report and construct validity. *Journal of Personality and Social Psychology, 82*, 642–662.

Giancola, F. (2006). The generation gap: More myth than reality. *Human Resource Planning, 29*, 32–37.

Giles, J., Wang, D., & Cai, W. (2011). The labor supply and retirement behavior of China's older workers and elderly in comparative perspective. *World Bank Policy Research Working Paper Series*. Retrieved April 7, 2017, from http://ssrn.com/abstract=1948991

Gilson, L. L., Maynard, M. T., Young, N. C. J., Vartiainen, M., & Hakonen, M. (2015). Virtual teams research: 10 years, 10 themes, and 10 opportunities. *Journal of Management, 41*, 1313–1337.

Goldberg, C. B., Finkelstein, L. M., Perry, E. L., & Konrad, A. M. (2004). Job and industry fit: The effects of age and gender matches on career progress outcomes. *Journal of Organizational Behavior, 25*, 807–829.

Goldberg, C. B., Perry, E. L., Finkelstein, L. M., & Shull, A. (2013). Antecedents and outcomes of targeting older applicants in recruitment. *European Journal of Work and Organizational Psychology, 22*, 265–278.

Goldstein, I. L., & Ford, J. K. (2002). *Training in organizations* (4th ed.). Belmont, CA: Wadsworth.

Gordon, R. A., & Arvey, R. D. (2004). Age bias in laboratory and field settings: A meta-analytic investigation. *Journal of Applied Psychology, 34*, 468–492.

Grant, A. M. (2012). Leading with meaning: Beneficiary contact, prosocial impact, and the performance effects of transformational leadership. *Academy of Management Journal, 55*, 458–476.

Gully, S. M., Phillips, J. M., & Kim, M. S. (2014). Strategic recruitment: A multilevel perspective. In K. Y. T. Yu & D. M. Cable (Eds.), *The Oxford handbook of recruitment* (pp. 161–183). New York, NY: Oxford University Press.

Hedge, J. W. (2008). Strategic human resource management and the older worker. *Journal of Workplace Behavioral Health, 23*, 109–123.

Hedge, J. W., Borman, W. C., & Lammlein, S. E. (2006). *The aging workforce: Realities, myths, and implications for organizations*. Washington, DC: American Psychological Association.

Heggestad, E. D., & Andrew, A. M. (2012). Aging, personality, and work attitudes. In J. W. Hedge & W. C. Borman (Eds.), *The Oxford handbook of work and aging* (p. 256–279). New York, NY: Oxford University Press.

Hendrickson, A. (2003). Human resource information systems: Backbone technology for contemporary human resources. *Journal of Labor Research, 24*, 381–394.

Hertel, G., Thielgen, M., Rauschenbach, C., Grube, A., Stamov-Roßnagel, C., & Krumm, C. (2013). Age differences in motivation and stress at work. In C. Schlick, E. Frieling, and J. Wegge (eds.) *Age-differentiated work systems* (pp. 113–147). Berlin: Springer.

Hertel, G., & Zacher, H. (in press). Managing the aging workforce. In C. Viswesvaran, N. Anderson, D. S. Ones, & H. K. Sinangil (Eds.), *The SAGE handbook of industrial, work, and organizational psychology* (2nd ed.), Vol. *3*.

Hoch, J. E., & Kozlowski, S. W. J. (2014). Leading virtual teams: Hierarchical leadership, structural supports, and shared team leadership. *Journal of Applied Psychology, 99*, 390–403.

Horwitz, S. K., & Horwitz, I. B. (2007). The effects of team diversity on team outcomes: A meta-analytic review of team demography. *Journal of Management, 33*, 987–1015.

Hughes, M. L., Geraci, L., & De Forrest, R. L. (2013). Aging 5 years in 5 minutes: The effect of taking a memory test on older adults' subjective age. *Psychological Science, 24*, 2481–2488.

Hurtz, G. M., & Donovan, J. J. (2000). Personality and job performance: The big five revisited. *Journal of Applied Psychology, 85*, 869–879.

Inceoglu, I., Segers, J., & Bartram, D. (2012). Age-related differences in work motivation. *Journal of Occupational and Organizational Psychology, 85*, 300–329.

Johnson, R., & Brown, K. (2017). E-learning. In G. Hertel, D. L. Stone, R. D. Johnson, & J. Passmore (Eds.), *The Wiley Blackwell handbook of the psychology of the internet at work* (pp. 369–400). Chichester: John Wiley & Sons, Ltd.

Judge, T. A., & Piccolo, R. F. (2004). Transformational and transactional leadership: A meta-analytic test of their relative validity. *Journal of Applied Psychology, 89*, 755–768.

Kahai, S., Avolio, B. J., & Sosik, J. (2017). E-leadership. In G. Hertel, D. L. Stone, R. D. Johnson, & J. Passmore (Eds.), *The Wiley Blackwell handbook of the psychology of the internet at work* (pp. 285–314). Chichester: John Wiley & Sons, Ltd.

Kanfer, R. (1990). Motivation theory and industrial and organizational psychology. In M. D. Dunette & L. Hough (Eds.), *Handbook of industrial and organizational psychology, Vol. 1* (pp. 75–170). Palo Alto, CA: Consulting Psychologists Press.

Kanfer, R., & Ackerman, P. L. (2004). Aging, adult development, and work motivation. *Academy of Management Review, 29*, 440–458.

Kanfer, R., Wanberg, C. R., & Kantrowitz, T. M. (2001). Job search and employment: A personality-motivational analysis and meta-analytic review. *Journal of Applied Psychology, 86*, 837–855.

Kearney, E., & Gebert, D. (2009). Managing diversity and enhancing team outcomes: The promise of transformational leadership. *Journal of Applied Psychology, 94*, 77–89.

Kearney, E., Gebert, D., & Voelpel, S. C. (2009). When and how diversity benefits teams: The importance of team members' need for cognition. *Academy of Management Journal, 52*, 581–598.

Kelloway, E. K., Barling, J., Kelley, E., Comtois, J., & Gatien, B. (2003). Remote transformational leadership. *Leadership & Organization Development Journal, 24*, 163–171.

Kennedy, K. M., Patridge, T., & Raz, N. (2008). Age-related differences in acquisition of perceptual-motor skills: Working memory as a mediator. *Aging, Neuropsychology, and Cognition: A Journal of Normal and Dysfunctional Development, 15*, 165–183.

Khalik, S., & Wei, A. C. (2016). More in Singapore remaining in workforce past 65. *Singapore Straits Times*. March 7, 2016.

Kim, S., & Feldman, D. C. (2000). Working in retirement: The antecedents of bridge employment and its consequences for quality of life in retirement. *Academy of Management Journal, 43*, 1195–1210.

Kite, M. E., Stockdale, G. D., Whitley, B. E., & Johnson, B. T. (2005). Attitudes toward younger and older adults: An updated meta-analytic review. *Journal of Social Issues, 61*, 241–266.

Klein, R. M., Dilchert, S., Ones, D. S., & Dages, K. D. (2015). Cognitive predictors and age-based adverse impact among business executives. *Journal of Applied Psychology, 100*, 1497–1510.

Kluger, A. N., & DeNisi, A. (1996). The effects of feedback interventions on performance: A historical review, a meta-analysis, and a preliminary feedback intervention theory. *Psychological Bulletin, 119*, 254–284.

Kooij, D. T. (2015). Successful aging at work: The active role of employees. *Work, Aging and Retirement, 1*, 309–319.

Kooij, D. T. A. M., de Lange, A. H., Jansen, P. G. W., Kanfer, R., and Dikkers, J. S. E. (2011). Age and work-related motives: Results of a meta-analysis. *Journal of Organizational Behavior, 32*, 197–225.

Kooij, D. T. A. M., Jansen, P. G. W., Dikkers, J. S. E., & de Lange, A. H. (2014). Managing aging workers: A mixed methods study on bundles of HR practices for aging workers. *International Journal of Human Resource Management, 25*, 2192–2212.

Kraimer, M. L., Takeuchi, R., & Frese, M. (2014). The global context and people at work: Special issue introduction. *Personnel Psychology, 67*, 5–21.

Kubeck, J. E., Delp, N. D., Haslett, T. K., & McDaniel, M. A. (1996). Does job-related training performance decline with age? *Psychology and Aging, 11*, 92–107.

Kunze, F., Boehm, S. A., & Bruch, H. (2011). Age diversity, age discrimination climate and performance consequences – a cross organizational study. *Journal of Organizational Behavior, 32*, 264–290.

Kupperschmidt, B. R. (2000). Multigeneration employees: Strategies for effective management. *Health Care Manager, 19*, 65–76.

Landy, F. J., & Farr, J. L. (1980). Performance rating. *Psychological Bulletin, 87*, 72–107.

Lee, I. (2005). The evolution of e-recruiting: A content analysis of Fortune 100 career web sites. *Journal of Commerce in Organizations, 3*, 57–68.

Liebermann, S. C., Wegge, J., Jungmann, F., & Schmidt, K. (2013). Age diversity and individual team member health: The moderating role of age and age stereotypes. *Journal of Occupational and Organizational Psychology, 86*, 184–202.

Leibold, M., & Voelpel, S. (2006). *Managing the aging workforce: Challenges and solutions*. Germany: Wiley and Publicis Corporate Publishing.

Liker, J. K., & Sindi, A. A. (1997). User acceptance of expert systems: A test of the theory of reasoned action. *Journal of Engineering and Technology Management, 14*, 147–173.

Liu, S., Courtenay, B. C., & Valentine, T. (2011). Managing older worker training: A literature review and conceptual framework. *Educational Gerontology, 37*, 1040–1062.

Liu, S., Huang, J. L., & Wang, M. (2014). Effectiveness of job search interventions: A meta-analytic review. *Psychological Bulletin, 140*, 1009–1041.

London, M. (2003). *Job feedback: Giving, seeking, and using feedback for performance improvement*. Mahwah, NJ: Lawrence Earlbaum.

Magidson, J. F., Roberts, B. W., Collado-Rodriguez, A., & Lejuez, C. W. (2014). Theory-driven intervention for changing personality: Expectancy value theory, behavioral activation, and conscientiousness. *Developmental Psychology, 50*, 1442–1450.

Major, D. A., Turner, J. E., & Fletcher, T. D. (2006). Linking proactive personality and the Big Five to motivation to learn and development activity. *Journal of Applied Psychology, 91*, 927–935.

Marler, J. H., & Fisher, S. L. (2013). An evidence-based review of e-HRM and strategic human resource management. *Human Resource Management Review, 23*, 18–36.

Martins, L. L., Gilson, L. L., & Maynard, M. T. (2004). Virtual teams: What do we know and where do we go from here? *Journal of Management, 30*, 805–835.

Maurer, S. D., & Liu, Y. (2007). Developing effective e-recruiting websites: Insights for managers from marketers. *Business Horizons, 50*, 305–314.

Maurer, T. J. (2001). Career-relevant learning and development, worker age, and beliefs about self-efficacy for development. *Journal of Management, 27*, 123–140.

Maurer, T. J., Shultz, K. S., & Adams, G. A. (2007). Employee development and training issues related to the aging workforce. In K. Schultz & G. Adams, *Aging and work in the 21st century* (pp. 163–178). Mahwah, NJ: Erlbaum.

Maurer, T. J., Weiss, E. M., & Barbeitte, F. G. (2003). A model of involvement in work-related learning and development activity: The effects of individual, situational, motivational, and age variables. *Journal of Applied Psychology, 88*, 707–724.

Mayer, R. E. (2005). *The Cambridge handbook of multimedia learning*. Cambridge, MA: Cambridge University Press.

Maynard, M. T., Gilson, L. L., Jones Young, N. C., & Vartiainen, M. (2017). Virtual teams. In G. Hertel, D. L. Stone, R. D. Johnson, & J. Passmore (Eds.). *The Wiley Blackwell handbook of the psychology of the internet at work* (pp. 315–346). Chichester: John Wiley & Sons, Ltd.

McAdams, D. P. (2001). Generativity in midlife. In M. E. Lachman (Ed.), *Handbook of midlife development: Wiley series on adulthood and aging* (pp. 395–443). Hoboken, NJ: John Wiley & Sons, Inc.

McAdams, D. P., de St Aubin, E., & Logan, R. L. (1993). Generativity among young, midlife, and older adults. *Psychology an Aging, 8*, 221–230.

McCausland, T. C., & King, E. B. (2014, May). *Chronological age and training outcomes: Examining psychological processes and cognitive ability*. Paper presented at the Annual Conference of the Society for Industrial and Organizational Psychology, Honolulu, HI, USA.

McCrae, R. R., Costa, P. T., de Lima, M. P., Simoes, A., Ostendorf, F., Angleitner, A., ... Piedmont, R. L. (1999). Age differences in personality across the adult life span: Parallels in five cultures. *Developmental Psychology, 35*, 466–477.

Mitzner, T. L., Boron, J. B., Fausset, C. B., Adams, A. E., Charness, N., Czaja, S. J., ... Sharit, J. (2010). Older adults talk technology: Technology usage and attitudes. *Computers in Human Behavior, 26*, 1710–1721.

Molleman, E. (2005). Diversity in demographic characteristics, abilities and personality traits: Do faultlines affect team functioning? *Group Decision and Negotiation, 14*, 173–193.

Morris, M. G., & Venkatesh, V. (2000). Age differences in technology adoption decisions: Implications for a changing work force. *Personnel Psychology, 53*, 375–403.

Muller, A., Heiden, B., Herbig, B., Poppe, F., & Angerer, P. (2015). Improving well-being at work: A randomized controlled intervention based on selection, optimization, and compensation. *Journal of Occupational Health Psychology, 21*, 169–181.

Murphy, K. R., & Cleveland, J. (1995). *Understanding performance appraisal: Social, organizational, and goal-based perspectives*. Thousand Oaks, CA: SAGE.

Myers, K. K., & Sadaghiani, K. (2010). Millennials in the workplace: A communication perspective on millennials' organizational relationships and performance. *Journal of Business and Psychology, 25*, 225–238.

Ng, T. W. H., & Feldman, D. C. (2008). The relationship of age to ten dimensions of job performance. *Journal of Applied Psychology, 93*, 392–423.

Ng, T. W. H., & Feldman, D. C. (2010). The relationships of age with job attitudes: A meta-analysis. *Personnel Psychology, 63*, 677–718.

Ng, T. W. H., & Feldman, D. C. (2012). Evaluating six common stereotypes about older workers with meta-analytic data. *Personnel Psychology, 65*, 821–858.

Ng, T. W. H., & Feldman, D. C. (2013). A meta-analysis of the relationships of age and tenure with innovation related behaviour. *Journal of Occupational and Organizational Psychology, 86*, 585–616.

Ng, T. W. H., & Feldman, D. C. (2014). The moderating effects of age in the relationships of job autonomy to work outcomes. *Work, Aging and Retirement*, DOI:https://dx.doi.org/10.1093/workar/wau003.

Olson, G. M., & Olson, J. S. (2000). Distance matters. *Journal of Human-Computer Interaction, 15*, 139–178.

Ouwehand, C., de Ridder, D. T. D., & Bensing, J. M. (2007). A review of successful aging models: Proposing proactive coping as an important additional strategy. *Clinical Psychology Review, 27*, 873–884.

Parida, S. K., & Nayak, S. (2012). The effectiveness of electronic human resource management (e-HRM) with special reference to National Thermal Power Corporation. *Training & Development Journal, 3*, 36–42.

Park, D. C., Lodi-Smith, J., Drew, L., Haber, S., Hebrank, A., Bischof, G. N., & Aamodt, W. (2014). The impact of sustained engagement on cognitive function in older adults: The Synapse Project. *Psychological Science, 25*, 103–112.

Parry, E., & Urwin, P. (2011). Generational differences in work values: A review of theory and evidence. *International Journal of Management Reviews, 13*, 79–96.

Perry, E. L., Kulik, C. T., & Bourhis, A. C. (1996). Moderating effects of personal and contextual factors in age discrimination. *Journal of Applied Psychology, 81*, 628–647.

Pew Research Center (2014, April). Internet Project Survey: Older adults and technology use. Retrieved April 7, 2017, from www.pewinternet.org/2014/04/03/older-adults-and-technology-use/

Porter, C. E., & Donthu, N. (2006). Using the technology acceptance model to explain how attitudes determine Internet usage: The role of perceived access barriers and demographics. *Journal of Business Research, 59*, 999–1007.

Posthuma, R. A., & Campion, M. A. (2009). Age stereotypes in the workplace: Common stereotypes, moderators, and future research directions. *Journal of Management, 35*, 158–188.

PricewaterhouseCoopers (2011). Millennials at work: Reshaping the workplace. Retrieved April 7, 2017, from https://www.pwc.com.

Purvanova, R. K., & Bono, J. E. (2009). Transformational leadership in context: Face-to-face and virtual teams. *The Leadership Quarterly, 20*, 343–357.

Rau, B. L., & Adams, G. A. (2005). Attracting retirees to apply: Desired organizational characteristics of bridge employment. *Journal of Organizational Behavior, 26*, 649–660.

Riach, K. (2007). "Othering" older worker identity in recruitment. *Human Relations, 60*, 1701–1726.

Richardson, R., Webb, J., Webber, L., & Smith, K. (2013). Age discrimination in the evaluation of job applicants. *Journal of Applied Social Psychology, 43*, 35–44.

Rico, R., & Cohen, S. G. (2005). Effects of task interdependence and type of communication on performance in virtual teams. *Journal of Managerial Psychology, 20*, 261–274.

Riordan, C. M. (2000). Relational demography within groups: Past developments, contradictions, and new directions. *Research in Personnel and Human Resources Management, 19*, 131–174.

Rizzuto, T. E. (2011). Age and technology innovation in the workplace: Does work context matter? *Computers in Human Behavior, 27*, 1612–1629.

Roberts, B. W., & DelVecchio, W. F. (2000). The rank-order consistency of personality traits from childhood to old age: A quantitative review of longitudinal studies. *Psychological Bulletin, 126*, 3–25.

Roberts, B. W., Walton, K. E., & Viechtbauer, W. (2006). Patterns of mean-level change in personality traits across the life course: A meta-analysis of longitudinal studies. *Psychological Bulletin, 132*, 1–25.

Salgado, J. F., & Moscoso, S. (2003). Internet-based personality testing: Equivalence of measures and assesses' perceptions and reactions. *International Journal of Selection and Assessment, 11*, 194–205.

Salmela-Aro, K., Mutanen, P., & Vuori, J. (2012). Promoting career preparedness and intrinsic work-goal motivation: RCT intervention. *Journal of Vocational Behavior, 80*, 67–75.

Salthouse, T. A. (1995). Differential age-related influences on memory for verbal-symbolic information and visual-spatial information? *The Journals of Gerontology, Series B, 50B*, 193–201.

Salthouse, T. A. (2011). Consequences of age-related cognitive declines. *Annual Review of Psychology, 63*, 5.1–5.26.

Shachaf, P. (2008). Cultural diversity and information and communication technology impacts on global virtual teams: An exploratory study. *Information & Management, 45*, 131–142.

Sharit, J., Hernandez, M. A., Czaja, S. J., & Pirolli, P. (2008). Investigating the roles of knowledge and cognitive abilities in older adult information seeking on the web. *ACM Transactions on Computer-Human Interaction, 15*, 3.1–3.25.

Shiu, E., Hassan, L. M., & Parry, S. (2015). The moderating effects of national age stereotyping on the relationships between job satisfaction and its determinants: A study of older workers across 26 countries. *British Journal of Management, 26*, 255–272.

Shore, L. M., Cleveland, J. N., & Goldberg, C. B. (2003). Work attitudes and decisions as a function of manager age and employee age. *Journal of Applied Psychology, 88*, 529–537.

Singh, P., & Finn, D. (2003). The effects of information technology on recruitment. *Journal of Labor Research, 24*, 395–408.

Smola, K. W., & Sutton, C. D. (2002). Generational differences: Revisiting generational work values for the new millennium. *Journal of Organizational Behavior, 23*, 363–382.

Soto, C. J. (2015). Is happiness good for your personality? Concurrent and prospective relations of the big five with subjective well-being. *Journal of Personality, 83*, 45–55.

Statista Dossier. (2014). Distribution of Internet users in Europe as of November 2014, by age group. Retrieved April 7, 2017, from www.statista.com/statistics/319905/age-distribution-of-internet-users-europe/

Stone, D. L., Deadrick, D. L., Lukaszewski, K. M., & Johnson, R. (2015). The influence of technology on the future of human resource management. *Human Resource Review, 25*, 216–231.

Stone, D. L., & Dulebohn, J. H. (2013). Emerging issues in theory and research on electronic human resource management (eHRM). *Human Resource Management Review, 23*, 1–5.

Strohmeier, S. (2009). Concepts of e-HRM consequences: A categorisation, review and suggestion. *International Journal of Human Resource Management, 20*, 528–543.

Sweller, J., van Merrienboer, J. J. G., & Paas, F. G. W. C. (1998). Cognitive architecture and instructional design. *Educational Psychology Review, 10*, 251–296.

Tacken, M., Marcellini, F., Mollenkopf, H., Ruoppila, I., & Szeman, Z. (2005). Use and acceptance of new technology by older people: Findings of the international MOBILATE survey: "Enhancing mobility in later life." *Gerontechnology, 3*, 126–137.

Taha, J., Czaja, S. J., & Sharit, J. (2015). Technology training for older job seeking adults: The efficacy of a program offered through a University-community collaboration. *Educational Gerontology, 42*, 276–287.

Tajfel, H. (1974). Social identity and intergroup behavior. *Social Science Information, 13*, 65–93.

Tarhini, A., Hone, K., & Liu, X. (2014). The effects of individual differences on e-learning users' behaviour in developing countries: A structural equation model. *Computers in Human Behavior, 41*, 153–163.

Timmerman, T. A. (2000). Racial diversity, age diversity, interdependence, and team performance. *Small Group Research, 31*, 592–606.

Toossi, M. (2012). Labor force projections to 2020: A more slowly growing workforce. *Monthly Labor Review, 135*, 43–64.

Trocchia, P. J., & Janda, S. (2000). A phenomenological investigation of Internet usage among older individuals. *Journal of Consumer Marketing, 17*, 605–616.

Truxillo, D. M., & Burlacu, G. B. (2016). Doe age matter to LMX and its outcomes? A review and future research directions. In T. N. Bauer & B. Erdogan (Eds.), *Oxford handbook of leader-member exchange* (pp. 397–214). New York, NY: Oxford University Press.

Truxillo, D. M., Cadiz, D. M., & Hammer, L. B. (2015). Supporting the aging workforce: A review and recommendations for workplace intervention research. *Annual Review of Organizational Psychology and Organizational Behavior, 2*, 351–381.

Truxillo, D. M., Cadiz, D. M., & Rineer, J. R. (2014). The aging workforce: Implications for human resource management research and practice. In *Oxford handbooks online*. New York, NY: Oxford University Press. DOI:10.1093/oxfordhb/9780199935406.013.004

Truxillo, D. M., McCune, E. A., Bertolino, M., & Fraccaroli, F. (2012). Perceptions of older versus younger workers in terms of Big Five facets, proactive personality, cognitive ability, and job performance. *Journal of Applied Social Psychology, 42*, 2607–2639.

Twenge, J. M. (2010). A review of the empirical evidence on generational differences in work attitudes. *Journal of Business and Psychology, 25*, 201–210.

Van Rooij, S. W. (2012). Training older workers: Lessons learned, unlearned, and relearned from the field of instructional design. *Human Resource Management, 51*, 281–298.

Venkatesh, R. (2013). Issues and challenges of e-recruitment in human resource management. *International Journal of Management, IT and Engineering, 3*, 228–239.

Venkatesh, V., & Bala, H. (2008). Technology acceptance model 3 and a research agenda on interventions. *Decision Sciences, 39*, 273–315.

Venkatesh, V., Morris, M. G., Davis, G. B., & Davis, F. D. (2003). User acceptance of information technology: Toward a unified view. *MIS Quarterly, 27*, 425–478.

Venkatesh, V., Thong, J. Y. L., & Xu, X. (2012). Consumer acceptance and use of information technology: Extending the unified theory of acceptance and use of technology. *MIS Quarterly, 36*, 157–178.

Verhaegen, P., & Salthouse, T. A. (1997). Meta-analyses of age-cognition relation in adulthood: Estimates of linear and non-linear age effects and structural models. *Psychological Bulletin, 12*, 231–249.

Voermans, M., & van Veldhoven, M. (2007). Attitude toward E-HRM: An empirical study at Philips. *Personnel Review, 36*, 887–902.

Vuori, J., Toppinen-Tanner, S., & Mutanen, P. (2012). Effects of resource-building group intervention on career management and mental health in work organizations: Randomized controlled field trial. *Journal of Applied Psychology, 97*, 273–286.

Walvoord, A. A. G., Redden, E. R., Elliott, L. R., & Coovert, M. D. (2008). Empowering followers in virtual teams: Guiding principles from theory and practice. *Computers in Human Behavior, 24*, 1884–1906.

Wanberg, C. R., Watt, J. D., & Rumsey, D. J. (1996). Individuals without jobs: An empirical study of job-seeking behavior and reemployment. *Journal of Applied Psychology, 81*, 76–87.

Wang, M., Adams, G. A., Beehr, T. A., & Shultz, K. S. (2009). Bridge employment and retirement: Issues and opportunities during the latter part of one's career. In S. G. Baugh & S. E. Sullivan (Eds.), *Maintaining focus, energy, and options over the career*. Charlotte, NC: Information Age Publishing.

Wang, M., Burlacu, G., Truxillo, D. M., James, K., & Yao, X. (2015). Age differences in feedback reactions: The roles of employee feedback orientation on social awareness and utility. *Journal of Applied Psychology, 100*, 1296–1308.

Wang, Y., Wang, H., & Shee, D. Y. (2007). Measuring e-learning systems success in an organizational context: Scale development and validation. *Computers in Human Behavior, 23*, 1792–1808.

Warr, P., Miles, A., & Platts, C. (2001). Age and personality in the British population between 16 and 64 years. *Journal of Occupational and Organizational Psychology, 74*, 165–199.

Weatherall, J. W. A. (2000). A grounded theory analysis of older adults and information technology. *Educational Gerontology, 26*, 371–386.

Webber, S. S., & Donahue, L. M. (2001). Impact of highly and less job-related diversity on work group cohesion and performance: A meta-analysis. *Journal of Management, 27*, 141–162.

Wegge, J., Roth, C., Neubach, B., Schmidt, K. H., & Kanfer, R. (2008). Age and gender diversity as determinants of performance and health in a public organization: The role of task complexity and group size. *Journal of Applied Psychology, 93*, 1301–1313.

Weise, B. S., Freund, A. M., & Baltes, P. B. (2002). Subjective career success and emotional well-being: Longitudinal predictive power of selection, optimization, and compensation. *Journal of Vocational Behavior, 60*, 321–335.

Weiss, E. M., & Maurer, T. J. (2004). Age discrimination in personnel decisions: A re-examination. *Journal of Applied Social Psychology, 34*, 1551–1562.

Wolfson, N., Cavanagh, T., & Kraiger, K. (2014). Older adults and technology-based instruction: Optimizing learning outcomes and transfer. *Academy of Management Learning & Education, 13*, 26–44.

Wolfson, N., & Kraiger, K. (2014a, May). *Shedding light on grey areas: The effect of technology-based collaboration on learning outcomes of older and younger adults*. Paper presented at the Annual Conference of the Society for Industrial and Organizational Psychology, Honolulu, HI, USA.

Wolfson, N., & Kraiger, K. (2014b). Cognitive aging and training: The role of instructional coherence and advance organizers. *Experimental Aging Research, 40*, 164–186.

Wolinsky, F. D., Weg, M. W. V., Howren, M. B., Jones, M. P., & Dotson, M. M. (2013). A randomized controlled trial of cognitive training using a visual speed of processing intervention in middle aged and older adults. *PLoS ONE 8*(5): e61624. DOI:10.1371/journal.pone.0061624

Yu, H., & Miller, P. (2005). Leadership style: The X generation and Baby Boomers compared in different cultural contexts. *Leadership & Organization Development Journal, 26*, 35–50.

Zacher, H. (2013). Older job seekers' job search intensity: The interplay of proactive personality, age and occupational future time perspective. *Ageing and Society, 33*, 1139–1166.

Zacher, H. (2015). The importance of a precise definition, comprehensive model, and critical discussion of successful aging at work. *Work, Aging and Retirement, 1*, 320–333.

Zacher, H., & Frese, M. (2011). Maintaining a focus on opportunities at work: The interplay between age, job complexity, and the use of selection, optimization, and compensation strategies. *Journal of Organizational Behavior, 32*, 291–318.

Zacher, H., Rosing, K., & Frese, M. (2011). Age and leadership: The moderating role of legacy beliefs. *Leadership Quarterly, 22*, 43–50.

Zacher, H., Rosing, K., Henning, T., & Frese, M. (2011). Establishing the new generation at work: Leader generativity as a moderator of the relationships between leader age, leader-member exchange, and leadership success. *Psychology and Aging, 26*, 241–252.

Zaniboni, S., Truxillo, D. M., & Fraccaroli, F. (2013). Differential effects of task variety and skill variety on burnout and turnover intentions for older and younger workers. *European Journal of Work and Organizational Psychology, 22*, 306–317.

Zigurs, I. (2003). Leadership in virtual teams: Oxymoron or opportunity? *Organizational Dynamics, 31*, 339–351.

Zinke, K., Zeintl, M., Rose, N. S., Putzmann, J., Pydde, A., & Kliegel, M. (2014). Working memory training and transfer in older adults: Effects of age, baseline performance, and training gains. *Developmental Psychology, 50*, 304–315.

21

The Future of Work

Stela Lupushor and Alex Fradera

Introduction

Dramatic shifts are influencing the world of work and are impacting corporations, non-profit institutions, governmental agencies, and independent agents alike. These shifts are changing the dynamics between work requesters, workers, and the workplace. They can be grouped in the following categories:

- *Datafication*: the conversion into bits and bytes of activities, interactions, and relationships between entities participating in the world of work, making it easier to use analytics to understand the workflow and work dynamics.
- *Digitization*: the transformation of work, knowledge distribution, and the workplace itself through the use of tools and apps, devices, sensors, robotics, artificial intelligence, etc. This transformation is moving work into the digital realm and creating opportunities to integrate the physical and digital workplace in new ways.
- *Disintermediation*: the decomposition of the value chain of work and removal of intermediaries driven by disruptions introduced by cross-chain players, reshaping how work gets done and reconfiguring the work ecosystem around the consumer's experience versus the traditional value chain.

These shifts demand that corporations rely more on technology, data, and analytics to make decisions about their business, workforce, and workplace strategies. Corporations have to adapt their internal processes and management systems, have to change the communication and knowledge distribution flows to make sense of and influence human behaviors in the workplace, and have to identify opportunities for learning and development to maximize workforce performance. The same shifts are also forcing workers to rethink their digital personas and reputations, the management of their networks, and the development

The Wiley Blackwell Handbook of the Psychology of the Internet at Work,
First Edition. Edited by Guido Hertel, Dianna L. Stone, Richard D. Johnson, and Jonathan Passmore.

of their skills and credentials. As computers take on a bigger role in managing information and intelligence, the value creation will shift toward innate human traits that are difficult to automate and digitize, such as creativity, expressiveness, intuition, emotional intelligence, and storytelling. These traits will influence how we connect and work with others, how we relate to brands, and how we make choices and purchasing decisions. Business models, products, and offerings will become less about their quality and price and more about the language of emotions used to describe them.

This chapter provides examples of disruptions for each of the three major shifts – datafication, digitization, and disintermediation, and articulates the potential impact on the traditional work processes, workplace, and workers from the angle of psychological impact.

Datafication

The digitization of our lives and work environment and the retranslation of work activity into bits and bytes has led to massive amounts of data that, through the use of analytics, provides us with the ability to form impressions about workers, work requesters' identities, and the value of their potential interaction. Insights from data also create a better understanding of the nature of business and the social interactions between the actors in this exchange. This understanding can be used to optimize future interactions, however, it also brings along a variety of risks. Such risks can be the introduction or amplification of biases, of misuse of data to make decisions that unilaterally benefit select actors, or the greater transparency that may make it extremely difficult to minimize the impact of past mistakes or external actions done with malicious intent. Below we discuss examples of datafication and implications of analytics use in the context of both corporate environment and personal life.

Data are everywhere

We leave a digital footprint with almost every step we take online or in the real world where we carry our devices. As we use these devices to manage our lives, our phones and laptops are continuously capturing data of all kinds; and as the convenience of smartphones is being bettered by wearable technologies that ensure that we are perpetually connected with our interests, this process is becoming ever-more invisible. Miniaturization makes it possible for this technology to be embedded in contact lenses, clipped or weaved into clothing, set into jewelry or earbuds, or even tattooed on skin. Our activity can be captured (even if anonymously) at a very detailed level – who did we talk to, who we exchanged emails with, where we have been, what music we like, what purchases we have made and what ratings we gave to whom, restaurants at which we ate, the videos we watched and the articles we read, and through wearables, our heart rate, skin conductance, and other subtle cues of which we ourselves may be unaware. And more parts of our environment are joining the chorus. Digital sensors that monitor and control disparate objects while connecting them to a common, typically Cloud-based software platform or network are also data conduits, shifting information about the patterns of heating our homes or when we switch off the lights. The intimacy we share with our devices is unprecedented.

Without realizing it, the data about our activity, when aggregated with similar information across many other users, allow companies to understand and detect patterns about human behaviors, personalities, and likes and dislikes. When combined with other information such as demographic details, location, life stages, and so forth, such data can help companies make precise conclusions about each and every individual's potential decision path and ways to interject with product recommendations at the perfect time through the perfect channel. As we gain the ability to decompose other personal information such as genome mapping or neuro-mapping, even more details about an individual can become part of the larger body of data subject to analysis. Such information can create vast opportunities to identify cures for illnesses. The danger is that, if misused, the same information can increase the risk of identity theft and other digital crimes, and become a way to damage an individual's reputation.

Possessing data is not enough. Making sense of them requires a good degree of analytics capabilities, which has been through its own evolutionary stages. Davenport (2013) has described the evolution of analytics in three phases. In Analytics 1.0, which was predominant for half a century until the mid-2000s, organizations aspired to gain an objective and deep understanding of business activities and to give leaders facts and data to make informed decisions, rather than relying on their intuition. The 2.0 phase of analytics was the era of big data that blossomed during the mid-2000s with the advent of Internet-based companies and social networks that began amassing significant amounts of information about products, consumers, interactions, transactions – a dataset that grew exponentially and evolved constantly. These types of data were generated by humans through the use of apps (reviews, blog posts, ratings, emails, surveys, clicks, social participation, videos, audios, etc.) or machines (devices, sensors), or sourced from the Internet or public data initiatives. Such vast amounts of data, mostly unstructured and complex, are processed by innovative methods that integrate sophisticated software, Cloud computing, in-memory analytics, and machine-learning methods (semi-automated model development and testing). These methods help us make sense out of the data and build the engines behind our ability to interpret public opinion and sentiment, to attract viewers to websites, to suggest new products to buy, or to recommend new connections.

Davis and Davidson (1991) noted that companies emit "information exhaust," which can then be analyzed and used to amplify their offerings. These authors suggested that companies should transform their businesses to build products and services based on that information. This launched the era of Analytics 3.0 – when every device, shipment, consumer, transaction, or activity leaves a digital trail, and now organizations have the ability to analyze those sets and use the patterns and insights gained to optimize business decisions at the frontline of the operations. This shift will only accelerate as the "Internet of Things" gains pace: this pervasive deployment of sensors that extends digitization and connectivity to previously analog tasks, processes, and machine and service operations. The information collected through the use of such devices can be analyzed and used by businesses to make informed choices about their products, supply chains, customer decision journeys, and their marketing messages. This is the era when organizations use analytics to provide ***intelligent*** customer offerings. As an example consider an online travel company. As soon as a user visits the website, an algorithm designed by economists analyzes the data, including the time between clicks and predicts whether the user is just browsing or is a potential buyer. Based on that information, the site is adjusted in milliseconds and provides the "browser" with more advertisements and the "buyer" with a simple experience to execute the purchase, as a result maximizing profit for the website. While targeted advertising can be taken as intrusive and reflecting corporations knowing and tracking too much, there is evidence that consumers are willing to share their data in return for more useful recommendations when the right circumstances are in place, such as a sense that the information collected is of an appropriate scope, and that the company can be trusted to take privacy concerns seriously (Leon, Rao, Schaub, Marsh, & Cranor, 2015).

In this era, a reconfigured work environment is relying on digital enablement to orchestrate the work. Analytics applied along the work lifecycle are used to identify the points where humans need to intervene. Both the technology and service delivery are wrapped around the end goal to minimize the effort and maximize the bottom line, be that a streamlined and information-rich workflow or an über-personalized consumer experience. Organizations are rethinking their processes and operations, and more importantly, they bring a different set of skills and expertise to build and manage the work in a digital and data-enabled way.

The abundant data are being used not only in the context of the enterprise, but also by users themselves in their personal and everyday lives. We take our devices everywhere and we generously share with them many personal details, including data about our activities, whereabouts, and communications. Our relationships with these devices and the associated applications create a significant digital exhaust about us and our lives too. Through the analysis of such data, one can understand the patterns of one's own behavior and the opportunities to influence a change. In a way, the devices can become the perfect "nagging" mechanism to remind us to take a walk around the block when we have been sitting for too long, to take a break when our pulse starts racing, to take the pill at the right time, or to call mom. We can now even schedule triggers that can motivate us to improve our health and our relationships. Such use of portable technology to nudge individual health behavior has been effective in helping managing health conditions such as diabetes, and facilitating cessation of smoking (Buhi et al., 2013), and contributes to weight loss and regular exercise (Stephens & Allen, 2013). It gives users a sense of satisfaction, achievement, and results based on the data collected and milestones achieved. The irony is that the same technology can also be disruptive and cause anxiety; one of the main interruptions is interference with sleep (Thomé, Härenstam, & Hagberg, 2011).

Companies are starting to understand the potential value of "digital nags" and are adopting the use of such devices to encourage employees to track their fitness activity in exchange for reduced healthcare premiums. They are reimbursing employees for the use of personal wearables to reduce the dependence on clunky devices and move around freely without feeling disconnected. Such devices are also used to monitor physical movement, such as the position of employees or soldiers in unique or potentially dangerous environments, to not only track location but also improve efficiency, manage resources more effectively, and even track the individual's heart rate, blood pressure, and hydration levels. The number of such uses continues to grow.

There is a darker side to having the power that data and analysis affords us. Individuals or organizations with malicious intent can find opportunities to manipulate the system, introduce biases, and misuse the data to enable decisions that unilaterally benefit them. Predictive models are increasingly being used in organizations to assess the risk of quitting in an employee or their potential for success based on personality traits. Similar models might assign scores of the likelihood of a person committing a crime. Although these measures have been crafted with the best of intentions, there are concerns that they inadvertently undermine the efforts to ensure individualized and equal opportunities or justice. These models may exacerbate unwarranted and unjust disparities that are already common in our organizations and society at large. Even unintended errors or biases in the analysis can drive leaders to make decisions that may negatively impact employees. There is a human being—with dreams and hopes, with aspirations and ambitions, with a family to feed, with a potential to be realized—behind every number and the power of analysis and the ensuing decisions should be used with great care.

Data at our fingertips – the changing human mind

There is an ongoing pursuit to develop technologies that help augment our minds and offer ways to offload cognitive demands on our environments. Most lucidly unpacked by Clark (1996), whenever we use a map to represent a route rather than relying on a picture in our head, it is not just us doing the thinking, but the map and us together. In the same vein, through the preponderance of technologies that organize our activities our minds are extended into our environment – for instance, by electronic reminders that enhance prospective memory (Sohn et al., 2005) serving as triggers when we enter a site to ensure we pick up a vital report; the power of deep diary/communication searches to track and recover meetings from even several years ago; or search and face recognition to remind us of names when we forget them. Older people often rely on calendars and diaries strategically to compensate for "raw" deficits relative to younger people (Ihle, Schnitzspahn, Rendell, Luong, & Kliegel, 2012), implying that they, in particular, could benefit from electronic scaffolding, and which may thus have a role in maintaining the effectiveness of the older workforce. It could also facilitate the return to the workforce of people with neurological memory deficits, among whom electronic companions have proved to have profound impact on quality of life (Van Den Broek, Downes, Johnson, Dayus, & Hilton, 2000). Although the augmentation of thinking in general is less remarked on, in aggregate, it is quite remarkable. As computing becomes even more smarter, this augmentation will continue.

Measured work, and the worker and her reputation

With more and more work components transitioning into the digital realm, both work and workers' effectiveness are becoming easier to measure. As transparency increases, new technology allows us to measure performance and the economic value of work with higher precision, and personal devices, wearables, and digital sensors are playing a significant part in the measurements-driven environment. It is important to distinguish here between the most intrusive, hierarchical use of technology, in the form of electronic surveillance, and its more imaginative uses. Surveillance can produce benefits for organizations, at least in the short-term. It is used to reduce risks, such as fraud or unethical behavior, and extract more from its human resources (HR). This is epitomized in a 1995 experiment by Aiello and Kolb, in which individuals monitored electronically were unlikely to engage in social loafing (expecting other group members to bear the brunt of the work); however, these individuals were also more stressed as a consequence. Indeed, the introduction of electronic monitoring may have led to more stress, fatigue, and health complaints (Smith, Carayon, Sanders, Lim, & LeGrande, 1992). Part of this may be because the electronic record can make accusations that are hard for a human to rebuke or even account for. In a qualitative study, a respondent reported how transitions between buildings were tracked using door keycard data (Stanton & Weiss, 2000). He feared that if he had been asked, "What were you doing in that part of the building for 30 minutes" some months back, "Most likely I wouldn't have any recollection why I was there." As insights from our computers become more subtle, such helpless reactions could easily become more common. Balancing this are upsides, as monitoring can promote fairness and ensure consistency of treatment (Niehoff & Moorman, 1993). However, it is clear that from an employee's perspective, a panopticon that counts your toilet breaks and personal emails is a bleak form of progress. Indeed, even performance benefits generated by draconian monitoring may be offset by invisible rebellions, such as workers shirking on an unmonitored aspect of the work (Belot & Schröder, 2013).

For many organizations there are smarter ways to use data. Rather than target individuals for discrepancies in productivity, the access to mass data allows looking at aggregate patterns and understanding how to make high-impact improvements. Moreover, when workers are given access to and a say about their data, the situation becomes brighter. First, they can address decrements in their own productivity, spotting and solving problems themselves. Second, access to data empowers them to secure more from the organization. However, productivity and quality of work are not the only factors becoming increasingly easier to evaluate – quality of relationships and individual character are following suit. Ratings, similar to those used across consumer shopping sites, are now used to evaluate an individual's output, especially if they participate in digital platforms connecting them to their work. Their skills and cognitive abilities can be subject to similar ratings. The size and quality of their networks can be quantified, and these data can elevate their ranking in recruiters' searches (Searle, 2006). The digital presence and the brand image projected by individuals online – which can be termed digital reputation – is becoming something that individuals carefully manage. They recognize the importance of intentionally building their digital reputation, since it can make a difference in their ability to find the next project, command a better price for their work, and form new and important relationships.

In addition, organizations are able to use people's data to make decisions about how they operate. Google's Project Aristotle revealed through extensive analysis that state factors (external factors that control the work environment) – specifically those that contribute to team psychological safety – were more important than individual talent factors in determining the success of a team (Duhigg, 2016). These kinds of insight empower organizations to determine what they need to do to facilitate success, and are likely to include a mix of universal factors (as the value of psychological safety is likely to be) and factors specific to that organization. Companies use individuals' information about their digital brand, performance, capabilities, and networks to make decisions about whether to recruit an individual and even predict his or her future performance and potential. The upside is that measuring value increases both the productivity and the ability of employers to realign incentives and optimally allocate work. The downside is that monitoring and feedback is not always enjoyable, especially when we cannot sustain high performance all the time. Receiving negative feedback, even if rare, can be discouraging and demotivating (Kluger & DeNisi, 1996), is seen as unjust (Baron, 1988), and can result in a backlash against managers (Geddes & Baron, 1997). Privacy and "second chances" in this transparent world can become more difficult to attain. An ideal solution is to create an approach where individuals have a say in the overall measurement system and the ability to correct errors to prevent any injustice; where younger individuals (still in their rebellious phase and apt to make risky online decisions regarding privacy, see Livingstone, 2008) have enough space to take risks and learn from their mistakes; where we as humans can preserve some degree of privacy.

Companies themselves are recognizing the importance of managing their employment brand and reputation on social networks and external media where current and past employees freely share their experiences and opinions (Hutton, Goodman, Alexander, Genest, 2001). In a world where both good and bad news can spread in an instant, it is becoming critical for companies to treat their employees with fairness and respect to avoid a public opinion that is skewed towards negative. Those who are unhappy are more likely to share their sentiment publicly, whether by spreading rumors (Bordia et al., 2014) or revealing information about the work environment that otherwise companies would prefer to not be publicly known, or that require them to address publicly to ensure the employment brand is not damaged. The more negativity there is in the comments, the more it points to systemic problems in the internal work environment, and the more difficult it becomes for a company to attract the individuals with desirable qualifications.

As, simultaneously, more able incumbents are likely to jump from a ship which is feeling increasingly shaky (Loi, Hang-Yue, & Foley, 2006), the organization may be left hemorrhaging talent without the ability to recoup it, leaving it with a hollowed-out, disengaged workforce – and in serious trouble. Put simply, digital reputation matters for individuals and companies alike.

Reflections: information revolution is about the speed and distance

Technological revolutions have been common over the millennia of world transformations. Gordon (2015) notes that in the nineteenth century the idea of speed was defined by horses and the rhythm of the days was dictated by the rise and fall of the sun. The invention of electricity brought light into the evenings, the construction of railroads reduced the idea of distance, with the invention of the telegraph and telephone further reduced the power of physical distance. When the prices of cars reduced, people were empowered to be in the driver's seat (literally) controlling the speed themselves. Over the years, innovations have continued to redefine the notion of speed, and now we are in an era where technology is changing, yet again, the speed of ***information*** travel. The ***information revolution*** is intensified by artificial intelligence, the Internet-of-Things, three-dimensional (3D) printing and other disrupting shifts described later in this chapter.

This revolution is spreading from the digital into the physical world. It breaks down geographic barriers and, for example, replaces long-distance travel with meetings conducted in the virtual world, where our avatars can interact in ways that resemble real life. The revolution is also breaking down speed barriers by creating alternative and faster routes for information to flow. Increasingly such information flow is freer of intermediaries or official sources, where knowledge used to be hoarded and dispensed in controlled manner. The revolution is democratizing knowledge and shrinking the world. In this world the analysis, recommendations, products, experiences – all revolve around a single individual, consumer, customer, employee. This world caters to a singular ***human*** with unique needs, wants, preferences, behaviors, experiences, a human on the information highway, who is yet again empowered to be in the driver's seat.

The digital platforms enabling the information revolution rely on the users to share their experiences and rank services and individuals or companies delivering them, the content they want to consume. It makes these opinions publicly available, and which are then used by search engines to offer the top-rated offerings back to their users, who are increasingly relying on such rating for decision-making. Combined with increasing understanding of user preferences and previously made choices, search engines are increasing their ability to monetize this free information flow and increase the concentration of their aggregation power. There are "side effects" of such relationships. Lack of diversity in the information being offered will further amplify the opinions that "fit" an individual's mindset or limit their networks to only those who share their opinions (Aral, 2014). Humans are susceptible to herd instincts, allowing our own thoughts and actions to be influenced by those around us (Cialdini, 1993; Festinger, 1954). Online, this means that when we see other people appreciate a certain book, enjoy a particular hotel or restaurant or like a certain doctor – and reward them with a high rating – it can lead us to feel the same positive feelings about the book, hotel, restaurant, or doctor and to likewise provide a similarly high online rating (Cosley et al., 2003). Such ratings can be easily manipulated and influenced, and the responsibility to short-circuit herding and biases is shared across parties – the designers of ratings sites, the marketing of online ratings, policy-makers, and each and every one of us who is sharing rating and consuming on their basis.

Digitization

A tide of changes are being brought about by the use of technology and the impact it has on work, knowledge, and the workplace. Some technology applications fall into one or more of the following groups:

- *Automation*: the use of software and applications to perform tasks, execute projects, connect expertise to work requests, convert knowledge into bits and bytes, and make it available at scale for further analysis, enhancement, and expansion
- *Integration*: the use of sensors and virtual or immersive reality to blur the boundaries between the digital and physical environments
- *Robotization*: the passing over of an increasing number of activities from human hands to technological systems (see also Steil & Maier, 2017, Chapter 18 in this volume). This includes manual, repetitive, and potentially dangerous activities that can be carried out by specialized robots and drones. It also extends – thanks to artificial intelligence and constantly-learning systems – to tasks that require massive processing capacity, cognitive abilities, or intelligence.

Digitization is breaking boundaries between the physical and digital workspaces, the way knowledge flows and is shared, and how the work gets executed. Additionally, digitization contributes to the massive amounts of data about the work itself and the interaction of all the actors involved. It creates new opportunities to design the worker's experience across physical and digital spaces in dramatically different ways.

Work away from work

Once upon a time, working from home seemed a romantic and highly exclusive option for a luxury creative class (Massey, 2016). With the decline of long-term employment and the increase in telecommuting options afforded by technology, the number of workers staying at home is rising. "The office" is a twentieth-century concept and the "architecture of offices, with their massed desks and corner offices, literalized the climb up the class ladder" (Mills as cited in Saval, 2014, pp. 284–286). Increasingly, companies are paying attention to the psychological impact of office life on their workers, and emphasis is now being placed on creating positive and engaging work "cultures" and "experiences," contributing to the departure from the notion that workers need to spend five days of the week at the office.

Companies reap cost benefits by reducing their investments in real estate and space management. There are also environmental benefits in reducing commuting time and involved energy spending. There is evidence that individuals perform better when they work from home, intermittently or permanently (Gajendran & Harrison, 2007). These individuals take less time for breaks and fewer sick days, and also experience a lower level of attrition, helping their company reduce overall workforce management costs (see, e.g., Bloom, Liang, Roberts, & Ying, 2015, E. L. Kelly et al., 2014). Teleworking is beneficial for individuals through a greater sense of autonomy (Gajendran & Harrison, 2007; see also Van Yperen & Wörtler, 2017, Chapter 8 in this volume), and freedom from typical interruptions (Haddad, Lyons, & Chatterjee, 2009) leading to reduced stress (Raghuram & Wiesenfeld, 2004) and more goal progress, as well as reduced commuting, an activity that reliably produces negative effect (Kahneman, Krueger, Schkade, Schwarz, & Stone, 2004).

Historically, higher levels of remote work were associated with poorer relationships with co-workers (Gajendran & Harrison, 2007). However, remote working may be particularly suitable for some workers. Teleworking allows participation for people whose physical disabilities make it difficult to have an office routine (West & Anderson 2005). It may also mitigate the impact of mental health issues such as mood disorders, which workplaces can aggravate, making "normal" engagement in a workplace challenging (Woo & Postolach, 2008). Less acute variation also seems to play a role: Anderson, Kaplan, and Vega (2014) demonstrated that the benefits of telecommuting were particularly strong for individuals higher in openness to experiences and weaker for individuals who seek social connection, or who experience higher levels of ruminative thought, more likely to be manifested when alone. Participants higher in neuroticism have more positive attitudes to remote work, which provides a buffer against the requirement for public performance (Clark, Karau, & Michalisin, 2012). Of course, developments in technology and work design may help modify the experience of remote work to optimize its suitability for an ever-wider range of personalities. An increasing range of features are making virtual work more effective, for example optimizing team-member avatars to accentuate the feeling of being one group (Van der Land, Schouten, Feldberg, Huysman, & van den Hooff, 2015). Unfortunately, individuals who work from home are less likely to be promoted, showing that leaders still value attendance over output (Elsbach, Cable, & Sherman, 2010). The difference is in the preference of managers for *passive face time*, which is the sheer presence and being observed at work without necessarily any evidence of doing work or the quality of work produced. To attain a balance between the wholesale abandonment of office space and five days at the office, a flexible definition of "employee" should be established with a possible solution being smaller, more permeable office spaces based on the co-working space models that were once exclusively used by freelancers and small companies. Work settings must be redefined such that they provide an integrated experience between the physical and digital office environments where employees can be equally productive, regardless of where they choose to perform their work. Design of digital environments should aim to optimize the experience of a wide range of users.

Fusion of the physical and digital

Today, our physical and digital worlds are inextricably woven together. It is possible to research a product online and read all the customer feedback, find the best price, then go to the store to see the product and perhaps even test it, before finally ordering it online. Many companies agonize over whether to invest in digital capabilities or focus primarily on their physical space and offerings. Those that have done the latter tend to run their digital operations as independent business units, a method which may be preferable for management, but does not align much with customer expectations. The innovations that fuse the digital and the physical open up vast new opportunities for companies to create a delightful experience for their customers. Fusion also extends to the products offered to customers. Publishers routinely offer bundle deals, with a PDF file arriving immediately–possibly before the physical book is even produced – and the physical version arriving at its own rate. Game companies provide electronic games attached to physicalized toys sold in stores or even printed using 3D technology (Carter, 2012). One reason for this approach is that touch, how a person senses "the world adjacent to his body by use of his body" (Gibson, 1966, p. 14), is a fundamental and effortless way of interacting with the world, and which grounds us in a way that is apparent by reference to neurological cases where body sensation is lost (see, e.g., Cole & Paillard, 1995). As perception draws on cross-modal processes – sound and touch inform vision, and vice versa – physical objects with haptic qualities maintain an attraction; they come with built-in intuition. Any person who is frustrated by a misbehaving water faucet with no discernable tap or valve, or who continues to use paper notepads (which have been shown to facilitate greater learning compared with computer note-taking;

Mueller & Oppenheimer, 2014) will appreciate this. Therefore, fusion solutions that retain physical perks together with the convenience of the virtual are a strong offer to consumers. The greatest barrier to adopting fusion strategies is inexperience with their execution. These strategies require a tightly integrated and collaborative effort between digital and physical worlds experts at every stage, from idea generation to development, testing, and launching, as well as the assembly of teams for every possible type of innovation project.

Fusion opportunities are seen across many fields and industries – education serves as an apt example. Augmentation of traditional lectures with student participation online or asynchronously has been done for some time. School are starting to experiment with highly individualized instruction, and integrating analytics and recommendation engines into their curriculums. With the advent of the Massive Online Open Courses (MOOC) platforms, a lecture itself is now taken online. Individuals around the world who might not have had access to such content before, can now take a class from a prestigious institution and get a certificate of completion and school credits. It is difficult to substitute the value of being present in the classroom and participating live in a brainstorming session or debate, but platforms such as MOOCs expand the access to education, enabling more individuals to participate in such conversations, and as a result democratizing knowledge that otherwise would be limited to those who can afford it. Although adoption of MOOCs is in the early stages and students' completion rates are uneven, they are gaining popularity with major educational institutions and some of the courses are being offered for a fee. At the same time, educational institutions are pondering on the impact MOOCs will have on their business models, class sizes, content delivery, and distribution channels. Again, the development of haptic qualities to complement the sight-and-sound that comes easily to the Internet will be critical in maximizing this learning potential (Hamam, Eid, & El Saddik, 2013).

With the rapid expansion of knowledge and the dwindling half-life of skills, we have to accelerate the pace of learning to stay ahead. The pressure to learn rapidly will continue to demand examination of educational degree standards and the number of years required to obtain one. Organizations will need to rethink their approach to traditional learning, and fuse digital learning and on-the-job development to keep pace with the dynamic environment. Fusion transformation does not imply replacing the physical with digital (e.g., the prediction about extinction of physical retail stores has not materialized), but about marrying the two to deliver an experience that oftentimes was previously unimaginable, while enabling better economics.

Virtual reality and holography facilitating the fusion

In 1965, Ivan Sutherland, a computer graphics pioneer, described virtual reality at an international meeting of techies as:

> *a room within which the computer can control the existence of matter. A chair displayed in such a room would be good enough to sit in. Handcuffs displayed in such a room would be confining, and a bullet displayed in such a room would be fatal. With appropriate programming, such a display could literally be the Wonderland into which Alice walked.*
>
> Sutherland (1965, p. 236)

While still in its early years, virtual reality has seen recent rapid advances, in the form of new devices, headsets, and body suits that have appeared on the market. Much of the excitement about virtual reality originates in the gaming community, a group eager to fully immerse themselves in the gaming experience. But gaming is just the start. Museums are integrating the visitor's experience and allow tourists to "visit" interactive exhibitions and engage with or imagine museum objects in their original context. Entertainment can now be consumed just as vividly without leaving the comfort of one's home. Galleries, interactive performances at theaters, virtual theme parks fall into the category "edutainment," in which the aim is to educate as well as entertain. Scientists are experiencing standing on Mars and attaining a more complete understanding of the planet. Many more opportunities for an integrated virtual experience are emerging.

Virtual reality opens the door to learning more about human behavior, as we become able to profoundly shift how we are represented to others. Researchers have put participants into virtual reality headsets that transform how they perceive their space, each other, and themselves via mirrors designed into the digitized virtual reality space. Child avatars can produce overestimation of sizes of objects (Banakou, Groten, & Slater, 2013). Those given an avatar taller than their true height tended to be more confident and made more aggressive offers in a negotiation ultimatum game, whereas those with particularly attractive avatars became more friendly, disclosing more personal information and reducing interpersonal distance with others in the space (Yee & Bailenson, 2007). Such Proteus effects – named after the Greek god's transformational capabilities – last even after one leaves the virtual realm. Virtual reality also offers benefits for remote working: collaboration between individuals over a distance is enhanced by a greater sense of presence when in virtual reality environments – of "being there" together (Van der Land, Schouten, Van den Hooff, & Feldberg, 2011)

Another interesting technology that allows fusing the physical and digital is the hologram, which has been popularized by science-fiction movies and television serials such as *Star Wars* and *Star Trek*. The *Star Trek* holodeck that displayed a 3D transparent image of someone one could interact with and ask questions is becoming a reality. The first holographic concert was held in 2012, when a virtual Tupac Shakur took the stage at the Coachella Valley Music and Arts Festival. This was the first time a hologram had been used in a live performance and the first time hologram technology was used to resurrect a dead singer. Since then holographic technology has been evolving and at the time of writing, devices are emerging that can transmit 3D images in close to real time, which could make the holographic telepresence a reality. Such technologies could enable speakers to deliver their keynotes to audiences on the other side of the world, or enable projecting a full-size image of the body of a professor into a classroom in front of a live audience thousands of miles away, with the image even able to make eye contact.

Returning to the importance of touch, researchers recognize that haptic information "is critically important for fast, accurate interaction with our environment" (Robles-De-La-Torre, 2006, p. 28). While still in experimental phases, there are prototypes of machines that render holograms touchable, adding to a growing body of telehaptic experiences. Such machines use ultrasonic radiation pressure to create a realistic tactile sensation (Makino, Furuyama, & Shinoda, 2015). These kinds of experience will most likely be limited to the gaming industry initially, but further developments might allow integration into experiential learning (Konnikova, 2015). This technology will also allow individuals to interact with a fused reality and learn new skills or practice them in a safe environment, eliminating the need for costly, labor-intensive efforts and potential dangerous settings that are otherwise required. Of course, as such technology goes through its teething phase, it may uncover new, unanticipated concerns of its own.

For the time being and until virtual reality and holograms are widely adopted, telepresence robots and technologies will continue to provide immersive experiences (Tsui, Desai, Yanco, & Uhlik, 2011). Most devices nowadays increasingly have built-in webcams to allow videoconferencing, tablets have been installed on roving robots in offices for remote employees to feel like they are physically there, part of the team. Members of distributed teams can just as effectively interact, participate in meetings and even "take a walk" together to get a cup of coffee.

Robotization, exoskeletons and artificial intelligence

Our world is starting to resemble images from "I, Robot" by Isaac Asimov where a whole metropolis is designed and run by an advanced robotic system. From ancient Greece onward, robots have fascinated engineers, inventors, and dreamers throughout history. The first *functioning* robot was built in 1738 by Jacques de Vaucanson and was equipped with the ability to play the flute. In modern days, humanoid robots are becoming affordable and available for purchase by the general public (Aldebaran Robotics, 2015). These humanoids are equipped with sensors and computers, and connected through the Cloud to a network of similar humanoids that constantly learn from each other, expanding their knowledge and constantly evolving. Later in this chapter we will describe how artificial intelligence (AI) combined with robotics is bound to impact many aspects of work environment.

The field of exoskeletal enhancements – ready-to-wear robotic prosthesis that function as replacements for limbs and other body parts – is seeing rapid developments. Exoskeletons and bionic limbs bring back hope and restore the quality of life for those with mobility-inhibiting injuries or lost strength. They have been designed by incorporating studies of normal human physiologies to mimic the natural movement of human limb muscle function, enabling propulsion very similar to that of natural walking. Many of these devices interact with the patient's nervous system through electrodes implanted on the residual limb to which the prosthetic is attached (Burke, 2015). Compared to more basic prosthetics, these new forms are experienced differently, with users describing them as transforming quality of life and being seen as "a part of me" (Lundberg, Hagberg & Bullington, 2011). Such enhancements redefine the notion of disability, and are extending lives and our ability to stay active. Robotic exoskeletons might compensate for spinal injuries or allow intensive walking training that helps children with neurological disorders maintain their mobility. Robotic arms and aids allow paralyzed individuals to extend their interaction with the world. Exoskeletal or bionic enhancements enable the replacement of deteriorating limbs giving them the ability to stay independent. This increases the opportunities for many individuals who otherwise would have no ability to stay gainfully employed. Corporations can take advantage of the growing segment of individuals who can benefit from assistive devices and keep them in employment longer. This requires rethinking the types of accommodations such individuals might need, including reconfiguration of the office space, assistive technologies and work redesign. Exoskeletal enhancements also open the door to whole new fields of opportunities for "cosmetic amplifications," where individuals augment their physical abilities with artificial and augmentative enhancements, and as a result gain advantage over "normally-abled" people. This is an emerging area and more research is needed, but the demand for this technology together with continuing investments in it to make it lighter, more affordable and usable, have the potential to lower healthcare costs and – more importantly – improve the quality of life of millions of people.

The physical world of robotics has been further enhanced by AI to the point where we are starting to close in on milestones such as the Turing test, proposed by computing visionary Alan Turing, in which an observer finds it impossible to tell between computer and human responses in a controlled conversation. Although AI intelligence is enabling robots to tighten their grasp of the syntax of the language (the grammatical rules and structures), it will take time to attain semantic understanding (the underlying meaning of the words and sentences). We already have computer programs that mimic conversational speech and can interact with humans, give directions, answer questions, and even interject some humor, albeit through fairly fixed repertoires, but the capability to "find something funny" still seems a little way off (Hernandez, 2016).

Machine learning algorithms are now used to train robots to perform tasks and associate different actions with commands given in everyday language. This learned behavior is stored in central repositories that can be accessed by other robots and together form a shared body of knowledge, an attempt to emulate one of the most central components of human culture – the ability to acquire and share knowledge. Furthermore, ongoing research is focusing on how robots can learn to perform the same task in entirely different settings, which will allow robots to adapt over time to perform new tasks or perform them in unfamiliar contexts. This will result in robots and systems that can learn and engage in ***autonomous*** behavior, all the more so as they gain access to more data, content, and knowledge via sources such as the Internet, and as the notion of digital shared knowledge becomes more feasible.

Deep-learning algorithms and methods attempt to emulate the way neurons and synapses in the brain change as new information and experiences are introduced. These methods are advancing the machine's ability to recognize faces in images, identify spoken words in audiorecordings of natural speech, and detect patterns in vast amounts of data. While these algorithms might enable machines to engage in simple conversations, according to cognitive scientists they requires the processing of countless transcripts of previous conversations, a process which differs from that used by the human brain to acquire language (Ernest & Marcus, 2015). There are many subtleties in the functioning of the human mind and reverse engineering of its properties will be a technological challenge for the years to come. The scientific community is still split over whether machines will evolve enough to emulate the human brain and ***think*** the way we do. Robots can "see" and "hear" and recognize patterns, but "understanding" them still remains a uniquely human ability. Early examples of technological developments, such as the IBM Watson cognitive computing system winning Jeopardy! or Google's DeepMind Alpha Go (Silver et al., 2016) winning "Go" against human players – once thought far beyond the reach of machines – have brought us to a reality where the solving of a given complex human-like intelligent activity should no longer be expected to happen in the decades to come but at "any time now" (Santens, 2016, n.p.).

There is fear that robots will replace many jobs in the future, a fear that is becoming real in the case of jobs that involve semi-repetitive or computational tasks (Vardi, 2012). A surplus in cheap labor is becoming less of a lure and companies increasingly make decisions to invest in industrial robots, investments that can be repaid in a few years. Scientists speculate that AI could drive global unemployment up to 50%, wiping out middle-class jobs and exacerbating inequality (Yuhas, 2016). In contrast to the industrial revolution, the "AI revolution" will not be a matter of physically powerful machines that outperform human laborers, but rather a contest between human wit and mechanical intelligence and strength (Vardi, 2012). Work that requires certain unique human capabilities is the most challenging to duplicate by machines, such as using common sense, having the ability to understand patterns, and being able to perform work that requires creativity and original thinking, complex social interactions, or a blend of analysis and dexterity. Will the machine one day take over? Or will machines offer us an opportunity to further extend the human mind, allowing still-greater human endeavors, such as the new insights and levels of grandmaster chess play enabled by human–AI teams (K. Kelly, 2014)?

Furthermore, can robots become so advanced that they will become the ultimate threat to our existence? Who will own these robots? Who will be responsible for mistakes made by robots? Will companies have to pay taxes for employing robots? What will be the definition of crime in this new digital world and is there possibility that the *enemy* will turn the robot into a device that changes its patterns and behaviors against its original intent? Many big questions will arise as we learn to coexist alongside these ever-evolving and constantly fine-tuning machines (see also Steil & Maier, 2017, Chapter 18 in this volume).

Reflections: machines doing our work

The jobs of tomorrow may be different, and skill needs will evolve, but human nature will continue to exercise its curiosity and sense of wonder about how things work, how they are built, and how they can be improved. In 1930, the economist John Maynard Keynes wrote a story (reproduced in Keynes, 1933) about a future society which has freed itself from prejudice, war, and every known form of cancer. And thanks to technology, it has also freed itself from work – the only problem left is the question of filling all the extra time. The last and only job left is quality control played by the protagonist. Ultimately, the human tendency to question, tinker, and create takes over and a major underground manufacturing ring develops, drawing in even the protagonist of the story, who is meant to be supervising this utopian system. The beauty of the story lies in the balance that is eventually restored in the world order – the balance that is attained as a result of human nature and the curiosity that drive us to want to use our skills, to aspire to grow, to continuously evolve, to feel competent. A balance that does not necessitate passing a Turing test.

Disintermediation of the Work Value Chain

Manual work has undergone massive transformation since the industrial revolution, thanks initially to factory production line development, and then through "scientific management," where figures such as Taylor and Gilbraith used studies of worker time and motion to systematically break apart and regularize tasks (Harper & Mousa, 2013). But digitization is taking this process into unprecedented domains. Work can be decomposed and atomized to a degree such that it can be easily disassembled and executed in a distributed fashion – that is, across extended distances, at speed. It can then be reassembled in a way that carries with it the information about all actors involved, the rating of quality of the output, and of the quality of relationships formed across the entire chain, including the formal and informal feedback obtained in the process of execution. It also carries the value assigned to each component and the transfer of that value between the worker and requester (monetary value, brand association, reputation enhancement, skills and knowledge, network connections, etc.)

Levy and Murnane (2004) describes how computers are changing the landscape through the example of the legal field, where machine learning has reached a point at which software can, with guidance from senior lawyers, perform the time-intensive task of legal discovery that requires combing through several documents to identify those that must be shown to the other side before trial. This work is typically performed by junior lawyers and paralegals, and now machines are stepping up to computerize the law or at least certain components of the work that do not require unique skills such as writing persuasive legal memos with novel arguments. This is also an example of work disintermediation, which is the replacement of intermediaries in the supply chain with automated steps that are becoming increasingly technologically sophisticated. That is, instead of going through an intermediary such as an agent, buyers and sellers can deal directly with each other and the interaction is enabled through technology solutions. Companies that are transforming their organizational processes to integrate the information on how the work gets done and to analyze data about each actor and the process of value creation have an advantage. These companies can use the insights gained about the work environment and work outcomes to optimize their internal processes and maximize impact on the bottom line.

Networked work

The disintermediation of work in the West begun much earlier, when labor-intensive and repetitive components were sent offshore for execution, mostly in pursuit of lower costs. Now, software and advanced algorithms are taking over those components and at a fast pace. It is a crucial step in enhancing productivity in many end-to-end work cycles, but the jobs that are expressed in programmable rules and require only moderate skills – typically involving middle-class wages in the developed world – are increasingly at risk.

The economy of the eighteenth century, described by Adam Smith (Smith & Garnier, 1838) was a genuine market economy of individuals engaging in commerce with one another. Over the next two centuries, the emergence of mass production and distribution yielded modern corporations. The entrepreneurs of Smith's time gave way to the salaried employees of the twentieth century. We are coming full circle and today, more and more individuals choose to make a living by working gigs. Small-scale entrepreneurs and small businesses are growing in numbers (Fairlie, Morelix, Reedy, & Russell, 2015) and giving rise to a new type of "networked work" offered through what is referred as the *independent*, or the *freelancers*' or the *gig* economy. Operating in such an economy comes with interesting challenges. For example, it is difficult to understand the true composition of the workforce, given the inability to measure and officially count the independent workforce; and the independent workforce does not have the security and protection (such as health benefits, unemployment insurance) that traditionally employed individuals have.

The disintermediation of work is amplified by digital and mobile technologies that change not only what we do and how we do it, but also where we do it (at home, at work, remotely). A multitude of digital platforms are now available to connect independent agents around the world to work (and the emphasis is on work, not necessarily jobs). Such platforms fall into three major groups: matching the individuals with regular jobs, online marketplaces for contingent work, and talent management marketplaces that allow assessment of skills, and recommend ways to optimize team formation and skills development. Such platforms bring radical transparency and efficiency to labor markets and could add 2% to the gross domestic product (GDP) by 2025 (Manyika, Lund, Robinson, Valentino, & Dobbs, 2015). Millions can benefit from these development by not only raising self-awareness and increase their skills, but also finding jobs faster, thereby reducing the unemployment or partial employment, as well as supplementing their income through freelancing. Adoption rates will vary, however, and economies with relatively fluid job markets such as the United States, will be among the early adopters, whereas countries with low unemployment and barriers to entry such as Japan and China will lag behind (Manyika et al., 2015).

The digital platforms connecting work with workers make it possible for companies to go beyond hiring of temporary workers; instead, work can be performed by one of a vast network of occasionally anonymous workers. This type of "networked work" may lead to commodification of some occupations. It also means a shift in the responsibility for personal development and many of the risks that employers once bore to individuals themselves. Individuals bid for work, and it is auctioned off to the lowest bidder, resulting in fewer stable jobs for tasks that can be performed remotely. It will also lead to a race to the bottom when it comes to wages and cost of labor. There are traits associated with people who choose to start businesses or become freelance, but as traditional work gives way to "networked work," more people will find themselves in these situations regardless of whether they are suited to it. In psychological terms, job insecurity erodes wellbeing, producing anxiety, distress, and depression, as well as higher levels of exhaustion, both mental and physical (DeWitte, 1993).

Empowerment of independents

Despite some of the downsides, the shift to independent and "networked work" can produce an interesting counter-effect. It creates an empowered worker, particularly in the case of those with higher and more specialized skillsets. Contrary to popular belief, most current independent workers are not forced into temporary work, but rather choose this route – and many of them do not want to return to traditional employment relationships (Johnson, 2013). Temperamentally, those who become and remain self-employed are more emotionally stable and open to experience (Patel & Thatcher, 2014), and those who are actively entrepreneurial are more extraverted, conscientious, and less agreeable (Rauch & Frese, 2007). Many individuals who strike out on their own are not "average" workers, at least in their own eyes. They are typically experts in their fields and possess advanced skills, experiences, and education. Companies employ them not as a low-cost solution, but as "guns for hire" who come to fix specific problems.

Workers are increasingly sitting in the driver's seat and will carefully choose the companies to which they bring their talent. They are inspired by success stories of those who struck out on their own and their freedom, self-reliance, and the possibility of becoming wealthy. Workers' decisions are informed by organizational reputation and many times they make choices whether or not to pursue work with a company based on public opinion and word of mouth, or based on the value of the organizational reputation that can be transferred to their own digital reputation. Working for or with a reputable brand allows individuals to add evidence of their skills and experience to their résumé (which is increasingly digital and at times specific to the platform they participate in), thereby increasing their value and ability to negotiate better terms with the next company that is considering employing them.

3D printing – new era in manufacturing

Another technology with the potential to further drive disintermediation is 3D printing, also known as additive manufacturing. 3D printing is the process through which many layers of material are "printed" as real 3D objects (hence "3D" printing). Digital files, drawn up by a designer or created by a 3D scan, communicate to the printer the dimensions of each required layer or "horizontal slice" to complete the object. 3D printers may range from small personal hobbyist machines to industrial printers. This technology utilizes a range of materials ("inks") that are most commonly plastic polymers and metals. In contrast with traditional subtractive manufacturing processes that are based on the removal of material to create new products, 3D printing has been used for rapid prototyping by manufacturers since the 1980s. It enabled sidestepping the often labor-intensive, expensive traditional processes such as cast production using molds and dies, milling and lathing and other machine work, and finally, the shipping of the object from a supplier. The advancements in speed, capabilities, and affordability of 3D printers have broadened their use and popularity. 3D printing is now moving into the mainstream and is being used in very different ways, including the printing of advanced prosthetics, and, potentially, even the printing of living tissue. This is, of course, dependent on the expanding list of new materials or "inks," to include ceramics and glass.

Widespread adoption of 3D printing has the potential to restructure the supply chains, and dramatically redefine manufacturing as we know it today. Instead of having to order and waiting for missing part of a water faucet to be delivered by a manufacturer, the part's specs could be obtained to simply print it at home. Such a scenario shrinks the supply chain and limits it to just the links between the individual, the specs and 3D printer – and in the process eliminating the links connecting development, prototyping, production, delivery, and warehousing of parts. We can imagine a future of "printer farms" full of printers humming away, overseen by just a handful of workers. What is "printed" would be limited only by the range of "inks" available. This future can shrink the distance between "global" and "local" by bridging the globally distributed free-flowing design and know-how with a localized method of production and distribution. A world in which those blessed with wild ideas but two left thumbs are invited back into the world of physical production. Moreover, beneficiaries of this model will not be just individuals, but also could include industries such as aerospace, electronics, transportation, power and utilities, energy, and industries with weak intellectual property protection. The competitive advantage will come from great design and marketing and less from production or even distribution and inventory management.

This future of controlled equipment, greater automation, and advanced robotics implies great opportunities, but also a potential for wide-scale loss of unskilled, factory floor manufacturing jobs. Job opportunities might gradually move to those with technical know-how or those who can or want to be trained to acquire it. However, one recent estimate suggests a net loss of 9 million jobs in the coming decade (Schiller, 2015). In the modern era, where employment is strongly associated with societal status and is the primary means to support oneself and one's dependents, being unemployed produces deeply negative effects, especially in the long term (McKee-Ryan, Song, Wanberg, & Kinicki, 2005). Even those who keep their jobs during this transitional period may be negatively affected by layoffs, experience "survivor syndrome," with guilt and concern over departed co-workers, and remain insecurity regarding their own job (Kim, 2003). A range of technology leaders have voiced concerns about the risks of an economy that depends on near-full employment, in a time where technology is accelerating in areas where it can make people redundant; many are suggesting political solutions such as a universal basic income (Santens, 2016).

Exchange of anything of value

Nowadays digital media allows taking a picture and sharing and replicating it over and over across devices or websites of our choosing. In the case of money, or in fact anything that carries value, the ability to replicate is not desirable – we instead need something that is immutable and enables trust between the parties who participate in the exchange. Crypto-currency (a form of money that can be transferred securely and anonymously across a widely distributed peer-to-peer network) and block-chain technology (a public ledger that tracks the assets and the exchange of legitimate transactions that have occurred on the network so far, maintained by collaborative efforts of all the nodes in the network) allow secure transfer of assets over the Internet. The block-chain network is a distributed peer-to-peer network where transactions are verified – thus avoiding double spending – by individual account-holders or "miners." These transactions are chunked into blocks, linked to each other in the "block-chain," which relies on specialized hardware and formidable cryptographic mathematics, rendering it practically impossible to replicate. Miners confirm transactions through a shared consensus system, and usually require several independent confirmations for the transaction to go through. This process guarantees random distribution and makes tampering difficult. It produces a "trustless" system in which no one has to put their faith in anyone else, because the network itself is guaranteed to keep a fair and accurate record of which asset belongs to whom, and as a result solving the Byzantine general's problem (Lamport, Shostak, & Pease, 1982). Such technological innovation will transform the role of intermediaries such as banks, currency exchange marketplaces, and other centralized institutions that keep track of payments and guarantee financial system. It has the potential to disrupt other centralized institutions in the "trust business," such as stock markets. With a simple framework where an individual's email is the address for the crypto-currency wallet, it increases the speed of communication and the number of transactions. It also removes friction from many areas, such as the cash-based economy, money distribution and transmission of currencies, informal employment, and smart contracts.

In 2015, the Honduras started building a registry title using block-chain technology. Kenya, Uruguay, and Panama are investing in block-chain technology to tackle regional payment problems. Similar explorations of the technology are spilling into many areas where property titles are involved, to enable owners to have true ownership of their physical and digital assets. Songs or books purchased via online platforms today are in reality leased, since their transfer to other devices or platforms is limited if not impossible. Likewise, artists, writers, and other producers of creative content are locked into the same platform's distribution channels. With technology such as block-chain, the ownership of intellectual property (digital rights management) can be captured and direct transactions with buyers executed, allowing artists and songwriters to escape the tyranny of record labels or publishers' distribution channels. They can use the same technology for royalty payments in a more transparent, disintermediated, and faster way.

Value is attributed not only to money but also to identity and reputation. Identity theft is a modern-day plague causing significant financial losses and reputational damage, and is mostly a result of giving out too much information to too many entities. Consider the amount of personal information one has to share when acquiring a mortgage, applying for a credit card, or even when purchasing a new device. Block-chain technology can digitize information about our identities, including data about our health, education, reputation, and assets, among others. It gives us the ability to not only share the basic necessary information but also control how that information is distributed, and how the receivers are using it. Similar to payment verification to prevent double-spending, block-chain technology can enable data distribution. The owner of the information (the individual and their personal identifiable information) can control what details they want to "send" to the requesting entity (bank, employer, shopping site, etc.) to inform decisions or actions without the need for the receiver to hold a copy of the data. For example, if there is a service that the requesting entity can ping to confirm that someone is over 18, they do not need to hold that fact on their own servers to process the credit card application. The owner of the data ensures the data are used for that single-use purpose only. The receiver's security paradigm thus changes dramatically. Thus, current models of privacy, data protection, and related regulation will have to evolve to adapt to this new paradigm shift as well.

Another arena in which such technology can eliminate friction is the distribution of social welfare. It reduces the cost of delivering money and the probability of theft, and increases the speed of transactions. Similarly, such technology can help the poor world's unbanked individuals to maintain a notarized transaction list that enables financial institutions to assess their creditworthiness, and grants them better access to insurance and other financial products. It provides underprivileged individuals with an easier path to financial inclusion. Lastly, such technology can become a "railroad," facilitating the exchange of value between work requesters and workers or those participating in the sharing economy (the market model of peer-to-peer exchange, also known as collaborative consumption, peer economy, and uberization). With block-chain technology, the identities of both work requesters and workers are securely captured. Similarly, the transaction that registers the value transfer from one individual to another can be securely and transparently captured. Work is done, money is paid for the work, the reputation of each party is updated with new details about the transaction, and all of it is securely stored and controlled by the involved parties. Such frictionless value exchange has the potential to disintermediate the internal corporate processes of managing employees, their records, compensation and payroll processes, and performance management systems. It also threatens traditional notions of employment models.

Value creation and migration

Slywotzky (1996) has argued that organizations are moving from chasing a pure market share to a new reality where business design is centered around customer experience, introducing dramatically different opportunities for capturing and protecting value. Traditional industries such as technology, telecommunications, media and advertising, and consumer electronics are witnessing their own value chains being disintermediated. Companies such as Apple, Amazon, Google, Tesla, and many others are disrupting these industries by moving away from capturing market share and optimizing supply chains and processes. Instead, these companies are thinking differently about how demand gets created and they start with the customer "hassle maps" to find elegant solutions, that is by simplifying the experience and the number of relationships a customer needs to manage. They analyze the end-to-end experience of the customer, the natural settings, and the environment where the experiences occur. They observe the challenges encountered by customers, and understand the decision path they have to take on the journey from thinking about the product that will help them solve the challenges, to wanting that product to finally buying it. These companies offer solutions to customers' challenges – be that a well-designed product, a delightful shopping experience, a fast delivery, or the creation of moments of happiness. They then build process and supply chains, as well as partnerships, to deliver better experiences, and they seek coalitions and alliances with parties that are highly motivated to help them delight the customer. These approaches are thus increasingly individualized, and based on building "persuasion profiles" (Kaptein & van Halteren, 2013) that utilize user data to respond to their key drivers. Subliminal emotional triggers are built into the experience and create visceral feelings that later translate into customer ratings, recommendations, and word-of-mouth spread of information. These customers' conversations translate into more sales, less spending on advertising, and most importantly, the engagement of employees (even if temporary), which subsequently leads to more and better product offerings. This virtuous cycle has been argued to create value for the customer, for the company, and for its employees (Slywotzky, 1996).

The cross-chain companies described above started as technology players disrupting telecommunications, media and consumer electronics value chains, and they are now moving to disrupt other industries such as retail, education, financial services, healthcare, auto, aerospace, and energy. However technology companies are not the only cross-chain players. Ride-sharing companies are investing in driverless cars, robotics, and drones. Large car manufacturers are partnering with Cloud providers to connect cars to sensor-laden smart homes. Neat and tidy supply chains are becoming a thing of the past, forcing companies to reinvent the business they are in and the competitive landscape they belong to.

Crossing the next frontier will be about finding solutions for employees' "hassle maps," by building a work environment that enables employees to be productive, perform their best, satisfying them both extrinsically in terms of status, material reward, and scope of influence, as well as supplying intrinsic motivators such as work that aligns with their interests, and compatible workmates and organizational culture (see, e.g., Ryan & Deci, 2000 on motivators), in order to ultimately realize their full potential. This will require organizations to implement technology solutions that allow employees to be more productive and gain access to information, knowledge and experts through analytics-driven recommendations and nudges. The work might have to be redesigned around solving problems or getting the work done versus following a process document or procedure. Companies will also have to rethink management systems and how the work is orchestrated, evaluated, and rewarded, and how experimentation and risk taking are encouraged. These are fundamentally ideas around job enlargement, especially enrichment, championed by Frederick Herzberg in the twentieth century (see, e.g., Herzberg, Mausner, & Snyderman, 2011), and the world is looking more and more ready for them. In addition, companies need to rethink internal processes that block motivation and superior performance. Such processes span the full spectrum of the traditional scope of HR, including programs such as traditional performance management, which in today's environment creates more harm than good through forced rankings, and infrequent and subjective, and at times demoralizing, feedback. Researchers have identified that performance management ratings are heavily colored by "idiosyncratic rater effects" (Scullen, Mount, & Goff, 2000, p. 957). A study rating 4492 managers on certain performance dimensions by two bosses, two peers, and two subordinates revealed that 62% of the variance in the ratings could be accounted for by individual raters' peculiarities of perception. Actual performance accounted for only 21% of the variance. This led the researchers to conclude:

> *Although it is implicitly assumed that the ratings measure the performance of the ratee, most of what is being measured by the ratings is the unique rating tendencies of the rater. Thus ratings reveal more about the rater than they do about the ratee."*
>
> Mount and Scullen as cited in Buckingham and Goodall (2015, n.p.)

Many companies, including Adobe, General Electric, IBM, Deloitte, and Goldman Sachs (Buckingham & Goodall, 2015), are starting to challenge their performance management approach and moving from an annual review process with forced rankings to more frequent check-ins and ongoing feedback from not only the manager but also from peers across the organization. Even the notion of "traditional career" is being revisited as individuals change jobs more frequently and tenures shrink. Companies are experimenting with a networked and flexible operating structure and work is being done by teams that inter-operate without the need for hierarchy (Bersin, 2016). It is an emerging practice in organizational design that facilitates innovation, rapid response to customer needs, and agility. People enjoy working in small groups, so such teams also have a positive impact on employee engagement and retention. In this context careers are not necessarily about upward movement but about skills and capabilities development through varied and challenging assignments, team leadership, mentoring, and opportunities to learn, moving toward a career model of growth in many directions. Similarly, obscure and seemingly unfair compensation models, promotions driven primarily by relationships, or by execution of tasks based on current quality definitions, could be among the first areas of opportunity for organizations to resolve with regard to their employees' "hassle maps."

Reflections: change is hard and is everyone's responsibility

The ever-evolving present requires significant amount of change at both individual and company levels. Although there is resistance to change, humans are often quite quick in adapting to changing environments. Their tolerance for change is critical to their ability to adapt to an ever-changing environment demanding new skills and behaviors. Change also impacts the organizational ability to rethink the work environment. After all, leaders are humans too, and they have own barriers for change. This might hinder their ability to make dramatically different decisions on behalf of the company on adapting to stay current and survive in the long term. At the organizational level, leaders need help to reimagine the future of their work environment in light of disruptions driven by technology, demographic shifts and work disintermediation. Such reflections and explorations will enable corporations transform their workplace and their relationships with their workforce, and accelerate adoption of organizational changes and the worker's personal evolution as a result of it.

The implications of datafication, digitization, and disintermediation

The three shifts described in this chapter are leading to wholesale change in the way we work, think about work, as well as how we engage in our wider personal life. Digital identity and reputation will form the basis for a new system to describe the value and worth of actors in the future world of work. Workers' digital imprints will enable them to carry their work and knowledge portfolios, profiles, networks, and expertise from one job, project, or experience to the next. Managing digital identities of corporations and institutions will be just as important, since these will determine attraction factors for workers, who will not only consider the monetary value of the relationship, but also the value the corporate brand will add to their own reputations.

The traditional roles primarily played by HR within corporate boundaries will have to shift to adapt to this new order and to the complexity of the relationships between work, the workplace, and workers. Such complexities will at first include organizational ability to manage the contractual aspects of employment, as well as to manage the considerable challenge of cross-cultural integration, which researchers now recognize – termed either cultural intelligence (Ang & Van Dyne, 2008) or cultural competence (Dolan & Kawamura, 2015) – as a distinct, quantifiable and developable individual difference. Organizations will also have to rethink their approach to employee engagement and evolving systems of workplace motivators, to align with changes due to role enrichment and worker independence. Furthermore, organizations will be required to reassess their approach to leadership and skills development, to the management of workplace diversity, and rethink performance management and recognition in this context. Over time, these complexities will be addressed by using a variety of applications, platforms, and service providers.

As many of the involved processes and tasks will be automated, space will be created for the role of HR to shift to three key areas:

- workplace design and shaping the experience of the workers across both the physical and digital realm to maximize performance and outcomes, and to amplify the drivers of creativity and general wellbeing
- orchestration of the relationships an organization has with its workforce versus being a mere administrator of processes and managing employment risks on behalf of the company; HR can adopt techniques and methods from marketing and customer

relationship management on how to build and manage relationships with its workforce by treating them as past, current and future customers, how to build the employment brand image such that it appeals to and attracts the candidates with the best cultural fit and critical skills, and how to build a better understanding of its workforce through sophisticated analytics. By employing such methods, HR will have the opportunity to create the stories necessary to match the emotional connections of workers with the organizational brand
- responsibility to humanize the workplace and prevent inequity, bias, and extremism.

At the human level, our intrinsic needs for self-actualization, love and belonging, and respect and dignity will remain constant. The digital world will change the vehicles through which we develop and express ourselves, how we connect with others, and how we define and attain our purposes. The organizations that build their models to work in harmony with these intrinsic needs of their workforce will have a chance to survive in the long term.

Our brains demonstrate neuroplasticity – they can rewrite themselves, on a structural, neuronal and network level, to an extent that we have discovered is greater than perhaps we can imagine (Joja, 2013). In the future that we face, we are lucky that we possess such flexibility. Organizations – in some ways analogous to brains, themselves collections of individuals in more or less stable networks – will also need to show such flexibility. Da Vinci questioned everything, and dismissed easy answers; Monet painted over 30 versions of Rouen Cathedral; it took Edison 1800 failed attempts to come up with the light bulb. We will need to show such character-exploring options, handling risk and embracing failure – and organizations need to ensure they are set up to encourage and handle it. After all, humans are animals, our deepest codes running unconsciously and without care for the long term. That is, we need to employ every trick we can to prioritize our highest mind and look to the future in the most rational and best possible way.

Future Research

This fast changing environment requires in-depth research and understanding of the implications that digitization, datafication, and disintermediation of work will have on our lives and organizational structures and norms. Many questions arise in relation to this future, to the opportunities to use the advancements in the fields of data, analytics, and neuroscience to better understand human behavioral patterns and ways to influence them. What is the impact of such opportunities on individual privacy and the ethics and morality of data use and analytics? How to balance wealth-building with solving weighty societal issues, and defining the steps governments must take in order to balance the needs of the global business environment and changes in domestic policies to stay competitive and continue to provide social protections to their citizens? What will be the impact on taxation? What will constitute crime in this increasingly digital world and how will security and protection of digital assets be ensured? How will organizations need to be structured, how should workflow be designed, and what is the notion of leadership and decision-making in the new era?

There are also specific psychological issues to consider. What will an epoch of mass remote work look like? If these trends follow through, it may be common to live in households composed of remote workers, with implications for home and work life. With regards to the former, we need to investigate how marital relations will be affected when couples spend the majority of days together from morning until night time.

With regards to the latter, how much will we be inclined to invest in the daily rapport with our virtual team mates when we have good friends or family right there, in our kitchen? How is remote working experienced for those who to date have not gravitated to roles where this is the norm, and how far can technology vanquish alienation? Should the office become a thing of the past for everyone? Furthermore, how will these concerns align with cohort effects, given that the newest cohort, sometimes termed Generation Z, enact more of their personal relationships online. Are they content for this to extend to the workplace, or is the face-to-face office an important buffer against the loss of physical human contact?

At the time of writing, developments in AI are beginning to penetrate mainstream awareness thanks to the salience of white-collar job loss (e.g., Dunkley, 2016). We need to understand changing attitudes to AI and machine-assisted work. Machines could enhance self-efficacy, but equally erode self-worth, depending on whether they are developed as appropriate technologies that help humans achieve more by extending their minds, or as the primary agent, with humans relegated to fact-checkers or flesh front-ends. Understanding this requires fine-grained analyses of the meaning and degree of rewards provided by different work activities within sectors facing a transition to AI-assisted work. It also necessitates monitoring whether the introduction of AI – and indeed other technology-related changes in work processes – enhances flow, creative self-efficacy, and gratification, or produces harmful outcomes. Changes in patterns of work merit steady exploration. In the future, it may be that "work is for robots" and traditional near-full employment will no longer make sense as a societal goal. Assuming we achieve this or other abundant futures, much of our understanding of the psychological consequences of employment and under-employment will become outdated. Currently it would be useful to engage in comparative work that contrasts cultures that assign different prestige to being employed and provide differing social security nets. Triangulating such research with work on the impact of volunteering and other unsalaried work – from editing Wikipedia to producing a blog – may help us understand the long-term human consequences of a life where structured activities give way to self-initiated goals.

Conclusion

This chapter explored several examples of datafication, digitization, and disintermediation. These paradigmatic shifts challenge the traditional norms that define a corporation/institution – from internal management systems, processes, physical footprints and culture to external partnerships that it builds to accomplish its business goals, companies it considers as competitors, and the positioning of the brand itself. As the organizational boundaries become more porous, work and the workplace are becoming increasingly digital and some work can now be done anywhere. Knowledge is growing more accessible, and the system of workplace motivators is shifting. The new order is disrupting how wealth is created, as well as how the relationships between work, the workplace, and workers are structured, and over time, it is reframing workplace norms (resource management, definition of leadership, definition and management of performance, brand value, etc.) The impact of such disruptions on the psychological wellbeing of individuals is not fully understood and will require much research in the years to come. One thing is clear – data and technology have a lot of power and with power comes great responsibility. It is the responsibility of all of us to innovate, design, build, and regulate in a way that minimizes the negatives and amplifies the positives of such an environment. Do no harm!

References

Aiello, J. R., & Kolb, K. J. (1995). Electronic performance monitoring and social context: Impact on productivity and stress. *Journal of Applied Psychology, 80*(3), 339–353. DOI:10.1037//0021-9010.80.3.339

Aldebaran Robotics (2015). SoftBank to Launch Sales of "Pepper." Retrieved July 7, 2015, from https://www.aldebaran.com/sites/aldebaran/files/press-releases/softbank_to_launch_sales_of_pepper_1.pdf

Anderson, A. J., Kaplan, S. A., & Vega, R. P. (2014). The impact of telework on emotional experience: When, and for whom, does telework improve daily affective well-being? *European Journal of Work and Organizational Psychology, 6*, 882–897. DOI:10.1080/1359432X.2014.966086

Ang, S., & Van Dyne, L. (2008). *Handbook of cultural intelligence*. London: Routledge.

Aral, S. (2014). The problem with online ratings. *MIT Sloan Management Review, Winter*. Retrieved April 23, 2017, from http://sloanreview.mit.edu/article/the-problem-with-online-ratings-2/

Banakou, D., Groten, R., & Slater, M. (2013). Illusory ownership of a virtual child body causes overestimation of object sizes and implicit attitude changes. *Proceedings of the National Academy of Sciences, 110*(31), 12846–12851. DOI:10.1073/pnas.1306779110

Baron, R. A. (1988). Negative effects of destructive criticism: Impact on conflict, self-efficacy, and task performance. *Journal of Applied Psychology, 73*(2), 199–207. DOI:10.1037/0021-9010.73.2.199

Belot, M., & Schröder, M. (2013). Does monitoring work?: A field experiment with multiple forms of counterproductive behaviour. Otto-von-Guericke-Universität Magdeburg, Retrieved 1 June, 2016, from www.fww.ovgu.de/fww_media/femm/femm_2013/2013_06.pdf

Bersin, J. (2016). Reinventing the career: What should organizations do? Retrieved April 17, 2017, from http://joshbersin.com/2016/04/reinventing-the-career-what-should-organizations-do/

Bloom, N., Liang, J., Roberts, J., & Ying, Z. J. (2015). Does working from home work? Evidence from a Chinese experiment. *Quarterly Journal of Economics, 130*(1), 165–218. DOI:10.3386/w18871

Bordia, P., Kiazad, K., Restubog, S. L. D., DiFonzo, N., Stenson, N., & Tang, R. L. (2014). Rumor as revenge in the workplace. *Group & Organization Management, 39*(4), 363–388. DOI:10.1177/1059601114540750

Buckingham, M., & Goodall, A. (2015, April). Reinventing performance management. *Harvard Business Review*. Retrieved April 7, 2016, from https://hbr.org/2015/04/reinventing-performance-management

Buhi, E. R., Trudnak, T. E., Martinasek, M. P., Oberne, A. B., Fuhrmann, H. J., & McDermott, R. J. (2013). Mobile phone-based behavioural interventions for health: A systematic review. *Health Education Journal, 72*(5), 564–583. DOI:10.1177/0017896912452071

Burke, E. (2015). The powered exoskeleton. *Biotech Primer Weekly*. Retrieved from http://weekly.biotechprimer.com/the-powered-exoskeleton/

Carter, M. (2012, June 21) Makie puts toy-making power in consumers' hands with 3-D printed dolls. Fast Company. Retrieved April 24, 2017, from https://www.fastcompany.com/1680956/makie-puts-toy-making-power-in-consumers-hands-with-3-d-printed-dolls

Cialdini, R. B. (1993). *Influence: Science and practice* (3rd ed.). New York: HarperCollins.

Clark, A. (1996). *Being there: Putting brain, body and world together again*. Boston, MA: MIT Press.

Clark, L. A., Karau, S. J., & Michalisin, M. D. (2012). Telecommuting attitudes and the "big five" personality dimensions. *Journal of Management Policy and Practice, 13*(3), 31–47.

Cole, J., & Paillard, J. (1995). Living without touch and peripheral information about body position and movement: Studies with deafferented subjects. In J. L. Bermudez, A. Marcel, & N. Eilan (Eds.), *The body and the self* (pp. 245–266). Boston, MA: MIT Press.

Cosley, D., Lam, S. K., Albert, I., Konstan, J. A., & Riedl, J. (2003). Is seeing believing?: How recommender system interfaces affect users' opinions. In *Proceedings of the SIGCHI Conference on Human Factors in Computing Systems* (pp. 585–592). New York, NY: ACM.

Davenport, T. H. (2013). Analytics 3.0. *Harvard Business Review*. Retrieved April 7, 2017, from https://hbr.org/2013/12/analytics-30

Davis, S., & Davidson, B. (1991). *2020 vision: Transform your business to succeed in tomorrow's economy.* New York, NY: Simon and Schuster.
Dewitte, H. (1993). Psychological consequences of long-term unemployment – review of the literature. *Psychologica Belgica, 33*(1), 1–35.
Dolan, S., & Kawamura, K. (2015). *Cross cultural competence: A field guide for developing global leaders and managers.* Bingley, UK: Emerald Group Publishing.
Duhigg, C. (2016). *Smarter faster better: The secrets of being productive in life and business.* New York, NY: Penguin Random House.
Dunkley, E. (2016, March 13). RBS Cuts face-to-face service and brings in. *The Financial Times.* London. Retrieved April 7, 2017, from www.ft.com/intl/cms/s/10df9f22-e90f-11e5-bb79-2303682345c8,Authorised=false.html?siteedition=intl&_i_location=http%3A%2F%2Fwww.ft.com%2Fcms%2Fs%2F0%2F10df9f22-e90f-11e5-bb79-2303682345c8.html%3Fsiteedition%3Dintl&_i_referer=&classification=conditio
Elsbach, K. D., Cable, D. M., & Sherman, J. W. (2010). How passive "face time" affects perceptions of employees: Evidence of spontaneous trait inference. *Human Relations, 63*(6), 735–760. http://doi.org/10.1177/0018726709353139
Ernest, E., & Marcus, G. (2015). Commonsense reasoning and commonsense knowledge in artificial intelligence. *Communications of the ACM. 58*, 92–103. Retrieved June, 1, 2016, from http://m.cacm.acm.org/magazines/2015/9/191169-commonsense-reasoning-and-commonsense-knowledge-in-artificial-intelligence/fulltext
Fairlie, R. W., Morelix, A., Reedy, E., & Russell, J. (2015). *2015 Kaufmann Index on startup activity: National trends.* Retrieved April 7, 2017, from www.kauffman.org/~/media/kauffman_org/research reports and covers/2015/05/kauffman_index_startup_activity_national_trends_2015.pdf
Festinger, L. (1954). A theory of social comparison processes. *Human Relations, 7*, 117–140. DOI:10.1177/001872675400700202
Gajendran, R. S., & Harrison, D. A. (2007). The good, the bad, and the unknown about telecommuting: meta-analysis of psychological mediators and individual consequences. *Journal of Applied Psychology, 92*(6), 1524–1541. DOI:10.1037/0021-9010.92.6.1524
Geddes, D., & Baron, R. A. (1997). Workplace aggression as a consequence of negative performance feedback. *Management Communication Quarterly, 10*(4), 433–454. DOI:10.1177/0893318997104002
Gibson, J. J. (1966). *The senses considered as perceptual systems.* Boston, MA: Houghton Mifflin.
Gordon, R. J. (2015). *The rise and fall of American growth: The U.S. standard of living since the civil war.* Princeton, NJ: Princeton University Press.
Haddad, H., Lyons, G., & Chatterjee, K. (2009). An examination of determinants influencing the desire for and frequency of part-day and whole-day homeworking. *Journal of Transport Geography, 17*(2), 124–133. DOI:10.1016/j.jtrangeo.2008.11.008
Hamam, A., Eid, M., & El Saddik, A. (2013). Effect of kinesthetic and tactile haptic feedback on the quality of experience of edutainment applications. *Multimedia Tools and Applications, 67*(2), 455–472. DOI:10.1007/s11042-012-0990-7
Harper, S., & Mousa, F. T. (2013). Time-and-motion studies. In R. Griffin (Ed.), *Oxford bibliographies in management.* New York, NY: Oxford University Press.
Hernandez, D. (2016). It's hard work being funny – especially for robots. Retrieved April 4, 2016, from http://fusion.net/story/251798/funny-robots/
Herzberg, F., Mausner, B., & Snyderman, B. B. (2011). *The motivation to work* (2nd ed.). Piscataway, NJ: Transaction Publishers.
Hutton, J. G., Goodman, M. B., Alexander, J. B., & Genest, C. M. (2001). Reputation management: The new face of corporate public relations? *Public Relations Review, 27*(3), 247–261. DOI:10.1016/S0363-8111(01)00085-6
Ihle, A., Schnitzspahn, K., Rendell, P. G., Luong, C., & Kliegel, M. (2012). Age benefits in everyday prospective memory: The influence of personal task importance, use of reminders and everyday stress. *Aging, Neuropsychology, and Cognition, 19*(1-2), 84–101. DOI:10.1080/13825585.2011.629288

Johnson, W. (2013). Independent work may be inevitable. *Harvard Business Review*. Retrieved April 7, 2017, from https://hbr.org/2013/02/independent-work-may-be-inevit/

Joja, D. O. (2013). Learning experience and neuroplasticity – a shifting paradigm. *Noema*, *7*, 159–170.

Kahneman, D., Krueger, A. B., Schkade, D. A., Schwarz, N., & Stone, A. A. (2004). A survey method for characterizing daily life experience: the day reconstruction method. *Science*, *306*(5702), 1776–1780. DOI:10.1126/science.1103572

Kaptein, M., & van Halteren, A. (2013). Adaptive persuasive messaging to increase service retention: Using persuasion profiles to increase the effectiveness of email reminders. *Personal and Ubiquitous Computing*, *17*(6), 1173–1185. DOI:10.1007/s00779-012-0585-3

Kelly, E. L., Moen, P., Oakes, J. M., Fan, W., Okechukwu, C., Davis, K. D., ... Casper, L. M. (2014). Changing work and work-family conflict: Evidence from the work, family, and health network. *American Sociological Review*, *79*(3), 485–516. doi:10.1177/0003122414531435

Kelly, K. (2014, December). The future of AI? Helping human beings think smarter. *Wired*. Retrieved April 23, 2017, from www.wired.co.uk/article/brain-power

Keynes, J. M. (1933). *Essays in persuasion*. London: Macmillan.

Kim, W.-B. (2003). Economic crisis, downsizing and "layoff survivor's syndrome." *Journal of Contemporary Asia*, *33*, 449–464. https://doi.org/10.1080/00472330380000281

Kluger, A. N., & DeNisi, A. (1996). The effects of feedback interventions on performance: A historical review, a meta-analysis, and a preliminary feedback intervention theory. *Psychological Bulletin*, *119*(2), 254–284. DOI:10.1037/0033-2909.119.2.254

Konnikova, M. (2015). Virtual reality gets real. *The Atlantic*. Retrieved April 7, 2017, from https://www.theatlantic.com/magazine/archive/2015/10/virtual-reality-gets-real/403225/

Lamport, L., Shostak, R., & Pease, M. (1982). The Byzantine generals problem. *ACM Transactions on Programming Languages and Systems*, *4*(3), 382–401. doi:10.1145/357172.357176

Leon, P., Rao, A., Schaub, F., Marsh, A., & Cranor, L. (2015). Why people are (Un) willing to share information with online advertisers. Technical Report CMU-ISR-15-106, Carnegie Mellon University. Retrieved April 7, 2017, from http://ra.adm.cs.cmu.edu/anon/usr0/ftp/home/anon/isr2015/CMU-ISR-15-106.pdf

Levy, F., & Murnane, R. J. (2004). *The new division of labor: how computers are creating the next job market*. Princeton, NJ: Princeton University Press.

Livingstone, S. (2008). Taking risky opportunities in youthful content creation: teenagers' use of social networking sites for intimacy, privacy and self-expression. *New Media & Society*, *10*(3), 393–411. DOI:10.1177/1461444808089415

Loi, R., Hang-Yue, N., & Foley, S. (2006). Linking employees' justice perceptions to organizational commitment and intention to leave: The mediating role of perceived organizational support. *Journal of Occupational and Organizational Psychology*, *79*(1), 101–120. DOI:10.1348/096317905X39657

Lundberg, M., Hagberg, K., & Bullington, J. (2011). My prosthesis as a part of me: A qualitative analysis of living with an osseointegrated prosthetic limb. *Prosthetics and Orthotics International*, *35*(2), 207–214. DOI:10.1177/0309364611409795

Makino, Y., Furuyama, Y., & Shinoda, H. (2015). HaptoClone (Haptic-Optical Clone). In *Proceedings of the 3rd ACM Symposium on Spatial User Interaction* (pp. 139–139). New York, NY: ACM Press. DOI:10.1145/2788940.2794354

Manyika, J., Lund, S., Robinson, K., Valentino, J., & Dobbs, R. (2015). *Connecting talent with opportunity in the digital age*. Retrieved April 7, 2017, from www.mckinsey.com/global-themes/employment-and-growth/connecting-talent-with-opportunity-in-the-digital-age

Massey, A. (2016). Why the future of work is at home. Retrieved April 4, 2016, from www.psmag.com/business-economics/office-youre-already-dead-to-me-leave-us-alone-gawd

McKee-Ryan, F., Song, Z., Wanberg, C. R., & Kinicki, A. J. (2005). Psychological and physical well-being during unemployment: a meta-analytic study. *Journal of Applied Psychology*, *90*(1), 53–76. DOI:10.1037/0021-9010.90.1.53

Mueller, P. A., & Oppenheimer, D. M. (2014). The pen is mightier than the keyboard: Advantages of longhand over laptop note taking. *Psychological Science*, *25*(6), 1159–1168. DOI:10.1177/0956797614524581

Niehoff, B., & Moorman, R. (1993). Justice as a mediator of the relationship between methods of monitoring and organizational citizenship behavior. *Academy of Management Journal, 36*(3), 527–556. DOI:10.2307/256591

Patel, P. C., & Thatcher, S. M. B. (2014). Sticking it out: Individual attributes and persistence in self-employment. *Journal of Management, 40*(7), 1932–1979. DOI:10.1177/0149206312446643

Raghuram, S., & Wiesenfeld, B. (2004). Work-nonwork conflict and job stress among virtual workers. *Human Resource Management.* DOI:10.1002/hrm.20019

Rauch, A., & Frese, M. (2007). Let's put the person back into entrepreneurship research: A meta-analysis on the relationship between business owners' personality traits, business creation, and success. *European Journal of Work and Organizational Psychology, 16*(4), 353–385. DOI:10.1080/13594320701595438

Robles-De-La-Torre, G. (2006). The importance of the sense of touch in virtual and real environments. *IEEE Multimedia, 13*(3), 24–30. DOI:10.1109/MMUL.2006.69

Ryan, R. M., & Deci, E. L. (2000). Intrinsic and extrinsic motivations: Classic definitions and new directions. *Contemporary Educational Psychology, 25*(1), 54–67. DOI:10.1006/ceps.1999.1020

Santens, S. (2016, March). Deep learning is going to teach us all the lesson of our lives: jobs are for machines. *Medium.* Retrieved April 7, 2017, from https://medium.com/basic-income/deep-learning-is-going-to-teach-us-all-the-lesson-of-our-lives-jobs-are-for-machines-7c6442e37a49

Saval, N. (2014). White-collar world. *Chronicle of Higher Education.* Retrieved April 7, 2017, from http://chronicle.com/article/Paper-Pushers/145839?cid=trend_right

Schiller, B. (2015, August). Robots will take your job, but first they'll be your annoying co-worker. *Fastcoexist.com.* Retrieved April 7, 2017, from http://www.fastcoexist.com/3050428/robots-will-take-your-job-but-first-theyll-be-your-annoying-co-worker

Scullen, S., Mount, M., & Goff, M. (2000). Understanding the latent structure of job performance ratings. *Journal of Applied Psychology, 85*(6), 956–970. DOI:10.1037/0021-9010.85.6.956

Searle, R. H. (2006). New technology: the potential impact of surveillance techniques in recruitment practices. *Personnel Review, 35*(3), 336–351. DOI:10.1108/00483480610656720

Silver, D., Huang, A., Maddison, C. J., Guez, A., Sifre, L., van den Driessche, G., ... Hassabis, D. (2016). Mastering the game of Go with deep neural networks and tree search. *Nature, 529*(7587), 484–489. DOI:10.1038/nature16961

Slywotzky, A. J. (1996). *Value migration: How to think several moves ahead of the competition.* Boston, MA: Harvard Business School Press.

Smith, A., & Garnier, M. (1838). An inquiry into the nature and causes of the wealth of nations. Edinburgh: T. Nelson.

Smith, M. J., Carayon, P., Sanders, K. J., Lim, S. Y., & LeGrande, D. (1992). Employee stress and health complaints in jobs with and without electronic performance monitoring. *Applied Ergonomics, 23*(1), 17–27. DOI:10.1016/0003-6870(92)90006-H

Sohn, T., Li, K. A., Lee, G., Smith, I., Scott, J., & Griswold, W. G. (2005). Place-its: A study of location-based reminders on mobile Phones. In M. Beigl, S. Intille, J. Rekimoto, & H. Tokuda (Eds.), *UbiComp 2005: Ubiquitous Computing. UbiComp 2005.* Lecture Notes in Computer Science, Vol. 3660 (pp. 232–250). Berlin: Springer Berlin. DOI:10.1007/11551201_14

Stanton, J. M., & Weiss, E. M. (2000). Electronic monitoring in their own words: An exploratory study of employees' experiences with new types of surveillance. *Computers in Human Behavior, 16*(4), 423–440. DOI:10.1016/S0747-5632(00)00018-2

Steil, J. J. & Maier, G. W. (2017). Robots in the digitalized workplace. In G. Hertel, D. L. Stone, R. D. Johnson, & J. Passmore (Eds.), *The Wiley Blackwell Handbook of the psychology of the internet at work* (pp. 403–422). Chichester: John Wiley & Sons, Ltd.

Stephens, J., & Allen, J. (2013). Mobile phone interventions to increase physical activity and reduce weight: A systematic review. *Journal of Cardiovascular Nursing, 28*(4), 320–329. DOI:10.1097/JCN.0b013e318250a3e7

Thomée, S., Härenstam, A., & Hagberg, M. (2011). Mobile phone use and stress, sleep disturbances, and symptoms of depression among young adults – a prospective cohort study. *BMC Public Health, 11*(1), 66. DOI:10.1186/1471-2458-11-66

Tsui, K. M., Desai, M., Yanco, H. A., & Uhlik, C. (2011, March). *Exploring use cases for telepresence robots*. Paper presented at the 6th ACMIEEE International Conference on Human-Robot Interaction (HRI 2011), Lausanne, Switzerland. Retrieved April 7, 2017, from http://ieeexplore.ieee.org/xpl/articleDetails.jsp?arnumber=6281254

Van den Broek, M. D., Downes, J., Johnson, Z., Dayus, B., & Hilton, N. (2000). Evaluation of an electronic memory aid in the neuropsychological rehabilitation of prospective memory deficits. *Brain Injury, 14*(5), 455–462. DOI:10.1080/026990500120556

Van der Land, S. F., Schouten, A. P., Feldberg, F., Huysman, M., & van den Hooff, B. (2015). Does avatar appearance matter? How team visual similarity and member-avatar similarity influence virtual team performance. *Human Communication Research, 41*(1), 128–153. DOI:10.1111/hcre.12044

Van der Land, S., Schouten, A. P., Van den Hooff, B., & Feldberg, F. (2011). Modeling the metaverse: A theoretical model of effective team collaboration in 3D virtual environments. *Journal For Virtual Worlds Research, 4*(3) DOI:10.4101/jvwr.v4i3.6126

Van Yperen, N. & Wörtler, B. (2017). Blended working. In G. Hertel, D. L. Stone, R. D. Johnson, & J. Passmore (Eds.), *The Wiley Blackwell handbook of the psychology of the internet at work* (pp. 157–174). Chichester: John Wiley & Sons, Ltd.

Vardi, M. Y. (2012). Artificial intelligence. *Communications of the ACM, 55*(1), 5–5. DOI:10.1145/2063176.2063177

West, M. D., & Anderson, J. (2005). Telework and employees with disabilities: Accommodation and funding options. *Journal of Vocational Rehabilitation, 23*(2), 115–122.

Woo, J.-M., & Postolach, T. T. (2008). The impact of work environment on mood disorders and suicide: Evidence and implications. *International Journal on Disability and Human Development, 7*(2), 185–200. DOI:10.1515/IJDHD.2008.7.2.185

Yee, N., & Bailenson, J. (2007). The Proteus effect: The effect of transformed self-representation on behavior. *Human Communication Research, 33*(3), 271–290. DOI:10.1111/j.1468-2958.2007.00299.x

Yuhas, A. (2016, February 13). Would you bet against sex robots? AI "could leave half of world unemployed." *The Guardian*. Retrieved April 7, 2017, from https://www.theguardian.com/technology/2016/feb/13/artificial-intelligence-ai-unemployment-jobs-moshe-vardi

Index

3D printing, *see* three-dimensional [3D] printing

A

abuse of computer access at work, *see* cyber-aggression; cyberdeviance; cyberloafing
access to technology, inequality of, *see* digital divide
accessibility of information 4
acquisition of information
  cyberdeviance focused on 133
action-centred study of work practices 41
Advanced Research Projects Agency Network (ARPANET) 2
aesthetics
  definition of 81
  employer websites 215
  measurement of 93
  user experience 81
affordances
  concept of 27
age differences at work
  cognition 454
  personality 455
  work motivation 456
  age bias by employers 247
  aging workforce 450
  blended working 164
  e-learning 382, 460, 467
  e-recruiting 457, 466
  e-selection 274, 459, 466
  future research 465
  human resources functions 457
  identification as older or younger worker 451
  Internet-based work tools 452
  introduction to 449
  job satisfaction 165
  leadership 462, 463
  meta-stereotypes concept 457
  societal context 457
  performance management 464
  relational demography concept 463
  summary overview 469
  virtual teams 462, 467
  within-person changes during career 454
agency
  of artifacts 120
  level of 89
  maximization 89
  power 40, 50
  relationships 50
  restoration of 41
  sense of 179
  theory 104
  trust 115
aggregators of job adverts 215
aggression, *see* cyber-aggression
AI, *see* artificial intelligence (AI)
Airbnb 3
Amazon 2, 117, 499
Amazon Mechanical Turk (AMT) 3, 51, 261
AMT, *see* Amazon Mechanical Turk (AMT)

The Wiley Blackwell Handbook of the Psychology of the Internet at Work,
First Edition. Edited by Guido Hertel, Dianna L. Stone, Richard D. Johnson, and Jonathan Passmore.

Apple, 3, 499
ARPANET 2
artificial intelligence (AI)
  digitization 481
  effects on work 3
  e-selection 275, 276
  future of work 492
  information revolution 487
  intelligent algorithms 4
  leadership 3
  robots 406, 415, 488
assistance from robots, *see* robots
automatization
  potential for 4
  psychological theories relevant to 6
autonomy
  blended working 162
  as core job characteristic 162
  enhancement of 63, 89, 185
  experience of 418, 458
  flexwork 181
  ICT use 184, 186
  Internet-based work 3
  lack of 162
  level of 137, 147, 162, 167, 181
  maintenance of 33, 437
  need for 160, 162, 165, 168, 222, 408, 418, 442
  older workers 458
  perception of 167, 488
  robots 408, 418
  telework 488
  value of 63, 437, 441

B

behavior, *see* psychology
Bezos, Jeff 51
bias
  age bias 247
  gender bias 247
  racial bias 247
  suppression 246
'big data'
  analysis 33
  articles on 260
  emergent technologies 118
  e-selection 275
  intelligent algorithms for handling 4
  mobile computing 204
  work performance measurement 453
blended working
  adoption of 166
  autonomy 162
  caregivers 164
  concept 10, 157, 168
  consequences
  future research 167
  intrinsic work motivation 162
  introduction to 157
  job satisfaction 158, 160, 163
  older workers 164
  perceived effectiveness 160
  perceived opportunity for (OBW) 162
  psychological factors 159
  self-determination theory (SDT) 160
  as solution for high job demands 161
  summary overview 168
  workers with physical disability 166
blogs
  benefits of 29
  'big data' 483
  digitized communication 27
  e-leadership 289
  enterprise social media 28
  ethical issues 236
  impact on work practices 41
  information sharing 48, 388
  persistence of data 47
  volunteering 503
boundary-crossing
  capabilities 5
  psychological theories relevant to 5
branding
  e-recruiting 220
breaktimes, *see* work breaks
bring your own device (BYOD), *see* mobile computing
browsing, *see* web browsing
bullying, *see* cyber-aggression
burnout, *see* stress

C

careers, *see* employment decision-making; e-recruiting; e-selection
caregivers
  blended working 164
CAT, *see* computerized adaptive testing (CAT)
cell phones, *see* 'smart' phones
channel expansion theory of computer-mediated communication 68
cloud computing
  mobile computing 200, 202
CMC, *see* computer-mediated communication (CMC)
cognition
  age 454
cognitive ability testing in e-selection 263
cohesion in virtual teams 332

collaborative work
collaboration tools 3
competences, *see* competences
interactive communication 4
Open Source Software projects 2
organization by computer-mediated communication 31
relationship depth 69
robots, with, *see* robots
shared understanding within 69
'sharing' communities 3
social impact theory 69
social loafing 69
trust 69
virtual teams, *see* virtual teams
commercial Internet, *see* e-commerce
communication
behaviors of online leaders 302
complementary successive technologies, use of 66
convergence processes 22, 66
conveyance processes 22, 66
cyberdeviance focused on 133
digitized, *see* digitized communication
email 52
enterprise wikis 53
e-selection 274
interactivity of 4
relationships 52
robots as embodied communication agents 416
technology, *see* information and communication technology
virtual teams 330, 334
work–life balance 183
communication visibility
enterprise social media 29
theory of 29
communities
democracy 44
embodiment 46
expertise finding within 48
leadership 302
Open Source Software projects 33, 302
of practice 45
relationships within 48
work practices 44
compensatory adaptation theory of computer-mediated communication 25
competences
competency modeling approaches 61
computer-mediated communication and related competencies, theories of 67
de-skilling effects of robot use 412
digital collaboration and related competencies, theories of 68
great eight competency framework 62
introduction to 61
research on
low convergence of studies 70
recommendations for structuring of future research 71
self-efficacy theory 64
summary overview 73
technology acceptance and perceived ease of use 64
technology choice and related competencies, theories of 66
theoretical basis for understanding relevance of 71
computer anxiety in e-learning 381
computer crime deterrance 145
computer game elements in business applications, *see* gamification
computer self-efficacy (CSE) 381
computerized adaptive testing (CAT) 262
computer-mediated communication (CMC)
benefits of 22
channel expansion theory 68
compensatory adaptation theory 25
competence model 64
competences 67
development phases 21
effects
adaptation theories of 24
affordances, concept of 27
constructivist and deterministic assumptions about 26
deficit theories of 23
electronic propinquity theory 68
hyperpersonal model 25
lack of social cues hypothesis 23
nonverbal cues, lack of 21, 67
organization of cooperation 31
selection of 22
social identity model of deindividuation effects (SIDE) 26
social information processing theory 24, 25, 32, 67
social media 27
social presence theory 67
Web 2.0 27
concepts, definitions, ideas and theories
aesthetics 81, 104
affordances 27
agency theory 104
blended working 10, 157, 168

concepts, definitions, ideas and theories (*cont'd*)
channel expansion theory 68
communication visibility theory 29
compensatory adaptation theory 25
competency 62
counterproductive work behavior 131
critical contact theory 219
crowdwork 3
cyberdeviance 132
distrust 114
e-learning 371
electronic propinquity theory 68
enterprise social media 28
e-recruiting 213
flexwork 176, 177
gamification, *see* gamification
Gemeinschaft 44
Gesellschaft 44
hybrid manufacturing 407
information and communication technologies (ICTs) 184
information technology artifact 117
Internet 2
item response theory 262
media richness theory 22
media synchronicity theory 22, 66
meta-stereotypes 457
mobile computing 195
online environments 120
organizational justice theory 238
organizational social media 28
privacy 432
processing fluency theory 92
psychological needs 160
reasoned action 64, 104
relational demography 463
self-determination theory (SDT) 160
self-efficacy 64
signaling theory 219
social capital 30, 31
social impact theory 69
social information processing theory 24, 25, 32, 67
social presence theory 67
succession theory 66
swift trust 64, 112
team virtuality 315, 320, 321, 323
technology acceptance 64
technology trust 64
telework 177
'the office' 488
user experience 79
virtual teams 315
work engagement 188
work team 452

constructivist perspectives
effects of computer-mediated communication 26
content, *see* information
convergence processes in communication 22, 66
conveyance processes in communication 22, 66
cooperation, *see* collaborative work
corporate intranets 27, 41
counterproductive work behavior, *see* cyberdeviance
critical contact theory
e-recruiting 219
cross-cultural virtual teams
shared understanding in 70
crowd labor relationships 51
crowdsourcing
birth of 2
labor relationships 51
crowdwork
concept of 3
embodiment of relationships 51
Internet-based 9
crypto-currency 497
CSE, *see* computer self-efficacy (CSE)
CUE model of user experience 80
cues-filtered-out theories, *see* nonverbal cues
cultural differences in privacy values 437
customer satisfaction 43, 106, 107, 108, 110
cyber-aggression
cyberdeviance 131, 133, 139, 140, 141, 142
risk of 8
cyberdeviance
counterproductive work behavior concept 131
cyber-aggression 131, 133, 139, 140, 141, 142
definition 132
focused on acquisition 133
focused on communication 133
future research 146
interpersonal (CD-I)
behavior-based 141
categorization 132
characteristics 139
demographic variables 141
future research 148
outcomes 142
perception-based 141
personality variables 142
reasons 141
research 139
intrapersonal (cyberloafing), *see* cyberloafing
introduction to 131

organizational (CD-O)
attitudinal-based 144
categorization 132
characteristics 143
deterrance measures, effect of 146
future research 149
individual difference variables 143
outcomes 146
reasons 143
research 143
summary overview 150
typology 132
cyberloafing
activities 133
blended working 159
categorization 132
in collaborative work 69
collaborative work 69
cyberdeviance 132
distinction from cyberdeviance 132
future research 147
mobile computing 200
outcomes 138
overview of studies 134
reasons
deterrance measures, effect of 135
facilitating conditions 137
individual differences 134
job/organizational characteristics 136
job-related stressors 137
perceptions of organizational justice 136
socially constructed norms and roles 136
work–life balance issues 137
'cyber-mobbing'
problem of 4
cyberslacking, *see* cyberloafing

D
data, *see* information
data analysis in online surveys 350
data collection in online surveys 350
definitions, *see* concepts, definitions, ideas and theories
deindividuation effects of computer-mediated communication
social identity model of 26
democracy and community engagement 44
demography
age differences at work 463
digital divide 427
relational demography concept 463
depersonalization effects of computer-mediated communication
social identity model of 26
de-skilling, *see* competences
deterministic perspectives
effects of computer-mediated communication 26
deterrance measures against computer misuse
effect on cyberdeviance 135, 146
device diversity issues
e-selection 273
digital collaboration, *see* collaborative work
digital divide
access to technology 426
cross-national differences 430
demographic factors 427
ethnic differences 428
future research 440
gender differences 427
problem of 424
theoretical bases 424
digitization
applications 488
implications 501
trend 481, 488
digitized communication
computer-mediated communication (CMC), *see* computer-mediated communication (CMC)
future research on 32
introduction to 21
knowledge building 29
organization of cooperation 31
pre-Internet 22
relationship-building 30
social media 27
summary overview of 33
visibility 29
disabled workers
blended working 166
e-selection 274
disintermediation
implications 501
trend 481
of work value chain 494
disruptive innovation via robots 416
distrust, *see* trust
diversity, *see* workforce diversity
'dotcom crisis' 2000 2

E
ease of use perceptions 64
eBay 2
e-commerce
birth of 2, 105
consolidation 2
'dotcom crisis' 2000 2
privacy concerns 106
trust 105, 108, 119

editability of social media content 47
education
  mobile computing uses 203
e-leadership, *see* leadership
e-learning
  age differences 460
  attitudes to 379
  communication with instructor 379
  computer anxiety 381
  computer self-efficacy (CSE) 381
  context of 372
  definition of 371
  dimensions of effectiveness 372
  effectiveness models 371
  future research 387
  gamification 389
  guidance to trainees 378
  individual differences and learner choices research model 372
  instructors
    communication with 379
    role of 379
  interaction 384
  introduction to 369
  knowledge repositories 388
  learner choices 386
  learner control 377
  learner task complexity 377
  learning climate 375
  metacognitive activity 386
  mobile learning 203, 387
  organizational context 374
  organizational support 374
  outcomes 373
  previous online course experience 382
  process of 384
  research inputs 374
  reviews of 370
  self-efficacy 379, 381
  social presence 385
  summary overview 390
  technology for
    media richness 376
    reliability 375
    useability and usefulness 375
  trainees
    age 382
    computer anxiety 381
    computer self-efficacy (CSE) 381
    motivation 383
    personality 380
    previous online course experience 382
    self-efficacy 379
  virtual realities 389
electronic propinquity theory of computer-mediated communication 68
Eluta 215
email
  access 3
  address, control over 434
  bring your own device (BYOD) 200
  communication 52
  compensatory adaptation theory 25
  crowd labor relationships 51
  crypto-currency 498
  cyberdeviance 132, 139, 141, 142
  cyberloafing 133, 134, 136, 137, 138, 147
  data on usage 482, 483
  data structuredness 276
  e-leadership 298, 300, 301, 306, 464
  e-learning 382
  e-recruiting 214
  e-selection 267
  ethnic differences in use 428
  expertise finding 49
  first email sent 2
  flexwork 188
  gender differences in use 427
  impact on on work practices 41
  mobile computing 195, 196, 199
  monitoring of use 485
  online surveys and feedback 349, 350, 354, 355, 356
  personal email websites 199
  restriction policy 147
  richness 22
  richness perception 68, 376
  social influence model of technology 23
  social networking 47
  spam 5
  stress 52
  usage data 482, 483
  usage monitoring 485
  videoconferencing 66
  virtual teams 318, 319, 324, 329, 334, 335, 338
  Web 2.0 41
  work simulations 267
email-free workday
employee monitoring by robots 413
employee surveys, *see* online surveys
employer websites
  aesthetics 215
  content 216
  effectiveness 215
  e-recruiting uses 221
  interactivity 216
  navigability 216

employment decision-making
  accuracy of information 240
  advanced notice of procedures 248
  age bias 247
  bias suppression 246
  consistency of procedure 239
  correctability 243
  e-recruiting, *see* e-recruiting
  ethical issues 236
  ethical values, correspondance with 245
  future research 249
  gender bias 247
  introduction to 231
  legal issues 236
  online applicant data usage
    advantages 233
    disadvantages 234
  organizational justice theory 238
  overview of studies 235
  privacy concerns 235
  procedural consistency 239
  racial bias 247
  social values, correspondance with 244
  summary overview 251
  validity of information 241
  verification of data 242
empowerment, *see* power
enterprise social media
  capabilities 28
  communication visibility 29
  definition of 28
  future research on 32
  knowledge-building uses 29
  organization of cooperation 31
  relationship-building uses 30
  visibility enhancement by 29
enterprise wikis
  knowledge sharing uses 53
e-recruiting
  age differences 457
  aggregators 215
  applicant quality 218
  applicant reactions 223
  branding 220
  critical contact theory 219
  current approaches 221
  definition of 213
  disadvantages of 223
  effectiveness of 226
  employer websites
    aesthetics 215
    content 216
    effectiveness 215
    interactivity 216
    navigability 216
  ethics 224
  first impressions, importance of 219
  gamification 222
  history of 214
  implementation issues 223
  interviews 216
  limitations of 223
  mobile recruiting 222
  networking 220
  novelty 220
  online job boards 214, 221
  organizational acceptance 224
  passive recruiting 219
  privacy concerns 224
  signaling theory 219
  social media 221
  social networking 217, 218, 219, 224
  summary overview 227
  theories applicable to 226
  volume management 217
ergonomics
  current study outline 41
  future research 54
  robots 414
  summary overview 54
  system-oriented approach to 40
  work practices, *see* work practices
e-selection
  administrators' resource requirements 271
  age differences 459
  artificial intelligence (AI) 275, 276
  audience for 269
  candidate reactions 269
  challenges 258
  cognitive ability, testing of 263
  communication issues 274
  computerized adaptive testing (CAT) 262
  data privacy and security 259, 272
  deployment 258
  device diversity issues 273
  employer involvement levels 270
  equivalence considerations in multiple test administration modes 260
  future research
    artificial intelligence (AI) 276
    'big data' 275
    collection of new types of data, opportunities for 276
    gamification 276
    machine learning 276
    simulations 276
    ubiquity of computing devices 276
    virtual reality 276

e-selection (*cont'd*)
game-like (gamified) assessments 268
global solutions 271
impact of computerization on survey assessments 265
implementation issues 268
international standards 271
interviews 265
introduction to 257
item response theory 262
knowledge testing 263
personality testing 264
protected groups, accommodation for 274
range of assessments and tests 263
security issues 273
situational judgment tests (SJTs) 266
summary overview 277
supervised testing 272
training issues 274
unsupervised testing 261, 272
work simulations 267
ethical issues
employment decision-making 236
e-recruiting 224
ethical values
correspondance of employment decisions with 245
ethnic minorities
digital divide 428
employer bias against 247
e-selection 274
privacy 438
exchange and transactions of value
innovation in 498
exoskeletons
future of work 492
expertise finding within organizations 48

F
Facebook
banning of 251
communities 47
cyberdeviance 139
development of 3
digitized communication 27
e-leadership 301
employment decision-making 231, 249, 250
e-recruiting 221
ethical issues 246
information technology artifact 117
legal issues 236, 237
organizational use 28
privacy 431
racial bias by employers 248
relationship-building uses 30
social capital 31
social networking 217
validity of information 241
verification of data 242
face-to-face (FtF) communication
competency modeling approaches 62
computer-mediated communication (CMC) compared 22
electronic propinquity theory 68
nonverbal cues 22, 67
richness 22
family, *see* work–life balance
File Transfer Protocol (FTP) 2
financial transaction technologies 497
flexwork
arrangements 178
categorization 177
concept 177
definition 176
future of work 488
future research 187
individual influences 179
information and communication technologies (ICTs)
flexwork in relation 176
ICT usage 175, 185
outcomes
conditional 186
negative 186
positive 185
summary overview 188
introduction to 175
offering of 177
organizational influences 179
outcomes
negative conditional outcomes 181
negative general outcomes 180
positive conditional outcomes 180
positive general outcomes 179
scope 177
suitability 178
summary overview 181, 188
telework, definition of 177
work–life balance in relation 176
'forgetting', *see* 'intentional forgetting'
FtF, *see* face-to-face (FtF) communication
FTP 2
future of work
3D printing 496
artificial intelligence 492
automation 488
change: challange and shared responsibility 501
datafication

implications 501
information revolution: shrinking of speed and distance 487
measurement of work 485
trend 481, 482
ubiquity of data 482
digitization
applications 488
implications 501
trend 481, 488
disintermediation
implications 501
trend 481
of work value chain 494
empowerment of independents 496
exchange of anything of value 497
exoskeletons 492
fusion of physical and digital 489
future research 502
holography 490
integration technologies 488
introduction to 481
machines doing human work 494
networked work 495
'the office', concept of 488
robotization 488, 492
summary overview 503
value creation and migration 499
virtual reality 490
work away from work 488

G
gamification
business applications 85
effects on work performance 86
e-learning 389
e-recruiting 222
e-selection 268, 276
industrial production case study
background to 89
breaktime management 90
collaboration optimization 91
results 91
team-oriented score system 90
team-working rhythm 91
overview of studies 88
Gemeinschaft concept of community 44
gender
bias by employers 247
digital divide 427
privacy 438
generation Y
mobile computing 199
Gesellschaft concept of community 44
Github 47
global positioning system (GPS)
effects on work 3
monitoring of workers 4
wearable technologies 360
GPS, *see* global positioning system (GPS)
Great Eight competency framework 62
group support system (GSS) 286, 293

H
hand-held devices, *see* mobile computing
harassment, *see* cyber-aggression
high job demands, blended working as solution 161
history of Internet at work 2
holography and future of work 490
homeworking, *see* blended working
human resource management, *see* age differences at work; employment decision-making; e-recruiting; e-selection
hybrid manufacturing concept 407, 408
hyperpersonal model of computer-mediated communication 25

I
ICTs, *see* information and communication technologies (ICTs)
ideas, *see* concepts, definitions, ideas and theories
implementation
e-recruiting 223
e-selection 268
leadership of IT implementation
complementary leadership processes 305
leadership role 304
senior leadership support 305
summary overview 307
incivility, *see* cyber-aggression
independence 496
individual differences and learner choices research model 372
individual reactions to Internet at work
blended working 157
competences 61
cyberdeviance 131
digitized communication 21
ergonomics 39
flexwork 175
mobile computing 195
overview of 9
privacy, *see* privacy
trust 103
user experience 79
work practices 41

industrial production
gamification case study, *see* gamification
hybrid manufacturing concept 407, 408
robots, *see* robots
three-dimensional [3D] printing 496
inequality of technology access, *see* digital divide
information
accessibility 4
accuracy 240
acquisition, cyberdeviance focused on 133
'big data' handling 4, 275, 414
collection of new types of data, opportunities for 276
content
associations with contributors 47
determination of 80
editability 47
employer websites 216
persistence (permanence) 47
user experience 80
visibility 47
correctability 243
data privacy and security
employment decision-making, *see* employment decision-making
e-selection 259, 273
data structuredness 276
datafication
implications 501
information revolution: shrinking of speed and distance 487
measurement of work 485
trend 481, 482
ubiquity of data 482
diversity 27
overload, psychological implications 5
personal information
authorization for access 434
authorization for release 435
checking of validity or accuracy 436
power and access to data 486
privacy 433
processing, theories of 25, 67
reprocessability 4
social information processing theory 25
technology, *see* information and communication technology
validity 241
verification 242
information and communication technologies (ICTs)
definition 184
in e-learning, *see* e-learning
flexwork
ICT in relation 176
ICT usage 175, 185
outcomes
conditional 186
negative 186
positive 185
summary overview 188
fusion of physical and digital 489
future research 187
leadership, *see* leadership
positive associations 184
succession theory 66
summary overview 187
virtual teams 318, 322, 323, 326, 334
information revolution, shrinking of speed and distance 487
information technology artifacts
agency of 120
conceptualization of 117
innovation
disruptive innovation via robots 416
enterprise social media as driver 30
e-recruiting 215
fusion of physical and digital 490
leadership 301
mobile computing as driver 195, 196, 201, 204
robots 403, 407, 416
by teams 500
in value exchange and transactions 498
instant messaging
boundary crossing 47
channel expansion theory 68
e-leadership 299, 300, 301
mobile computing 196
virtual teams 318, 324, 349, 355
integration technologies and future of work 488
integrative model of leadership 287, 288
intelligent algorithms
for 'big data' handling 4
intelligent systems
trust in humans 115
'intentional forgetting'
in knowledge management 4
interaction in e-learning 384
interactivity
of communication 4
employer websites 216
increase in 3
psychological theories relevant to 6
Internet
birth of 2
collaborative work via 2
commercial Internet, *see* e-commerce

definition of 2
firsts 2
globally commercialization 2
interactivity 3
sensor technologies 3
Web 2.0 3
Internet at work
automatization potential 4
blended working, *see* blended working
boundary-crossing capabilities 5
brief history 2
collaborative work, *see* collaborative work
communication capabilities 4
competences, *see* competences
core characteristics of Internet-based work 3
cyberdeviance, *see* cyberdeviance
digitized communication, *see* computer-mediated communication (CMC); digitized communication
e-leadership, *see* leadership
e-learning, *see* e-learning
employment decision-making, *see* employment decision-making
e-recruiting, *see* e-recruiting
ergonomics, *see* ergonomics
e-selection, *see* e-selection
flexwork, *see* flexwork
impact of Internet on work 1
individual reactions 9
information capabilities 4
introduction to 1
mobile computing, *see* mobile computing
'the office', concept of 488
online surveys, *see* online surveys
organizational implications 11
potential benefits 7
psychological implications 5
psychological research into 8
robots, *see* robots
societal implications 13
summary overview of 15
trust, *see* trust
user experience, *see* user experience
virtual teams, *see* virtual teams
Internet of Things (IoT)
boundary crossing 5
impact 3
mobile computing 197
Internet-based recruiting
psychological implications of 6
Internet-based technologies
labels associated with 2
Internet-based work
potential risks 8
'Internet-of-Things' 3
interpersonal cyberdeviance (CD-I), *see* cyberdeviance
interviews, *see* online interviews
intranets 27, 41
IoT, *see* Internet of Things (IoT)
iPhone 3
item response theory 262

J
job advert aggregators 215
job satisfaction, *see* motivation and satisfaction
job stress, *see* stress
justice, *see* organizational justice

K
knowledge
building 29
meta-knowledge 29
visibility 29
knowledge management
'big data' handling 4
'intentional forgetting' 4
knowledge repositories in e-learning 388
knowledge sharing
enterprise wikis 53
visibility 30
knowledge testing in e-selection 263

L
labor
crowd relationships 51
labor decisions, *see* employment decision-making
labor rights
'sharing' communities 3
law and legal issues
employment decision-making 236
Facebook 236, 237
LinkedIn 236
personal data use in employment decision-making 236
privacy laws 439
leadership
age differences 462, 463
effect of information technology
consensus on 287
question of 285
future research 307
integrative model 287, 288
introduction to 285
IT implementation
complementary leadership processes 305
leadership role 304
senior leadership support 305
summary overview 307

leadership (*cont'd*)
  power 287, 289, 297, 300, 306
  same-place IT-supported meetings
    development of 289
    group support system (GSS) facilitation 286, 293
    studies that did not manipulate leadership behavior 289
    studies that manipulated leadership behavior 291
    summary overview 294
  in Social Media Age
    communication behaviors of online leaders 302
    dangers of social media 302
    impact of Internet 301
    online communities 302
    summary overview 309
  virtual teams
    increased relevance of leadership 296
    more powerful leadership mechanisms 298
    reduced impact of leadership 295
    research on leadership 329, 335
    summary overview 300
    use of 295
learning
  e-learning, *see* e-learning
  machine learning in e-selection 276
  mobile computing uses 203
learning performance
  user experience 82, 84
legal issues, *see* law
LinkedIn
  communities 46
  development of 3
  e-leadership 301
  employment decision-making 231, 233, 236, 249, 250
  e-recruiting 217, 219, 220, 221, 225, 226
  legal issues 236
Linux 2, 302
live stream
  first live stream 2
location data
  e-recruiting uses 225

M

machine learning in e-selection 276
machines doing human work 494
manufacturing, *see* industrial production
measurement of work performance, *see* work performance, measurement and monitoring
media
  richness theory 22
  selection for tasks 22
  social influence model of technology 23
  synchronicity theory 22, 66
meetings, leadership in, *see* leadership
metacognitive activity in e-learning 386
meta-knowledge
  building of 29
  relationship-building 30
meta-stereotypes concept 457
micro-blogging, *see* blogs
millennials (generation Y)
  mobile computing 199
misuse of computer access at work, *see* cyberdeviance; cyberloafing
mobile computing
  bring your own device (BYOD)
    advantages 198
    birth of 198
    control of 199
    device heterogeneity 199
    disadvantages 199
    infrastructure issues 199
    organizational responses 200
    security of 199
  cloud computing 200, 202
  cyberloafing 200
  definition 195
  devices 195, 196
  e-recruiting 222
  evolution 197, 198
  future research 204
  future trends 202
  global spread 201
  Internet of Things (IoT) 197, 203
  introduction to 195
  mobile learning 203, 387
  mobility and mobile use 196
  organizational responses 200
  overlap of private and professional contexts 197
  ownership issues 195
  research on mobile computing 196
  rise of 195
  summary overview 205
  technologies 195, 196
  training role 203
  ubiquity 196
  work-life balance 198
mobile phones, *see* 'smart' phones
monitoring of workers, *see* work performance, measurement and monitoring
Monster.com 214

motivation and satisfaction
aesthetics 85
age differences 456
blended working 158, 160, 162, 163
bring your own device (BYOD) 198, 199
competence 64
cooperation 32
customers 43, 106, 107, 108, 110
cyberdeviance 141, 142, 143, 144
e-learning 383
flexwork 180, 181, 182, 186
gamification 91
mobile computing 198
older workers 165
online surveys and feedback 356, 359, 361
psychological needs concept 160
usability 81

N
navigability of employer websites 216
networking, e-recruiting via 220
nonverbal cues
lack of social cues hypothesis 23, 67
presence of 22
norms and roles at work
cyberloafers 136

O
'the office', concept of 488
older workers, *see* age differences at work
online applicant data usage, *see* employment decision-making
Online Career Center 214
online communities, *see* communities
online environments
conceptualization 120
distrust 114
e-commerce, *see* e-commerce
trust repair 114
virtual teams, *see* virtual teams
online interviews
e-recruiting 216
e-selection 265
online job boards 214, 221
online leaders, *see* leadership
online surveys
automated employee survey processes 361
benchmarks for results 358
communication of results 350
data analysis 350
data collection 350
dynamic filtering 357
employee surveys 347
future research 362
future technologies 360
impact of computerization on survey assessments 265
introduction to 347
layout and design, importance of 356
planning phase 349
process, impact of Internet on 349
required fields 357
response rates 353
responsive layout 357
summary overview 363
wearable technologies 360
online trust, *see* trust
Open Source Software projects
collaborative work 2
communities 33, 302
crowdwork 3
Github 47
private implementation 28
relationships 50
organizational behavior 160
organizational cooperation, *see* cooperation
organizational cyberdeviance, *see* cyberdeviance; cyberloafing
organizational implications of Internet at work 11
organizational justice
cyberloafers' perceptions of 136
employment decision-making 238
organizational responses to Internet at work
e-learning 369
employment decision-making 231
e-recruiting 213
e-selection 257
leadership 285
online surveys 347
virtual teams 315
organizational social media, definition of 28
own device at work, *see* mobile computing

P
participatory design approach to work practices 41, 50
perceived ease of use
technology acceptance and 64
performance, *see* task performance; work performance
persistence (permanence) of social media content 47
personal data use by employers, *see* employment decision-making
personal mobile devices, *see* mobile computing

personality age and 455
personality testing in e-selection 264
personalized assistance from robots 413
power
access to data 486
agency 40, 50
independence 496
information revolution 487
leadership 287, 289, 297, 298, 300, 306
perceptions of empowerment 183
relationships 50
virtual teams 337
privacy
ability to check validity or accuracy of data 436
authorization for access to information 434
authorization for release of personal information 435
cultural and national differences in privacy values 437
definition of 432
e-commerce 106
employment decision-making 235
e-recruiting 224
ethnic differences 438
gender differences 438
global privacy laws 439
individual factors 438
informational factors 433
model of 432
personality differences 439
procedural factors 434
sociocultural factors 436
violation of 431
processing fluency theory 92
productivity
blended working 158, 159, 164, 165
community 44, 47, 48
cyberloafing effects, *see* web browsing
data about 348
e-learning 374
enhancement by enterprise social media 30
e-selection 269
flexwork 179, 180
gamification 94
impact of Internet-based tools 449
measurement of 485
mobile computing 196, 197, 198, 199, 203, 205
networked work 495
online surveys 361, 362, 363
social media 33
visibility 30
wearable technologies 361
psychological needs concept 160
psychological research
agenda for 8
combination with human-computer interaction research 362
psychology
computer anxiety 381
of computer misuse, *see* cyberdeviance; cyberloafing
computer self-efficacy (CSE) 381
implications of Internet at work 5
metacognitive activity in e-learning 386
motivation and satisfaction, *see* motivation and satisfaction
motivation in e-learning 383
psychological needs concept 160

R

race, *see* ethnic minorities
reasoned action
technology acceptance 64
trust 104
recruitment, *see* Internet-based recruiting
relational demography concept 463
relationship-building
meta-knowledge 30
social capital, definition of 30
relationships
agency 50
communication 52
crowd labor relationships 51
depth in collaborative work 69
email communication 52
embodiment 50
power 50
trust 49
remote working, *see* flexwork
richness, *see* media richness theory
risk
emergent technologies 118
Internet-based work 8
trust 104
robots
acceptance of 410
application development 414
artificial intelligence (AI) 406, 415, 488
big data 413
de-skilling effects 412
disruptive innovation 416
as embodied communication agents 416
flexible adaptation to user requirements 413
future applications 416
future of work 488, 492
future research 417

human-like appearance and cognition 409
hybrid manufacturing concept 407, 408
integration into work systems 414
introduction to 403
long-term effects 412
monitoring of users 413
personalization 413
robotic assistant systems 416
summary overview 418
technological capabilities 404
usability 410
user adaptation 412
user experience 410
user perspectives 408

S
SAP community network 87
satisfaction, *see* motivation and satisfaction
search engines
employment decision-making uses, *see* employment decision-making
e-recruiting uses 225
security, *see* information, data privacy and security
self-determination theory (SDT)
blended working 160
self-efficacy concept 64
Computer self-efficacy 391
in e-learning 379, 381
sensor technologies 3
sexual harassment, *see* cyber-aggression
shared understanding in collaborative work 69
'sharing' communities
birth of 3
labor rights 3
SIDE, *see* social identity model of deindividuation effects (SIDE)
signaling theory
e-recruiting 219
simulations in e-selection 276
situational judgment tests (SJTs) 266
skills, *see* competences
Skype 3, 217, 265
'smart' phones
access to 424
blended working 157
digital divide 424
e-leadership 301, 302, 307, 308
e-selection 262, 273, 274, 276
future of work 482
introduction of 3
millennials 203
mobile computing 195, 196, 198, 199, 200, 201
mobile learning 388
online surveys and feedback 356, 358, 359
robot development 411
technology trust 116
SNS, *see* social networking
social capital, definition of 30, 31
social context for older workers 457
social cues, *see* nonverbal cues
social identity model of deindividuation effects (SIDE) 26
social impact theory in collaborative work 69
social influence model of technology 23
social information processing theory 24, 25, 32, 67
social issues of Internet at work
digital divide
access to technology 426
cross-national differences 430
demographic factors 427
ethnic differences 428
future research 440
gender differences 427
problem of 424
theoretical bases 424
future research 440
introduction to 423
privacy
ability to check validity or accuracy of data 436
authorization for access to information 434
authorization for release of personal information 435
cultural and national differences in privacy values 437
definition of 432
ethnic differences 438
future research 441
gender differences 438
global privacy laws 439
individual factors 438
informational factors 433
model of 432
personality differences 439
procedural factors 434
sociocultural factors 436
violation of 431
summary overview 442
social loafing (cyberloafing), *see* cyberdeviance
social media
affordances 47
associations between users and content 47
computer-mediated communication (CMC) 27

Social media (*cont'd*)
development of 3
digitized communication 27
distinctiveness from earlier technologies 28
editability of content 47
enterprise social media 28
e-recruiting 221
as ICT 184
leadership, *see* leadership
organizational 28
persistence (permanence) of user contributions 47
relationship-building uses 28
visibility of user presence 47
social networking
communities 47
e-leadership 289
email 47
employment decision-making uses 231
e-recruiting 217, 218, 219, 224
e-selection 259, 275
gamification 87
impact on work practices 41
intranets 41
trust 118
use of 21, 27
user motivations 46
virtual teams 318
social presence in e-learning 385
social presence theory of computer-mediated communication 67
social values
correspondance of employment decisions with 244
socially constructed work norms and roles
cyberloafers and 136
societal and cross-sectorial perspectives
employee age differences in using Internet-based tools at work 449
future of work 481
robots 403
social issues of Internet at work 423
societal implications of Internet at work 13
stress
burnout 141, 184, 186, 361
cyberloafing 137
high job demands 161
performance under 83, 84
surveys, *see* online surveys; online surveys and feedback
swift trust model 64, 112
synchronicity theory 22, 66
systems perspectives
ergonomics 40

T
task analysis of work practices 41
task complexity in e-learning 377
task interdependence in virtual teams 334
task performance
cyberdeviance 142
trust 109
user experience 83, 84
tasks
selection of media for 22
testing tasks, user experience 83, 84
teams, *see* virtual teams; work teams
technology
acceptance and perceived ease of use 64
future applications, *see* future of work
ICT, *see* information and communication technologies (ICTs)
machines doing human work 494
robots, *see* robots
social influence model 23
technology trust concept 116
trust 115
technology choice and related competencies
theories of 66
telecommuting, *see* flexwork
telephone communication
computer-mediated communication (CMC) compared 22
richness 22
'smart' phones, *see* 'smart' phones
telework, *see* flexwork
testing
cognitive ability, testing of 263
computerized adaptive testing (CAT) 262
equivalence considerations in multiple test administration modes 260
item response theory 262
knowledge testing 263
performance in testing tasks 83, 84
personality testing 264
range of e-selection assessments and tests 263
situational judgment tests (SJTs) 266
supervised testing 272
unsupervised testing 261, 272
theories, *see* concepts, definitions, ideas and theories
three-dimensional [3D] printing
information revolution 487
interactive communication 4
in manufacturing 496
time constraints in virtual teams 334, 337

Tocqueville, Alexis de 44
training
  e-selection 274
  mobile computing uses 203
transactions of value, innovation in 498
transparency, *see* visibility
Trivago 215
trust
  agency 115
  avoidance of negative consequences 120
  balance of trust and distrust 120
  broader conceptualization of information technology artifact 117
  collaborative work 69
  contextualized models 118
  contextualized understanding of 121
  definition 104
  development view 111
  distinguishing trust in technological actors from social actors 116
  distrust in online environments 114
  e-commerce 105
  e-commerce trust research overview 108
  emergent technologies and risks 118
  future research 119
  importance 104
  initial trust model 112
  integrative view 111
  by intelligent systems in humans 115
  introduction to 103
  models of 110, 113
  online trust research 106
  relationships 49
  repair 114
  summary overview 122
  swift trust model 64, 112
  technology trust concept 116
  transference 113
  virtual teams 107, 331
Twitter
  big-data 275
  communities 48
  development of 3
  e-leadership 301
  employment decision-making 242
  e-recruiting 221
  e-selection 275
  social networking 217
  verification of data 242

U
Uber 3
usability
  user experience 81
user experience
  aesthetics 81, 93
  case study 85
  content 80
  CUE model 80
  definition of 79
  effects on work performance 81
  gamification, *see* gamification
  introduction to 79
  learning performance 82, 84
  measurement of 92
  overview of studies 82
  performance in testing tasks 83, 84
  performance under stress 83, 84
  processing fluency theory 92
  psychological processes affecting user performance 91
  research
    future 94
    practical implications of 92
  summary overview 94
  task performance 83, 84
  usability 81

V
value exchange and transactions
  innovation in 498
Viadeo 3
videoconferencing
  choice of 66
  effects on work 23
  e-leadership 299, 300
  email and 66
  E-recruiting interviews 216
  future of work 491
  gamification 90
  Internet of Things (IoT) 203
  job applicant reactions to 224
  uses 21
  virtual teams 324, 331, 335
virtual collaboration, *see* collaborative work
virtual online environments, *see* online environments
virtual reality
  e-learning 389
  e-selection 276
  future of work 490
virtual teams
  action processes 330
  affective reactions 332, 337
  age differences 462
  cohesion 332

communication media availability 334
communication processes 330
definition 315
diversity 322
effectiveness measures 337
emergent mentalities 331, 337
future research
inputs
compositional factors 336
contextual factors 336
leadership 335
technology 334
mediators
emergent mentalities 337
group potency 337
team empowerment 337
team processes 336
transition processes 336
moderators, time constraints as 337
opportunities 334
outcomes
effectiveness measures 337
members' affective reactions 337
performance measures 337
geographic dispersion 322
group potency 337
interactive communication 4
interpersonal processes 331
introduction to 315
leadership
increased relevance of 296
more powerful leadership mechanisms 298
reduced impact of 295
research on 329, 335
summary overview 300
use of virtual teams 295
performance measures 337
psychological implications of 5
research on
cross-sectional design 317
exemplar studies 327
input variables
compositional factors 326
leadership 329
primary inputs 326
technology 326
virtuality 326
laboratory studies 317
level of analysis 318
mediators
action processes 330
cohesion 332
communication processes 330
emergent mentalities 331
interpersonal processes 331
shared mental models 331
team identification and commitment 332
team processes 330
transition processes 330
trust 331
methodological considerations 317
moderators
communication media availability 334
importance 333
task interdependence 334
team virtuality 333
time constraints 334
outcomes
affective outcomes 332
team effectiveness 333
student samples 317
theoretical considerations 316
shared mental models 331
shared understanding in cross-cultural teams 70
social loafing 69
social network analysis of 318
task interdependence 334
team effectiveness 333
team empowerment 337
team identification and commitment 332
team processes 330, 336
team virtuality
concept 315
definitions 320
dimensions
composite measures 324
conceptualizations 321, 323
diversity 322
geographic dispersion 322
multiple dimensions 324
operationalization 322, 323
technology usage 322
types of technology used 323
as input 326
research on 319, 333
technology types used 323
technology use 318, 322, 326, 334
time constraints 334, 337
transition processes 330, 336
trust 107, 331
visibility
communication visibility theory 29
enhancement by enterprise social media 29
knowledge sharing 30
social media affordance 47
volunteering
future of work 503

W
wearable technologies in online surveys 360
Web 2.0
 birth of 3, 21
 computer-mediated communication (CMC) 27
 email 41
 enterprise wikis 53
 impact on work practices 41
web browsing
 cyberloafing 138
 future of work 483
web conferencing
 interactive communication 4
web page, first 2
web searching automatization 4
Web-based Work, *see* Internet at work
websites
 aesthetics of 93
 automated scanning of 4
 bring your own device (BYOD) 199
 cyberloafing 138
 employer websites
  aesthetics 215
  content 216
  e-recruiting 215
  interactivity 216
  navigability 216
 e-recruiting 213, 215
 future of work 483, 497
 perception of 91
 privacy issues 434, 435
 social networking 87
 task performance on 84
 trust building 107, 116
 trust transference 113
 user experience of 80
Wikipedia
 birth of 3
 community 32, 303
 enterprise wikis 53
 volunteering 503
wikis
 impact on work practices 41
women
 email use 427
 employer bias against 247
work breaks
 cyberloafing 138
 gamification and breaktime management 90
work communities, *see* communities
work engagement concept 188
work motivation, *see* motivation and satisfaction
work norms and roles
 cyberloafers and 136
work performance
 age differences 464
 measurement and monitoring
  datafication and future of work 485
  by robots 413
  virtual teams 337
 psychological processes affecting 91
 trust 109
 user experience 81
 virtual teams 337
work practices
 action-centred study 41
 communities 44
 email impact 41
 embodiment of 43
 participatory design approach 41, 50
 research
  current 41
  future 54
 summary overview 54
 task analysis 41
work relationships, *see* relationships
work simulations in e-selection 267
work stress, *see* stress
work teams
 concept 452
 gamification and team-working rhythm 91
 innovation by 500
 virtual teams, *see* virtual teams
worker monitoring by robots 413
workers with physical disability, *see* disabled workers
workforce diversity
 e-selection 274
 virtual teams 322
working from home, *see* blended working
work-life balance
 blurring of boundaries 181
 boundary crossing 182
 cross-boundary communication 183
 cyberloafing 137
 flexwork, *see* flexwork
 ICTs 183
 mobile computing 198
 outcomes of work–family integration 183
workplace aggression, *see* cyber-aggression

Y
YouTube 3

Printed and bound by CPI Group (UK) Ltd, Croydon, CR0 4YY
17/09/2026
14964302-0006